Student Solutions Manual

CINDY TRIMBLE

Intermediate Algebra

THIRD EDITION

Elayn Martin-Gay

PEARSON
Prentice Hall

Upper Saddle River, NJ 07458

Editor-in-Chief: Chris Hoag
Executive Editor: Paul Murphy
Project Manager: Mary Beckwith
Editorial Assistant: Abigail Rethore
Executive Managing Editor: Kathleen Schiaparelli
Senior Managing Editor: Nicole M. Jackson
Assistant Managing Editor: Karen Bosch Petrov
Production Editor: Traci Douglas
Supplement Cover Manager: Paul Gourhan
Supplement Cover Designer: Christopher Kossa
Manufacturing Buyer: Ilene Kahn
Manufacturing Manager: Alexis Heydt-Long

Pearson Prentice Hall
Pearson Education, Inc.
Upper Saddle River, NJ 07458

Printed in the United States of America

10 9 8 7 6 5 4 3 2 1

ISBN 0-13-219595-X Standalone
0-13-219594-1 Component

Pearson Education Ltd., *London*
Pearson Education Australia Pty. Ltd., *Sydney*
Pearson Education Singapore, Pte. Ltd.
Pearson Education North Asia Ltd., *Hong Kong*
Pearson Education Canada, Inc., *Toronto*
Pearson Educación de Mexico, S.A. de C.V.
Pearson Education—Japan, *Tokyo*
Pearson Education Malaysia, Pte. Ltd.

Contents

Chapter 1	1
Chapter 2	37
Chapter 3	98
Chapter 4	156
Chapter 5	218
Chapter 6	279
Chapter 7	361
Chapter 8	412
Chapter 9	527
Chapter 10	590
Appendices	645

Chapter 1

Section 1.2 Practice Problems

1. $\{x|x$ is a whole number between 0 and 4$\}$ $= \{1, 2, 3\}$

2. $\{x|x$ is a natural number greater than 80$\}$ $= \{81, 82, 83, ...\}$

3. Since 0 is not a natural number the statement is false.

4. Since 9 is not an element of $\{4, 6, 8, 10\}$ the statement is true.

5. True; every whole number is a real number.

6. True; every integer is a rational number.

7. False; $\sqrt{3}$ is an irrational number.

8. Seventeen times a number is written as $17 \cdot x$ or $17x$.

9. Five more than six times a number is written as $6x + 5$.

10. The quotient of six and a number is written as $\frac{6}{x}$ or $6 \div x$.

11. One-fourth subtracted from three times a number is written as $3x - \frac{1}{4}$.

12. Eleven less than a number is written as $x - 11$.

13. Three times the difference of a number and ten is written as $3(x - 10)$.

Exercise Set 1.2

1. $\{x|x$ is a natural number less than 6$\}$ $= \{1, 2, 3, 4, 5\}$

3. $\{x|x$ is a natural number between 10 and 17$\}$ $= \{11, 12, 13, 14, 15, 16\}$

5. $\{x|x$ is a whole number that is not a natural number$\} = 0$

7. $\{x|x$ is an even whole number less than 9$\}$ $= \{0, 2, 4, 6, 8\}$

9. The whole numbers in the set are $3, 0,$ and $\sqrt{36}$.

11. The natural numbers in the set are 3 and $\sqrt{36}$.

13. The irrational number in the set is $\sqrt{7}$.

15. $-11 \in \{x|x$ is an integer$\}$

17. $0 \notin \{x|x$ is a positive integer$\}$

19. $12 \notin \{1, 3, 5, ...\}$

21. Since the set of whole numbers is contained in the set of real numbers, the statement is true.

23. Since real numbers can be irrational, the statement is true.

25. Since the number 0 is a whole number but not a natural number, the statement is false.

27. Since $\frac{1}{2}$ is a rational number and every rational number is a real number, the statement is true.

29. Since 0 is an integer, the statement is true.

31. Since $\sqrt{7}$ is an irrational number, the statement is false.

33. Since there are no days of the week which start with the letter *B*, the statement is true.

35. Since real numbers can be integers, the statement is true.

37. Since irrational numbers are numbers which cannot be written as rational numbers, the statement is false.

39. Twice a number is written as $2x$.

41. Ten less than a number is written as $x - 10$.

43. The sum of a number and two is written as $x + 2$.

45. A number divided by 11 is written as $\frac{x}{11}$ or $x \div 11$.

47. Four subtracted from a number is written as $x - 4$.

49. A number plus two and three tenths is written as $x + 2.3$ or $x + 2\frac{3}{10}$.

51. A number less than one and one third is written as $1\frac{1}{3} - x$.

53. Nine times a number is written as $9x$.

55. Nine added to a number is written as $x + 9$.

57. Five more than twice a number is written as $2x + 5$.

59. Twelve minus three times a number is written as $12 - 3x$.

61. One plus twice a number is written as $1 + 2x$.

63. Ten subtracted from five times a number is written as $5x - 10$.

65. The quotient of five and the difference of four and a number is written as $\frac{5}{4-x}$.

67. Twice the sum of a number and three is written as $2(x + 3)$.

69. answers may vary

71. answers may vary

73. The only whole number that is not a natural number is the number 0.

75.

Year	Increase in Life Expectancy (in years)
1950	$71.1 - 65.7 = 5.4$
1960	$73.2 - 71.1 = 2.1$
1970	$74.9 - 73.2 = 1.7$
1980	$77.5 - 74.9 = 2.6$
1990	$78.9 - 77.5 = 1.4$
2000	$79.6 - 78.9 = 0.7$

77. answers may vary

Section 1.3 Practice Problems

1. The difference of x and 7 is 45 is written as $x - 7 = 45$.

2. The product of 5 and x amounts to the sum of x and 14 is written as $5x = x + 14$.

3. The quotient of y and 23 is the same as 20 subtracted from y is written as $\frac{y}{23} = y - 20$.

4. 7 is to the right of -7 on a number line so $7 > -7$.

5. -1 is to the left of 11 on a number line so $-1 < 11$.

6. -10 is to the right of -12 on a number line so $-10 > -12$.

7. -3.25 is to the left of -3.025 on a number line so $-3.25 < -3.025$.

8. The true statement is $7.206 = 7.2060$.

9. 18.6 is to the right of –14.2 on a number line so $18.6 > -14.2$.

10. The denominators are the same, so $\frac{4}{7} < \frac{5}{7}$ since $4 < 5$.

11. By dividing, we see that $\frac{3}{8} = 0.375$ and $\frac{1}{3} = 0.333\ldots$. Thus $\frac{3}{8} > \frac{1}{3}$ since $0.375 > 0.333\ldots$.

12. True, since $-11 < 16$ is true.

13. True, since $-7 = -7$ is true.

14. True, since $-7 = -7$ is true.

15. False, since neither $-25 < -30$ nor $-25 = -30$ is true.

16. The opposite of 7 is –7.

17. The opposite of $\frac{2}{13}$ is $-\frac{2}{13}$.

18. The opposite of $-\frac{5}{7}$ is $-\left(-\frac{5}{7}\right)$ or $\frac{5}{7}$.

19. The opposite of –4.7 is –(–4.7) or 4.7.

20. The reciprocal of 13 is $\frac{1}{13}$.

21. The reciprocal of –5 is $-\frac{1}{5}$.

22. The reciprocal of $\frac{2}{3}$ is $\frac{3}{2}$ since $\frac{2}{3} \cdot \frac{3}{2} = 1$.

23. By the commutative property of addition $9 + 4x = 4x + 9$.

24. By the associative property of multiplication $5 \cdot (6x) = (5 \cdot 6)x = 30x$.

25. $7(4x - y) = 7 \cdot 4x - 7 \cdot y = 28x - 7y$

26. $8(3 + x) = 8 \cdot 3 + 8 \cdot x = 24 + 8x$

27. $5x(y - 4) = 5x \cdot y - 5x \cdot 4 = 5xy - 20x$

Exercise Set 1.3

1. The sum of 10 and x is –12 is written as $10 + x = -12$.

3. Twice x plus 5 is the same as –14 is written as $2x + 5 = -14$.

5. The quotient of n and 5 is 4 times n is written as $\frac{n}{5} = 4n$.

7. The difference of z and one-half is the same as the product of z and one-half is written as $z - \frac{1}{2} = \frac{1}{2}z$.

9. 0 is to the right of –2, so $0 > -2$.

11. –16 is to the right of –17, so $-16 > -17$.

13. $\frac{12}{3} = 4$, $\frac{12}{2} = 6$, and 4 is to the left of 6, so $\frac{12}{3} < \frac{12}{2}$.

15. The true statement is $7.4 = 7.40$.

17. 8.6 is to the right of –3.5, so $8.6 > -3.5$.

19. The denominators are the same, so $\frac{7}{11} < \frac{9}{11}$ since $7 < 9$.

21. By dividing, we see that $\frac{1}{2} = 0.5$ and $\frac{5}{8} = 0.625$. Thus, $\frac{1}{2} < \frac{5}{8}$, since $0.5 < 0.625$.

23. -7.9 is to the left of -7.09, so $-7.9 < -7.09$.

25. True, since $-6 < 0$ is true.

27. True, since $-3 = -3$ is true.

29. False, since neither $-14 > -1$ nor $-14 = -1$ is true.

31. True, since $-3 = -3$ is true.

33. The opposite of 9 is -9.

35. The opposite of -6.2 is $-(-6.2)$ or 6.2.

37. The opposite of $\frac{4}{7}$ is $-\frac{4}{7}$.

39. The opposite of $-\frac{5}{11}$ is $-\left(-\frac{5}{11}\right)$ or $\frac{5}{11}$.

41. The opposite of 0 is -0 or 0.

43. The reciprocal of 5 is $\frac{1}{5}$.

45. The reciprocal of -8 is $-\frac{1}{8}$.

47. The reciprocal of $-\frac{1}{4}$ is -4.

49. The reciprocal of 0 is undefined, since there is no number that when multiplied by 0 gives a product of 1.

51. The reciprocal of $\frac{7}{8}$ is $\frac{8}{7}$ since $\frac{7}{8} \cdot \frac{8}{7} = 1$.

	Number	Opposite	Reciprocal
53.	25	-25	$\frac{1}{25}$
55.	-10	10	$-\frac{1}{10}$
57.	$-\frac{1}{7}$	$\frac{1}{7}$	-7
59.	0	0	undefined
61.	$\frac{16}{19}$	$-\frac{16}{19}$	$\frac{19}{16}$

63. $7x + y = y + 7x$ by the commutative property of addition.

65. $z \cdot w = w \cdot z$ by the commutative property of multiplication.

67. $\frac{1}{3} \cdot \frac{x}{5} = \frac{x}{5} \cdot \frac{1}{3}$ by the commutative property of multiplication.

69. $5 \cdot (7x) = (5 \cdot 7)x$ by the associative property of multiplication.

71. $(x + 1.2) + y = x + (1.2 + y)$ by the associative property of addition.

73. $(14z) \cdot y = 14(z \cdot y)$ by the associative property of multiplication.

75. $3(x + 5) = 3 \cdot x + 3 \cdot 5 = 3x + 15$

77. $4(z - 6) = 4 \cdot z - 4 \cdot 6 = 4z - 24$

79. $8(2a + b) = 8 \cdot 2a + 8 \cdot b = 16a + 8b$

81. $6x(y - 4) = 6x \cdot y - 6x \cdot 4 = 6xy - 24x$

83. $0.4(2x + 5y) = 0.4 \cdot 2x + 0.4 \cdot 5y = 0.8x + 2y$

85. $\frac{1}{2}(4x - 9y) = \frac{1}{2} \cdot 4x - \frac{1}{2} \cdot 9y = 2x - \frac{9}{2}y$

87. $2(6x+5y+2z)=2\cdot 6x+2\cdot 5y+2\cdot 2z$
$=12x+10y+4z$

89. $3x + 6 = 6 + 3x$ by the commutative property of addition.

91. $\frac{2}{3}+\left(-\frac{2}{3}\right)=0$ by the additive inverse property.

93. $7 \cdot 1 = 7$ by the multiplicative identity property.

95. $10(2y) = (10 \cdot 2)y$ by the associative property.

97. $3(x + 4) = 3 \cdot x + 3 \cdot 4 = 3x + 12$

99. $4 + 8y = 8y + 4$

101. zero; answers may vary

103. no; answers may vary

105. answers may vary

Section 1.4 Practice Problems

1. $|7| = 7$

2. $|-20| = -(-20) = 20$

3. $\left|-\frac{4}{9}\right|=-\left(-\frac{4}{9}\right)=\frac{4}{9}$

4. $-|5.9| = -5.9$

5. $-|-3| = -3$

6. $-7 + (-10) = -(7 + 10) = -17$

7. $8 + (-12) = -4$

8. $-14 + 20 = 6$

9. $-4.6 + (-1.9) = -6.5$

10. $-\frac{2}{3}+\frac{1}{6}=-\frac{2}{3}\cdot\frac{2}{2}+\frac{1}{6}=-\frac{4}{6}+\frac{1}{6}=-\frac{3}{6}=-\frac{1}{2}$

11. $-\frac{1}{7}+\frac{1}{2}=-\frac{1}{7}\cdot\frac{2}{2}+\frac{1}{2}\cdot\frac{7}{7}=-\frac{2}{14}+\frac{7}{14}=\frac{5}{14}$

12. $7 - 14 = 7 + (-14) = -7$

13. $-10 - (-2) = -10 + 2 = -8$

14. $13.3 - (-8.9) = 13.3 + 8.9 = 22 2$

15. $-\frac{1}{3}-\frac{1}{2}=-\frac{1}{3}+\left(-\frac{1}{2}\right)$
$=-\frac{1}{3}\cdot\frac{2}{2}+\left(-\frac{1}{2}\cdot\frac{3}{3}\right)$
$=-\frac{2}{6}+\left(-\frac{3}{6}\right)$
$=-\frac{5}{6}$

16. $18 + 3 - 4 = 21 - 4 = 17$

17. $-3-11+7=-3+(-11)+7$
$=-14+7$
$=-7$

18. $-4(-2) = 8$

19. $5\left(-\frac{1}{10}\right)=\frac{5}{1}\cdot\left(-\frac{1}{10}\right)=-\frac{5}{10}=-\frac{1}{2}$

20. $-3.2(0.1) = -0.32$

21. $8(-6) = -48$

22. $-\frac{2}{5}\left(-\frac{1}{3}\right)=\frac{2}{15}$

23. $(-1.3)(-1.5) = 1.95$

24. $0(-10) = 0$

25. $\frac{45}{-9}=-5$

26. $\frac{-16}{-4} = 4$

27. $\frac{25}{-5} = -5$

28. $\frac{-3}{0}$ is undefined.

29. $\frac{0}{-3} = 0$

30. $\frac{-1}{-4} = \frac{1}{4}$

31. $-\frac{3}{4} \div \left(-\frac{3}{8}\right) = -\frac{3}{4} \cdot \left(-\frac{8}{3}\right) = \frac{24}{12} = 2$

32. $-\frac{1}{11} \div \frac{2}{7} = -\frac{1}{11} \cdot \left(\frac{7}{2}\right) = -\frac{7}{22}$

33. $4^2 = 4 \cdot 4 = 16$

34. $-3^2 = -(3 \cdot 3) = -9$

35. $-3^3 = -(3 \cdot 3 \cdot 3) = -27$

36. $\left(\frac{1}{3}\right)^4 = \left(\frac{1}{3}\right)\left(\frac{1}{3}\right)\left(\frac{1}{3}\right)\left(\frac{1}{3}\right) = \frac{1}{81}$

37. $(-3)^2 = (-3)(-3) = 9$

38. $(-3)^3 = (-3)(-3)(-3) = -27$

39. $\sqrt{36} = 6$, since 6 is positive and $6^2 = 36$.

40. $\sqrt{4} = 2$, since 2 is positive and $2^2 = 4$.

41. $\sqrt{\frac{1}{49}} = \frac{1}{7}$, since $\frac{1}{7}$ is positive and $\left(\frac{1}{7}\right)^2 = \frac{1}{49}$.

42. $\sqrt[3]{64} = 4$, since $4^3 = 64$.

43. $\sqrt[4]{1} = 1$, since 1 is positive and $1^4 = 1$.

44. $\sqrt[5]{243} = 3$, since $3^5 = 243$.

Exercise Set 1.4

1. $|2| = 2$

3. $|-4| = 4$

5. $|0| = 0$

7. $-|3| = -3$

9. $-\left|-\frac{2}{9}\right| = -\frac{2}{9}$

11. $-3 + 8 = 5$

13. $-14 + (-10) = -24$

15. $-4.3 - 6.7 = -4.3 + (-6.7) = -11$

17. $13 - 17 = 13 + (-17) = -4$

19. $$\begin{aligned} \frac{11}{15} - \left(-\frac{3}{5}\right) &= \frac{11}{15} + \frac{3}{5} \\ &= \frac{11}{15} + \frac{3}{5} \cdot \frac{3}{3} \\ &= \frac{11}{15} + \frac{9}{15} \\ &= \frac{20}{15} \\ &= \frac{4}{3} \end{aligned}$$

21. $19-10-11=19+(-10)+(-11)$
$=9+(-11)$
$=-2$

23. $-14-7=-14+(-7)=-21$

25. $-\frac{4}{5}-\left(-\frac{3}{10}\right)=-\frac{4}{5}\cdot\frac{2}{2}+\frac{3}{10}$
$=-\frac{8}{10}+\frac{3}{10}$
$=-\frac{5}{10}$
$=-\frac{1}{2}$

27. $8-14=8+(-14)=-6$

29. $-5\cdot 12=-60$

31. $-8(-10)=80$

33. $-17\cdot 0=0$

35. $0(-1)=0$

37. $\frac{-9}{3}=-3$

39. $\frac{-12}{-4}=3$

41. $3\left(-\frac{1}{18}\right)=-\frac{1}{6}$

43. $(-0.7)(-0.8)=0.56$

45. $9.1\div(-1.3)=-7$

47. $-4(-2)(-1)=8(-1)=-8$

49. $-7^2=-(7\cdot 7)=-49$

51. $(-6)^2=(-6)(-6)=36$

53. $(-2)^3=(-2)(-2)(-2)=-8$

55. $\left(-\frac{1}{3}\right)^3=\left(-\frac{1}{3}\right)\left(-\frac{1}{3}\right)\left(-\frac{1}{3}\right)=-\frac{1}{27}$

57. $\sqrt{49}=7$, since $7^2=49$.

59. $\sqrt{64}=8$, since $8^2=64$.

61. $\sqrt{\frac{1}{9}}=\frac{1}{3}$, since $\left(\frac{1}{3}\right)^2=\frac{1}{9}$.

63. $\sqrt[3]{64}=4$, since $4^3=64$.

65. $\sqrt{\frac{4}{25}}=\frac{2}{5}$, since $\left(\frac{2}{5}\right)^2=\frac{4}{25}$.

67. $-4+7=3$

69. $-9+(-3)=-12$

71. $6(-3)=-18$

73. $-9\cdot 8=-72$

75. $(-11)^2=(-11)(-11)=121$

77. $\frac{16}{-2}=-8$

79. $-18\div 6=-3$

81. $-\frac{2}{7}\cdot\left(-\frac{1}{6}\right)=\frac{2}{42}=\frac{1}{21}$

83. $-2(-3.6)=7.2$

85. $-4-(-19)=-4+19=15$

87. $6.3-18.5=6.3+(-18.5)=-12.2$

89. $\frac{0}{-5}=0$

91. $\frac{-18}{0}$ is undefined.

93. $\sqrt{\frac{1}{16}} = \frac{1}{4}$

95. $\frac{-6}{7} \div 2 = \frac{-6}{7} \cdot \frac{1}{2} = \frac{-6}{14} = -\frac{3}{7}$

97. $\frac{-5.2}{-1.3} = 4$

99. $-25 \div (-5) = 5$

101. $\sqrt[4]{81} = 3$, since $3^4 = 81$.

103. $\sqrt[3]{8} = 2$, since $2^3 = 8$.

105. $-\frac{1}{6} \div \left(\frac{9}{10}\right) = -\frac{1}{6} \cdot \left(\frac{10}{9}\right) = -\frac{10}{54} = -\frac{5}{27}$

107. $-\frac{2}{3} \cdot \left(\frac{6}{4}\right) = -\frac{12}{12} = -1$

109. $\frac{3}{5} \div \left(-\frac{2}{5}\right) = \frac{3}{5} \cdot \left(-\frac{5}{2}\right) = -\frac{15}{10} = -\frac{3}{2}$

111.
$$\begin{aligned} 16 - 8 - 9 &= 16 + (-8) - 9 \\ &= 8 - 9 \\ &= 8 + (-9) \\ &= -1 \end{aligned}$$

113. $-5 + (-7) - 10 = -12 + (-10) = -22$

115. $-7(-1)(5) = 7(5) = 35$

117. $-6(-5)(0) = 30(0) = 0$

119. $-4^4 = -(4 \cdot 4 \cdot 4 \cdot 4) = -256$

121. answers may vary

123. True; $-7 - 1 = -7 + (-1) = -8$ and $-7(-1) = 7$ and $-8 < 7$, since -8 is to the left of 7.

125. $1 - \frac{1}{5} - \frac{3}{7} = \frac{35}{35} - \frac{7}{35} - \frac{15}{35} = \frac{28}{35} - \frac{15}{35} = \frac{13}{35}$

127.
$$\begin{aligned} 10{,}203 - 5998 &= 10{,}203 + (-5998) \\ &= 4205 \text{ m} \end{aligned}$$
The volcano rises to 4205 meters above sea level.

129. Since each color has an equal sector of the spinner in b, spinner b would lead to a fair game.

131. For player 1 to lose, he would want to choose spinner d since spinner d has the smallest yellow area.

133. Yes, a two-way tie could occur if two players have 6 points each (the third player has 0 points); or two players have 5 points each (the third player has 2 points).

135. $\sqrt{273} \approx 16.5227$

137. $\sqrt{19.6} \approx 4.4272$

139. After 5 years, a person investing in common stocks may expect to lose at most an average of 17% per year.

141. After 5 years, the highest average annual return is 23% and the lowest average annual return is −17%.
$23 - (-17) = 23 + 17 = 40$
The difference is 40%.

Integrated Review

1. $\{x|x$ is a natural number less than $4\}$
$= \{1, 2, 3\}$

2. $\{x|x$ is an odd whole number less than $6\}$
$= \{1, 3, 5\}$

3. $\{x|x$ is an even natural number greater than $7\} = \{8, 10, 12, \ldots\}$

4. $\{x | x \text{ is a whole number between 10 and 15}\} = \{11, 12, 13, 14\}$

5. Twice the difference of a number and three is written as $2(x - 3)$.

6. The quotient of six and the sum of a number and ten is written as $\frac{6}{x+10}$.

7. -4 is to the right of -6 so $-4 > -6$.

8. The true statement is $8.6 = 8.600$.

9. The denominators are the same so $\frac{9}{10} < \frac{11}{10}$ since $9 < 11$.

10. -6.1 is to the left of -6.01 so $-6.1 < -6.01$.

11. The product of 5 and x is the same as 20 is written as $5x = 20$.

12. The sum of a and 12 amounts to 14 is written as $a + 12 = 14$.

13. The quotient of y and 10 is the same as the product of y and 10 is written as $\frac{y}{10} = y \cdot 10$.

14. The sum of x and 1 equals the difference of x and 1 is written as $x + 1 = x - 1$.

15. $-4 + 7 = 3$

16. $-11 + 20 = 9$

17. $-4(7) = -28$

18. $-11(20) = -220$

19. $-8 - (-13) = -8 + 13 = 5$

20. $-12 - 16 = -12 + (-16) = -28$

21. $\frac{-20}{-4} = 5$

22. $\frac{-18}{6} = -3$

23. $-5^2 = -(5 \cdot 5) = -25$

24. $(-5)^2 = (-5)(-5) = 25$

25. $-6 - 1 + 20 = -6 + (-1) + 20 = -7 + 20 = 13$

26.
$$\begin{aligned} 18 - 4 - 19 &= 18 + (-4) - 19 \\ &= 14 - 19 \\ &= 14 + (-19) \\ &= -5 \end{aligned}$$

27. $\frac{0}{-3} = 0$

28. $\frac{5}{0}$ is undefined.

29. $-4(3)(2) = -12(2) = -24$

30. $-5(-1)(6) = 5(6) = 30$

31. $-\frac{1}{2} \cdot \frac{6}{7} = -\frac{6}{14} = -\frac{3}{7}$

32. $\frac{4}{5} \cdot \left(-\frac{1}{8}\right) = -\frac{4}{40} = -\frac{1}{10}$

33.
$$\begin{aligned} \frac{3}{10} - \frac{4}{5} &= \frac{3}{10} + \left(-\frac{4}{5}\right) \\ &= \frac{3}{10} + \left(-\frac{4}{5} \cdot \frac{2}{2}\right) \\ &= \frac{3}{10} + \left(-\frac{8}{10}\right) \\ &= -\frac{5}{10} \\ &= -\frac{1}{2} \end{aligned}$$

34. $-\frac{2}{3}-\frac{1}{4}=-\frac{2}{3}+\left(-\frac{1}{4}\right)$
$=-\frac{2}{3}\cdot\frac{4}{4}+\left(-\frac{1}{4}\cdot\frac{3}{3}\right)$
$=-\frac{8}{12}+\left(-\frac{3}{12}\right)$
$=-\frac{11}{12}$

35. $\frac{1.6}{-0.2}=-8$

36. $-\frac{4.8}{16}=-0.3$

37. $6.7-(-1.3)=6.7+1.3=8$

38. $-4.6+9=4.4$

39. $\frac{1}{2}+\left(-\frac{7}{8}\right)=\frac{1}{2}\cdot\frac{4}{4}+\left(-\frac{7}{8}\right)=\frac{4}{8}+\left(-\frac{7}{8}\right)=-\frac{3}{8}$

40. $\frac{1}{2}\div\left(-\frac{7}{8}\right)=\frac{1}{2}\cdot\left(-\frac{8}{7}\right)=-\frac{8}{14}=-\frac{4}{7}$

41. $\sqrt{49}=7$, since $7^2=49$.

42. $\sqrt[3]{27}=3$, since $3^3=27$.

43. $-2^2=-(2\cdot 2)=-4$

44. $(-2)^3=(-2)(-2)(-2)=-8$

45. $\sqrt{\frac{1}{81}}=\frac{1}{9}$, since $\left(\frac{1}{9}\right)^2=\frac{1}{81}$.

46. $\sqrt{\frac{1}{100}}=\frac{1}{10}$, since $\left(\frac{1}{10}\right)^2=\frac{1}{100}$.

	Number	Opposite	Reciprocal
47.	-6	6	$-\frac{1}{6}$
48.	4	-4	$\frac{1}{4}$
49.	$\frac{5}{7}$	$-\frac{5}{7}$	$\frac{7}{5}$
50.	$-\frac{7}{30}$	$\frac{7}{30}$	$-\frac{30}{7}$

51. $9(m+5)=9\cdot m+9\cdot 5=9m+45$

52. $11(7+r)=11\cdot 7+11\cdot r=77+11r$

53. $3(2y-3x)=3\cdot 2y-3\cdot 3x=6y-9x$

54. $8(4m-7n)=8\cdot 4m-8\cdot 7n=32m-56n$

55. $0.2(3a+7)=0.2\cdot 3a+0.2\cdot 7=0.6a+1.4$

56. $0.6(2n+5)=0.6\cdot 2n+0.6\cdot 5=1.2n+3$

57. $\frac{1}{5}(10x-19y+20)=\frac{1}{5}\cdot 10x-\frac{1}{5}\cdot 19y+\frac{1}{5}\cdot 20$
$=2x-\frac{19}{5}y+4$

Section 1.5 Practice Problems

1. $15-2\cdot 5=15-10=5$

2. $2+5(2-6)^2=2+5(-4)^2$
$=2+5(16)$
$=2+80$
$=82$

3. $\frac{|-3|^2+5}{\sqrt{9}-10}=\frac{3^2+5}{3-10}=\frac{9+5}{-7}=\frac{14}{-7}=-2$

4. $7-[2(1-3)+5(10-12)]$
$=7-[2(-2)+5(-2)]$
$=7-[-4+(-10)]$
$=7-[-14]$
$=7+14$
$=21$

5. $\dfrac{|9-16|+5^2}{-3\sqrt{8+1}+(-4)^2}=\dfrac{|-7|+5^2}{-3\sqrt{9}+(-4)^2}$
$=\dfrac{7+25}{-3\cdot 3+16}$
$=\dfrac{32}{-9+16}$
$=\dfrac{32}{7}$

6. a. Replace x with 5 and y with –4.
$2x-6y=2\cdot 5-6(-4)$
$=10-(-24)$
$=10+24$
$=34$

b. Replace y with –4.
$-3y^2=-3(-4)^2=-3(16)=-48$

c. Replace x with 5 and y with –4.
$\dfrac{\sqrt{x}}{y}+\dfrac{y}{x}=\dfrac{\sqrt{5}}{-4}+\left(-\dfrac{4}{5}\right)$
$=-\dfrac{\sqrt{5}}{4}\cdot\dfrac{5}{5}+\left(-\dfrac{4}{5}\cdot\dfrac{4}{4}\right)$
$=-\dfrac{5\sqrt{5}}{20}+\left(-\dfrac{16}{20}\right)$
$=-\dfrac{5\sqrt{5}+16}{20}$

7. To complete the table, we evaluate $\dfrac{5(x-32)}{9}$ at each given replacement value.

a. When $x=-13$,
$\dfrac{5(-13-32)}{9}=\dfrac{5(-45)}{9}=-25.$

b. When $x=0$,
$\dfrac{5(0-32)}{9}=\dfrac{5(-32)}{9}=\dfrac{-160}{9}=-17\dfrac{7}{9}.$

c. When $x=41$, $\dfrac{5(41-32)}{9}=\dfrac{5(9)}{9}=5.$

8. $9x-15x+7=(9-15)x+7=-6x+7$

9. $8y+y=8y+1y=(8+1)y=9y$

10. $4x+12x-9-10=(4+12)x+(-9-10)$
$=16x-19$

11. $-4x+7-5x-8=-4x-5x+7-8$
$=(-4-5)x+(7-8)$
$=-9x-1$

12. $5y-6y+2-11+y=5y-6y+1y+2-11$
$=(5-6+1)y+(2-11)$
$=0y-9$
$=-9$

13. $-3(y+1)+4=-3(y)+(-3)(1)-4$
$=-3y-3+4$
$=-3y+1$

14. $8x+2-4(x-9)=8x+2-4x+36$
$=4x+38$

15. $(3.2x-4.1)-(-x+7.6)$
$=(3.2x-4.1)-1(-x+7.6)$
$=3.2x-4.1+1x-7.6$
$=4.2x-11.7$

16. $\dfrac{1}{5}(15m-40n)-\dfrac{1}{4}(8m-4n+1)+\dfrac{1}{5}$
$=3m-8n-2m+1n-\dfrac{1}{4}+\dfrac{1}{5}$
$=m-7n-\dfrac{1}{20}$

Exercise Set 1.5

1. $3(5-7)^4=3(-2)^4=3\cdot 16=48$

3. $-3^2+2^3=-9+8=-1$

5. $\dfrac{3.1-(-1.4)}{-0.5}=\dfrac{4.5}{-0.5}=-9$

7. $|3.6-7.2|+|3.6+7.2|=|-3.6|+|10.8|$
$=3.6+10.8$
$=14.4$

9. $(-3)^2+2^3=9+8=17$

11. $-8\div 4\cdot 2=-2\cdot 2=-4$

13. $4[8-(2-4)]=4[8-(-2)]=4[10]=40$

15. $-8\left(-\dfrac{3}{4}\right)-8=6-8=-2$

17. $2-[(7-6)+(9-19)]=2-[1+(-10)]$
$=2-[-9]$
$=11$

19. $\dfrac{(-9+6)(-1^2)}{-2-2}=\dfrac{(-3)(-1)}{-4}=-\dfrac{3}{4}$

21. $(\sqrt[3]{8})(-4)-(\sqrt{9})(-5)=(2)(-4)-(3)(-5)$
$=-8-(-15)$
$=7$

23. $12+\{6-[5-2(-5)]\}=12+\{6-[5-(-10)]\}$
$=12+\{6-[15]\}$
$=12+\{-9\}$
$=3$

25. $25-[(3-5)+(14-18)]^2$
$=25-[(-2)+(-4)]^2$
$=25-[-6]^2$
$=25-36$
$=-11$

27. $\dfrac{\left(3-\sqrt{9}\right)-(-5-1.3)}{-3}=\dfrac{(3-3)-(-5-1.3)}{-3}$
$=\dfrac{0-(-6.3)}{-3}$
$=\dfrac{6.3}{-3}$
$=-2.1$

29. $\dfrac{|3-9|-|-5|}{-3}=\dfrac{|-6|-|-5|}{-3}$
$=\dfrac{6-5}{-3}$
$=\dfrac{1}{-3}$
$=-\dfrac{1}{3}$

31. $\dfrac{3(-2+1)}{5}-\dfrac{-7(2-4)}{1-(-2)}=\dfrac{3(-1)}{5}-\dfrac{-7(-2)}{3}$
$=-\dfrac{3}{5}-\dfrac{14}{3}$
$=-\dfrac{3}{5}\cdot\dfrac{3}{3}-\dfrac{14}{3}\cdot\dfrac{5}{5}$
$=-\dfrac{9}{15}-\dfrac{70}{15}$
$=-\dfrac{79}{15}$

33. $\dfrac{\frac{1}{3}\cdot 9-7}{3+\frac{1}{2}\cdot 4}=\dfrac{3-7}{3+2}=\dfrac{-4}{5}=-\dfrac{4}{5}$

35. $3\{-2+5[1-2(-2+5)]\}=3\{-2+5[1-2(3)]\}$
$=3\{-2+5[1-6]\}$
$=3\{-2+5[-5]\}$
$=3\{-2+[-25]\}$
$=3\{-27\}$
$=-81$

37. $\dfrac{-4\sqrt{80+1}+(-4)^2}{3^3+|-2(3)|} = \dfrac{-4\sqrt{81}+(-4)^2}{3^3+|-6|}$

$= \dfrac{-4\cdot 9+16}{27+6}$

$= \dfrac{-36+16}{33}$

$= \dfrac{-20}{33}$

$= -\dfrac{20}{33}$

39. $-150(3.25 - 1.68) = -150(1.57) = -235.5$

41. $\left(\dfrac{5.6-8.4}{1.9-2.7}\right)^2 = \left(\dfrac{-2.8}{-0.8}\right)^2 = (3.5)^2 = 12.25$

43. Replace x with 9 and y with -2.

$$\begin{aligned}9x-6y &= 9(9)-6(-2)\\ &= 81-(-12)\\ &= 81+12\\ &= 93\end{aligned}$$

45. Replace y with -2.

$-3y^2 = -3(-2)^2 = -3\cdot 4 = -12$

47. Replace x with 9 and y with -2.

$$\begin{aligned}\frac{\sqrt{x}}{y}-\frac{y}{x} &= \frac{\sqrt{9}}{-2}-\frac{-2}{9}\\ &= -\frac{3}{2}+\frac{2}{9}\\ &= -\frac{3}{2}\cdot\frac{9}{9}+\frac{2}{9}\cdot\frac{2}{2}\\ &= -\frac{27}{18}+\frac{4}{18}\\ &= -\frac{23}{18}\end{aligned}$$

49. Replace x with 9 and y with −2.

$$\frac{3+2|x-y|}{x+2y} = \frac{3+2|9-(-2)|}{9+2(-2)}$$
$$= \frac{3+2|11|}{9+(-4)}$$
$$= \frac{3+22}{5}$$
$$= \frac{25}{5}$$
$$= 5$$

51. Replace x with 9 and y with −2.

$$\frac{y^3+\sqrt{x-5}}{|4x-y|} = \frac{(-2)^3+\sqrt{9-5}}{|4(9)-(-2)|}$$
$$= \frac{-8+\sqrt{4}}{|36+2|}$$
$$= \frac{-8+2}{|38|}$$
$$= -\frac{6}{38}$$
$$= -\frac{3}{19}$$

53. a.

Length, y	**Perimeter, $8 + 2y$**
5	$8 + 2(5) = 8 + 10 = 18$
7	$8 + 2(7) = 8 + 14 = 22$
10	$8 + 2(10) = 8 + 20 = 28$
100	$8 + 2(100) = 8 + 200 = 208$

b. As the length increases, the perimeter increases; answers may vary.

55. a.

Number of Bookshelves, x	**Cost per bookshelf, $\frac{100x+5000}{x}$**
10	$\frac{100(10)+5000}{10} = \frac{1000+5000}{10} = \frac{6000}{10} = 600$
100	$\frac{100(100)+5000}{100} = \frac{10,000+5000}{100} = \frac{15,000}{100} = 150$
1000	$\frac{100(1000)+5000}{1000} = \frac{100,000+5000}{1000} = \frac{105,000}{1000} = 105$

b. As the number of bookshelves manufactured increases, the cost per bookshelf decreases; answers may vary.

57. Replace z with 8.4.
$439.39z = 439.39 \cdot 8.4 = 3690.876$
The average cost of 8.4 ounces of gold at that time was \$3690.88.

59. $6x + 2x = (6 + 2)x = 8x$

61. $19y - y = 19y - 1y = (19 - 1)y = 18y$

63. $$\begin{aligned} 9x - 8 - 10x &= 9x - 10x - 8 \\ &= (9 - 10)x - 8 \\ &= -1x - 8 \\ &= -x - 8 \end{aligned}$$

65. $$\begin{aligned} -9 + 4x + 18 - 10x &= 4x - 10x - 9 + 18 \\ &= -6x + 9 \end{aligned}$$

67. $$\begin{aligned} 3a - 4b + a - 9b &= 3a + 1a - 4b - 9b \\ &= (3+1)a + (-4-9)b \\ &= 4a - 13b \end{aligned}$$

69. $x - y + x - y = x + x - y - y = 2x - 2y$

71. $$\begin{aligned} &1.5x + 2.3 - 0.7x - 5.9 \\ &= 1.5x - 0.7x + 2.3 - 5.9 \\ &= (1.5 - 0.7)x + 2.3 - 5.9 \\ &= 0.8x - 3.6 \end{aligned}$$

73. $$\begin{aligned} \frac{3}{4}b - \frac{1}{2} + \frac{1}{6}b - \frac{2}{3} &= \frac{3}{4}b + \frac{1}{6}b - \frac{1}{2} - \frac{2}{3} \\ &= \frac{9}{12}b + \frac{2}{12}b - \frac{3}{6} - \frac{4}{6} \\ &= \frac{11}{12}b - \frac{7}{6} \end{aligned}$$

75. $$\begin{aligned} &8ab - 4.6 - 11ab - 8.2 \\ &= 8ab - 11ab - 4.6 - 8.2 \\ &= -3ab - 12.8 \end{aligned}$$

77. $2(3x + 7) = 2 \cdot 3x + 2 \cdot 7 = 6x + 14$

79. $5k - (3k - 10) = 5k - 3k + 10 = 2k + 10$

81. $$\begin{aligned} (3x+4) - (6x-1) &= 3x + 4 - 6x + 1 \\ &= 3x - 6x + 4 + 1 \\ &= -3x + 5 \end{aligned}$$

83. $$\begin{aligned} 3(x-2) + x + 15 &= 3x - 6 + 1x + 15 \\ &= 3x + 1x - 6 + 15 \\ &= 4x + 9 \end{aligned}$$

85. $$\begin{aligned} -(n+5) + (5n-3) &= -n - 5 + 5n - 3 \\ &= -n + 5n - 5 - 3 \\ &= 4n - 8 \end{aligned}$$

87. $$\begin{aligned} 4(6n-3) - 3(8n+4) &= 24n - 12 - 24n - 12 \\ &= 24n - 24n - 12 - 12 \\ &= -24 \end{aligned}$$

89. $$\begin{aligned} \frac{1}{4}(8x-4) - \frac{1}{5}(20x - 6y) &= 2x - 1 - 4x + \frac{6}{5}y \\ &= 2x - 4x + \frac{6}{5}y - 1 \\ &= -2x + \frac{6}{5}y - 1 \end{aligned}$$

91. $$\begin{aligned} 3x - 2(x-5) + x &= 3x - 2x + 10 + x \\ &= 3x - 2x + x + 10 \\ &= 2x + 10 \end{aligned}$$

93. $$\begin{aligned} &\frac{1}{6}(24a - 18b) - \frac{1}{7}(7a - 21b - 2) - \frac{1}{5} \\ &= 4a - 3b - a + 3b + \frac{2}{7} - \frac{1}{5} \\ &= 4a - a - 3b + 3b + \frac{10}{35} - \frac{7}{35} \\ &= 3a + \frac{3}{35} \end{aligned}$$

95. $$\begin{aligned} &5.7a + 1.7 - 3(2.1a - 0.6) \\ &= 5.7a + 1.7 - 6.3a + 1.8 \\ &= 5.7a - 6.3a + 1.7 + 1.8 \\ &= -0.6a + 3.5 \end{aligned}$$

97. $-4[6(2t+1)-(9+10t)]$
$=-4[12t+6-9-10t]$
$=-4[12t-10t+6-9]$
$=-4[2t-3]$
$=-8t+12$

99. $-1.2(5.7x-3.6)+8.75x$
$=-6.84x+4.32+8.75x$
$=-6.84x+8.75x+4.32$
$=1.91x+4.32$

101. $8.1z+7.3(z+5.2)-6.85$
$=8.1z+7.3z+37.96-6.85$
$=15.4z+31.11$

103. $(2+7)\cdot(1+3)=(9)\cdot(4)=36$

105. answers may vary

107. $(3x+2)-(5x+7)=3x+2-5x-7$
$=3x-5x+2-7$
$=-2x-5$

109. From the graph the population that was over 65 in 1970 was about 20 million.

111. From the graph, the predicted population over 65 in 2030 will be 70 million.

113. The population over 65 is increasing; answers may vary.

115. $\dfrac{-1.682-17.895}{(-7.102)(-4.691)}\approx\dfrac{-19.577}{33.3155}\approx-0.5876$

Section 1.6 Practice Problems

1. $5^2\cdot5^6=5^{2+6}=5^8$

2. $x^5\cdot x^9=x^{5+9}=x^{14}$

3. $y\cdot y^4\cdot y^3=(y^1\cdot y^4)\cdot y^3=y^5\cdot y^3=y^8$

4. $(7y^5)(6y)=7(6)y^5y^1=42y^6$

5. $(-3x^2y^7)(5xy^6)=-3(5)x^2x^1y^7y^6$
$=-15x^3y^{13}$

6. $8^0=1$

7. $-14^0=-(14)^0=-1$

8. $(y-3)^0=1$

9. $5x^0=5(1)=5$

10. $\dfrac{6^{10}}{6^2}=6^{10-2}=6^8$

11. $\dfrac{y^6}{y^2}=y^{6-2}=y^4$

12. $\dfrac{36x^5}{9x}=4x^{5-1}=4x^4$

13. $\dfrac{10a^7b^9}{15a^5b^9}=\dfrac{2}{3}a^{7-5}\cdot b^{9-9}$
$=\dfrac{2}{3}a^2\cdot b^0$
$=\dfrac{2}{3}a^2(1)$
$=\dfrac{2}{3}a^2$

14. $7^{-2}=\dfrac{1}{7^2}=\dfrac{1}{49}$

15. $5x^{-4}=5\cdot\dfrac{1}{x^4}=\dfrac{5}{x^4}$

16. $(2x)^{-1}=\dfrac{1}{(2x)^1}=\dfrac{1}{2x}$

17. $3^{-1}+2^{-2}=\dfrac{1}{3^1}+\dfrac{1}{2^2}=\dfrac{1}{3}+\dfrac{1}{4}=\dfrac{4}{12}+\dfrac{3}{12}=\dfrac{7}{12}$

18. $\frac{1}{y^{-4}} = \frac{1}{\frac{1}{y^4}} = 1 \div \frac{1}{y^4} = 1 \cdot \frac{y^4}{1} = y^4$

19. $\frac{x^3}{x^{10}} = x^{3-10} = x^{-7} = \frac{1}{x^7}$

20. $\frac{4^2}{4^5} = 4^{2-5} = 4^{-3} = \frac{1}{4^3} = \frac{1}{64}$

21. $y^{-10} \cdot y^3 = y^{-10+3} = y^{-7} = \frac{1}{y^7}$

22. $\frac{q^5}{q^{-4}} = q^{5-(-4)} = q^9$

23. $\frac{5^{-4}}{5^{-2}} = 5^{-4-(-2)} = 5^{-2} = \frac{1}{5^2} = \frac{1}{25}$

24. $\frac{10x^{-8}y^5}{20xy^{-5}} = \frac{x^{-8-1} \cdot y^{5-(-5)}}{2}$

$= \frac{x^{-9} \cdot y^{10}}{2}$

$= \frac{y^{10}}{2x^9}$

25. $\frac{(4x^{-1})(x^5)}{x^7} = \frac{4x^{-1+5}}{x^7}$

$= \frac{4x^4}{x^7}$

$= 4x^{4-7}$

$= 4x^{-3}$

$= \frac{4}{x^3}$

26. $x^{3m} \cdot x^n = x^{3m+n}$

27. $\frac{x^{2m-2}}{x^{m-6}} = x^{(2m-2)-(m-6)}$

$= x^{2m-2-m+6}$

$= x^{m+4}$

28. $1,760,000 = 1.76 \times 10^6$

29. $0.00028 = 2.8 \times 10^{-4}$

30. $8.6 \times 10^7 = 86,000,000$

31. $3.022 \times 10^{-4} = 0.0003022$

Calculator Explorations

1. $(3 \times 10^{11}) \cdot (2 \times 10^{32}) = 6 \times 10^{43}$
2. $(6 \times 10^{14}) \div (3 \times 10^9) = 2 \times 10^5$
3. $(5.2 \times 10^{23}) \cdot (7.3 \times 10^4) = 3.796 \times 10^{28}$
4. $(4.38 \times 10^{41}) \div (3 \times 10^{17}) = 1.46 \times 10^{24}$

Mental Math

1. $5x^{-1}y^{-2} = \frac{5}{x^1y^2} = \frac{5}{xy^2}$
2. $7xy^{-4} = \frac{7x}{y^4}$
3. $a^2b^{-1}c^{-5} = \frac{a^2}{bc^5}$
4. $a^{-4}b^2c^{-6} = \frac{b^2}{a^4c^6}$
5. $\frac{y^{-2}}{x^{-4}} = \frac{x^4}{y^2}$

6. $\dfrac{x^{-7}}{z^{-3}} = \dfrac{z^3}{x^7}$

Exercise Set 1.6

1. $4^2 \cdot 4^3 = 4^{2+3} = 4^5$

3. $x^5 \cdot x^3 = x^{5+3} = x^8$

5. $m \cdot m^7 \cdot m^6 = m^1 \cdot m^7 \cdot m^6 = m^{1+7+6} = m^{14}$

7. $(4xy)(-5x) = 4(-5)x^1x^1y^1$
$= -20x^{1+1}y^1$
$= -20x^2y$

9. $(-4x^3p^2)(4y^3x^3) = -4(4)x^3x^3y^3p^2$
$= -16x^{3+3}y^3p^2$
$= -16x^6y^3p^2$

11. $-8^0 = -(8)^0 = -1$

13. $(4x+5)^0 = 1$

15. $-x^0 = -(x)^0 = -1$

17. $4x^0 + 5 = 4 \cdot 1 + 5 = 4 + 5 = 9$

19. $\dfrac{a^5}{a^2} = a^{5-2} = a^3$

21. $-\dfrac{26z^{11}}{2z^7} = -\dfrac{26}{2}z^{11-7} = -13z^4$

23. $\dfrac{x^9y^6}{x^8y^6} = x^{9-8}y^{6-6} = x^1y^0 = x^1(1) = x$

25. $\dfrac{12x^4y^7}{9xy^5} = \dfrac{12}{9}x^{4-1}y^{7-5} = \dfrac{4}{3}x^3y^2$

27. $\dfrac{-36a^5b^7c^{10}}{6ab^3c^4} = \dfrac{-36}{6}a^{5-1}b^{7-3}c^{10-4}$
$= -6a^4b^4c^6$

29. $4^{-2} = \dfrac{1}{4^2} = \dfrac{1}{16}$

31. $(-3)^{-3} = \dfrac{1}{(-3)^3} = -\dfrac{1}{27}$

33. $\dfrac{x^7}{x^{15}} = x^{7-15} = x^{-8} = \dfrac{1}{x^8}$

35. $5a^{-4} = 5 \cdot \dfrac{1}{a^4} = \dfrac{5}{a^4}$

37. $\dfrac{x^{-7}}{y^{-2}} = x^{-7} \cdot \dfrac{1}{y^{-2}} = \dfrac{1}{x^7} \cdot y^2 = \dfrac{y^2}{x^7}$

39. $\dfrac{x^{-2}}{x^5} = x^{-2-5} = x^{-7} = \dfrac{1}{x^7}$

41. $\dfrac{8r^4}{2r^{-4}} = \dfrac{8}{2}r^{4-(-4)} = 4r^8$

43. $\dfrac{x^{-9}x^4}{x^{-5}} = \dfrac{x^{-9+4}}{x^{-5}} = \dfrac{x^{-5}}{x^{-5}} = x^{-5-(-5)} = x^0 = 1$

45. $\dfrac{2a^{-6}b^2}{18ab^{-5}} = \dfrac{2}{18}a^{-6-1}b^{2-(-5)} = \dfrac{1}{9}a^{-7}b^7 = \dfrac{b^7}{9a^7}$

47. $\frac{(24x^8)(x)}{20x^{-7}} = \frac{24x^{8+1}}{20x^{-7}}$
$= \frac{24x^9}{20x^{-7}}$
$= \frac{24}{20}x^{9-(-7)}$
$= \frac{6}{5}x^{16}$
$= \frac{6x^{16}}{5}$

49. $-7x^3 \cdot 20x^9 = -7 \cdot 20x^{3+9} = -140x^{12}$

51. $x^7 \cdot x^8 \cdot x = x^{7+8+1} = x^{16}$

53. $2x^3 \cdot 5x^7 = 2 \cdot 5x^{3+7} = 10x^{10}$

55. $(5x)^0 + 5x^0 = 1 + 5(1) = 1 + 5 = 6$

57. $\frac{z^{12}}{z^{15}} = z^{12-15} = z^{-3} = \frac{1}{z^3}$

59. $3^0 - 3t^0 = 1 - 3(1) = 1 - 3 = -2$

61. $\frac{y^{-3}}{y^{-7}} = y^{-3-(-7)} = y^4$

63. $4^{-1} + 3^{-2} = \frac{1}{4} + \frac{1}{3^2} = \frac{1}{4} + \frac{1}{9} = \frac{9}{36} + \frac{4}{36} = \frac{13}{36}$

65. $3x^{-1} = 3 \cdot \frac{1}{x} = \frac{3}{x}$

67. $\frac{r^4}{r^{-4}} = r^{4-(-4)} = r^8$

69. $\frac{x^{-7}y^{-2}}{x^2y^2} = x^{-7-2}y^{-2-2}$
$= x^{-9}y^{-4}$
$= \frac{1}{x^9} \cdot \frac{1}{y^4}$
$= \frac{1}{x^9y^4}$

71. $(-4x^2y)(3x^4)(-2xy^5)$
$= -4(3)(-2)x^2x^4x^1y^1y^5$
$= 24x^{2+4+1}y^{1+5}$
$= 24x^7y^6$

73. $2^{-4}x = \frac{1}{2^4} \cdot x = \frac{1}{16}x = \frac{x}{16}$

75. $\frac{5^{17}}{5^{13}} = 5^{17-13} = 5^4 = 625$

77. $\frac{8^{-7}}{8^{-6}} = 8^{-7-(-6)} = 8^{-1} = \frac{1}{8}$

79. $\frac{9^{-5}a^4}{9^{-3}a^{-1}} = 9^{-5-(-3)}a^{4-(-1)}$
$= 9^{-2}a^5$
$= \frac{1}{9^2} \cdot a^5$
$= \frac{a^5}{81}$

81. $\frac{14x^{-2}yz^{-4}}{2xyz} = \frac{14}{2}x^{-2-1}y^{1-1}z^{-4-1}$
$= 7x^{-3}y^0z^{-5}$
$= \frac{7}{x^3z^5}$

83. $x^5 \cdot x^{7a} = x^{5+7a}$

85. $\dfrac{x^{3t-1}}{x^t} = x^{3t-1-t} = x^{2t-1}$

87. $x^{4a} \cdot x^7 = x^{4a+7}$

89. $\dfrac{z^{6x}}{z^7} = z^{6x-7}$

91. $\dfrac{x^{3t} \cdot x^{4t-1}}{x^t} = \dfrac{x^{3t+4t-1}}{x^t}$
$= \dfrac{x^{7t-1}}{x^t}$
$= x^{7t-1-1t}$
$= x^{6t-1}$

93. $31,250,000 = 3.125 \times 10^7$

95. $0.016 = 1.6 \times 10^{-2}$

97. $67,413 = 6.7413 \times 10^4$

99. $0.0125 = 1.25 \times 10^{-2}$

101. $0.000053 = 5.3 \times 10^{-5}$

103. $778,300,000 = 7.783 \times 10^8$

105. $6,404,000,000 = 6.404 \times 10^9$

107. $124,000,000,000 = 1.24 \times 10^{11}$

109. $0.001 = 1.0 \times 10^{-3}$

111. $3.6 \times 10^{-9} = 0.0000000036$

113. $9.3 \times 10^7 = 93,000,000$

115. $1.278 \times 10^6 = 1,278,000$

117. $7.35 \times 10^{12} = 7,350,000,000,000$

119. $4.03 \times 10^{-7} = 0.000000403$

121. $3.0 \times 10^8 = 300,000,000$

123. $4.9 \times 10^9 = 4,900,000,000$

125. answers may vary

127. answers may vary

Section 1.7 Practice Problems

1. $(y^2)^8 = y^{2 \cdot 8} = y^{16}$

2. $(3^3)^2 = 3^{3 \cdot 2} = 3^6 = 729$

3. $(6^2)^{-1} = 6^{2 \cdot (-1)} = 6^{-2} = \dfrac{1}{6^2} = \dfrac{1}{36}$

4. $(x^{-5})^{-7} = x^{-5(-7)} = x^{35}$

5. $(3x^4)^3 = 3^3 \cdot (x^4)^3 = 3^3 \cdot x^{4 \cdot 3} = 27x^{12}$

6. $\left(\dfrac{4}{5}\right)^2 = \dfrac{4^2}{5^2} = \dfrac{16}{25}$

7. $\left(\dfrac{4m^5}{n^3}\right)^3 = \dfrac{(4m^5)^3}{(n^3)^3} = \dfrac{4^3 \cdot (m^5)^3}{(n^3)^3} = \dfrac{64m^{15}}{n^9}$

8. $\left(\dfrac{2^{-1}}{y}\right)^{-3} = \dfrac{(2^{-1})^{-3}}{y^{-3}} = \dfrac{2^3}{y^{-3}} = 8y^3$

9. $(a^{-4}b^3c^{-2})^6 = (a^{-4})^6 \cdot (b^3)^6 \cdot (c^{-2})^6$
$= a^{-24}b^{18}c^{-12}$
$= \dfrac{b^{18}}{a^{24}c^{12}}$

10. $(7xy^{-2})^{-2} = 7^{-2}(x)^{-2}(y^{-2})^{-2}$
$= 7^{-2}x^{-2}y^{4}$
$= \dfrac{y^4}{7^2x^2}$
$= \dfrac{y^4}{49x^2}$

11. $\left(\dfrac{y^{-7}}{y^{-10}}\right)^{-3} = \dfrac{(y^{-7})^{-3}}{(y^{-10})^{-3}}$
$= \dfrac{y^{21}}{y^{30}}$
$= y^{21-30}$
$= y^{-9}$
$= \dfrac{1}{y^9}$

12. $\left(\dfrac{3}{5}\right)^{-2} = \dfrac{3^{-2}}{5^{-2}} = \dfrac{5^2}{3^2} = \dfrac{25}{9}$

13. $\dfrac{6^{-2}x^{-4}y^{10}}{x^2y^{-6}} = 6^{-2}x^{-4-2}y^{10-(-6)}$
$= 6^{-2}x^{-6}y^{16}$
$= \dfrac{y^{16}}{6^2x^6}$
$= \dfrac{y^{16}}{36x^6}$

14. $\left(\dfrac{4a^3b^2}{b^{-6}c}\right)^{-2} = \left(\dfrac{4a^3b^8}{c}\right)^{-2}$
$= \dfrac{4^{-2}a^{-6}b^{-16}}{c^{-2}}$
$= \dfrac{c^2}{4^2a^6b^{16}}$
$= \dfrac{c^2}{16a^6b^{16}}$

15. $\left(\dfrac{4x^3}{3y^{-1}}\right)^3\left(\dfrac{y^{-2}}{3x^{-1}}\right)^{-1} = \dfrac{64x^9}{27y^{-3}} \cdot \dfrac{y^2}{3^{-1}x^1}$
$= \dfrac{64 \cdot 3 \cdot x^9 \cdot y^2 \cdot y^3}{27x^1}$
$= \dfrac{64x^8y^5}{9}$

16. $x^{-n}(3x^n)^2 = x^{-n}3^2x^{2n} = 9x^{-n-2n} = 9x^n$

17. $\dfrac{(y^{2m})^2}{y^{m-3}} = \dfrac{y^{2m\cdot 2}}{y^{m-3}}$
$= \dfrac{y^{4m}}{y^{m-3}}$
$= y^{4m-(m-3)}$
$= y^{4m-m+3}$
$= y^{3m+3}$

18. $(9.6\times10^6)(4\times10^{-8}) = 9.6\times4\times10^6\times10^{-8}$
$= 38.4\times10^{-2}$
$= (3.84\times10^1)\times10^{-2}$
$= 3.84\times10^{-1}$

19. $\dfrac{4.2\times10^7}{7\times10^{-3}} = \left(\dfrac{4.2}{7}\right)\left(\dfrac{10^7}{10^{-3}}\right)$
$= 0.6\times10^{7-(-3)}$
$= 0.6\times10^{10}$
$= (6\times10^{-1})(10^{10})$
$= 6\times10^9$

20. $\dfrac{3000\times0.000012}{400} = \dfrac{(3\times10^3)(1.2\times10^{-5})}{4\times10^2}$
$= \dfrac{3(1.2)}{4}\cdot\dfrac{10^3\ 10^{-5}}{10^2}$
$= 0.9\times10^{-4}$
$= (9\times10^{-1})\times10^{-4}$
$= 9\times10^{-5}$

Mental Math

1. $(x^4)^5 = x^{4\cdot 5} = x^{20}$

2. $(5^6)^2 = 5^{6\cdot 2} = 5^{12}$

3. $x^4 \cdot x^5 = x^{4+5} = x^9$

4. $x^7 \cdot x^8 = x^{7+8} = x^{15}$

5. $(y^6)^7 = y^{6\cdot 7} = y^{42}$

6. $(x^3)^4 = x^{3\cdot 4} = x^{12}$

7. $(z^4)^9 = z^{4\cdot 9} = z^{36}$

8. $(z^3)^7 = z^{3\cdot 7} = z^{21}$

9. $(z^{-6})^{-3} = z^{-6(-3)} = z^{18}$

10. $(y^{-4})^{-2} = y^{-4(-2)} = y^8$

Exercise Set 1.7

1. $(3^{-1})^2 = 3^{-2} = \dfrac{1}{3^2} = \dfrac{1}{9}$

3. $(x^4)^{-9} = x^{-36} = \dfrac{1}{x^{36}}$

5. $(3x^2y^3)^2 = 3^2(x^2)^2(y^3)^2 = 9x^4y^6$

7. $$\left(\frac{2x^5}{y^{-3}}\right)^4 = (2x^5y^3)^4 = 2^4(x^5)^4(y^3)^4 = 16x^{20}y^{12}$$

9. $$\begin{aligned}(2a^2bc^{-3})^{-6} &= 2^{-6}(a^2)^{-6}b^{-6}(c^{-3})^{-6}\\ &= 2^{-6}a^{-12}b^{-6}c^{18}\\ &= \frac{c^{18}}{2^6a^{12}b^6}\\ &= \frac{c^{18}}{64a^{12}b^6}\end{aligned}$$

11. $$\left(\frac{7}{8}\right)^3 = \frac{7^3}{8^3} = \frac{343}{512}$$

13. $$\begin{aligned}(-2^{-2}y^{-1})^{-3} &= [-1\cdot(2)^{-2}y^{-1}]^{-3}\\ &= (-1)^{-3}(2^{-2})^{-3}(y^{-1})^{-3}\\ &= (-1)\cdot 2^6y^3\\ &= -64y^3\end{aligned}$$

15. $$\begin{aligned}\left(\frac{a^{-4}}{a^{-5}}\right)^{-2} &= (a^{-4-(-5)})^{-2}\\ &= (a^1)^{-2}\\ &= a^{-2}\\ &= \frac{1}{a^2}\end{aligned}$$

17. $$\begin{aligned}\left(\frac{6p^6}{p^{12}}\right)^2 &= (6p^{6-12})^2\\ &= (6p^{-6})^2\\ &= 6^2p^{-12}\\ &= \frac{36}{p^{12}}\end{aligned}$$

19. $$\begin{aligned}(-8y^3xa^{-2})^{-3} &= [-1\cdot 8y^3xa^{-2}]^{-3}\\ &= (-1)^{-3}\cdot 8^{-3}y^{-9}x^{-3}a^6\\ &= -\frac{a^6}{8^3x^3y^9}\\ &= -\frac{a^6}{512x^3y^9}\end{aligned}$$

21. $\left(\frac{3}{4}\right)^{-3} = \frac{3^{-3}}{4^{-3}} = \frac{4^3}{3^3} = \frac{64}{27}$

23. $\left(\frac{2a^{-2}b^5}{4a^2b^7}\right)^{-2} = \left(\frac{1}{2}a^{-4}b^{-2}\right)^{-2}$

$= \left(\frac{1}{2}\right)^{-2} a^8b^4$

$= \frac{1^{-2}a^8b^4}{2^{-2}}$

$= \frac{2^2a^8b^4}{1^2}$

$= 4a^8b^4$

25. $\left(\frac{x^{-2}y^{-2}}{a^{-3}}\right)^{-7} = \frac{(x^{-2})^{-7}(y^{-2})^{-7}}{(a^{-3})^{-7}} = \frac{x^{14}y^{14}}{a^{21}}$

27. $(y^{-5})^2 = y^{-5\cdot 2} = y^{-10} = \frac{1}{y^{10}}$

29. $(5^{-1})^3 = 5^{-3} = \frac{1}{5^3} = \frac{1}{125}$

31. $(x^7)^{-9} = x^{7\cdot(-9)} = x^{-63} = \frac{1}{x^{63}}$

33. $\left(\frac{x^7y^{-3}}{z^{-4}}\right)^{-5} = \frac{(x^7)^{-5}(y^{-3})^{-5}}{(z^{-4})^{-5}}$

$= \frac{x^{-35}y^{15}}{z^{20}}$

$= \frac{y^{15}}{x^{35}z^{20}}$

35. $(4x^2)^2 = 4^2(x^2)^2 = 16x^4$

37. $\left(\frac{4^{-4}}{y^3x}\right)^{-2} = \frac{(4^{-4})^{-2}}{(y^3)^{-2}x^{-2}} = \frac{4^8}{y^{-6}x^{-2}} = 4^8x^2y^6$

39. $\left(\frac{2x^{-3}}{y^{-1}}\right)^{-3} = \frac{2^{-3}(x^{-3})^{-3}}{(y^{-1})^{-3}}$

$= \frac{2^{-3}x^9}{y^3}$

$= \frac{x^9}{2^3y^3}$

$= \frac{x^9}{8y^3}$

41. $\frac{4^{-1}x^2yz}{x^{-2}yz^3} = 4^{-1}x^{2-(-2)}y^{1-1}z^{1-3}$

$= \frac{1}{4}x^4y^0z^{-2}$

$= \frac{x^4}{4z^2}$

43. $\left(\frac{3x^5}{6x^4}\right)^4 = \left(\frac{1}{2}x\right)^4$

$= \left(\frac{1}{2}\right)^4 x^4$

$= \frac{1^4}{2^4}x^4$

$= \frac{1}{16}x^4$

$= \frac{x^4}{16}$

45. $\frac{(y^3)^{-4}}{y^3} = \frac{y^{-12}}{y^3} = y^{-12-3} = y^{-15} = \frac{1}{y^{15}}$

47. $\frac{3^{-2}a^{-5}b^6}{4^{-2}a^{-7}b^{-3}} = \left(\frac{3^{-2}}{4^{-2}}\right)a^{-5-(-7)}b^{6-(-3)}$

$= \frac{4^2}{3^2}a^2b^9$

$= \frac{16a^2b^9}{9}$

49. $(4x^6y^5)^{-2}(6x^4y^3)$
$= 4^{-2}(x^6)^{-2}(y^5)^{-2} \cdot 6x^4y^3$
$= \frac{1}{4^2}x^{-12}y^{-10} \cdot 6x^4y^3$
$= \frac{6}{16}x^{-12+4}y^{-10+3}$
$= \frac{3}{8}x^{-8}y^{-7}$
$= \frac{3}{8x^8y^7}$

51. $x^6(x^6bc)^{-6} = x^6 \cdot (x^6)^{-6}(b)^{-6}(c)^{-6}$
$= x^6x^{-36}b^{-6}c^{-6}$
$= x^{6-36}b^{-6}c^{-6}$
$= x^{-30}b^{-6}c^{-6}$
$= \frac{1}{x^{30}b^6c^6}$

53. $\frac{2^{-3}x^2y^{-5}}{5^{-2}x^7y^{-1}} = \frac{5^2}{2^3}x^{2-7}y^{-5-(-1)}$
$= \frac{25}{8}x^{-5}y^{-4}$
$= \frac{25}{8x^5y^4}$

55. $\left(\frac{2x^2}{y^4}\right)^3\left(\frac{2x^5}{y}\right)^{-2} = \frac{2^3x^6}{y^{12}} \cdot \frac{2^{-2}x^{-10}}{y^{-2}}$
$= \frac{2^{3+(-2)}x^{6+(-10)}}{y^{12+(-2)}}$
$= \frac{2^1x^{-4}}{y^{10}}$
$= \frac{2}{x^4y^{10}}$

57. $(x^{3a+6})^3 = x^{(3a+6)\cdot 3} = x^{9a+18}$

59. $\frac{x^{4a}(x^{4a})^3}{x} = \frac{x^{4a}(x^{12a})}{x} = \frac{x^{16a}}{x} = x^{16a-1}$

61. $(b^{5x-2})^2 = b^{(5x-2)\cdot 2} = b^{10x-4}$

63. $\frac{(y^{2a})^8}{y^{a-3}} = \frac{y^{16a}}{y^{a-3}} = y^{16a-(a-3)} = y^{15a+3}$

65. $\left(\frac{2x^{3t}}{x^{2t-1}}\right)^4 = (2x^{3t-(2t-1)})^4$
$= (2x^{t+1})^4$
$= 2^4x^{(t+1)\cdot 4}$
$= 16x^{4t+4}$

67. $\frac{25x^{2a+b}y^{2a-b}}{5x^{a-b}y^{a+b}}$
$= \left(\frac{25}{5}\right)x^{2a+b-(a-b)}y^{2a-b-(a+b)}$
$= 5x^{a+2b}y^{a-2b}$

69. $(5\times10^{11})(2.9\times10^{-3}) = 5\times2.9\times10^{11}\times10^{-3}$
$= 14.5\times10^8$
$= (1.45\times10^1)\times10^8$
$= 1.45\times10^9$

71. $(2\times10^5)^3 = 2^3\times(10^5)^3 = 8\times10^{15}$

73. $\frac{3.6\times10^{-4}}{9\times10^2} = \frac{3.6}{9}\cdot\frac{10^{-4}}{10^2}$
$= 0.4\times10^{-6}$
$= (4\times10^{-1})\times10^{-6}$
$= 4\times10^{-7}$

75. $\frac{0.0069}{0.023} = \frac{6.9\times10^{-3}}{2.3\times10^{-2}} = \frac{6.9}{2.3}\cdot\frac{10^{-3}}{10^{-2}} = 3\times10^{-1}$

77. $\frac{18,200\times 100}{91,000}=\frac{1.82\times 10^4\times 10^2}{9.1\times 10^4}$
$=\frac{1.82}{9.1}\cdot\frac{10^6}{10^4}$
$=0.2\times 10^2$
$=(2\times 10^{-1})\times 10^2$
$=2\times 10^1$

79. $\frac{6000\times 0.006}{0.009\times 400}=\frac{6\times 10^3\times 6\times 10^{-3}}{9\times 10^{-3}\times 4\times 10^2}$
$=\frac{36\times 10^0}{36\times 10^{-1}}$
$=\frac{36}{36}\cdot\frac{10^0}{10^{-1}}$
$=1\times 10^1$

81. $\frac{0.00064\times 2000}{16,000}=\frac{6.4\times 10^{-4}\times 2\times 10^3}{1.6\times 10^4}$
$=\frac{12.8\times 10^{-1}}{1.6\times 10^4}$
$=\frac{12.8}{1.6}\cdot\frac{10^{-1}}{10^4}$
$=8\times 10^{-5}$

83. $\frac{66,000\times 0.001}{0.002\times 0.003}=\frac{6.6\times 10^4\times 1\times 10^{-3}}{2\times 10^{-3}\times 3\times 10^{-3}}$
$=\frac{6.6\times 10^1}{6\times 10^{-6}}$
$=\frac{6.6}{6}\cdot\frac{10^1}{10^{-6}}$
$=1.1\times 10^7$

85. $\frac{9.24\times 10^{15}}{(2.2\times 10^{-2})(1.2\times 10^{-5})}$
$=\frac{9.24}{(2.2)(1.2)}\cdot\frac{10^{15}}{10^{-2}\cdot 10^{-5}}$
$=\frac{9.24}{2.64}\times 10^{22}$
$=3.5\times 10^{22}$

87. $200,000\times 10^{-8}=(2\times 10^5)\times 10^{-8}=2\times 10^{-3}$

The computer would take 2×10^{-3} second to add two numbers together 200,000 times.

89. $(3.8\times 10^{-6})(1.64\times 10^{-5})$
$=3.8\times 1.64\times 10^{-6}\times 10^{-5}$
$=6.232\times 10^{-11}$

The average size of a grain of salt is 6.232×10^{-11} cubic meter.

91. The area of a parallelogram is base · height. The area of the given parallelogram is

$\left(\frac{3x^{-1}}{y^{-3}}\right)(5x^{-7})=\frac{15x^{-8}}{y^{-3}}=\frac{15y^3}{x^8}$ square feet.

93. Since $D=\frac{M}{V}$, then $D\cdot V=M$.

$(3.12\times 10^{-2})(4.269\times 10^{14})$
$=3.12\times 4.269\times 10^{-2}\times 10^{14}$
$=13.31928\times 10^{12}$
$=(1.331928\times 10^1)\times 10^{12}$
$=1.331928\times 10^{13}$

The mass of water in Lake Superior is 1.331928×10^{13} tons.

95. No

97. $\frac{2.95\times 10^8}{3.536\times 10^6}=\frac{2.95}{3.536}\cdot\frac{10^8}{10^6}$
$\approx 0.834276\times 10^2$
≈ 83

The population density of the U.S. in 2004 was approximately 83 people per square mile.

99. $\frac{1.276\times 10^8}{3.3\times 10^7}=\frac{1.276}{3.3}\cdot\frac{10^8}{10^7}$
$\approx 0.3866\times 10^1$
$=3.866$

In 2004 the population of Japan was approximately 3.9 times greater than the population of Oceania.

Chapter 1 Vocabulary Check

1. A(n) algebraic expression is formed by numbers and variables connected by the operations of addition, subtraction, multiplication, division, raising to powers, and/or taking roots.

2. The opposite of a number a is $-a$.

3. $3(x - 6) = 3x - 18$ by the distributive property.

4. The absolute value of a number is the distance between that number and 0 on the number line.

5. A(n) exponent is a shorthand notation for repeated multiplication of the same factor.

6. A letter that represents a number is called a variable.

7. The symbols < and > are called inequality symbols.

8. If a is not 0, then a and $\frac{1}{a}$ are called reciprocals.

9. $A + B = B + A$ by the commutative property of addition.

10. $(A + B) + C = A + (B + C)$ by the associative property of addition.

11. The numbers 0, 1, 2, 3, ... are called whole numbers.

12. If a number corresponds to a point on the number line, we know that number is a real number.

Chapter 1 Review

1. The quotient of a number and seven is written as $\frac{x}{7}$.

2. The product of a number and seven is written as $7x$.

3. Four times the sum of a number and ten is written as $4(x + 10)$.

4. The difference of three times a number and nine is written as $3x - 9$.

5. $\{x|x$ is an odd integer between -2 and $4\}$ $= \{-1, 1, 3\}$

6. $\{x|x$ is an even integer between -3 and $7\}$ $= \{-2, 0, 2, 4, 6\}$

7. $\{x|x$ is a negative whole number$\} = \{\ \}$

8. $\{x|x$ is a natural number that is not a rational number$\} = \{\ \}$

9. $\{x|x$ is a whole number greater than $5\}$ $= \{6, 7, 8, ...\}$

10. $\{x|x$ is an integer less than $3\}$ $= \{..., -1, 0, 1, 2\}$

11. Since 10 is an element of D, $10 \in D$ is true.

12. Since 59 is not an element of B, $59 \in B$ is false.

13. Since $\sqrt{169} = 13$ is not an element of A, $\sqrt{169} \notin A$ is true.

14. Since 0 is not an element of F, $0 \notin F$ is true.

15. Since π is not an element of E, $\pi \in E$ is false.

16. Since π is an element of H, $\pi \in H$ is true.

17. Since $\sqrt{4} = 2$ is not an element of G, $\sqrt{4} \in G$ is false.

18. Since -9 is an element of C, $-9 \in C$ is true.

19. The whole numbers in the set are $\left\{5, \frac{8}{2}, \sqrt{9}\right\}$.

20. The natural numbers in the set are $\left\{5, \frac{8}{2}, \sqrt{9}\right\}$.

21. The rational numbers in the set are $\left\{5, -\frac{2}{3}, \frac{8}{2}, \sqrt{9}, 0.3, 1\frac{5}{8}, -1\right\}$.

22. The irrational numbers in the set are $\left\{\sqrt{7}, \pi\right\}$.

23. The real numbers in the set are $\left\{5, -\frac{2}{3}, \frac{8}{2}, \sqrt{9}, 0.3, \sqrt{7}, 1\frac{5}{8}, -1, \pi\right\}$.

24. The integers in the set are $\left\{5, \frac{8}{2}, \sqrt{9}, -1\right\}$.

25. Twelve is the product of x and negative 4 is written as $12 = -4x$.

26. The sum of n and twice n is negative fifteen is written as $n + 2n = -15$.

27. Four times the sum of y and three is -1 is written as $4(y + 3) = -1$.

28. The difference of t and five, multiplied by six is four is written as $6(t - 5) = 4$.

29. Seven subtracted from z is six is written as $z - 7 = 6$.

30. Ten less than the product of x and nine is five is written as $9x - 10 = 5$.

31. The difference of x and 5 is the same as 12 is written as $x - 5 = 12$.

32. The opposite of four is equal to the product of y and seven is written as $-4 = 7y$.

33. Two-thirds is equal to twice the sum of n and one-fourth is written as $\frac{2}{3} = 2\left(n + \frac{1}{4}\right)$.

34. The sum of t and six amounts to negative twelve is written as $t + 6 = -12$.

35. The opposite of $-\frac{3}{4}$ is $\frac{3}{4}$.

36. The opposite of 0.6 is -0.6.

37. The opposite of 0 is 0.

38. The opposite of -1 is 1.

39. The reciprocal of $-\frac{3}{4}$ is $-\frac{4}{3}$.

40. The reciprocal of $\frac{1}{5}$ is 5.

41. The reciprocal of 0 is undefined.

42. The reciprocal of -1 is -1.

43. $(M + 5) + P = M + (5 + P)$ by the associative property of addition.

44. $5(3x - 4) = 15x - 20$ by the distributive property.

45. $(-4) + 4 = 0$ by the additive inverse property.

46. $(3 + x) + 7 = 7 + (3 + x)$ by the commutative property of addition.

47. $(XY)Z = X(YZ)$ by the associative property of multiplication.

48. $\left(-\frac{3}{5}\right) \cdot \left(-\frac{5}{3}\right) = 1$ by the multiplicative inverse property.

49. $T \cdot 1 = T$ by the multiplicative identity property.

50. $(x + y) + z = (y + x) + z$ by the commutative property of addition.

51. $A + 0 = A$ by the additive identity property.

52. $8 \cdot 1 = 8$ by the multiplicative identity property.

53. $5(x - 3z) = 5x - 15z$ by the distributive property.

54. $(7 + y) + (3 + x) = (3 + x) + (7 + y)$ by the commutative property.

55. $0 = 2 + (-2)$, for example, by the additive inverse property.

56. $1 = 2 \cdot \frac{1}{2}$, for example, by the multiplicative inverse property.

57. $[(3.4)(0.7)]5 = (3.4)[(0.7)5]$ by the associative property.

58. $7 = 7 + 0$ by the additive identity property.

59. $-9 > -12$, since -9 is to the right of -12.

60. $0 > -6$, since 0 is to the right of -6.

61. $-3 < -1$, since -3 is to the left of -1.

62. $7 = |-7|$, since $|-7| = 7$ and $7 = 7$.

63. $-5 < -(-5)$, since $-(-5) = 5$ and -5 is to the left of 5.

64. $-(-2) > -2$, since $-(-2) = 2$ and 2 is to the right of -2.

65. $$\begin{aligned}-3(2x-7y) &= -3\cdot 2x-(-3)\cdot 7y\\ &= -6x-(-21y)\\ &= -6x+21y\end{aligned}$$

66. $$\begin{aligned}-9(10a+4b) &= -9\cdot 10a+(-9)\cdot 4b\\ &= -90a+(-36b)\\ &= -90a-36b\end{aligned}$$

67. $$\begin{aligned}\frac{1}{2}(18m-8n+1) &= \frac{1}{2}\cdot 18m-\frac{1}{2}\cdot 8n+\frac{1}{2}\cdot 1\\ &= 9m-4n+\frac{1}{2}\end{aligned}$$

68. $$\begin{aligned}\frac{1}{3}(12x-33y+2) &= \frac{1}{3}\cdot 12x-\frac{1}{3}\cdot 33y+\frac{1}{3}\cdot 2\\ &= 4x-11y+\frac{2}{3}\end{aligned}$$

69. $-7 + 3 = -4$

70. $-10 + (-25) = -35$

71. $5(-0.4) = -2$

72. $(-3.1)(-0.1) = 0.31$

73. $-7 - (-15) = -7 + 15 = 8$

74. $9 - (-4.3) = 9 + 4.3 = 13.3$

75. $\sqrt{16}-2^3 = 4-8 = 4+(-8) = -4$

76. $\sqrt[3]{27}-5^2 = 3-25 = 3+(-25) = -22$

77. $-24 \div 0$ is undefined.

78. $0 \div (-45) = 0$

79. $(-36) \div (-9) = 4$

80. $60 \div (-12) = -5$

81. $$\begin{aligned}-\frac{4}{5}-\left(-\frac{2}{3}\right) &= -\frac{4}{5}+\frac{2}{3}\\ &= -\frac{4}{5}\cdot\frac{3}{3}+\frac{2}{3}\cdot\frac{5}{5}\\ &= -\frac{12}{15}+\frac{10}{15}\\ &= -\frac{2}{15}\end{aligned}$$

82. $\frac{5}{4}-\left(-2\frac{3}{4}\right)=\frac{5}{4}+2\frac{3}{4}=\frac{5}{4}+\frac{11}{4}=\frac{16}{4}=4$

83. $1-\frac{1}{4}-\frac{1}{3}=\frac{12}{12}-\frac{3}{12}-\frac{4}{12}=\frac{5}{12}$

84. 31,441 – 1589 = 29,852 feet below sea level

85.
$$\begin{aligned}-5+7-3-(-10)&=2-3-(-10)\\&=-1-(-10)\\&=-1+10\\&=9\end{aligned}$$

86.
$$\begin{aligned}8-(-3)+(-4)+6&=8+3+(-4)+6\\&=11+(-4)+6\\&=7+6\\&=13\end{aligned}$$

87. $3(4-5)^4=3(-1)^4=3(1)=3$

88. $6(7-10)^2=6(-3)^2=6(9)=54$

89. $\left(-\frac{8}{15}\right)\cdot\left(-\frac{2}{3}\right)^2=\left(-\frac{8}{15}\right)\left(\frac{4}{9}\right)=-\frac{32}{135}$

90.
$$\begin{aligned}\left(-\frac{3}{4}\right)^2\cdot\left(-\frac{10}{21}\right)&=\left(\frac{9}{16}\right)\left(-\frac{10}{21}\right)\\&=-\frac{90}{336}\\&=-\frac{15}{56}\end{aligned}$$

91. $-\frac{6}{15}\div\frac{8}{25}=-\frac{6}{15}\cdot\frac{25}{8}=-\frac{150}{120}=-\frac{5}{4}$

92. $\frac{4}{9}\div\left(-\frac{8}{45}\right)=\frac{4}{9}\cdot\left(-\frac{45}{8}\right)=-\frac{180}{72}=-\frac{5}{2}$

93.
$$\begin{aligned}-\frac{3}{8}+3(2)\div 6&=-\frac{3}{8}+6\div 6\\&=-\frac{3}{8}+1\\&=-\frac{3}{8}+\frac{8}{8}\\&=\frac{5}{8}\end{aligned}$$

94.
$$\begin{aligned}5(-2)-(-3)-\frac{1}{6}+\frac{2}{3}&=-10-(-3)-\frac{1}{6}+\frac{2}{3}\\&=-10+3-\frac{1}{6}+\frac{2}{3}\\&=-7-\frac{1}{6}+\frac{2}{3}\\&=-\frac{42}{6}-\frac{1}{6}+\frac{4}{6}\\&=-\frac{39}{6}\\&=-6\frac{1}{2}\end{aligned}$$

95.
$$\begin{aligned}\left|2^3-3^2\right|-|5-7|&=|8-9|-|5-7|\\&=|-1|-|-2|\\&=1-2\\&=-1\end{aligned}$$

96.
$$\begin{aligned}\left|5^2-2^2\right|+|9\div(-3)|&=|25-4|+|9\div(-3)|\\&=|21|+|-3|\\&=21+3\\&=24\end{aligned}$$

97.
$$\begin{aligned}(2^3-3^2)-(5-7)&=(8-9)-(5-7)\\&=-1-(-2)\\&=-1+2\\&=1\end{aligned}$$

98.
$$\begin{aligned}(5^2-2^2)+[9\div(-3)]&=(25-4)-[9\div(-3)]\\&=21+[-3]\\&=18\end{aligned}$$

99.
$$\begin{aligned}\frac{(8-10)^3-(-4)^2}{2+8(2)\div 4}&=\frac{(-2)^3-16}{2+16\div 4}\\&=\frac{-8-16}{2+4}\\&=\frac{-24}{6}\\&=-4\end{aligned}$$

100. $\dfrac{(2+4)^2+(-1)^5}{12\div 2\cdot 3-3}=\dfrac{6^2+(-1)}{6\cdot 3-3}$
$=\dfrac{36+(-1)}{18-3}$
$=\dfrac{35}{15}$
$=\dfrac{7}{3}$

101. $\dfrac{(4-9)+4-9}{10-12\div 4\cdot 8}=\dfrac{-5+4-9}{10-3\cdot 8}$
$=\dfrac{-1-9}{10-24}$
$=\dfrac{-10}{-14}$
$=\dfrac{5}{7}$

102. $\dfrac{3-7-(7-3)}{15+30\div 6\cdot 2}=\dfrac{3-7-4}{15+5\cdot 2}=\dfrac{-4-4}{15+10}=-\dfrac{8}{25}$

103. $\dfrac{\sqrt{25}}{4+3\cdot 7}=\dfrac{5}{4+21}=\dfrac{5}{25}=\dfrac{1}{5}$

104. $\dfrac{\sqrt{64}}{24-8\cdot 2}=\dfrac{8}{24-16}=\dfrac{8}{8}=1$

105.

Radius, *r*	**Circumference, $2\pi r$**
1	2(3.14)(1) = 6.28
10	2(3.14)(10) = 62.8
100	2(3.14)(100) = 628

106. As the radius of the circle increases, the circumference increases.

107. $14x-3-11x-10=14x-11x-3-10$
$=3x-13$

108. $81y+19-y-20=81y-y+19-20$
$=80y-1$

109. $7a-3(2a-y)+4y-6$
$=7a-6a+3y+4y-6$
$=a+7y-6$

110. $9b-4(8b-x)+9x-10$
$=9b-32b+4x+9x-10$
$=-23b+13x-10$

111. $\dfrac{1}{5}(15m-5n)-(3m-5)+2n$
$=3m-n-3m+5+2n$
$=3m-3m-n+2n+5$
$=n+5$

112. $-(2x-9)+\dfrac{1}{4}(8x+4y)-13$
$=-2x+9+2x+y-13$
$=-2x+2x+y+9-13$
$=y-4$

113. $(-2)^2=(-2)(-2)=4$

114. $(-3)^4=(-3)(-3)(-3)(-3)=81$

115. $-2^2=-1(2\cdot 2)=-4$

116. $-3^4=-1(3\cdot 3\cdot 3\cdot 3)=-81$

117. $8^0=1$

118. $-9^0=-(9^0)=-1$

119. $-4^{-2}=-(4^{-2})=-\left(\dfrac{1}{4^2}\right)=-\left(\dfrac{1}{16}\right)=-\dfrac{1}{16}$

120. $(-4)^{-2}=\dfrac{1}{(-4)^2}=\dfrac{1}{16}$

121. $-xy^2\cdot y^3\cdot xy^2z=-1x^1x^1y^2y^3y^2z$
$=-1\cdot x^{1+1}\cdot y^{2+3+2}\cdot z$
$=-x^2y^7z$

122. $(-4xy)(-3xy^2b) = (-4)(-3)x^1x^1y^1y^2b$
$= 12x^{1+1}y^{1+2}b$
$= 12x^2y^3b$

123. $a^{-14}\cdot a^5 = a^{-14+5} = a^{-9} = \frac{1}{a^9}$

124. $\frac{a^{16}}{a^{17}} = a^{16-17} = a^{-1} = \frac{1}{a^1} = \frac{1}{a}$

125. $\frac{x^{-7}}{x^4} = x^{-7-4} = x^{-11} = \frac{1}{x^{11}}$

126. $\frac{9a(a^{-3})}{18a^{15}} = \frac{9}{18}\cdot\frac{a^1\cdot a^{-3}}{a^{15}}$
$= \frac{1}{2}\cdot\frac{a^{-2}}{a^{15}}$
$= \frac{1}{2}a^{-17}$
$= \frac{1}{2a^{17}}$

127. $\frac{y^{6p-3}}{y^{6p+2}} = y^{6p-3-(6p+2)} = y^{-5} = \frac{1}{y^5}$

128. $(3x^{2a+b}y^{-3b})^2 = 3^2(x^{2a+b})^2(y^{-3b})^2$
$= 9x^{4a+2b}y^{-6b}$
$= \frac{9x^{4a+2b}}{y^{6b}}$

129. $36,890,000 = 3.689\times10^7$

130. $0.000362 = 3.62\times10^{-4}$

131. $1.678\times10^{-6} = 0.000001678$

132. $4.1\times10^5 = 410,000$

133. $(8^5)^3 = 8^{5\times3} = 8^{15}$

134. $\left(\frac{a}{4}\right)^2 = \frac{a^2}{4^2} = \frac{a^2}{16}$

135. $(3x)^3 = 3^3x^3 = 27x^3$

136. $(-4x)^{-2} = \frac{1}{(-4x)^2} = \frac{1}{(-1)^2(4)^2x^2} = \frac{1}{16x^2}$

137. $\left(\frac{6x}{5}\right)^2 = \frac{(6x)^2}{5^2} = \frac{6^2x^2}{5^2} = \frac{36x^2}{25}$

138. $(8^6)^{-3} = 8^{6\cdot-3} = 8^{-18} = \frac{1}{8^{18}}$

139. $\left(\frac{4}{3}\right)^{-2} = \frac{4^{-2}}{3^{-2}} = \frac{3^2}{4^2} = \frac{9}{16}$

140. $(-2x^3)^{-3} = (-1\cdot2x^3)^{-3}$
$= (-1)^{-3}2^{-3}(x^3)^{-3}$
$= \frac{1}{(-1)^3}\cdot\frac{1}{2^3}\cdot x^{-9}$
$= -1\cdot\frac{1}{8}\cdot x^{-9}$
$= -\frac{1}{8x^9}$

141. $\left(\frac{8p^6}{4p^4}\right)^{-2} = \left(\frac{8}{4}\cdot p^{6-4}\right)^{-2}$
$= (2p^2)^{-2}$
$= 2^{-2}(p^2)^{-2}$
$= \frac{1}{2^2}p^{-4}$
$= \frac{1}{2^2p^4}$
$= \frac{1}{4p^4}$

142. $(-3x^{-2}y^2)^3 = (-1\cdot 3x^{-2}y^2)^3$
$= (-1)^3(3)^3(x^{-2})^3(y^2)^3$
$= -1\cdot 27x^{-6}y^6$
$= -\frac{27y^6}{x^6}$

143. $\left(\frac{x^{-5}y^{-3}}{z^3}\right)^{-5} = \frac{(x^{-5})^{-5}(y^{-3})^{-5}}{(z^3)^{-5}}$
$= \frac{x^{25}y^{15}}{z^{-15}}$
$= x^{25}y^{15}z^{15}$

144. $\frac{4^{-1}x^3yz}{x^{-2}yx^4} = \frac{4^{-1}x^3yz}{x^2y}$
$= 4^{-1}x^{3-2}y^{1-1}z$
$= 4^{-1}x^1y^0z$
$= \frac{xz}{4}$

145. $(5xyz)^{-4}(x^{-2})^{-3} = 5^{-4}x^{-4}y^{-4}z^{-4}x^6$
$= 5^{-4}x^2y^{-4}z^{-4}$
$= \frac{x^2}{5^4y^4z^4}$
$= \frac{x^2}{625y^4z^4}$

146. $\frac{2(3yz)^{-3}}{y^{-3}} = \frac{2\cdot 3^{-3}y^{-3}z^{-3}}{y^{-3}}$
$= 2\cdot 3^{-3}\cdot y^{-3-(-3)}\cdot z^{-3}$
$= 2\cdot 3^{-3}\cdot y^0\cdot z^{-3}$
$= \frac{2}{3^3z^3}$
$= \frac{2}{27z^3}$

147. $x^{4a}(3x^{5a})^3 = x^{4a}\cdot 3^3(x^{5a})^3$
$= x^{4a}\cdot 3^3\cdot x^{15a}$
$= x^{4a+15a}\cdot 3^3$
$= 27x^{19a}$

148. $\frac{4y^{3x-3}}{2y^{2x+4}} = \frac{4}{2}\cdot y^{3x-3-(2x+4)} = 2y^{x-7}$

	Number	Opposite of Number	Reciprocal of Number
149.	$-\frac{3}{4}$	$\frac{3}{4}$	$-\frac{4}{3}$
150.	5	-5	$\frac{1}{5}$

151. $-2\left(5x+\frac{1}{2}\right)+7.1 = -2\cdot 5x+(-2)\cdot\frac{1}{2}+7.1$
$= -10x+(-1)+7.1$
$= -10x+6.1$

152. $\sqrt{36}\div 2\cdot 3 = 6\div 2\cdot 3 = 3\cdot 3 = 9$

153. $-\frac{7}{11}-\left(-\frac{1}{11}\right) = -\frac{7}{11}+\frac{1}{11} = -\frac{6}{11}$

154. $10-(-1)+(-2)+6 = 10+1+(-2)+6$
$= 11+(-2)+6$
$= 9+6$
$= 15$

155. $\left(-\frac{2}{3}\right)^3\div\frac{10}{9} = (-1)^3\cdot\frac{2^3}{3^3}\div\frac{10}{9}$
$= -\frac{8}{27}\div\frac{10}{9}$
$= -\frac{8}{27}\cdot\frac{9}{10}$
$= -\frac{72}{270}$
$= -\frac{4}{15}$

156. $$\frac{(3-5)^2+(-1)^3}{1+2(3-(-1))^2} = \frac{(-2)^2+(-1)}{1+2(3+1)^2} = \frac{4+(-1)}{1+2(4)^2} = \frac{3}{1+2\cdot 16} = \frac{3}{1+32} = \frac{3}{33} = \frac{1}{11}$$

157. $$\frac{1}{3}(9x-3y)-(4x-1)+4y = 3x-y-4x+1+4y = 3x-4x-y+4y+1 = -x+3y+1$$

158. $$-5^{-2} = -1\cdot 5^{-2} = -1\cdot\frac{1}{5^2} = -\frac{1}{25}$$

159. $$\left(\frac{5x^7}{10x^{-3}}\right)^{-3} = \left(\frac{1}{2}x^{10}\right)^{-3} = \left(\frac{1}{2}\right)^{-3}(x^{10})^{-3} = \frac{1^{-3}}{2^{-3}}x^{-30} = \frac{2^3}{1^3}x^{-30} = \frac{8}{x^{30}}$$

160. $$\begin{aligned}(-5a^{-2}bc)^{-2}\cdot(3a^{-3}b^2c^3)^2 &= (-1\cdot 5a^{-2}bc)^{-2}(3a^{-3}b^2c^3)^2\\ &= (-1)^{-2}5^{-2}(a^{-2})^{-2}b^{-2}c^{-2}\cdot 3^2(a^{-3})^2(b^2)^2(c^3)^2\\ &= \frac{1}{(-1)^2}\cdot\frac{1}{5^2}a^4b^{-2}c^{-2}\cdot 3^2a^{-6}b^4c^6\\ &= \frac{3^2}{5^2}a^{4+(-6)}b^{-2+4}c^{-2+6}\\ &= \frac{9}{25}a^{-2}b^2c^4\\ &= \frac{9b^2c^4}{25a^2}\end{aligned}$$

Chapter 1 Test

1. Since –2.3 is to the left of 2.33, $-2.3 < 2.33$, so the statement is false.

2. Since $-6^2 = -36$ and $(-6)^2 = 36$, $-6^2 \neq (-6)^2$ so the statement is false.

3. Since $(-2)(-3)(0) = 0$ and $\frac{(-4)}{0}$ is undefined, $(-2)(-3)(0) \neq \frac{(-4)}{0}$, so the statement is false.

4. $\{x|x$ is a whole number less than $2\} = \{0, 1\}$

5. Since the set of natural numbers is contained in the set of integers, the statement is true.

6. Since not all rational numbers are integers, the statement is false.

7. $5 - 12 \div 3(2) = 5 - 4(2) = 5 - 8 = -3$

8. $$\begin{aligned}(4-9)^3 - |-4-6|^2 &= (-5)^3 - |-10|^2\\ &= -125 - 10^2\\ &= -125 - 100\\ &= -225\end{aligned}$$

9. $$\begin{aligned}[3|4-5|^5 - (-9)] \div (-6) &= [3|-1|^5 - (-9)] \div (-6)\\ &= [3(1)^5 - (-9)] \div (-6)\\ &= [3\cdot 1 - (-9)] \div (-6)\\ &= [3 - (-9)] \div (-6)\\ &= [3 + 9] \div (-6)\\ &= [12] \div (-6)\\ &= -2\end{aligned}$$

10. $$\frac{6(7-9)^3+(-2)}{(-2)(-5)(-5)} = \frac{6(-2)^3+(-2)}{(10)(-5)} = \frac{6(-8)+(-2)}{-50} = \frac{-48+(-2)}{-50} = \frac{-50}{-50} = 1$$

11. $$\frac{\left(4-\sqrt{16}\right)-(-7-20)}{-2(1-4)^2} = \frac{(4-4)-(-27)}{-2(-3)^2} = \frac{0+27}{-2(9)} = \frac{27}{-18} = -\frac{3}{2}$$

12. Replace q with 4 and r with -2.
$$q^2 - r^2 = 4^2 - (-2)^2 = 16 - 4 = 12$$

13. Replace q with 4, r with -2, and t with 1.
$$\frac{5t-3q}{3r-1} = \frac{5(1)-3(4)}{3(-2)-1} = \frac{5-12}{-6-1} = \frac{-7}{-7} = 1$$

14. a.

Adults, x	Total Cost, $5.75x$
1	$5.75(1) = 5.75$
3	$5.75(3) = 17.25$
10	$5.75(10) = 57.50$
20	$5.75(20) = 115.00$

b. As the number of adults increases, the total cost increases.

15. Three times the quotient of n and five is the opposite of n is written as $3\left(\frac{n}{5}\right) = -n.$

16. Twenty is equal to six subtracted from twice x is written as $20 = 2x - 6$.

17. Negative two is equal to x divided by the sum of x and 5 is written as $-2 = \frac{x}{x+5}$.

18. $6(x - 4) = 6x - 24$ by the distributive property.

19. $(4 + x) + z = 4 + (x + z)$ by the associative property of addition.

20. $(-7) + 7 = 0$ by the additive inverse property.

21. The reciprocal of $-\frac{7}{11}$ is $-\frac{11}{7}$; the opposite of $-\frac{7}{11}$ is $\frac{7}{11}$.

22. $$9x+12y-3x-6.2+20 = 9x-3x+12y-6.2+20 = 6x+12y+13.8$$

23. $$\frac{1}{3}(15x-27y)-2(3x-y)-4y = 5x-9y-6x+2y-4y = 5x-6x-9y+2y-4y = -x-11y$$

24. $$(-9x)^{-2} = (-1\cdot 9x)^{-2} = (-1)^{-2}(9)^{-2}x^{-2} = \frac{1}{(-1)^2(9)^2x^2} = \frac{1}{81x^2}$$

25. $$\frac{6^{-1}a^2b^{-3}}{3^{-2}a^{-5}b^2} = \frac{3^2}{6^1}a^{2-(-5)}b^{-3-2} = \frac{9}{6}a^7b^{-5} = \frac{3a^7}{2b^5}$$

26. $\left(\dfrac{-xy^{-5}z}{xy^3}\right)^{-5} = (-1x^{1-1}y^{-5-3}z^1)^{-5}$

$= (-1x^0y^{-8}z^1)^{-5}$

$= (-1y^{-8}z^1)^{-5}$

$= (-1)^{-5}y^{40}z^{-5}$

$= \dfrac{1}{(-1)^5} \cdot y^{40}z^{-5}$

$= -1y^{40}z^{-5}$

$= -\dfrac{y^{40}}{z^5}$

27. $(-6a^{-5}b^{12})^{-2}(3ab^5)^2$

$= (-1 \cdot 6a^{-5}b^{12})^{-2}(3ab^5)^2$

$= (-1)^{-2}6^{-2}a^{10}b^{-24}3^2a^2b^{10}$

$= \dfrac{1}{(-1)^2} \cdot \dfrac{3^2}{6^2} \cdot a^{10+2}b^{-24+10}$

$= \dfrac{9}{36}a^{12}b^{-14}$

$= \dfrac{a^{12}}{4b^{14}}$

28. $\dfrac{(x^w)^2}{(x^{4w})^{-2}} = \dfrac{x^{2w}}{x^{-8w}} = x^{2w-(-8w)} = x^{10w}$

29. $630{,}000{,}000 = 6.3 \times 10^8$

30. $0.01200 = 1.2 \times 10^{-2}$

31. $5.0 \times 10^{-6} = 0.000005$

$\dfrac{(0.00012)(144{,}000)}{0.0003}$

$= \dfrac{(1.2 \times 10^{-4})(1.44 \times 10^5)}{3 \times 10^{-4}}$

$= \dfrac{1.728 \times 10^1}{3 \times 10^{-4}}$

$= \dfrac{1.728}{3} \cdot \dfrac{10^1}{10^{-4}}$

$= 0.576 \times 10^5$

32. $= (5.76 \times 10^{-1}) \times 10^5$

$= 5.76 \times 10^4$

33. $\dfrac{(0.0024)(0.00012)}{0.00032}$

$= \dfrac{(2.4 \times 10^{-3})(1.2 \times 10^{-4})}{3.2 \times 10^{-4}}$

$= \dfrac{(2.4)(1.2)}{3.2} \cdot \dfrac{(10^{-3})(10^{-4})}{10^{-4}}$

$= 0.9 \times 10^{-3}$

$= (9 \times 10^{-1}) \times 10^{-3}$

$= 9 \times 10^{-4}$

34. $848{,}600{,}000 = 8.486 \times 10^8$

Chapter 2

Section 2.1 Practice Problems

1.
$$\begin{aligned}14-x&=21\\14-(-7)&\stackrel{?}{=}21\\14+7&\stackrel{?}{=}21\\21&=21\quad\text{True}\end{aligned}$$
Since a true statement results, −7 is a solution.

2.
$$\begin{aligned}x-10&=2x-14\\(8)-10&\stackrel{?}{=}2(8)-14\\-2&\stackrel{?}{=}16-14\\-2&=2\quad\text{False}\end{aligned}$$
Since a false statement results, 8 is not a solution.

3.
$$\begin{aligned}3x+6&=21\\3x+6-6&=21-6\\3x&=15\\\frac{3x}{3}&=\frac{15}{3}\\x&=5\end{aligned}$$
The solution set is $\{5\}$.

4.
$$\begin{aligned}4.5&=3+2.5x\\4.5-3&=3+2.5x-3\\1.5&=2.5x\\\frac{1.5}{2.5}&=\frac{2.5x}{2.5}\\0.6&=x\end{aligned}$$
The solution set is $\{0.6\}$.

5.
$$\begin{aligned}-2x+2-3x&=8x+20-7x\\-5x+2&=1x+20\\-5x+2-1x&=1x+20-1x\\-6x+2&=20\\-6x+2-2&=20-2\\-6x&=18\\\frac{-6x}{-6}&=\frac{18}{-6}\\x&=-3\end{aligned}$$
The solution set is $\{-3\}$.

6.
$$\begin{aligned}4(x-2)&=6x-10\\4x-8&=6x-10\\4x-8-4x&=6x-10-4x\\-8&=2x-10\\-8+10&=2x-10+10\\2&=2x\\\frac{2}{2}&=\frac{2x}{2}\\1&=x\end{aligned}$$
The solution set is $\{1\}$.

7.
$$\begin{aligned}\frac{x}{6}-\frac{x}{8}&=\frac{1}{8}\\24\left(\frac{x}{6}-\frac{x}{8}\right)&=24\left(\frac{1}{8}\right)\\24\left(\frac{x}{6}\right)-24\left(\frac{x}{8}\right)&=3\\4x-3x&=3\\x&=3\end{aligned}$$
The solution set is $\{3\}$.

8.
$$\begin{aligned}\frac{x-1}{3}+\frac{2}{3}&=x-\frac{2x+3}{9}\\9\left(\frac{x-1}{3}+\frac{2}{3}\right)&=9\left(x-\frac{2x+3}{9}\right)\\9\left(\frac{x-1}{3}\right)+9\left(\frac{2}{3}\right)&=9x-9\left(\frac{2x+3}{9}\right)\\3(x-1)+3(2)&=9x-(2x+3)\\3x-3+6&=9x-2x-3\\3x+3&=7x-3\\3x+3-3x&=7x-3-3x\\3&=4x-3\\3+3&=4x-3+3\\6&=4x\\\frac{6}{4}&=\frac{4x}{4}\\\frac{3}{2}&=x\end{aligned}$$
The solution set is $\left\{\frac{3}{2}\right\}$.

9.
$$\begin{aligned} 0.2x+0.1 &= 0.12x-0.06 \\ 100(0.2x+0.1) &= 100(0.12x-0.06) \\ 100(0.2x)+100(0.1) &= 100(0.12x)-100(0.06) \\ 20x+10 &= 12x-6 \\ 20x+10-12x &= 12x-6-12x \\ 8x+10 &= -6 \\ 8x+10-10 &= -6-10 \\ 8x &= -16 \\ \frac{8x}{8} &= \frac{-16}{8} \\ x &= -2 \end{aligned}$$

The solution set is $\{-2\}$.

10.
$$\begin{aligned} 5x-1 &= 5(x+3) \\ 5x-1 &= 5x+15 \\ 5x-1-5x &= 5x+15-5x \\ -1 &= 15 \quad \text{False} \end{aligned}$$

The solution set is $\varnothing$ or $\{\ \}$.

11.
$$\begin{aligned} -4(x-1) &= -4x-9+13 \\ -4x+4 &= -4x+4 \\ -4x+4+4x &= -4x+4+4x \\ 4 &= 4 \\ 4-4 &= 4-4 \\ 0 &= 0 \quad \text{True} \end{aligned}$$

The solution set is $\{x|x \text{ is a real number}\}$.

Mental Math

1.
$$\begin{aligned} 3x &= 18 \\ \frac{3x}{3} &= \frac{18}{3} \\ x &= 6 \end{aligned}$$

$\{6\}$

2.
$$\begin{aligned} 2x &= 60 \\ \frac{2x}{2} &= \frac{60}{2} \\ x &= 30 \end{aligned}$$

$\{30\}$

3.
$$\begin{aligned} x-7 &= 10 \\ x-7+7 &= 10+7 \\ x &= 17 \end{aligned}$$

$\{17\}$

4. $x-2=15$
$x-2+2=15+2$
$x=17$
$\{17\}$

5. $\frac{x}{2}=4$
$2\cdot\frac{x}{2}=4\cdot 2$
$x=8$
$\{8\}$

6. $\frac{x}{3}=5$
$3\cdot\frac{x}{3}=5\cdot 3$
$x=15$
$\{15\}$

7. $x+1=11$
$x+1-1=11-1$
$x=10$
$\{10\}$

8. $x+4=20$
$x+4-4=20-4$
$x=16$
$\{16\}$

Exercise Set 2.1

1. $\frac{x}{-6}=4$
$\frac{-24}{-6}\stackrel{?}{=}4$
$4=4$ True
Yes, –24 is a solution.

3. $x-17=20$
$-3-17\stackrel{?}{=}20$
$-20=20$ False
No, –3 is not a solution.

5. $5+3x=-1$
$5+3(-2)\stackrel{?}{=}-1$
$-1=-1$ True
Yes, –2 is a solution.

7. $x-7=x+2$
$5-7\stackrel{?}{=}5+2$
$-2=7$ False
No, 5 is not a solution.

9. $4(x-3)=12$
$4(5-3)\stackrel{?}{=}12$
$8=12$ False
No, 5 is not a solution.

11. $4x-2=5x+6$
$4(-8)-2\stackrel{?}{=}5(-8)+6$
$-32-2\stackrel{?}{=}-40+6$
$-34=-34$ True
Yes, –8 is a solution.

13. $-5x=-30$
$\frac{-5x}{-5}=\frac{-30}{-5}$
$x=6$
Check: $-5x=-30$
$-5(6)\stackrel{?}{=}-30$
$-30=-30$
The solution set is $\{6\}$.

15. $-10=x+12$
$-10-12=x+12-12$
$-22=x$
Check: $-10=x+12$
$-10\stackrel{?}{=}-22+12$
$-10=-10$ True
The solution set is $\{-22\}$.

17. $x-2.8=1.9$
$x-2.8+2.8=1.9+2.8$
$x=4.7$
Check: $x-2.8=1.9$
$4.7-2.8\stackrel{?}{=}1.9$
$1.9=1.9$ True
The solution set is $\{4.7\}$.

19.
$$\begin{aligned}5x-4&=26+2x\\5x-4-2x&=26+2x-2x\\3x-4&=26\\3x-4+4&=26+4\\3x&=30\\\frac{3x}{3}&=\frac{30}{3}\\x&=10\end{aligned}$$

Check:
$$\begin{aligned}5x-4&=26+2x\\5(10)-4&\stackrel{?}{=}26+2(10)\\50-4&\stackrel{?}{=}26+20\\46&=46 \quad \text{True}\end{aligned}$$

The solution set is {10}.

21.
$$\begin{aligned}-4.1-7z&=3.6\\-4.1-7z+4.1&=3.6+4.1\\-7z&=7.7\\\frac{-7z}{-7}&=\frac{7.7}{-7}\\z&=-1.1\end{aligned}$$

Check:
$$\begin{aligned}-4.1-7z&=3.6\\-4.1-7(-1.1)&\stackrel{?}{=}3.6\\-4.1+7.7&\stackrel{?}{=}3.6\\3.6&=3.6 \quad \text{True}\end{aligned}$$

The solution set is {−1.1}.

23.
$$\begin{aligned}5y+12&=2y-3\\5y+12-2y&=2y-3-2y\\3y+12&=-3\\3y+12-12&=-3-12\\3y&=-15\\\frac{3y}{3}&=\frac{-15}{3}\\y&=-5\end{aligned}$$

Check:
$$\begin{aligned}5y+12&=2y-3\\5(-5)+12&\stackrel{?}{=}2(-5)-3\\-25+12&\stackrel{?}{=}-10-3\\-13&=-13 \quad \text{True}\end{aligned}$$

The solution set is {−5}.

25.
$$\begin{aligned}3x-4-5x&=x+4+x\\-2x-4&=2x+4\\-2x-4+4&=2x+4+4\\-2x&=2x+8\\-2x-2x&=2x+8-2x\\-4x&=8\\\frac{-4x}{-4}&=\frac{8}{-4}\\x&=-2\end{aligned}$$

Check:
$$\begin{aligned}3x-4-5x&=x+4+x\\3(-2)-4-5(-2)&\stackrel{?}{=}-2+4+(-2)\\-6-4+10&\stackrel{?}{=}0\\0&=0 \quad \text{True}\end{aligned}$$

The solution set is {−2}.

27.
$$\begin{aligned}8x-5x+3&=x-7+10\\3x+3&=x+3\\3x+3-3&=x+3-3\\3x&=x\\3x-x&=x-x\\2x&=0\\\frac{2x}{2}&=\frac{0}{2}\\x&=0\end{aligned}$$

Check:
$$\begin{aligned}8x-5x+3&=x-7+10\\8(0)-5(0)+3&\stackrel{?}{=}(0)-7+10\\0-0+3&\stackrel{?}{=}0-7+10\\3&=3 \quad \text{True}\end{aligned}$$

The solution set is {0}.

29.
$$\begin{aligned}5x+12&=2(2x+7)\\5x+12&=4x+14\\5x+12-4x&=4x+14-4x\\x+12&=14\\x+12-12&=14-12\\x&=2\end{aligned}$$

Check:
$$\begin{aligned}5x+12&=2(2x+7)\\5(2)+12&\stackrel{?}{=}2(2\cdot 2+7)\\10+12&\stackrel{?}{=}2(11)\\22&=22 \quad \text{True}\end{aligned}$$

The solution set is {2}.

31.

$$\begin{aligned} 3(x-6) &= 5x \\ 3x-18 &= 5x \\ 3x-18-3x &= 5x-3x \\ -18 &= 2x \\ \frac{-18}{2} &= \frac{2x}{2} \\ -9 &= x \end{aligned}$$

Check:

$$\begin{aligned} 3(x-6) &= 5x \\ 3[(-9)-6] &\stackrel{?}{=} 5(-9) \\ 3(-15) &\stackrel{?}{=} -45 \\ -45 &= -45 \quad \text{True} \end{aligned}$$

The solution set is $\{-9\}$.

33.

$$\begin{aligned} -2(5y-1)-y &= -4(y-3) \\ -10y+2-y &= -4y+12 \\ -11y+2 &= -4y+12 \\ -11y+2+11y &= -4y+12+11y \\ 2 &= 7y+12 \\ 2-12 &= 7y+12-12 \\ -10 &= 7y \\ \frac{-10}{7} &= \frac{7y}{7} \\ -\frac{10}{7} &= y \end{aligned}$$

Check:

$$\begin{aligned} -2(5y-1)-y &= -4(y-3) \\ -2\left[5\left(-\frac{10}{7}\right)-1\right]-\left(-\frac{10}{7}\right) &\stackrel{?}{=} -4\left(-\frac{10}{7}-3\right) \\ -2\left(-\frac{57}{7}\right)+\frac{10}{7} &\stackrel{?}{=} -4\left(-\frac{31}{7}\right) \\ \frac{114}{7}+\frac{10}{7} &\stackrel{?}{=} \frac{124}{7} \\ \frac{124}{7} &= \frac{124}{7} \quad \text{True} \end{aligned}$$

The solution set is $\left\{-\frac{10}{7}\right\}$.

35.

$$\begin{aligned} \frac{x}{2}+\frac{x}{3} &= \frac{3}{4} \\ 12\left(\frac{x}{2}+\frac{x}{3}\right) &= 12\left(\frac{3}{4}\right) \\ 12\left(\frac{x}{2}\right)+12\left(\frac{x}{3}\right) &= 9 \\ 6x+4x &= 9 \\ 10x &= 9 \\ \frac{10x}{10} &= \frac{9}{10} \\ x &= \frac{9}{10} \end{aligned}$$

Check:

$$\begin{aligned} \frac{x}{2}+\frac{x}{3} &= \frac{3}{4} \\ \frac{\left(\frac{9}{10}\right)}{2}+\frac{\left(\frac{9}{10}\right)}{3} &\stackrel{?}{=} \frac{3}{4} \\ \frac{9}{20}+\frac{3}{10} &\stackrel{?}{=} \frac{3}{4} \\ \frac{9}{20}+\frac{6}{20} &\stackrel{?}{=} \frac{3}{4} \\ \frac{15}{20} &\stackrel{?}{=} \frac{3}{4} \\ \frac{3}{4} &= \frac{3}{4} \quad \text{True} \end{aligned}$$

The solution set is $\left\{\frac{9}{10}\right\}$.

37.

$$\begin{aligned} \frac{3t}{4}-\frac{t}{2} &= 1 \\ 4\left(\frac{3t}{4}-\frac{t}{2}\right) &= 4(1) \\ 4\left(\frac{3t}{4}\right)-4\left(\frac{t}{2}\right) &= 4 \\ 3t-2t &= 4 \\ t &= 4 \end{aligned}$$

Check:

$$\begin{aligned} \frac{3t}{4}-\frac{t}{2} &= 1 \\ \frac{3(4)}{4}-\frac{4}{2} &\stackrel{?}{=} 1 \\ 3-2 &\stackrel{?}{=} 1 \\ 1 &= 1 \quad \text{True} \end{aligned}$$

The solution set is $\{4\}$.

39.
$$\frac{n-3}{4}+\frac{n+5}{7}=\frac{5}{14}$$
$$28\left(\frac{n-3}{4}+\frac{n+5}{7}\right)=28\left(\frac{5}{14}\right)$$
$$28\left(\frac{n-3}{4}\right)+28\left(\frac{n+5}{7}\right)=2(5)$$
$$7(n-3)+4(n+5)=10$$
$$7n-21+4n+20=10$$
$$11n-1=10$$
$$11n-1+1=10+1$$
$$11n=11$$
$$\frac{11n}{11}=\frac{11}{11}$$
$$n=1$$

Check:
$$\frac{n-3}{4}+\frac{n+5}{7}=\frac{5}{14}$$
$$\frac{(1)-3}{4}+\frac{(1)+5}{7}\stackrel{?}{=}\frac{5}{14}$$
$$\frac{-2}{4}+\frac{6}{7}\stackrel{?}{=}\frac{5}{14}$$
$$\frac{-14}{28}+\frac{24}{28}\stackrel{?}{=}\frac{5}{14}$$
$$\frac{10}{28}\stackrel{?}{=}\frac{5}{14}$$
$$\frac{5}{14}=\frac{5}{14}\quad\text{True}$$

The solution set is $\{1\}$.

41.
$$0.6x-10=1.4x-14$$
$$10(0.6x-10)=10(1.4x-14)$$
$$10(0.6x)-10(10)=10(1.4x)-10(14)$$
$$6x-100=14x-140$$
$$6x-100-6x=14x-140-6x$$
$$-100=8x-140$$
$$-100+140=8x-140+140$$
$$40=8x$$
$$\frac{40}{8}=\frac{8x}{8}$$
$$5=x$$

Check:
$$0.6x-10=1.4x-14$$
$$0.6(5)-10\stackrel{?}{=}1.4(5)-14$$
$$3-10\stackrel{?}{=}7-14$$
$$-7=-7\quad\text{True}$$

The solution set is $\{5\}$.

43.
$$\frac{3x-1}{9}+x=\frac{3x+1}{3}+4$$
$$9\left(\frac{3x-1}{9}+x\right)=9\left(\frac{3x+1}{3}+4\right)$$
$$9\left(\frac{3x-1}{9}\right)+9(x)=9\left(\frac{3x+1}{3}\right)+9(4)$$
$$3x-1+9x=3(3x+1)+36$$
$$12x-1=9x+3+36$$
$$12x-1=9x+39$$
$$12x-1+1=9x+39+1$$
$$12x=9x+40$$
$$12x-9x=9x+40-9x$$
$$3x=40$$
$$\frac{3x}{3}=\frac{40}{3}$$
$$x=\frac{40}{3}$$

Check:
$$\frac{3x-1}{9}+x=\frac{3x+1}{3}+4$$
$$\frac{3\left(\frac{40}{3}\right)-1}{9}+\left(\frac{40}{3}\right)\stackrel{?}{=}\frac{3\left(\frac{40}{3}\right)+1}{3}+4$$
$$\frac{40-1}{9}+\frac{40}{3}\stackrel{?}{=}\frac{40+1}{3}+4$$
$$\frac{39}{9}+\frac{120}{9}\stackrel{?}{=}\frac{41}{3}+4$$
$$\frac{159}{9}\stackrel{?}{=}\frac{41}{3}+\frac{12}{3}$$
$$\frac{159}{9}\stackrel{?}{=}\frac{53}{3}$$
$$\frac{159}{9}=\frac{159}{9}\quad\text{True}$$

The solution set is $\left\{\frac{40}{3}\right\}$.

45.
$$\begin{aligned}1.5(4-x)&=1.3(2-x)\\10[1.5(4-x)]&=10[1.3(2-x)]\\15(4-x)&=13(2-x)\\60-15x&=26-13x\\60-15x+15x&=26-13x+15x\\60&=26+2x\\60-26&=26+2x-26\\34&=2x\\\frac{34}{2}&=\frac{2x}{2}\\17&=x\end{aligned}$$
Check:
$$\begin{aligned}1.5(4-x)&=1.3(2-x)\\1.5(4-17)&\stackrel{?}{=}1.3(2-17)\\1.5(-13)&\stackrel{?}{=}1.3(-15)\\-19.5&=-19.5\quad\text{True}\end{aligned}$$
The solution set is {17}.

47.
$$\begin{aligned}4(n+3)&=2(6+2n)\\4n+12&=12+4n\\4n+12-12&=12+4n-12\\4n&=4n\\4n-4n&=4n-4n\\0&=0\end{aligned}$$
Since 0 = 0 is a true statement for every value of n, the solution set is the set of all real numbers. The solution set is $\{n|n$ is a real number.$\}$.

49.
$$\begin{aligned}3(x+1)+5&=3x+2\\3x+3+5&=3x+2\\3x+8&=3x+2\\3x+8-3x&=3x+2-3x\\8&=2\quad\text{False}\end{aligned}$$
The equation 8 = 2 is a false statement, so the original equation has no solution. The solution set is ∅.

51.
$$\begin{aligned}2(x-8)+x&=3(x-6)+2\\2x-16+x&=3x-18+2\\3x-16&=3x-16\\3x-16+16&=3x-16+16\\3x&=3x\\3x-3x&=3x-3x\\0&=0\end{aligned}$$
Since 0 = 0 is a true statement for every value of x, the solution set is the set of all real numbers. The solution set is $\{x|x$ is a real number$\}$.

53.
$$\begin{aligned}4(x+5)&=3(x-4)+x\\4x+20&=3x-12+x\\4x+20&=4x-12\\4x+20-4x&=4x-12-4x\\20&=-12\quad\text{False}\end{aligned}$$
The equation 20 = −12 is a false statement, so the original equation has no solution. The solution set is ∅.

55.
$$\begin{aligned}\frac{3}{8}+\frac{b}{3}&=\frac{5}{12}\\24\left(\frac{3}{8}+\frac{b}{3}\right)&=24\left(\frac{5}{12}\right)\\24\left(\frac{3}{8}\right)+24\left(\frac{b}{3}\right)&=24\left(\frac{5}{12}\right)\\9+8b&=10\\9+8b-9&=10-9\\8b&=1\\\frac{8b}{8}&=\frac{1}{8}\\b&=\frac{1}{8}\end{aligned}$$
The solution set is $\left\{\frac{1}{8}\right\}$.

57.
$$\begin{aligned}x-10&=-6x-10\\x-10+6x&=-6x-10+6x\\7x-10&=-10\\7x-10+10&=-10+10\\7x&=0\\\frac{7x}{7}&=\frac{0}{7}\\x&=0\end{aligned}$$
The solution set is {0}.

59. $$\begin{aligned} 5(x-2)+2x &= 7(x+4)-38 \\ 5x-10+2x &= 7x+28-38 \\ 7x-10 &= 7x-10 \\ 7x-10-7x &= 7x-10-7x \\ -10 &= -10 \quad \text{True} \end{aligned}$$

The solution set is $\{x|x \text{ is a real number}\}$.

61. $$\begin{aligned} y+0.2 &= 0.6(y+3) \\ 10(y+0.2) &= 10[0.6(y+3)] \\ 10y+2 &= 6(y+3) \\ 10y+2 &= 6y+18 \\ 10y+2-6y &= 6y+18-6y \\ 4y+2 &= 18 \\ 4y+2-2 &= 18-2 \\ 4y &= 16 \\ \frac{4y}{4} &= \frac{16}{4} \\ y &= 4 \end{aligned}$$

The solution set is $\{4\}$.

63. $$\begin{aligned} \frac{1}{4}(a+2) &= \frac{1}{6}(5-a) \\ 12\left[\frac{1}{4}(a+2)\right] &= 12\left[\frac{1}{6}(5-a)\right] \\ 3(a+2) &= 2(5-a) \\ 3a+6 &= 10-2a \\ 3a+6-6 &= 10-2a-6 \\ 3a &= 4-2a \\ 3a+2a &= 4-2a+2a \\ 5a &= 4 \\ \frac{5a}{5} &= \frac{4}{5} \\ a &= \frac{4}{5} \end{aligned}$$

The solution set is $\left\{\frac{4}{5}\right\}$.

65. $$\begin{aligned} 2y+5(y-4) &= 4y-2(y-10) \\ 2y+5y-20 &= 4y-2y+20 \\ 7y-20 &= 2y+20 \\ 7y-20-2y &= 2y+20-2y \\ 5y-20 &= 20 \\ 5y-20+20 &= 20+20 \\ 5y &= 40 \\ \frac{5y}{5} &= \frac{40}{5} \\ y &= 8 \end{aligned}$$

The solution set is $\{8\}$.

67. $$\begin{aligned} 6x-2(x-3) &= 4(x+1)+4 \\ 6x-2x+6 &= 4x+4+4 \\ 4x+6 &= 4x+8 \\ 4x+6-4x &= 4x+8-4x \\ 6 &= 8 \quad \text{False} \end{aligned}$$

The solution set is $\varnothing$.

69. $$\begin{aligned} \frac{m-4}{3}-\frac{3m-1}{5} &= 1 \\ 15\left(\frac{m-4}{3}-\frac{3m-1}{5}\right) &= 15(1) \\ 15\left(\frac{m-4}{3}\right)-15\left(\frac{3m-1}{5}\right) &= 15 \\ 5(m-4)-3(3m-1) &= 15 \\ 5m-20-9m+3 &= 15 \\ -4m-17 &= 15 \\ -4m-17+17 &= 15+17 \\ -4m &= 32 \\ \frac{-4m}{-4} &= \frac{32}{-4} \\ m &= -8 \end{aligned}$$

The solution set is $\{-8\}$.

71.
$$\begin{aligned}
8x-12-3x&=9x-7\\
5x-12&=9x-7\\
5x-12+12&=9x-7+12\\
5x&=9x+5\\
5x-9x&=9x+5-9x\\
-4x&=5\\
\frac{-4x}{-4}&=\frac{5}{-4}\\
x&=-\frac{5}{4}
\end{aligned}$$

The solution set is $\left\{-\frac{5}{4}\right\}$.

73.
$$\begin{aligned}
-(3x-5)-(2x-6)+1&=-5(x-1)-(3x+2)+3\\
-3x+5-2x+6+1&=-5x+5-3x-2+3\\
-5x+12&=-8x+6\\
-5x+12+8x&=-8x+6+8x\\
3x+12&=6\\
3x+12-12&=6-12\\
3x&=-6\\
\frac{3x}{3}&=\frac{-6}{3}\\
x&=-2
\end{aligned}$$

The solution set is $\{-2\}$.

75.
$$\begin{aligned}
\frac{1}{3}(y+4)+6&=\frac{1}{4}(3y-1)-2\\
12\left[\frac{1}{3}(y+4)+6\right]&=12\left[\frac{1}{4}(3y-1)-2\right]\\
12\left[\frac{1}{3}(y+4)\right]+12(6)&=12\left[\frac{1}{4}(3y-1)\right]-12(2)\\
4(y+4)+72&=3(3y-1)-24\\
4y+16+72&=9y-3-24\\
4y+88&=9y-27\\
4y+88-88&=9y-27-88\\
4y&=9y-115\\
4y-9y&=9y-115-9y\\
-5y&=-115\\
\frac{-5y}{-5}&=\frac{-115}{-5}\\
y&=23
\end{aligned}$$

The solution set is $\{23\}$.

77. $$\begin{aligned} 2[7-5(1-n)]+8n &= -16+3[6(n+1)-3n] \\ 2[7-5+5n]+8n &= -16+3[6n+6-3n] \\ 2[2+5n]+8n &= -16+3[3n+6] \\ 4+10n+8n &= -16+9n+18 \\ 4+18n &= 9n+2 \\ 4+18n-9n &= 9n+2-9n \\ 4+9n &= 2 \\ 4+9n-4 &= 2-4 \\ 9n &= -2 \\ \frac{9n}{9} &= \frac{-2}{9} \\ n &= -\frac{2}{9} \end{aligned}$$

The solution set is $\left\{-\frac{2}{9}\right\}$.

79. The quotient of 8 and a number is written as $\frac{8}{x}$.

81. The product of 8 and a number is written as $8x$.

83. Five subtracted from twice a number is written as $2x - 5$.

85. $$\begin{aligned} 2x+19 &= 13 \\ 2x+19-19 &= 13-19 \\ 2x &= -6 \\ \frac{2x}{2} &= \frac{-6}{2} \\ x &= -3 \end{aligned}$$

87. $$\begin{aligned} 9x+1.6 &= 4x+0.4 \\ 9x+1.6-1.6 &= 4x+0.4-1.6 \\ 9x &= 4x-1.2 \\ 9x-4x &= 4x-1.2-4x \\ 5x &= -1.2 \\ \frac{5x}{5} &= \frac{-1.2}{5} \\ x &= -0.24 \end{aligned}$$

89. **a.** $4(x + 1) + 1 = 4x + 4 + 1 = 4x + 5$

b. $$\begin{aligned} 4(x+1)+1 &= -7 \\ 4x+4+1 &= -7 \\ 4x+5 &= -7 \\ 4x &= -7-5 \\ 4x &= -12 \\ x &= \frac{-12}{4} \\ x &= -3 \end{aligned}$$

The solution set is $\{-3\}$.

c. answers may vary

91. answers may vary

93. $$\begin{aligned} 3.2x+4 &= 5.4x-7 \\ 3.2x+4-4 &= 5.4x-7-4 \\ 3.2x &= 5.4x-11 \end{aligned}$$

The equations are equivalent when $K = -11$.

95. $$\begin{aligned} \frac{7}{11}x+9 &= \frac{3}{11}x-14 \\ \frac{7}{11}x+9-9 &= \frac{3}{11}x-14-9 \\ \frac{7}{11}x &= \frac{3}{11}x-23 \end{aligned}$$

The equations are equivalent when $K = -23$.

97. answers may vary

99. $$\begin{aligned} -9.112y &= -47.537304 \\ \frac{-9.112y}{-9.112} &= \frac{-47.537304}{-9.112} \\ y &= 5.217 \end{aligned}$$

Check: $$\begin{aligned} -9.112y &= -47.537304 \\ -9.112(5.217) &\stackrel{?}{=} -47.537304 \\ -47.537304 &= -47.537304 \quad \text{True} \end{aligned}$$

The solution set is $\{5.217\}$.

101.
$$x(x-6)+7 = x(x+1)$$
$$x^2-6x+7 = x^2+x$$
$$x^2-6x+7-x^2 = x^2+x-x^2$$
$$-6x+7 = x$$
$$-6x+7+6x = x+6x$$
$$7 = 7x$$
$$\frac{7}{7} = \frac{7x}{7}$$
$$1 = x$$

Check: $x(x-6)+7 = x(x+1)$
$$1(1-6)+7 \stackrel{?}{=} 1(1+1)$$
$$-5+7 \stackrel{?}{=} 2$$
$$2 = 2 \quad \text{True}$$

The solution set is {1}.

Section 2.2 Practice Problems

1. a. x is the first even integer, $x + 2$ is the next consecutive even integer, and $x + 4$ is the third consecutive even integer. The sum of three consecutive even integers is $(x) + (x + 2) + (x + 4) = 3x + 6$.

b. The perimeter of a triangle is the sum of the lengths of the sides.
$$(x)+(2x+1)+(4x) = x+2x+1+4x = 7x+1$$

2. Let x be the number of Microsoft's MSN Radio and windowsMedia.com subscribers. Then $x + 1359.2$ is the number of listeners of AOL radio Network and $4x + 230.1$ is the number of listeners of Yahoo!'s LAUNCHcast.
$x + (x + 1359.2) + (4x + 230.1)$
$= 6x + 1589.3$

3. Let x be the first number. The other two numbers are $3x$ and $x + 50$.
$$x+3x+x+50 = 235$$
$$5x+50 = 235$$
$$5x = 185$$
$$x = 37$$
$3x = 3(37) = 111$
$x + 50 = (37) + 50 = 87$
The numbers are 37, 87, and 111.

4. Let x be the original price of the home.
$$x-0.06x = 83,660$$
$$0.94x = 83,660$$
$$x = 89,000$$
The original price of the home was $89,000.

5. Let x be the width of the rectangle. The length of the rectangle is $2x + 5$. The perimeter is the sum of the lengths of the sides.
$$(x)+(2x+5)+(x)+(2x+5) = 106$$
$$6x+10 = 106$$
$$6x = 96$$
$$x = 16$$
$2x + 5 = 2(16) + 5 = 37$
The length is 37 m; the width is 16 m.

6. Let x be the first integer. Then $x + 1$ is the next consecutive integer and $x + 2$ is the next consecutive integer after that.
$$(x)+(x+1)+(x+2) = 378$$
$$3x+3 = 378$$
$$3x = 375$$
$$x = 125$$
$x + 1 = 126$
$x + 2 = 127$
The three consecutive integers whose sum is 378 are 125, 126, and 127.

Exercise Set 2.2

1. The perimeter is the sum of the lengths of the sides.
$y + y + y + y = 4y$

3. Let z be the first of three consecutive integers. Then, $z + 1$ and $z + 2$ are the other two integers.
$(z) + (z + 1) + (z + 2) = 3z + 3$

5. x nickels have a value of $5x$ cents. $(x + 3)$ dimes have a value of $10(x + 3)$ cents, and $2x$ quarters have a value of $25(2x)$ cents.
$$5x+10(x+3)+25(2x) = 5x+10x+30+50x = 65x+30$$
The total amount of money is $(65x + 30)$ cents.

7. Each of the four vertical fences has a dimension of x. Each of the three horizontal fences has a dimension of $2x + 1$.
$4(x)+3(2x+1)=4x+6x+3=10x+3$
The total amount of fencing needed so that each rectangle has the same dimensions is $10x + 3$.

9. The length of the unknown vertical side is $10 - 2 = 8$. The length of the unknown horizontal side is
$x - 3 - (x - 10) = x - 3 - x + 10 = 7$.
$10 + (x - 10) + 2 + 7 + 8 + (x - 3) = 2x + 14$
The perimeter is $2x + 14$.

11. Let x be the number.
$$\begin{aligned} 4(x-2) &= 2+4x+2x \\ 4x-8 &= 2+6x \\ 4x &= 10+6x \\ -2x &= 10 \\ x &= -5 \end{aligned}$$
The number is −5.

13. Let x be the first number. The other two numbers are then $5x$ and $100 + x$.
$$\begin{aligned} x+5x+100+x &= 415 \\ 7x+100 &= 415 \\ 7x &= 315 \\ x &= 45 \end{aligned}$$
$5x = 5(45) = 225$
$x + 100 = 45 + 100 = 145$
The numbers are 45, 225, and 145.

15. Let x be the number of acres of land that is federally owned. Then x is 29% of 2271.
$x = 0.29(2271) = 658.59$
Let $2271 - x$ be the number of acres of land that is not federally owned.
$2271 - (658.59) = 1612.41$
1612.41 million acres of land in the US is not federally owned.

17. Let x be the number of minor earthquakes in 2004. Then x is 91% of 3456.
$x = 0.91(3456) = 3144.96$
There were 3145 minor earthquakes in the United States in 2004.

19. Let x be the number of online shoppers who prefer to do business with only large, well-known retailers. Then x is 15% of 1500.
$x = 0.15(1500) = 225$
Let $1500 - x$ be the number of shoppers who are willing to do business with any size retailers.
$1500 - (225) = 1275$
1275 shoppers are willing to do business with any size retailers.

21. Let x be the percent of e-mail users that spend less than 15 minutes on e-mail at work per day.
$$\begin{aligned} x+50+8+10+9 &= 100 \\ x+77 &= 100 \\ x &= 23 \end{aligned}$$
23% of e-mail users spend less than 15 minutes on e-mail each day.

23. Let x be the number of employees who use e-mail for more than 3 hours per day. Then x is 9% of 4633.
$x = 0.09(4633) = 416.97$
417 employees use e-mail for more than 3 hours per day.

25. Let x be the number of seats at Heinz Field. Then INVESCO Field has $x + 11{,}675$ seats.
$$\begin{aligned} x+(x+11{,}675) &= 140{,}575 \\ 2x+11{,}675 &= 140{,}575 \\ 2x &= 128{,}900 \\ x &= 64{,}450 \end{aligned}$$
$x + 11{,}675 = 64{,}450 + 11{,}675 = 76{,}125$
INVESCO Field at Mile High has 76,125 seats, while Heinz Field has 64,450 seats.

27. The perimeter is the sum of the lengths of the sides.
$$\begin{aligned} (x)+(5x)+(6x-3) &= 483 \\ 12x-3 &= 483 \\ 12x &= 486 \\ x &= 40.5 \end{aligned}$$
$5x = 5(40.5) = 202.5$
$6x - 3 = 6(40.5) - 3 = 243 - 3 = 240$
The lengths of the sides are 40.5 ft, 202.5 ft, and 240 ft.

29. Let x be the population of the New York metropolitan region. Then Mexico City's population is $x + 0.4$, and Tokyo's population is $2x - 1.6$.

$$\begin{aligned} x+(x+0.4)+2x-1.6 &= 72 \\ 4x-1.2 &= 72 \\ 4x &= 73.2 \\ x &= 18.3 \end{aligned}$$

$x + 0.4 = 18.3 + 0.4 = 18.7$

$2x - 1.6 = 2(18.3) - 1.6 = 36.6 - 1.6 = 35.0$

The population of New York is 18.3 million, the population of Mexico City is 18.7 million, and the population of Tokyo is 35.0 million.

31.
$$\begin{aligned} (x)+(4x)+(x+6) &= 180 \\ 6x+6 &= 180 \\ 6x &= 174 \\ x &= 29 \end{aligned}$$

$4x = 4(29) = 116$

$x + 6 = (29) + 6 = 35$

The angles have measures of 29°, 35°, and 116°.

33.
$$\begin{aligned} 4x+(5x+1)+(5x+3) &= 102 \\ 14x+4 &= 102 \\ 14x &= 98 \\ x &= 7 \end{aligned}$$

$4x = 4(7) = 28$

$5x + 1 = 5(7) + 1 = 35 + 1 = 36$

$5x + 3 = 5(7) + 3 = 35 + 3 = 38$

The sides have lengths 28 meters, 36 meters, and 38 meters.

35.
$$\begin{aligned} (x)+(1.5x)+(x)+(2.5x-9) &= 99 \\ 6x-9 &= 99 \\ 6x &= 108 \\ x &= 18 \end{aligned}$$

$1.5x = 1.5(18) = 27$

$2.5x - 9 = 2.5(18) - 9 = 45 - 9 = 36$

The sides have lengths of 18 in., 18 in., 27 in., and 36 in.

37.
$$\begin{aligned} (2x-51)+\left(\frac{3}{2}x+3\right)+x &= 780 \\ 2(2x-51)+2\left(\frac{3}{2}x+3\right)+2(x) &= 2(780) \\ 4x-102+3x+6+2x &= 1560 \\ 9x-96 &= 1560 \\ 9x &= 1656 \\ x &= 184 \end{aligned}$$

$2x - 51 = 2(184) - 51 = 368 - 51 = 317$

$\frac{3}{2}x+3=\frac{3}{2}(184)+3=276+3=279$

Occupation	Increase in number of jobs
Security guards	317 thousand
Home Health aides	279 thousand
Computer systems analysts	184 thousand
Total	780 thousand

39. Let x be the number of medical assistant jobs. Then $2x + 173$ is the number of postsecondary teacher jobs, and the number of registered nurse jobs is $3x - 22$.

$$\begin{aligned} x+(2x+173)+(3x-22) &= 1441 \\ 6x+151 &= 1441 \\ 6x &= 1290 \\ x &= 215 \end{aligned}$$

$2x + 173 = 2(215) + 173 = 430 + 173 = 603$

$3x - 22 = 3(215) - 22 = 645 - 22 = 623$

Medical assistant jobs will grow by 215 thousand, post-secondary teacher jobs will grow by 603 thousand, and registered nurse jobs will grow by 623 thousand.

41. Let x be the number of seats in a B737-200 aircraft. Then the number of seats in a B767-300ER is $x + 88$, and the number of seats in a F-100 is $x - 32$.

$$x+(x+88)+(x-32)=413$$
$$3x+56=413$$
$$3x=357$$
$$x=119$$

$x + 88 = 119 + 88 = 207$
$x - 32 = 119 - 32 = 87$
The B737-200 has 119 seats, the B767-300ER has 207 seats, and the F-100 has 87 seats.

43. Let x be the price of the fax machine before tax.

$$x+0.08x=464.40$$
$$1.08x=464.40$$
$$x=430$$

The fax machine costs $430 before tax.

45. Let x be the closing price on August 19, 2004. Then the closing price on December 28, 2004 is x increased by 92.1% of x.

$$x+0.921x=192.76$$
$$1.921x=192.76$$
$$x\approx 100.34$$

The closing price on August 19, 2004 was $100.34.

47. Let x be the first integer. Then $x + 1$ is the next consecutive integer and $x + 2$ is the next consecutive integer after that.

$$x+(x+1)+(x+2)=228$$
$$3x+3=228$$
$$3x=225$$
$$x=75$$

$x + 1 = (75) + 1 = 76$
$x + 2 = (75) + 2 = 77$
The three consecutive integers are 75, 76, and 77.

49. Let $x°$ be the measure of the second angle. Then the first angle measures $(2x)°$, and the third angle measures $(3x - 12)°$.

$$x+2x+(3x-12)=180$$
$$6x-12=180$$
$$6x=192$$
$$x=32$$

$2x = 2(32) = 64$
$3x - 12 = 3(32) - 12 = 96 - 12 = 84$
The angles measure 64°, 32°, and 84°.

51. Let x be the length of one side of the square frame. Then $x + 6$ is the length of one side of the triangular frame.

$$4x=3(x+6)$$
$$4x=3x+18$$
$$x=18$$

$x + 6 = (18) + 6 = 24$
The length of one side of the square frame is 18 cm. The length of one side of the triangular frame is 24 cm.

53. Let x be the expected population of South Africa in 2050. Then x is 46,900,000 less the decrease of 11% of 46,900,000.

$$x=46,900,000-0.11(46,900,000)$$
$$=46,900,000-5,159,000$$
$$=41,741,000$$

The population of South Africa is expected to be 41,741,000 in 2050.

55. Let x be Fallon's zip code. Then $x + 2$ is Fernley's zip code and $x + 4$ is the zip code of Gardnerville Ranchos.

$$2x+(x+4)=268,222$$
$$3x+4=268,222$$
$$3x=268,218$$
$$x=89,406$$

$x + 2 = (89,406) + 2 = 89,408$
$x + 4 = (89,406) + 4 = 89,410$
Fallon's zip code is 89406. Fernley's zip code is 89408. Gardnerville Ranchos' zip code is 89410.

57. Let $x°$ be the measure of the first angle. Then its supplement measures $(3x + 20)°$.

$$x+(3x+20)=180$$
$$4x+20=180$$
$$4x=160$$
$$x=40$$

$3x + 20 = 3(40) + 20 = 120 + 20 = 140$
The angles measure 40° and 140°.

59. Let x be the number of bulb hours for a halogen bulb. Then $25x$ is the number of bulb hours for a fluorescent bulb, and

$x - 2500$ is the number of bulb hours for an incandescent bulb.

$$x + 25x + (x - 2500) = 105{,}500$$
$$27x - 2500 = 105{,}500$$
$$27x = 108{,}000$$
$$x = 4000$$

$25x = 25(4000) = 100{,}000$

$x - 2500 = (4000) - 2500 = 1500$

A halogen bulb lasts 4000 bulb hours. A fluorescent bulb lasts 100,000 bulb hours. An incandescent bulb lasts 1500 bulb hours.

61. Let x be the number of home runs that Derek Jeter hit. Then Mark Mulder hit $(x + 2)$ home runs, and Geoff Jenkins hit $(x + 4)$ home runs.

$$x + (x + 2) + (x + 4) = 75$$
$$3x + 6 = 75$$
$$3x = 69$$
$$x = 23$$

$x + 2 = 23 + 2 = 25$

$x + 4 = 23 + 4 = 27$

Jeter hit 23 home runs, Mulder hit 25 home runs, and Jenkins hit 27 home runs in 2004.

63. $4ab - 3bc = 4(-5)(-8) - 3(-8)(2)$
$= 160 + 48$
$= 208$

65. $n^2 - m^2 = (-3)^2 - (-8)^2 = 9 - 64 = -55$

67. $P + PRT = (3000) + (3000)(0.0325)(2)$
$= 3000 + 195$
$= 3195$

69. yes; answers may vary

71. answers may vary

73. a.

$$y = -72.8x + 2843.5$$
$$0 = -72.8x + 2843.5$$
$$72.8x = 2843.5$$
$$x \approx 39.06$$

The average number of cigarettes smoked will be 0 39 years after 1990, or in 2029.

b. $y = -72.8x + 2843.5$

$y = -72.8(20) + 2843.5$
$= -1456 + 2843.5$
$= 1387.5$

The average number of cigarettes smoked by an American adult is predicted to be 1387.5 in 2010.

c. The average number of cigarettes smoked daily in 2010 is predicted to be $\frac{1387.5}{365} \approx 4$.

This does not represent the average number of cigarettes smoked by an American smoker, because it is the average for *all* Americans, both smokers and non-smokers.

75. Let x be the first odd integer. Then $x + 2$ is the next consecutive odd integer.

$$7x = 5(x + 2) + 54$$
$$7x = 5x + 10 + 54$$
$$7x = 5x + 64$$
$$2x = 64$$
$$x = 32$$

No such odd integers exist.

77.

$$R = C$$
$$60x = 50x + 5000$$
$$10x = 5000$$
$$x = 500$$

$50x + 5000 = 50(500) + 5000$
$= 25{,}000 + 5000$
$= 30{,}000$

500 computer boards must be sold to break even. It costs $30,000 to produce the 500 boards.

79. The company makes a profit if it makes and sells more products than the break-even number.

Section 2.3 Practice Problems

1. $I = Prt$
$$\frac{I}{rt} = \frac{Prt}{rt}$$
$$\frac{I}{rt} = P$$

2. $2y + 5x = 10$
$$2y + 5x - 5x = 10 - 5x$$
$$2y = 10 - 5x$$
$$\frac{2y}{2} = \frac{10-5x}{2}$$
$$y = \frac{10-5x}{2} \text{ or } y = 5 - \frac{5}{2}x$$

3. $A = \frac{1}{2}(B+b)h$
$$2A = 2\left[\frac{1}{2}(B+b)h\right]$$
$$2A = (B+b)h$$
$$2A = Bh + bh$$
$$2A - bh = Bh$$
$$B = \frac{2A - bh}{h} \text{ or } B = \frac{2A}{h} - b$$

4. Use $A = P\left(1+\frac{r}{n}\right)^{nt}$ with $P = 5000$, $r = 4\% = 0.04$, $n = 12$, and $t = 2$.
$$A = 5000\left(1+\frac{0.04}{12}\right)^{12\cdot 2}$$
$$A = 5000(1.0033333)^{24}$$
$$A \approx 5000(1.083142959)$$
$$A \approx \$5415.71$$

Mental Math

1. $2x + y = 5$
$$y = 5 - 2x$$

2. $7x - y = 3$
$$-y = 3 - 7x$$
$$y = -3 + 7x \text{ or } y = 7x - 3$$

3. $a - 5b = 8$
$$a = 5b + 8$$

4. $7r + s = 10$
$$s = 10 - 7r$$

5. $5j + k - h = 6$
$$5j + k = h + 6$$
$$k = h - 5j + 6$$

6. $w - 4y + z = 0$
$$w + z = 4y$$
$$z = 4y - w$$

Exercise Set 2.3

1. $d = rt$
$$\frac{d}{r} = \frac{rt}{r}$$
$$\frac{d}{r} = t$$

3. $I = Prt$
$$\frac{I}{Pt} = \frac{Prt}{Pt}$$
$$\frac{I}{Pt} = r$$

5. $P = a + b + c$
$$P - a - b = c$$

7. $9x - 4y = 16$
$$-4y = -9x + 16$$
$$y = \frac{-9x+16}{-4}$$
$$y = \frac{9x-16}{4} \text{ or } y = \frac{9}{4}x - 4$$

9. $P = 2l + 2w$
$$P - 2w = 2l$$
$$\frac{P-2w}{2} = l \text{ or } l = \frac{P}{2} - w$$

11.
$$E = I(r+R)$$
$$\frac{E}{I} = r+R$$
$$\frac{E}{I} - R = r$$

13. $5x + 4y = 20$
$$4y = 20 - 5x$$
$$y = \frac{20-5x}{4} \text{ or } y = 5 - \frac{5}{4}x$$

15.
$$S = 2LW + 2LH + 2WH$$
$$S - 2LW = 2LH + 2WH$$
$$S - 2LW = H(2L + 2W)$$
$$\frac{S-2LW}{2L+2W} = H$$

17. $C = 2\pi r$
$$\frac{C}{2\pi} = \frac{2\pi r}{2\pi}$$
$$\frac{C}{2\pi} = r$$

19.
$$C = \frac{5}{9}(F-32)$$
$$\frac{9}{5}C = \frac{9}{5}\left[\frac{5}{9}(F-32)\right]$$
$$\frac{9}{5}C = F - 32$$
$$\frac{9}{5}C + 32 = F$$

21. Use $A = P\left(1+\frac{r}{n}\right)^{nt}$ with $P = 3500$, $r = 3\% = 0.03$, and $t = 10$.

$n = 1$: $A = 3500\left(1+\frac{0.03}{1}\right)^{1\cdot 10}$
$$A = 3500(1.03)^{10}$$
$$A \approx \$4703.71$$

$n = 2$: $A = 3500\left(1+\frac{0.03}{2}\right)^{2\cdot 10}$
$$A = 3500(1.015)^{20}$$
$$A \approx \$4713.99$$

$n = 4$: $A = 3500\left(1+\frac{0.03}{4}\right)^{4\cdot 10}$
$$A = 3500(1.0075)^{40}$$
$$A \approx \$4719.22$$

$n = 12$: $A = 3500\left(1+\frac{0.03}{12}\right)^{12\cdot 10}$
$$A = 3500(1.0025)^{120}$$
$$A \approx \$4722.74$$

$n = 365$: $A = 3500\left(1+\frac{0.03}{365}\right)^{365\cdot 10}$
$$A \approx \$4724.45$$

23. Use $A = P\left(1+\frac{r}{n}\right)^{nt}$ with $P = 6000$, $r = 4\% = 0.04$, and $t = 5$.

a. semiannually: $n = 2$:
$$A = 6000\left(1+\frac{0.04}{2}\right)^{2\cdot 5}$$
$$A = 6000(1.02)^{10}$$
$$A \approx \$7313.97$$

b. quarterly: $n = 4$: $A = 6000\left(1+\frac{0.04}{4}\right)^{4\cdot 5}$
$$A = 6000(1.01)^{20}$$
$$A \approx \$7321.14$$

c. monthly: $n = 12$:
$$A = 6000\left(1+\frac{0.04}{12}\right)^{12\cdot 5}$$
$$A \approx \$7325.98$$

25. $C = \frac{5}{9}(F-32)$
$$C = \frac{5}{9}(104-32)$$
$$C = \frac{5}{9}(72)$$
$$C = 40$$
104°F is 40°C.

27. Roundtrip distance = 90 + 90 = 180 miles

$D = r \cdot t$

$180 = 50t$

$\frac{180}{50} = t$

$t = 3.6$ hours = 3 hours 36 minutes

29. $A = l \cdot w$

$A = 64 \cdot 64$

$A = 4096$ sq ft

Packages $= \frac{4096}{24} \approx 170.6$

Buy 171 packages of tiles.

31. $A = \frac{1}{2} \cdot b \cdot h$

$18 = \frac{1}{2} \cdot 4 \cdot h$

$18 = 2h$

$9 = h$

The height of the kite is 9 ft.

33. $A = 2(l \cdot h) + 2(w \cdot h)$

$A = 2(14 \cdot 8) + 2(16 \cdot 8)$

$A = 224 + 256 = 480$ sq ft

Two coats $= 2 \cdot 480 = 960$ sq ft

Gallons $= \frac{960}{500} = 1.92$

Buy 2 gallons of paint.

35. **a.** $V = \pi r^2 h$

$V = \pi(4.2)^2(21.2)$

$V = 373.968\pi$

$V = 1174.86$ cubic meters

b. $V = \frac{4}{3}\pi r^3$

$V = \frac{4}{3}\pi(4.2)^3$

$V \approx 310.34$ cubic meters

c. 1174.86 + 310.34 = 1485.20 cubic meters

37. Radius of satellite orbit $= 22{,}248 + 4000$
$= 26{,}248$ mi

$C = 2\pi r = 2 \cdot \pi \cdot 26{,}248 \approx 164{,}921$ mi

39. $1.3 \text{ miles} \times 5280 \frac{\text{ft}}{\text{mile}} = 6864$ ft

$V = \pi r^2 h$

$3800 = \pi r^2(6864)$

$\frac{3800}{6864\pi} = \frac{\pi r^2(6864)}{6864\pi}$

$0.17622 = r^2$

$0.42 \approx r$

The radius of the hole is approximately 0.42 ft.

41. $C = \pi d = \pi \cdot 41{,}125 = 41.125\pi$ ft
$\approx 41.125 \cdot 3.14$
$= 129.1325$ ft

43. $C = 4h + 9f + 4p$

$C - 4h - 4p = 9f$

$\frac{C - 4h - 4p}{9} = f$

45. $C = 4h + 9f + 4p$

$C = 4 \cdot 7 + 9 \cdot 14 + 4 \cdot 6$

$C = 28 + 126 + 24 = 178$

A serving of cashews contains 178 calories.

47. $C = 4h + 9f + 4p$

$130 = 4(31) + 9(0) + 4p$

$130 = 124 + 0 + 4p$

$6 = 4p$

$1.5 = p$

1.5 grams of protein

49. {–3, –2, –1}

51. $x + 5 \le 6$

$x \le 1$

{–3, –2, –1, 0, 1}

53. answers may vary

55.

Planet	AU from Sun
Mercury	$\frac{36}{92.9}=0.388$
Venus	$\frac{67.2}{92.9}=0.723$
Earth	$\frac{92.9}{92.9}=1.0$
Mars	$\frac{141.5}{92.9}=1.523$
Jupiter	$\frac{483.3}{92.9}=5.202$
Saturn	$\frac{886.1}{92.9}=9.538$
Uranus	$\frac{1783}{92.9}=19.193$
Neptune	$\frac{2793}{92.9}=30.065$
Pluto	$\frac{3670}{92.9}=39.505$

57. $C=\dfrac{1,700,000,000}{250,000,000}=6.8$

Cost is \$6.80 per person.

59. answers may vary

61. $N=R^{*}\times f_p\times n_e\times f_l\times f_i\times f_c\times L$

$$\frac{N}{R^{*}\times f_p\times f_l\times f_i\times f_c\times L}=\frac{R^{*}\times f_p\times n_e\times f_l\times f_i\times f_c\times L}{R^{*}\times f_p\times f_l\times f_i\times f_c\times L}$$

$$\frac{N}{R^{*}\times f_p\times f_l\times f_i\times f_c\times L}=n_e$$

Section 2.4 Practice Problems

1. 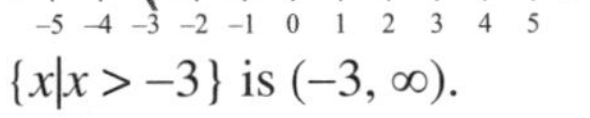

$\{x|x>-3\}$ is $(-3,\infty)$.

2. 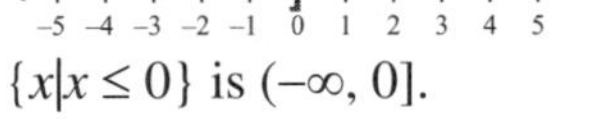

$\{x|x\le 0\}$ is $(-\infty, 0]$.

3. 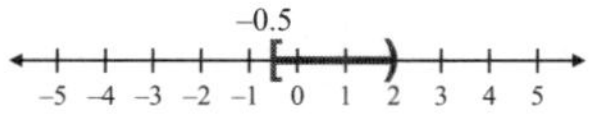

$\{x|-0.5\le x<2\}$ is $[-0.5, 2)$.

4.
$$\begin{aligned} x+3&<1\\ x+3-3&<1-3\\ x&<-2\end{aligned}$$

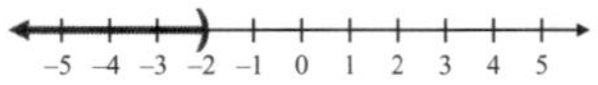

$\{x|x<-2\}$ is $(-\infty,-2)$.

5.
$$\begin{aligned} 3x-4&<4x\\ 3x-4-3x&<4x-3x\\ -4&<x\\ x&>-4\end{aligned}$$

$\{x|x>-4\}$ is $(-4,\infty)$.

6.
$$\begin{aligned} 5x-1&\ge 4x+4\\ 5x-1+1&\ge 4x+4+1\\ 5x&\ge 4x+5\\ 5x-4x&\ge 4x+5-4x\\ x&\ge 5\end{aligned}$$

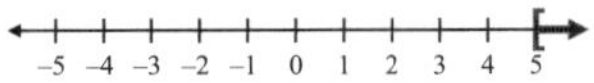

$\{x|x\ge 5\}$ is $[5,\infty)$.

7.
$$\begin{aligned} \frac{1}{6}x&\le\frac{2}{3}\\ 6\cdot\frac{1}{6}x&\le 6\cdot\frac{2}{3}\\ x&\le 4\end{aligned}$$

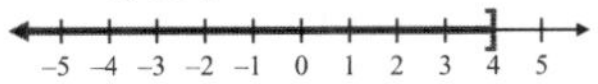

$\{x|x\le 4\}$ is $(-\infty, 4]$.

8.
$$\begin{aligned} -1.1x&<5.5\\ \frac{-1.1x}{-1.1}&>\frac{5.5}{-1.1}\\ x&>-5\end{aligned}$$

$\{x|x>-5\}$ is $(-5,\infty)$.

9.
$$\begin{aligned}
-(2x-6) &\le 4(2x-4)+2\\
-2x+6 &\le 8x-16+2\\
-2x+6 &\le 8x-14\\
-2x+6-6 &\le 8x-14-6\\
-2x &\le 8x-20\\
-2x-8x &\le 8x-20-8x\\
-10x &\le -20\\
\frac{-10x}{-10} &\ge \frac{-20}{-10}\\
x &\ge 2
\end{aligned}$$
The solution set is $[2, \infty)$.

10.
$$\begin{aligned}
\frac{3}{4}(x+2) &\ge x-6\\
4\left[\frac{3}{4}(x+2)\right] &\ge 4(x-6)\\
3(x+2) &\ge 4x-24\\
3x+6 &\ge 4x-24\\
3x+6-6 &\ge 4x-24-6\\
3x &\ge 4x-30\\
3x-4x &\ge 4x-30-4x\\
-x &\ge -30\\
\frac{-x}{-1} &\le \frac{-30}{-1}\\
x &\le 30
\end{aligned}$$
The solution set is $(-\infty, 30]$.

11.
$$\begin{aligned}
5(x-3) &< 5x+2\\
5x-15 &< 5x+2\\
5x-15-5x &< 5x+2-5x\\
-15 &< 2 \quad \text{True}
\end{aligned}$$
The solution set is $(-\infty, \infty)$.

12. Let x be the amount of sales.
$$\begin{aligned}
1000+0.15x &\ge 4000\\
0.15x &\ge 3000\\
x &\ge 20,000
\end{aligned}$$
The minimum amount of sales needed is \$20,000.

13.
$$\begin{aligned}
c &> -9.2t+527.33\\
100 &> -9.2t+527.33\\
-427.33 &> -9.2t\\
\frac{-427.33}{-9.2} &< \frac{-9.2t}{-9.2}\\
46.45 &< t
\end{aligned}$$
$1990 + 47 = 2037$
Consumption will be less than 100 billion per year after the year 2037.

Mental Math

1. $x-2<4$
$x<6$
$\{x|x<6\}$

2. $x-1>6$
$x>7$
$\{x|x>7\}$

3. $x+5\ge 15$
$x\ge 10$
$\{x|x\ge 10\}$

4. $x+1\le 8$
$x\le 7$
$\{x|x\le 7\}$

5. $3x>12$
$x>4$
$\{x|x>4\}$

6. $5x<20$
$x<4$
$\{x|x<4\}$

7. $\frac{x}{2}\le 1$
$x\le 2$
$\{x|x\le 2\}$

8. $\frac{x}{4}\ge 2$
$x\ge 8$
$\{x|x\ge 8\}$

Exercise Set 2.4

1. 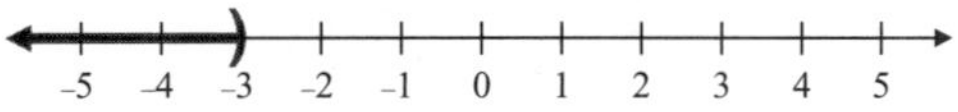

$\{x|x < -3\}$ is $(-\infty, -3)$.

3. 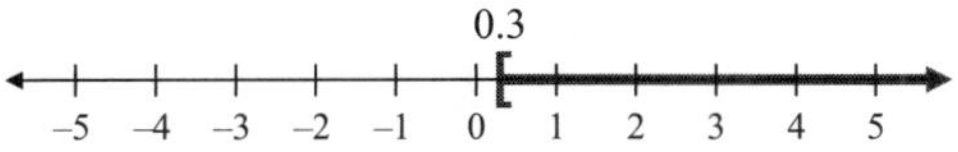

$\{x|x \geq 0.3\}$ is $[0.3, \infty)$.

5. 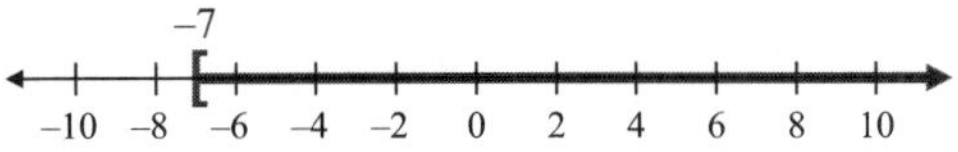

$\{x|-7 \leq x\}$ is $[-7, \infty)$.

7. 

$\{x|-2 < x < 5\}$ is $(-2, 5)$.

9. 

$\{x|5 \geq x > -1\}$ is $(-1, 5]$.

11.
$$x - 7 \geq -9$$
$$x - 7 + 7 \geq -9 + 7$$
$$x \geq -2$$
The solution set is $[-2, \infty)$.

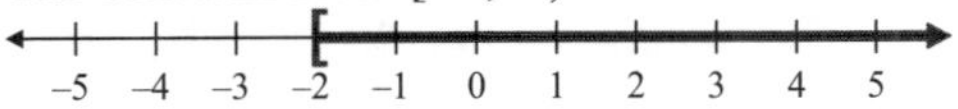

13.
$$7x < 6x + 1$$
$$7x - 6x < 6x + 1 - 6x$$
$$x < 1$$
The solution set is $(-\infty, 1)$.

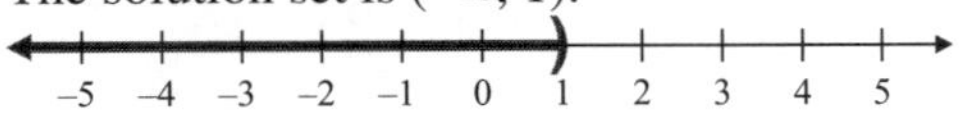

15.
$$8x - 7 \leq 7x - 5$$
$$8x - 7 + 7 \leq 7x - 5 + 7$$
$$8x \leq 7x + 2$$
$$8x - 7x \leq 7x + 2 - 7x$$
$$x \leq 2$$
The solution set is $(-\infty, 2]$.

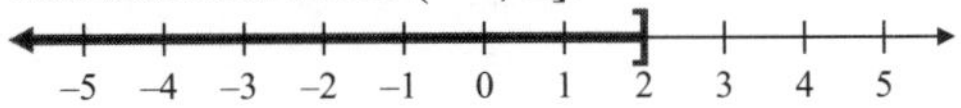

17.
$$\frac{3}{4}x \geq 6$$
$$\frac{4}{3}\left(\frac{3}{4}x\right) \geq \frac{4}{3}(6)$$
$$x \geq 8$$
The solution set is $[8, \infty)$.

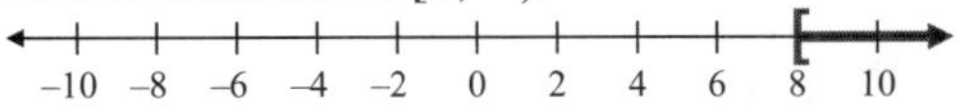

19.
$$5x < -23.5$$
$$\frac{5x}{5} < \frac{-23.5}{5}$$
$$x < -4.7$$
The solution set is $(-\infty, -4.7)$.

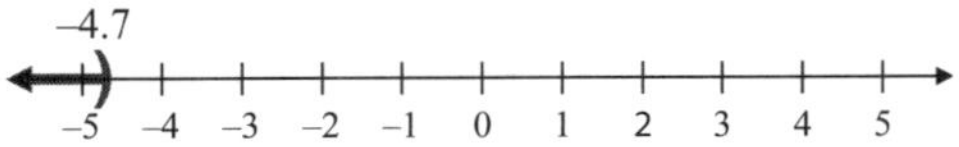

21.
$$-3x \geq 9$$
$$\frac{-3x}{-3} \leq \frac{9}{-3}$$
$$x \leq -3$$
The solution set is $(-\infty, -3]$.

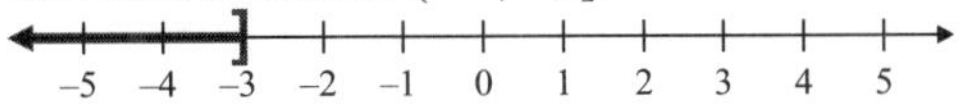

23.
$$-2x + 7 \geq 9$$
$$-2x + 7 - 7 \geq 9 - 7$$
$$-2x \geq 2$$
$$\frac{-2x}{-2} \leq \frac{2}{-2}$$
$$x \leq -1$$
The solution set is $(-\infty, -1]$.

25.
$$15 + 2x \geq 4x - 7$$
$$15 + 2x - 4x \geq 4x - 7 - 4x$$
$$15 - 2x \geq -7$$
$$15 - 2x - 15 \geq -7 - 15$$
$$-2x \geq -22$$
$$\frac{-2x}{-2} \leq \frac{-22}{-2}$$
$$x \leq 11$$
The solution set is $(-\infty, 11]$.

27.
$$\begin{aligned} 4(2x+1) &> 4 \\ 8x+4 &> 4 \\ 8x+4-4 &> 4-4 \\ 8x &> 0 \\ \frac{8x}{8} &> \frac{0}{8} \\ x &> 0 \end{aligned}$$
The solution set is $(0, \infty)$.

29.
$$\begin{aligned} 3(x-5) &< 2(2x-1) \\ 3x-15 &< 4x-2 \\ 3x-15-3x &< 4x-2-3x \\ -15 &< x-2 \\ -15+2 &< x-2+2 \\ -13 &< x \end{aligned}$$
The solution set is $(-13, \infty)$.

31.
$$\begin{aligned} \frac{5x+1}{7}-\frac{2x-6}{4} &\ge -4 \\ 28\left[\frac{5x+1}{7}-\frac{2x-6}{4}\right] &\ge 28(-4) \\ 28\left(\frac{5x+1}{7}\right)-28\left(\frac{2x-6}{4}\right) &\ge -112 \\ 4(5x+1)-7(2x-6) &\ge -112 \\ 20x+4-14x+42 &\ge -112 \\ 6x+46 &\ge -112 \\ 6x+46-46 &\ge -112-46 \\ 6x &\ge -158 \\ \frac{6x}{6} &\ge \frac{-158}{6} \\ x &\ge -\frac{79}{3} \end{aligned}$$
The solution set is $\left[-\frac{79}{3}, \infty\right)$.

33.
$$\begin{aligned} -3(2x-1) &< -4[2+3(x+2)] \\ -6x+3 &< -4[2+3x+6] \\ -6x+3 &< -4[3x+8] \\ -6x+3 &< -12x-32 \\ -6x+3+12x &< -12x-32+12x \\ 6x+3 &< -32 \\ 6x+3-3 &< -32-3 \\ 6x &< -35 \\ \frac{6x}{6} &< \frac{-35}{6} \\ x &< -\frac{35}{6} \end{aligned}$$
The solution set is $\left(-\infty, -\frac{35}{6}\right)$.

35.
$$\begin{aligned} x+9 &< 3 \\ x+9-9 &< 3-9 \\ x &< -6 \end{aligned}$$
The solution set is $(-\infty, -6)$.

37.
$$\begin{aligned} -x &< -4 \\ \frac{-x}{-1} &> \frac{-4}{-1} \\ x &> 4 \end{aligned}$$
The solution set is $(4, \infty)$.

39.
$$\begin{aligned} -7x &\le 3.5 \\ \frac{-7x}{-7} &\ge \frac{3.5}{-7} \\ x &\ge -0.5 \end{aligned}$$
The solution set is $[-0.5, \infty)$.

41.
$$\begin{aligned} \frac{1}{2}+\frac{2}{3} &\ge \frac{x}{6} \\ 6\left(\frac{1}{2}+\frac{2}{3}\right) &\ge 6\left(\frac{x}{6}\right) \\ 6\left(\frac{1}{2}\right)+6\left(\frac{2}{3}\right) &\ge x \\ 3+4 &\ge x \\ 7 &\ge x \end{aligned}$$
The solution set is $(-\infty, 7]$.

43.
$$\begin{aligned} -5x+4 &\le -4(x-1) \\ -5x+4 &\le -4x+4 \\ -5x+4-4 &\le -4x+4-4 \\ -5x &\le -4x \\ -5x+4x &\le -4x+4x \\ -x &\le 0 \\ \frac{-x}{-1} &\ge \frac{0}{-1} \\ x &\ge 0 \end{aligned}$$
The solution set is $[0, \infty)$.

45.
$$\begin{aligned} \frac{3}{4}(x-7) &\ge x+2 \\ 4\left[\frac{3}{4}(x-7)\right] &\ge 4(x+2) \\ 3(x-7) &\ge 4(x+2) \\ 3x-21 &\ge 4x+8 \\ 3x-21-3x &\ge 4x+8-3x \\ -21 &\ge x+8 \\ -21-8 &\ge x+8-8 \\ -29 &\ge x \end{aligned}$$
The solution set is $(-\infty, -29]$.

47.
$$\begin{aligned} 0.8x+0.6x &\ge 4.2 \\ 1.4x &\ge 4.2 \\ \frac{1.4x}{1.4} &\ge \frac{4.2}{1.4} \\ x &\ge 3 \end{aligned}$$
The solution set is $[3, \infty)$.

49.
$$\begin{aligned} 4(x-6)+2x-4 &\ge 3(x-7)+10x \\ 4x-24+2x-4 &\ge 3x-21+10x \\ 6x-28 &\ge 13x-21 \\ 6x-28-6x &\ge 13x-21-6x \\ -28 &\ge 7x-21 \\ -28+21 &\ge 7x-21+21 \\ -7 &\ge 7x \\ \frac{-7}{7} &\ge \frac{7x}{7} \\ -1 &\ge x \end{aligned}$$
The solution set is $(-\infty, -1]$.

51.
$$\begin{aligned} 14-(5x-6) &\ge -6(x+1)-5 \\ 14-5x+6 &\ge -6x-6-5 \\ -5x+20 &\ge -6x-11 \\ -5x+20-20 &\ge -6x-11-20 \\ -5x &\ge -6x-31 \\ -5x+6x &\ge -6x-31+6x \\ x &\ge -31 \end{aligned}$$
The solution set is $[-31, \infty)$.

53.
$$\begin{aligned} \frac{1}{2}(3x-4) &\le \frac{3}{4}(x-6)+1 \\ 4\left[\frac{1}{2}(3x-4)\right] &\le 4\left[\frac{3}{4}(x-6)+1\right] \\ 2(3x-4) &\le 3(x-6)+4 \\ 6x-8 &\le 3x-18+4 \\ 6x-8 &\le 3x-14 \\ 6x-8-3x &\le 3x-14-3x \\ 3x-8 &\le -14 \\ 3x-8+8 &\le -14+8 \\ 3x &\le -6 \\ \frac{3x}{3} &\le \frac{-6}{3} \\ x &\le -2 \end{aligned}$$
The solution set is $(-\infty, -2]$.

55.
$$\begin{aligned} 0.4(4x-3) &< 1.2(x+2) \\ 10[0.4(4x-3)] &< 10[1.2(x+2)] \\ 4(4x-3) &< 12(x+2) \\ 16x-12 &< 12x+24 \\ 16x-12+12 &< 12x+24+12 \\ 16x &< 12x+36 \\ 16x-12x &< 12x+36-12x \\ 4x &< 36 \\ \frac{4x}{4} &< \frac{36}{4} \\ x &< 9 \end{aligned}$$
The solution set is $(-\infty, 9)$.

57.

$$\frac{2}{5}x-\frac{1}{4}\le\frac{3}{10}x-\frac{4}{5}$$
$$20\left(\frac{2}{5}x-\frac{1}{4}\right)\le 20\left(\frac{3}{10}x-\frac{4}{5}\right)$$
$$20\left(\frac{2}{5}x\right)-20\left(\frac{1}{4}\right)\le 20\left(\frac{3}{10}x\right)-20\left(\frac{4}{5}\right)$$
$$8x-5\le 6x-16$$
$$8x-5-6x\le 6x-16-6x$$
$$2x-5\le -16$$
$$2x-5+5\le -16+5$$
$$2x\le -11$$
$$\frac{2x}{2}\le\frac{-11}{2}$$
$$x\le -\frac{11}{2}$$

The solution set is $\left(-\infty, -\frac{11}{2}\right]$.

59. a. Let x be the final exam score.

$$\frac{72+67+82+79+x+x}{6}\ge 77$$
$$6\left(\frac{72+67+82+79+2x}{6}\right)\ge 6(77)$$
$$72+67+82+79+2x\ge 462$$
$$300+2x\ge 462$$
$$300+2x-300\ge 462-300$$
$$2x\ge 162$$
$$\frac{2x}{2}\ge\frac{162}{2}$$
$$x\ge 81$$

The solution is $\{x|x\ge 81\}$.

b. A final exam grade of 81 or higher will result in an average of 77 or higher.

61. a. Let x be the weight of luggage and cargo. The six passengers have a total weight of 6(160) pounds.

$$x+6(160)\le 2000$$
$$x+960\le 2000$$
$$x\le 1040$$

The solution is $\{x|x\le 1040\}$.

b. The luggage and cargo must weigh no more than 1040 pounds.

63. a. Let x be the number of whole boxes on the elevator.

$$66x+147\le 1500$$
$$66x+147-147\le 1500-147$$
$$66x\le 1353$$
$$x\le 20.5$$

The solution is $\{x|x\le 20\}$.

b. She can move at most 20 whole boxes at one time.

65. a. Let x be the number of monthly calls. The monthly charge for plan 2 is $\$(13 + 0.06x)$, and plan 1 is more economical when the monthly charge for plan 2 is more than \$25.

$$13+0.06x>25$$
$$0.06x>12$$
$$x>200$$

The solution is $\{x|x>200\}$.

b. Plan 1 is more economical for more than 200 monthly calls.

67. Glass does not become a true liquid until temperatures are greater than or equal to 500°C. Use $F=\frac{9}{5}C+32.$ Replace C with 500.

$$F=\frac{9}{5}C+32$$
$$F=\frac{9}{5}(500)+32$$
$$F=9(100)+32$$
$$F=900+32$$
$$F=932$$

500°C = 932°F

So glass does not become a true liquid until temperatures are greater than or equal to 932°F. The solution set is $\{F|F\ge 932°\}$.

69. a.

$$s>35{,}000$$
$$651.2t+27{,}821>35{,}000$$
$$651.2t>7179$$
$$t>\frac{7179}{651.2}\approx 11$$

Beginning salaries will be greater than \$35,000 11 years after 1989, which is in the year 2000.

b. answers may vary

71. Consumption of whole milk is decreasing over time; answers may vary.

73. 2010 is 14 years after 1996, so 2010 corresponds to $t = 14$.

$w = -1.9t + 70.1$
$w = -1.9(14) + 70.1$
$w = -26.6 + 70.1$
$w = 43.5$

The average consumption of whole milk is expected to be 43.5 pounds per person per year in 2010.

75.

$$\begin{aligned} -1.9t + 70.1 &< 50 \\ -1.9t + 70.1 - 70.1 &< 50 - 70.1 \\ -1.9t &< -20.1 \\ \frac{-1.9t}{-1.9} &> \frac{-20.1}{-1.9} \\ t &> 10.579 \end{aligned}$$

10.579 years after 1996 corresponds to some time during the year 2006. During 2006, the consumption of whole milk will be less than 50 pounds per person per year.

77. answers may vary

79. a.

$$\begin{aligned} -0.8t + 34.3 &> -1.9t + 70.1 \\ -0.8t + 34.3 - 34.3 &> -1.9t + 70.1 - 34.3 \\ -0.8t &> -1.9t + 35.8 \\ -0.8t + 1.9t &> -1.9t + 35.8 + 1.9t \\ 1.1t &> 35.8 \\ \frac{1.1t}{1.1} &> \frac{35.8}{1.1} \\ t &> 32.\overline{54} \end{aligned}$$

b. $1996 + 33 = 2029$

2029 is the first year when the consumption of skim milk is greater than the consumption of whole milk.

81. The integers that are both greater than or equal to 0 and less than or equal to 7 are {0, 1, 2, 3, 4, 5, 6, 7}.

83. The integers that are less than 6 and less than −5 are {..., −9, −8, −7, −6}.

85.

$$\begin{aligned} 3x - 12 &= 3 \\ 3x - 12 + 12 &= 3 + 12 \\ 3x &= 15 \\ \frac{3x}{3} &= \frac{15}{3} \\ x &= 5 \end{aligned}$$

The solution set is {5}.

87.

$$\begin{aligned} -5x - 4 &= -x - 4 \\ -5x - 4 + 4 &= -x - 4 + 4 \\ -5x &= -x \\ -5x + x &= -x + x \\ -4x &= 0 \\ \frac{-4x}{-4} &= \frac{0}{-4} \\ x &= 0 \end{aligned}$$

The solution set is {0}.

89. To solve $-3x \le 14$, both sides must be divided by −3, so the inequality symbol must be reversed.

91.

$$\begin{aligned} -x - 17 &\ge -23 \\ -x - 17 + 17 &\ge -23 + 17 \\ -x &\ge -6 \end{aligned}$$

To solve this inequality, both sides must be divided by −1, so the inequality symbol must be reversed.

93.

$$\begin{aligned} 2x - 3 &= 5 \\ 2x &= 8 \\ x &= 4 \end{aligned}$$

The solution set is {4}.

95.

$$\begin{aligned} 2x - 3 &> 5 \\ 2x &> 8 \\ x &> 4 \end{aligned}$$

The solution set is $(4, \infty)$.

97.

$$\begin{aligned} 4(x - 1) &\ge 4x - 8 \\ 4x - 4 &\ge 4x - 8 \\ 4x - 4 - 4x &\ge 4x - 8 - 4x \\ -4 &\ge -8 \end{aligned}$$

Since the statement $-4 \ge -8$ is true regardless of the value of x, every real number is a solution.

The solution set is $(-\infty, \infty)$.

99. $7x < 7(x-2)$
$7x < 7x - 14$
$7x - 7x < 7x - 14 - 7x$
$0 < -14$
Since the statement $0 < -14$ is false, the original inequality has no solution. The solution set is $\varnothing$.

101. answers may vary

103. answers may vary

Integrated Review

1. $-4x = 20$
$x = -5$
The solution set is $\{-5\}$.

2. $-4x < 20$
$x > -5$
The solution set is $(-5, \infty)$.

3. $\frac{3}{4}x \ge 2$
$3x \ge 8$
$x \ge \frac{8}{3}$
The solution set is $\left[\frac{8}{3}, \infty\right)$.

4. $5x + 3 \ge 2 + 4x$
$x + 3 \ge 2$
$x \ge -1$
The solution set is $[-1, \infty)$.

5. $6(y-4) = 3(y-8)$
$6y - 24 = 3y - 24$
$6y = 3y$
$3y = 0$
$y = 0$
The solution set is $\{0\}$.

6. $-4x \le \frac{2}{5}$
$x \ge -\frac{1}{10}$
The solution set is $\left[-\frac{1}{10}, \infty\right)$.

7. $-3x \ge \frac{1}{2}$
$x \le -\frac{1}{6}$
The solution set is $\left(-\infty, -\frac{1}{6}\right]$.

8. $5(y+4) = 4(y+5)$
$5y + 20 = 4y + 20$
$5y = 4y$
$y = 0$
The solution set is $\{0\}$.

9. $7x < 7(x-2)$
$7x < 7x - 14$
$0 < -14$
False; the solution set is $\varnothing$.

10. $\frac{-5x+11}{2} \le 7$
$-5x + 11 \le 14$
$-5x \le 3$
$x \ge -\frac{3}{5}$
The solution set is $\left[-\frac{3}{5}, \infty\right)$.

11. $-5x + 1.5 = -19.5$
$-5x = -21$
$x = \frac{21}{5} = 4.2$
The solution set is $\{4.2\}$.

12. $-5x + 4 = -26$
$-5x = -30$
$x = 6$
The solution set is $\{6\}$.

13. $$\begin{aligned} 5+2x-x &= -x+3-14 \\ 5+x &= -x-11 \\ x &= -x-16 \\ 2x &= -16 \\ x &= -8 \end{aligned}$$
The solution set is $\{-8\}$.

14. $$\begin{aligned} 12x+14 &< 11x-2 \\ 12x &< 11x-16 \\ x &< -16 \end{aligned}$$
The solution set is $(-\infty, -16)$.

15. $$\begin{aligned} \frac{x}{5}-\frac{x}{4} &= \frac{x-2}{2} \\ 20\left[\frac{x}{5}-\frac{x}{4}\right] &= 20\left(\frac{x-2}{2}\right) \\ 20\left(\frac{x}{5}\right)-20\left(\frac{x}{4}\right) &= 10(x-2) \\ 4x-5x &= 10x-20 \\ -x &= 10x-20 \\ -11x &= -20 \\ x &= \frac{20}{11} \end{aligned}$$
The solution set is $\left\{\frac{20}{11}\right\}$.

16. $$\begin{aligned} 12x-12 &= 8(x-1) \\ 12x-12 &= 8x-8 \\ 12x &= 8x+4 \\ 4x &= 4 \\ x &= 1 \end{aligned}$$
The solution set is $\{1\}$.

17. $$\begin{aligned} 2(x-3) &> 70 \\ 2x-6 &> 70 \\ 2x &> 76 \\ x &> 38 \end{aligned}$$
The solution set is $(38, \infty)$.

18. $$\begin{aligned} -3x-4.7 &= 11.8 \\ -3x &= 16.5 \\ x &= -5.5 \end{aligned}$$
The solution set is $\{-5.5\}$.

19. $$\begin{aligned} -2(b-4)-(3b-1) &= 5b+3 \\ -2b+8-3b+1 &= 5b+3 \\ -5b+9 &= 5b+3 \\ -5b &= 5b-6 \\ -10b &= -6 \\ b &= \frac{-6}{-10}=\frac{3}{5} \end{aligned}$$
The solution set is $\left\{\frac{3}{5}\right\}$.

20. $$\begin{aligned} 8(x+3) &< 7(x+5)+x \\ 8x+24 &< 7x+35+x \\ 8x+24 &< 8x+35 \\ 24 &< 35 \quad \text{True} \end{aligned}$$
The solution set is $(-\infty, \infty)$.

21. $$\begin{aligned} \frac{3t+1}{8} &= \frac{5+2t}{7}+2 \\ 56\left(\frac{3t+1}{8}\right) &= 56\left[\frac{5+2t}{7}+2\right] \\ 7(3t+1) &= 56\left(\frac{5+2t}{7}\right)+56(2) \\ 21t+7 &= 8(5+2t)+112 \\ 21t+7 &= 40+16t+112 \\ 21t+7 &= 16t+152 \\ 21t &= 16t+145 \\ 5t &= 145 \\ t &= 29 \end{aligned}$$
The solution set is $\{29\}$.

22. $$\begin{aligned} 4(x-6)-x &= 8(x-3)-5x \\ 4x-24-x &= 8x-24-5x \\ 3x-24 &= 3x-24 \\ -24 &= -24 \quad \text{True} \end{aligned}$$
The solution set is $\{x|x \text{ is a real number}\}$.

23. $\frac{x+3}{12}+\frac{x-5}{15}<\frac{2}{3}$

$60\left[\frac{x+3}{12}+\frac{x-5}{15}\right]<60\left(\frac{2}{3}\right)$

$60\left(\frac{x+3}{12}\right)+60\left(\frac{x-5}{15}\right)<40$

$5(x+3)+4(x-5)<40$

$5x+15+4x-20<40$

$9x-5<40$

$9x<45$

$x<5$

The solution set is $(-\infty, 5)$.

24. $\frac{y}{3}+\frac{y}{5}=\frac{y+3}{10}$

$30\left[\frac{y}{3}+\frac{y}{5}\right]=30\left[\frac{y+3}{10}\right]$

$30\left(\frac{y}{3}\right)+30\left(\frac{y}{5}\right)=3(y+3)$

$10y+6y=3y+9$

$16y=3y+9$

$13y=9$

$y=\frac{9}{13}$

The solution set is $\left\{\frac{9}{13}\right\}$.

25. $5(x-6)+2x>3(2x-1)-4$

$5x-30+2x>6x-3-4$

$7x-30>6x-7$

$7x>6x+23$

$x>23$

The solution set is $(23, \infty)$.

26. $14(x-1)-7x\le 2(3x-6)+4$

$14x-14-7x\le 6x-12+4$

$7x-14\le 6x-8$

$7x\le 6x+6$

$x\le 6$

The solution set is $(-\infty, 6]$.

27. $\frac{1}{4}(3x+2)-x\ge\frac{3}{8}(x-5)+2$

$8\left[\frac{1}{4}(3x+2)-x\right]\ge 8\left[\frac{3}{8}(x-5)+2\right]$

$8\left[\frac{1}{4}(3x+2)\right]-8(x)\ge 8\left[\frac{3}{8}(x-5)\right]+8(2)$

$2(3x+2)-8x\ge 3(x-5)+16$

$6x+4-8x\ge 3x-15+16$

$-2x+4\ge 3x+1$

$-2x\ge 3x-3$

$-5x\ge -3$

$x\le\frac{3}{5}$

The solution set is $\left(-\infty, \frac{3}{5}\right]$.

28. $\frac{1}{3}(x-10)-4x>\frac{5}{6}(2x+1)-1$

$6\left[\frac{1}{3}(x-10)-4x\right]>6\left[\frac{5}{6}(2x+1)-1\right]$

$6\left[\frac{1}{3}(x-10)\right]-6(4x)>6\left[\frac{5}{6}(2x+1)\right]-6(1)$

$2(x-10)-24x>5(2x+1)-6$

$2x-20-24x>10x+5-6$

$-22x-20>10x-1$

$-22x>10x+19$

$-32x>19$

$x<-\frac{19}{32}$

The solution set is $\left(-\infty, -\frac{19}{32}\right)$.

Section 2.5 Practice Problems

1. The numbers 3, 4, and 5 are in both sets. The intersection is {3, 4, 5}.

2. $x+5<9$ and $3x-1<2$

$x<4$ and $3x<3$

$x<4$ and $x<1$

The solution set is

$(-\infty, 4)\cap(-\infty, 1)=(-\infty, 1)$.

3. $4x \geq 0$ and $2x + 4 \leq 2$
$x \geq 0$ and $2x \leq -2$
$x \geq 0$ and $x \leq -1$
There is no number greater than or equal to 0 and less than or equal to −1. The solution set is ∅.

4. $5 < 1 - x < 9$
$5 - 1 < 1 - x - 1 < 9 - 1$
$4 < -x < 8$
$-4 > x > -8$
The solution set is (−8, −4).

5. $-2 < \frac{3}{4}x + 2 \leq 5$
$-2 - 2 < \frac{3}{4}x + 2 - 2 \leq 5 - 2$
$-4 < \frac{3}{4}x \leq 3$
$4(-4) < 4\left(\frac{3}{4}x\right) \leq 4(3)$
$-16 < 3x \leq 12$
$\frac{-16}{3} < \frac{3x}{3} \leq \frac{12}{3}$
$-\frac{16}{3} < x \leq 4$
The solution set is $\left(-\frac{16}{3}, 4\right]$.

6. The numbers that are in either set are {1, 2, 3, 4, 5, 6}. This set is the union.

7. $3x - 2 \geq 10$ or $x - 6 \leq -4$
$3x \geq 12$ or $x \leq 2$
$x \geq 4$ or $x \leq 2$
The solution set is $(-\infty, 2] \cup [4, \infty)$.

8. $x - 7 \leq -1$ or $2x - 6 \geq 2$
$x \leq 6$ or $2x \geq 8$
$x \leq 6$ or $x \geq 4$
The solution set is $(-\infty, \infty)$.

Exercise Set 2.5

1. $C \cup D = \{2, 3, 4, 5\} \cup \{4, 5, 6, 7\}$
$= \{2, 3, 4, 5, 6, 7\}$

3. $A \cap D = \{x | x \text{ is an even integer}\} \cap \{4, 5, 6, 7\}$
$= \{4, 6\}$

5. $A \cup B = \{x \mid x \text{ is an even integer}\} \cup \{x \mid x \text{ is an odd integer}\}$
$= \{x \mid x \text{ is an integer}\}$
$= \{\ldots, -2, -1, 0, 1, 2, \ldots\}$

7. $B \cap D = \{x | x \text{ is an odd integer}\} \cap \{4, 5, 6, 7\}$
$= \{5, 7\}$

9. $B \cup C = \{x \mid x \text{ is an odd integer}\} \cup \{2, 3, 4, 5\}$
$= \{x \mid x \text{ is an odd integer or } x = 2 \text{ or } x = 4\}$

11. $A \cap C = \{x | x \text{ is an even integer}\} \cap \{2, 3, 4, 5\}$
$= \{2, 4\}$

13.
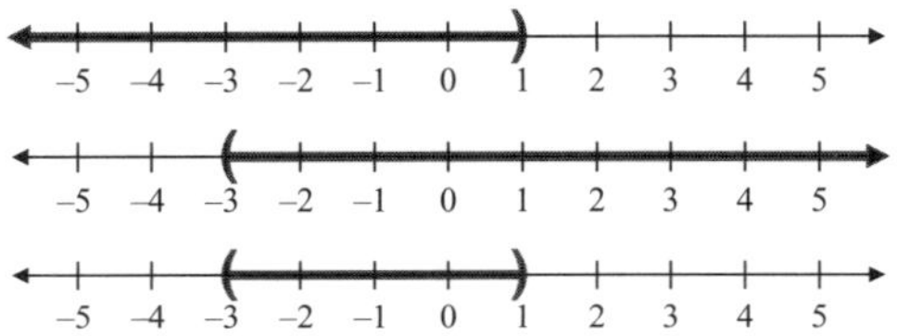

The solution set is $(-3, 1)$.

15.
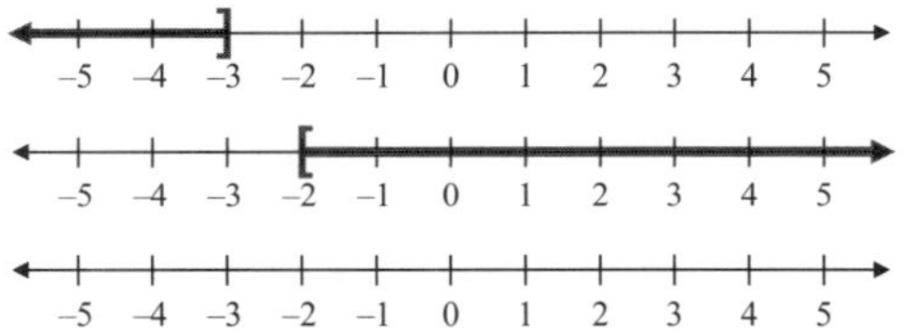

The solution set is $\varnothing$.

17.
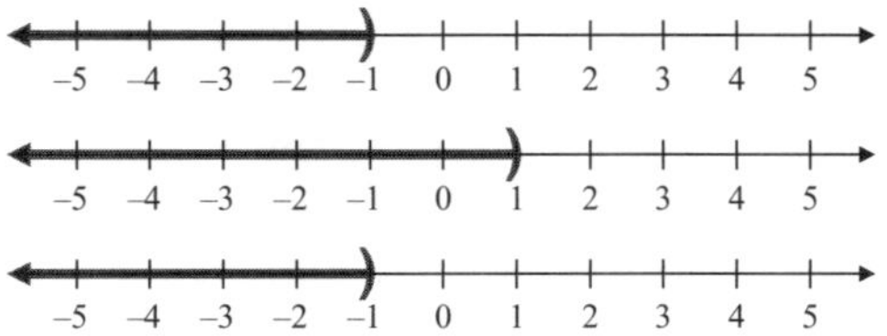

The solution set is $(-\infty, -1)$.

19. $x + 1 \geq 7$ and $3x - 1 \geq 5$
$x \geq 6$ and $3x \geq 6$
$x \geq 6$ and $x \geq 2$

The solution set is $[6, \infty)$.

21. $4x + 2 \le -10$ and $2x \le 0$
$4x \le -12$ and $x \le 0$
$x \le -3$ and $x \le 0$
The solution set is
$(-\infty, -3] \cap (-\infty, 0] = (-\infty, -3]$.

23. $-2x < -8$ and $x - 5 < 5$
$x > 4$ and $x < 10$
The solution set is $(4, 10)$.

25.
$$5 < x - 6 < 11$$
$$5 + 6 < x - 6 + 6 < 11 + 6$$
$$11 < x < 17$$
The solution set is $(11, 17)$.

27.
$$-2 \le 3x - 5 \le 7$$
$$-2 + 5 \le 3x - 5 + 5 \le 7 + 5$$
$$3 \le 3x \le 12$$
$$\frac{3}{3} \le \frac{3x}{3} \le \frac{12}{3}$$
$$1 \le x \le 4$$
The solution set is $[1, 4]$.

29.
$$1 \le \frac{2}{3}x + 3 \le 4$$
$$1 - 3 \le \frac{2}{3}x + 3 - 3 \le 4 - 3$$
$$-2 \le \frac{2}{3}x \le 1$$
$$\frac{3}{2}(-2) \le \frac{3}{2}\left(\frac{2}{3}x\right) \le \frac{3}{2}(1)$$
$$-3 \le x \le \frac{3}{2}$$
The solution set is $\left[-3, \frac{3}{2}\right]$.

31.
$$-5 \le \frac{-3x+1}{4} \le 2$$
$$4(-5) \le 4\left(\frac{-3x+1}{4}\right) \le 4(2)$$
$$-20 \le -3x + 1 \le 8$$
$$-20 - 1 \le -3x + 1 - 1 \le 8 - 1$$
$$-21 \le -3x \le 7$$
$$\frac{-21}{-3} \ge \frac{-3}{-3}x \ge \frac{7}{-3}$$
$$7 \ge x \ge -\frac{7}{3}$$
The solution set is $\left[-\frac{7}{3}, 7\right]$.

33.

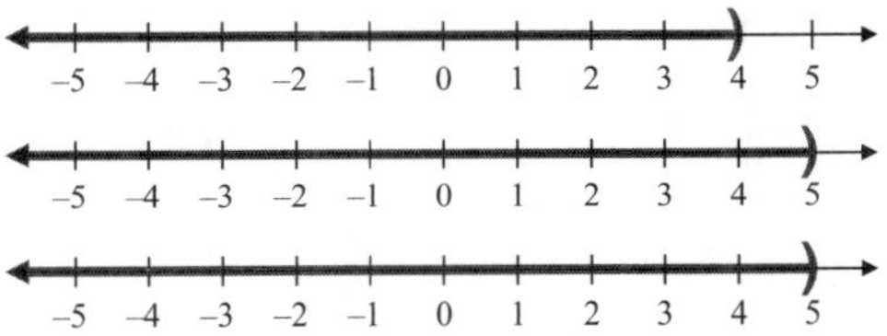

The solution set is $(-\infty, 5)$.

35.

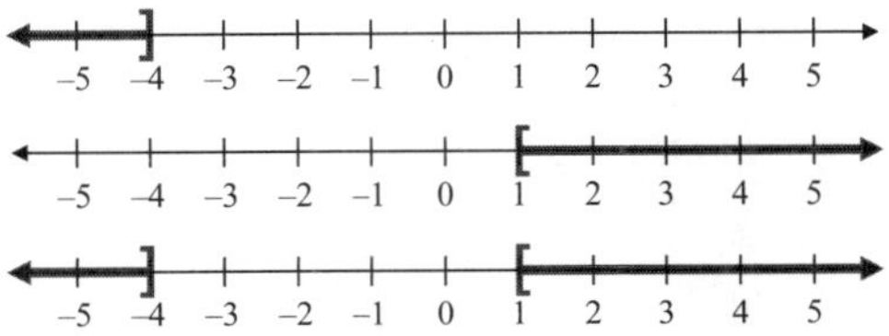

The solution set is $(-\infty, -4] \cup [1, \infty)$.

37.

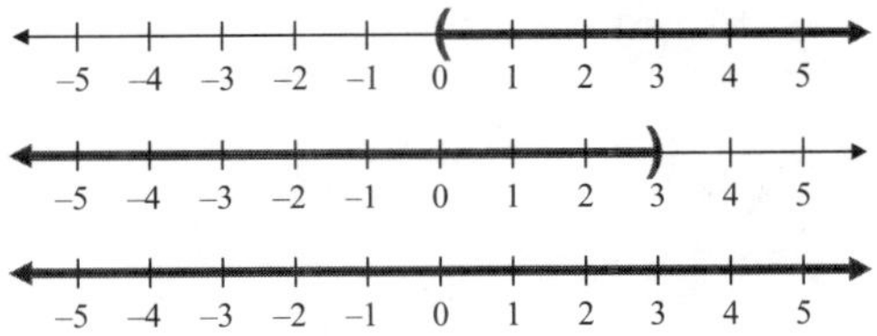

The solution set is $(-\infty, \infty)$.

39. $-2x \le -4$ or $5x - 20 \ge 5$
$x \ge 2$ or $5x \ge 25$
$x \ge 2$ or $x \ge 5$
The solution set is $[2, \infty)$.

41. $x + 4 < 0$ *or* $6x > -12$
$x < -4$ *or* $x > -2$
The solution set is $(-\infty, -4) \cup (-2, \infty)$.

43. $3(x-1) < 12$ or $x+7 > 10$
$3x-3 < 12$ or $x > 3$
$3x < 15$ or $x > 3$
$x < 5$ or $x > 3$
The solution set is $(-\infty, \infty)$.

45. $x < \frac{2}{3}$ and $x > -\frac{1}{2}$
The solution set is
$\left(-\infty, \frac{2}{3}\right) \cap \left(-\frac{1}{2}, \infty\right) = \left(-\frac{1}{2}, \frac{2}{3}\right)$.

47. $x < \frac{2}{3}$ or $x > -\frac{1}{2}$
The solution set is
$\left(-\infty, \frac{2}{3}\right) \cup \left(-\frac{1}{2}, \infty\right) = (-\infty, \infty)$.

49.
$$0 \le 2x-3 \le 9$$
$$0+3 \le 2x-3+3 \le 9+3$$
$$3 \le 2x \le 12$$
$$\frac{3}{2} \le \frac{2x}{2} \le \frac{12}{2}$$
$$\frac{3}{2} \le x \le 6$$
The solution set is $\left[\frac{3}{2}, 6\right]$.

51.
$$\frac{1}{2} < x - \frac{3}{4} < 2$$
$$4\left(\frac{1}{2}\right) < 4\left(x - \frac{3}{4}\right) < 4(2)$$
$$2 < 4(x) - 4\left(\frac{3}{4}\right) < 8$$
$$2 < 4x - 3 < 8$$
$$2+3 < 4x-3+3 < 8+3$$
$$5 < 4x < 11$$
$$\frac{5}{4} < \frac{4x}{4} < \frac{11}{4}$$
$$\frac{5}{4} < x < \frac{11}{4}$$
The solution set is $\left(\frac{5}{4}, \frac{11}{4}\right)$.

53. $x+3 \ge 3$ and $x+3 \le 2$
$x \ge 0$ and $x \le -1$
The solution set is $[0, \infty) \cap (-\infty, -1] = \varnothing$.

55. $3x \ge 5$ or $-\frac{5}{8}x - 6 > 1$
$x \ge \frac{5}{3}$ or $-\frac{5}{8}x > 7$
$x \ge \frac{5}{3}$ or $-\frac{8}{5}\left(\frac{-5x}{8}\right) < \frac{-8}{5}(7)$
$x \ge \frac{5}{3}$ or $x < -\frac{56}{5}$
The solution set is $\left(-\infty, -\frac{56}{5}\right) \cup \left[\frac{5}{3}, \infty\right)$.

57.
$$0 < \frac{5-2x}{3} < 5$$
$$3(0) < 3\left(\frac{5-2x}{3}\right) < 3(5)$$
$$0 < 5-2x < 15$$
$$0-5 < 5-2x-5 < 15-5$$
$$-5 < -2x < 10$$
$$\frac{-5}{-2} > \frac{-2x}{-2} > \frac{10}{-2}$$
$$\frac{5}{2} > x > -5$$
$$-5 < x < \frac{5}{2}$$
The solution set is $\left(-5, \frac{5}{2}\right)$.

59.
$$-6 < 3(x-2) \le 8$$
$$-6 < 3x-6 \le 8$$
$$-6+6 < 3x-6+6 \le 8+6$$
$$0 < 3x \le 14$$
$$\frac{0}{3} < \frac{3x}{3} \le \frac{14}{3}$$
$$0 < x \le \frac{14}{3}$$
The solution set is $\left(0, \frac{14}{3}\right]$.

61. $-x+5>6$ and $1+2x\le -5$
$-x>1$ and $2x\le -6$
$x<-1$ and $x\le -3$
The solution set is
$(-\infty, -1)\cap(-\infty, -3] = (-\infty, -3]$.

63. $3x+2\le 5$ or $7x>29$
$3x\le 3$ or $x>\frac{29}{7}$
$x\le 1$ or $x>\frac{29}{7}$
The solution set is $(-\infty, 1]\cup\left(\frac{29}{7}, \infty\right)$.

65. $5-x>7$ and $2x+3\ge 13$
$-x>2$ and $2x\ge 10$
$x<-2$ and $x\ge 5$
The solution set is $(-\infty, -2)\cap[5, \infty) = \varnothing$.

67.
$$-\frac{1}{2}\le\frac{4x-1}{6}<\frac{5}{6}$$
$$6\left(-\frac{1}{2}\right)\le 6\left(\frac{4x-1}{6}\right)<6\left(\frac{5}{6}\right)$$
$$-3\le 4x-1<5$$
$$-3+1\le 4x-1+1<5+1$$
$$-2\le 4x<6$$
$$-\frac{2}{4}\le\frac{4x}{4}<\frac{6}{4}$$
$$-\frac{1}{2}\le x<\frac{3}{2}$$
The solution set is $\left[-\frac{1}{2}, \frac{3}{2}\right)$.

69.
$$\frac{1}{15}<\frac{8-3x}{15}<\frac{4}{5}$$
$$15\left(\frac{1}{15}\right)<15\left(\frac{8-3x}{15}\right)<15\left(\frac{4}{5}\right)$$
$$1<8-3x<12$$
$$1-8<8-3x-8<12-8$$
$$-7<-3x<4$$
$$\frac{-7}{-3}>\frac{-3x}{-3}>\frac{4}{-3}$$
$$\frac{7}{3}>x>-\frac{4}{3}$$
$$-\frac{4}{3}<x<\frac{7}{3}$$
The solution set is $\left(-\frac{4}{3}, \frac{7}{3}\right)$.

71.
$$0.3<0.2x-0.9<1.5$$
$$10(0.3)<10(0.2x-0.9)<10(1.5)$$
$$3<10(0.2x)-10(0.9)<15$$
$$3<2x-9<15$$
$$3+9<2x-9+9<15+9$$
$$12<2x<24$$
$$\frac{12}{2}<\frac{2x}{2}<\frac{24}{2}$$
$$6<x<12$$
The solution set is (6, 12).

73. $|-7|-|19| = 7-19 = -12$

75. $-(-6)-|-10| = 6-10 = -4$

77. $|x| = 7$ when $x = 7$ or $x = -7$.
The solution set is $\{-7, 7\}$.

79. $|x| = 0$ when $x = 0$.
The solution set is $\{0\}$.

81. Both lines are above the level representing 48 pounds per person for the years 1993, 1994, 1995, 1998, 1999, and 2002.

83. answers may vary

85. $x - 6 < 3x < 2x + 5$

$$\begin{aligned} x-6<3x \quad &\text{and} \quad 3x<2x+5 \\ -6<2x \quad &\text{and} \quad x<5 \\ -3<x \quad &\text{and} \quad x<5 \end{aligned}$$

The solution set is $(-3, \infty) \cap (-\infty, 5) = (-3, 5)$.

87. $x + 3 < 2x + 1 < 4x + 6$

$$\begin{aligned} x+3<2x+1 \quad &\text{and} \quad 2x+1<4x+6 \\ x<2x-2 \quad &\text{and} \quad 2x<4x+5 \\ -x<-2 \quad &\text{and} \quad -2x<5 \\ x>2 \quad &\text{and} \quad x>-\frac{5}{2} \end{aligned}$$

The solution set is $(2, \infty) \cap \left(-\frac{5}{2}, \infty\right) = (2, \infty)$.

89.
$$\begin{aligned} -29 \le C &\le 35 \\ -29 \le \frac{5}{9}(F-32) &\le 35 \\ \frac{9}{5}(-29) \le \frac{9}{5}\left[\frac{5}{9}(F-32)\right] &\le \frac{9}{5}(35) \\ -52.2 \le F-32 &\le 63 \\ -52.2+32 \le F-32+32 &\le 63+32 \\ -20.2 \le F &\le 95 \end{aligned}$$

The temperatures ranged from −20.2°F to 95°F.

91. Let x be his final exam score.

$$\begin{aligned} 70 \le \frac{68+65+75+78+2x}{6} &\le 79 \\ 6(70) \le 6\left(\frac{68+65+75+78+2x}{6}\right) &\le 6(79) \\ 420 \le 68+65+75+78+2x &\le 474 \\ 420 \le 286+2x &\le 474 \\ 420-286 \le 286+2x-286 &\le 474-286 \\ 134 \le 2x &\le 188 \\ \frac{134}{2} \le \frac{2x}{2} &\le \frac{188}{2} \\ 67 \le x &\le 94 \end{aligned}$$

To earn a C in the course, his final exam grade must be between 67 and 94.

Section 2.6 Practice Problems

1. $|y| = 5$

$y = 5$ or $y = -5$

The solution set is $\{-5, 5\}$.

2. $|p| = -4$

The absolute value of a number is never negative, so the equation has no solution. The solution set is $\varnothing$.

3. $|x| = 0$

We are looking for all numbers whose distance from 0 is zero units. The only number is 0. The solution set is $\{0\}$.

4. $|4x + 2| = 6$

$$\begin{aligned} 4x+2=6 \quad &\text{or} \quad 4x+2=-6 \\ 4x=4 \quad &\quad\;\; 4x=-8 \\ x=1 \quad &\quad\;\; x=-2 \end{aligned}$$

The solution set is $\{1, -2\}$.

5. $\left|\frac{x}{3}+4\right| = 1$

$$\begin{aligned} \frac{x}{3}+4=1 \quad &\text{or} \quad \frac{x}{3}+4=-1 \\ \frac{x}{3}=-3 \quad &\text{or} \quad \frac{x}{3}=-5 \\ 3\left(\frac{x}{3}\right)=3(-3) \quad &\text{or} \quad 3\left(\frac{x}{3}\right)=3(-5) \\ x=-9 \quad &\text{or} \quad x=-15 \end{aligned}$$

The solution set is $\{-9, -15\}$.

6. $|4x+2|+1=7$

$|4x+2|=6$

$$\begin{aligned} 4x+2=6 \quad &\text{or} \quad 4x+2=-6 \\ 4x=4 \quad &\text{or} \quad 4x=-8 \\ x=1 \quad &\text{or} \quad x=-2 \end{aligned}$$

The solution set is $\{1, -2\}$.

7. $|4x-5|=|3x+5|$

$$\begin{array}{lcl} 4x-5=3x+5 & \text{or} & 4x-5=-(3x+5) \\ 4x-5=3x+5 & \text{or} & 4x-5=-3x-5 \\ 4x=3x+10 & \text{or} & 4x=-3x \\ x=10 & \text{or} & 7x=0 \\ x=10 & \text{or} & x=0 \end{array}$$

The solution set is $\{0, 10\}$.

8. $|x+2|=|4-x|$

$$\begin{array}{lcl} x+2=4-x & \text{or} & x+2=-(4-x) \\ x+2=4-x & \text{or} & x+2=-4+x \\ x=2-x & \text{or} & x=-6+x \\ 2x=2 & \text{or} & 0=-6 \\ x=1 & & \text{False} \end{array}$$

The solution set is $\{1\}$.

9. $|x|<4$

$-4<x<4$

The solution set is $(-4, 4)$.

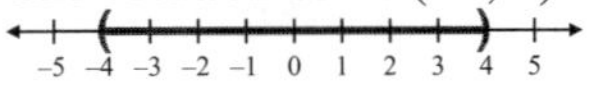

10. $|x|\ge 5$

$x\ge 5 \cup x\le -5$

The solution set is $(-\infty, -5] \cup [5, \infty)$.

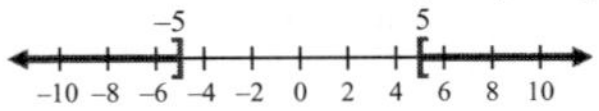

11. $|x+2|>4$

$$\begin{array}{lcl} x+2<-4 & \text{or} & x+2>4 \\ x<-6 & \text{or} & x>2 \end{array}$$

The solution set is $(-\infty, -6) \cup (2, \infty)$.

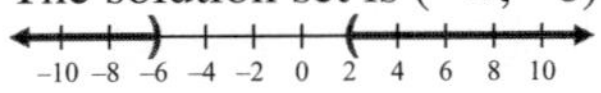

12. $|x-3|=5$

$$\begin{array}{lcl} x-3=5 & \text{or} & x-3=-5 \\ x=8 & \text{or} & x=-2 \end{array}$$

The solution set is $\{-2, 8\}$.

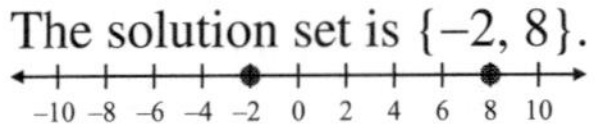

13. $|x-2|\le 1$

$$\begin{aligned} -1\le x-2&\le 1 \\ -1+2\le x-2+2&\le 1+2 \\ 1\le x&\le 3 \end{aligned}$$

The solution set is $[1, 3]$.

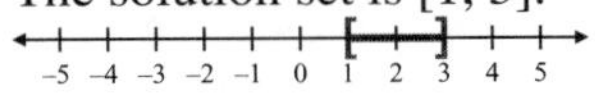

14. $|2x-5|+2\le 9$

$$\begin{aligned} |2x-5|&\le 7 \\ -7\le 2x-5&\le 7 \\ -7+5\le 2x-5+5&\le 7+5 \\ -2\le 2x&\le 12 \\ -1\le x&\le 6 \end{aligned}$$

The solution set is $[-1, 6]$.

15. $|x|<-1$

The absolute value of a number is never negative. Thus it will then never be less than -1. The solution set is $\varnothing$.

16. $|x+1|\ge -3$

The absolute value of a number is always nonnegative. Thus it will always be greater than -3. The solution set contains all real numbers, or $(-\infty, \infty)$.

Mental Math

1. $|2x+1|=3$

$2x+1=3$ or $2x+1=-3$ d

2. $|2x+1|\le 3$

$-3\le 2x+1<3$ e

3. $|2x+1|<3$

$-3<2x+1<3$ c

4. $|2x+1|\ge 3$

$2x+1\ge 3$ or $2x+1\le -3$ b

5. $|2x+1|>3$

$2x+1>3$ or $2x+1<-3$ a

Exercise Set 2.6

1. $|x|=7$

$x=7$ or $x=-7$

The solution set is $\{7, -7\}$.

3. $|x|=-4$

The absolute value of a number is never negative. The solution set is $\varnothing$.

5. $|3x| = 12.6$
$3x = 12.6$ or $3x = -12.6$
$x = 4.2$ or $x = -4.2$
The solution set is $\{4.2, -4.2\}$.

7. $3|x| - 5 = 7$
$3|x| = 12$
$|x| = 4$
$x = 4$ or $x = -4$
The solution set is $\{-4, 4\}$.

9. $-6|x| + 44 = -10$
$-6|x| = -54$
$|x| = 9$
$x = 9$ or $x = -9$
The solution set is $\{-9, 9\}$.

11. $|x - 9| = 14$
$x - 9 = 14$ or $x - 9 = -14$
$x = 23$ or $x = -5$
The solution set is $\{-5, 23\}$.

13. $|2x - 5| = 9$
$2x - 5 = 9$ or $2x - 5 = -9$
$2x = 14$ or $2x = -4$
$x = 7$ or $x = -2$
The solution set is $\{7, -2\}$.

15. $\left|\frac{x}{2} - 3\right| = 1$
$\frac{x}{2} - 3 = 1$ or $\frac{x}{2} - 3 = -1$
$\frac{x}{2} = 4$ or $\frac{x}{2} = 2$
$x = 8$ or $x = 4$
The solution set is $\{4, 8\}$.

17. $|z| + 4 = 9$
$|z| = 5$
$z = 5$ or $z = -5$
The solution set is $\{5, -5\}$.

19. $|3x| + 5 = 14$
$|3x| = 9$
$3x = 9$ or $3x = -9$
$x = 3$ or $x = -3$
The solution set is $\{-3, 3\}$.

21. $\left|\frac{4x - 6}{3}\right| = 6$
$\frac{4x - 6}{3} = 6$ or $\frac{4x - 6}{3} = -6$
$4x - 6 = 18$ or $4x - 6 = -18$
$4x = 24$ or $4x = -12$
$x = 6$ or $x = -3$
The solution set is $\{-3, 6\}$.

23. $|2x| = 0$
$2x = 0$
$x = 0$
The solution set is $\{0\}$.

25. $|4n + 1| + 10 = 4$
$|4n + 1| = -6$
Since $|4n + 1| \geq 0$, there are no values satisfying $|4n + 1| = -6$.
The solution set is $\varnothing$.

27. $3|x - 1| + 19 = 23$
$3|x - 1| = 4$
$|x - 1| = \frac{4}{3}$
$x - 1 = \frac{4}{3}$ or $x - 1 = -\frac{4}{3}$
$x = \frac{7}{3}$ or $x = -\frac{1}{3}$
The solution set is $\left\{-\frac{1}{3}, \frac{7}{3}\right\}$.

29. $|5x - 7| = |3x + 11|$

$$\begin{aligned} 5x-7 &= 3x+11 &\text{ or }&& 5x-7 &= -(3x+11) \\ 2x-7 &= 11 &\text{ or }&& 5x-7 &= -3x-11 \\ 2x &= 18 &\text{ or }&& 8x-7 &= -11 \\ x &= 9 &\text{ or }&& 8x &= -4 \\ &&&& x &= -\frac{4}{8} = -\frac{1}{2} \end{aligned}$$

The solution set is $\left\{-\frac{1}{2}, 9\right\}$.

31. $|z + 8| = |z - 3|$

$$\begin{aligned} z+8 &= z-3 &\text{ or }&& z+8 &= -(z-3) \\ 8 &= -3 &\text{ or }&& z+8 &= -z+3 \\ &\text{False} &\text{ or }&& z &= -z-5 \\ &\text{False} &\text{ or }&& 2z &= -5 \\ &\text{False} &\text{ or }&& z &= -\frac{5}{2} \end{aligned}$$

The solution set is $\left\{-\frac{5}{2}\right\}$.

33. $|2y - 3| = |9 - 4y|$

$$\begin{aligned} 2y-3 &= 9-4y &\text{ or }&& 2y-3 &= -(9-4y) \\ 6y &= 12 &\text{ or }&& 2y-3 &= -9+4y \\ y &= 2 &\text{ or }&& 6 &= 2y \\ y &= 2 &\text{ or }&& 3 &= y \end{aligned}$$

The solution set is {2, 3}.

35. $\left|\frac{3}{4}x - 2\right| = \left|\frac{1}{4}x + 6\right|$

$$\begin{aligned} \frac{3}{4}x-2 &= \frac{1}{4}x+6 &\text{ or }&& \frac{3}{4}x-2 &= -\left(\frac{1}{4}x+6\right) \\ \frac{3}{4}x &= \frac{1}{4}x+8 &\text{ or }&& \frac{3}{4}x-2 &= -\frac{1}{4}x-6 \\ \frac{2}{4}x &= 8 &\text{ or }&& \frac{3}{4}x &= -\frac{1}{4}x-4 \\ 2x &= 32 &\text{ or }&& x &= -4 \\ x &= 16 &\text{ or }&& x &= -4 \end{aligned}$$

The solution set is {−4, 16}.

37. $|2x - 6| = |10 - 2x|$

$2x-6=10-2x$ or $2x-6=-(10-2x)$
$4x-6=10$ or $2x-6=-10+2x$
$4x=16$ or $-6=-10$ False
$x=4$

The solution set is $\{4\}$.

39. $|x + 4| = |7 - x|$

$x+4=7-x$ or $x+4=-(7-x)$
$x=3-x$ or $x+4=-7+x$
$2x=3$ or $4=-7$
$x=\frac{3}{2}$ or False

The solution set is $\left\{\frac{3}{2}\right\}$.

41. $\left|\frac{2x+1}{5}\right|=\left|\frac{3x-7}{3}\right|$

$\frac{2x+1}{5}=\frac{3x-7}{3}$ or $\frac{2x+1}{5}=-\left(\frac{3x-7}{3}\right)$
$15\cdot\frac{2x+1}{5}=15\cdot\frac{3x-7}{3}$ or $\frac{2x+1}{5}=\frac{-3x+7}{3}$
$3(2x+1)=5(3x-7)$ or $15\cdot\frac{2x+1}{5}=15\cdot\frac{-3x+7}{3}$
$6x+3=15x-35$ or $3(2x+1)=5(-3x+7)$
$-9x+3=-35$ or $6x+3=-15x+35$
$-9x=-38$ or $21x+3=35$
$x=\frac{-38}{-9}=\frac{38}{9}$ or $21x=32$
$x=\frac{32}{21}$

The solution set is $\left\{\frac{32}{21}, \frac{38}{9}\right\}$.

43. $|5x + 1| = |4x - 7|$

$5x+1=4x-7$ or $5x+1=-(4x-7)$
$5x=4x-8$ or $5x+1=-4x+7$
$x=-8$ or $5x=-4x+6$
$x=-8$ or $9x=6$
$x=-8$ or $x=\frac{6}{9}=\frac{2}{3}$

The solution set is $\left\{-8, \frac{2}{3}\right\}$.

45. $|x| \le 4$
$-4 \le x \le 4$
The solution set is $[-4, 4]$.

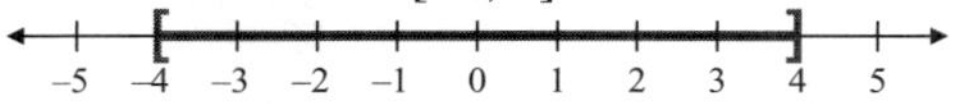

47. $|x| > 3$
$x < -3$ or $x > 3$
The solution set is $(-\infty, -3) \cup (3, \infty)$.

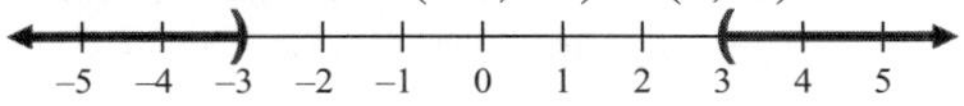

49. $|x + 3| < 2$
$-2 < x + 3 < 2$
$-5 < x < -1$
The solution set is $(-5, -1)$.

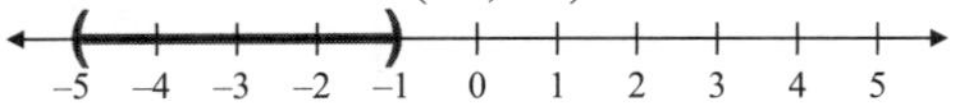

51. $|y - 6| \ge 7$
$y - 6 \le -7$ or $y - 6 \ge 7$
$y \le -1$ or $y \ge 13$
The solution set is $(-\infty, -1] \cup [13, \infty)$.

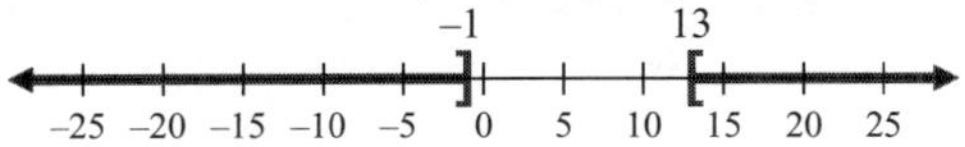

53. $\left|\frac{x+2}{3}\right| < 1$

$-1 < \frac{x+2}{3} < 1$
$-3 < x + 2 < 3$
$-5 < x < 1$
The solution set is $(-5, 1)$.

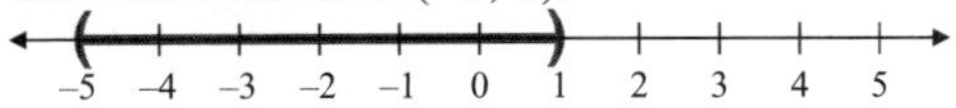

55. $|x| + 7 \le 12$
$|x| \le 5$
$-5 \le x \le 5$
The solution set is $[-5, 5]$.

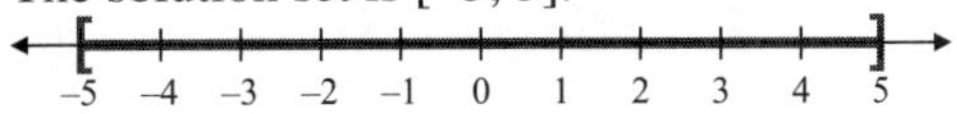

57. $|2x + 3| \le 0$
Since absolute value can never be negative, the only solution is when $|2x + 3| = 0$.
$|2x + 3| = 0$
$2x + 3 = 0$
$2x = -3$
$x = -\frac{3}{2}$

The solution set is $\left\{-\frac{3}{2}\right\}$.

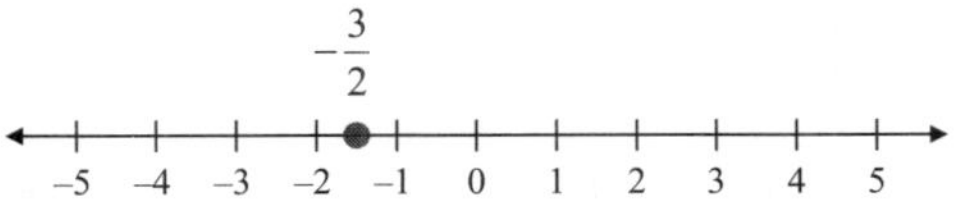

59. $|x| + 2 > 6$
$|x| > 4$
$x < -4$ or $x > 4$
The solution set is $(-\infty, -4) \cup (4, \infty)$.

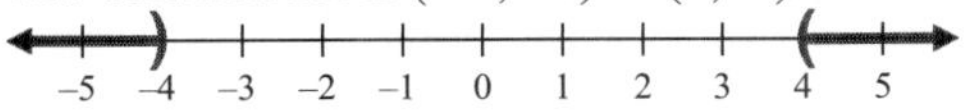

61. $|2x + 7| \le 13$
$-13 \le 2x + 7 \le 13$
$-20 \le 2x \le 6$
$-10 \le x \le 3$
The solution set is $[-10, 3]$.

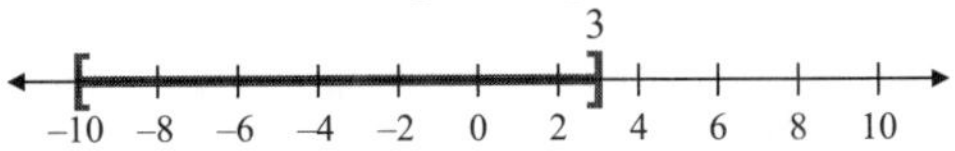

63. $|8 - 3x| < 5$
$-5 < 8 - 3x < 5$
$-13 < -3x < -3$
$\frac{13}{3} > x > 1$

The solution set is $\left(1, \frac{13}{3}\right)$.

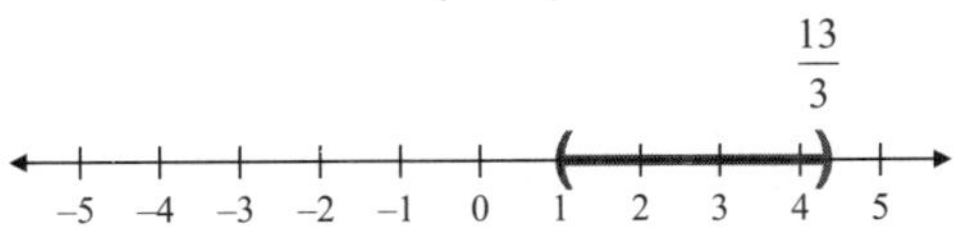

65. $|x+10| \ge 14$

$x+10 \le -14$ or $x+10 \ge 14$

$x \le -24$ or $x \ge 4$

The solution set is $(-\infty, -24] \cup [4, \infty)$.

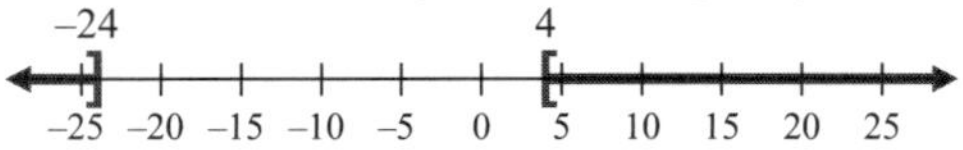

67. $|2x-7| \le 11$

$-11 \le 2x-7 \le 11$

$-4 \le 2x \le 18$

$-2 \le x \le 9$

The solution set is $[-2, 9]$.

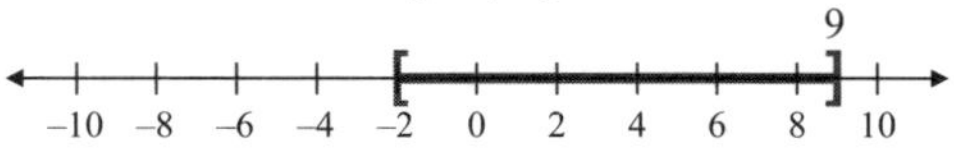

69. $|x| > -4$

The absolute value of a number is always nonnegative. Thus, it will always be greater than −4. The solution set is all real numbers or $(-\infty, \infty)$.

71. $6+|4x-1| \le 9$

$|4x-1| \le 3$

$-3 \le 4x-1 \le 3$

$-2 \le 4x \le 4$

$-\frac{2}{4} \le x \le 1$

$-\frac{1}{2} \le x \le 1$

The solution set is $\left[-\frac{1}{2}, 1\right]$.

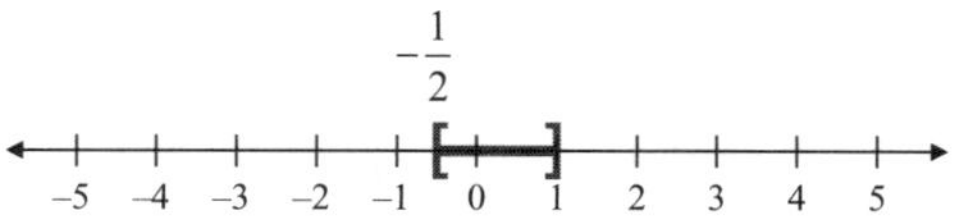

73. $|6x-8|-7 > -3$

$|6x-8| > 4$

$6x-8 < -4$ or $6x-8 > 4$

$6x < 4$ or $6x > 12$

$x < \frac{2}{3}$ or $x > 2$

The solution set is $\left(-\infty, \frac{2}{3}\right) \cup (2, \infty)$.

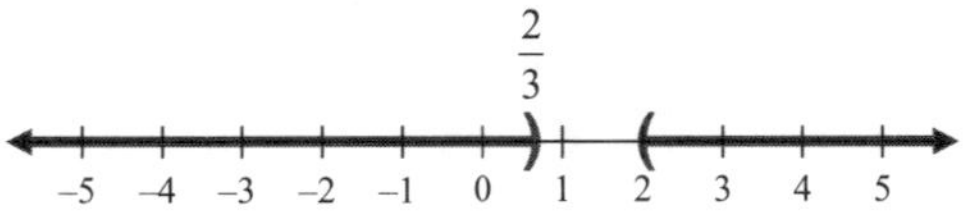

75. $|5x+3| < -6$

The absolute value of a number is never negative. Thus it will never be less than −6. The solution set is ∅.

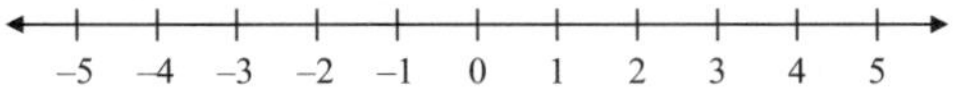

77. $\left|\frac{x+6}{3}\right| > 2$

$\frac{x+6}{3} < -2$ or $\frac{x+6}{3} > 2$

$x+6 < -6$ or $x+6 > 6$

$x < -12$ or $x > 0$

The solution set is $(-\infty, -12) \cup (0, \infty)$.

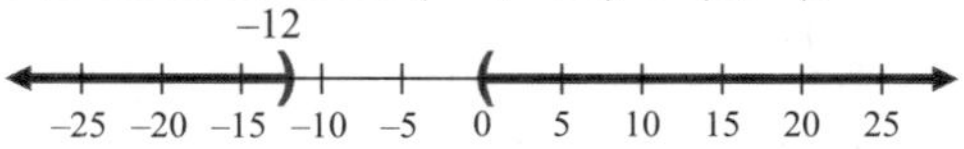

79. $|x| = 13$

$x = 13$ or $x = -13$

The solution set is $\{-13, 13\}$.

81. $|x| < 13$

$-13 < x < 13$

The solution set is $(-13, 13)$.

83. $|x|+12 = 9$

$|x| = -3$

The absolute value of a number is never negative. Thus it will never be equal to −3. The solution set is ∅.

85. $2|x|-9 \le 11$

$2|x| \le 20$

$|x| \le 10$

$-10 \le x \le 10$

The solution set is $[-10, 10]$.

87. $|2x-3|=7$

$2x-3=7$ or $2x-3=-7$

$2x=10$ or $2x=-4$

$x=5$ or $x=-2$

The solution set is $\{5, -2\}$.

89. $|x-5|\ge 12$

$x-5\le -12$ or $x-5\ge 12$

$x\le -7$ or $x\ge 17$

The solution set is $(-\infty, -7] \cup [17, \infty)$.

91. $|9+4x|=0$

$9+4x=0$

$4x=-9$

$x=-\frac{9}{4}$

The solution set is $\left\{-\frac{9}{4}\right\}$.

93. $|2x+1|-7<-4$

$|2x+1|<3$

$-3<2x+1<3$

$-4<2x<2$

$-2<x<1$

The solution set is $(-2, 1)$.

95. $\left|\frac{1}{3}x+1\right|>5$

$\frac{1}{3}x+1<-5$ or $\frac{1}{3}x+1>5$

$\frac{1}{3}x<-6$ or $\frac{1}{3}x>4$

$x<-18$ or $x>12$

The solution set is $(-\infty, -18) \cup (12, \infty)$.

97. $|3x-5|+4=5$

$|3x-5|=1$

$3x-5=-1$ or $3x-5=1$

$3x=4$ or $3x=6$

$x=\frac{4}{3}$ or $x=2$

The solution set is $\left\{\frac{4}{3}, 2\right\}$.

99. $|6x+11|=-1$

Since the absolute value is never negative, the solution set is $\varnothing$.

101. $\left|\frac{1-2x}{3}\right|=6$

$\frac{1-2x}{3}=-6$ or $\frac{1-2x}{3}=6$

$1-2x=-18$ or $1-2x=18$

$-2x=-19$ or $-2x=17$

$x=\frac{19}{2}$ or $x=-\frac{17}{2}$

The solution set is $\left\{-\frac{17}{2}, \frac{19}{2}\right\}$.

103. $\left|\frac{3x-5}{6}\right|>5$

$\frac{3x-5}{6}<-5$ or $\frac{3x-5}{6}>5$

$3x-5<-30$ or $3x-5>30$

$3x<-25$ or $3x>35$

$x<-\frac{25}{3}$ or $x>\frac{35}{3}$

The solution set is $\left(-\infty, -\frac{25}{3}\right) \cup \left(\frac{35}{3}, \infty\right)$.

105. $|6x - 3| = |4x + 5|$

$$\begin{aligned} 6x-3 &= 4x+5 &\text{ or }&& 6x-3 &= -(4x+5) \\ 6x-3 &= 4x+5 &\text{ or }&& 6x-3 &= -4x-5 \\ 2x-3 &= 5 &\text{ or }&& 10x-3 &= -5 \\ 2x &= 8 &\text{ or }&& 10x &= -2 \\ x &= 4 &\text{ or }&& x &= -\frac{1}{5} \end{aligned}$$

The solution set is $\left\{-\frac{1}{5}, 4\right\}$.

107. $\left|\frac{1+3n}{4}\right| = |-4|$

$$\left|\frac{1+3n}{4}\right| = 4$$

$$\begin{aligned} \frac{1+3n}{4} &= 4 &\text{ or }&& \frac{1+3n}{4} &= -4 \\ 1+3n &= 16 &\text{ or }&& 1+3n &= -16 \\ 3n &= 15 &\text{ or }&& 3n &= -17 \\ n &= 5 &\text{ or }&& n &= -\frac{17}{3} \end{aligned}$$

The solution set is $\left\{-\frac{17}{3}, 5\right\}$.

109. From the circle graph, cheddar cheese made up 32% of U.S. cheese production in 2003.

111. 8% of 360° is 0.08(360°) = 28.8°.

113. Since absolute value is never negative, the solution set is ∅.

115. The absolute value of a number is always nonnegative. Thus it will always be greater than –4. The solution set is all real numbers or $\{x|x$ is a real number$\}$.

117. All numbers whose distance from 0 is 5 units is written as $|x| = 5$.

119. All numbers whose distance from 0 is less than 7 units is written as $|x| < 7$.

121. $-5 \le x \le 5$ is $|x| \le 5$. Answers may vary.

123. $|3.5 - x| < 0.05$

$$\begin{aligned} -0.05 < 3.5 - x &< 0.05 \\ -3.55 < -x &< -3.45 \\ 3.55 > x &> 3.45 \\ 3.45 < x &< 3.55 \end{aligned}$$

The acceptable measured values are between 3.45 and 3.55.

The Bigger Picture

1. $9x - 14 = 11x + 2$

$$\begin{aligned} 9x &= 11x + 16 \\ -2x &= 16 \\ x &= -8 \end{aligned}$$

The solution set is $\{-8\}$.

2. $|x - 4| = 17$

$$\begin{aligned} x-4 &= 17 &\text{ or }&& x-4 &= -17 \\ x &= 21 &\text{ or }&& x &= -13 \end{aligned}$$

The solution set is $\{-13, 21\}$.

3.

$$\begin{aligned} x-1 &\le 5 &\text{ or }&& 3x-2 &\le 10 \\ x &\le 6 &\text{ or }&& 3x &\le 12 \\ x &\le 6 &\text{ or }&& x &\le 4 \end{aligned}$$

The solution set is $(-\infty, 6] \cup (-\infty, 4] = (-\infty, 6]$.

4.

$$\begin{aligned} -x &< 7 &\text{ and }&& 4x &\le 20 \\ x &> -7 &\text{ and }&& x &\le 5 \end{aligned}$$

The solution set is $(-7, \infty) \cap (-\infty, 5] = (-7, 5]$.

5. $|x - 2| = |x + 15|$

$$\begin{aligned} x-2 &= x+15 &\text{ or }&& x-2 &= -(x+15) \\ -2 &= 15 &\text{ or }&& x-2 &= -x-15 \\ &\text{False} &\text{ or }&& x &= -x-13 \\ &\text{False} &\text{ or }&& 2x &= -13 \\ &\text{False} &\text{ or }&& x &= -\frac{13}{2} \end{aligned}$$

The solution set is $\left\{-\frac{13}{2}\right\}$.

6. $9y-6y+1=4y+10-y+3$
$$3y+1=3y+13$$
$$1=13$$
False; the solution set is $\varnothing$.

7.
$$1.5x-3=1.2x-18$$
$$10[1.5x-3]=10[1.2x-18]$$
$$15x-30=12x-180$$
$$15x=12x-150$$
$$3x=-150$$
$$x=-50$$
The solution set is $\{-50\}$.

8.
$$\frac{7x+1}{8}-3=x+\frac{2x+1}{4}$$
$$8\left[\frac{7x+1}{8}-3\right]=8\left[x+\frac{2x+1}{4}\right]$$
$$7x+1-8(3)=8x+2(2x+1)$$
$$7x+1-24=8x+4x+2$$
$$7x-23=12x+2$$
$$7x=12x+25$$
$$-5x=25$$
$$x=-5$$
The solution set is $\{-5\}$.

9. $|5x+2|-10\le -3$
$$|5x+2|\le 7$$
$$-7\le 5x+2\le 7$$
$$-9\le 5x\le 5$$
$$-\frac{9}{5}\le x\le 1$$
The solution set is $\left[-\frac{9}{5}, 1\right]$.

10. $|x+11|>2$
$$x+11>2 \quad \text{or} \quad x+11<-2$$
$$x>-9 \quad \text{or} \quad x<-13$$
The solution set is $(-\infty, -13)\cup(-9, \infty)$.

11. $|9x+2|-1=24$
$$|9x+2|=25$$
$$9x+2=25 \quad \text{or} \quad 9x+2=-25$$
$$9x=23 \quad \text{or} \quad 9x=-27$$
$$x=\frac{23}{9} \quad \text{or} \quad x=-3$$
The solution set is $\left\{-3, \frac{23}{9}\right\}$.

12. $\left|\frac{3x-1}{2}\right|=|2x+5|$
$$\frac{3x-1}{2}=2x+5 \quad \text{or} \quad \frac{3x-1}{2}=-(2x+5)$$
$$3x-1=2(2x+5) \quad \text{or} \quad 3x-1=-2(2x+5)$$
$$3x-1=4x+10 \quad \text{or} \quad 3x-1=-4x-10$$
$$3x=4x+11 \quad \text{or} \quad 3x=-4x-9$$
$$-x=11 \quad \text{or} \quad 7x=-9$$
$$x=-11 \quad \text{or} \quad x=-\frac{9}{7}$$
The solution set is $\left\{-11, -\frac{9}{7}\right\}$.

Chapter 2 Vocabulary Check

1. The statement "$x<5$ or $x>7$" is called a(n) compound inequality.

2. An equation in one variable that has no solution is called a(n) contradiction.

3. The intersection of two sets is the set of all elements common to both sets.

4. The union of two sets is the set of all elements that belong to either of the sets.

5. An equation in one variable that has every number (for which the equation is defined) as a solution is called a(n) identity.

6. The equation $d = rt$ is also called a(n) formula.

7. A number's distance from 0 is called its absolute value.

8. When a variable in an equation is replaced by a number and the resulting equation is true, then that number is called a(n) solution of the equation.

9. The integers 17, 18, 19 are examples of consecutive integers.

10. The statement $5x - 0.2 < 7$ is an example of a(n) linear inequality in one variable.

11. The statement $5x - 0.2 = 7$ is an example of a(n) linear equation in one variable.

Chapter 2 Review

1. $4(x-5) = 2x-14$
$4x-20 = 2x-14$
$4x = 2x+6$
$2x = 6$
$x = 3$
The solution set is $\{3\}$.

2. $x+7 = -2(x+8)$
$x+7 = -2x-16$
$x = -2x-23$
$3x = -23$
$x = -\frac{23}{3}$
The solution set is $\left\{-\frac{23}{3}\right\}$.

3. $3(2y-1) = -8(6+y)$
$6y-3 = -48-8y$
$6y = -45-8y$
$14y = -45$
$y = -\frac{45}{14}$
The solution set is $\left\{-\frac{45}{14}\right\}$.

4. $-(z+12) = 5(2z-1)$
$-z-12 = 10z-5$
$-z = 10z+7$
$-11z = 7$
$z = -\frac{7}{11}$
The solution set is $\left\{-\frac{7}{11}\right\}$.

5. $n-(8+4n) = 2(3n-4)$
$n-8-4n = 6n-8$
$-8-3n = 6n-8$
$-3n = 6n$
$-9n = 0$
$n = 0$
The solution set is $\{0\}$.

6. $4(9v+2) = 6(1+6v)-10$
$36v+8 = 6+36v-10$
$36v+8 = 36v-4$
$8 = -4$
False; the solution set is $\varnothing$.

7. $0.3(x-2) = 1.2$
$10[0.3(x-2)] = 10(1.2)$
$3(x-2) = 12$
$3x-6 = 12$
$3x = 18$
$x = 6$
The solution set is $\{6\}$.

8. $1.5 = 0.2(c-0.3)$
$10(1.5) = 10[0.2(c-0.3)]$
$15 = 2(c-0.3)$
$15 = 2c-0.6$
$15.6 = 2c$
$7.8 = c$
The solution set is $\{7.8\}$.

9. $-4(2-3x) = 2(3x-4)+6x$
$-8+12x = 6x-8+6x$
$-8+12x = 12x-8$
$-8 = -8$ True
The solution set is $\{x | x \text{ is a real number}\}$.

10. $6(m-1)+3(2-m)=0$
$$6m-6+6-3m=0$$
$$3m=0$$
$$m=0$$
The solution set is $\{0\}$.

11. $6-3(2g+4)-4g=5(1-2g)$
$$6-6g-12-4g=5-10g$$
$$-6-10g=5-10g$$
$$-6=5 \quad \text{False}$$
The solution set is $\varnothing$.

12. $20-5(p+1)+3p=-(2p-15)$
$$20-5p-5+3p=-2p+15$$
$$15-2p=-2p+15$$
$$15=15 \quad \text{True}$$
The solution set is $\{p|p \text{ is a real number}\}$.

13. $\frac{x}{3}-4=x-2$
$$3\left(\frac{x}{3}-4\right)=3(x-2)$$
$$x-12=3x-6$$
$$x=3x+6$$
$$-2x=6$$
$$x=-3$$
The solution set is $\{-3\}$.

14. $\frac{9}{4}y=\frac{2}{3}y$
$$12\left(\frac{9}{4}y\right)=12\left(\frac{2}{3}y\right)$$
$$27y=8y$$
$$19y=0$$
$$y=0$$
The solution set is $\{0\}$.

15. $\frac{3n}{8}-1=3+\frac{n}{6}$
$$24\left(\frac{3n}{8}-1\right)=24\left(3+\frac{n}{6}\right)$$
$$24\left(\frac{3n}{8}\right)-24(1)=24(3)+24\left(\frac{n}{6}\right)$$
$$9n-24=72+4n$$
$$9n=96+4n$$
$$5n=96$$
$$n=\frac{96}{5}$$
The solution set is $\left\{\frac{96}{5}\right\}$.

16. $\frac{z}{6}+1=\frac{z}{2}+2$
$$6\left(\frac{z}{6}+1\right)=6\left(\frac{z}{2}+2\right)$$
$$6\left(\frac{z}{6}\right)+6(1)=6\left(\frac{z}{2}\right)+6(2)$$
$$z+6=3z+12$$
$$z=3z+6$$
$$-2z=6$$
$$z=-3$$
The solution set is $\{-3\}$.

17. $\frac{y}{4}-\frac{y}{2}=-8$
$$4\left(\frac{y}{4}-\frac{y}{2}\right)=4(-8)$$
$$4\left(\frac{y}{4}\right)-4\left(\frac{y}{2}\right)=-32$$
$$y-2y=-32$$
$$-y=-32$$
$$y=32$$
The solution set is $\{32\}$.

18. $$\frac{2x}{3}-\frac{8}{3}=x$$
$$3\left(\frac{2x}{3}-\frac{8}{3}\right)=3(x)$$
$$3\left(\frac{2x}{3}\right)-3\left(\frac{8}{3}\right)=3x$$
$$2x-8=3x$$
$$-8=x$$
The solution set is $\{-8\}$.

19. $$\frac{b-2}{3}=\frac{b+2}{5}$$
$$15\left(\frac{b-2}{3}\right)=15\left(\frac{b+2}{5}\right)$$
$$5(b-2)=3(b+2)$$
$$5b-10=3b+6$$
$$5b=3b+16$$
$$2b=16$$
$$b=8$$
The solution set is $\{8\}$.

20. $$\frac{2t-1}{3}=\frac{3t+2}{15}$$
$$15\left(\frac{2t-1}{3}\right)=15\left(\frac{3t+2}{15}\right)$$
$$5(2t-1)=3t+2$$
$$10t-5=3t+2$$
$$10t=3t+7$$
$$7t=7$$
$$t=1$$
The solution set is $\{1\}$.

21. $$\frac{2(t+1)}{3}=\frac{2(t-1)}{3}$$
$$3\left[\frac{2(t+1)}{3}\right]=3\left[\frac{2(t-1)}{3}\right]$$
$$2(t+1)=2(t-1)$$
$$2t+2=2t-2$$
$$2=-2 \quad \text{False}$$
The solution set is $\varnothing$.

22. $$\frac{3a-3}{6}=\frac{4a+1}{15}+2$$
$$30\left(\frac{3a-3}{6}\right)=30\left(\frac{4a+1}{15}+2\right)$$
$$5(3a-3)=30\left(\frac{4a+1}{15}\right)+30(2)$$
$$15a-15=2(4a+1)+60$$
$$15a-15=8a+2+60$$
$$15a-15=8a+62$$
$$15a=8a+77$$
$$7a=77$$
$$a=11$$
The solution set is $\{11\}$.

23. Let x be the number.
$$2(x-3)=1+3x$$
$$2x-6=1+3x$$
$$2x=7+3x$$
$$-x=7$$
$$x=-7$$
The number is -7.

24. Let x be the first number and $x + 5$ be the second number.
$$x+(x+5)=285$$
$$2x+5=285$$
$$2x=280$$
$$x=140$$
$$x + 5 = (140) + 5 = 145$$
The two numbers are 140 and 145.

25. 40% of 130 = 40% · 130 = 0.40 · 130 = 52

26. 1.5% of 8 = 1.5% · 8 = 0.015 · 8 = 0.12

27. Let x be the average annual earnings for a worker with a bachelor's degree in 2000.
$$x+0.1745x=60,665$$
$$1.1745x=60,665$$
$$x=51,651.77$$
The average annual earnings for a worker with a bachelor's degree in 2000 was \$51,652.

28. Four consecutive integers are x, $x+1$, $x+2$, and $x+3$.

$$\begin{aligned}(x+1)+(x+2)+(x+3)-2x&=16\\3x+6-2x&=16\\x+6&=16\\x&=10\end{aligned}$$

$x+1=(10)+1=11$
$x+2=(10)+2=12$
$x+3=(10)+3=13$
The integers are 10, 11, 12, and 13.

29. Let x be the first odd integer and $x+2$ is the next consecutive odd integer.

$$\begin{aligned}5(x)&=3(x+2)+54\\5x&=3x+6+54\\5x&=3x+60\\2x&=60\\x&=30\end{aligned}$$

Since x is not odd, no such integers exist.

30. Let x be the width of a playing field and $2x-5$ is the length of this field.

$$\begin{aligned}x+(2x-5)+x+(2x-5)&=230\\6x-10&=230\\6x&=240\\x&=40\end{aligned}$$

$2x-5=2(40)-5=80-5=75$
The playing field has a width of 40 m and a length of 75 m.

31. Let x be the number of miles driven per day. Then $0.12(x-100)$ represents the cost of driving more than 100 miles per day.

$$\begin{aligned}2[19.95+0.12(x-100)]&=46.86\\2(19.95)+2[0.12(x-100)]&=46.86\\39.90+0.24(x-100)&=46.86\\39.90+0.24x-24&=46.86\\15.90+0.24x&=46.86\\0.24x&=30.96\\x&=129\end{aligned}$$

Therefore in two days, Mr. Woo drives 258 miles.

32. Recall that

$$\begin{aligned}R&=C\\16.5x&=4.5x+3000\\12.0x&=3000\\x&=250\end{aligned}$$

250 calculators need to be sold to break even.

33.

$$\begin{aligned}V&=LWH\\\frac{V}{LH}&=\frac{LWH}{LH}\\\frac{V}{LH}&=W\end{aligned}$$

34.

$$\begin{aligned}C&=2\pi r\\\frac{C}{2\pi}&=\frac{2\pi r}{2\pi}\\\frac{C}{2\pi}&=r\end{aligned}$$

35.

$$\begin{aligned}5x-4y&=-12\\-4y&=-5x-12\\4y&=5x+12\\y&=\frac{5x+12}{4}\end{aligned}$$

36.

$$\begin{aligned}5x-4y&=-12\\5x&=4y-12\\x&=\frac{4y-12}{5}\end{aligned}$$

37.

$$\begin{aligned}y-y_1&=m(x-x_1)\\\frac{y-y_1}{x-x_1}&=\frac{m(x-x_1)}{x-x_1}\\\frac{y-y_1}{x-x_1}&=m\end{aligned}$$

38.

$$\begin{aligned}y-y_1&=m(x-x_1)\\y-y_1&=mx-mx_1\\y-y_1+mx_1&=mx\\\frac{y-y_1+mx_1}{m}&=\frac{mx}{m}\\\frac{y-y_1+mx_1}{m}&=x\end{aligned}$$

39.
$$E = I(R+r)$$
$$E = IR + Ir$$
$$E - IR = Ir$$
$$\frac{E-IR}{I} = \frac{Ir}{I}$$
$$\frac{E-IR}{I} = r$$

40.
$$S = vt + gt^2$$
$$S - vt = gt^2$$
$$\frac{S-vt}{t^2} = \frac{gt^2}{t^2}$$
$$\frac{S-vt}{t^2} = g$$

41.
$$T = gr + gvt$$
$$T = g(r+vt)$$
$$\frac{T}{r+vt} = \frac{g(r+vt)}{r+vt}$$
$$\frac{T}{r+vt} = g$$

42.
$$I = Prt + P$$
$$I = P(rt+1)$$
$$\frac{I}{rt+1} = \frac{P(rt+1)}{rt+1}$$
$$\frac{I}{rt+1} = P$$

43. Use $A = P\left(1+\frac{r}{n}\right)^{nt}$ with $P = 3000$, $r = 3\% = 0.03$, and $t = 7$.

a. $n = 2$: $A = 3000\left(1+\frac{0.03}{2}\right)^{2\cdot 7}$
$$A = 3000(1.015)^{14}$$
$$A \approx \$3695.27$$

b. $n = 52$: $A = 3000\left(1+\frac{0.03}{52}\right)^{52\cdot 7}$
$$A \approx \$3700.81$$

44. Recall that $C = \frac{5}{9}(F-32)$. Replace F with 90.
$$C = \frac{5}{9}(90-32)$$
$$C = \frac{5}{9}(58)$$
$$C = 32.\overline{2}$$
$$90°\text{F} = 32\frac{2}{9}°\text{C}$$

45. Let x be the original width of the photograph and $x + 2$ be the original length of the photograph. Then $x + 4$ and $x + 6$ are the new dimensions of width and length, respectively.
$$(x+4)(x+6) = x(x+2) + 88$$
$$x^2 + 6x + 4x + 24 = x^2 + 2x + 88$$
$$x^2 + 10x + 24 = x^2 + 2x + 88$$
$$10x + 24 = 2x + 88$$
$$10x = 2x + 64$$
$$8x = 64$$
$$x = 8$$
$x + 2 = (8) + 2 = 10$
The original width was 8 in. and original length was 10 in.

46. $A = l \cdot w$
$A = 18 \cdot 21$
$A = 378$ sq ft
$$\text{Packages} = \frac{378}{24} = 15.75$$
Buy 16 packages of tiles.

47. $3(x-5) > -(x+3)$
$$3x - 15 > -x - 3$$
$$3x > -x + 12$$
$$4x > 12$$
$$x > 3$$
The solution set is $(3, \infty)$.

48. $-2(x+7) \ge 3(x+2)$
$-2x-14 \ge 3x+6$
$-2x \ge 3x+20$
$-5x \ge 20$
$x \le -4$
The solution set is $(-\infty, -4]$.

49. $4x-(5+2x) < 3x-1$
$4x-5-2x < 3x-1$
$2x-5 < 3x-1$
$2x < 3x+4$
$-x < 4$
$x > -4$
The solution set is $(-4, \infty)$.

50. $3(x-8) < 7x+2(5-x)$
$3x-24 < 7x+10-2x$
$3x-24 < 5x+10$
$3x < 5x+34$
$-2x < 34$
$x > -17$
The solution set is $(-17, \infty)$.

51. $24 \ge 6x-2(3x-5)+2x$
$24 \ge 6x-6x+10+2x$
$24 \ge 10+2x$
$14 \ge 2x$
$7 \ge x$
$x \le 7$
The solution set is $(-\infty, 7]$.

52. $\frac{x}{3}+\frac{1}{2} > \frac{2}{3}$
$6\left(\frac{x}{3}+\frac{1}{2}\right) > 6\left(\frac{2}{3}\right)$
$2x+3 > 4$
$2x > 1$
$x > \frac{1}{2}$
The solution set is $\left(\frac{1}{2}, \infty\right)$.

53. $x+\frac{3}{4} < -\frac{x}{2}+\frac{9}{4}$
$4\left(x+\frac{3}{4}\right) < 4\left(-\frac{x}{2}+\frac{9}{4}\right)$
$4(x)+4\left(\frac{3}{4}\right) < 4\left(-\frac{x}{2}\right)+4\left(\frac{9}{4}\right)$
$4x+3 < -2x+9$
$4x < -2x+6$
$6x < 6$
$x < 1$
The solution set is $(-\infty, 1)$.

54. $\frac{x-5}{2} \le \frac{3}{8}(2x+6)$
$8\left(\frac{x-5}{2}\right) \le 8\left[\frac{3}{8}(2x+6)\right]$
$4(x-5) \le 3(2x+6)$
$4x-20 \le 6x+18$
$4x \le 6x+38$
$-2x \le 38$
$x \ge -19$
The solution set is $[-19, \infty)$.

55. Let x be the number of pounds of laundry. The cost to do it at the laundromat is $0.5(10) + 0.4(x - 10)$. It is more economical to use the housekeeper when the cost at the laundromat exceeds \$15.
$0.5(10)+0.4(x-10) > 15$
$5+0.4x-4 > 15$
$0.4x+1 > 15$
$0.4x > 14$
$x > 35$
It is more economical to use the housekeeper for more than 35 pounds of laundry per week.

56.
$$500 \le F \le 1000$$
$$500 \le \frac{9}{5}C + 32 \le 1000$$
$$468 \le \frac{9}{5}C \le 968$$
$$2340 \le 9C \le 4840$$
$$260 \le C \le 537\frac{7}{9}$$
$$260° \le C \le 538°$$

57. Let x be the score from the eighth judge.
$$\frac{9.5+9.7+9.9+9.7+9.7+9.6+9.5+x}{8} \ge 9.65$$
$$\frac{67.6+x}{8} \ge 9.65$$
$$8\left(\frac{67.6+x}{8}\right) \ge 8(9.65)$$
$$67.6+x \ge 77.2$$
$$x \ge 9.6$$
To win the silver medal, Nana must score at least 9.6 points from the judge.

58. Let x be the amount of money she must save each summer.
$$4000 \le 500 + 2x \le 8000$$
$$3500 \le 2x \le 7500$$
$$1750 \le x \le 3750$$
Carol must earn between \$1750 and \$3750 each of the next two summers to buy the car.

59. $1 \le 4x - 7 \le 3$
$$8 \le 4x \le 10$$
$$2 \le x \le \frac{10}{4}$$
$$2 \le x \le \frac{5}{2}$$
The solution set is $\left[2, \frac{5}{2}\right]$.

60. $-2 \le 8 + 5x < -1$
$$-10 \le 5x < -9$$
$$-2 \le x < -\frac{9}{5}$$
The solution set is $\left[-2, -\frac{9}{5}\right)$.

61. $-3 < 4(2x-1) < 12$
$-3 < 8x-4 < 12$
$1 < 8x < 16$
$\frac{1}{8} < x < 2$
The solution set $\left(\frac{1}{8}, 2\right)$.

62. $-6 < x-(3-4x) < -3$
$-6 < x-3+4x < -3$
$-6 < 5x-3 < -3$
$-3 < 5x < 0$
$-\frac{3}{5} < x < 0$
The solution set is $\left(-\frac{3}{5}, 0\right)$.

63. $\frac{1}{6} < \frac{4x-3}{3} \le \frac{4}{5}$
$\frac{1}{2} < 4x-3 \le \frac{12}{5}$
$\frac{1}{2}+3 < 4x \le \frac{12}{5}+3$
$\frac{7}{2} < 4x \le \frac{27}{5}$
$\frac{7}{8} < x \le \frac{27}{20}$
The solution set is $\left(\frac{7}{8}, \frac{27}{20}\right]$.

64. $x \le 2$ and $x > -5$
The solution set is
$(-\infty, 2] \cap (-5, \infty) = (-5, 2]$.

65. $3x-5 > 6$ or $-x < -5$
$3x > 11$ or $x > 5$
$x > \frac{11}{3}$ or $x > 5$
The solution set is $\left(\frac{11}{3}, \infty\right)$.

66. $|x-7| = 9$
$x-7=9$ or $x-7=-9$
$x=16$ or $x=-2$
The solution set is $\{16, -2\}$.

67. $|8-x| = 3$
$8-x=3$ or $8-x=-3$
$-x=-5$ or $-x=-11$
$x=5$ or $x=11$
The solution set is $\{5, 11\}$.

68. $|2x+9| = 9$
$2x+9=9$ or $2x+9=-9$
$2x=0$ or $2x=-18$
$x=0$ or $x=-9$
The solution set is $\{0, -9\}$.

69. $|-3x+4| = 7$
$-3x+4=7$ or $-3x+4=-7$
$-3x=3$ or $-3x=-11$
$x=-1$ or $x=\frac{11}{3}$
The solution set is $\left\{-1, \frac{11}{3}\right\}$.

70. $|3x-2|+6=10$
$|3x-2|=4$
$3x-2=4$ or $3x-2=-4$
$3x=6$ or $3x=-2$
$x=2$ or $x=-\frac{2}{3}$
The solution set is $\left\{2, -\frac{2}{3}\right\}$.

71. $5+|6x+1|=5$
$|6x+1|=0$
$6x+1=0$
$6x=-1$
$x=-\frac{1}{6}$
The solution set is $\left\{-\frac{1}{6}\right\}$.

72. $-5 = |4x - 3|$
The absolute value of a number is always nonnegative. Therefore it can never equal −5. The solution set is ∅.

73. $|5-6x|+8=3$
$|5-6x|=-5$
The absolute value of a number is always nonnegative. Therefore it can never equal −5. The solution set is ∅.

74. $-8=|x-3|-10$
$2=|x-3|$
$x-3=2$ or $x-3=-2$
$x=5$ or $x=1$
The solution set is {1, 5}.

75. $\left|\frac{3x-7}{4}\right|=2$
$\frac{3x-7}{4}=2$ or $\frac{3x-7}{4}=-2$
$3x-7=8$ or $3x-7=-8$
$3x=15$ or $3x=-1$
$x=5$ or $x=-\frac{1}{3}$
The solution set is $\left\{-\frac{1}{3}, 5\right\}$.

76. $|6x + 1| = |15 + 4x|$
$6x+1=15+4x$ or $6x+1=-(15+4x)$
$6x=14+4x$ or $6x+1=-15-4x$
$2x=14$ or $6x=-16-4x$
$x=7$ or $10x=-16$
$x=7$ or $x=-\frac{16}{10}=-\frac{8}{5}$
The solution set is $\left\{7, -\frac{8}{5}\right\}$.

77. $|5x - 1| < 9$
$-9<5x-1<9$
$-8<5x<10$
$-\frac{8}{5}<x<2$
The solution set is $\left(-\frac{8}{5}, 2\right)$.

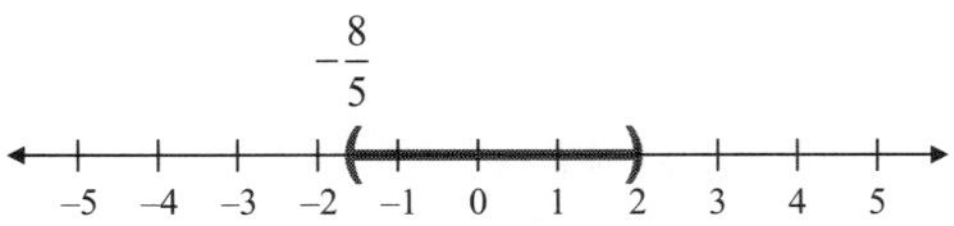

78. $|6 + 4x| \ge 10$
$6+4x\le-10$ or $6+4x\ge10$
$4x\le-16$ or $4x\ge4$
$x\le-4$ or $x\ge1$
The solution set is $(-\infty, -4] \cup [1, \infty)$.

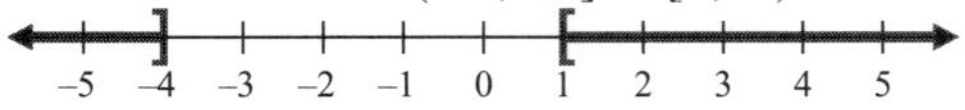

79. $|3x|-8>1$
$|3x|>9$
$3x<-9$ or $3x>9$
$x<-3$ or $x>3$
The solution set is $(-\infty, -3) \cup (3, \infty)$.

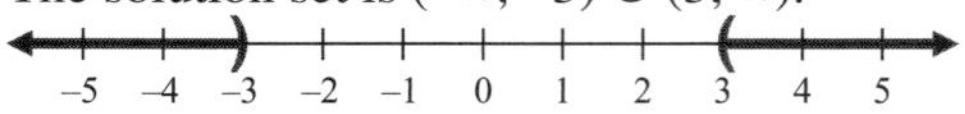

80. $9+|5x|<24$
$|5x|<15$
$-15<5x<15$
$-3<x<3$
The solution set is (−3, 3).

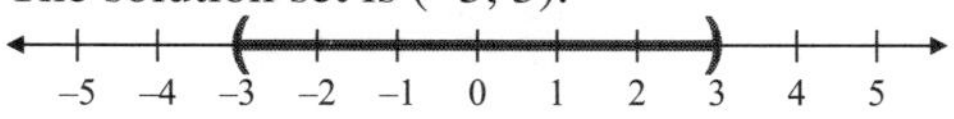

81. $|6x - 5| \le -1$
The absolute value of a number is never negative. Thus it will never be less than or equal to −1. The solution set is ∅.

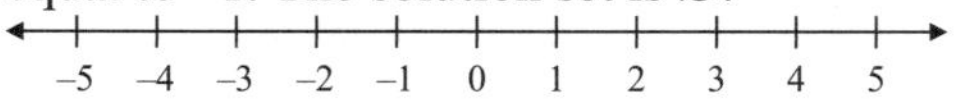

82. $\left|3x+\frac{2}{5}\right| \ge 4$

$$3x+\frac{2}{5} \ge 4 \quad \text{or} \quad 3x+\frac{2}{5} \le -4$$
$$5\left(3x+\frac{2}{5}\right) \ge 5(4) \quad \text{or} \quad 5\left(3x+\frac{2}{5}\right) \le 5(-4)$$
$$15x+2 \ge 20 \quad \text{or} \quad 15x+2 \le -20$$
$$15x \ge 18 \quad \text{or} \quad 15x \le -22$$
$$x \ge \frac{18}{15} \quad \text{or} \quad x \le -\frac{22}{15}$$
$$x \ge \frac{6}{5} \quad \text{or} \quad x \le -\frac{22}{15}$$

The solution set is $\left(-\infty, -\frac{22}{5}\right] \cup \left[\frac{6}{5}, \infty\right)$.

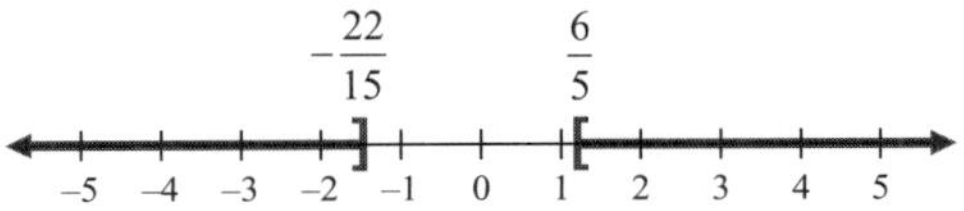

83. $\left|\frac{x}{3}+6\right| - 8 > -5$

$$\left|\frac{x}{3}+6\right| > 3$$
$$\frac{x}{3}+6 < -3 \quad \text{or} \quad \frac{x}{3}+6 > 3$$
$$\frac{x}{3} < -9 \quad \text{or} \quad \frac{x}{3} > -3$$
$$x < -27 \quad \text{or} \quad x > -9$$

The solution set is $(-\infty, -27) \cup (-9, \infty)$.

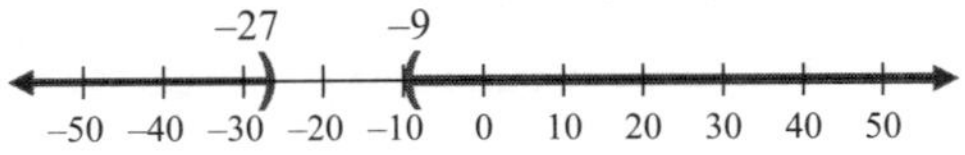

84. $\left|\frac{4(x-1)}{7}\right| + 10 < 2$

$$\left|\frac{4(x-1)}{7}\right| < -8$$

The absolute value of a number is never negative. Thus it will never be less than −8. The solution set is ∅.

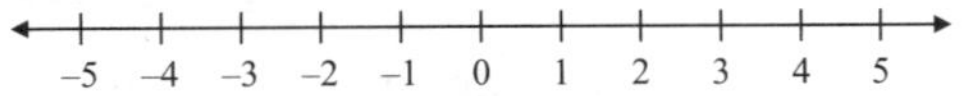

85. $\frac{x-2}{5}+\frac{x+2}{2}=\frac{x+4}{3}$

$$30\left(\frac{x-2}{5}+\frac{x+2}{2}\right)=30\left(\frac{x+4}{3}\right)$$
$$30\left(\frac{x-2}{5}\right)+30\left(\frac{x+2}{2}\right)=10(x-4)$$
$$6(x-2)+15(x+2)=10x+40$$
$$6x-12+15x+30=10x+40$$
$$21x+18=10x+40$$
$$21x=10x+22$$
$$11x=22$$
$$x=2$$

The solution set is {2}.

86. $\frac{2z-3}{4}-\frac{4-z}{2}=\frac{z+1}{3}$

$$12\left(\frac{2z-3}{4}-\frac{4-z}{2}\right)=12\left(\frac{z+1}{3}\right)$$
$$12\left(\frac{2z-3}{4}\right)-12\left(\frac{4-z}{2}\right)=4(z+1)$$
$$3(2z-3)-6(4-z)=4z+4$$
$$6z-9-24+6z=4z+4$$
$$12z-33=4z+4$$
$$12z=4z+37$$
$$8z=37$$
$$z=\frac{37}{8}$$

The solution set is $\left\{\frac{37}{8}\right\}$.

87. Let x be the number of tourists in France. Then $x + 9$ is the number of tourists in the USA, and $x + 44$ is the number of tourists in China.

$$x+(x+9)+(x+44)=332$$
$$3x+53=332$$
$$3x=279$$
$$x=93$$

$x + 9 = (93) + 9 = 102$
$x + 44 = (93) + 44 = 137$
In 2020, France will have 93 million tourists, the USA will have 102 million tourists, and China will have 137 million tourists.

88.
$$A = \frac{h}{2}(B+b)$$
$$2A = h(B+b)$$
$$2A = hB + hb$$
$$2A - hb = hB$$
$$\frac{2A - hb}{h} = B$$

89.
$$V = \frac{1}{3}\pi r^2 h$$
$$3V = \pi r^2 h$$
$$\frac{3V}{\pi r^2} = h$$

90. Volume of Box $= l \cdot w \cdot h$
$= 8 \cdot 5 \cdot 3$
$= 120$ cubic inches

Volume of cylinder $= \pi r^2 h$
$= \pi \cdot 3^2 \cdot 6$
$= 54\pi$
≈ 169.65 cubic inches

The cylinder is larger and holds more ice cream.

91.
$$D = r \cdot t$$
$$130 = r \cdot 2.25$$
$$57.\overline{7} = r$$
His average speed was 58 mph.

92.
$$48 + x \ge 5(2x+4) - 2x$$
$$48 + x \ge 10x + 20 - 2x$$
$$48 + x \ge 8x + 20$$
$$x \ge 8x - 28$$
$$-7x \ge -28$$
$$x \le 4$$
The solution set is $(-\infty, 4]$.

93.
$$\frac{3(x-2)}{5} > -\frac{5(x-2)}{3}$$
$$15\left[\frac{3(x-2)}{5}\right] > 15\left[-\frac{5(x-2)}{3}\right]$$
$$9(x-2) > -25(x-2)$$
$$9x - 18 > -25x + 50$$
$$9x > -25x + 68$$
$$34x > 68$$
$$x > 2$$
The solution set is $(2, \infty)$.

94.
$$0 \le \frac{2(3x+4)}{5} \le 3$$
$$0 \le 2(3x+4) \le 15$$
$$0 \le 6x + 8 \le 15$$
$$-8 \le 6x \le 7$$
$$-\frac{8}{6} \le x \le \frac{7}{6}$$
$$-\frac{4}{3} \le x \le \frac{7}{6}$$
The solution set is $\left[-\frac{4}{3}, \frac{7}{6}\right]$.

95. $x \le 2$ or $x > -5$
The solution set is
$(-\infty, 2] \cup (-5, \infty) = (-\infty, \infty)$.

96.
$-2x \le 6$ and $-2x + 3 < -7$
$x \ge -3$ and $-2x < -10$
$x \ge -3$ and $x > 5$

The solution set is $[-3, \infty) \cap (5, \infty) = (5, \infty)$.

97.
$$|7x| - 26 = -5$$
$$|7x| = 21$$
$7x = 21$ or $7x = -21$
$x = 3$ or $x = -3$

The solution set is $\{-3, 3\}$.

98.
$$\left|\frac{9-2x}{5}\right| = -3$$
The absolute value of a number is never negative. Thus it will never be equal to -3. The solution set is $\varnothing$.

99. $|x - 3| = |7 + 2x|$

$$x - 3 = 7 + 2x \quad \text{or} \quad x - 3 = -(7 + 2x)$$
$$x = 10 + 2x \quad \text{or} \quad x - 3 = -7 - 2x$$
$$-x = 10 \quad \text{or} \quad x = -4 - 2x$$
$$x = -10 \quad \text{or} \quad 3x = -4$$
$$x = -10 \quad \text{or} \quad x = -\frac{4}{3}$$

The solution set is $\left\{-10, -\frac{4}{3}\right\}$.

100. $|6x - 5| \ge -1$

The absolute value of a number is always nonnegative. Thus it will always be greater than −1. The solution set is all real numbers or $(-\infty, \infty)$.

101. $\left|\frac{4x-3}{5}\right| < 1$

$$-1 < \frac{4x-3}{5} < 1$$
$$-1 < \frac{4x-3}{5} < 1$$
$$-5 < 4x - 3 < 5$$
$$-2 < 4x < 8$$
$$\frac{-2}{4} < x < 2$$
$$-\frac{1}{2} < x < 2$$

The solution set is $\left(-\frac{1}{2}, 2\right)$.

Chapter 2 Test

1.
$$8x + 14 = 5x + 44$$
$$8x + 14 - 5x = 5x + 44 - 5x$$
$$3x + 14 = 44$$
$$3x + 14 - 14 = 44 - 14$$
$$3x = 30$$
$$\frac{3x}{3} = \frac{30}{3}$$
$$x = 10$$

The solution set is $\{10\}$.

2. $9(x + 2) = 5[11 - 2(2 - x) + 3]$
$$9x + 18 = 5[11 - 4 + 2x + 3]$$
$$9x + 18 = 5[2x + 10]$$
$$9x + 18 = 10x + 50$$
$$9x = 10x + 32$$
$$-x = 32$$
$$x = -32$$

The solution set is $\{-32\}$.

3. $3(y - 4) + y = 2(6 + 2y)$
$$3y - 12 + y = 12 + 4y$$
$$4y - 12 = 12 + 4y$$
$$-12 = 12 \quad \text{False}$$

The solution set is $\varnothing$.

4. $7n - 6 + n = 2(4n - 3)$
$$8n - 6 = 8n - 6$$
$$-6 = -6 \quad \text{True}$$

The solution set is $\{n|n \text{ is a real number}\}$.

5.
$$\frac{7w}{4} + 5 = \frac{3w}{10} + 1$$
$$20\left(\frac{7w}{4} + 5\right) = 20\left(\frac{3w}{10} + 1\right)$$
$$20\left(\frac{7w}{4}\right) + 20(5) = 20\left(\frac{3w}{10}\right) + 20(1)$$
$$35w + 100 = 6w + 20$$
$$35w + 100 - 6w = 6w + 20 - 6w$$
$$29w + 100 = 20$$
$$29w + 100 - 100 = 20 - 100$$
$$29w = -80$$
$$\frac{29w}{29} = \frac{-80}{29}$$
$$w = -\frac{80}{29}$$

The solution set is $\left\{-\frac{80}{29}\right\}$.

6. $$\frac{z+7}{9}+1=\frac{2z+1}{6}$$
$$18\left(\frac{z+7}{9}+1\right)=18\left(\frac{2z+1}{6}\right)$$
$$18\left(\frac{z+7}{9}\right)+18(1)=3(2z+1)$$
$$2(z+7)+18=6z+3$$
$$2z+14+18=6z+3$$
$$2z+32=6z+3$$
$$2z=6z-29$$
$$-4z=-29$$
$$z=\frac{29}{4}$$
The solution set is $\left\{\frac{29}{4}\right\}$.

7. $|6x-5|-3=-2$
$|6x-5|=1$
$6x-5=1$ or $6x-5=-1$
$6x=6$ or $6x=4$
$x=1$ or $x=\frac{4}{6}=\frac{2}{3}$
The solution set is $\left\{1, \frac{2}{3}\right\}$.

8. $|8-2t|=-6$
Since the absolute value is never negative, the solution set is $\varnothing$.

9. $|2x-3|=|4x+5|$
$2x-3=4x+5$ or $2x-3=-(4x+5)$
$2x-3=4x+5$ or $2x-3=-4x-5$
$-2x-3=5$ or $6x-3=-5$
$-2x=8$ or $6x=-2$
$x=-4$ or $x=-\frac{1}{3}$
The solution set is $\left\{-4, -\frac{1}{3}\right\}$.

10. $|x-5|=|x+2|$
$x-5=x+2$ or $x-5=-(x+2)$
$-5=2$ or $x-5=-x-2$
False or $x=-x+3$
False or $2x=3$
False or $x=\frac{3}{2}$
The solution set is $\left\{\frac{3}{2}\right\}$.

11. $3x-4y=8$
$$-4y=-3x+8$$
$$4y=3x-8$$
$$y=\frac{3x-8}{4}$$

12. $$S=gt^2+gvt$$
$$S=g(t^2+vt)$$
$$\frac{S}{t^2+vt}=g$$

13. $$F=\frac{9}{5}C+32$$
$$F-32=\frac{9}{5}C$$
$$\frac{5}{9}(F-32)=C$$

14. $3(2x-7)-4x>-(x+6)$
$$6x-21-4x>-x-6$$
$$2x-21>-x-6$$
$$2x>-x+15$$
$$3x>15$$
$$x>5$$
The solution set is $(5, \infty)$.

15. $\frac{3x-2}{3}-\frac{5x+1}{4} \ge 0$

$12\left(\frac{3x-2}{3}-\frac{5x+1}{4}\right) \ge 12(0)$

$12\left(\frac{3x-2}{3}\right)-12\left(\frac{5x+1}{4}\right) \ge 0$

$4(3x-2)-3(5x+1) \ge 0$

$12x-8-15x-3 \ge 0$

$-3x-11 \ge 0$

$-3x \ge 11$

$x \le -\frac{11}{3}$

The solution set is $\left(-\infty, -\frac{11}{3}\right]$.

16. $-3 < 2(x-3) \le 4$

$-3 < 2x-6 \le 4$

$3 < 2x \le 10$

$\frac{3}{2} < x \le 5$

The solution set is $\left(\frac{3}{2}, 5\right]$.

17. $|3x+1| > 5$

$3x+1 < -5$ or $3x+1 > 5$

$3x < -6$ or $3x > 4$

$x < -2$ or $x > \frac{4}{3}$

The solution set is $(-\infty, -2) \cup \left(\frac{4}{3}, \infty\right)$.

18. $|x-5|-4 < -2$

$|x-5| < 2$

$-2 < x-5 < 2$

$3 < x < 7$

The solution set is (3, 7).

19. $x \ge 5$ and $x \ge 4$

The solution set is $[5, \infty) \cap [4, \infty) = [5, \infty)$.

20. $x \ge 5$ or $x \ge 4$

The solution set is $[5, \infty) \cup [4, \infty) = [4, \infty)$.

21. $-1 \le \frac{2x-5}{3} < 2$

$3(-1) \le 3\left(\frac{2x-5}{3}\right) < 3(2)$

$-3 \le 2x-5 < 6$

$-3+5 \le 2x < 6+5$

$2 \le 2x < 11$

$\frac{2}{2} \le \frac{2x}{2} < \frac{11}{2}$

$1 \le x < \frac{11}{2}$

The solution set is $\left[1, \frac{11}{2}\right)$.

22. $6x+1 > 5x+4$ or $1-x > -4$

$6x > 5x+3$ or $-x > -5$

$x > 3$ or $x < 5$

The solution set is

$(3, \infty) \cup (-\infty, 5) = (-\infty, \infty)$.

23. 12% of $80 = 0.12 \cdot 80 = 9.6$

24. Let x be the number of new vehicles sold by Ford in 2000. The number of new vehicles sold in 2003 is x decreased by 9.48%.

$x-0.0948x = 6,720,000$

$0.9052x = 6,720,000$

$x = 7,423,773.752$

Ford sold 7,424,000 new vehicles in 2000.

25. Recall that $C = 2\pi r$. Here $C = 78.5$.

$78.5 = 2\pi r$

$r = \frac{78.5}{2\pi} = \frac{39.25}{\pi}$

Also recall that $A = \pi r^2$.

$A = \pi\left(\frac{39.25}{\pi}\right)^2 = \frac{39.25^2}{\pi} \approx \frac{39.25^2}{3.14} \approx 491$

The area of the pen is about 491 square feet. Each dog requires at least 60 square feet of space, and $\frac{491}{60} \approx 8.18$. At most 8 dogs could be kept in the pen.

26. Let x be the number of people employed as database administrators, computer support specialists, and all other computer scientists in 1996. The number of people employed in these fields in 2006 is x increased by 118%.

$$x + 1.18x = 461,000$$
$$2.18x = 461,000$$
$$x = 211,467.8899$$

In 1996 211,468 people were employed as database administrators, computer support specialists, and all other computer scientists.

27. Use $A = P\left(1+\frac{r}{n}\right)^{nt}$ where $P = 2500$, $r = 3.5\% = 0.035$, $t = 10$, and $n = 4$.

$$A = 2500\left(1+\frac{0.035}{4}\right)^{4\cdot 10}$$
$$A = 2500(1.00875)^{40}$$
$$A \approx \$3542.27$$

28. Let x be the amount of money international travelers spent in New York. Then $x + 4$ is the amount of money international travelers spent in California and $2x - 1$ is the amount of money international travelers spent in Florida.

$$x + (x+4) + (2x-1) = 39$$
$$4x + 3 = 39$$
$$4x = 36$$
$$x = 9$$

$x + 4 = (9) + 4 = 13$

$2x - 1 = 2(9) - 1 = 18 - 1 = 17$

International travelers spent \$9 billion in New York, \$13 billion in California, and \$17 billion in Florida.

Cumulative Review Chapters 1–2

1. $\{x|x$ is a natural number greater than $100\}$
$= \{101, 102, 103, ...\}$

2. **a.** $\{x|x$ is an integer between -3 and $5\}$
$= \{-2, -1, 0, 1, 2, 3, 4\}$

b. $\{x|x$ is a whole number between 3 and $5\}$
$= \{4\}$

3. The sum of x and 5 is 20 is written as $x + 5 = 20$.

4. Subtract 19 from x and the difference is the product of 4 and x is written as $x - 19 = 4x$.

5. The quotient of z and 9 amounts to 9 plus z is written as $\frac{z}{9} = 9 + z$.

6. **a.** Because -3 is to the right of -5, $-3 > -5$.

b. Because $\frac{-12}{-4} = 3$, $3 = 3$.

c. Because 0 is to the right of -2, $0 > -2$.

7. The opposite of 8 is -8.

8. The opposite of 0 is 0.

9. The opposite of $-\frac{1}{5}$ is $\frac{1}{5}$.

10. The opposite of -7.3 is 7.3.

11. $-3 + (-11) = -14$

12. $-20.2 + 7.8 = -12.4$

13. $-10 + 15 = 5$

14. $-\frac{1}{2} + \frac{7}{2} = \frac{-1+7}{2} = \frac{6}{2} = 3$

15. $-\frac{2}{3} + \frac{3}{7}$

$$-\frac{2}{3}\cdot\frac{7}{7} + \frac{3}{7}\cdot\frac{3}{3}$$
$$-\frac{14}{21} + \frac{9}{21} = \frac{-14+9}{21} = -\frac{5}{21}$$

16. $1.7 - 8.9 = 1.7 + (-8.9) = -7.2$

17. $\frac{20}{-4} = -5$

18. $\frac{30}{0} = \text{undefined}$

19. $\frac{0}{-8} = 0$

20. $-\frac{3}{4} \div \frac{9}{4} = -\frac{3}{4} \cdot \frac{4}{9} = -\frac{3}{9} = -\frac{1}{3}$

21. $\frac{-10}{-80} = \frac{1}{8}$

22. $-\frac{3}{4}\left(-\frac{4}{7}\right) = \frac{3}{7}$

23. $3 + 2 \cdot 30 = 3 + 60 = 63$

24. $\frac{\sqrt{9+40} - 2^2}{20 \div 2 \cdot 2} = \frac{\sqrt{49} - 4}{10 \cdot 2} = \frac{7-4}{20} = \frac{3}{20}$

25. $3x - 5x + 4 = -2x + 4$

26. $-3(4+y) + \frac{1}{3}(6 - 12y) = -12 - 3y + 2 - 4y$
$= -7y - 10$

27. $y + 3y = 4y$

28. $6 + 2(7x - 1) + 4 = 6 + 14x - 2 + 4 = 14x + 8$

29. $\frac{x^7}{x^4} = x^{7-4} = x^3$

30.
$$\begin{aligned}\frac{a^{-2} \cdot b^{-8}}{b^{-7}} &= a^{-2} \cdot b^{-8-(-7)} \\ &= a^{-2} \cdot b^{-8+7} \\ &= a^{-2} \cdot b^{-1} \\ &= \frac{1}{a^2 b}\end{aligned}$$

31. $\frac{20x^6}{4x^5} = 5x^{6-5} = 5x$

32.
$$\begin{aligned}\frac{(-3)^9 z^7 \cdot z^2}{(-3)^{11} z^{-11}} &= \frac{(-3)^9 z^9}{(-3)^{11} z^{-11}} \\ &= (-3)^{9-11} z^{9-(-11)} \\ &= (-3)^{-2} z^{9+11} \\ &= (-3)^{-2} z^{20} \\ &= \frac{z^{20}}{(-3)^2} \\ &= \frac{z^{20}}{9}\end{aligned}$$

33.
$$\begin{aligned}\left(\frac{3x^2 y}{y^{-9} z}\right)^{-2} &= \left(\frac{3x^2 y^{10}}{z}\right)^{-2} \\ &= \frac{3^{-2} x^{-4} y^{-20}}{z^{-2}} \\ &= \frac{z^2}{3^2 x^4 y^{20}} \\ &= \frac{z^2}{9x^4 y^{20}}\end{aligned}$$

34.
$$\begin{aligned}&(-4a^{-3} b^{-4})(3a^{-2} b)^2 \\ &= (-4a^{-3} b^{-4})(3^2 a^{-4} b^2) \\ &= (-4)(3^2) \cdot a^{-3} a^{-4} b^{-4} b^2 \\ &= (-4)(9) a^{-7} b^{-2} \\ &= \frac{-36}{a^7 b^2}\end{aligned}$$

35. $\left(\frac{3a^2}{2x^{-1}}\right)^3\left(\frac{x^{-3}}{4a^{-2}}\right)^{-1} = \left(\frac{3^3a^6}{2^3x^{-3}}\right)\left(\frac{x^3}{4^{-1}a^2}\right)$

$= \frac{27a^6x^3}{8(4^{-1})a^2x^{-3}}$

$= \frac{4 \cdot 27a^4x^6}{8}$

$= \frac{27a^4x^6}{2}$

36. $11.2 = 1.2 - 5x$

$10 = -5x$

$-2 = x$

The solution set is $\{-2\}$.

37. $2x + 5 = 9$

$2x = 4$

$x = 2$

The solution set is $\{2\}$.

38. $2x + 1.5 = -0.2 + 1.6x$

$10(2x + 1.5) = 10(-0.2 + 1.6x)$

$20x + 15 = -2 + 16x$

$20x = -17 + 16x$

$4x = -17$

$x = -4.25$

The solution set is $\{-4.25\}$.

39. $6x - 4 = 2 + 6(x - 1)$

$6x - 4 = 2 + 6x - 6$

$6x - 4 = 6x - 4$

$-4 = -4$ True

The solution set is $\{x | x \text{ is a real number}\}$.

40. a. The sum of three consecutive integers if x is the first consecutive integer is written as $x + (x + 1) + (x + 2) = 3x + 3$.

b. The perimeter of a square with side length $3x + 1$ is written as $4(3x + 1) = 12x + 4$.

41. Let x be the original price of a particular computer model. The selling price is x decreased by 8%.

$x - 0.08x = 2162$

$0.92x = 2162$

$x = 2350$

The original price was \$2350.

42. Let x be the first number and $3x + 2$ be the second number.

$(3x + 2) - x = 24$

$2x + 2 = 24$

$2x = 22$

$x = 11$

$3x + 2 = 3(11) + 2 = 33 + 2 = 35$

The two numbers are 11 and 35.

43. $V = lwh$

$\frac{V}{lw} = \frac{lwh}{lw}$

$\frac{V}{lw} = h$

44. $7x - 4y = 10$

$7x = 4y + 10$

$x = \frac{4y + 10}{7}$

45. $\frac{1}{4}x \le \frac{3}{2}$

$4\left(\frac{1}{4}x\right) \le 4\left(\frac{3}{2}\right)$

$x \le 6$

The solution set is $(-\infty, 6]$.

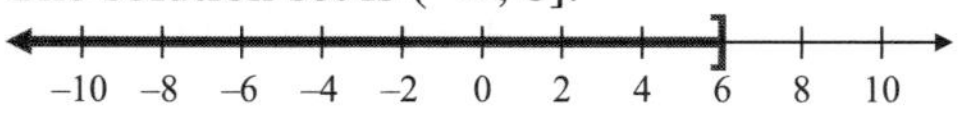

46. $2(7x - 1) - 5x > -(-7x) + 4$

$14x - 2 - 5x > 7x + 4$

$9x - 2 > 7x + 4$

$9x > 7x + 6$

$2x > 6$

$x > 3$

The solution set is $(3, \infty)$.

47. $-2x-5<-3$ or $6x<0$
$-2x<2$ or $x<0$
$x>-1$ or $x<0$
The solution set is
$(-1, \infty) \cup (-\infty, 0) = (-\infty, \infty)$.

48. $4(x+1)-3<4x+1$
$4x+4-3<4x+1$
$4x+1<4x+1$
$1<1$ False
The solution set is $\varnothing$.

49. $|x|=3$
$x=3$ or $x=-3$
The solution set is $\{-3, 3\}$.

50. $|x+3|=|7-x|$
$x+3=7-x$ or $x+3=-(7-x)$
$x=4-x$ or $x+3=-7+x$
$2x=4$ or $3=-7$
$x=2$ or False
The solution set is $\{2\}$.

Chapter 3

Section 3.1 Practice Problems

1. **a.** (3, –2) is in quadrant IV.

   b. (3.5, 4.5) is in quadrant I.

   c. (0, 3) is not in any quadrant.

   d. (–4, 1) is in quadrant II.

   e. (–1, 0) is not in any quadrant.

   f. $\left(-2\frac{1}{2}, -3\right)$ is in quadrant III.

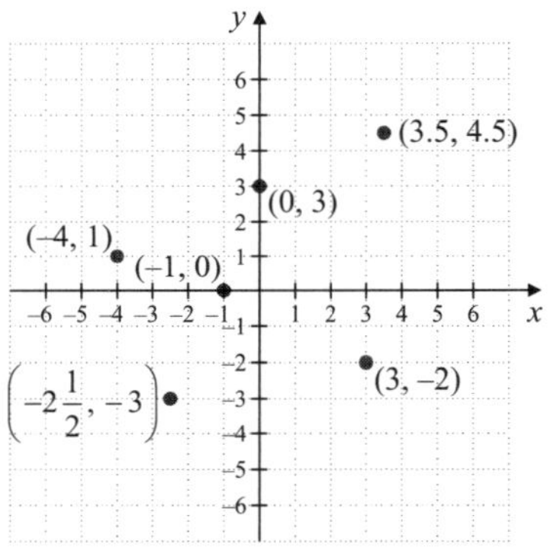

2.

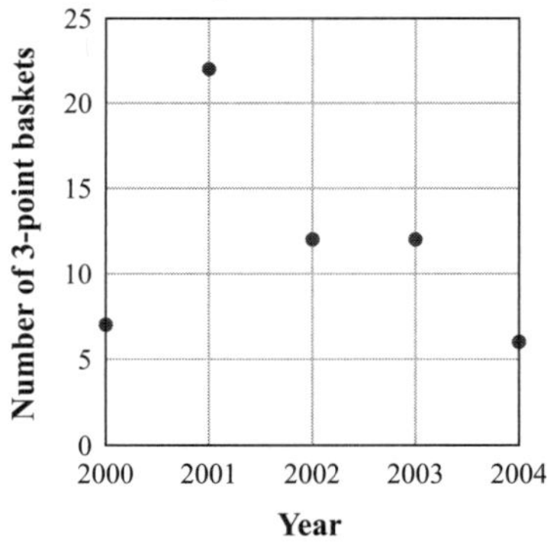

3. (0, –6):
$$2x + y = -6$$
$$2(0) + (-6) \stackrel{?}{=} -6$$
$$-6 = -6 \quad \text{True}$$
(0, –6) is a solution.
(1, 4):
$$2x + y = -6$$
$$2(1) + 4 \stackrel{?}{=} -6$$
$$2 + 4 \stackrel{?}{=} -6$$
$$6 = -6 \quad \text{False}$$
(1, 4) is not a solution.
(–1, –4):
$$2x + y = -6$$
$$2(-1) + (-4) \stackrel{?}{=} -6$$
$$-2 - 4 \stackrel{?}{=} -6$$
$$-6 = -6 \quad \text{True}$$
(–1, –4) is a solution

4. $y = 4x - 3$
Find three ordered pair solutions.

x	y
0	–3
1	1
2	5

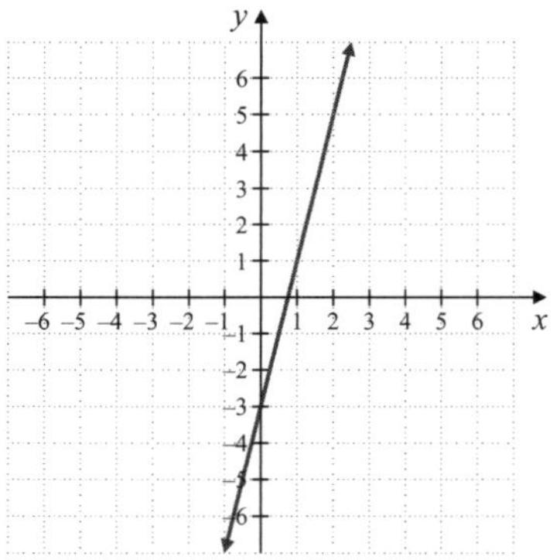

5. $2x + 3y = -6$
x-intercept:
$$2x + 3y = -6$$
$$2x + 3(0) = -6$$
$$2x = -6$$
$$x = -3$$
x-intercept (–3, 0)

y-intercept: $$\begin{aligned} 2x+3y&=-6\\ 2(0)+3y&=-6\\ 3y&=-6\\ y&=-2 \end{aligned}$$

y-intercept $(0, -2)$

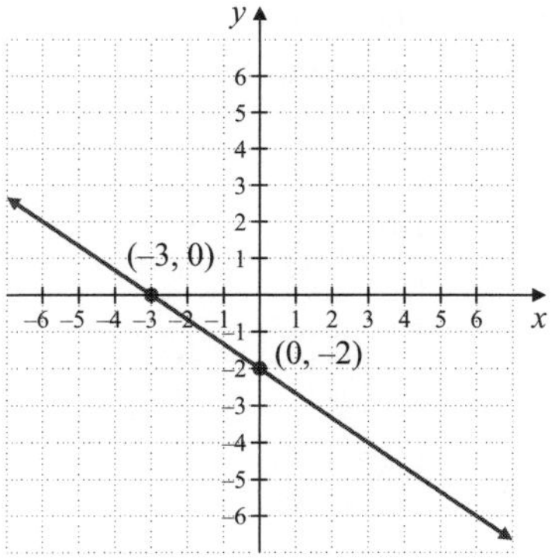

6. $y=\dfrac{1}{4}x$

Find three ordered pair solutions.

x	y
0	0
1	$\frac{1}{4}$
2	$\frac{1}{2}$

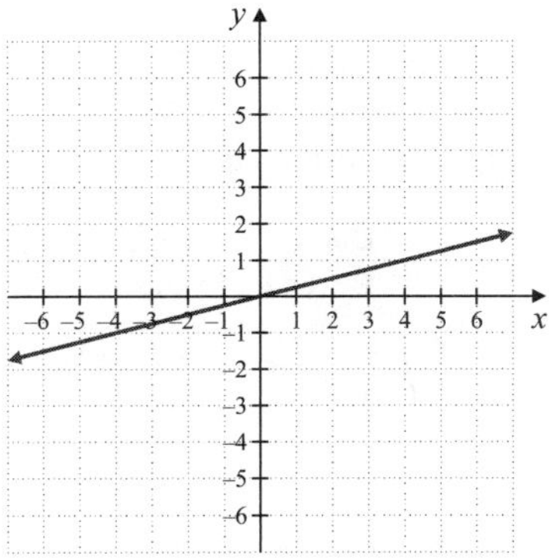

7. $x=-1$

This is a vertical line with an x-intercept at $(-1, 0)$ and no y-intercept.

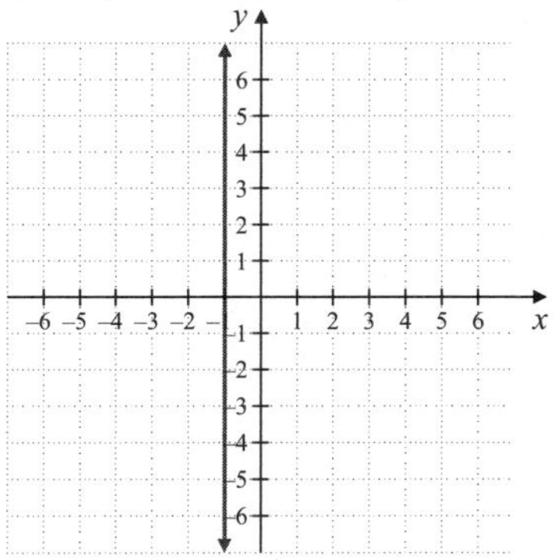

8. $y=2$

This is a horizontal line with a y-intercept at $(0, 2)$ and no x-intercept.

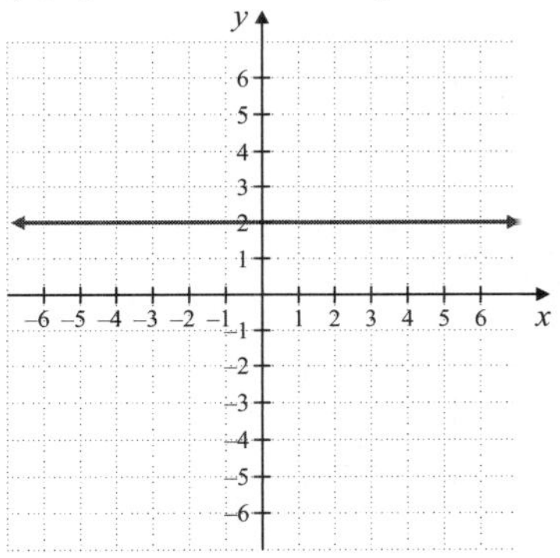

Calculator Exploration

1. $y = 6x - 1$

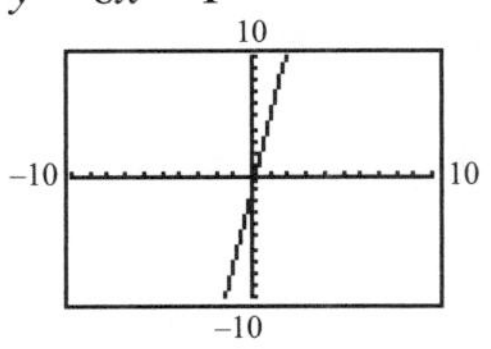

2. $y = 3x - 2$

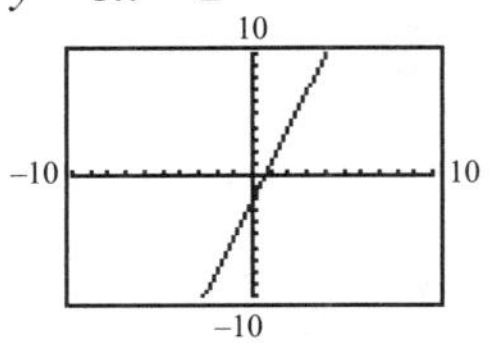

3. $y = -3.2x + 7.9$

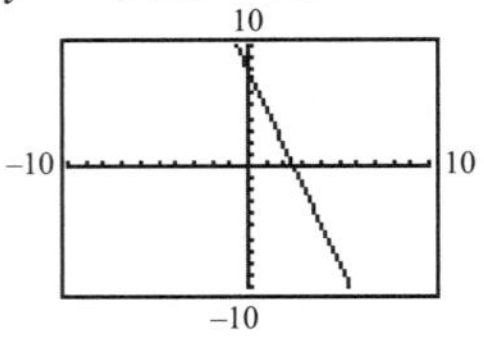

4. $y = -x + 5.85$

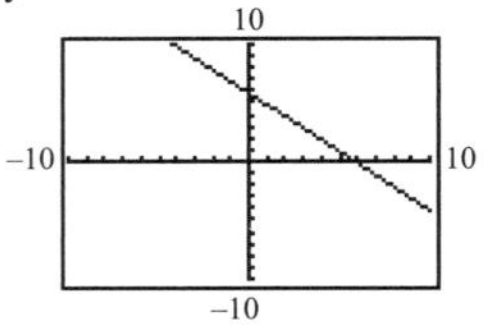

5. $y = \frac{1}{4}x - \frac{2}{3}$

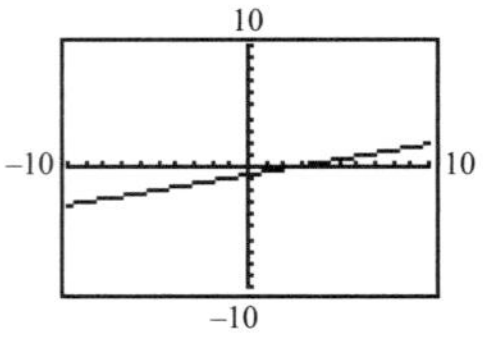

6. $y = \frac{2}{3}x - \frac{1}{5}$

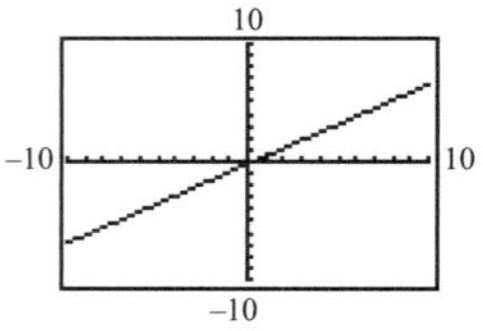

Mental Math

1. Point A is at (5, 2).
2. Point B is at (2, 5).
3. Point C is at (3, −1).
4. Point D is at (−1, 3).
5. Point E is at (−5, −2).
6. Point F is at (−3, 5).
7. Point G is at (−1, 0).
8. Point H is at (0, −3).

Exercise Set 3.1

1. (3, 2) is in quadrant I.
   (−5, 3) is in quadrant II.
   $\left(5\frac{1}{2}, -4\right)$ is in quadrant IV.
   (0, 3.5) is on the y-axis.
   (−2, −4) is in quadrant III.

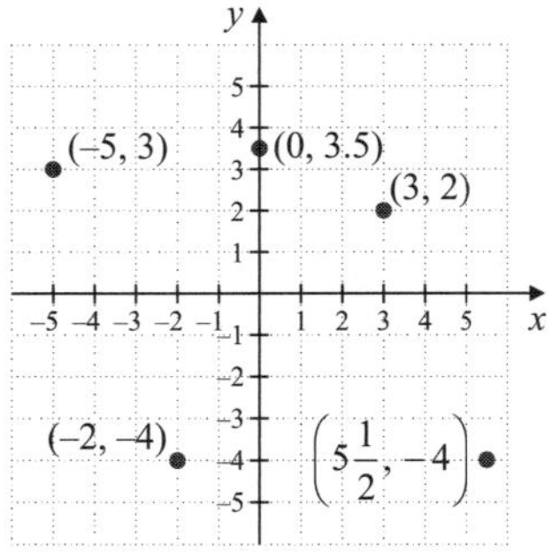

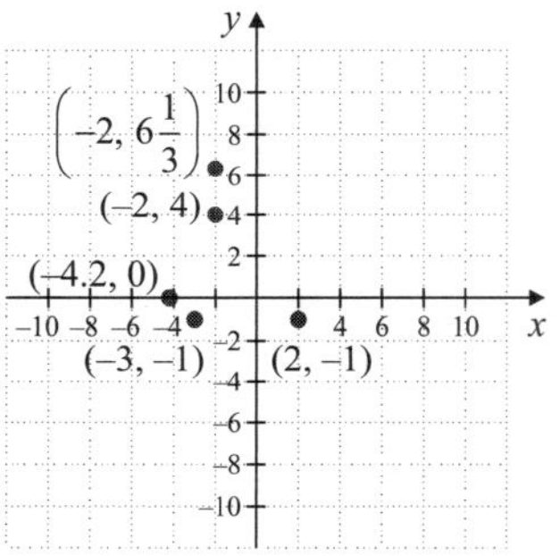

3. $(x, -y)$ is in quadrant IV.

5. $(x, 0)$ is on the x-axis.

7. $(-x, -y)$ is in quadrant III.

9. 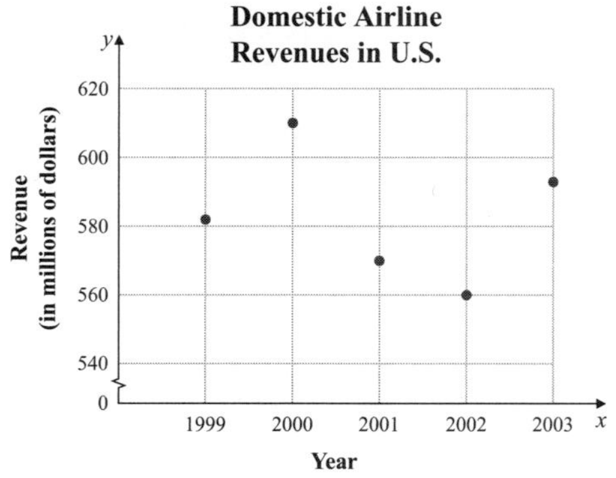

11. (0, 5): $y = 3x - 5$

$5 \stackrel{?}{=} 3(0) - 5$

$5 = -5$ False

(0, 5) is not a solution.

(−1, −8): $y = 3x - 5$

$-8 \stackrel{?}{=} 3(-1) - 5$

$-8 \stackrel{?}{=} -3 - 5$

$-8 = -8$ True

(−1, −8) is a solution.

13. (1, 0): $-6x + 5y = -6$

$-6(1) + 5(0) \stackrel{?}{=} -6$

$-6 + 0 \stackrel{?}{=} -6$

$-6 = -6$ True

(1, 0) is a solution.

$\left(2, \frac{6}{5}\right)$: $-6x + 5y = -6$

$-6(2) + 5\left(\frac{6}{5}\right) \stackrel{?}{=} -6$

$-12 + 6 \stackrel{?}{=} -6$

$-6 = -6$ True

$\left(2, \frac{6}{5}\right)$ is a solution.

15. (1, −3): $y = -3$

$-3 = -3$ True

(1, −3) is a solution.

(−3, 6): $y = -3$

$6 = -3$ False

(−3, 6) is not a solution.

17. $x - 2y = 4$

Find three ordered pair solutions.

x	y
4	0
0	−2
−2	−3

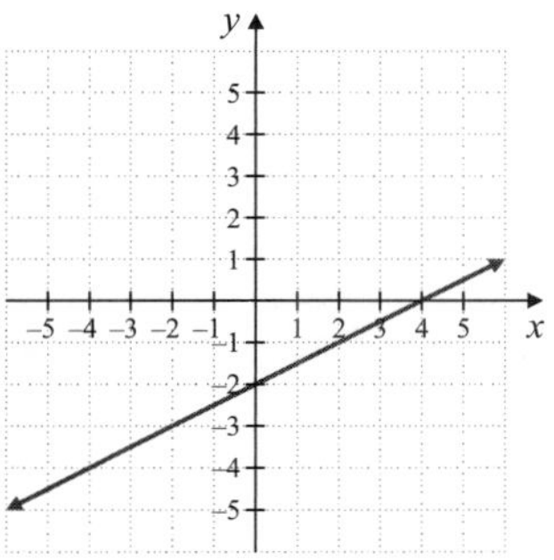

19. $3x + 2y = 6$

Find three ordered pair solutions.

x	y
0	3
2	0
4	−3

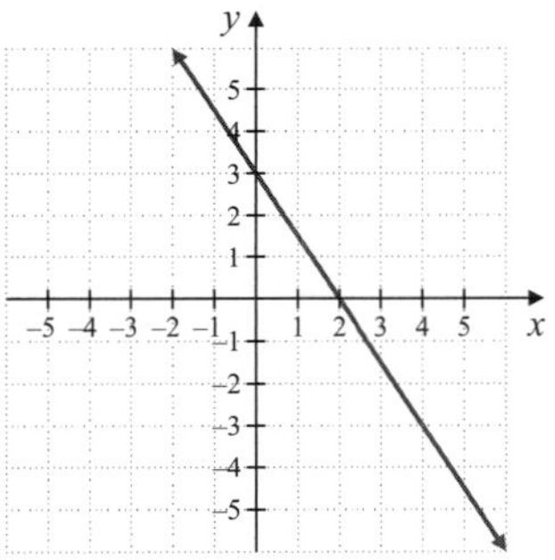

21. $x = 4$

Any ordered pair with an x-coordinate of 4 is a solution to $x = 4$.

x	y
4	−1
4	0
4	1

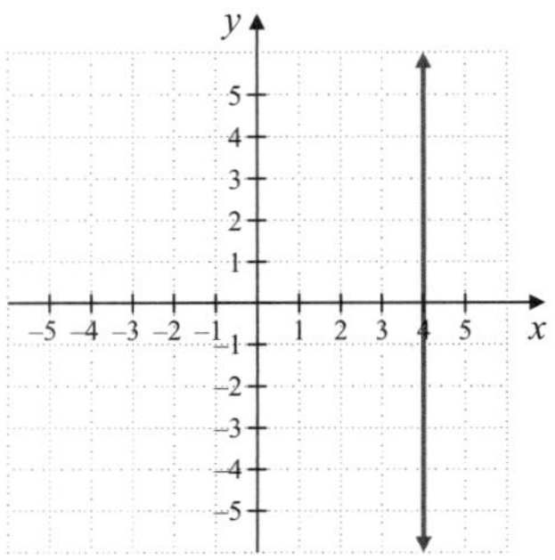

23. $x - 3y = 6$
Find three ordered pair solutions.

x	y
0	−2
6	0
−3	−3

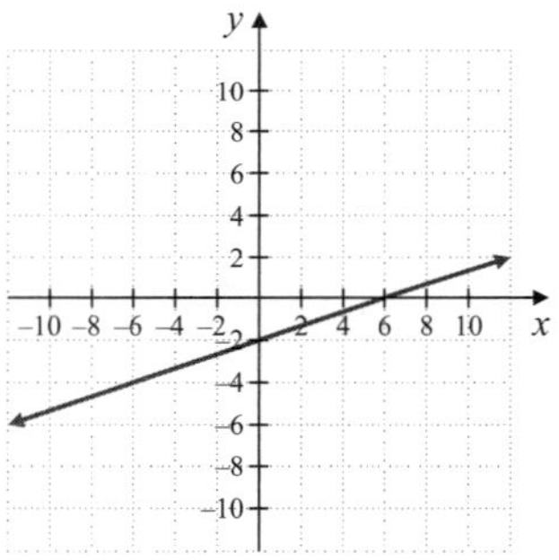

25. $y = 3x$
Find three ordered pair solutions.

x	y
0	0
1	3
−1	−3

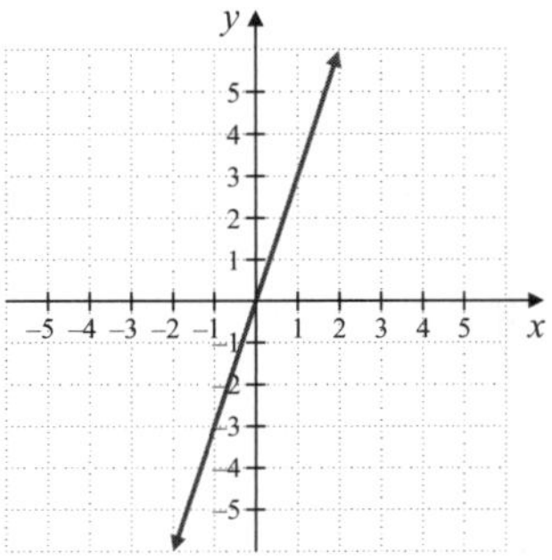

27. $y = -2$
Any ordered pair with a y-coordinate of −2 is a solution to $y = -2$.

x	y
−1	−2
0	−2
1	−2

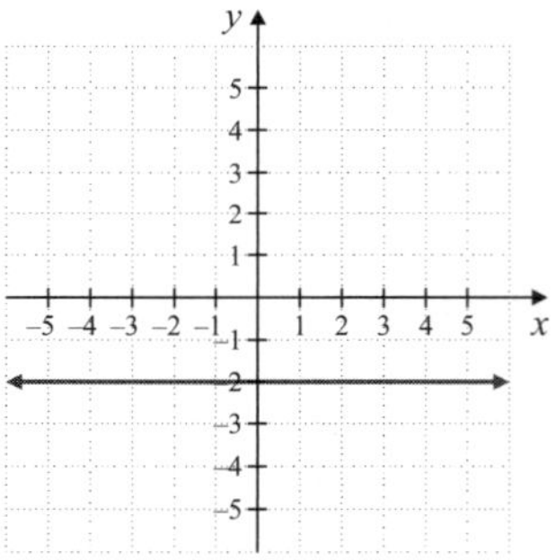

29. $4x + 5y = 15$
Find three ordered pair solutions.

x	y
0	3
5	−1
−5	7

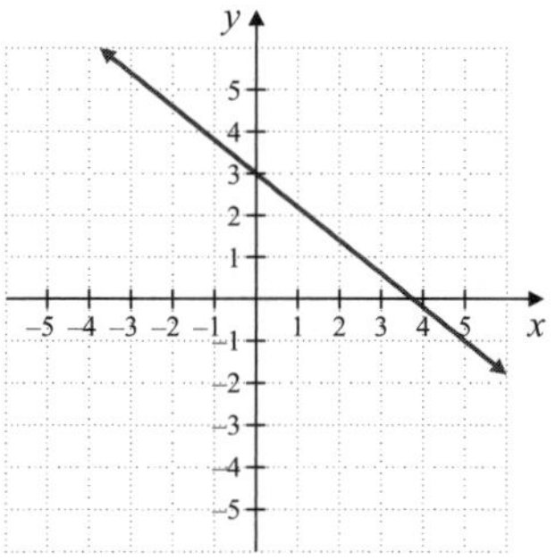

31. $5y = x - 10$
Find three ordered pair solutions.

x	y
0	–2
10	0
5	–1

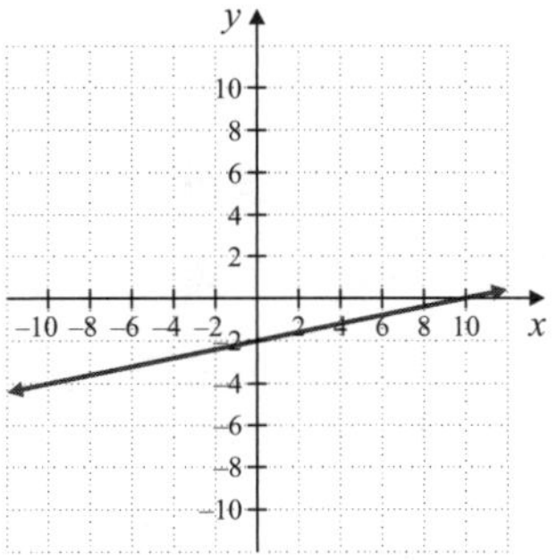

33. $x = \frac{1}{2}$
This is a vertical line with an x-intercept at $\left(\frac{1}{2}, 0\right)$ and no y-intercept.

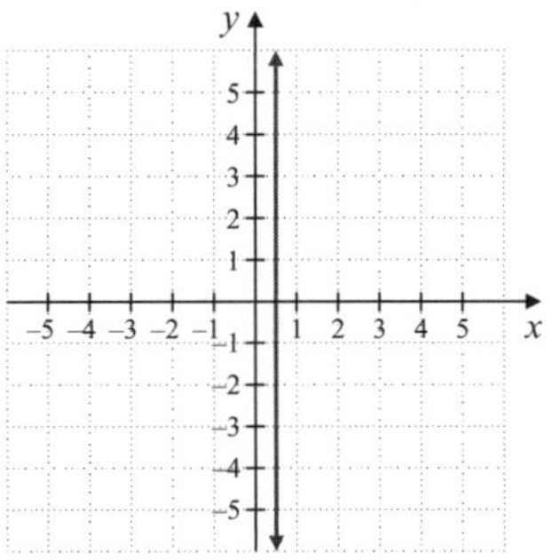

35. $y = 0.5x$
Find three ordered pair solutions.

x	y
0	0
1	0.5
2	1

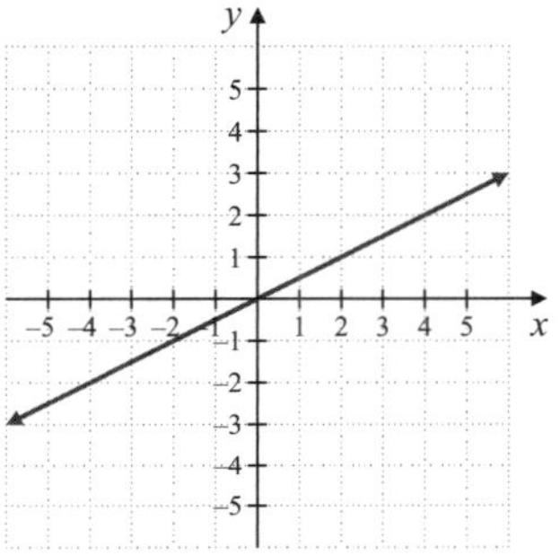

37. $y = -4x + 1$
Find three ordered pair solutions.

x	y
0	1
1	–3
–1	5

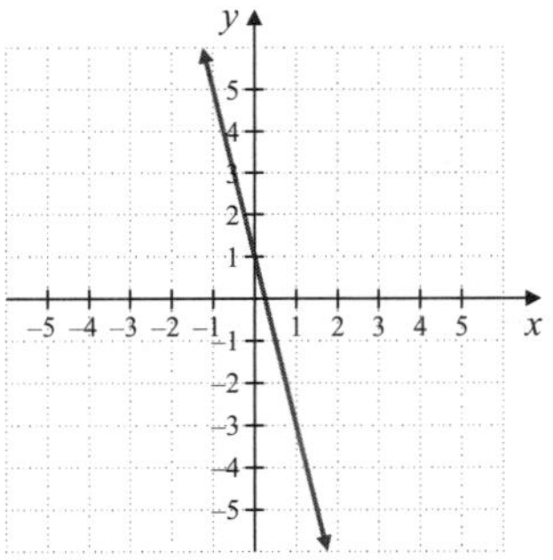

39.
$$2y - 6 = 0$$
$$2y = 6$$
$$y = 3$$
This is a horizontal line with a y-intercept at (0, 3) and no x-intercept.

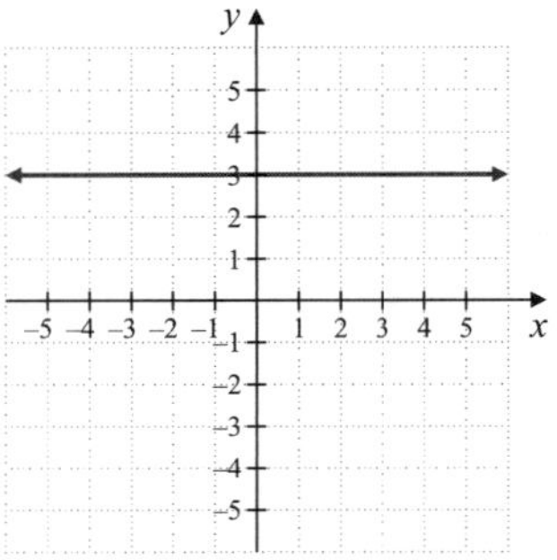

41. $y = -\frac{2}{3}x - 4$
Find three ordered pair solutions.

x	y
−3	−2
0	−4
3	−6

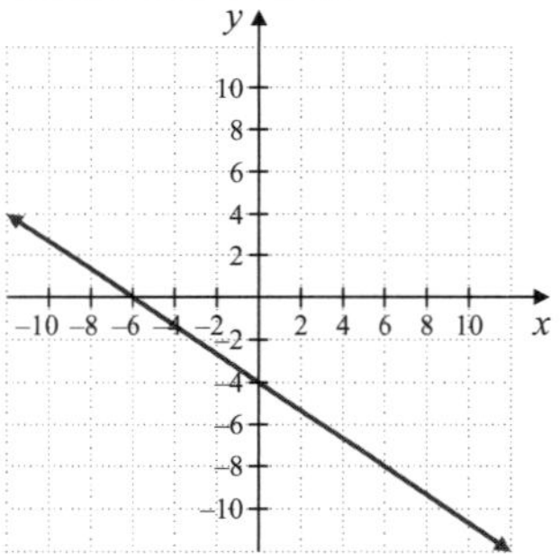

43. $5x - 2y = -10$
Find three ordered pair solutions.

x	y
0	5
−2	0
2	10

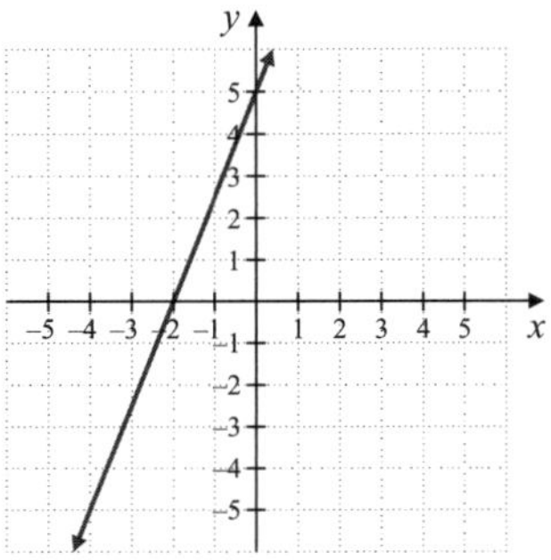

45. $\frac{-6-3}{2-8} = \frac{-9}{-6} = \frac{3}{2}$

47. $\frac{-8-(-2)}{-3-(-2)} = \frac{-6}{-1} = 6$

49. $\frac{0-6}{5-0} = \frac{-6}{5} = -\frac{6}{5}$

51. Choice b correctly describes the location of the point.

53. The description matches graph B.

55. The description matches graph C.

57. From the graph, the minimum hourly wage went from \$3.80 to \$4.25 in 1991.

59. answers may vary

61. **a.** $y = 2x + 6$
Find three ordered pair solutions.

x	y
0	6
1	8
2	10

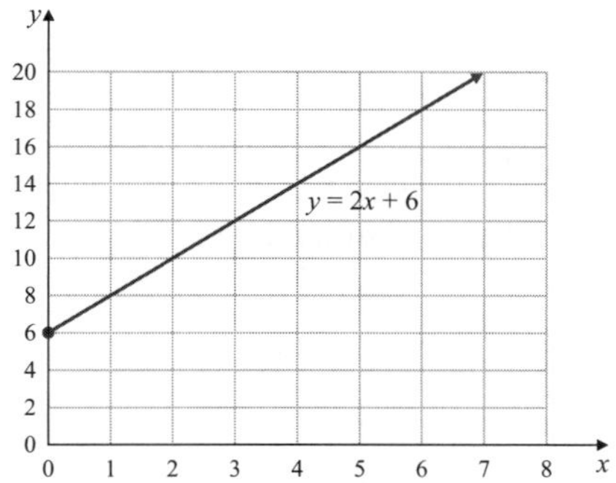

b. The line goes through the point (4, 14), so the perimeter is 14 inches when the length is 4 inches.

63. Since the point (0, 7000) is on the graph, the copy machine was purchased at \$7000.

65. The value was originally \$7000. After one year, the value was \$6500, so the loss in value during the first year was \$500.

67. This method is called straight-line depreciation because depreciation is the same from year to year.

69. a. $2x + 3y = 1500$
Let $x = 0$.
$$2(0)+3y=1500$$
$$3y=1500$$
$$y=500$$
The ordered pair is (0, 500). This corresponds to manufacturing 0 tables and 500 chairs.

b. $2x + 3y = 1500$
Let $y = 0$.
$$2x+3(0)=1500$$
$$2x=1500$$
$$x=750$$
The ordered pair is (750, 0). This corresponds to manufacturing 750 tables and 0 chairs.

c.
$$2x+3y=1500$$
$$2(50)+3y=1500$$
$$100+3y=1500$$
$$3y=1400$$
$$y=466$$
466 chairs are produced.

71. answers may vary

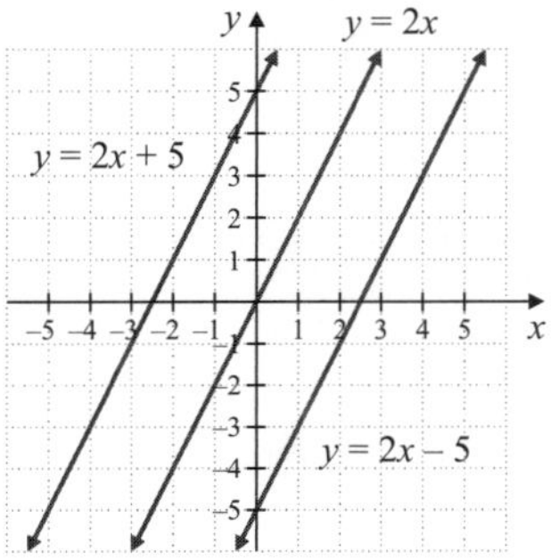

73. The vertical line $x = 0$ has y-intercepts.

75.

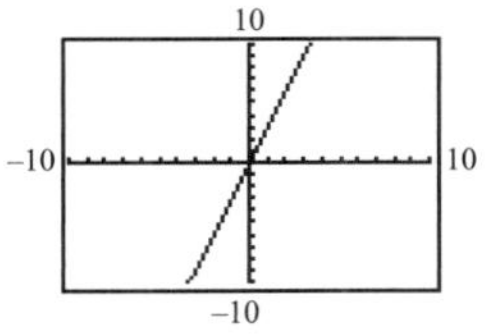

77.

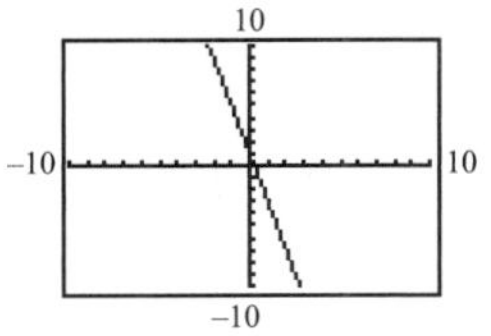

Section 3.2 Practice Problems

1. $(x_1, y_1) = (-1, -2)$
$(x_2, y_2) = (2, 5)$
$$m = \frac{y_2 - y_1}{x_2 - x_1} = \frac{5-(-2)}{2-(-1)} = \frac{7}{3}$$

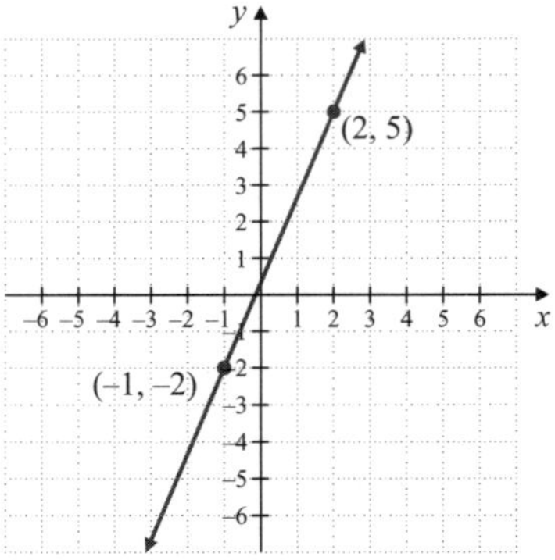

2. $(x_1, y_1) = (1, -1)$
$(x_2, y_2) = (-2, 4)$
$$m = \frac{y_2 - y_1}{x_2 - x_1} = \frac{4-(-1)}{-2-1} = -\frac{5}{3}$$

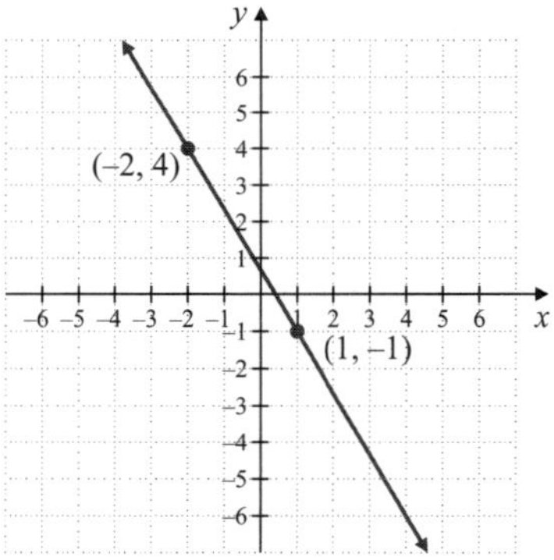

3. $y = 2x + 4$
Find two points on the line.
Let $x = 0$: $y = 2x + 4$
$y = 2 \cdot 0 + 4$
$y = 4$
Let $x = 1$: $y = 2x + 4$
$y = 2 \cdot 1 + 4$
$y = 6$

$(x_1, y_1) = (0, 4)$
$(x_2, y_2) = (1, 6)$
$$m = \frac{6-4}{1-0} = \frac{2}{1} = 2$$
The slope of the line is 2.

4. $2x - 4y = 8$
$-4y = -2x + 8$
$y = \frac{1}{2}x - 2$

The slope of the line is $\frac{1}{2}$ and the y-intercept is (0, −2).

5. $x = 3$
Find two points.
$(x_1, y_1) = (3, 0)$
$(x_2, y_2) = (3, 4)$
$$m = \frac{y_2 - y_1}{x_2 - x_1} = \frac{4-0}{3-3} = \frac{4}{0}$$
The slope of the vertical line $x = 3$ is undefined.

6. $y = -3$
Find two points.
$(x_1, y_1) = (0, -3)$
$(x_2, y_2) = (1, -3)$
$$m = \frac{y_2 - y_1}{x_2 - x_1} = \frac{-3-(-3)}{1-0} = \frac{0}{1} = 0$$
The slope of the horizontal line $y = -3$ is 0.

7. $(x_1, y_1) = (-2, 5)$
$(x_2, y_2) = (-4, 2)$
$$m = \frac{y_2 - y_1}{x_2 - x_1} = \frac{2-5}{-4-(-2)} = \frac{-3}{-2} = \frac{3}{2}$$

The slope of the line is $\frac{3}{2}$.

8. $2x+5y=1$

$$5y=-2x+1$$
$$y=-\frac{2}{5}x+\frac{1}{5}$$

$4x+10y=3$

$$10y=-4x+3$$
$$y=-\frac{4}{10}x+\frac{3}{10}$$
$$y=-\frac{2}{5}x+\frac{3}{10}$$

The two lines have the same slope, therefore the two lines are parallel.

9. $x-4y=3$

$$-4y=-x+3$$
$$y=\frac{1}{4}x-\frac{3}{4}$$

$3x+12y=7$

$$12y=-3x+7$$
$$y=\frac{-3}{12}x+\frac{7}{12}$$
$$y=-\frac{1}{4}x+\frac{7}{12}$$

The slopes are not the same and their product is not -1, $\left[\left(\frac{1}{4}\right)\cdot\left(-\frac{1}{4}\right)=-\frac{1}{16}\right]$.

Therefore, the lines are neither parallel nor perpendicular.

Calculator Explorations

1. $y = 3x$, $y = 3x + 4$

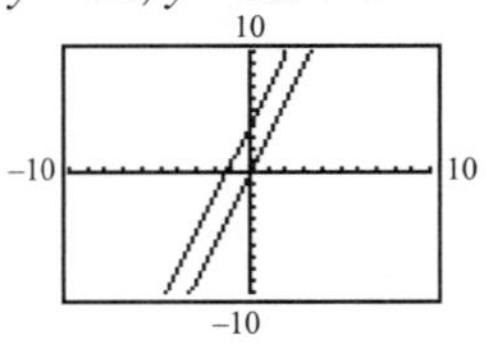

answers may vary

2. $y = 5x$, $y = 5x - 2$

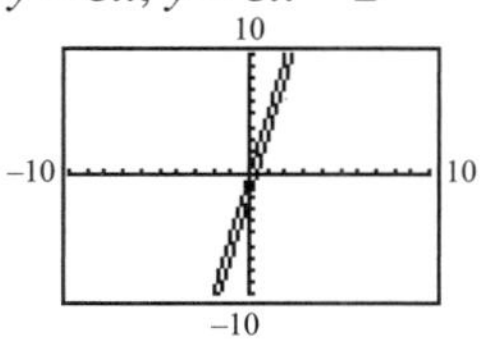

answers may vary

3. $y=-\frac{2}{3}x+1$, $y=-\frac{2}{3}x+6$

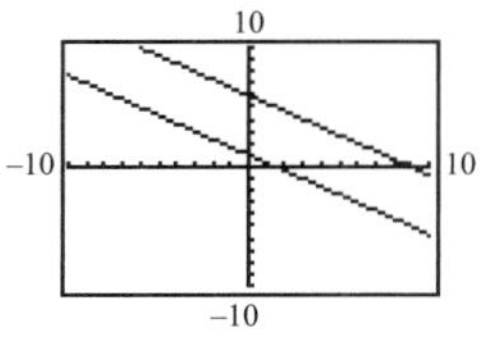

answers may vary

4. $y=-\frac{1}{4}x-3$, $y=-\frac{1}{4}x+6$

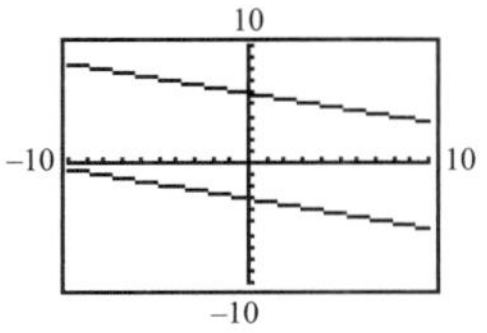

answers may vary

5. $y = 4.61x - 1.86$, $y = 4.61x + 2.11$

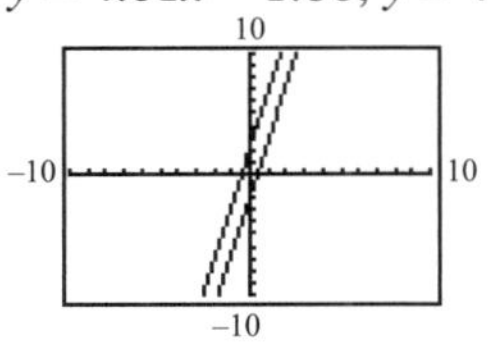

answers may vary

6. $y = 3.78x + 1.92$, $y = 3.78x + 8.08$

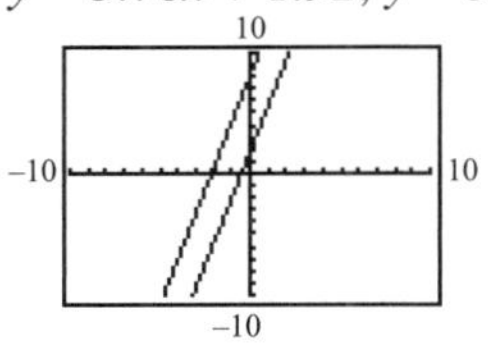

answers may vary

Mental Math

1. $m=\frac{7}{6}$; the line slants upward.

2. $m = -3$; the line slants downward.

3. $m = 0$; the line slants horizontally.

4. m is undefined; the line slants vertically.

Exercise Set 3.2

1. $m = \frac{11-2}{8-3} = \frac{9}{5}$

3. $m = \frac{8-1}{1-3} = -\frac{7}{2}$

5. $m = \frac{3-8}{4-(-2)} = -\frac{5}{6}$

7. $m = \frac{-4-(-6)}{4-(-2)} = \frac{2}{6} = \frac{1}{3}$

9. $m = \frac{11-(-1)}{-12-(-3)} = \frac{12}{-9} = -\frac{4}{3}$

11. $m = \frac{5-5}{3-(-2)} = \frac{0}{5} = 0$

13. $m = \frac{2-(-2)}{2-0} = \frac{4}{2} = 2$

15. $m = \frac{3-(-1)}{-2-3} = -\frac{4}{5}$

17. $m = \frac{\text{rise}}{\text{run}} = \frac{8}{12} = \frac{2}{3}$

The pitch of the roof is $\frac{2}{3}$.

19. $m = \frac{\text{rise}}{\text{run}} = \frac{15}{100} = \frac{3}{20}$

The grade of the road is $\frac{3}{20}$.

21. $y = -x + 5$
$m = -1$; y-intercept = $(0, 5)$

23. $y = 5x - 2$
$m = 5$; y-intercept = $(0, -2)$

25. $2x + y = 7$
$y = -2x + 7$
$m = -2$; y-intercept = $(0, 7)$

27. $2x - 3y = 10$
$-3y = -2x + 10$
$y = \frac{2}{3}x - \frac{10}{3}$
$m = \frac{2}{3}$; y-intercept = $\left(0, -\frac{10}{3}\right)$

29. $x = 4$
This is a vertical line, so the slope is undefined. Since the line is not the y-axis, there is no y-intercept.

31. $y = -2$
$y = 0x - 2$
$m = 0$; y-intercept = $(0, -2)$

33. $y = \frac{1}{2}x$
$y = \frac{1}{2}x + 0$
$m = \frac{1}{2}$; y-intercept = $(0, 0)$

35. $3y + 8 = x$
$3y = x - 8$
$y = \frac{1}{3}x - \frac{8}{3}$
$m = \frac{1}{3}$; y-intercept = $\left(0, -\frac{8}{3}\right)$

37. $-6x + 5y = 30$
$5y = 6x + 30$
$y = \frac{6}{5}x + 6$
$m = \frac{6}{5}$; y-intercept = $(0, 6)$

39. $y = 7x$
$y = 7x + 0$
$m = 7$; y-intercept = (0, 0)

41. $x + 2 = 0$
$x = -2$
This is a vertical line, so the slope is undefined. Since the line is not the y-axis, there is no y-intercept.

43. $y = 2x + 3$
$m = 2$; y-intercept = (0, 3)
The graph that matches these conditions is graph A.

45. $y = -2x + 3$
$m = -2$; y-intercept = (0, 3)
The graph that matches these conditions is graph B.

47. l_2 has the greater slope because the slope of l_2 is positive and the slope of l_1 is negative.

49. l_2 has the greater slope because the slope of l_2 is 0 and the slope of l_1 is negative.

51. l_2 has the greater slope because the slope of l_2 is a larger positive number than the slope of l_1.

53. $y = -3x + 6$ $\quad$ $y = 3x + 5$
$m = -3$ $\quad$ $m = 3$
y-int = (0, 6) $\quad$ y-int = (0, 5)
Neither, since their slopes are not equal, nor does their product equal −1.

55. $y = -9x + 3$ $\quad$ $y = \frac{3}{2}x - 7$
$m = -9$ $\quad$ $m = \frac{3}{2}$
y-int = (0, 3) $\quad$ y-int = (0, −7)
Neither, since their slopes are not equal, nor does their product equal −1.

57. $2x - y = -10$
$-y = -2x - 10$
$y = 2x + 10$

$2x + 4y = 2$
$4y = -2x + 2$
$y = -\frac{1}{2}x + \frac{1}{2}$

$m = 2$ $\quad$ $m = -\frac{1}{2}$

y-int = (0, 10) $\quad$ $y\text{-int} = \left(0, \frac{1}{2}\right)$

The lines are perpendicular since the product of their slopes is −1.

59. $y = 12x + 6$ $\quad$ $y = 12x - 2$
$m = 12$ $\quad$ $m = 12$
y-int = (0, 6) $\quad$ y-int = (0, −2)
The lines are parallel since their slopes are equal, but their y-intercepts are different.

61. $-4x + 2y = 5$
$2y = 4x + 5$
$y = 2x + \frac{5}{2}$

$2x - y = 7$
$-y = -2x + 7$
$y = 2x - 7$

$m = 2$ $\quad$ $m = 2$

$y\text{-int} = \left(0, \frac{5}{2}\right)$ $\quad$ y-int = (0, −7)

The lines are parallel, since they have the same slope, but different y-intercepts.

63. $7x - y = 1$
$-y = -7x + 1$
$y = 7x - 1$

$x + 7y = 3$
$7y = -x + 3$
$y = -\frac{1}{7}x + \frac{3}{7}$

$m = 7$ $\quad$ $m = -\frac{1}{7}$

y-int = (0, −1) $\quad$ $y\text{-int} = \left(0, \frac{3}{7}\right)$

The lines are perpendicular since the product of their slopes is −1.

65. $m = \frac{0-5}{10-0} = \frac{-5}{10} = -\frac{1}{2}$ $m = \frac{0-(-10)}{5-0} = \frac{10}{5} = 2$

The lines are perpendicular since the product of their slopes is −1.

67. $m = \frac{-9-8}{1-(-3)} = -\frac{17}{4}$ $m = \frac{16-(-1)}{-1-(-2)} = \frac{17}{1}$

Neither, since their slopes are not equal, nor does their product equal −1.

69. $|x-3| = 6$

$x-3=-6$ or $x-3=6$

$x=-3$ or $x=9$

The solution set is $\{-3, 9\}$.

71. $|2x+5| > 3$

$2x+5>3$ or $2x+5<-3$

$2x>-2$ or $2x<-8$

$x>-1$ or $x<-4$

The solution set is $(-\infty, -4) \cup (-1, \infty)$.

73. $|3x-4| \le 2$

$-2 \le 3x-4 \le 2$

$2 \le 3x \le 6$

$\frac{2}{3} \le x \le 2$

The solution set is $\left\{x \middle| \frac{2}{3} \le x \le 2\right\}$ or $\left[\frac{2}{3}, 2\right]$.

75. The calculation is incorrect.

$m = \frac{-14-6}{7-(-2)} = -\frac{20}{9}$

77. The calculation is correct.

79. $y = -\frac{7}{2}x - 6$

$m = -\frac{7}{2}$

The slope of a parallel line is $-\frac{7}{2}$.

81. $y = -\frac{7}{2}x - 6$

$m = -\frac{7}{2}$

The slope of a perpendicular line is $-\left(-\frac{2}{7}\right) = \frac{2}{7}$.

83. $5x - 2y = 6$

$-2y = -5x + 6$

$y = \frac{5}{2}x - 3$

$m = \frac{5}{2}$

The slope of a parallel line is $\frac{5}{2}$.

85. $5x - 2y = 6$

$-2y = -5x + 6$

$y = \frac{5}{2}x - 3$

The slope of a perpendicular line is $-\left(\frac{2}{5}\right) = -\frac{2}{5}$.

87. The coordinates of point *C* is (10, 13).

89. *F*(22, 2) and *G*(26, 8)

$m = \frac{8-2}{26-22} = \frac{6}{4} = \frac{3}{2}$

The rate of change in altitude is $\frac{3}{2}$ yards per second.

91. answers may vary

93. a. $l_1: m = \dfrac{-2-4}{2-(-1)} = \dfrac{-6}{3} = -2$

$l_2: m = \dfrac{6-2}{-8-(-4)} = \dfrac{4}{-4} = -1$

$l_3: m = \dfrac{-4-0}{0-(-6)} = \dfrac{-4}{6} = -\dfrac{2}{3}$

b. Since l_1 is the steepest line, for lines with negative slopes, the steeper line has the lesser slope.

Section 3.3 Practice Exercises

1. $y = \dfrac{2}{3}x + 1$

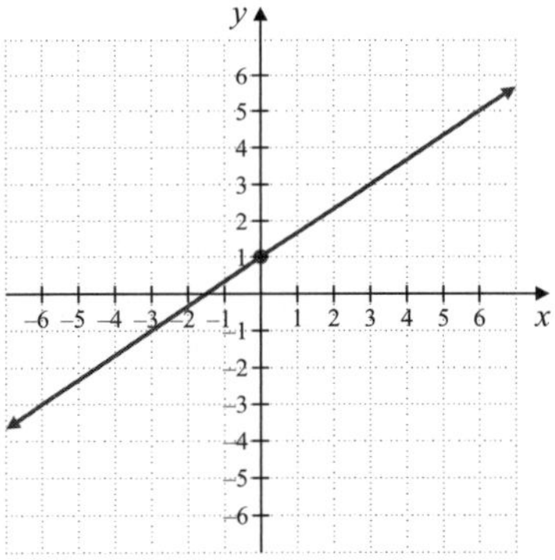

2. $y = 3x + y = -2$

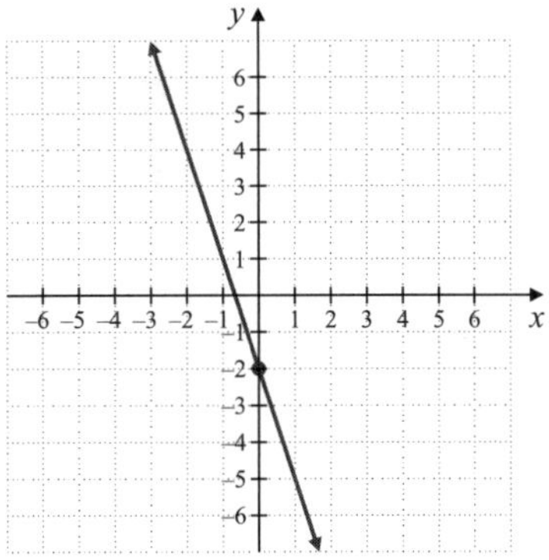

3. $y = mx + b$

$y = \dfrac{1}{7}x + (-5)$

$y = \dfrac{1}{7}x - 5$

4. a. $x = 2010 - 1980 = 30$
$y = 110{,}520(30) + 2{,}127{,}400$
$y = 5{,}443{,}000$

b. The number of people age 85 or older in the United States increases at a rate of 110,520 per year.

c. At year $x = 0$, or 1980, there were 2,127,400 people age 85 or older in the United States.

Calculator Explorations

1. $x = 3.5y$

$y = \dfrac{1}{3.5}x$

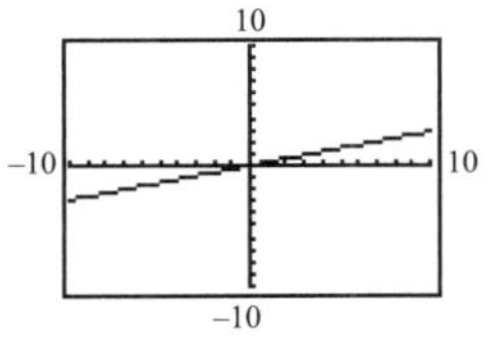

2. $-2.7y = x$

$y = -\dfrac{1}{2.7}x$

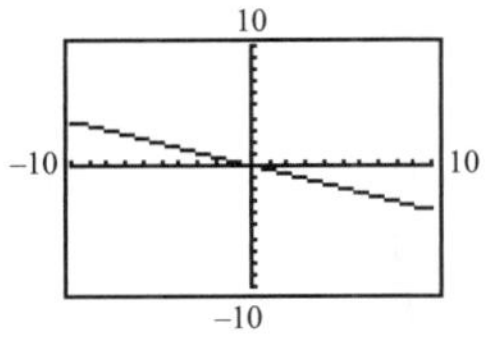

3. $5.78x + 2.31y = 10.98$

$$2.31y = -5.78x + 10.98$$

$$y = -\frac{5.78}{2.31}x + \frac{10.98}{2.31}$$

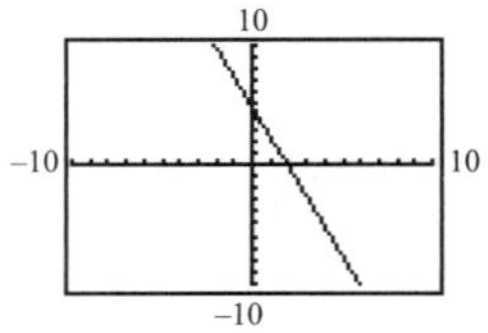

4. $-7.22x + 3.89y = 12.57$

$$3.89y = 7.22x + 12.57$$

$$y = \frac{7.22}{3.89}x + \frac{12.57}{3.89}$$

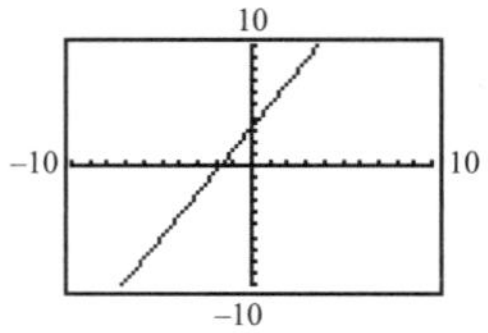

5. $y - x = 3.78$

$$y = x + 3.78$$

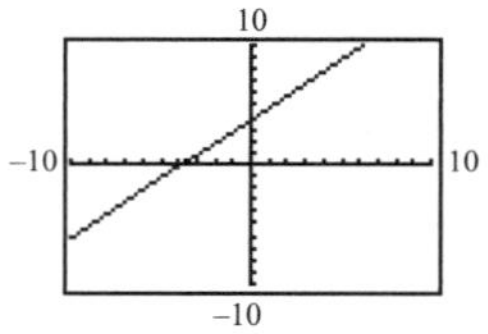

6. $3y - 5x = 6x - 4$

$$3y = 11x - 4$$

$$y = \frac{11}{3}x - \frac{4}{3}$$

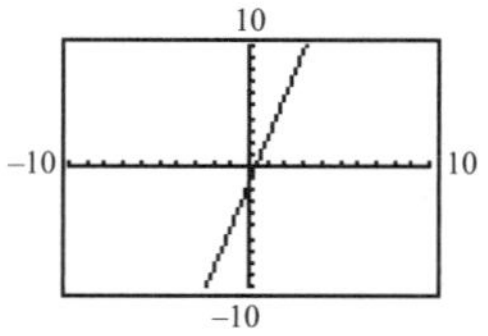

7. $y - 5.6x = 7.7x + 1.5$

$$y = 13.3x + 1.5$$

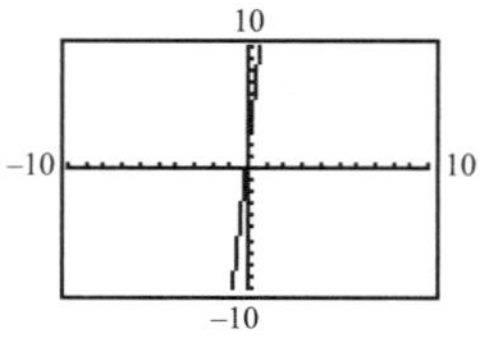

8. $y + 2.6x = -3.2$

$$y = -2.6x - 3.2$$

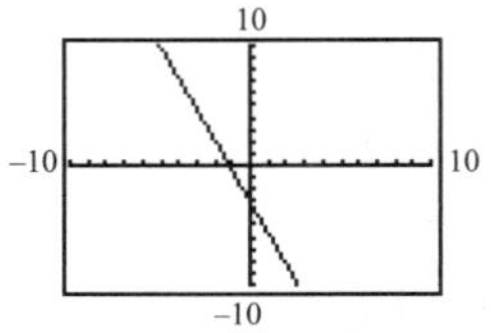

Mental Math

1. $y = -4x + 12$
$m = -4$
$y\text{-int} = (0, 12)$

2. $y = \frac{2}{3}x - \frac{7}{2}$

$m = \frac{2}{3}$

$y\text{-int} = \left(0, -\frac{7}{2}\right)$

3. $y = 5x$
$m = 5$
$y\text{-int} = (0, 0)$

4. $y = -x$
$m = -1$
$y\text{-int} = (0, 0)$

5. $y = \frac{1}{2}x + 6$

$m = \frac{1}{2}$

$y\text{-int} = (0, 6)$

6. $y = -\frac{2}{3}x + 5$

$m = -\frac{2}{3}$

$y\text{-int} = (0, 5)$

Exercise Set 3.3

1.

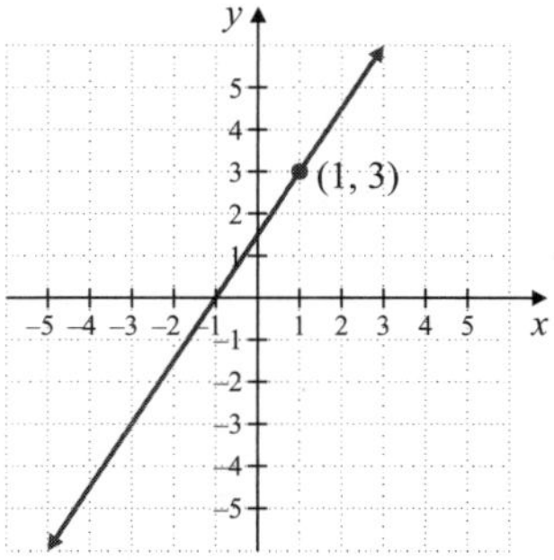

3.

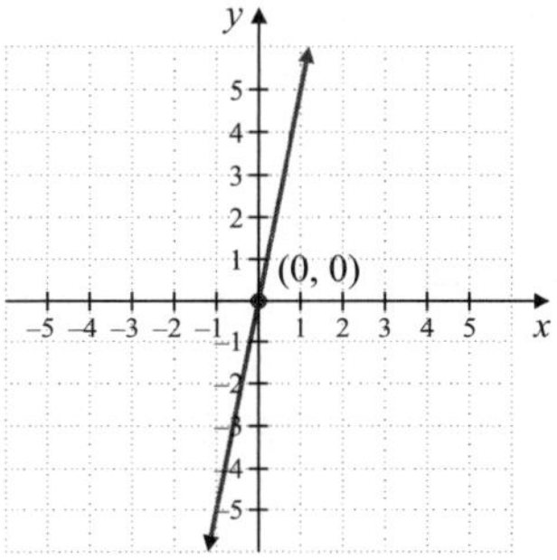

5.

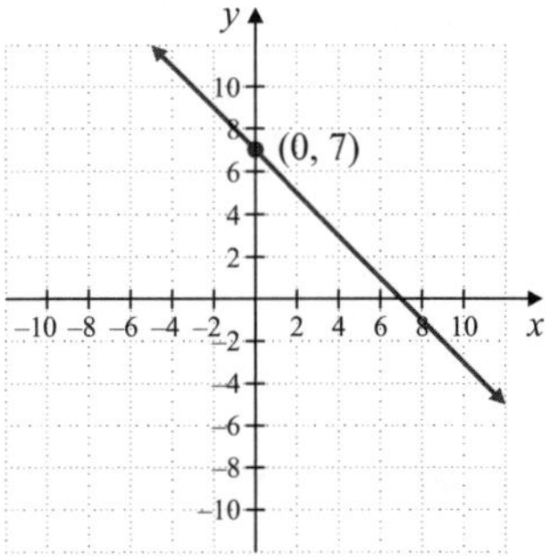

7. $y = -2x$
$m = -2$, y-int = (0, 0)

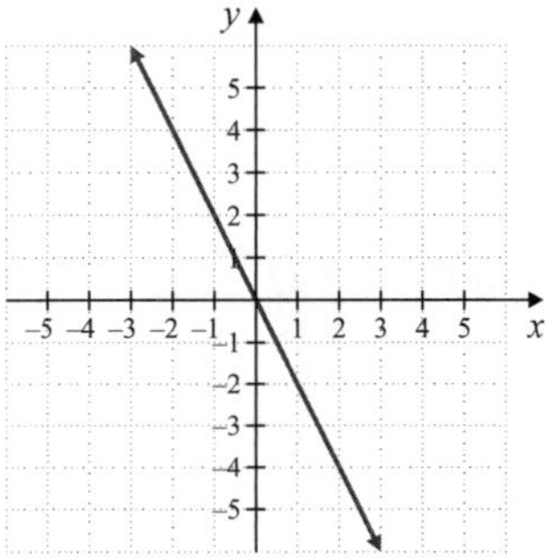

9. $y = -2x + 3$
$m = -2$, y-int = (0, 3)

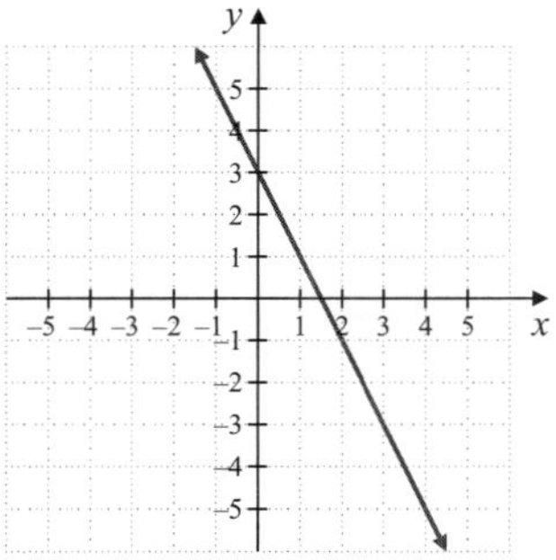

11. $y = \frac{1}{2}x$

$m = \frac{1}{2}$, y-int = (0, 0)

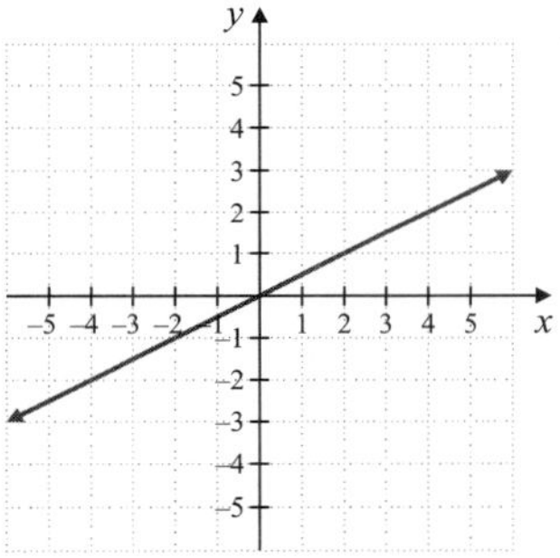

13. $y = \frac{1}{2}x - 4$

$m = \frac{1}{2}$, y-int = (0, −4)

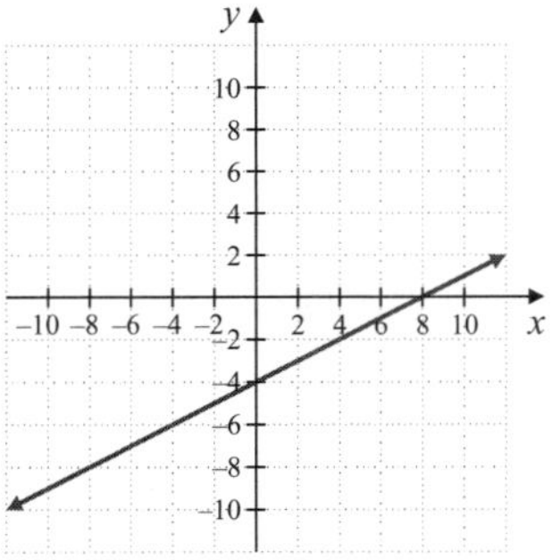

15. $7x - 2y = 10$

$$-2y = -7x + 10$$
$$y = \frac{7}{2}x - 5$$

$m = \frac{7}{2}$, y-int = (0, −5)

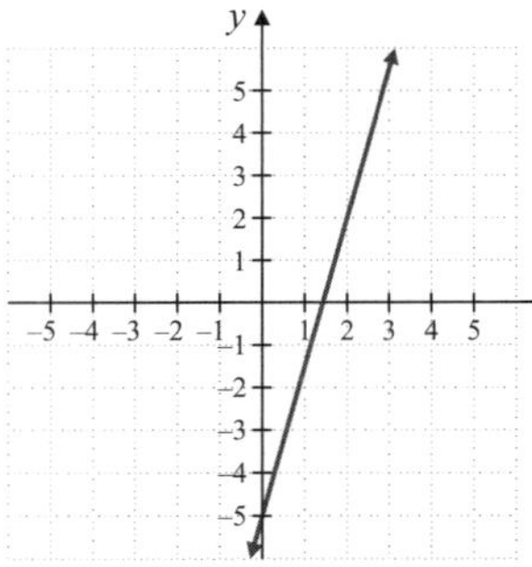

17. $x + 2y = 8$

$$2y = -x + 8$$
$$y = -\frac{1}{2}x + 4$$

$m = -\frac{1}{2}$, y-int = (0, 4)

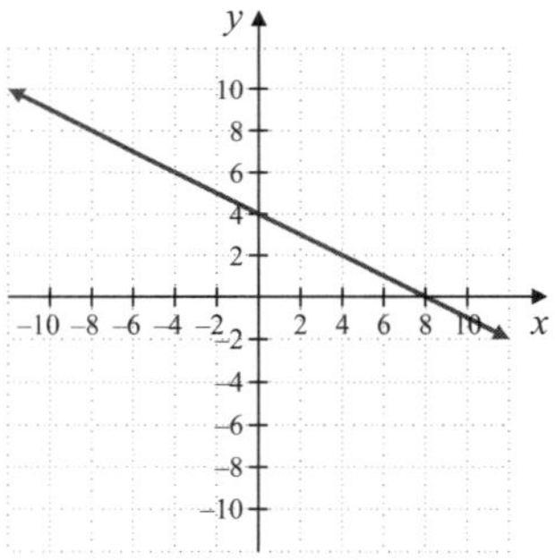

19. $y = \frac{1}{2}x - 1$

$m = \frac{1}{2}$; y-int = (0, −1)

This is graph C.

21. $y = -2x + 3$

$m = -2$; y-intercept = (0, 3)

This is graph D.

23. $y = mx + b$

$y = -x + 1$

25. $y = mx + b$

$$y = 2x + \frac{3}{4}$$

27. $y = mx + b$

$$y = \frac{2}{7}x + 0$$
$$y = \frac{2}{7}x$$

29. a. Use $y = 1983x + 42{,}972$ with $x = 1$, since 2000 is one year after 1999.
$y = 1983(1) + 42{,}972 = 44{,}955$
The average annual income is $44,955.

b. From $y = 1983x + 42{,}972$, $m = 1983$. The average annual income increases by $1983 each year.

c. From $y = 1983x + 42{,}972$, the y-intercept is (0, 42,972). At year 0, which is 1999, the average annual income was $42,972.

31. a. $53x - 5y = -930$

$$-5y = -53x - 930$$
$$y = 10.6x + 186$$

$m = 10.6$
y-int = (0, 186)

b. From $y = 10.6x + 186$, $m = 10.6$. The number of people employed as network systems and data communications analysts increases 10.6 thousand for every 1 year.

c. From $y = 10.6x + 186$, the y-intercept is (0, 186). At year 0, which is 2002, 186 thousand network systems and data communications analysts were employed.

33. a. Use $y = 17.7x + 44.9$ with $x = 7$, since 2004 is 7 years after 1997.
$y = 17.7(7) + 44.9 = 168.8$
In 2004, there were approximately 168.8 million cellular subscribers in the U.S.

b. Let $y = 220$ and solve for x.

$$220 = 17.7x + 44.9$$
$$175.1 = 17.7x$$
$$9.89 \approx x$$

The number of cellular subscribers will exceed 220 million about 9.9 years after 1997, or in 2006.

c. answers may vary

35.
$$y - 2 = 5(x + 6)$$
$$y - 2 = 5x + 30$$
$$y = 5x + 32$$

37.
$$y - (-1) = 2(x - 0)$$
$$y + 1 = 2x$$
$$y = 2x - 1$$

39. The equation is incorrect, the correct equations is $y = 3x + 1.7$.

41. The equation is correct.

43. $y = 42{,}000x + 2{,}900{,}000$, where y is revenue and x is number of years since 1995.

Section 3.4 Practice Problems

1.
$$y - y_1 = m(x - x_1)$$
$$y - (-4) = -2(x - 2)$$
$$y + 4 = -2x + 4$$
$$y = -2x$$

2. $m = \dfrac{4-0}{-2-3} = \dfrac{4}{-5}$

$$y - y_1 = m(x - x_1)$$

(3, 0): $y - 0 = -\dfrac{4}{5}(x - 3)$

$$y = -\frac{4}{5}x + \frac{12}{5}$$
$$5y = -4x + 12$$
$$4x + 5y = 12$$

3. $y = b$

Since the line contains the point (−1, 6), the equation is $y = 6$.

4. $x = c$

Since the line contains the point (4, 7), the equation is $x = 4$.

5.
$$3x + y = 5$$
$$y = -3x + 5$$

$m = -3$

m of parallel line = −3

$$y - y_1 = m(x - x_1)$$
$$y - 2 = -3(x - (-1))$$
$$y - 2 = -3x - 3$$
$$y = -3x - 1$$

6.
$$2x + 4y = 5$$
$$4y = -2x + 5$$
$$y = -\frac{2}{4}x + \frac{5}{4}$$
$$y = -\frac{1}{2}x + \frac{5}{4}$$

$m = -\dfrac{1}{2}$

m of perpendicular line = 2

$$y - y_1 = m(x - x_1)$$
$$y - 4 = 2(x - 3)$$
$$y - 4 = 2x - 6$$
$$y = 2x - 2$$

7. Let x = the number of years after 1995. Then y = the number of contracts obtained in the year corresponding to x.

(2, 15) and (8, 33)

$$m = \frac{33-15}{8-2} = \frac{18}{6} = 3$$
$$y - y_1 = m(x - x_1)$$
$$y - 15 = 3(x - 2)$$
$$y - 15 = 3x - 6$$
$$y = 3x + 9$$

Then predict the number of contracts expected in 2011.

$x = 2011 - 1995 = 16$

Use $y = 3x + 9$.

$y = 3(16) + 9 = 57$

The company can expect 57 new contracts in 2011.

Calculator Explorations

1. $y = 2.3x + 6.7$
   $x = 5.1,\ y = 18.4$

2. $y = -4.8x + 2.9$
   $x = -1.8,\ y = 11.5$

3. $y = -5.9x - 1.6$
   $x = -1.5,\ y = 7.2$

4. $y = 0.4x - 8.6$
   $x = 10.5,\ y = -4.4$

5. $y = 5.2x - 3.3$
   $x = 2.3,\ y = 8.7$
   $x = 7.6,\ y = 36$

6. $y = -6.2x - 8.3$
   $x = 3.2,\ y = -28.1$
   $x = -3.3,\ y = 12$

Mental Math

1. $y - 4 = -2(x - 1)$
   $m = -2;\ (1, 4)$

2. $y - 6 = -3(x - 4)$
   $m = -3;\ (4, 6)$

3. $y - 0 = \frac{1}{4}(x - 2)$

   $m = \frac{1}{4};\ (2, 0)$

4. $y - 1 = -\frac{2}{3}(x - 0)$

   $m = -\frac{2}{3};\ (0, 1)$

5. $y + 2 = 5(x - 3)$
   $m = 5;\ (3, -2)$

6. $y - 7 = 4(x + 6)$
   $m = 4;\ (-6, 7)$

Exercise Set 3.4

1. $$\begin{aligned} y - y_1 &= m(x - x_1) \\ y - 2 &= 3(x - 1) \\ y - 2 &= 3x - 3 \\ y &= 3x - 1 \end{aligned}$$

3. $$\begin{aligned} y - y_1 &= m(x - x_1) \\ y - (-3) &= -2(x - 1) \\ y + 3 &= -2x + 2 \\ y &= -2x - 1 \end{aligned}$$

5. $$\begin{aligned} y - y_1 &= m(x - x_1) \\ y - 2 &= \frac{1}{2}[x - (-6)] \\ y - 2 &= \frac{1}{2}(x + 6) \\ y - 2 &= \frac{1}{2}x + 3 \\ y &= \frac{1}{2}x + 5 \end{aligned}$$

7. $$\begin{aligned} y - y_1 &= m(x - x_1) \\ y - 0 &= -\frac{9}{10}(x - (-3)) \\ y &= -\frac{9}{10}(x + 3) \\ y &= -\frac{9}{10}x - \frac{27}{10} \end{aligned}$$

9. $m = \frac{6 - 0}{4 - 2} = \frac{6}{2} = 3$
   $$\begin{aligned} y - y_1 &= m(x - x_1) \\ y - 0 &= 3(x - 2) \\ y &= 3x - 6 \\ 6 &= 3x - y \\ 3x - y &= 6 \end{aligned}$$

11. $m = \frac{13-5}{-6-(-2)} = \frac{8}{-4} = -2$

$$\begin{aligned} y - y_1 &= m(x - x_1) \\ y - 5 &= -2(x - (-2)) \\ y - 5 &= -2(x + 2) \\ y - 5 &= -2x - 4 \\ y &= -2x + 1 \\ 2x + y &= 1 \end{aligned}$$

13. $m = \frac{-3-(-4)}{-4-(-2)} = \frac{1}{-2} = -\frac{1}{2}$

$$\begin{aligned} y - y_1 &= m(x - x_1) \\ y - (-4) &= -\frac{1}{2}[x - (-2)] \\ y + 4 &= -\frac{1}{2}(x + 2) \\ -2y - 8 &= x + 2 \\ -10 &= x + 2y \\ x + 2y &= -10 \end{aligned}$$

15. $m = \frac{-9-(-8)}{-6-(-3)} = \frac{-1}{-3} = \frac{1}{3}$

$$\begin{aligned} y - y_1 &= m(x - x_1) \\ y - (-8) &= \frac{1}{3}(x - (-3)) \\ y + 8 &= \frac{1}{3}(x + 3) \\ y + 8 &= \frac{1}{3}x + 1 \\ y &= \frac{1}{3}x - 7 \\ 3y &= x - 21 \\ x - 3y &= 21 \end{aligned}$$

17. $m = \frac{\frac{7}{10} - \frac{4}{10}}{-\frac{1}{5} - \frac{3}{5}} = \frac{\frac{3}{10}}{-\frac{4}{5}} = \frac{3}{10}\left(-\frac{5}{4}\right) = -\frac{3}{8}$

$$\begin{aligned} y - y_1 &= m(x - x_1) \\ y - \frac{4}{10} &= -\frac{3}{8}\left(x - \frac{3}{5}\right) \\ y - \frac{4}{10} &= -\frac{3}{8}x + \frac{9}{40} \\ y &= -\frac{3}{8}x + \frac{5}{8} \\ 8y &= -3x + 5 \\ 3x + 8y &= 5 \end{aligned}$$

19. A vertical line has an equation of the form $x = c$. Since the line passes through the point (2, 6) its equation is $x = 2$.

21. A horizontal line has an equation of the form $y = b$. Since the line passes through the point (–3, 1), its equation is $y = 1$.

23. A line with an undefined slope is a vertical line. A vertical line has an equation of the form $x = c$. Since the line passes through the point (0, 5), its equation is $x = 0$.

25. $y = 4x - 2$; $m = 4$
A parallel line will have slope 4

$$\begin{aligned} y - y_1 &= m(x - x_1) \\ y - 8 &= 4(x - 3) \\ y - 8 &= 4x - 12 \\ y &= 4x - 4 \end{aligned}$$

27. $3y = x - 6$

$y = \frac{1}{3}x - 2,\ m = \frac{1}{3}$

A perpendicular line will have slope –3.

$$\begin{aligned} y - y_1 &= m(x - x_1) \\ y - (-5) &= -3(x - 2) \\ y + 5 &= -3x + 6 \\ y &= -3x + 1 \end{aligned}$$

29. The line $y = 7$ is horizontal, so a parallel line will also be horizontal, with equation $y = 4$.

31. $3x+2y=5$

$$2y=-3x+5$$

$$y=-\frac{3}{2}x+\frac{5}{2};\ m=-\frac{3}{2}$$

A parallel line will have slope $-\frac{3}{2}$.

$$y-y_1=m(x-x_1)$$

$$y-(-3)=-\frac{3}{2}(x-(-2))$$

$$y+3=-\frac{3}{2}x-\frac{6}{2}$$

$$y+3=-\frac{3}{2}x-3$$

$$y=-\frac{3}{2}x-6$$

33. The line $x = 3$ is vertical, so a perpendicular line will be horizontal, with equation $y = -5$.

35. $x-4y=4$

$$x-4=4y$$

$$\frac{1}{4}x-1=y;\ m=\frac{1}{4}$$

A perpendicular line will have slope of -4.

$$y-y_1=m(x-x_1)$$

$$y-5=-4(x-(-1))$$

$$y-5=-4x-4$$

$$y=-4x+1$$

37. $y-y_1=m(x-x_1)$

$$y-3=2[x-(-2)]$$

$$y-3=2(x+2)$$

$$y-3=2x+4$$

$$-7=2x-y$$

$$2x-y=-7$$

39. $m=\frac{y_2-y_1}{x_2-x_1}=\frac{2-6}{5-1}=\frac{-4}{4}=-1$

$$y-y_1=m(x-x_1)$$

$$y-6=-1(x-1)$$

$$y-6=-x+1$$

$$y=-x+7$$

41. $y=mx+b$

$$y=-\frac{1}{2}x+\frac{3}{8}$$

43. $m=\frac{y_2-y_1}{x_2-x_1}=\frac{-6-(-4)}{0-(-7)}=-\frac{2}{7}$

$$y-y_1=m(x-x_1)$$

$$y-(-6)=-\frac{2}{7}(x-0)$$

$$y+6=-\frac{2}{7}x$$

$$7y+42=-2x$$

$$2x+7y=-42$$

45. $y-y_1=m(x-x_1)$

$$y-0=-\frac{4}{3}[x-(-5)]$$

$$y=-\frac{4}{3}(x+5)$$

$$3y=-4(x+5)$$

$$3y=-4x-20$$

$$4x+3y=-20$$

47. A vertical line has an equation of the form $x = c$. Since the line passes through the point $(-2, -10)$, its equation is $x = -2$.

49. $2x+4y=9$

$$4y=-2x+9$$

$$y=-\frac{1}{2}x+\frac{9}{4}$$

$$m=-\frac{1}{2}$$

A parallel line will have slope $-\frac{1}{2}$.

$$y-y_1=m(x-x_1)$$

$$y-(-2)=-\frac{1}{2}(x-6)$$

$$y+2=-\frac{1}{2}x+3$$

$$y=-\frac{1}{2}x+1$$

$$2y=-x+2$$

$$x+2y=2$$

51. $y - y_1 = m(x - x_1)$
$y - 12 = 0(x - (-9))$
$y - 12 = 0$
$y = 12$

53. $8x - y = 9$
$-y = -8x + 9$
$y = 8x - 9$
$m = 8$
A parallel line will have slope 8.
$y - y_1 = m(x - x_1)$
$y - 1 = 8(x - 6)$
$y = 8x - 47$
$47 = 8x - y$
$8x - y = 47$

55. The line $y = 9$ is horizontal, so a perpendicular line will be vertical, with equation $x = 5$.

57. $m = \dfrac{-5-(-8)}{-6-2} = \dfrac{3}{-8} = -\dfrac{3}{8}$
$y - y_1 = m(x - x_1)$
$y - (-8) = -\frac{3}{8}(x - 2)$
$y + 8 = -\frac{3}{8}x + \frac{3}{4}$
$y = -\frac{3}{8}x - \frac{29}{4}$

59. a. $m = \dfrac{96-32}{3-1} = \dfrac{64}{2} = 32$
$y - y_1 = m(x - x_1)$
$y - 32 = 32(x - 1)$
$y - 32 = 32x - 32$
$y = 32x$

b. Let $x = 4$. $y = 32x = 32(4) = 128$ feet per second. The rock would have traveled 128 feet per second at 4 seconds.

61. a. Use the points (1, 30,000) and (4, 66,000).
$m = \dfrac{66,000 - 30,000}{4-1}$
$= \dfrac{36,000}{3}$
$= 12,000$
$y - y_1 = m(x - x_1)$
$y - 30,000 = 12,000(x - 1)$
$y - 30,000 = 12,000x - 12,000$
$y = 12,000x + 18,000$

b. Let $x = 7$.
$y = 12,000(7) + 18,000$
$= 84,0000 + 18,000$
$= 102,000$
The profit at the end of the seventh year is expected to be $102,000.

c. Find x when $y = 126,000$.
$126,000 = 12,000x + 18,000$
$108,000 = 12,000x$
$9 = x$
The profit should reach $126,000 at the end of the ninth year.

63. a. Use the points (0, 142,200) and (3, 170,000).
$m = \dfrac{170,000 - 142,200}{3-0}$
$= \dfrac{278,000}{3}$
$= 9266.7$
$y - y_1 = m(x - x_1)$
$y - 142,200 = 9266.7(x - 0)$
$y = 9266.7x + 142,200$

b. Let $x = 8$.
$y = 9266.7x + 142,200$
$y = 9266.7(8) + 142,200$
$y = 216,334$
The median existing home price for 2008 is $216,334.

65. a. Use the points (0, 70) and (3, 70.3).

$$m = \frac{70.3 - 70}{3 - 0} = \frac{0.3}{3} = 0.1$$

$$y - y_1 = m(x - x_1)$$
$$y - 70 = 0.1(x - 0)$$
$$y = 0.1x + 70$$

b. Let $x = 7$, since 2008 is 7 years after 2001.
$y = 0.1(7) + 70 = 0.7 + 70 = 70.7$
There will be approximately 70.7 million basic cable subscribers in 2008.

67. $y = 7x + 3$
$y = 7(4) + 3$
$y = 31$
(4, 31)

69. $y = 4.2x$
$y = 4.2(-2)$
$y = -8.4$
(−2, −8.4)

71. $y = x^2 + 2x + 1$
$y = 1^2 + 2(1) + 1$
$y = 4$
(1, 4)

73. $m = \frac{1-3}{1-0} = \frac{-2}{1} = -2$

$$y - y_1 = m(x - x_1)$$
$$y - 3 = -2(x - 0)$$
$$y - 3 = -2x$$
$$2x + y = 3$$

75. $m = \frac{5-1}{4-(-2)} = \frac{4}{6} = \frac{2}{3}$

$$y - 5 = \frac{2}{3}(x - 4)$$
$$y - 5 = \frac{2}{3}x - \frac{8}{3}$$
$$3y - 15 = 2x - 8$$
$$2x - 3y = -7$$

77. True

79. answers may vary

81.

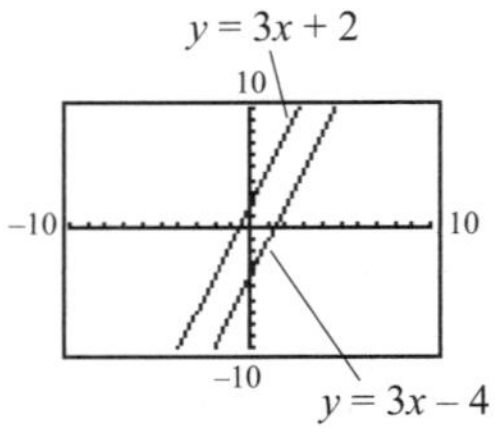

Integrated Review

1. $y = -2x$

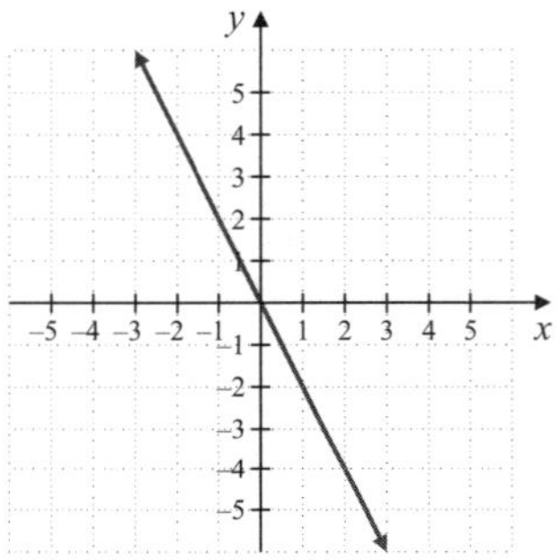

2. $3x - 2y = 6$

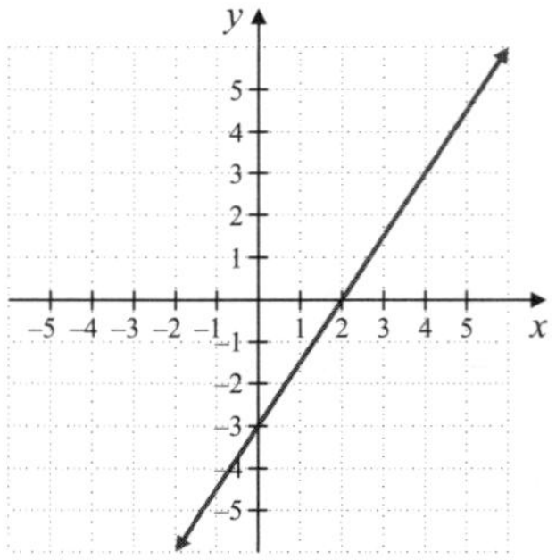

3. $x = -1\frac{1}{2}$

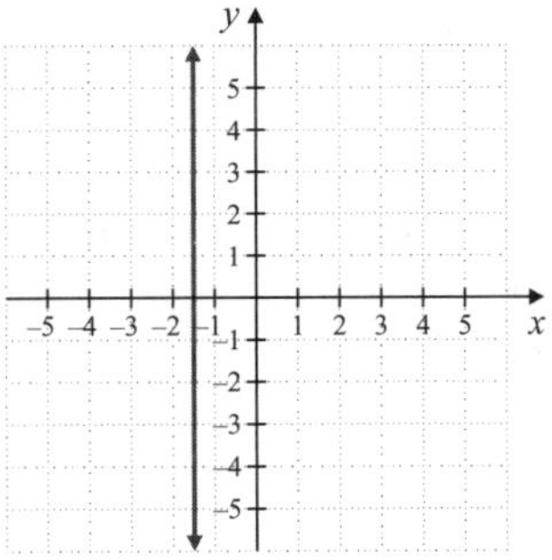

4. $y = 1.5$

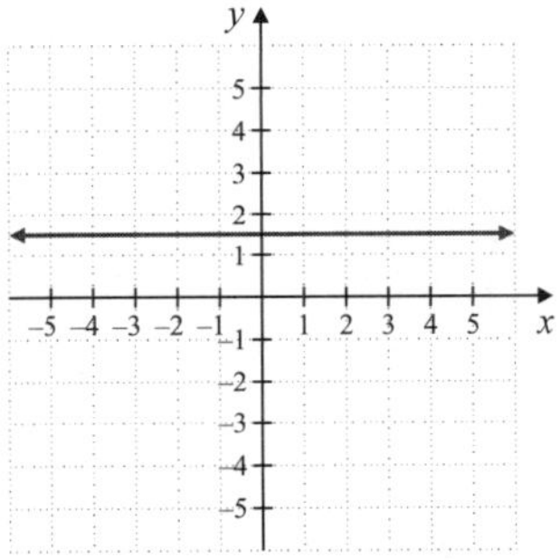

5. $m = \frac{-5-(-5)}{3-(-2)} = \frac{0}{5} = 0$

6. $m = \frac{5-2}{0-5} = -\frac{3}{5}$

7. $y = 3x - 5$
$m = 3$; y-int = $(0, -5)$

8. $5x - 2y = 7$
$5x - 7 = 2y$
$\frac{5}{2}x - \frac{7}{2} = y$
$m = \frac{5}{2}$; y-int $= \left(0, -\frac{7}{2}\right)$

9. $y = 8x - 6$ $\quad$ $y = 8x + 6$
$m = 8$; y-int = $(0, -6)$ $\quad$ $m = 8$; y-int: $(0, 6)$
The lines are parallel, since they have the same slope.

10. $y = \frac{2}{3}x + 1$ $\quad$ $2y + 3x = 1$
$2y = -3x + 1$
$y = -\frac{3}{2}x + \frac{1}{2}$

$m = \frac{2}{3}$; y-int = $(0, 1)$ $\quad$ $m = -\frac{3}{2}$;
y-int $= \left(0, \frac{1}{2}\right)$

The lines are perpendicular since the product of their slopes is -1.

11. $m = \frac{2-6}{5-1} = \frac{-4}{4} = -1$
$y - y_1 = m(x - x_1)$
$y - 6 = -1(x - 1)$
$y = -x + 7$

12. $x = c$
$x = -2$

13. $y = b$
$y = 0$

14. $m = \frac{-5-(-8)}{-6-2} = \frac{3}{-8}$
$y - y_1 = m(x - x_1)$
$y - (-8) = -\frac{3}{8}(x - 2)$
$y + 8 = -\frac{3}{8}x + \frac{6}{8}$
$y = -\frac{3}{8}x - \frac{29}{4}$

15. $y - y_1 = m(x - x_1)$
$y - 4 = -5(x - (-2))$
$y - 4 = -5x - 10$
$y = -5x - 6$

16. $y = mx + b$
$y = -4x + \frac{1}{3}$

17. $y = mx + b$

$y = \frac{1}{2}x - 1$

18. $y - y_1 = m(x - x_1)$

$y - 0 = 3\left(x - \frac{1}{2}\right)$

$y = 3x - \frac{3}{2}$

19. $3x - y = 5$

$3x - 5 = y$

$m = 3$

A parallel line will have slope of 3.

$y - y_1 = m(x - x_1)$

$y - (-5) = 3(x - (-1))$

$y + 5 = 3x + 3$

$y = 3x - 2$

20. $4x - 5y = 10$

$4x - 10 = 5y$

$\frac{4}{5}x - 2 = y$

$m = \frac{4}{5}$

A perpendicular line will have slope of $-\frac{5}{4}$.

$y - y_1 = m(x - x_1)$

$y - 4 = -\frac{5}{4}(x - 0)$

$y = -\frac{5}{4}x + 4$

21. $4x + y = \frac{2}{3}$

$y = -4x + \frac{2}{3}$

$m = -4$

A perpendicular line will have slope of $\frac{1}{4}$.

$y - y_1 = m(x - x_1)$

$y - (-3) = \frac{1}{4}(x - 2)$

$y + 3 = \frac{1}{4}x - \frac{1}{2}$

$y = \frac{1}{4}x - \frac{7}{2}$

22. $5x + 2y = 2$

$2y = -5x + 2$

$y = -\frac{5}{2}x + 1$

$m = -\frac{5}{2}$

A parallel line will have slope of $-\frac{5}{2}$.

$y - y_1 = m(x - x_1)$

$y - 0 = -\frac{5}{2}(x - (-1))$

$y = -\frac{5}{2}x - \frac{5}{2}$

23. $x = c$

$x = -1$

24. $y - y_1 = m(x - x_1)$

$y - 3 = 0(x - (-1))$

$y = 3$

Section 3.5 Practice Problems

1. $x + 3y > 4$

Test (0, 0)

$0 + 3(0) > 4$

$0 > 4$ False

Shade the half-plane that does not contain (0, 0).

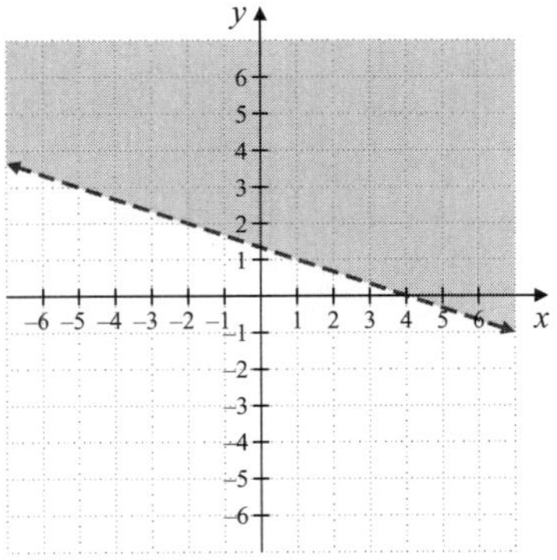

2. $x \le 2y$
Test (0, 1)
$0 \le 2(1)$
$0 \le 2$ True
Shade the half-plane that contains (0, 1).

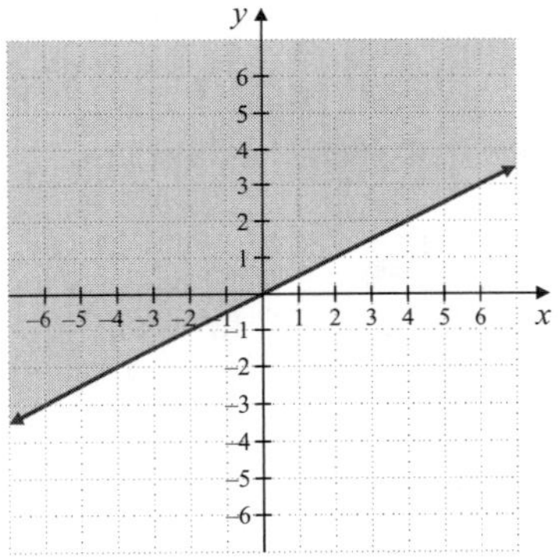

3. $x \le 2$ and $y \ge x + 1$

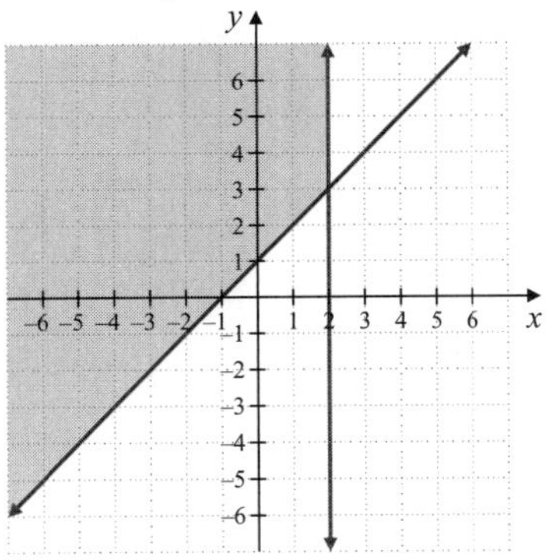

4. $x + 2y \le 4$ or $y \ge -1$

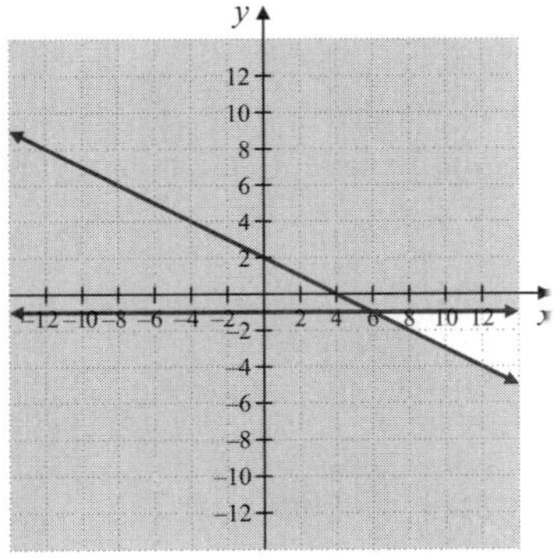

Exercise Set 3.5

1. $x < 2$

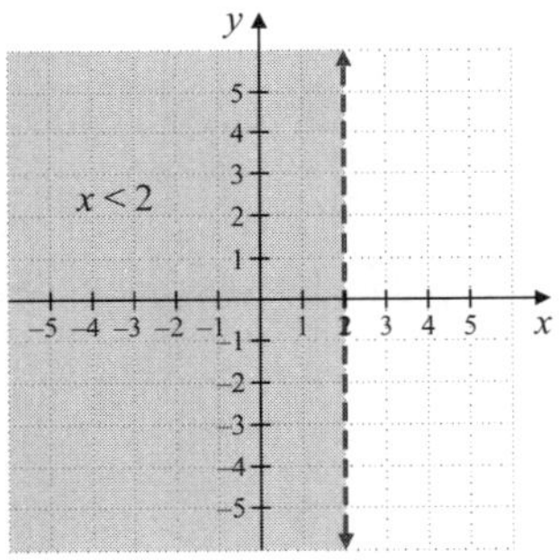

3. $x - y \ge 7$
Test (0, 0)
$0 - 0 \ge 7$
$0 \ge 7$ False
Shade the half-plane that does not contain (0, 0).

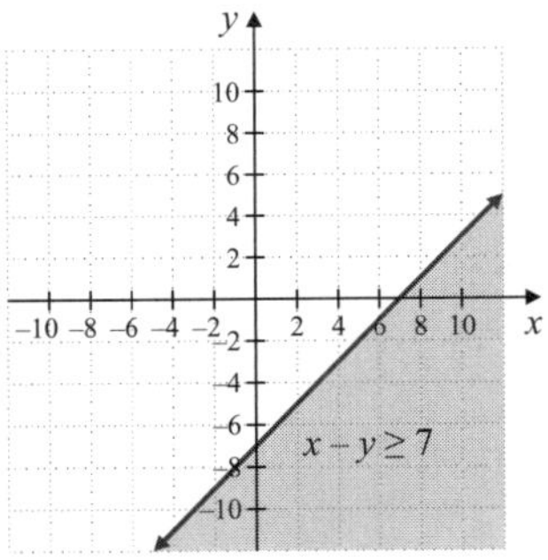

5. $3x + y > 6$
Test (0, 0)
$3(0) + 0 > 6$
$0 > 6$ False
Shade the half-plane that does not contain (0, 0).

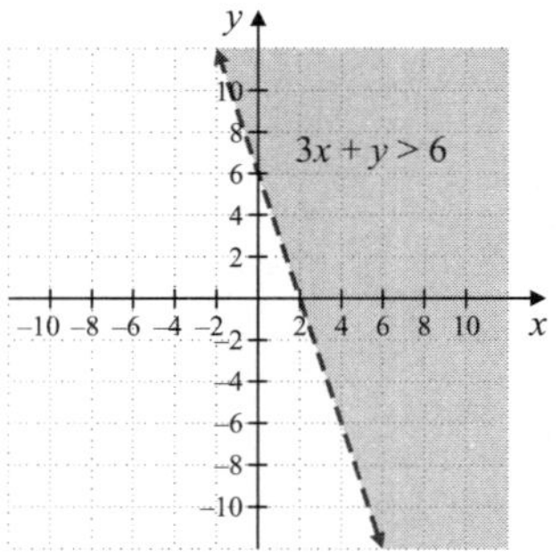

7. $y \le -2x$
Test (1, 0)
$0 \le -2(1)$
$0 \le -2$ False
Shade the half-plane that does not contain (1, 0).

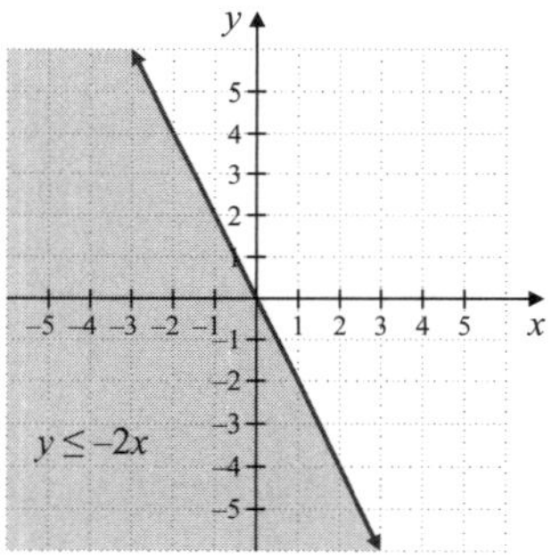

9. $2x + 4y \ge 8$
Test (0, 0)
$2(0) + 4(0) \ge 8$
$0 \ge 8$ False
Shade the half-plane that does not contain the point (0, 0).

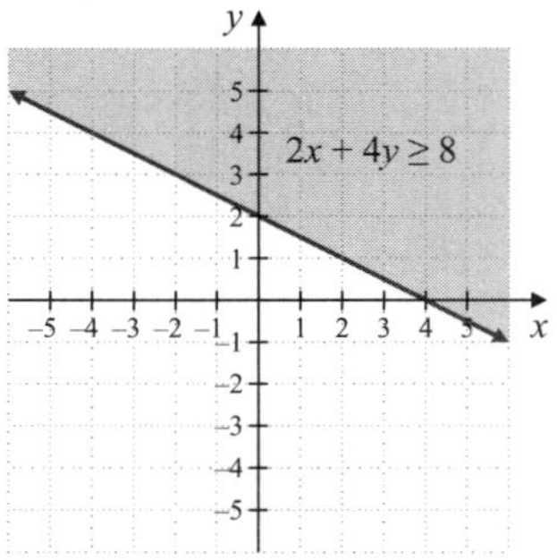

11. $5x + 3y > -15$
Test (0, 0)
$5(0) + 3(0) > -15$
$0 > -15$ True
Shade the half-plane that contains (0, 0).

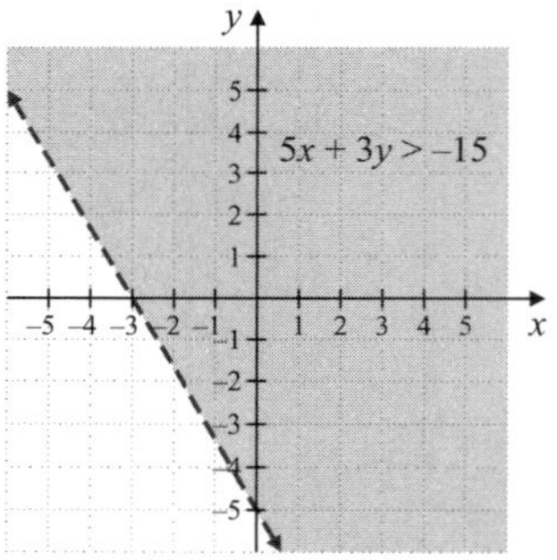

13. $y \le -4.5$

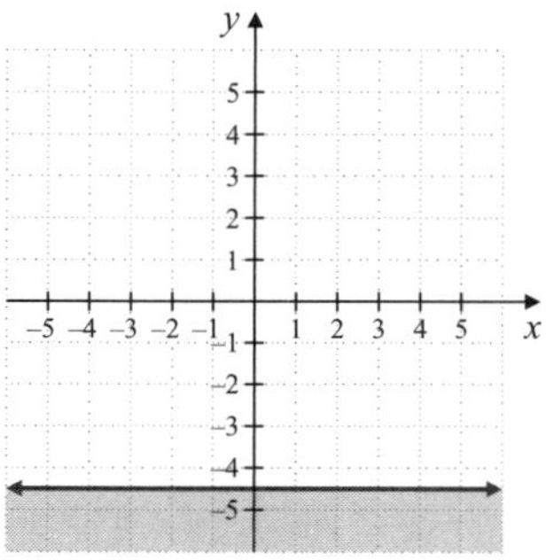

15. $x > 4y$

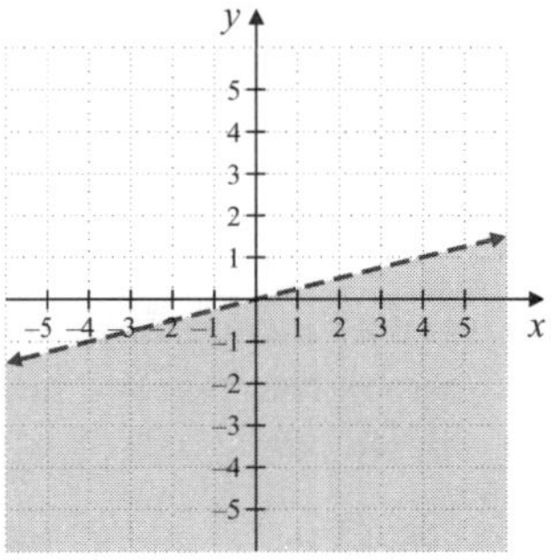

17. $x \geq 3$ and $y \leq -2$

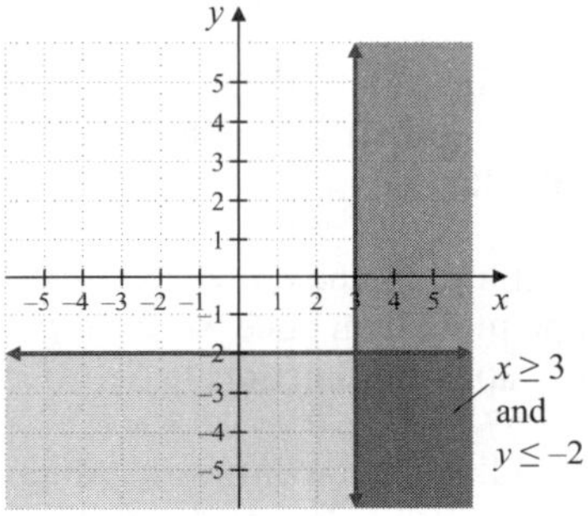

19. $x \leq -2$ or $y \geq 4$

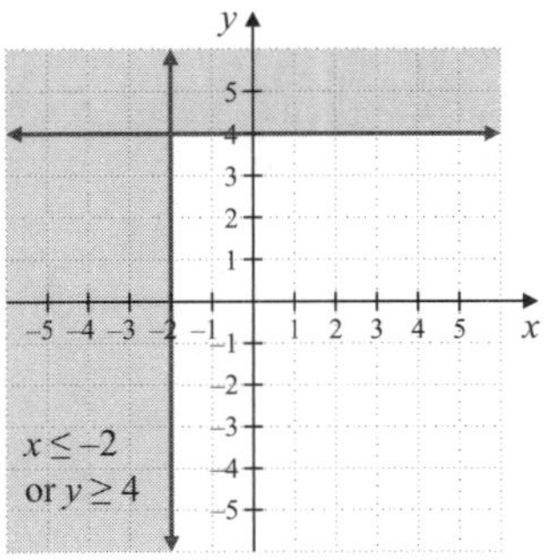

21. $x - y < 3$ and $x > 4$

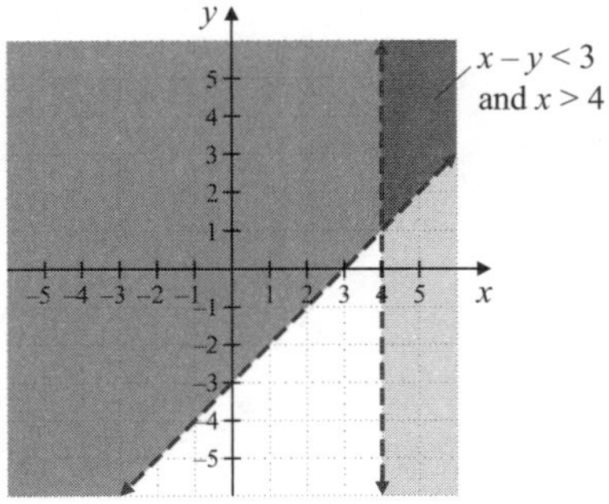

23. $x + y \leq 3$ or $x - y \geq 5$

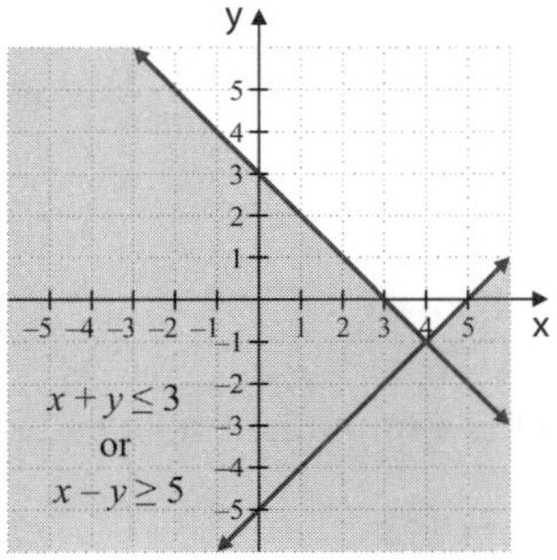

25. $x - y \geq 2$ or $y < 5$

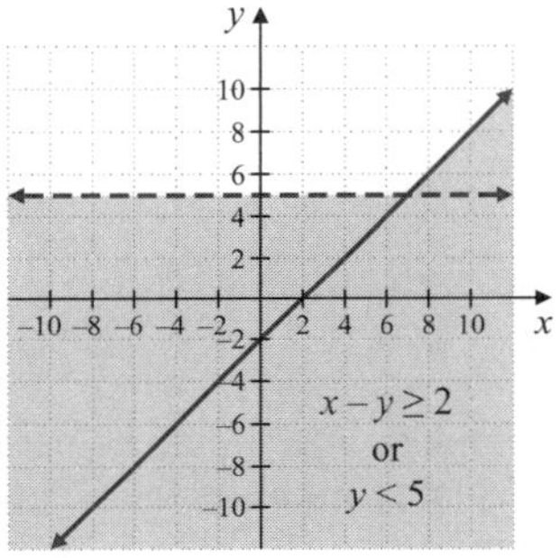

27. $2x > y$ and $3x - 9y > 12$

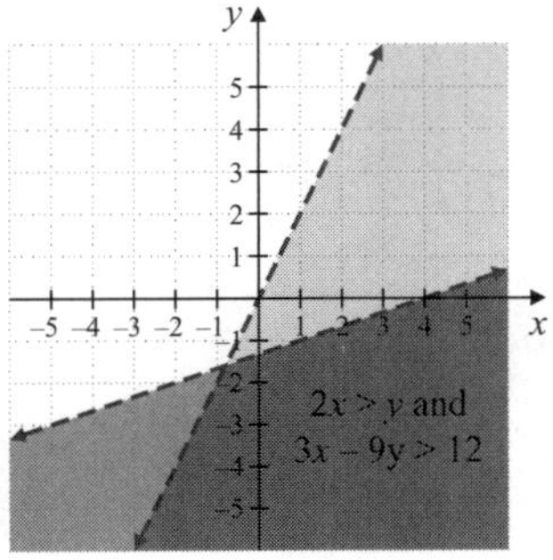

29. $y \leq 2x + 3$
The graph will have a solid line. Testing (0, 0) results in a true inequality, so the graph is graph D.

31. $y > 2x + 3$
The graph will have a dashed line. Testing (0, 0) results in a false inequality, so the graph is graph A.

33. $x - y = 4$
$3-(-1) \stackrel{?}{=} 4$
$4 = 4$ True

$x + 2y = 1$
$3 + 2(-1) \stackrel{?}{=} 1$
$3 - 2 \stackrel{?}{=} 1$
$1 = 1$ True
Yes, (3, −1) is a solution of both equations.

35. $3x + 2y = -12$
$3(-4) + 2(0) \stackrel{?}{=} -12$
$-12 \stackrel{?}{=} -12$ True

$x = 4y$
$-4 \stackrel{?}{=} 4(0)$
$-4 = 0$ False
No, (−4, 0) is not a solution of both equations.

37. A dashed boundary line should be used when the inequality involves the > symbol or the < symbol.

39.

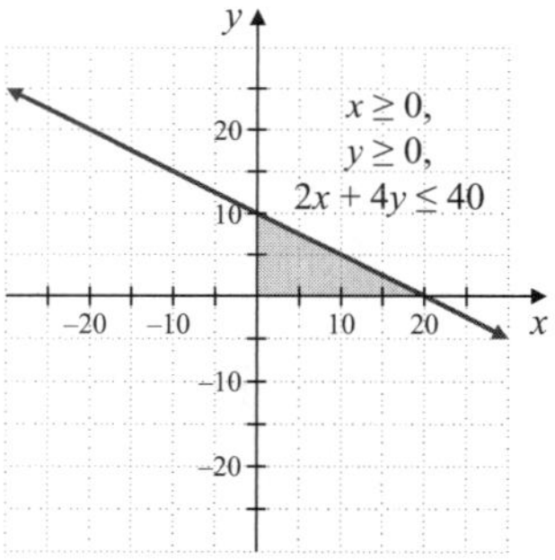

Section 3.6 Practice Problems

1. Domain = {1, 2, 0}
Range = {6, 8, 3, −2}

2. Domain = {1}
Range = {−1, 0, 1, 2, 3}

3. Domain
= {Arkansas, Texas, Oklahoma, Oregon}
Range = {4, 5, 32}

4. The relation is a function.

5. The relation is not a function since 0 is paired with −2 and 3.

6. The relation is a function.

7. The relation is a function.

8. The relation is not a function since the x-value 10 corresponds to two y-values, −3 and 3, in the ordered pairs (10, −3), (10, 3).

9. A vertical line can be drawn that intersects the graph exactly once, so the graph is the graph of a function.

10. A vertical line can be drawn that intersects the graph exactly once, so the graph is the graph of a function.

11. A vertical line can be drawn that intersects the graph more than once, so the graph is not the graph of a function.

12. A vertical line can be drawn that intersects the graph exactly once, so the graph is the graph of a function.

13. A vertical line can be drawn that intersects the graph more than once, so the graph is not the graph of a function.

14. $g(x) = 4x + 5$
$g(0) = 4(0) + 5 = 5$

15. $g(x) = 4x + 5$
$g(-5) = 4(-5) + 5 = -20 + 5 = -15$

16. $f(x) = 3x^2 - x + 2$
$f(2) = 3(2)^2 - 2 + 2 = 12$

17. $f(x) = 3x^2 - x + 2$
$f(-1) = 3(-1)^2 - (-1) + 2 = 6$

18.

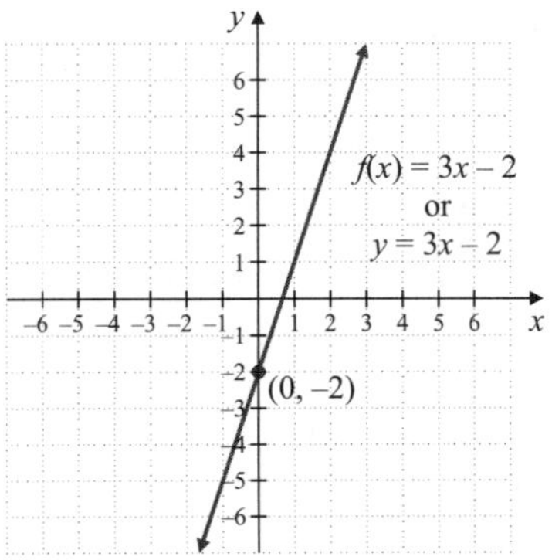

Exercise Set 3.6

1. Domain = {−1, 0, −2, 5}
Range = {7, 6, 2}
The relation is a function.

3. Domain = {−2, 6, −7}
Range = {4, −3, −8}
The relation is not a function since −2 is paired with 4 and −3.

5. Domain = {1}
Range = {1, 2, 3, 4}
The relation is not a function since 1 is paired with 1, 2, 3 and 4.

7. Domain $= \left\{\frac{3}{2}, 0\right\}$

Range $= \left\{\frac{1}{2}, -7, \frac{4}{5}\right\}$

The relation is not a function since $\frac{3}{2}$ is paired with $\frac{1}{2}$ and −7.

9. Domain = {−3, 0, 3}
Range = {−3, 0, 3}
The relation is a function.

11. Domain = {−1, 1, 2, 3}
Range = {2, 1}
The relation is a function.

13. Domain = {Colorado, Alaska, Massachusetts, Delaware, Illinois, Pennsylvania}
Range = {1, 7, 10, 19}
The relation is a function.

15. Domain = {32°, 104°, 212°, 50°}
Range = {0°, 40°, 10°, 100°}
The relation is a function.

17. Domain = {−1, 2, 5, 100}
Range = {0}
The relation is a function.

19. The relation is a function.

21. Since each x-value corresponds to exactly one y-value, the relation is a function.

23. Since each x-value corresponds to more than one y-value, the relation is not a function.

25. $y - x = 7$
$y = x + 7$
Since each x-value corresponds to exactly one y-value, the relation is a function.

27. A vertical line can be drawn that intersects the graph exactly once, so the graph is the graph of a function.

29. A vertical line can be drawn that intersects the graph more than once, so the graph is not the graph of a function.

31. A vertical line can be drawn that intersects the graph exactly once, so the graph is the graph of a function.

33. A vertical line can be drawn that intersects the graph more than once, so the graph is not the graph of a function.

35. A vertical line can be drawn that intersects the graph more than once, so the graph is not the graph of a function.

37. A vertical line can be drawn that intersects the graph more than once, so the graph is not the graph of a function.

39. A vertical line can be drawn that intersects the graph more than once, so the graph is not the graph of a function.

41. A vertical line can be drawn that intersects the graph more than once, so the graph is not the graph of a function.

43. A vertical line can be drawn that intersects the graph exactly once, so the graph is the graph of a function.

45. $f(x) = 3x + 3$
$f(4) = 3(4) + 3 = 12 + 3 = 15$

47. $h(x) = 5x^2 - 7$
$h(-3) = 5(-3)^2 - 7 = 45 - 7 = 38$

49. $g(x) = 4x^2 - 6x + 3$

$$\begin{aligned} g(2) &= 4(2)^2 - 6(2) + 3 \\ &= 4(4) - 12 + 3 \\ &= 16 - 12 + 3 \\ &= 7 \end{aligned}$$

51. $g(x) = 4x^2 - 6x + 3$
$g(0) = 4(0)^2 - 6(0) + 3 = 0 - 0 + 3 = 3$

53. $f(x) = \frac{1}{2}x$

a. $f(0) = \frac{1}{2}(0) = 0$

b. $f(2) = \frac{1}{2}(2) = 1$

c. $f(-2) = \frac{1}{2}(-2) = -1$

55. $f(x) = -5$

a. $f(2) = -5$

b. $f(0) = -5$

c. $f(606) = -5$

57. $A(r) = \pi r^2$
$A(5) = \pi(5)^2 = 25\pi$
The area is 25π square centimeters.

59. $V(x) = x^3$
$V(14) = (14)^3 = 2744$
The volume is 2744 cubic inches.

61. Use $H(f) = 2.59f = 47.24$ with $f = 46$.

$$\begin{aligned} H(46) &= 2.59(46) + 47.24 \\ &= 119.14 + 47.24 \\ &= 166.38 \end{aligned}$$

The height of the woman is 166.38 centimeters.

63. $D(x) = \frac{136}{25}x$

$D(30) = \frac{136}{25}(30) = 163.2$
The dose needed is 163.2 milligrams.

65. a. $C(x) = 2.28x + 94.86$

$$\begin{aligned} C(2) &= 2.28(2) + 94.86) \\ &= 4.56 + 94.86 \\ &= 99.42 \end{aligned}$$

The per capita consumption of poultry was 99.42 pounds in 2003.

b. Use $x = 6$ since 2007 is 6 years after 2001.

$$\begin{aligned} C(6) &= 2.28(6) + 94.86 \\ &= 13.68 + 94.86 \\ &= 108.54 \end{aligned}$$

The per capita consumption of poultry is expected to be 108.54 pounds in 2007.

67. $f(x) = 2x + 3$ or $y = 2x + 3$
$m = 2$; y-int = (0, 3)

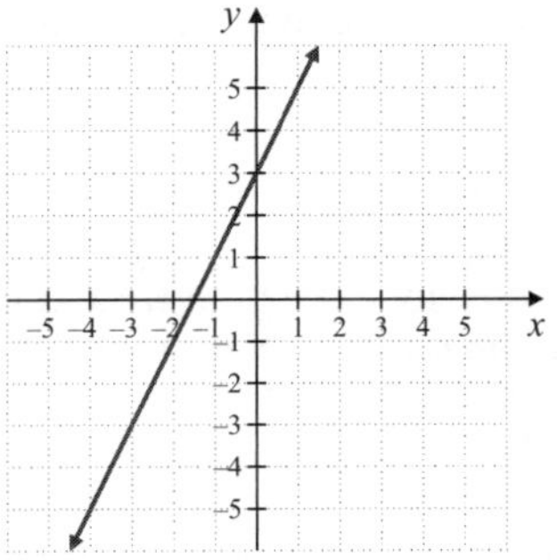

69. $f(x) = -3x$ or $y = -3x = -3x + 0$
$m = -3$; y-intercept = (0, 0)

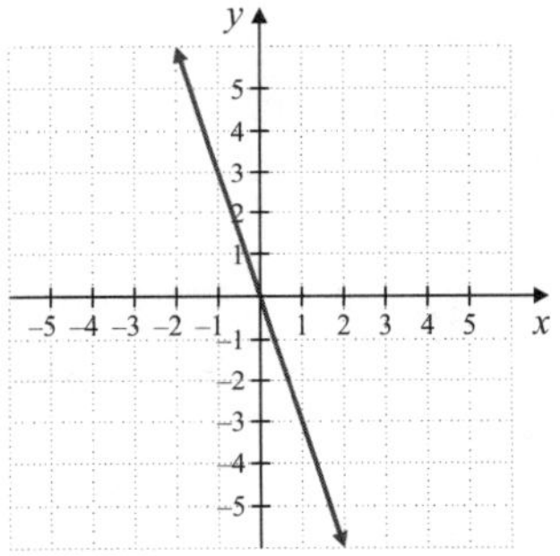

71. $f(x) = -x + 2$ or $y = -x + 2$
$m = -1$; y-int = (0, 2)

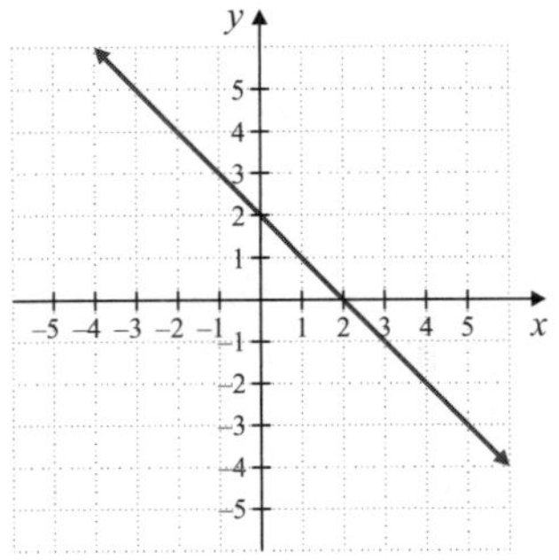

73. $2x - 7 \le 21$
$2x \le 28$
$x \le 14$
The solution set is $(-\infty, 14]$.

75. $5(x-2) \ge 3(x-1)$
$5x - 10 \ge 3x - 3$
$2x \ge 7$
$x \ge \frac{7}{2}$
The solution set is $\left[\frac{7}{2}, \infty\right)$.

77. $\frac{x}{2} + \frac{1}{4} < \frac{1}{8}$
$\frac{x}{2} < \frac{1}{8} - \frac{1}{4}$
$\frac{x}{2} < -\frac{1}{8}$
$x < -\frac{1}{4}$
The solution set is $\left(-\infty, -\frac{1}{4}\right)$.

79. no; answers may vary

81. yes; answers may vary

83. true; $x = 7$ and $y = 50$

85. $f(x) = x^2 + 1$
$f(7) = (7)^2 + 1 = 49 + 1 = 50$
Since $f(7) = 50$ when $f(x) = x^2 + 1$, a possible function is $f(x) = x^2 + 1$; true.

87. answers may vary

89. $y = x + 7$ can be written as $f(x) = x + 7$.

91. answers may vary

93. $f(x) = x^2 - 12$

a. $f(12) = (12)^2 - 12 = 144 - 12 = 132$

b. $f(a) = (a)^2 - 12 = a^2 - 12$

c. $f(-x) = (-x)^2 - 12 = x^2 - 12$

d. $f(x+h) = (x+h)^2 - 12$
$= x^2 + 2xh + h^2 - 12$

Section 3.7 Practice Problems

1. Domain = [−2, 4]
   Range = [−3, 4]

2. Domain = [0, ∞)
   Range = (−∞, ∞)

3. Domain = (−∞, ∞)
   Range = (−∞, ∞)

4. Domain = [−2, 2]
   Range = [−2, 2]

5. $f(x) = \begin{cases} 3x+4 & \text{if } x < 0 \\ -x+2 & \text{if } x \geq 0 \end{cases}$

   $f(-4) = 3(-4) + 4 = -12 + 4 = -8$
   (−4, −8)
   $f(3) = -3 + 2 = -1$
   (3, −1)
   $f(0) = -0 + 2 = 2$
   (0, 2)

6. 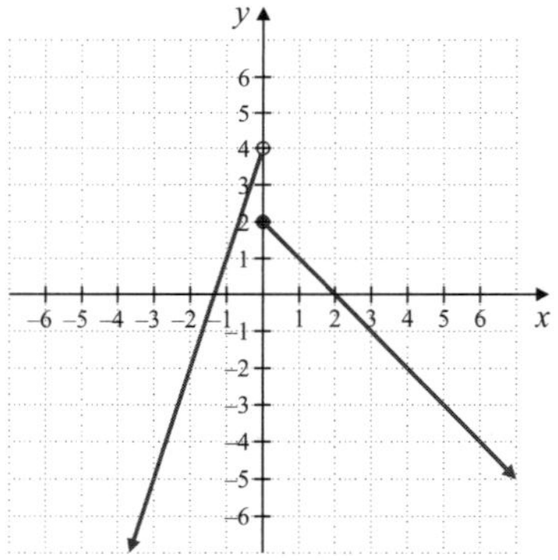

Exercise Set 3.7

1. Domain = [0, ∞)
   Range = (−∞, ∞)
   The relation is not a function since it fails the vertical line test (try $x = 1$).

3. Domain = (−∞, ∞)
   Range = [0, ∞)
   The relation is a function since it passes the vertical line test.

5. Domain = (−∞, ∞)
   Range = (−∞, −3] ∪ [3, ∞)
   The relation is not a function since it fails the vertical line test (try $x = 3$).

7. Domain = [1, 7]
   Range = [1, 7]
   The relation is not a function since it fails the vertical line test.

9. Domain = {−2}
   Range = (−∞, ∞)
   The relation is not a function since it fails the vertical line test.

11. Domain = (−∞, ∞)
    Range = (−∞, 3]
    The relation is a function since it passes the vertical line test.

13. Domain = (−∞, ∞)
    Range = (−∞, 3]
    The relation is a function since it passes the vertical line test.

15. Domain = [2, ∞)
    Range = [3, ∞)
    The relation is a function since it passes the vertical line test.

17. $f(x) = \begin{cases} 2x & \text{if } x < 0 \\ x+1 & \text{if } x \geq 0 \end{cases}$

    Graph $y = 2x$ for $x < 0$, with an open circle at $x = 0$.
    Graph $y = x + 1$ for $x \geq 0$, with a closed circle at $x = 0$.

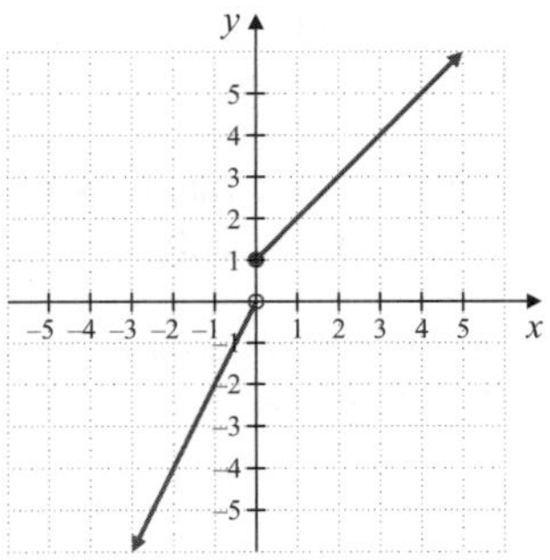

19. $f(x) = \begin{cases} 4x+5 & \text{if} \quad x \le 0 \\ \frac{1}{4}x+2 & \text{if} \quad x > 0 \end{cases}$

Graph $y = 4x + 5$ for $x \le 0$, with a closed circle at $x = 0$. Graph $y = \frac{1}{4}x + 2$ for $x > 0$, with an open circle at $x = 0$.

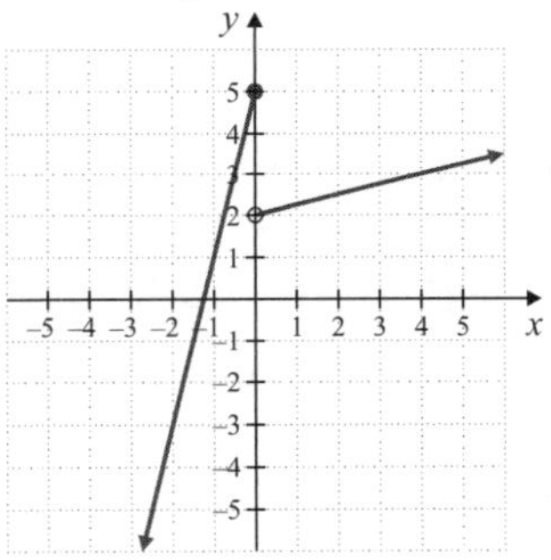

21. $g(x) = \begin{cases} -x & \text{if} \quad x \le 1 \\ 2x+1 & \text{if} \quad x > 1 \end{cases}$

Graph $y = -x$ for $x \le 1$, with a closed circle at $x = 1$. Graph $y = 2x + 1$ for $x > 1$, with an open circle at $x = 1$.

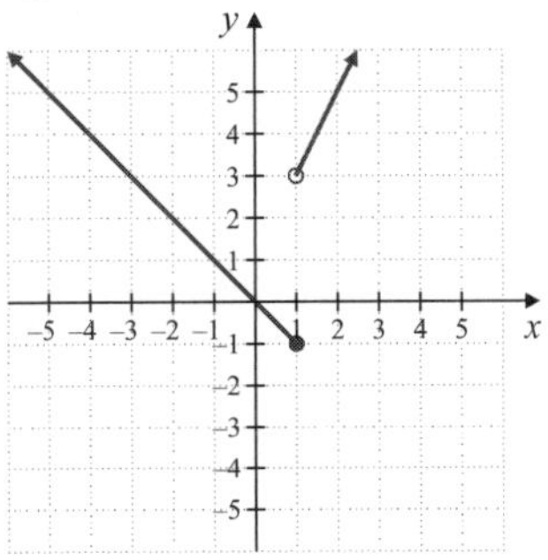

23. $f(x) = \begin{cases} 5 & \text{if} \quad x < -2 \\ 3 & \text{if} \quad x \ge -2 \end{cases}$

Graph $y = 5$ for $x < -2$, with an open circle at $x = -2$. Graph $y = 3$ for $x \ge -2$, with a closed circle at $x = -2$.

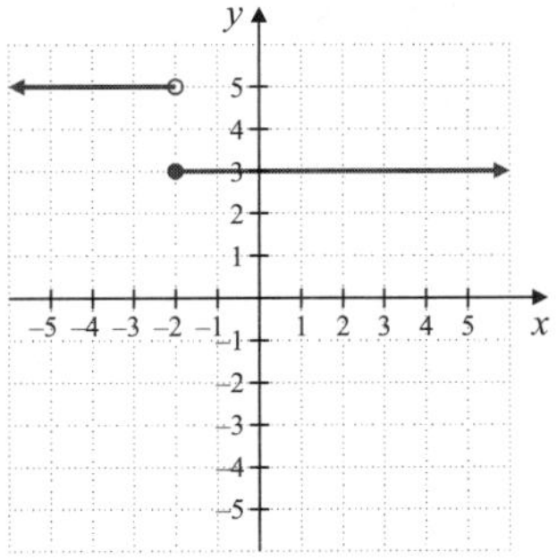

25. $f(x) = \begin{cases} -2x & \text{if} \quad x \le 0 \\ 2x+1 & \text{if} \quad x > 0 \end{cases}$

Graph $y = -2x$ for $x \le 0$, with a closed circle at $x = 0$. Graph $y = 2x + 1$ for $x > 0$, with an open circle at $x = 0$.

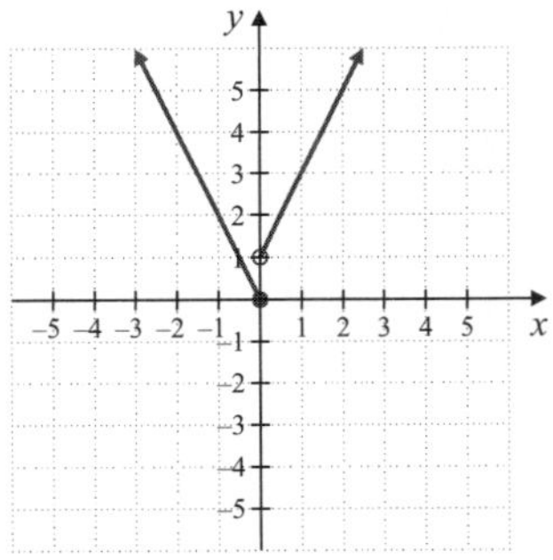

Domain = $(-\infty, \infty)$
Range = $[0, \infty)$

27. $h(x) = \begin{cases} 5x-5 & \text{if} \quad x < 2 \\ -x+3 & \text{if} \quad x \ge 2 \end{cases}$

Graph $y = 5x - 5$ for $x < 2$ with an open circle at $x = 2$. Graph $y = -x + 3$ for $x \ge 2$ with a closed circle at $x = 2$.

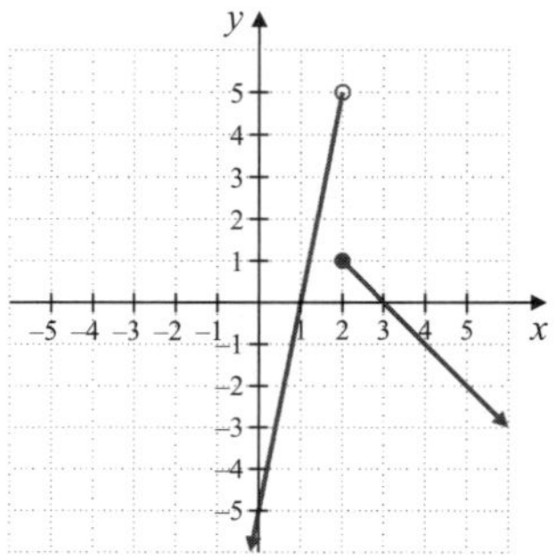

Domain = $(-\infty, \infty)$
Range = $(-\infty, 5)$

29. $f(x) = \begin{cases} x+3 & \text{if } x < -1 \\ -2x+4 & \text{if } x \geq -1 \end{cases}$

Graph $y = x + 3$ for $x < -1$, with an open circle at $x = -1$. Graph $y = -2x + 4$ for $x \geq -1$, with a closed circle at $x = -1$.

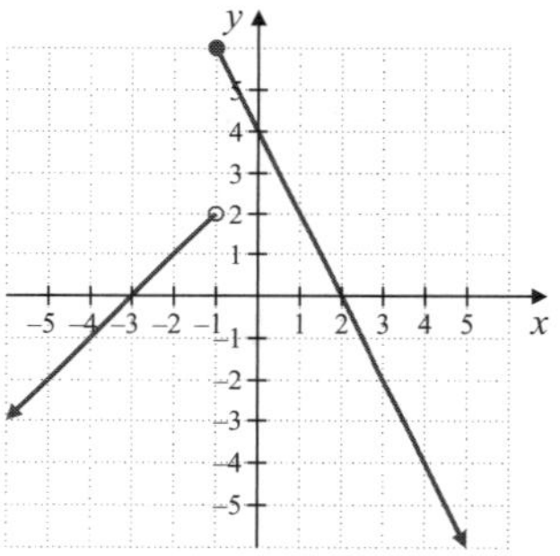

Domain = $(-\infty, \infty)$
Range = $(-\infty, 6]$

31. $g(x) = \begin{cases} -2 & \text{if } x \leq 0 \\ -4 & \text{if } x \geq 1 \end{cases}$

Graph $y = -2$ for $x \leq 0$, with a closed circle at $x = 0$. Graph $y = -4$ for $x \geq 1$, with a closed circle at $x = 1$.

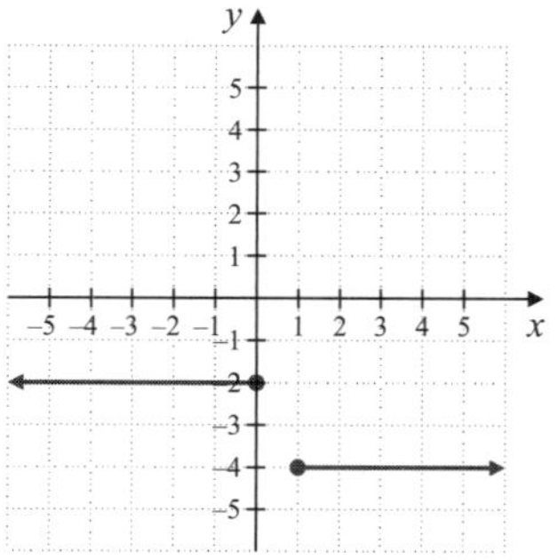

Domain = $(-\infty, 0] \cup [1, \infty)$
Range = $[-4, -2]$

33. $y = -1$ is a horizontal line that crosses the y-axis at $(0, -1)$. This is a graph A.

35. $x = 3$ is a vertical line that crosses the x-axis at $(3, 0)$. This is graph D.

37. answers may vary

39.

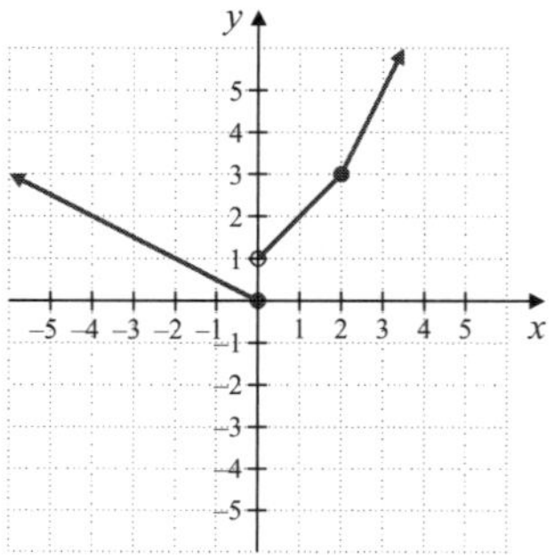

Section 3.8 Practice Problems

1. The graph of $g(x) = x^2 - 1$ is the graph of $f(x) = x^2$ shifted 1 unit down.

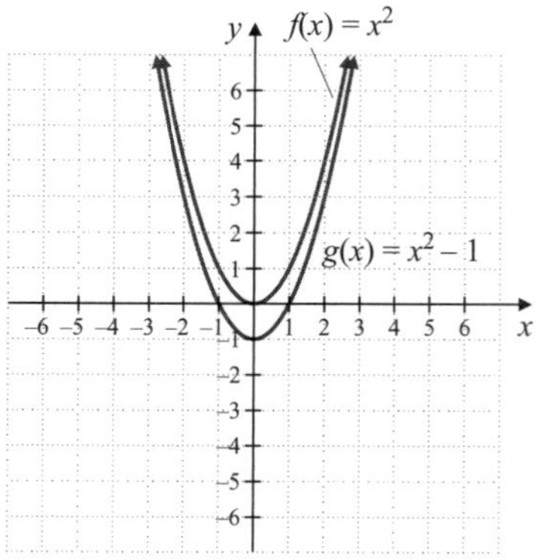

2. The graph of $g(x) = \sqrt{x} + 2$ is the graph of $f(x) = \sqrt{x}$ shifted 2 units up.

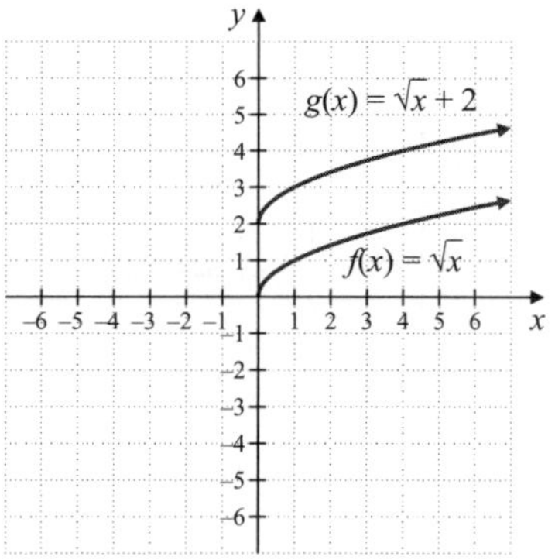

3. The graph of $g(x) = |x + 2|$ is the graph of $f(x) = |x|$ shifted 2 units to the left.

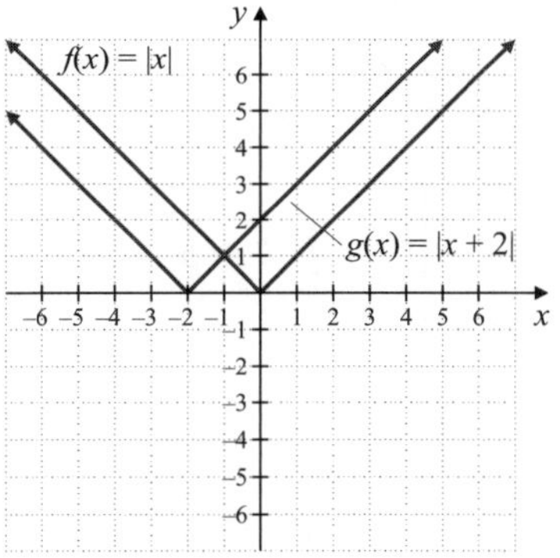

4. The graph of $g(x) = (x+2)^2 - 1$ is the graph of $f(x) = x^2$ shifted 2 units to the left and 1 unit down.

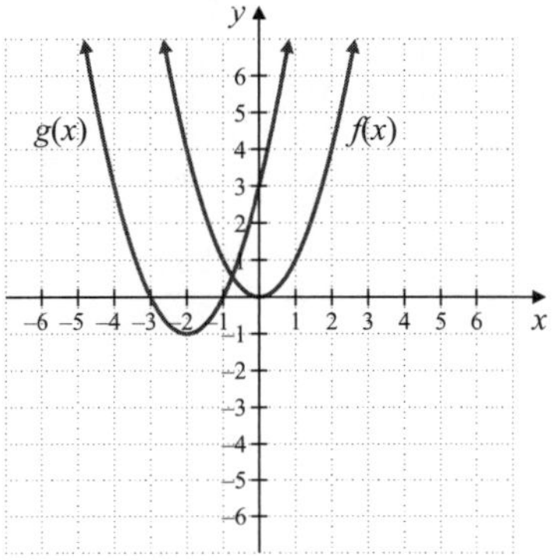

5. The graph of $h(x) = -(x-3)^2 + 2$ is the graph of $f(x)$ reflected about the x-axis, then moved three units to the right and two units upward.

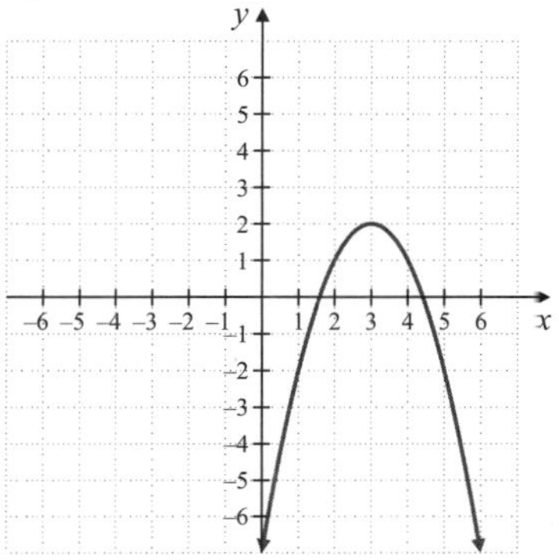

$h(x) = -(x - 3)^2 + 2$

Mental Math

1. $y = \sqrt{x}$ matches with graph c.

2. $y = x^2$ matches with graph b.

3. $y = x$ matches with graph d.

4. $y = |x|$ matches with graph a.

Exercise Set 3.8

1. The graph of $f(x) = |x| + 3$ is the graph of $y = |x|$ shifted 3 units up.

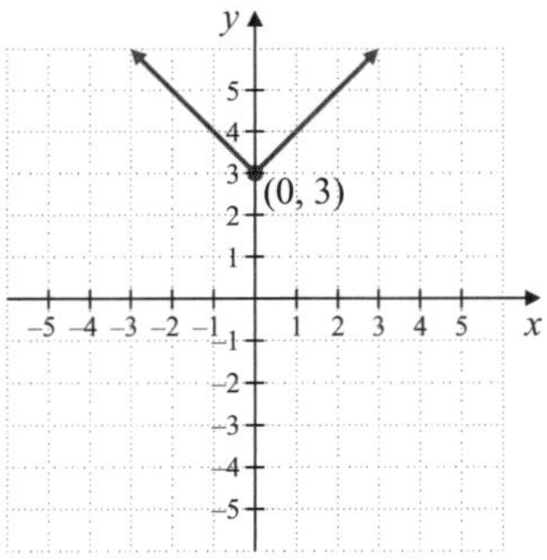

3. The graph of $f(x) = \sqrt{x} - 2$ is the graph of $y = \sqrt{x}$ shifted 2 units down.

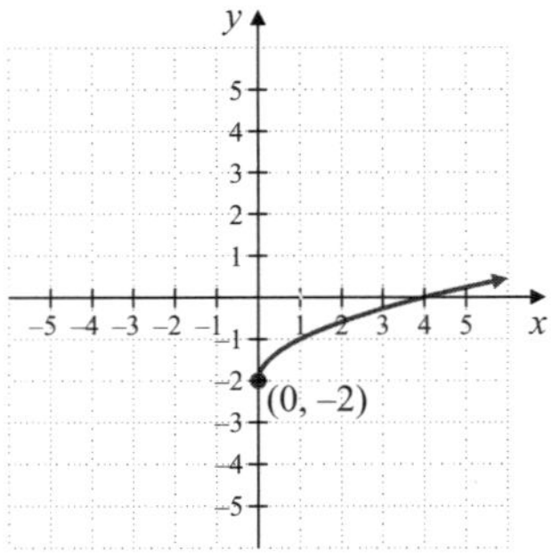

5. The graph of $f(x) = |x - 4|$ is the graph of $y = |x|$ shifted 4 units to the right.

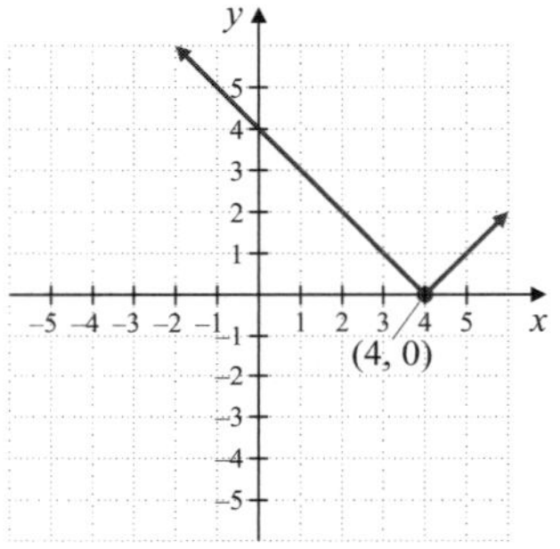

7. The graph of $f(x) = \sqrt{x+2}$ is the graph of $y = \sqrt{x}$ shifted 2 units to the left.

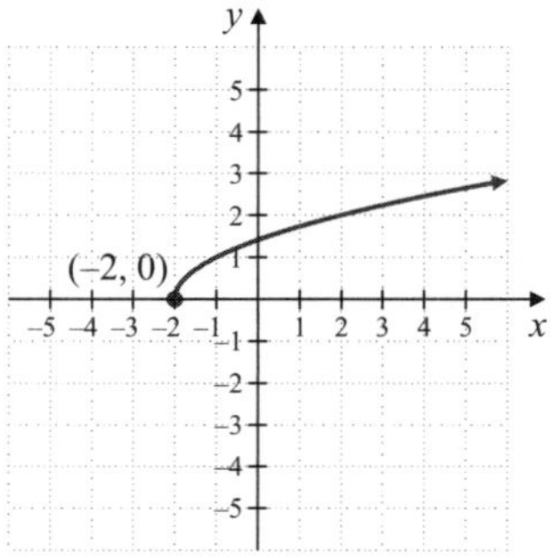

9. The graph of $y = (x-4)^2$ is the graph of $y = x^2$ shifted 4 units to the right.

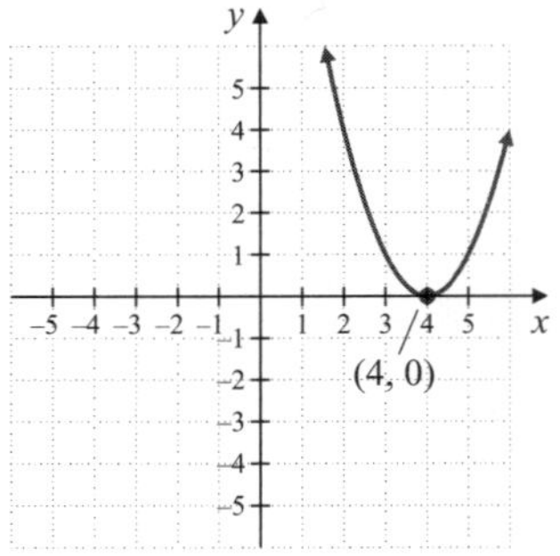

11. The graph of $f(x) = x^2 + 4$ is the graph of $y = x^2$ shifted 4 units up.

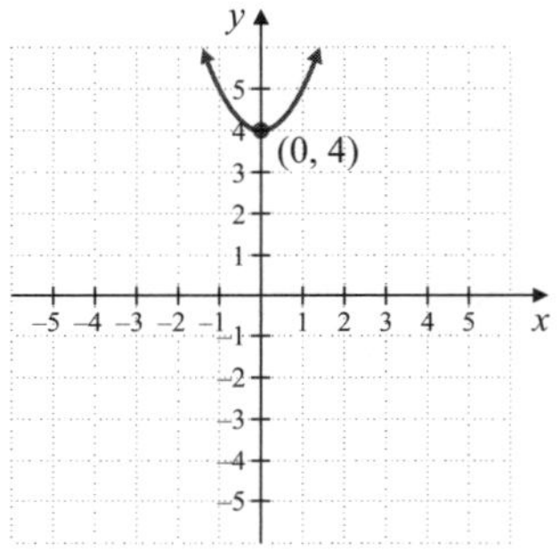

13. The graph of $f(x) = \sqrt{x-2} + 3$ is the graph of $y = \sqrt{x}$ shifted 2 units to the right and 3 units up.

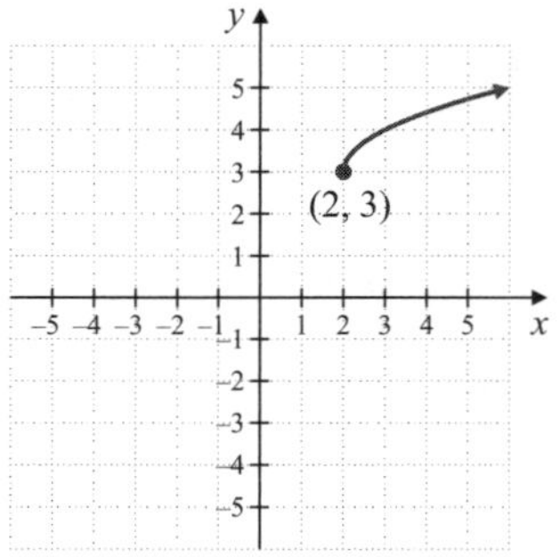

15. The graph of $f(x) = |x - 1| + 5$ is the graph of $y = |x|$ shifted 1 unit to the right and 5 units up.

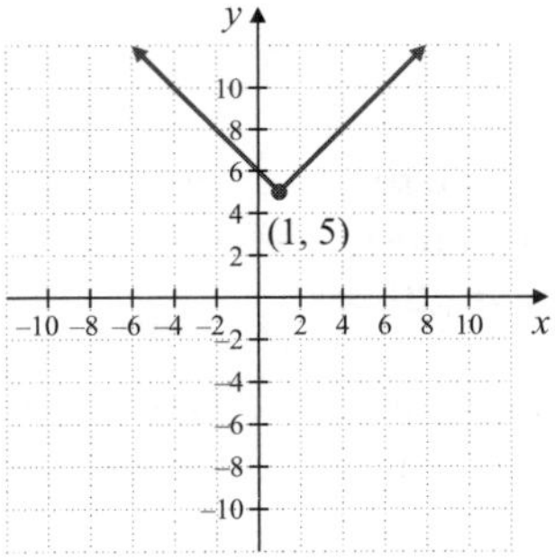

17. The graph of $f(x) = \sqrt{x+1} + 1$ is the graph of $y = \sqrt{x}$ shifted 1 unit to the left and 1 unit up.

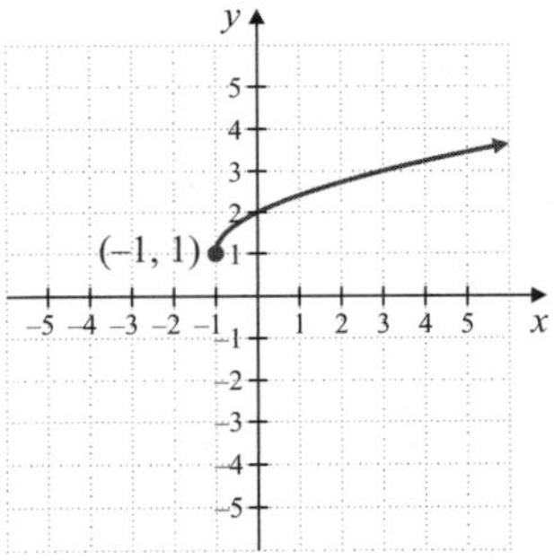

19. The graph of $f(x) = |x + 3| - 1$ is the graph of $y = |x|$ shifted 3 units to the left and 1 unit down.

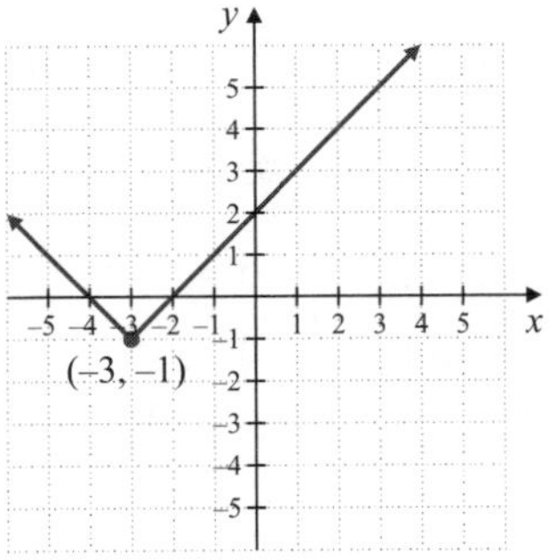

21. The graph of $g(x) = (x-1)^2 - 1$ is the graph of $y = x^2$ shifted 1 unit to the right and 1 unit down.

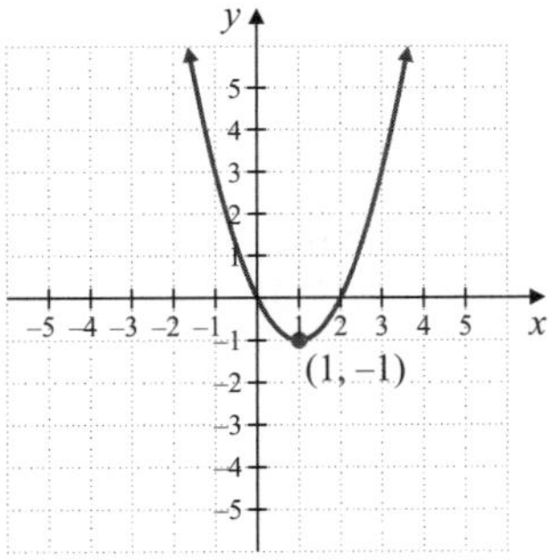

23. The graph of $f(x) = (x+3)^2 - 2$ is the graph of $y = x^2$ shifted 3 units to the left and 2 units down.

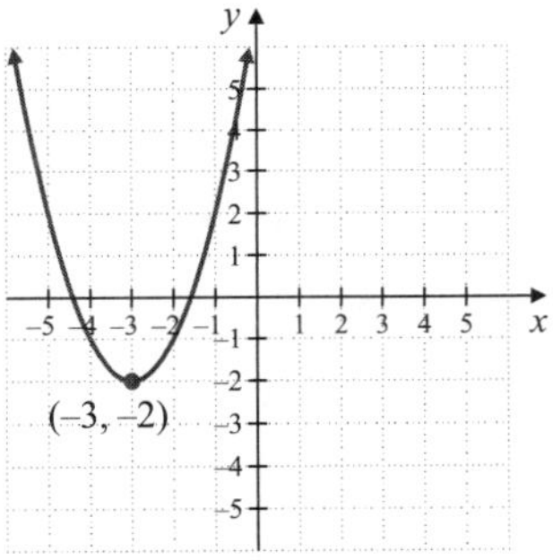

25. The graph of $f(x) = -(x-1)^2$ is the graph of $y = x^2$ reflected about the *x*-axis and shifted 1 unit to the right.

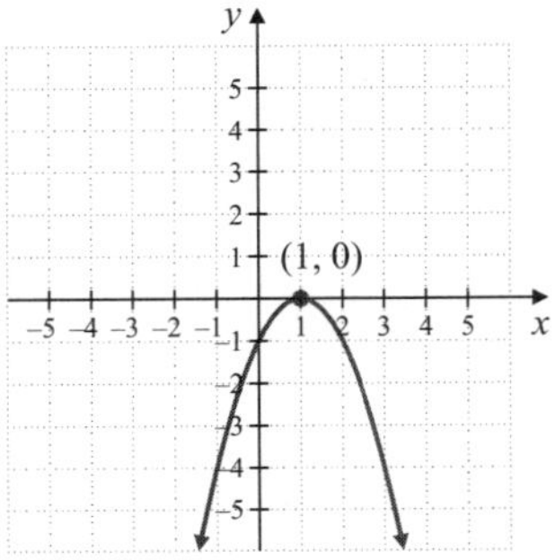

27. The graph of $h(x) = -\sqrt{x} + 3$ is the graph of $y = \sqrt{x}$ reflected about the x-axis and shifted 3 units up.

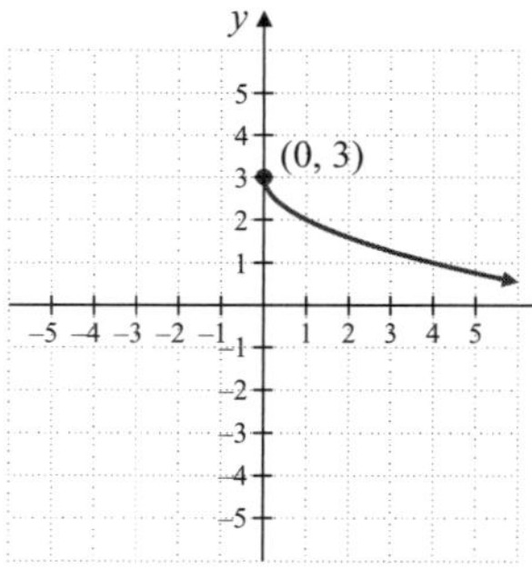

29. The graph of $h(x) = -|x + 2| + 3$ is the graph of $y = |x|$ reflected about the x-axis then shifted 2 units to the left and 3 units up.

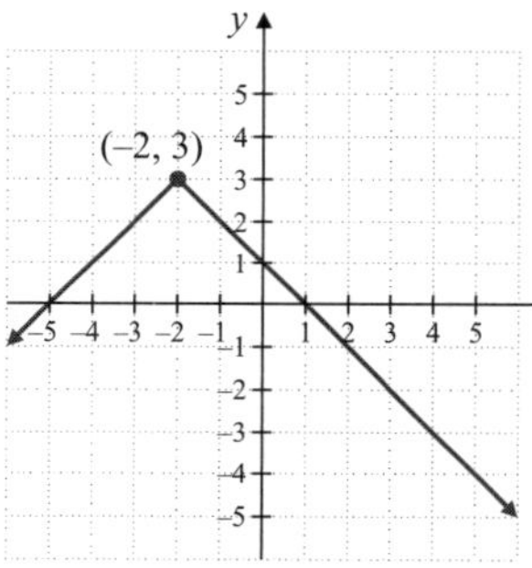

31. The graph of $f(x) = (x - 3) + 2$ is the graph of $y = x$ shifted 3 units to the right and 2 units up.

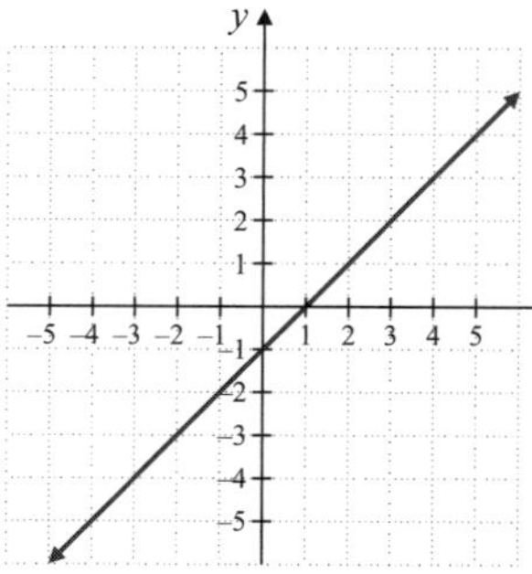

33. $-3x^4 \cdot 5x^4 = -3 \cdot 5 \cdot x^{4+4} = -15x^8$

35. $y^7 \cdot y^{11} = y^{7+11} = y^{18}$

37. Domain = $[2, \infty)$
Range = $[3, \infty)$

39. Domain = $(-\infty, \infty)$
Range = $(-\infty, 3]$

41. Since the square root of a negative number is not a real number, $x - 20 \geq 0$ or $x \geq 20$. The domain is $[20, \infty)$.

43. Since the absolute value of any real number will be positive, the domain is $(-\infty, \infty)$.

45. Since the square root of a negative number is not a real number, $x + 103 \geq 0$ or $x \geq -103$. The domain is $[-103, \infty)$.

47. $f(x) = \begin{cases} |x| & \text{if} \quad x \leq 0 \\ x^2 & \text{if} \quad x > 0 \end{cases}$

Graph $y = |x|$ for $x \leq 0$ with a closed circle at $x = 0$. Graph $y = x^2$ for $x > 0$ with an open circle at $x = 0$.

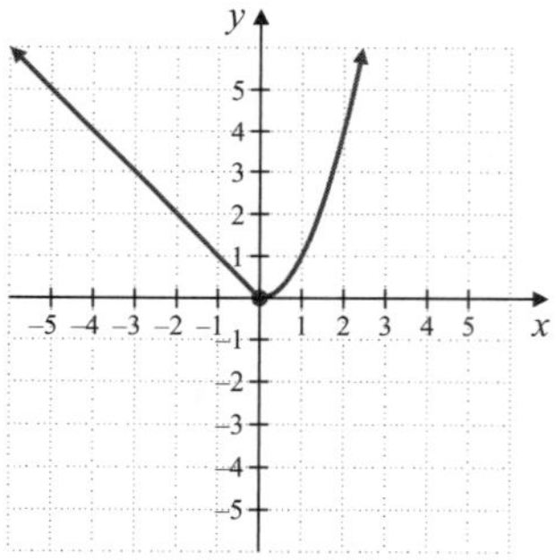

Domain = $(-\infty, \infty)$
Range = $[0, \infty)$

49. $g(x) = \begin{cases} |x - 2| & \text{if} \quad x < 0 \\ -x^2 & \text{if} \quad x \geq 0 \end{cases}$

Graph $y = |x - 2|$ for $x < 0$, with an open circle at $x = 0$.

Graph $y = -x^2$ for $x \geq 0$, with a closed circle at $x = 0$.

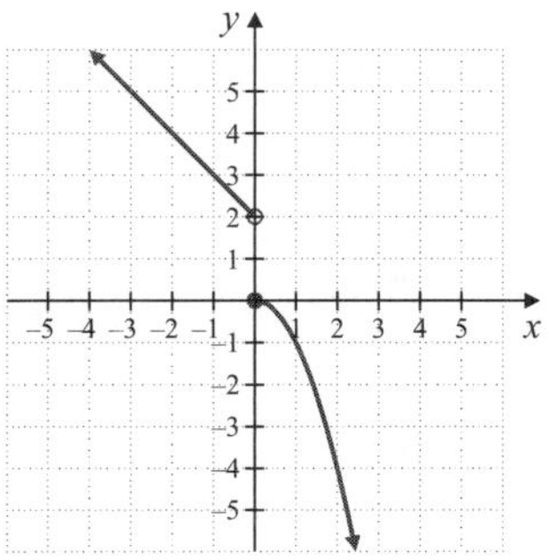

Domain = $(-\infty, \infty)$
Range = $(-\infty, 0] \cup (2, \infty)$

Chapter 3 Vocabulary Check

1. A relation is a set of ordered pairs.

2. The graph of every linear equation in two variables is a line.

3. The statement $-x + 2y > 0$ is called a linear inequality in two variables.

4. Standard form of a linear equation in two variables is $Ax + By = C$.

5. The range of a relation is the set of all second components of the ordered pairs of the relation.

6. Parallel lines have the same slope and different y-intercepts.

7. Slope-intercept form of a linear equation in two variables is $y = mx + b$.

8. A function is a relation in which each first component in the ordered pairs corresponds to exactly one second component.

9. In the equation $y = 4x - 2$, the coefficient of x is the slope of its corresponding graph.

10. Two lines are perpendicular if the product of their slopes is -1.

11. To find the x-intercept of a linear equation, let $y = 0$ and solve for the other variable.

12. The domain of a relation is the set of all first components of the ordered pairs of the relation.

13. A linear function is a function that can be written in the form $f(x) = mx + b$.

14. To find the y-intercept of a linear equation, let $x = 0$ and solve for the other variable.

15. The equation $y - 8 = -5(x + 1)$ is written in point-slope form.

Chapter 3 Review

1. $A(2, -1)$ is in quadrant IV.
$B(-2, 1)$ is in quadrant II.
$C(0, 3)$ is on the y-axis.
$D(-3, -5)$ is in quadrant III.

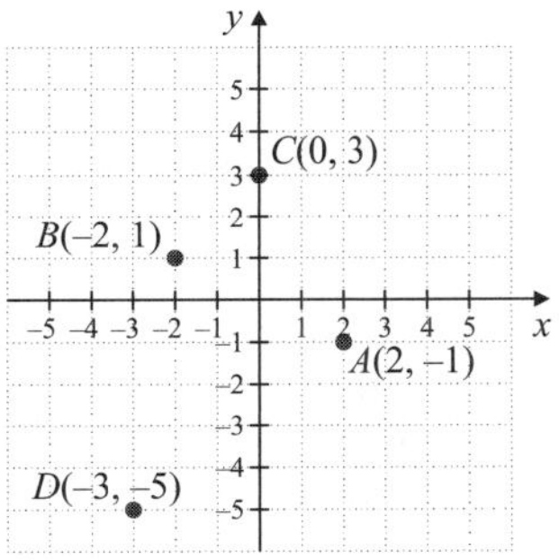

2. $A(-3, 4)$ is in quadrant II.
$B(4, -3)$ is in quadrant IV.
$C(-2, 0)$ is on the x-axis.
$D(-4, 1)$ is in quadrant II.

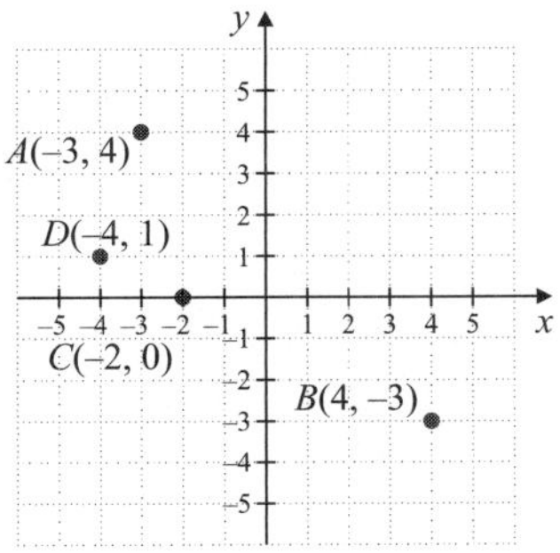

3.

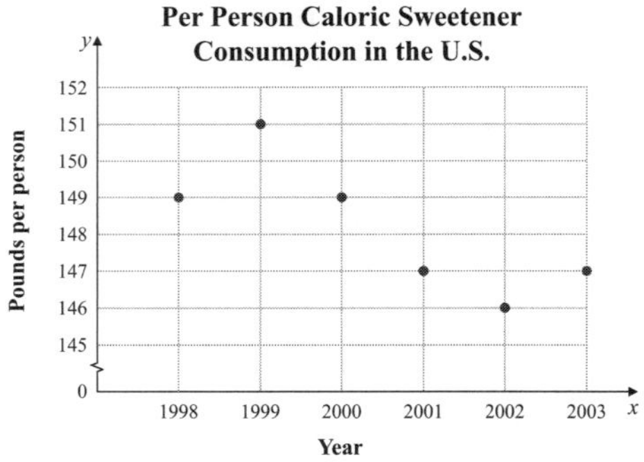

4.

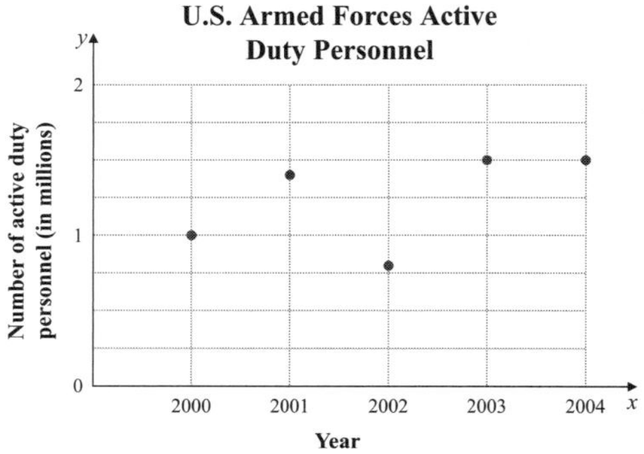

5. $7x - 8y = 56$

(0, 56): $7(0) - 8(56) \stackrel{?}{=} 56$

$-448 = 56$ False

(0, 56) is not a solution.

(8, 0): $7(8) - 8(0) \stackrel{?}{=} 56$

$56 = 56$ True

(8, 0) is a solution.

6. $-2x + 5y = 10$

(−5, 0): $-2(-5) + 5(0) \stackrel{?}{=} 10$

$10 = 10$ True

(−5, 0) is a solution.

(1, 1): $-2(1) + 5(1) \stackrel{?}{=} 10$

$-2 + 5 \stackrel{?}{=} 10$

$3 = 10$ False

(1, 1) is not a solution.

7. $x = 13$

(13, 5): $13 = 13$ True

(13, 5) is a solution.

(13, 13): $13 = 13$ True

(13, 13) is a solution.

8. $y = 2$

(7, 2): $2 = 2$ True

(7, 2) is a solution.

(2, 7): $7 \neq 2$ False

(2, 7) is not a solution.

9. $3x - y = 3$

Find three ordered pair solutions.

x	y
0	−3
1	0
2	3

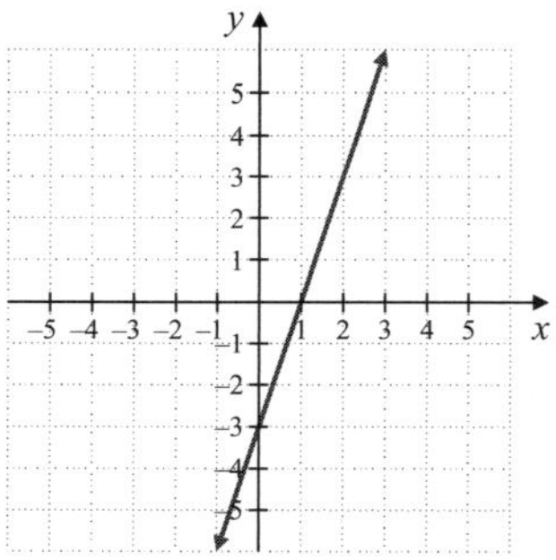

10. $2x - y = 4$

Find three ordered pair solutions.

x	y
0	−4
1	−2
3	2

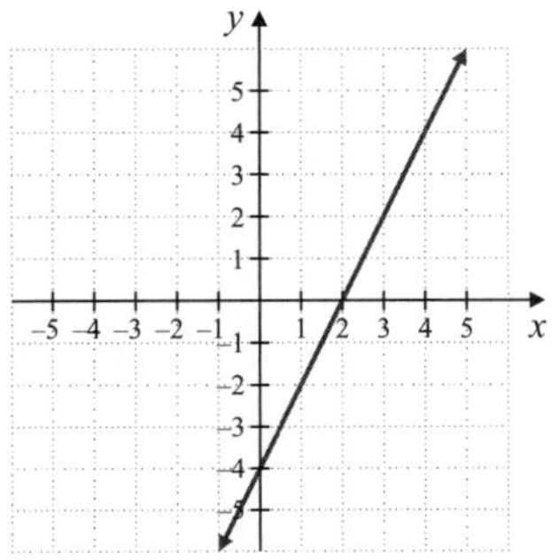

11. $4x + 5y = 20$
Find three ordered pair solutions.

x	y
0	4
5	0
10	−4

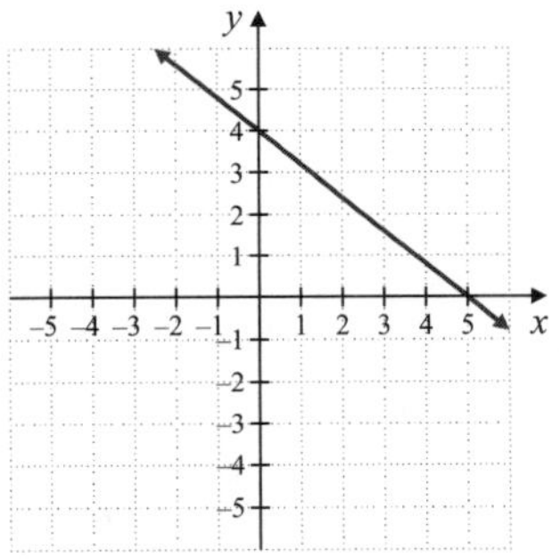

12. $y = 5$
This is a horizontal line with a y-intercept at (0, 5) and no x-intercept.

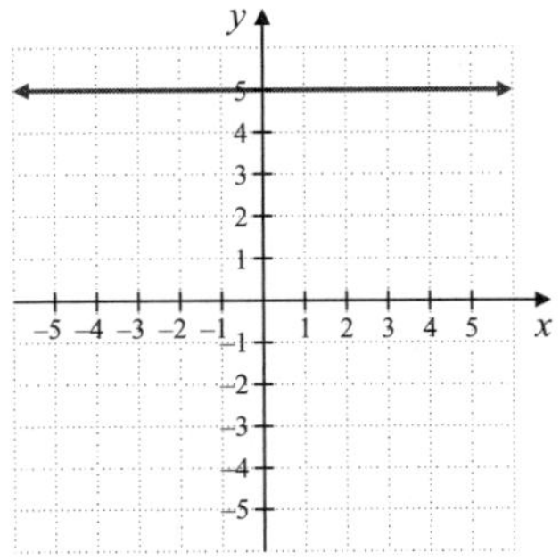

13. $x = -2$
This is a vertical line with an x-intercept at (−2, 0) and no y-intercept.

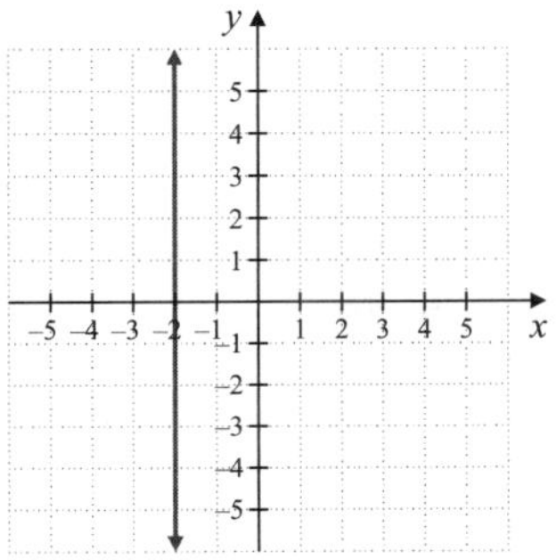

14. $y = \frac{1}{3}x$
Find three ordered pair solutions.

x	y
−3	−1
0	0
3	1

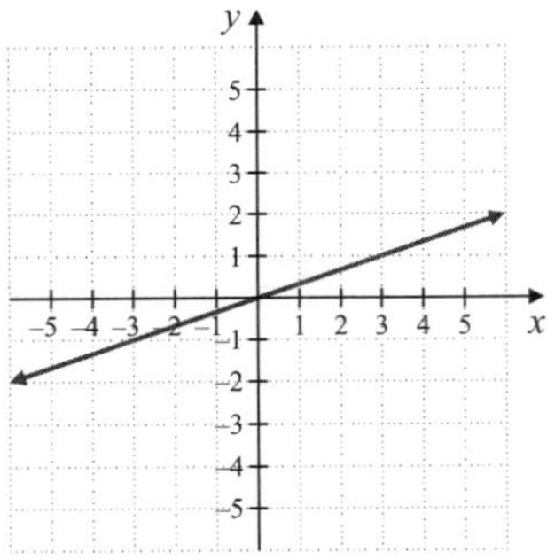

15. $m = \frac{y_2 - y_1}{x_2 - x_1} = \frac{-4-8}{6-2} = \frac{-12}{4} = -3$

16. $m = \frac{y_2 - y_1}{x_2 - x_1} = \frac{13-9}{5-(-3)} = \frac{4}{8} = \frac{1}{2}$

17. $m = \frac{y_2 - y_1}{x_2 - x_1} = \frac{6-(-4)}{-3-(-7)} = \frac{10}{4} = \frac{5}{2}$

18. $m = \frac{y_2 - y_1}{x_2 - x_1} = \frac{7-(-2)}{-5-7} = \frac{9}{-12} = -\frac{3}{4}$

19. $m = \frac{y_2 - y_1}{x_2 - x_1} = \frac{3-0}{2-0} = \frac{3}{2}$

20. $m = \frac{y_2 - y_1}{x_2 - x_1} = \frac{3-0}{0-1} = -3$

21. $m = \frac{y_2 - y_1}{x_2 - x_1} = \frac{2-0}{0-4} = -\frac{2}{4} = -\frac{1}{2}$

22. $m = \frac{y_2 - y_1}{x_2 - x_1} = \frac{4-(-1)}{2-(-3)} = \frac{5}{5} = 1$

23. l_2 has the greater slope because the slope of l_2 is positive and the slope of l_1 is negative.

24. l_2 has the greater slope because the slope of l_2 is positive and the slope of l_1 is 0.

25. l_2 has the greater slope because the slope of l_2 is steeper than l_1 therefore the slope of l_2 is a larger positive number than l_1.

26. l_1 has the greater slope because the slope of l_1 is zero and the slope of l_2 is negative.

27. $y = -3x + \frac{1}{2}$

$m = -3;\ y\text{-int} = \left(0, \frac{1}{2}\right)$

28. $y = 2x + 4$
$m = 2;\ y\text{-int} = (0, 4)$

29. $6x - 15y = 20$

$6x - 20 = 15y$

$\frac{6}{15}x - \frac{20}{15} = y$

$\frac{2}{5}x - \frac{4}{3} = y$

$m = \frac{2}{5};\ y\text{-int} = \left(0, -\frac{4}{3}\right)$

30. $4x + 14y = 21$

$14y = -4x + 21$

$y = -\frac{4}{14}x + \frac{21}{14}$

$y = -\frac{2}{7}x + \frac{3}{2}$

$m = -\frac{2}{7};\ y\text{-int} = \left(0, \frac{3}{2}\right)$

31. $y - 3 = 0$

$y = 3$

$m = 0$

32. $x = -5$
m = undefined

33. $y = -2x + 6$ $\quad$ $y = 2x - 1$
$m = -2;\ y\text{-int} = (0, 6)$ $\quad$ $m = 2;\ y\text{-int} = (0, -1)$
Neither, since their slopes are not equal, nor does their product equal -1.

34. $-x + 3y = 2$ $\quad$ $6x - 18y = 3$

$3y = x + 2$ $\quad$ $6x - 3 = 18y$

$y = \frac{1}{3}x + \frac{2}{3}$ $\quad$ $\frac{1}{3}x - \frac{1}{6} = y$

$m = \frac{1}{3};\ y\text{-int} = \left(0, \frac{2}{3}\right)$

$m = \frac{1}{3};\ y\text{-int} = \left(0, -\frac{1}{6}\right)$

The lines are parallel since they have the same slopes, but different y-intercepts.

35. $y=\frac{3}{4}x+1$ $\qquad$ $y=-\frac{4}{3}x+1$

$m=\frac{3}{4}$; y-int = (0, 1)

$m=-\frac{4}{3}$; y-int = (0, 1)

The lines are perpendicular since the product of their slopes is −1.

36. $x-2y=6$ $\qquad$ $4x+y=8$

$x-6=2y$ $\qquad$ $y=-4x+8$

$\frac{1}{2}x-3=y$

$m=\frac{1}{2}$; y-int = (0, −3) $\quad$ $m=-4$;

y-int = (0, 8)

Neither, since their slopes are not equal, nor does their product equal −1.

37.

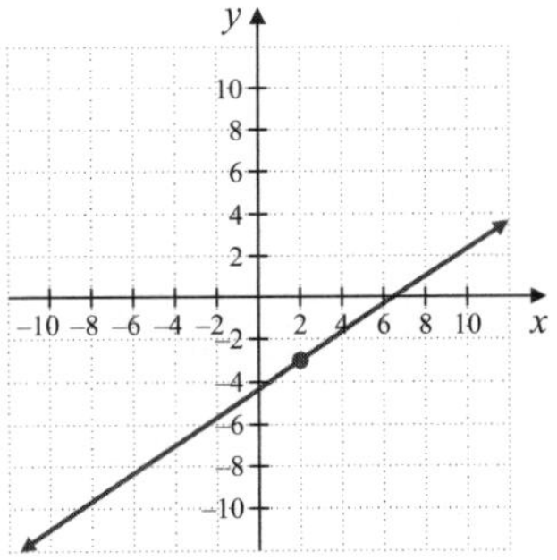

38.

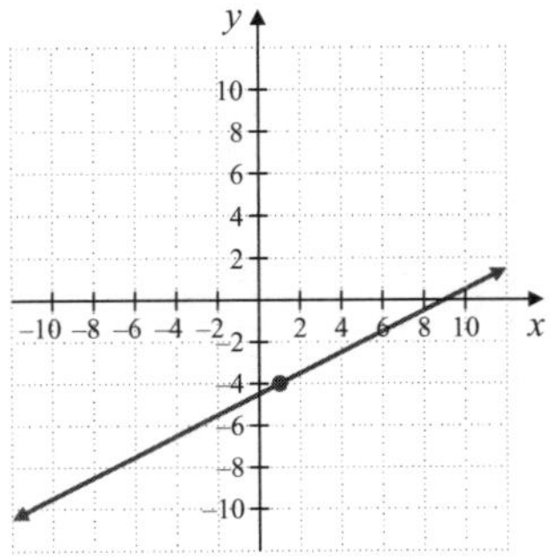

39.

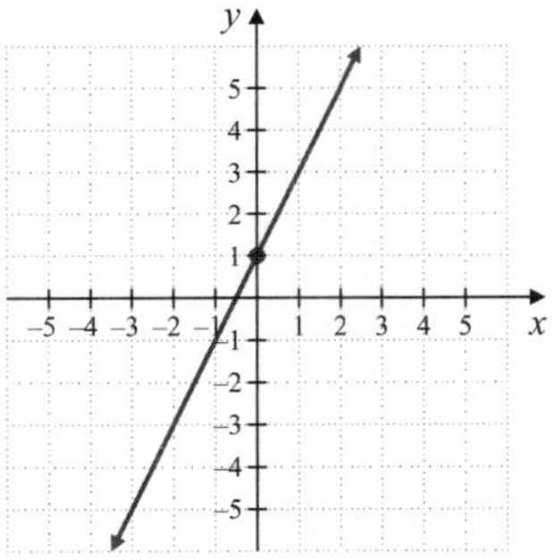

40.

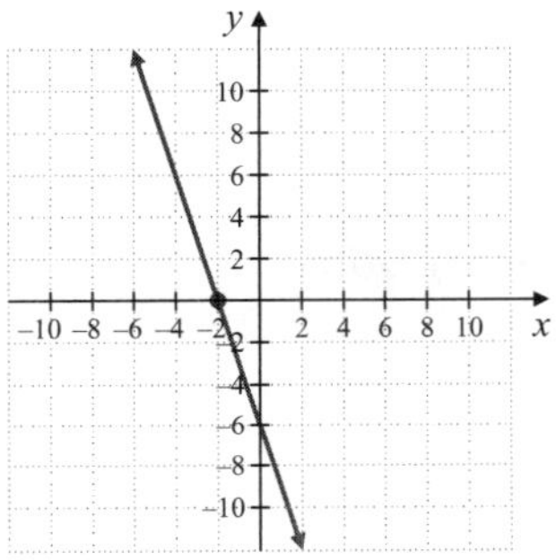

41. $y=-x+1$

$m=-1$; y-int = (0, 1)

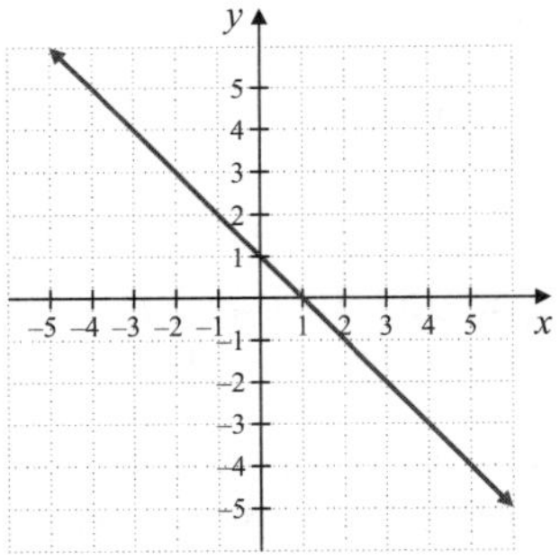

42. $y=4x-3$

$m=4$; y-int = (0, −3)

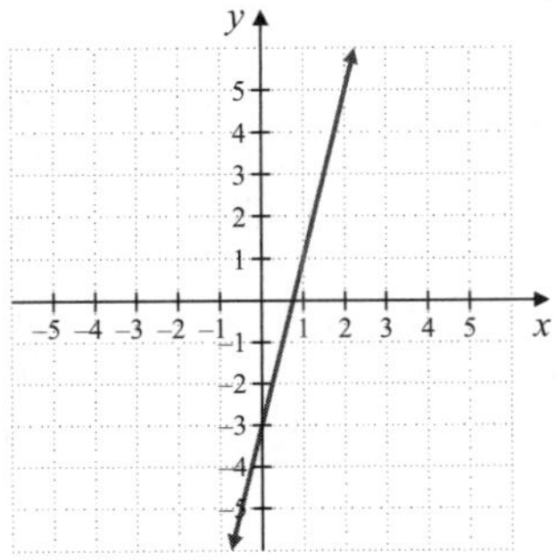

43. $3x - y = 6$
$3x - 6 = y$
$m = 3$; y-int = (0, −6)

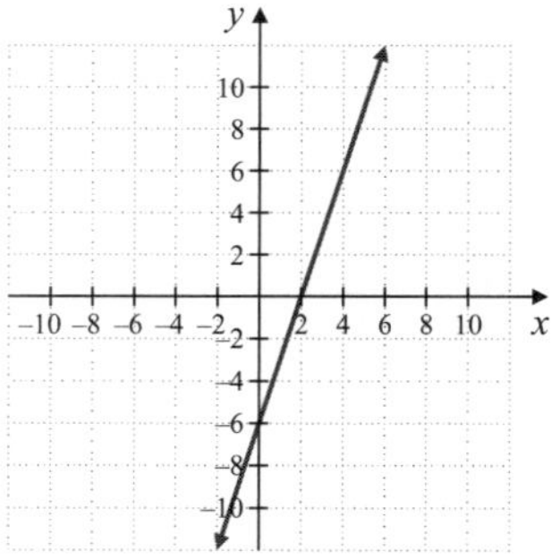

44. $y = -5x$
$m = -5$; y-int = (0, 0)

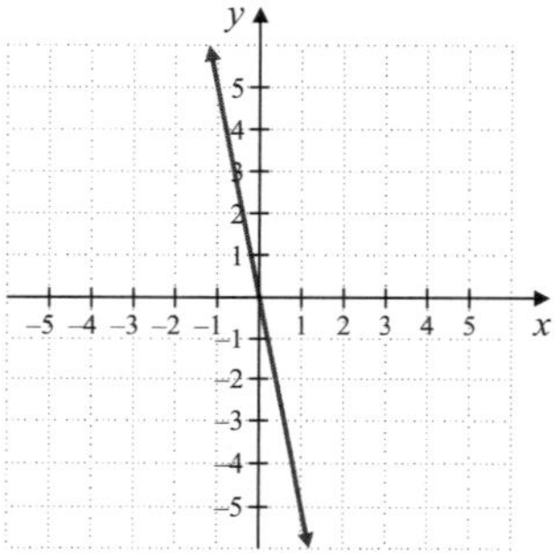

45. $C = 0.3x + 42$

- **a.** $C(150) = 0.3(150) + 42 = 87$
  It would cost $87 to rent the minivan.

- **b.** The slope is 0.3 which means the cost of renting increases by $0.30 for each additional mile driven.

- **c.** The y-intercept is (0, 42). The cost of renting for 0 miles driven is $42.

46. A horizontal line has an equation of the form $y = b$. Since the line passes through the point (3, −1), its equation is $y = -1$.

47. A line with undefined slope is a vertical line which has the form $x = c$. Since the line passes through (−4, −3), its equation is $x = -4$.

48.
$$\begin{aligned} y - y_1 &= m(x - x_1) \\ y - 5 &= 3(x - (-3)) \\ y - 5 &= 3x + 9 \\ y &= 3x + 14 \end{aligned}$$

49. $m = \dfrac{-2-(-1)}{-4-(-6)} = \dfrac{-1}{2}$
$$\begin{aligned} y - y_1 &= m(x - x_1) \\ y - (-2) &= -\frac{1}{2}(x - (-4)) \\ y + 2 &= -\frac{1}{2}(x + 4) \\ y + 2 &= -\frac{1}{2}x - 2 \\ y &= -\frac{1}{2}x - 4 \end{aligned}$$

50. $y = -2x + 3$; $m = -2$
A parallel line will have slope −2.
$$\begin{aligned} y - y_1 &= m(x - x_1) \\ y - (-6) &= -2(x - 2) \\ y + 6 &= -2x + 4 \\ y &= -2x - 2 \end{aligned}$$

51.
$$\begin{aligned} 4x + 3y &= 5 \\ 3y &= -4x + 5 \\ y &= -\frac{4}{3}x + \frac{5}{3} \end{aligned}$$

A perpendicular line will have slope $\frac{3}{4}$.
$$\begin{aligned} y - y_1 &- m(x - x_1) \\ y - (-1) &= \frac{3}{4}(x - (-6)) \\ y + 1 &= \frac{3}{4}x + \frac{9}{2} \\ y &= \frac{3}{4}x + \frac{7}{2} \end{aligned}$$

52. Use (2, 14,300) and (4, 10,200).

a. $m = \frac{14,300 - 10,200}{2 - 4} = \frac{4100}{-2} = -2050$

$y - y_1 = m(x - x_1)$
$y - 10,200 = -2050(x - 4)$
$y - 10,200 = -2050x + 8200$
$y = -2050x + 18,400$

b. Let $x = 6$, since 2006 is 6 years after 2000.
$y = -2050x + 18,400$
$y = -2050(6) + 18,400$
$y = -12,300 + 18,400$
$y = 6100$
In 2006 the value of the automobile will be $6100.

53. a. Use (7, 210,000) and (12, 270,000).

$m = \frac{270,000 - 210,000}{12 - 7}$
$= \frac{60,000}{5}$
$= 12,000$

$y - y_1 = m(x - x_1)$
$y - 210,000 = 12,000(x - 7)$
$y - 210,000 = 12,000x - 84,000$
$y = 12,000x + 126,000$

b. Let $x = 18$ since 2008 is 18 years after 1990.
$y = 12,000x + 126,000$
$y = 12,000(18) + 126,000$
$y = 216,000 + 126,000$
$y = 342,000$
In 2008, the estimated value of the building is $342,000.

54. $3x + y > 4$
Test (0, 0)
$3(0) + 0 > 4$
$0 > 4$ False
Shade the half-plane that does not contain the point (0, 0).

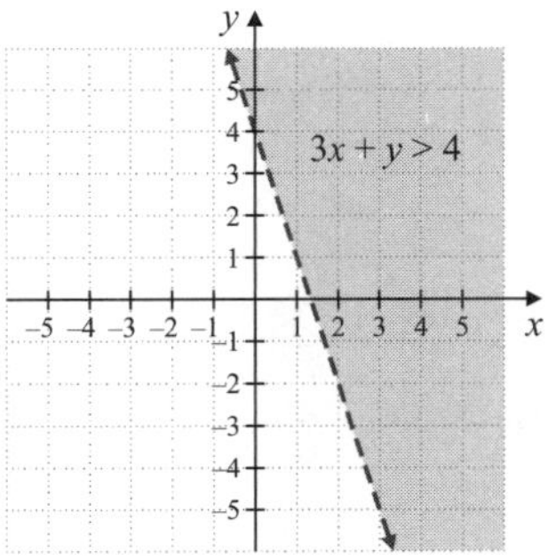

55. $\frac{1}{2}x - y < 2$
Test (0, 0)
$\frac{1}{2}(0) - 0 < 2$
$0 < 2$ True
Shade the half-plane that contains the point (0, 0).

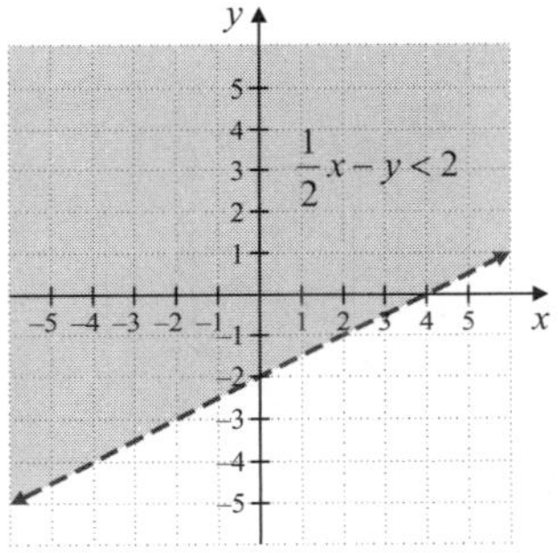

56. $5x - 2y \le 9$
Test (0, 0)
$5(0) - 2(0) \le 9$
$0 \le 9$ True
Shade the half-plane that contains the point (0, 0).

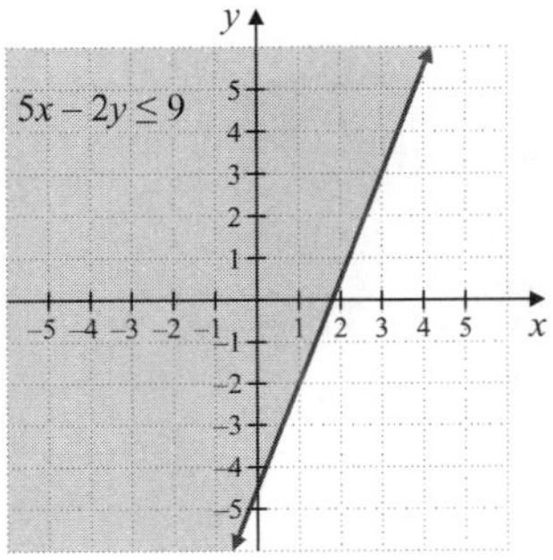

57. $y < 1$

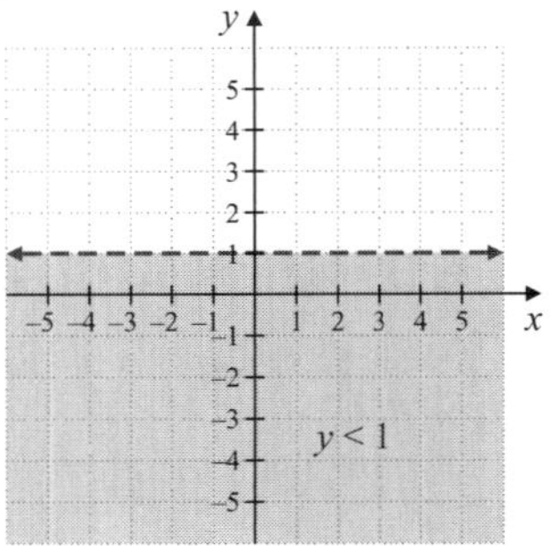

58. $x > -2$

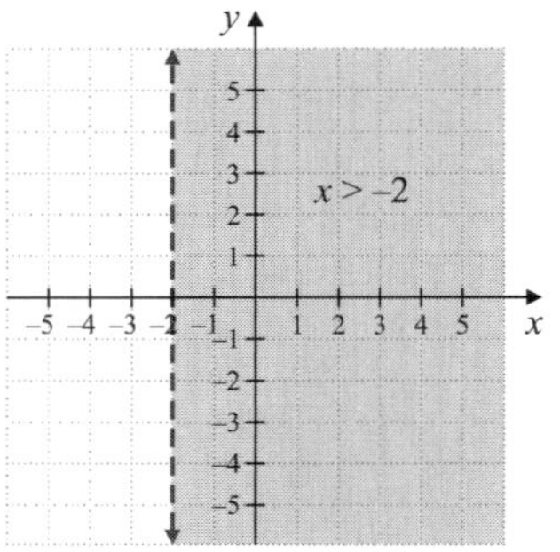

59. $y > 2x + 3$ or $x \le -3$

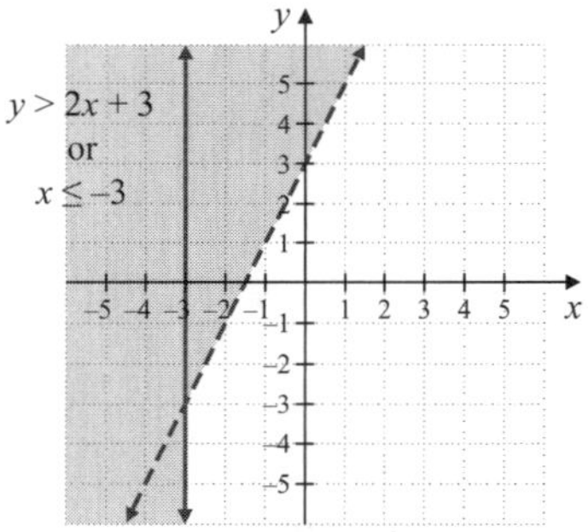

60. $2x < 3y + 8$ and $y \ge -2$

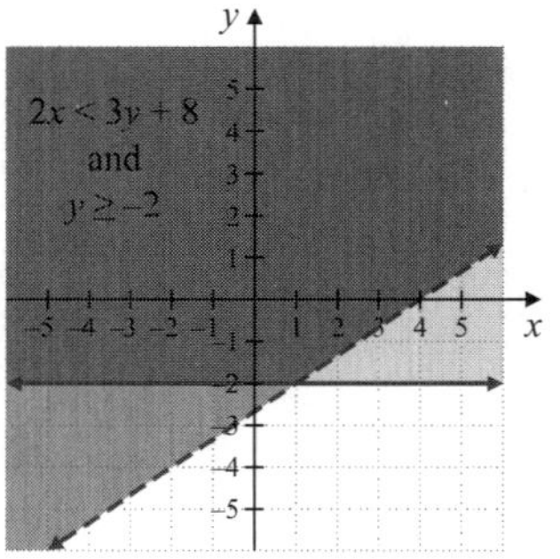

61. Domain $= \left\{-\frac{1}{2}, 6, 0, 25\right\}$

Range $= \left\{\frac{3}{4}, 0.65, -12, 25\right\}$

The relation is a function.

62. Domain $= \left\{\frac{3}{4}, 0.65, -12, 25\right\}$

Range $= \left\{-\frac{1}{2}, 6, 0, 25\right\}$

The relation is a function.

63. Domain = {2, 4, 6, 8}
Range = {2, 4, 5, 6}
The relation is not a function since 2 is paired with 2 and 4.

64. Domain = {triangle, square, rectangle, parallelogram}
Range = {3, 4}
The relation is a function.

65. Domain = $(-\infty, \infty)$
Range = $(-\infty, -1] \cup [1, \infty)$
The relation is not a function since it fails the vertical line test.

66. Domain = {−3}
Range = $(-\infty, \infty)$
The relation is not a function since it fails the vertical line test.

67. Domain = $(-\infty, \infty)$
Range = {4}
The relation is a function since it passes the vertical line test.

68. Domain = $[-1, 1]$
Range = $[-1, 1]$
The relation is not a function since it fails the vertical line test.

69. $f(x) = x - 5$
$f(2) = 2 - 5 = -3$

70. $g(x) = -3x$
$g(0) = -3(0) = 0$

71. $g(x) = -3x$
$g(-6) = -3(-6) = 18$

72. $h(x) = 2x^2 - 6x + 1$
$h(-1) = 2(-1)^2 - 6(-1) + 1 = 2 + 6 + 1 = 9$

73. $h(x) = 2x^2 - 6x + 1$
$h(1) = 2(1)^2 - 6(1) + 1 = 2 - 6 + 1 = -3$

74. $f(x) = x - 5$
$f(5) = 5 - 5 = 0$

75. $J(x) = 2.54x$
$J(150) = 2.54(150) = 381$
The person would weight 381 pounds on Jupiter.

76. $J(x) = 2.54x$
$J(2000) = 2.54(2000) = 5080$
The probe would weigh 5080 pounds on Jupiter.

77. $f(x) = x + 2$ or $y = x + 2$
$m = 1$; y-int = (0, 2)

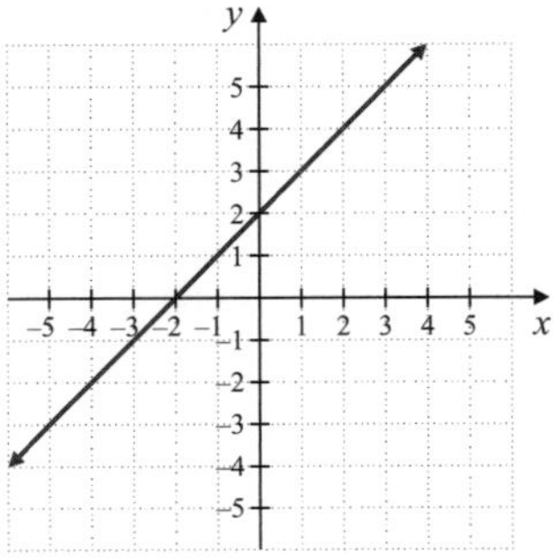

78. $f(x) = -\frac{1}{2}x + 3$ or $y = -\frac{1}{2}x - 3$

$m = -\frac{1}{2}$; y-int = (0, 3)

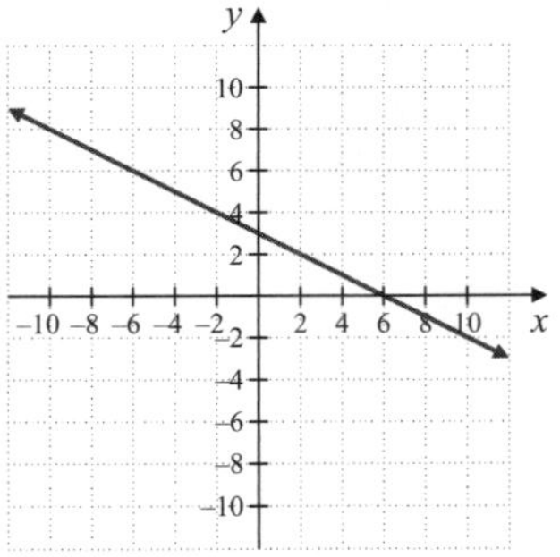

79. Domain = {2}
Range = $(-\infty, \infty)$

80. Domain = $(-\infty, \infty)$
Range = $(-\infty, \infty)$

81. Domain = [−4, 4]
Range = [−1, 5]

82. Domain = $(-\infty, \infty)$
Range = [−5]

83. $f(x) = \begin{cases} -3x & \text{if } x < 0 \\ x - 3 & \text{if } x \geq 0 \end{cases}$

Graph $y = -3x$ for $x < 0$, with an open circle at $x = 0$. Graph $y = x - 3$ for $x \geq 0$, with a closed circle at $x = 0$.

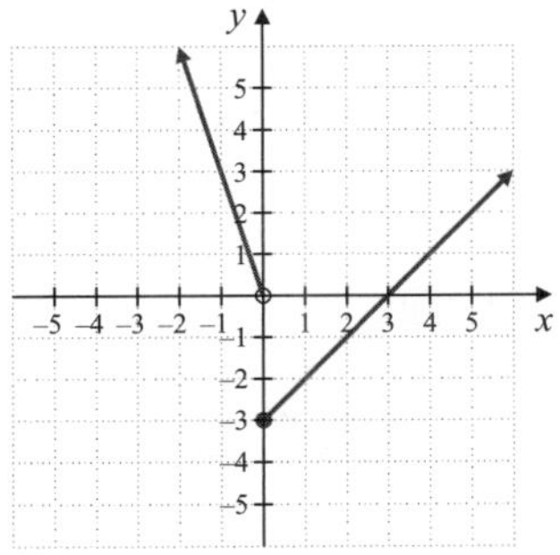

84. $g(x) = \begin{cases} -\frac{1}{5}x & \text{if } x \le -1 \\ -4x+2 & \text{if } x > -1 \end{cases}$

Graph $y = -\frac{1}{5}x$ for $x \le -1$ with a closed circle at $x = -1$. Graph $y = -4x + 2$ for $x > -1$ with an open circle at $x = -1$.

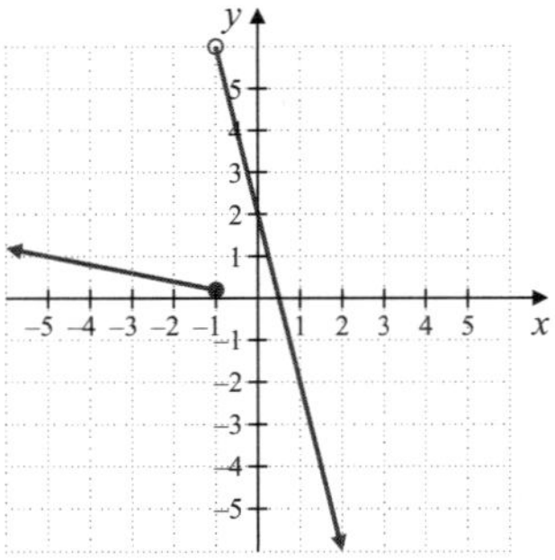

85. The graph of $y = \sqrt{x} - 4$ is the graph of $y = \sqrt{x}$ shifted 4 units down.

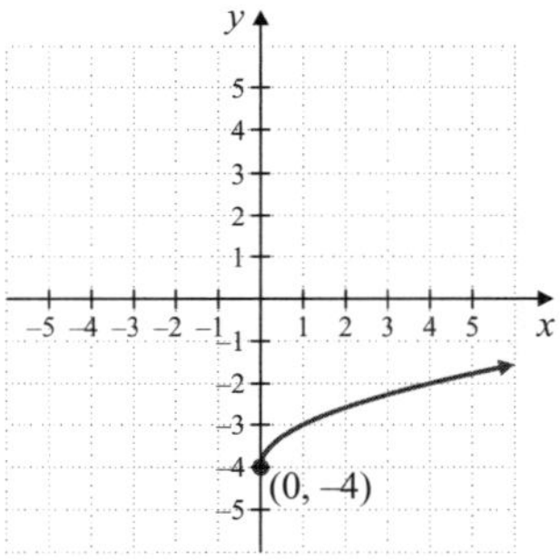

86. The graph of $f(x) = \sqrt{x-4}$ is the graph of $y = \sqrt{x}$ shifted 4 units to the right.

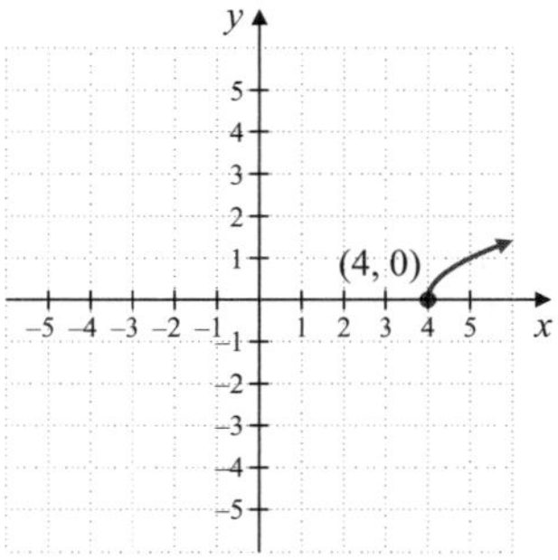

87. The graph of $g(x) = |x - 2| - 2$ is the graph of $y = |x|$ shifted 2 units to the right and 2 units down.

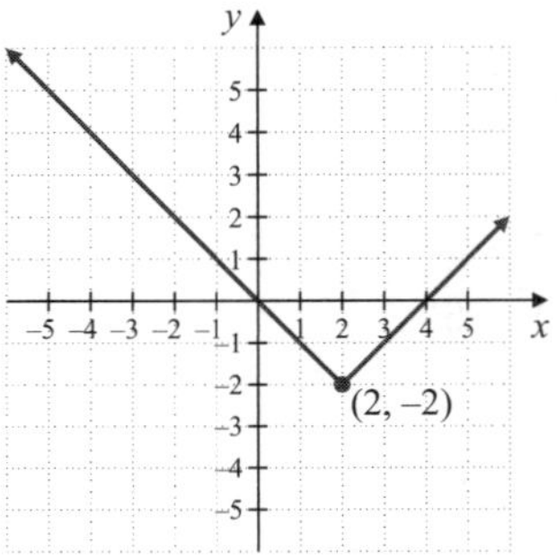

88. The graph of $h(x) = -(x+3)^2 - 1$ is the graph of $y = x^2$ reflected about the x-axis then shifted 3 units to the left and 1 unit down.

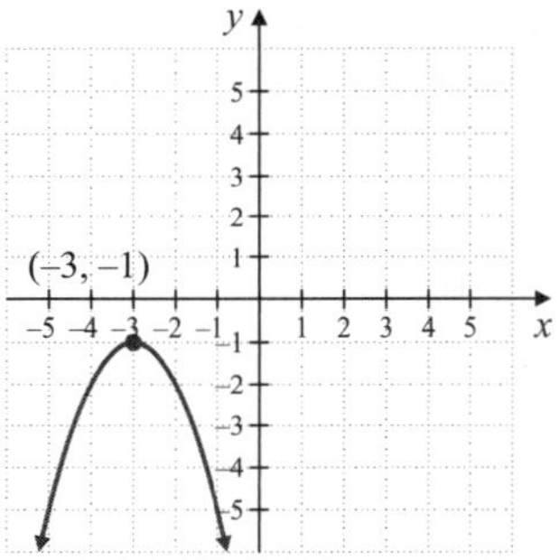

89. $3x - 2y = -9$

$3x + 9 = 2y$

$\frac{3}{2}x + \frac{9}{2} = y$

$m = \frac{3}{2}$; y-int $= \left(0, \frac{9}{2}\right)$

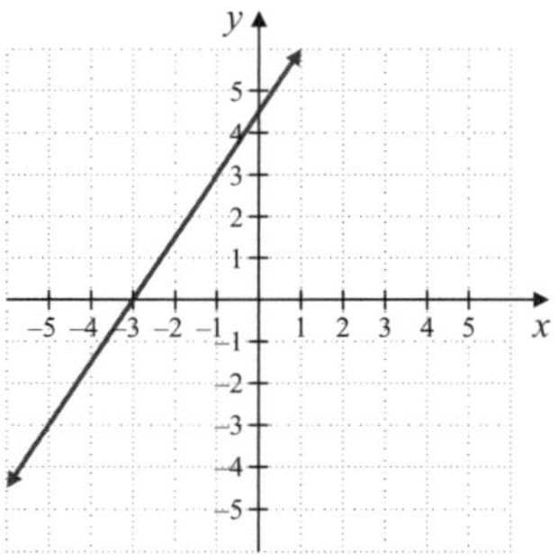

90. $x = -4y$

$-\frac{1}{4}x = y$

$m = -\frac{1}{4}$; y-int $= (0, 0)$

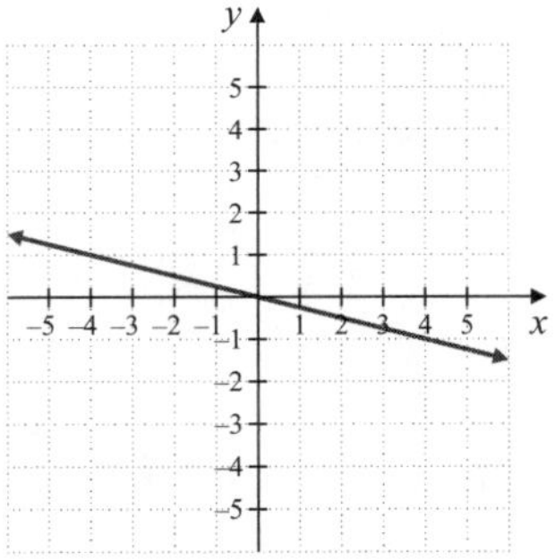

91. $3y \ge x$

Test (0, 1)

$3(1) \ge 0$

$3 \ge 0$ True

Shade the half-plane containing (0, 1).

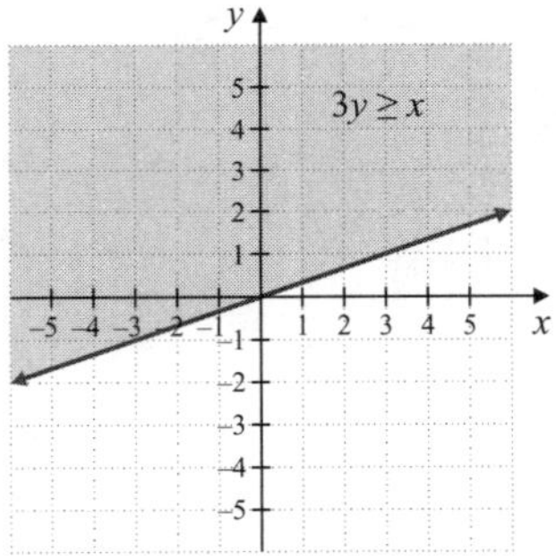

92. A vertical line has an undefined slope and the form of $x = c$. Since the line goes through the point (–2, –4), the equation of the line is $x = -2$.

93.
$$y - y_1 = m(x - x_1)$$
$$y - 5 = 0(x - 2)$$
$$y - 5 = 0$$
$$y = 5$$

94.
$$y - y_1 = m(x - x_1)$$
$$y - (-2) = 2(x - 5)$$
$$y + 2 = 2x - 10$$
$$y = 2x - 12$$

95. $m = \frac{-8-3}{-4-(-5)} = \frac{-11}{1} = -11$
$$y - y_1 = m(x - x_1)$$
$$y - 3 = -11(x - (-5))$$
$$y - 3 = -11x - 55$$
$$y = -11x - 52$$

96. $y = -\frac{3}{2}x + 1$

$m = -\frac{3}{2}$; y-int $= (0, 1)$

A parallel line will have a slope of $-\frac{3}{2}$.
$$y - y_1 = m(x - x_1)$$
$$y - (-2) = -\frac{3}{2}(x - (-4))$$
$$y + 2 = -\frac{3}{2}(x + 4)$$
$$y + 2 = -\frac{3}{2}x - 6$$
$$y = -\frac{3}{2}x - 8$$

97. $2x - 3y = 6$
$$2x - 6 = 3y$$
$$\frac{2}{3}x - 2 = y$$
A perpendicular line will have a slope of $-\frac{3}{2}$.
$$y - y_1 = m(x - x_1)$$
$$y - 5 = -\frac{3}{2}(x - (-4))$$
$$y - 5 = -\frac{3}{2}(x + 4)$$
$$y - 5 = -\frac{3}{2}x - 6$$
$$y = -\frac{3}{2}x - 1$$

98. Domain = $(-\infty, \infty)$

Range = $(-\infty, 0]$

99. Domain = $(-\infty, \infty)$
Range = $(-\infty, \infty)$

100. $f(x)=\begin{cases} x-2 & \text{if } x\le 0 \\ -\dfrac{x}{3} & \text{if } x\ge 3 \end{cases}$

Graph $y = x - 2$ for $x \le 0$, with a closed circle at $x = 0$. Graph $y=-\dfrac{x}{3}$ for $x \ge 3$, with a closed circle at $x = 3$.

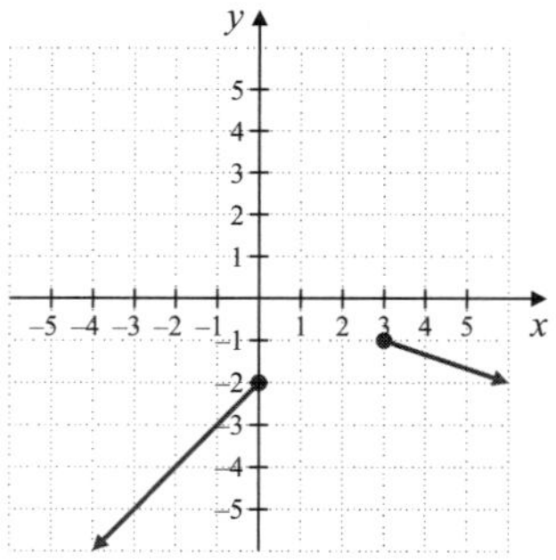

101. $g(x)=\begin{cases} 4x-3 & \text{if } x\le 1 \\ 2x & \text{if } x>1 \end{cases}$

Graph $y = 4x - 3$ for $x \le 1$, with a closed circle at $x = 1$. Graph $y = 2x$ for $x > 1$, with an open circle at $x = 1$.

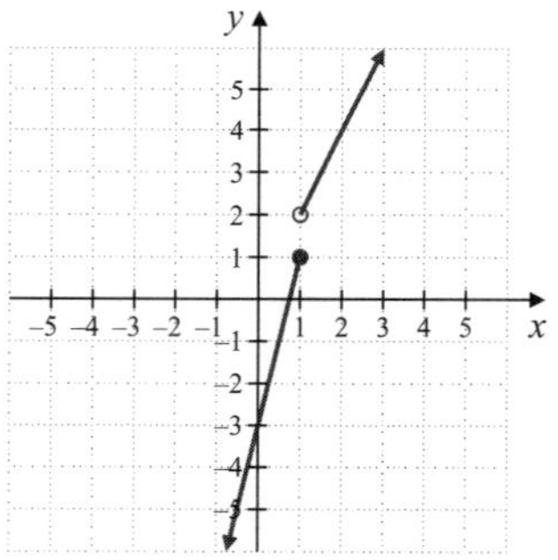

102. The graph $f(x)=\sqrt{x-2}$ is the graph of $y=\sqrt{x}$ shifted 2 units to the right.

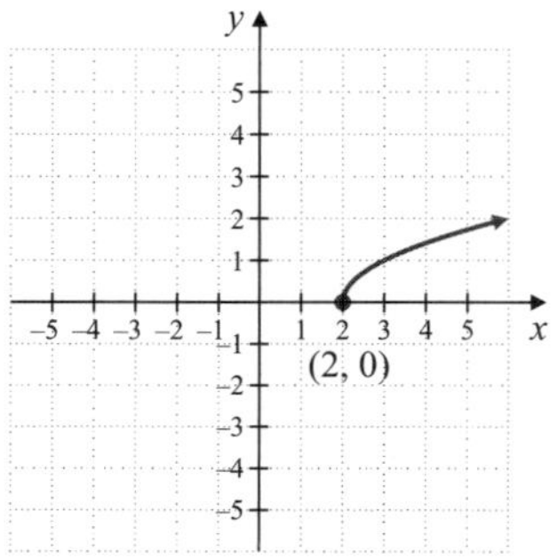

103. The graph if $f(x) = |x + 1| - 3$ in the graph of $y = |x|$ shifted 1 unit to the left and 3 units down.

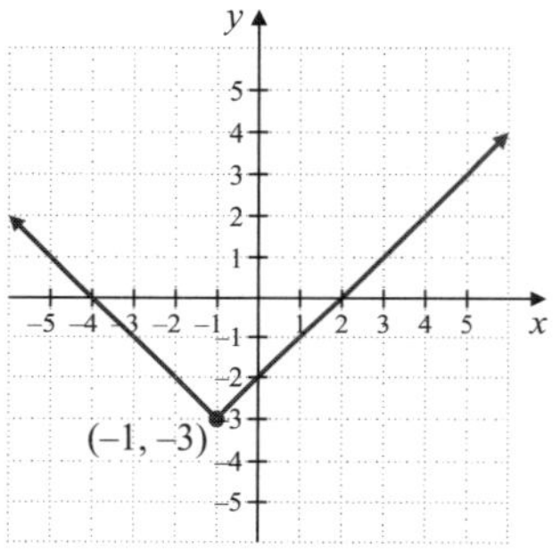

Chapter 3 Test

1.

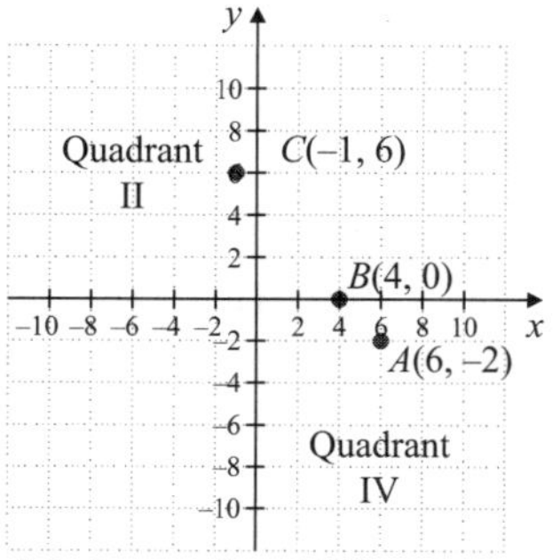

A is in quadrant IV.
B is on the *x*-axis, no quadrant.
C is in quadrant II.

2.

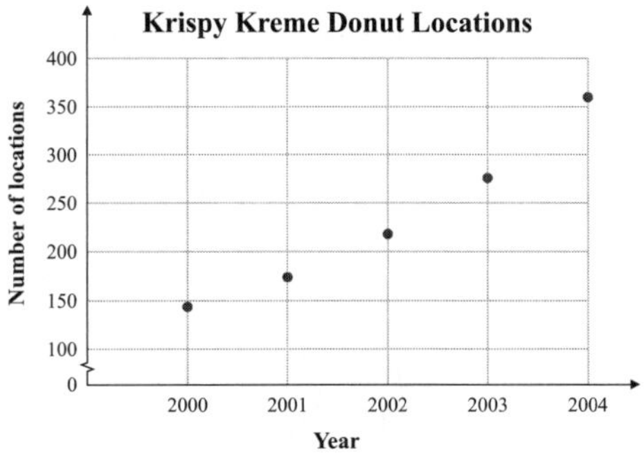

3. $-3x + y = -3$
Find three ordered pair solutions.

x	y
0	-3
1	0
2	3

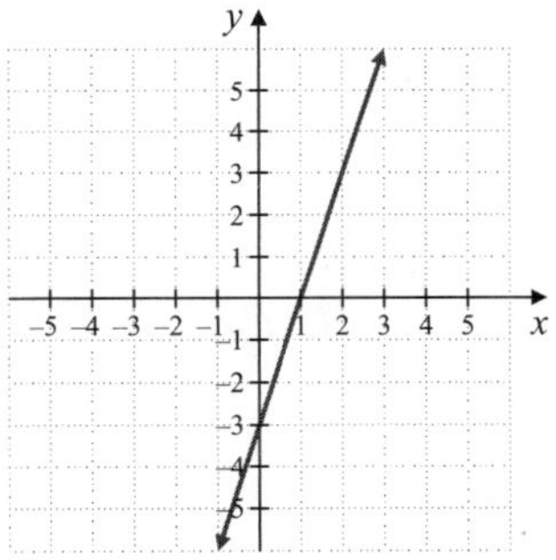

4. $2x - 3y = -6$
Find three ordered pair solutions.

x	y
-3	0
0	2
3	4

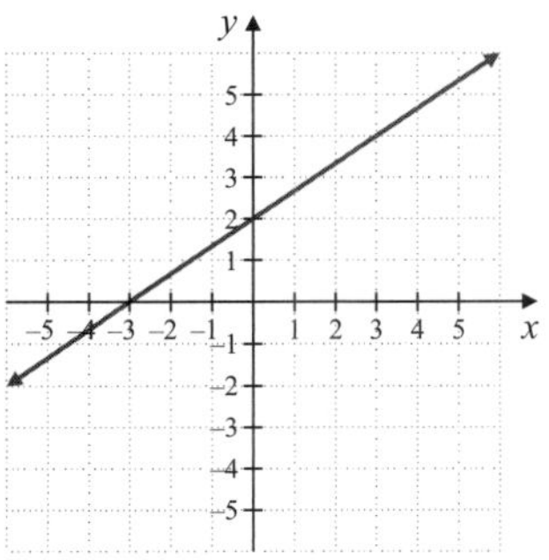

5. $4x + 6y = 7$
Find three ordered pair solutions.

x	y
-1	2
2	0
5	-2

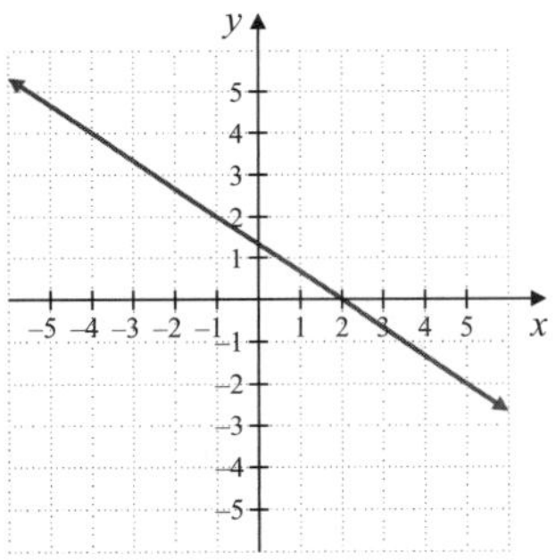

6. $y = -3$
This is a horizontal line with a y-intercept of $(0, -3)$ and no x-intercept.

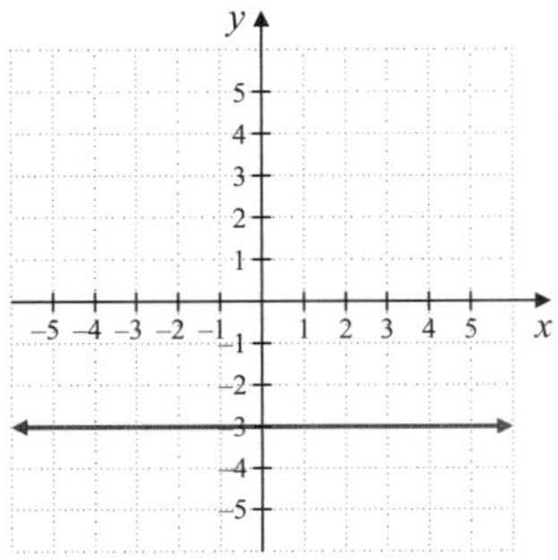

7. $m = \frac{10-(-8)}{-7-5} = \frac{18}{-12} = -\frac{3}{2}$

8. $3x + 12y = 8$
$12y = -3x + 8$
$y = -\frac{3}{12}x + \frac{8}{12}$
$y = -\frac{1}{4}x + \frac{2}{3}$
$m = -\frac{1}{4}$; $y\text{-int} = \left(0, \frac{2}{3}\right)$

9. $f(x) = 3x + 1$ or $y = 3x + 1$
$m = 3$; y-intercept = $(0, 1)$
The equation matches graph C.

10. $f(x) = 3x - 2$ or $y = 3x - 2$
$m = 3$; y-int = $(0, -2)$
The equation matches graph A.

11. $f(x) = 3x + 2$ or $y = 3x + 2$
$m = 3$; y-int = $(0, 2)$
The equation matches graph B.

12. $f(x) = 3x - 5$ or $y = 3x - 5$
$m = 3$; y-int = $(0, -5)$
The equation matches graph D.

13. A horizontal line has equation $y = b$. Since the line passes through $(2, -8)$, its equation is $y = -8$.

14. A vertical line has equation $x = a$. Since the line passes through $(-4, -3)$, its equation is $x = -4$.

15. A line perpendicular to $x = 5$ is a vertical line which has equation $y = b$. Since the line passes through $(3, -2)$, its equation is $y = -2$.

16. $y - y_1 = m(x - x_1)$
$y - (-1) = -3(x - 4)$
$y + 1 = -3x + 12$
$3x + y = 11$

17. $y = mx + b$
$y = 5x + (-2)$
$y = 5x - 2$
$2 = 5x - y$
$5x - y = 2$

18. $m = \frac{-3-(-2)}{6-4} = \frac{-1}{2}$
$y - y_1 = m(x - x_1)$
$y - (-2) = -\frac{1}{2}(x - 4)$
$y + 2 = -\frac{1}{2}x + 2$
$y = -\frac{1}{2}x$

19. $3x - y = 4$
$3x - 4 = y$
$m = 3$

A perpendicular line will have slope of $-\frac{1}{3}$.

$y - y_1 = m(x - x_1)$
$y - 2 = -\frac{1}{3}(x - (-1))$
$y - 2 = -\frac{1}{3}x - \frac{1}{3}$
$y = -\frac{1}{3}x + \frac{5}{3}$

20. $2y + x = 3$
$2y = -x + 3$
$y = -\frac{1}{2}x + 3$

$m = -\frac{1}{2}$

A parallel line will have slope of $-\frac{1}{2}$.

$$y - y_1 = m(x - x_1)$$
$$y - (-2) = -\frac{1}{2}(x - 3)$$
$$y + 2 = -\frac{1}{2}x + \frac{3}{2}$$
$$y = -\frac{1}{2}x - \frac{1}{2}$$

21. L_1: $2x - 5y = 8$
$$-5y = -2x + 8$$
$$y = \frac{2}{5}x - \frac{8}{5}$$

$$m = \frac{2}{5}$$

$$y\text{-int} = \left(0, -\frac{8}{5}\right)$$

L_2: $m = \frac{-1-4}{-1-1} = \frac{-5}{-2} = \frac{5}{2}$

Since the slopes of the lines are different, and the product of the slopes is not −1, the lines are neither parallel nor perpendicular.

22. a. $y = 0.232x + 63.953$
$y(60) = 0.232(60) + 63.953$
$y = 77.873$ or 78
78 games were won in the 2004 season.

b. $y = 0.232x + 63.95$
$y(93) = 0.232(93) + 63.953$
$y = 85.529$ or 86
86 games were won in the 2004 season.

c.
$$y = 0.232x + 63.953$$
$$95 = 0.232x + 63.953$$
$$31.047 = 0.232x$$
$$133.823 = x$$
A payroll of \$134 million would be necessary.

d. The slope is 0.232. Every million dollars spent on payroll increases winnings by 0.232 games.

23. $x \le -4$
Test (0, 0)
$0 \le -4$ False
Shade the half plane not containing (0, 0).

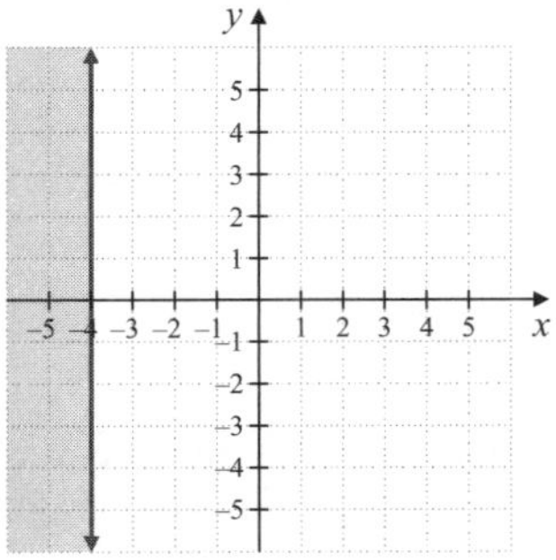

24. $2x - y > 5$
Test (0, 0)
$2(0) - 0 > 5$
$0 > 5$ False
Shade the half plane not containing (0, 0).

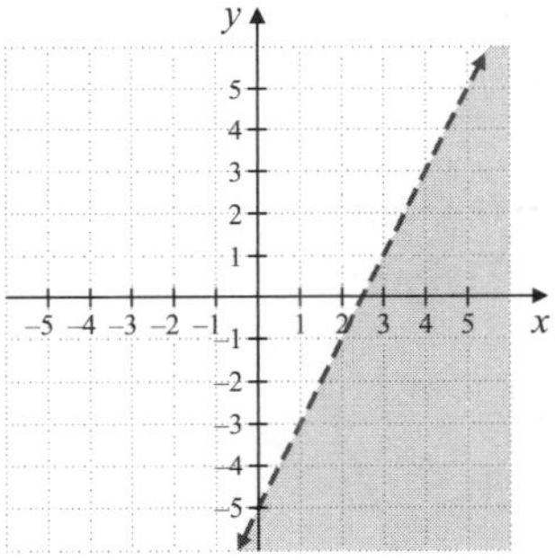

25. $2x + 4y < 6$ and $y \le -4$

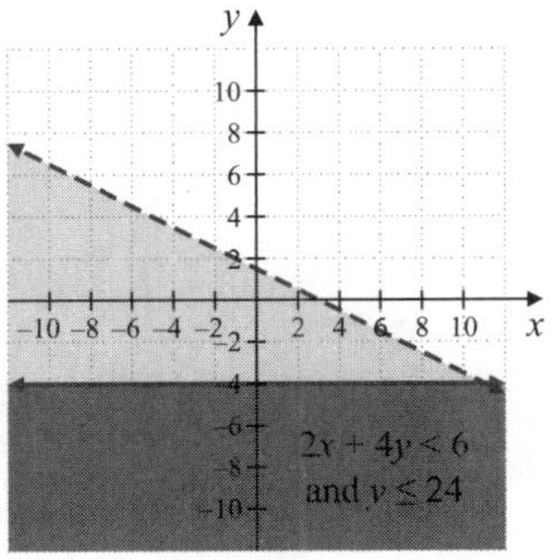

26. Domain = $(-\infty, \infty)$
Range = [5]
The relation is a function since it passes the vertical line test.

27. Domain = $\{-2\}$
Range = $(-\infty, \infty)$
The relation is not a function since it fails the vertical line test.

28. Domain = $(-\infty, \infty)$
Range = $[0, \infty)$
The relation is a function since it passes the vertical line test.

29. Domain = $(-\infty, \infty)$
Range = $(-\infty, \infty)$
Since no vertical line can intersect the graph more than once, the graph is the graph of a function.

30. $f(x) = \begin{cases} -\frac{1}{2}x & \text{if} \quad x \le 0 \\ 2x - 3 & \text{if} \quad x > 0 \end{cases}$

Domain = $(-\infty, \infty)$
Range = $(-3, \infty)$
Graph $y = -\frac{1}{2}x$ for $x \le 0$ with a closed circle at $x = 0$. Graph $y = 2x - 3$ for $x > 0$ with an open circle at $x = 0$.

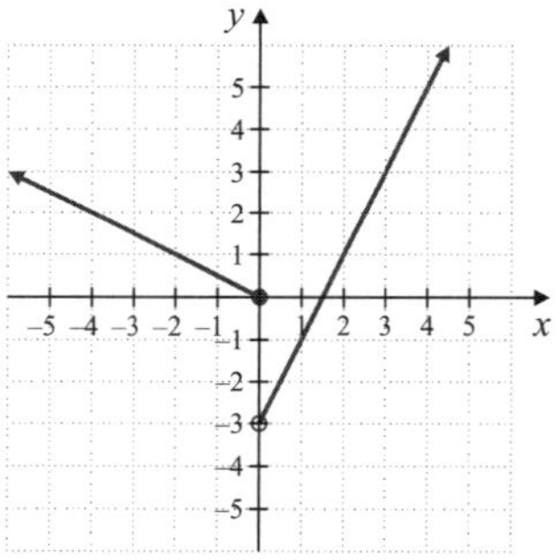

31. The graph of $f(x) = (x-4)^2$ is the graph of $y = x^2$ shifted 4 units to the right.

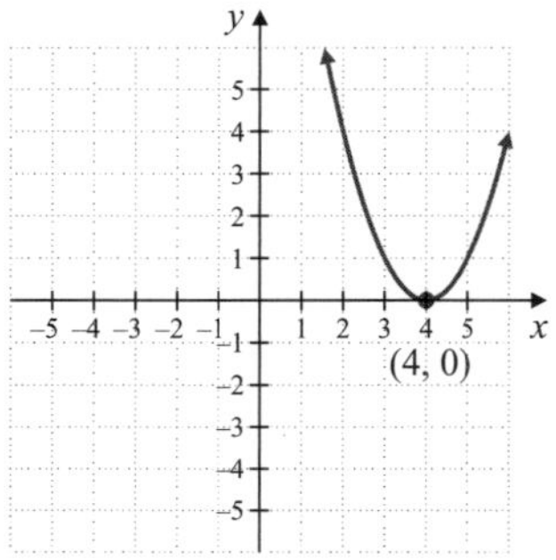

32. The graph of $g(x) = -|x + 2| - 1$ is the graph of $y = |x|$ reflected about the x-axis shifted 2 units to the left and 1 unit down.

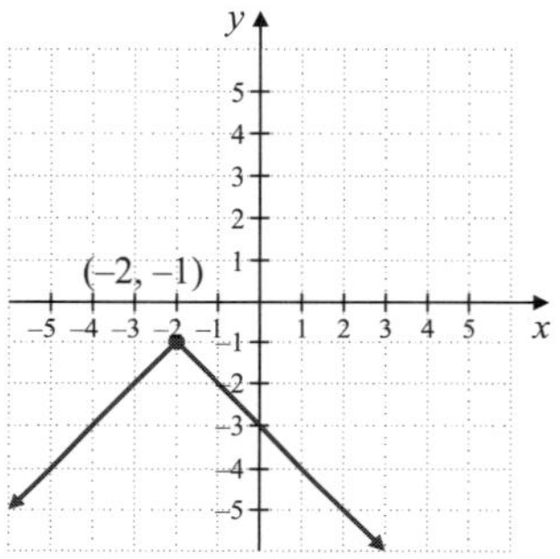

Domain = $(-\infty, \infty)$
Range = $(-\infty, -1]$

33. The graph of $h(x) = \sqrt{x} - 1$ is the graph of $y = \sqrt{x}$ shifted 1 unit down.

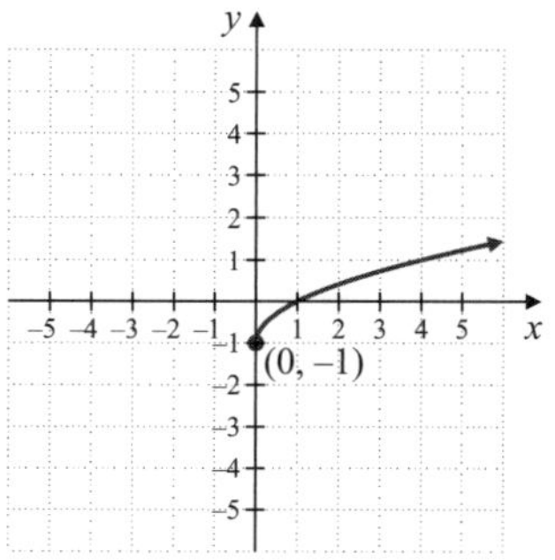

Cumulative Review Chapters 1–3

1. $\{2, 3, 4, 5\}$

2. $\{-1, 0, 1, 2\}$

3. $\{101, 102, 103, \ldots\}$

4. a. $\frac{-42}{-6} = 7$

b. $\frac{0}{14} = 0$

c. $-1(-5)(-2) = -10$

5. $-\frac{1}{9}$

6. $\frac{1}{10.2} = \frac{10}{102} = \frac{5}{51}$

7. $\frac{1}{\frac{7}{4}} = 1 \cdot \frac{4}{7} = \frac{4}{7}$

8. a. $2x^2 = 2(7)^2 = 2 \cdot 49 = 98$

b. $2x^2 = 2(-7)^2 = 2 \cdot 49 = 98$

9. $|3| = 3$

10. $\left|-\frac{7}{15}\right| = \frac{7}{15}$

11. $-|2.7| = -2.7$

12. $-|-102.86| = -102.86$

13.
$$\begin{aligned} 3-[6(4-6)+2(5-9)] &= 3-[6(-2)+2(-4)] \\ &= 3-(-12-8) \\ &= 3-(-20) \\ &= 3+20 \\ &= 23 \end{aligned}$$

14.
$$\begin{aligned} -2+3[5-(7-10)] &= -2+3[5-(-3)] \\ &= -2+3[8] \\ &= -2+24 \\ &= 22 \end{aligned}$$

15. $2^2 \cdot 2^5 = 2^{2+5} = 2^7$

16. $-7x^9 \cdot \frac{3}{14}x^4 = -\frac{21}{14}x^{9+4} = -\frac{3}{2}x^{13}$

17. $y \cdot y^2 \cdot y^4 = y^{1+2+4} = y^7$

18.
$$\begin{aligned} &(-2ab)(-0.3a^7b^{10})(8.1b^2) \\ &= 4.86a^{1+7}b^{1+10+2} \\ &= 4.86a^8b^{13} \end{aligned}$$

19. $(5x^2)^3 = 5^3 \cdot x^{2 \cdot 3} = 125x^6$

20. $(3^{-2}y^4)^2 = 3^{-4}y^{4 \cdot 2} = \frac{y^8}{81}$

21. $\left(\frac{2^{-3}}{y}\right)^{-2} = \frac{2^{-3 \cdot -2}}{y^{-2}} = 2^6 y^2 = 64y^2$

22. $\left(\frac{5a^3}{b^{-2}c^5}\right)^{-3} = \frac{5^{-3}a^{-9}}{b^{-2 \cdot -3}c^{5 \cdot -3}} = \frac{c^{15}}{125a^9b^6}$

23.
$$\begin{aligned} 0.6 &= 2-3.5c \\ -1.4 &= -3.5c \\ 0.4 &= c \end{aligned}$$
The solution set is $\{0.4\}$.

24.
$$\begin{aligned} 5(x-7) &= 4x-35+x \\ 5x-35 &= 5x-35 \\ 5x &= 5x \\ 0 &= 0 \end{aligned}$$
The solution set is $\{x|x \text{ is a real number}\}$.

25. Let x be the first number.

$$x+2x+3+4x=164$$
$$7x=161$$
$$x=23$$

$2x + 3 = 2(23) + 3 = 49$

$4x = 4(23) = 92$

The three numbers are 23, 49, 92.

26. Let x be the first number. Then $x + 2$ and $x + 4$ are the next consecutive odd integers.

$$x+x+2+x+4=213$$
$$3x+6=213$$
$$3x=207$$
$$x=69$$

$x + 2 = 69 + 2 = 71$

$x + 4 = 69 + 4 = 73$

The three consecutive odd integers are 69, 71, 73.

27. $A=P\left(1+\frac{r}{n}\right)^{nt}$

$$A=10,000\left(1+\frac{0.05}{4}\right)^{4\cdot 3}$$
$$A=10,000(1.0125)^{12}$$
$$A=10,000(1.1607545)$$
$$A=11,607.55$$

In 3 years she will have \$11,607.55 in the account.

28. **a.** $-4 + (-3) = -7$

b. $\frac{1}{2}-\left(-\frac{1}{3}\right)=\frac{1}{2}+\frac{1}{3}=\frac{3}{6}+\frac{2}{6}=\frac{5}{6}$

c. $7 - 20 = 7 + (-20) = -13$

29. $\{x|x \geq 2\}$

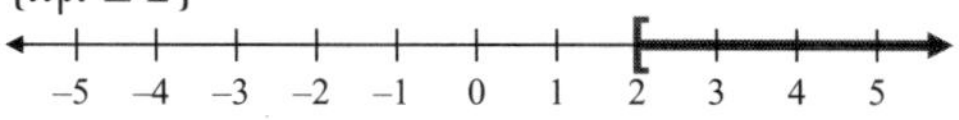

$[2, \infty)$

30. $\left\{x\middle|x<-\frac{3}{2}\right\}$

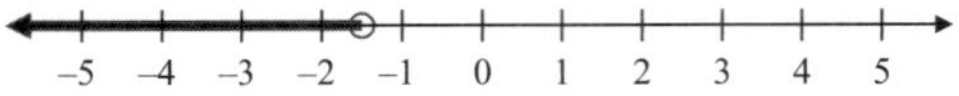

$\left(-\infty, -\frac{3}{2}\right)$

31. $\{x|0.5 < x \leq 3\}$

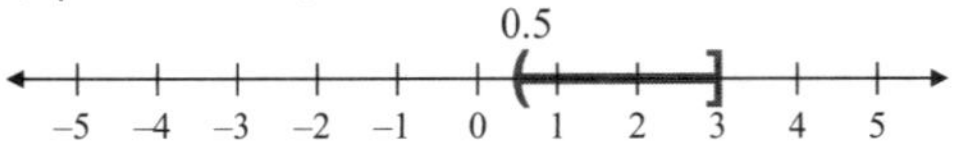

$(0.5, 3]$

32. $3x+10>\frac{5}{2}(x-1)$

$$3x+10>\frac{5}{2}x-\frac{5}{2}$$
$$6x+20>5x-5$$
$$x>-25$$

The solution set is $(-25, \infty)$.

33. $x-7<2$ and $2x+1<9$

$x<9$ and $2x<8$

$x<4$

The solution set is $(-\infty, 4)$.

34. $x-2<6$ or $3x+1>1$

$x<8$ or $3x>0$

$x>0$

The solution set is $(-\infty, \infty)$.

35. $|x-3|=|5-x|$

$$x-3=5-x$$
$$2x=8$$
$$x=4$$

The solution set is $\{4\}$.

36. $|5x-2|-7=-4$

$|5x-2|=3$

$5x-2=3$ or $5x-2=-3$

$5x=5$ $\quad$ $5x=-1$

$x=1$ $\quad$ $x=-\frac{1}{5}$

The solution set is $\left\{1, -\frac{1}{5}\right\}$.

37. $|x-3| > 7$

$x-3<-7$ or $x-3>7$

$x<-4$ or $x>10$

The solution set is $(-\infty, -4) \cup (10, \infty)$.

38. $|-x+8|-2 \le 8$

$|-x+8| \le 10$

$-10 \le -x+8 \le 10$

$-18 \le -x \le 2$

$-18 \ge x \ge 2$

The solution set is $[-2, 18]$.

39. $3x + 4y = -12$

Let $x = 0$.

$3x+4y=-12$

$3(0)+4y=-12$

$4y=-12$

$y=-3$

y-int: $(0, -3)$

Let $y = 0$.

$3x+4y=-12$

$3x+4(0)=-12$

$3x=-12$

$x=-4$

x-int: $(-4, 0)$

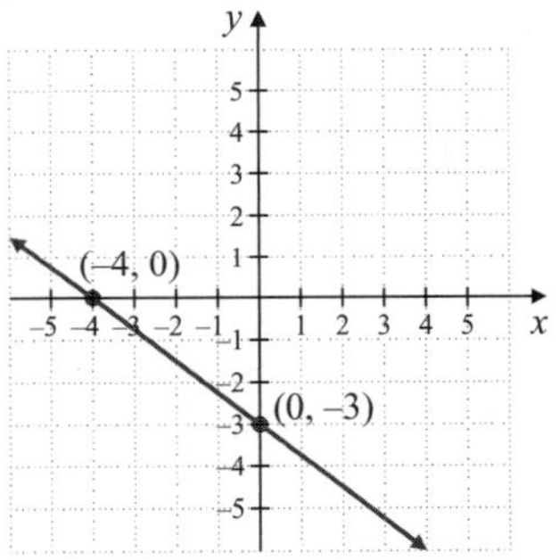

40. $7x+2y=10$

$2y=-7x+10$

$y=-\frac{7}{2}x+5$

$m=-\frac{7}{2}$

y-int = $(0, 5)$

41. $y = mx + b$

$y = 3x + 2$

$m = 3$

42. $m=\frac{y_2-y_1}{x_2-x_1}=\frac{9-6}{0-(-1)}=\frac{3}{1}=3$

43. $y = mx + b$

$y=\frac{2}{3}x-7$

44. $m=\frac{y_2-y_1}{x_2-x_1}=\frac{7-5}{-4-(-2)}=\frac{2}{-2}=-1$

$y-y_1=m(x-x_1)$

$y-7=-1(x-(-4))$

$y-7=-x-4$

$x+y=3$

45. A horizontal line with the form $y = b$ going through the point (2, 3) will have an equation $y = 3$.

46. A vertical line with the form $x = c$ going through the point $\left(-2, -\frac{3}{4}\right)$ will have an equation $x = -2$.

47. $f(x)=7x^2-3x+1$

$f(1)=7(1)^2-3(1)+1=7-3+1=5$

48. $f(x) = -2.3x - 6$

$y = -2.3x - 6$

$m = -2.3$

49. $g(x) = 3x - 2$

$g(0) = 3(0) - 2 = -2$

50. A vertical line can be drawn that intersects the graph more than once, so the graph is not the graph of a function.

Chapter 4

Section 4.1 Practice Problems

1. $\begin{cases} x - y = 3 \\ 2x - 3y = 5 \end{cases}$

$x - y = 3$
$4 - 1 \stackrel{?}{=} 3$
$3 = 3$ True
$2x - 3y = 5$
$2(4) - 3(1) \stackrel{?}{=} 5$
$8 - 3 \stackrel{?}{=} 5$
$5 = 5$ True

Yes, (4, 1) is a solution.

2. $\begin{cases} 3x - y = -12 \\ x - y = 0 \end{cases}$

$3x - y = -12$ $\quad$ $x - y = 0$
$3(-3) - 3 \stackrel{?}{=} -12$ $\quad$ $-3 - 3 \stackrel{?}{=} 0$
$-9 - 3 \stackrel{?}{=} -12$ $\quad$ $-6 = 0$ False
$-12 = -12$ True

No, (–3, 3) is not a solution.

3. $\begin{cases} x - y = 2 \\ x + 3y = 6 \end{cases}$

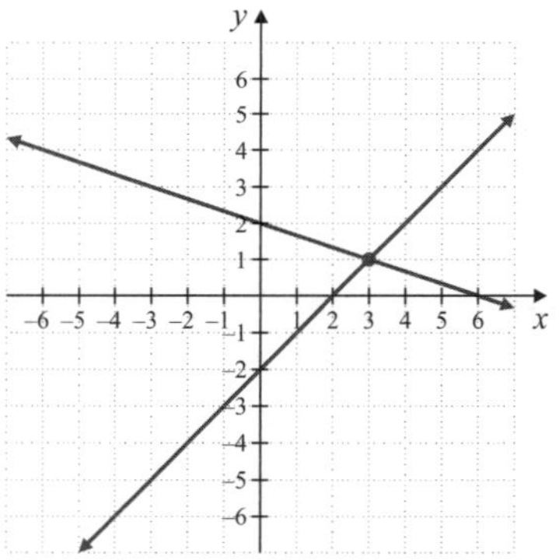

The solution of the system is (3, 1).

4. $\begin{cases} y = -3x \\ 6x + 2y = 4 \end{cases}$

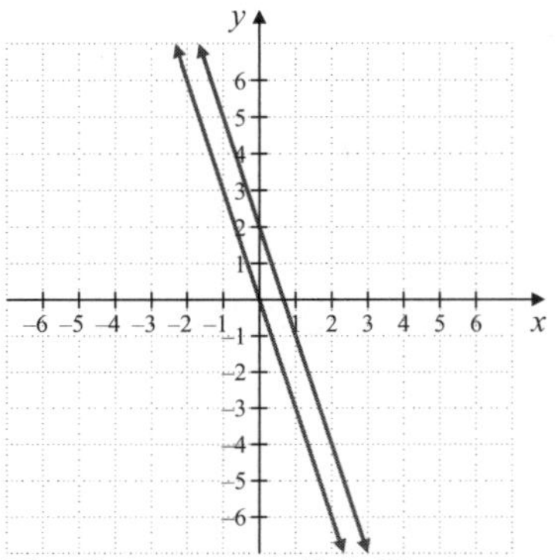

The lines are parallel. The system has no solution.

5. $\begin{cases} -2x + y = 1 \\ 4x - 2y = -2 \end{cases}$

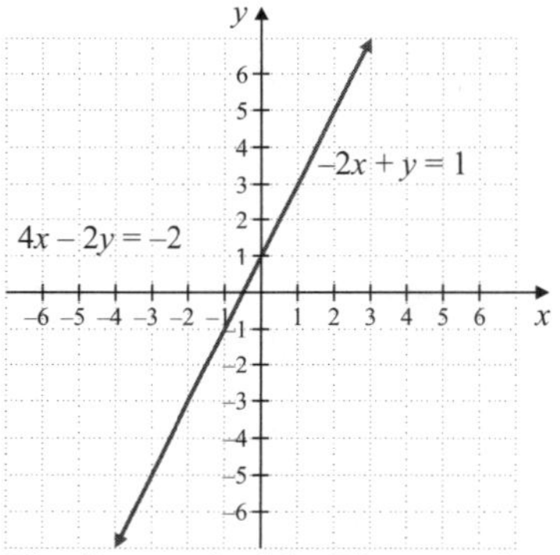

The lines coincide. The equations are dependent. The solution set is $\{(x, y) | -2x + y = 1\}$.

6. $\begin{cases} 6x - 4y = 10 \\ y = 3x - 3 \end{cases}$

Replace y with $3x - 3$ in the first equation.

$$6x - 4y = 10$$
$$6x - 4(3x - 3) = 10$$
$$6x - 12x + 12 = 10$$
$$-6x = -2$$
$$x = \frac{1}{3}$$

Replace x with $\frac{1}{3}$ in the second equation.

$$y = 3x - 3$$
$$y = 3\left(\frac{1}{3}\right) - 3$$
$$y = 1 - 3$$
$$y = -2$$

The solution of the system is $\left(\frac{1}{3}, -2\right)$.

7. $\begin{cases} -\frac{x}{2} + \frac{y}{4} = \frac{1}{2} \\ \frac{x}{2} + \frac{y}{2} = -\frac{1}{8} \end{cases}$

Multiply the first equation by 4 and the second equation by 8 to clear fractions.

$\begin{cases} -2x + y = 2 \\ 4x + 4y = -1 \end{cases}$

Solve the first equation for y.

$$-2x + y = 2$$
$$y = 2x + 2$$

Replace y with $2x + 2$ in the second equation.

$$4x + 4y = -1$$
$$4x + 4(2x + 2) = -1$$
$$4x + 8x + 8 = -1$$
$$12x = -9$$
$$x = -\frac{3}{4}$$

Replace x with $-\frac{3}{4}$ in the equation $y = 2x + 2$.

$$y = 2x + 2$$
$$y = 2\left(-\frac{3}{4}\right) + 2$$
$$y = -\frac{3}{2} + \frac{4}{2}$$
$$y = \frac{1}{2}$$

The solution of the system is $\left(-\frac{3}{4}, \frac{1}{2}\right)$.

8. $\begin{cases} 3x - y = 1 \\ 4x + y = 6 \end{cases}$

Add the equations.

$$\begin{array}{r} 3x - y = 1 \\ \underline{4x + y = 6} \\ 7x \quad = 7 \\ x = 1 \end{array}$$

Replace x with 1 in the first equation.

$$3x - y = 1$$
$$3(1) - y = 1$$
$$3 - y = 1$$
$$-y = -2$$
$$y = 2$$

The solution of the system is (1, 2).

9. $\begin{cases} \frac{x}{3} + 2y = -1 \\ x + 6y = 2 \end{cases}$

Multiply the first equation by −3.

$\begin{cases} -x - 6y = 3 \\ x + 6y = 2 \end{cases}$

Add the equations.

$$\begin{array}{r} -x - 6y = 3 \\ \underline{x + 6y = 2} \\ 0 = 5 \end{array}$$

The statement $0 = 5$ is false, so the system has no solution.

10. $\begin{cases} 2x - 5y = 6 \\ 3x - 4y = 9 \end{cases}$

Multiply the first equation by 3 and the second equation by −2.

$\begin{cases} 6x - 15y = 18 \\ -6x + 8y = -18 \end{cases}$

Add the equations.

$$\begin{array}{r} 6x - 15y = 18 \\ \underline{-6x + 8y = -18} \\ -7y = 0 \\ y = 0 \end{array}$$

Replace y with 0 in the first equation.

$$2x - 5y = 6$$
$$2x - 5(0) = 6$$
$$2x = 6$$
$$x = 3$$

The solution of the system is (3, 0).

11. $\begin{cases} 4x-7y=10 \\ -8x+14y=-20 \end{cases}$

Multiply the first equation by 2.

$\begin{cases} 8x-14y=20 \\ -8x+14y=-20 \end{cases}$

Add the equations.

$$\begin{array}{r} 8x-14y=20 \\ \underline{-8x+14y=-20} \\ 0=0 \end{array}$$

The statement 0 = 0 is true, so the system has infinitely many solutions. The solution set is $\{(x, y)|4x-7y=10\}$.

Calculator Explorations

1. $\begin{cases} y=-1.65x+3.65 \\ y=4.56x-9.44 \end{cases}$

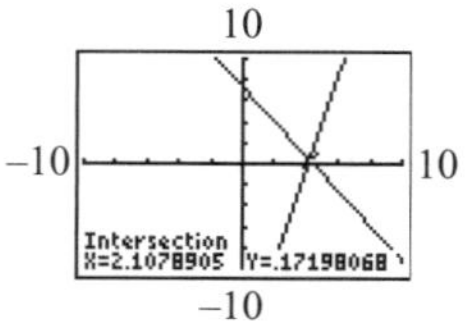

The solution of the system is approximately (2.11, 0.17).

2. $\begin{cases} y=7.61x+3.48 \\ y=-1.26x-6.43 \end{cases}$

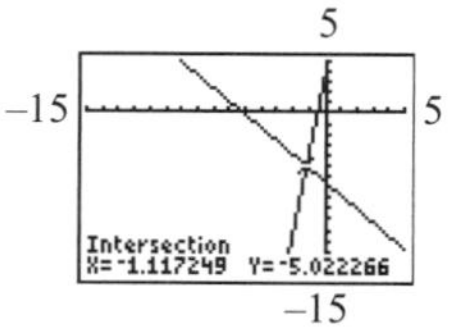

The solution of the system is approximately (−1.12, −5.02).

3. $\begin{cases} 2.33x-4.72y=10.61 \\ 5.86x-6.22y=-8.89 \end{cases}$

Solve each equation for *y*.

$\begin{cases} y=\dfrac{2.33x-10.61}{4.72} \\ y=\dfrac{5.86x+8.89}{6.22} \end{cases}$

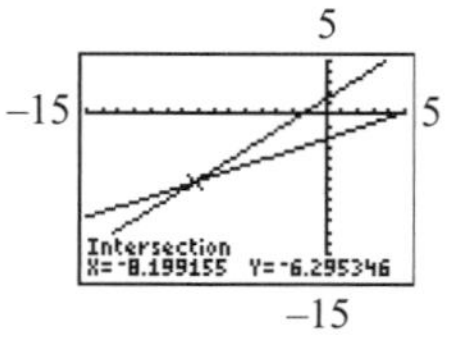

The solution of the system is approximately (−8.20, −6.30).

4. $\begin{cases} -7.89x-5.68y=3.26 \\ -3.65x+4.98y=11.77 \end{cases}$

Solve each equation for *y*.

$\begin{cases} y=\dfrac{-7.89x-3.26}{5.68} \\ y=\dfrac{3.65x+11.77}{4.98} \end{cases}$

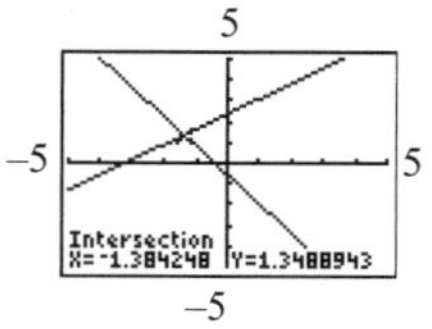

The solution of the system is approximately (−1.38, 1.35).

Mental Math

1. Choose the graph with parallel lines, B.
2. Choose the graph with coinciding lines, C.
3. Choose the graph with intersection (1, −2), A.
4. Choose the graph with intersection (−3, 0), D.

Exercise Set 4.1

1. $\begin{cases} x-y=3 \\ 2x-4y=8 \end{cases}$

$$\begin{array}{rl} x-y=3 & \\ 2-(-1)\stackrel{?}{=}3 & \\ 2+1\stackrel{?}{=}3 & \\ 3=3 & \text{True} \end{array} \qquad \begin{array}{rl} 2x-4y=8 & \\ 2(2)-4(-1)\stackrel{?}{=}8 & \\ 4+4\stackrel{?}{=}8 & \\ 8=8 & \text{True} \end{array}$$

Yes, (2, −1) is a solution of the system.

3. $\begin{cases} 2x-3y=-9 \\ 4x+2y=-2 \end{cases}$

$$\begin{aligned} 2x-3y &= -9 \\ 2(3)-3(5) &\stackrel{?}{=} -9 \\ 6-15 &\stackrel{?}{=} -9 \\ -9 &= -9 \quad \text{True} \\ 4x+2y &= -2 \\ 4(3)+2(5) &\stackrel{?}{=} -2 \\ 12+10 &\stackrel{?}{=} -2 \\ 22 &= -2 \quad \text{False} \end{aligned}$$

No, (3, 5) is not a solution of the system.

5. $\begin{cases} y=-5x \\ x=-2 \end{cases}$

$$\begin{aligned} y &= -5x \\ 10 &\stackrel{?}{=} -5(-2) \\ 10 &= 10 \quad \text{True} \end{aligned} \qquad \begin{aligned} x &= -2 \\ -2 &= -2 \quad \text{True} \end{aligned}$$

Yes, (−2, 10) is a solution of the system.

7. $\begin{cases} 3x+7y=-19 \\ -6x=5y+8 \end{cases}$

$$\begin{aligned} 3x+7y &= -19 \\ 3\left(\frac{2}{3}\right)+7(-3) &\stackrel{?}{=} -19 \\ 2+(-21) &\stackrel{?}{=} -19 \\ -19 &= -19 \quad \text{True} \\ -6x &= 5y+8 \\ -6\left(\frac{2}{3}\right) &\stackrel{?}{=} 5(-3)+8 \\ -4 &\stackrel{?}{=} -15+8 \\ -4 &= -7 \quad \text{False} \end{aligned}$$

No, $\left(\frac{2}{3}, -3\right)$ is not a solution of the system.

9. $\begin{cases} x+y=1 \\ x-2y=4 \end{cases}$

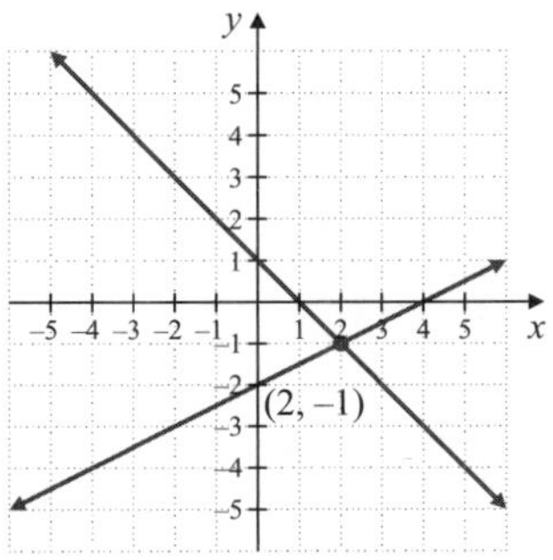

The solution of the system is (2, −1).

11. $\begin{cases} 2y-4=0 \\ x+2y=5 \end{cases}$

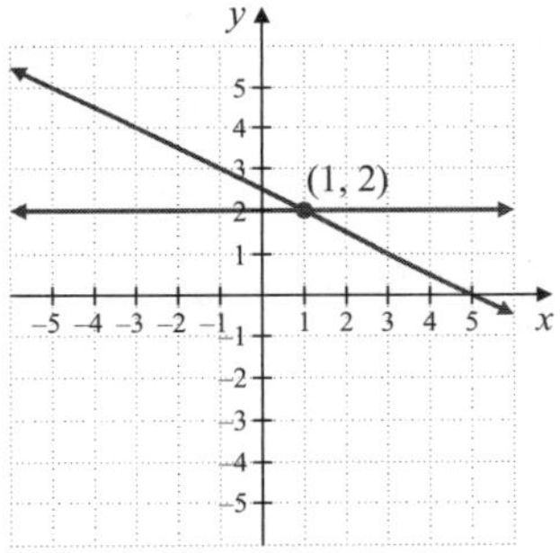

The solution of the system is (1, 2).

13. $\begin{cases} 3x-y=4 \\ 6x-2y=4 \end{cases}$

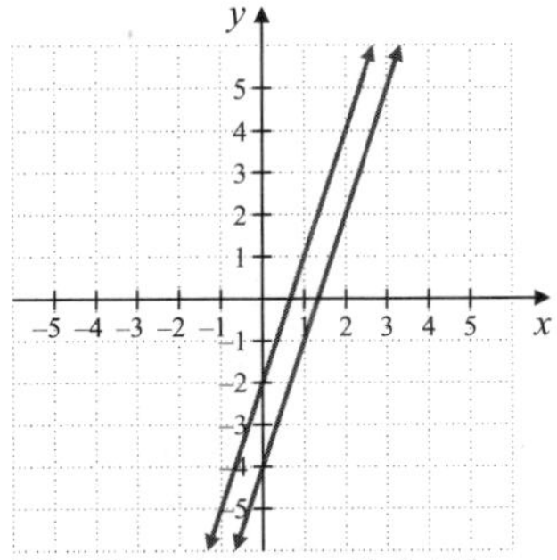

The lines are parallel. The system has no solution.

15. $\begin{cases} y = -3x \\ 2x - y = -5 \end{cases}$

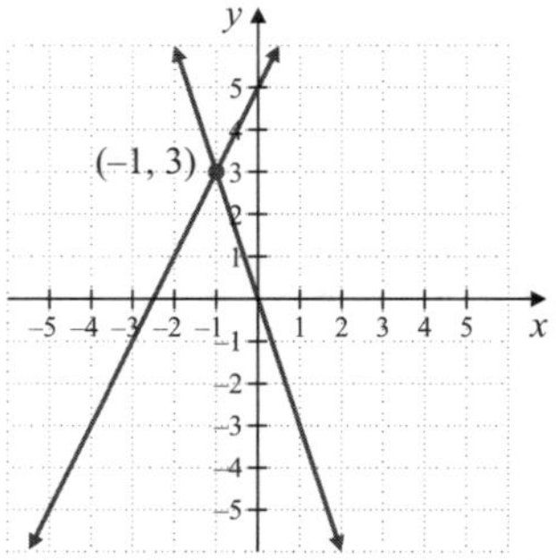

The solution of the system is (–1, 3).

17. $\begin{cases} x + y = 10 \\ y = 4x \end{cases}$

Replace y by $4x$ in the first equation.

$$x + y = 10$$
$$x + 4x = 10$$
$$5x = 10$$
$$x = 2$$

Replace x with 2 in the second equation.
$y = 4x = 4(2) = 8$
The solution of the system is (2, 8).

19. $\begin{cases} 4x - y = 9 \\ 2x + 3y = -27 \end{cases}$

Solve the first equation for y.

$$4x - y = 9$$
$$4x - 9 = y$$

Replace y with $4x - 9$ in the second equation.

$$2x + 3y = -27$$
$$2x + 3(4x - 9) = -27$$
$$2x + 12x - 27 = -27$$
$$14x = 0$$
$$x = 0$$

Replace x with 0 in the equation $y = 4x - 9$.
$y = 4x - 9 = 4(0) - 9 = 0 - 9 = -9$
The solution of the system is (0, –9).

21. $\begin{cases} \frac{1}{2}x + \frac{3}{4}y = -\frac{1}{4} \\ \frac{3}{4}x - \frac{1}{4}y = 1 \end{cases}$

Multiply both equations by 4 to clear the fractions.

$\begin{cases} 2x + 3y = -1 \\ 3x - y = 4 \end{cases}$

Solve the second equation for y.

$$3x - y = 4$$
$$-y = -3x + 4$$
$$y = 3x - 4$$

Replace y with $3x - 4$ in the (simplified) first equation.

$$2x + 3y = -1$$
$$2x + 3(3x - 4) = -1$$
$$2x + 9x - 12 = -1$$
$$11x = 11$$
$$x = 1$$

Replace x by 1 in the equation $y = 3x - 4$.
$y = 3x - 4 = 3(1) - 4 = 3 - 4 = -1$
The solution of the system is (1, –1).

23. $\begin{cases} x = -3y + 4 \\ 3x + 9y = 12 \end{cases}$

Replace x with $-3y + 4$ in the second equation.

$$3x + 9y = 12$$
$$3(-3y + 4) + 9y = 12$$
$$-9y + 12 + 9y = 12$$
$$0 = 0$$

The statement $0 = 0$ is true, so the system has infinitely many solutions. The solution set is $\{x|x = -3y + 4\}$.

25. $\begin{cases} 2x - 4y = 0 \\ x + 2y = 5 \end{cases}$

Multiply the second equation by 2.

$\begin{cases} 2x - 4y = 0 \\ 2x + 4y = 10 \end{cases}$

Add the equations.

$$\begin{array}{r} 2x-4y=0 \\ 2x+4y=10 \\ \hline 4x=10 \\ x=\frac{5}{2} \end{array}$$

Replace x with $\frac{5}{2}$ in the first equation.

$$\begin{aligned} 2x-4y&=0 \\ 2\left(\frac{5}{2}\right)-4y&=0 \\ 5-4y&=0 \\ 5&=4y \\ \frac{5}{4}&=y \end{aligned}$$

The solution of the system is $\left(\frac{5}{2}, \frac{5}{4}\right)$.

27. $\begin{cases} 5x+2y=1 \\ x-3y=7 \end{cases}$

Multiply the second equation by –5.

$\begin{cases} 5x+2y=1 \\ -5x+15y=-35 \end{cases}$

Add the equations.

$$\begin{array}{r} 5x+2y=1 \\ -5x+15y=-35 \\ \hline 17y=-34 \\ y=-2 \end{array}$$

Replace y with –2 in the second equation.

$$\begin{aligned} x-3y&=7 \\ x-3(-2)&=7 \\ x+6&=7 \\ x&=1 \end{aligned}$$

The solution of the system is (1, –2).

29. $\begin{cases} 5x-2y=27 \\ -3x+5y=18 \end{cases}$

Multiply the first equation by 3 and the second equation by 5.

$\begin{cases} 15x-6y=81 \\ -15x+25y=90 \end{cases}$

Add the equations.

$$\begin{array}{r} 15x-6y=81 \\ -15x+25y=90 \\ \hline 19y=171 \\ y=9 \end{array}$$

Replace y with 9 in the first equation.

$$\begin{aligned} 5x-2y&=27 \\ 5x-2(9)&=27 \\ 5x-18&=27 \\ 5x&=45 \\ x&=9 \end{aligned}$$

The solution of the system is (9, 9).

31. $\begin{cases} 3x=5y+11 \\ 2x=6y+2 \end{cases}$

Multiply the first equation by –6 and the second equation by 5.

$\begin{cases} -18x=-30y-66 \\ 10x=30y+10 \end{cases}$

Add the equations.

$$\begin{array}{r} -18x=-30y-66 \\ 10x=30y+10 \\ \hline -8x=-56 \\ x=7 \end{array}$$

Replace x with 7 in the first equation.

$$\begin{aligned} 3x&=5y+11 \\ 3(7)&=5y+11 \\ 21&=5y+11 \\ 10&=5y \\ 2&=y \end{aligned}$$

The solution of the system is (7, 2).

33. $\begin{cases} x-2y=4 \\ 2x-4y=4 \end{cases}$

Multiply the first equation by –2.

$\begin{cases} -2x+4y=-8 \\ 2x-4y=4 \end{cases}$

Add the equations.

$$\begin{array}{r} -2x+4y=-8 \\ 2x-4y=4 \\ \hline 0=-4 \end{array}$$

The statement 0 = –4 is false, so the system has no solution.

35. $\begin{cases} 3x + y = 1 \\ 2y = 2 - 6x \end{cases}$

Multiply the first equation by −2 and rearrange the second equation.

$\begin{cases} -6x - 2y = -2 \\ 6x + 2y = 2 \end{cases}$

Add the equations.

$$\begin{array}{r} -6x - 2y = -2 \\ 6x + 2y = 2 \\ \hline 0 = 0 \end{array}$$

The statement 0 = 0 is true, so the system has infinitely many solutions. The solution set is $\{(x, y) | 3x + y = 1\}$.

37. $\begin{cases} 2x + 5y = 8 \\ 6x + y = 10 \end{cases}$

Multiply the first equation by −3, then add the resulting equations.

$$\begin{array}{r} -6x - 15y = -24 \\ 6x + y = 10 \\ \hline -14y = -14 \\ y = 1 \end{array}$$

Replace y with 1 in the second equation.

$$\begin{aligned} 6x + y &= 10 \\ 6x + 1 &= 10 \\ 6x &= 9 \\ x &= \frac{3}{2} \end{aligned}$$

The solution of the system is $\left(\frac{3}{2}, 1\right)$.

39. $\begin{cases} x + y = 1 \\ x - 2y = 4 \end{cases}$

Multiply the first equation by −1, then add the resulting equations.

$$\begin{array}{r} -x - y = -1 \\ x - 2y = 4 \\ \hline -3y = 3 \\ y = -1 \end{array}$$

Replace y with −1 in the first equation.

$$\begin{aligned} x + y &= 1 \\ x - 1 &= 1 \\ x &= 2 \end{aligned}$$

The solution of the system is (2, −1).

41. $\begin{cases} \frac{1}{3}x + y = \frac{4}{3} \\ -\frac{1}{4}x - \frac{1}{2}y = -\frac{1}{4} \end{cases}$

Multiply the first equation by 3 and the second equation by 4 to clear fractions, then add the resulting equations.

$$\begin{array}{r} x + 3y = 4 \\ -x - 2y = -1 \\ \hline y = 3 \end{array}$$

Replace y with 3 in the equation $x + 3y = 4$.

$$\begin{aligned} x + 3y &= 4 \\ x + 3(3) &= 4 \\ x + 9 &= 4 \\ x &= -5 \end{aligned}$$

The solution of the system is (−5, 3).

43. $\begin{cases} 4x + 2y = 5 \\ 2x + y = -1 \end{cases}$

Multiply the second equation by −2, then add the resulting equations.

$$\begin{array}{r} 4x + 2y = 5 \\ -4x - 2y = 2 \\ \hline 0 = 7 \end{array}$$

The statement 0 = 7 is false, so the system has no solution.

45. $\begin{cases} 10y - 2x = 1 \\ 5y = 4 - 6x \end{cases}$

$\begin{cases} 10y - 2x = 1 \\ 5y + 6x = 4 \end{cases}$

Multiply the first equation by 3, then add the resulting equations.

$$\begin{array}{r} 30y - 6x = 3 \\ 5y + 6x = 4 \\ \hline 35y \quad\quad = 7 \\ y = \frac{1}{5} \end{array}$$

Replace y with $\frac{1}{5}$ in the first equation.

$$10y - 2x = 1$$
$$10\left(\frac{1}{5}\right) - 2x = 1$$
$$2 - 2x = 1$$
$$-2x = -1$$
$$x = \frac{1}{2}$$

The solution of the system is $\left(\frac{1}{2}, \frac{1}{5}\right)$.

47. $\begin{cases} \frac{3}{4}x + \frac{5}{2}y = 11 \\ \frac{1}{16}x - \frac{3}{4}y = -1 \end{cases}$

Multiply the first equation by 4 and the second equation by 16 to clear fractions.

$\begin{cases} 3x + 10y = 44 \\ x - 12y = -16 \end{cases}$

Solve the second equation for x.

$$x - 12y = -16$$
$$x = 12y - 16$$

Replace x with $12y - 16$ in the first equation.

$$3x + 10y = 44$$
$$3(12y - 16) + 10y = 44$$
$$36y - 48 + 10y = 44$$
$$46y = 92$$
$$y = 2$$

Replace y with 2 in the equation $x = 12y - 16$.

$x = 12y - 16 = 12(2) - 16 = 24 - 16 = 8$

The solution of the system is (8, 2).

49. $\begin{cases} x = 3y + 2 \\ 5x - 15y = 10 \end{cases}$

Replace x by $3y + 2$ in the second equation.

$$5x - 15y = 10$$
$$5(3y + 2) - 15y = 10$$
$$15y + 10 - 15y = 10$$
$$10 = 10$$

The statement 10 = 10 is true, so the system has infinitely many solutions. The solution set is $\{(x, y)|x = 3y + 2\}$.

51. $\begin{cases} \frac{x}{3} + y = \frac{4}{3} \\ -x + 2y = 11 \end{cases}$

Multiply the first equation by 3 to clear fractions, then add the resulting equations.

$$\begin{array}{r} x + 3y = 4 \\ \underline{-x + 2y = 11} \\ 5y = 15 \\ y = 3 \end{array}$$

Replace y with 3 in the second equation.

$$-x + 2y = 11$$
$$-x + 2(3) = 11$$
$$-x + 6 = 11$$
$$-x = 5$$
$$x = -5$$

The solution of the system is (−5, 3).

53. $\begin{cases} 2x = 6 \\ y = 5 - x \end{cases}$

Solve the first equation for x.

$$2x = 6$$
$$x = 3$$

Replace x with 3 in the second equation.

$y = 5 - x = 5 - 3 = 2$

The solution of the system is (3, 2).

55. $\begin{cases} \frac{x+5}{2} = \frac{6-4y}{3} \\ \frac{3x}{5} = \frac{21-7y}{10} \end{cases}$

Multiply the first equation by 6 and the second equation by 10 to clear fractions.

$\begin{cases} 3x + 15 = 12 - 8y \\ 6x = 21 - 7y \end{cases}$

$\begin{cases} 3x + 8y = -3 \\ 6x + 7y = 21 \end{cases}$

Multiply the first equation by −2, then add the resulting equations.

$$\begin{array}{r} -6x - 16y = 6 \\ \underline{6x + 7y = 21} \\ -9y = 27 \\ y = -3 \end{array}$$

Replace y with −3 in the equation $3x + 8y = -3$.

$$\begin{aligned} 3x+8y&=-3\\ 3x+8(-3)&=-3\\ 3x-24&=-3\\ 3x&=21\\ x&=7 \end{aligned}$$

The solution of the system is (7, –3).

57. $\begin{cases} 4x-7y=7 \\ 12x-21y=24 \end{cases}$

Multiply the first equation by –3, then add the resulting equations.

$$\begin{aligned} -12x+21y&=-21\\ 12x-21y&=24\\ \hline 0&=3 \end{aligned}$$

The statement 0 = 3 is false, so the system has no solution.

59. $\begin{cases} \frac{2}{3}x-\frac{3}{4}y=-1 \\ -\frac{1}{6}x+\frac{3}{8}y=1 \end{cases}$

Multiply the first equation by 12 and the second equation by 24 to clear fractions, then add the resulting equations.

$$\begin{aligned} 8x-9y&=-12\\ -4x+9y&=24\\ \hline 4x\quad&=12\\ x&=3 \end{aligned}$$

Replace x with 3 in the equation $8x - 9y = -12$.

$$\begin{aligned} 8x-9y&=-12\\ 8(3)-9y&=-12\\ 24-9y&=-12\\ -9y&=-36\\ y&=4 \end{aligned}$$

The solution of the system is (3, 4).

61. $\begin{cases} 2x-y=-1 \\ y=-2x \end{cases}$

Replace y by $-2x$ in the first equation:

$$\begin{aligned} 2x-y&=-1\\ 2x-(-2x)&=-1\\ 4x&=-1\\ x&=-\frac{1}{4} \end{aligned}$$

Replace x with $-\frac{1}{4}$ in the second equation.

$$y=-2x=-2\left(-\frac{1}{4}\right)=\frac{1}{2}$$

The solution of the system is $\left(-\frac{1}{4}, \frac{1}{2}\right)$.

63. $\begin{cases} 0.7x-0.2y=-1.6 \\ 0.2x-y=-1.4 \end{cases}$

Solve the second equation for y.

$$\begin{aligned} 0.2x-y&=-1.4\\ y&=0.2x+1.4 \end{aligned}$$

Replace y with $0.2x + 1.4$ in the first equation.

$$\begin{aligned} 0.7x-0.2y&=-1.6\\ 0.7x-0.2(0.2x+1.4)&=-1.6\\ 0.7x-0.04x-0.28&=-1.6\\ 0.66x&=-1.32\\ x&=-2 \end{aligned}$$

Replace x with –2 in the equation $y = 0.2x + 1.4$.

$$\begin{aligned} y&=0.2x+1.4\\ &=0.2(-2)+1.4\\ &=-0.4+1.4\\ &=1 \end{aligned}$$

The solution of the system is (–2, 1).

65. $\begin{cases} 4x-1.5y=10.2 \\ 2x+7.8y=-25.68 \end{cases}$

Multiply the second equation by –2 and add the resulting equations.

$$\begin{aligned} 4x-1.5y&=10.2\\ -4x-15.6y&=51.36\\ \hline -17.1y&=61.56\\ y&=-3.6 \end{aligned}$$

Replace y with –3.6 in the first equation.

$$\begin{aligned} 4x-1.5y&=10.2\\ 4x-1.5(-3.6)&=10.2\\ 4x+5.4&=10.2\\ 4x&=4.8\\ x&=1.2 \end{aligned}$$

The solution of the system is (1.2, –3.6).

67. Replace x with 1, y with 2, and z with 5.
$$\begin{aligned} 3x-4y+2z &= 5 \\ 3(1)-4(2)+2(5) &\stackrel{?}{=} 5 \\ 3-8+10 &\stackrel{?}{=} 5 \\ 5 &= 5 \quad \text{True} \end{aligned}$$
The given replacement values make the equation true.

69. Replace x with 0, y with -1, and z with 5.
$$\begin{aligned} -x-5y+3z &= 15 \\ -0-5(-1)+3(5) &\stackrel{?}{=} 15 \\ 5+15 &\stackrel{?}{=} 15 \\ 20 &= 15 \quad \text{False} \end{aligned}$$
The given replacement values make the equation false.

71.
$$\begin{array}{r} 3x+2y-5z=10 \\ \underline{-3x+4y+z=15} \\ 0x+6y-4z=25 \\ 6y-4z=25 \end{array}$$

73.
$$\begin{array}{r} 10x+5y+6z=14 \\ \underline{-9x+5y-6z=-12} \\ x+10y+0z=2 \\ x+10y=2 \end{array}$$

75. $\begin{cases} y=2x-5 \\ y=2x+1 \end{cases}$

The lines have the same slope and different y-intercepts, so they are parallel. There is no solution.

77. $\begin{cases} x+y=3 \\ 5x+5y=15 \end{cases}$

The result of multiplying the first equation by 5 is $5x + 5y = 15$, which is the second equation. The system is dependent, so it has an infinite number of solutions.

79. No; two lines cannot intersect at exactly two points.

81. The lines intersect at the point (5, 21), which corresponds to 5000 DVDs at a price of \$21 per DVD.

83. The graph for supply is higher than that for demand when x is greater than 6. Therefore, supply is greater than demand.

85. $\begin{cases} y=2.5x \\ y=0.9x+3000 \end{cases}$

Replace y with $2.5x$ in the second equation.
$$\begin{aligned} y &= 0.9x+3000 \\ 2.5x &= 0.9x+3000 \\ 1.6x &= 3000 \\ x &= 1875 \end{aligned}$$
Replace x with 1875 in the first equation.
$y = 2.5x = 2.5(1875) = 4687.5$
The break-even point is (1875, 4687.5).

87. At $x = 2000$, the graph for revenue is higher than that for cost, so the company makes money.

89. The revenue graph is above the cost graph to the right of the intersection point. Thus, for x-values greater than 1875, the company will make a profit.

91. answers may vary; one possibility:
$\begin{cases} -2x+y=1 \\ x-2y=-8 \end{cases}$

93. **a.** The slope of the red meat equation is negative, so the consumption of red meat is decreasing. The slope of the poultry line is positive, so the consumption of poultry is increasing.

b. $\begin{cases} y=-0.56x+114 \\ y=0.98x+67 \end{cases}$

Replace y with $-0.56x + 114$ in the second equation.
$$\begin{aligned} y &= 0.98x+67 \\ -0.56x+114 &= 0.98x+67 \\ -0.56x+47 &= 0.98x \\ 47 &= 1.54x \\ 30.5 &\approx x \end{aligned}$$
Replace x with 30.5 in the first equation.

$$\begin{aligned} y &= -0.56x + 114 \\ &= -0.56(30.5) + 114 \\ &\approx 97 \end{aligned}$$

The solution of the system is approximately (31, 97).

c. The meaning of the intersection point (31, 97) is that 31 years after 1997, or in 2028, the red meat and poultry consumption will both be about 97 pounds per person per year.

95. $\begin{cases} \frac{1}{x} + y = 12 \\ \frac{3}{x} - y = 4 \end{cases}$

Let $a = \frac{1}{x}$ to get the system:

$\begin{cases} a + y = 12 \\ 3a - y = 4 \end{cases}$

Add the equations.

$$\begin{array}{r} a + y = 12 \\ 3a - y = 4 \\ \hline 4a \quad = 16 \\ a = 4 \end{array}$$

Replace a with 4 in $a + y = 12$.

$$\begin{aligned} a + y &= 12 \\ 4 + y &= 12 \\ y &= 8 \end{aligned}$$

Since $a = \frac{1}{x}$, then $\frac{1}{x} = 4$, and $x = \frac{1}{4}$. The solution of the system is $\left(\frac{1}{4}, 8\right)$.

97. $\begin{cases} \frac{1}{x} + \frac{1}{y} = 5 \\ \frac{1}{x} - \frac{1}{y} = 1 \end{cases}$

Let $a = \frac{1}{x}$ and $b = \frac{1}{y}$ to get the system:

$\begin{cases} a + b = 5 \\ a - b = 1 \end{cases}$

Add the equations.

$$\begin{array}{r} a + b = 5 \\ a - b = 1 \\ \hline 2a \quad = 6 \\ a = 3 \end{array}$$

Replace a with 3 in $a + b = 5$.

$$\begin{aligned} a + b &= 5 \\ 3 + b &= 5 \\ b &= 2 \end{aligned}$$

Since $a = \frac{1}{x}$, then $\frac{1}{x} = 3$, and $x = \frac{1}{3}$.

Since $b = \frac{1}{y}$, then $\frac{1}{y} = 2$, and $y = \frac{1}{2}$.

The solution of the system is $\left(\frac{1}{3}, \frac{1}{2}\right)$.

99. $\begin{cases} \frac{2}{x} + \frac{3}{y} = -1 \\ \frac{3}{x} - \frac{2}{y} = 18 \end{cases}$

Let $a = \frac{1}{x}$ and $b = \frac{1}{y}$ to get the system:

$\begin{cases} 2a + 3b = -1 \\ 3a - 2b = 18 \end{cases}$

Multiply the first equation by 2 and the second equation by 3, and add the resulting equations.

$$\begin{array}{r} 4a + 6b = -2 \\ 9a - 6b = 54 \\ \hline 13a \quad = 52 \\ a = 4 \end{array}$$

Replace a with 4 in the equation $2a + 3b = -1$.

$$\begin{aligned} 2a + 3b &= -1 \\ 2(4) + 3b &= -1 \\ 8 + 3b &= -1 \\ 3b &= -9 \\ b &= -3 \end{aligned}$$

Since $a = \frac{1}{x}$, then $\frac{1}{x} = 4$, and $x = \frac{1}{4}$. Since

$b = \frac{1}{y}$, then $\frac{1}{y} = -3$, and $y = -\frac{1}{3}$.

The solution of the system is $\left(\frac{1}{4}, -\frac{1}{3}\right)$.

101. $\begin{cases} \frac{2}{x} - \frac{4}{y} = 5 \\ \frac{1}{x} - \frac{2}{y} = \frac{3}{2} \end{cases}$

Let $a = \frac{1}{x}$ and $b = \frac{1}{y}$ to get the system:

$\begin{cases} 2a - 4b = 5 \\ a - 2b = \frac{3}{2} \end{cases}$

Multiply the second equation by –2 and add the resulting equations.

$$\begin{array}{r} 2a - 4b = 5 \\ \underline{-2a + 4b = -3} \\ 0 = 2 \end{array}$$

The statement 0 = 2 is false, so the system has no solution.

Section 4.2 Practice Problems

1. $\begin{cases} 2x - y + 3z = 13 & (1) \\ x + y - z = -2 & (2) \\ 3x + 2y + 2z = 13 & (3) \end{cases}$

Add equation (1) and (2) to eliminate y.

$$\begin{array}{l} 2x - y + 3z = 13 \\ \underline{\ \ x + y - z = -2} \\ 3x \quad + 2z = 11 \quad (4) \end{array}$$

Multiply equation (1) by 2 and add it to equation (3) to eliminate y.

$$\begin{array}{l} 4x - 2y + 6z = 26 \\ \underline{3x + 2y + 2z = 13} \\ 7x \quad + 8z = 39 \quad (5) \end{array}$$

Multiply equation (4) by –4 and add it to equation (5).

$$\begin{array}{r} -12x - 8z = -44 \\ \underline{7x + 8z = 39} \\ -5x \quad = -5 \\ x = 1 \end{array}$$

Replace x with 1 in equation (4).

$$\begin{aligned} 3x + 2z &= 11 \\ 3(1) + 2z &= 11 \\ 3 + 2z &= 11 \\ 2z &= 8 \\ z &= 4 \end{aligned}$$

Replace x with 1 and z with 4 in equation (2).

$$\begin{aligned} x + y - z &= -2 \\ 1 + y - 4 &= -2 \\ y - 3 &= -2 \\ y &= 1 \end{aligned}$$

The solution of the system is (1, 1, 4).

2. $\begin{cases} 2x + 4y - 2z = 3 & (1) \\ -x + y - z = 6 & (2) \\ x + 2y - z = 1 & (3) \end{cases}$

Multiply equation (2) by 2 and add it to equation (1) to eliminate x.

$$\begin{array}{r} 2x + 4y - 2z = 3 \\ \underline{-2x + 2y - 2z = 12} \\ 6y - 4z = 15 \quad (4) \end{array}$$

Add equations (2) and (3) to eliminate x.

$$\begin{array}{r} -x + y - z = 6 \\ \underline{x + 2y - z = 1} \\ 3y - 2z = 7 \quad (5) \end{array}$$

Multiply equation (5) by –2 and add it to equation (4).

$$\begin{array}{r} 6y - 4z = 15 \\ \underline{-6y + 4z = -14} \\ 0 = 1 \end{array}$$

The statement 0 = 1 is false, so the system has no solution.

3. $\begin{cases} 3x + 2y \quad = -1 & (1) \\ 6x \quad - 2z = 4 & (2) \\ \quad y - 3z = 2 & (3) \end{cases}$

Multiply equation (1) by –2 and add it to equation (2) to eliminate x.

$$\begin{array}{r} -6x - 4y \quad = 2 \\ \underline{6x \quad - 2z = 4} \\ -4y - 2z = 6 \quad (4) \end{array}$$

Multiply equation (3) by 4 and add it to equation (4) to eliminate y.

$$\begin{array}{r} 4y-12z=8 \\ \underline{-4y-2z=6} \\ -14z=14 \\ z=-1 \end{array}$$

Replace z with -1 in equation (2).

$$\begin{aligned} 6x-2z&=4 \\ 6x-2(-1)&=4 \\ 6x+2&=4 \\ 6x&=2 \\ x&=\frac{1}{3} \end{aligned}$$

Replace z with -1 in equation (3).

$$\begin{aligned} y-3z&=2 \\ y-3(-1)&=2 \\ y+3&=2 \\ y&=-1 \end{aligned}$$

The solution of the system is $\left(\frac{1}{3}, -1, -1\right)$.

4. $\begin{cases} x-3y+4z=2 & (1) \\ -2x+6y-8z=-4 & (2) \\ \frac{1}{2}x-\frac{3}{2}y+2z=1 & (3) \end{cases}$

Multiply equation (2) by $-\frac{1}{2}$ and equation (3) by 2.

$$\begin{cases} x-3y+4z=2 \\ x-3y+4z=2 \\ x-3y+4z=2 \end{cases}$$

All three equations are identical. The system is dependent and the solution set is $\{(x, y, z) | x-3y+4z=2\}$.

5. $\begin{cases} 2x+5y-3z=30 & (1) \\ x+y=-3 & (2) \\ 2x-z=0 & (3) \end{cases}$

Solve equation (2) for y and equation (3) for z.

$$\begin{aligned} x+y&=-3 \\ y&=-x-3 \end{aligned} \qquad \begin{aligned} 2x-z&=0 \\ z&=2x \end{aligned}$$

Substitute $-x-3$ for y and $2x$ for z in equation (1).

$$\begin{aligned} 2x+5y-3z&=30 \\ 2x+5(-x-3)-3(2x)&=30 \\ 2x-5x-15-6x&=30 \\ -9x&=45 \\ x&=-5 \end{aligned}$$

Replace x with -5 in the equation $y=-x-3$.

$y=-x-3=-(-5)-3=5-3=2$

Replace x with -5 in the equation $z=2x$.

$z=2x=2(-5)=-10$

The solution of the system is $(-5, 2, -10)$.

Exercise Set 4.2

1. a.
$$\begin{aligned} x+y+z&=3 \\ -1+3+1&\stackrel{?}{=}3 \\ 3&=3 \quad \text{True} \end{aligned}$$

b.
$$\begin{aligned} -x+y+z&=5 \\ -(-1)+3+1&\stackrel{?}{=}5 \\ 1+3+1&\stackrel{?}{=}5 \\ 5&=5 \quad \text{True} \end{aligned}$$

c.
$$\begin{aligned} -x+y+2z&=0 \\ -(-1)+3+2(1)&\stackrel{?}{=}0 \\ 1+3+2&\stackrel{?}{=}0 \\ 6&=0 \quad \text{False} \end{aligned}$$

d.
$$\begin{aligned} x+2y-3z&=2 \\ -1+2(3)-3(1)&\stackrel{?}{=}2 \\ -1+6-3&\stackrel{?}{=}2 \\ 2&=2 \quad \text{True} \end{aligned}$$

$(-1, 3, 1)$ is a solution to the equations a, b, and d.

3. yes; answers may vary

5. $\begin{cases} x-y+z=-4 & (1) \\ 3x+2y-z=5 & (2) \\ -2x+3y-z=15 & (3) \end{cases}$

Add the equations (1) and (2) to eliminate z.

$$\begin{array}{r} x-y+z=-4 \\ \underline{3x+2y-z=5} \\ 4x+y=1 \quad (4) \end{array}$$

Add equations (1) and (3) to eliminate z.

$$\begin{array}{r} x - y + z = -4 \\ -2x + 3y - z = 15 \\ \hline -x + 2y \quad = 11 \end{array} \quad (5)$$

Multiply equation (5) by 4 and add it to equation (4).

$$\begin{array}{r} 4x + y = 1 \\ -4x + 8y = 44 \\ \hline 9y = 45 \\ y = 5 \end{array}$$

Replace y with 5 in equation (4).

$$\begin{aligned} 4x + y &= 1 \\ 4x + 5 &= 1 \\ 4x &= -4 \\ x &= -1 \end{aligned}$$

Replace x with -1 and y with 5 in equation (1).

$$\begin{aligned} x - y + z &= -4 \\ -1 - 5 + z &= -4 \\ -6 + z &= -4 \\ z &= 2 \end{aligned}$$

The solution of the system is (–1, 5, 2).

7. $\begin{cases} x + y \quad = 3 & (1) \\ 2y \quad = 10 & (2) \\ 3x + 2y - 3z = 1 & (3) \end{cases}$

Solve equation (2) for y.

$$\begin{aligned} 2y &= 10 \\ y &= 5 \end{aligned}$$

Replace y with 5 in equation (1).

$$\begin{aligned} x + y &= 3 \\ x + 5 &= 3 \\ x &= -2 \end{aligned}$$

Replace x with –2 and y with 5 in equation (3).

$$\begin{aligned} 3x + 2y - 3z &= 1 \\ 3(-2) + 2(5) - 3z &= 1 \\ -6 + 10 - 3z &= 1 \\ 4 - 3z &= 1 \\ -3z &= -3 \\ z &= 1 \end{aligned}$$

The solution of the system is (–2, 5, 1).

9. $\begin{cases} 2x + 2y + z = 1 & (1) \\ -x + y + 2z = 3 & (2) \\ x + 2y + 4z = 0 & (3) \end{cases}$

Add equations (2) and (3) to eliminate x.

$$\begin{array}{r} -x + y + 2z = 3 \\ x + 2y + 4z = 0 \\ \hline 3y + 6z = 3 \\ y + 2z = 1 \end{array} \quad (4)$$

Multiply equation (2) by 2 and add it to equation (1) to eliminate x.

$$\begin{array}{r} 2x + 2y + z = 1 \\ -2x + 2y + 4z = 6 \\ \hline 4y + 5z = 7 \end{array} \quad (5)$$

Multiply equation (4) by –4 and add it to equation (5).

$$\begin{array}{r} -4y - 8z = -4 \\ 4y + 5z = 7 \\ \hline -3z = 3 \\ z = -1 \end{array}$$

Replace z with –1 in equation (4).

$$\begin{aligned} y + 2z &= 1 \\ y + 2(-1) &= 1 \\ y - 2 &= 1 \\ y &= 3 \end{aligned}$$

Replace y with 3 and z with –1 in equation (3).

$$\begin{aligned} x + 2y + 4z &= 0 \\ x + 2(3) + 4(-1) &= 0 \\ x + 6 - 4 &= 0 \\ x + 2 &= 0 \\ x &= -2 \end{aligned}$$

The solution of the system is (–2, 3, –1).

11. $\begin{cases} x - 2y \ + z = -5 & (1) \\ -3x + 6y - 3z = 15 & (2) \\ 2x - 4y + 2z = -10 & (3) \end{cases}$

Multiply equation (1) by –3.

$-3x + 6y - 3z = 15$

This is equation (2).

Multiply equation (1) by 2.

$2x - 4y + 2z = -10$

This is equation (3).

The system is dependent and the solution set is $\{(x, y, z) | x - 2y + z = -5\}$.

13. $\begin{cases} 4x - y + 2z = 5 & (1) \\ 2y + z = 4 & (2) \\ 4x + y + 3z = 10 & (3) \end{cases}$

Multiply equation (1) by −1 and add it to equation (3) to eliminate x.

$$\begin{array}{r} -4x + y - 2z = -5 \\ \underline{4x + y + 3z = 10} \\ 2y + z = 5 \quad (4) \end{array}$$

Multiply equation (2) by −1 and add it to equation (4).

$$\begin{array}{r} -2y - z = -4 \\ \underline{2y + z = 5} \\ 0 = 1 \end{array}$$

The statement 0 = 1 is false, so the system has no solution.

15. $\begin{cases} x + 5z = 0 & (1) \\ 5x + y = 0 & (2) \\ y - 3z = 0 & (3) \end{cases}$

Solve equation (1) for x.

$$x + 5z = 0$$
$$x = -5z$$

Solve equation (3) for y.

$$y - 3z = 0$$
$$y = 3z$$

Replace x with $-5z$ and y with $3z$ in equation (2).

$$5x + y = 0$$
$$5(-5z) + 3z = 0$$
$$-25z + 3z = 0$$
$$-22z = 0$$
$$z = 0$$

Replace z with 0 in equation (1).

$$x + 5z = 0$$
$$x + 5(0) = 0$$
$$x + 0 = 0$$
$$x = 0$$

Replace x with 0 in equation (2).

$$5x + y = 0$$
$$5(0) + y = 0$$
$$0 + y = 0$$
$$y = 0$$

The solution of the system is (0, 0, 0).

17. $\begin{cases} 6x - 5z = 17 & (1) \\ 5x - y + 3z = -1 & (2) \\ 2x + y = -41 & (3) \end{cases}$

Add equations (2) and (3) to eliminate y.

$$\begin{array}{r} 5x - y + 3z = -1 \\ \underline{2x + y = -41} \\ 7x + 3z = -42 \quad (4) \end{array}$$

Multiply equation (1) by 3 and equation (4) by 5, then add the resulting equations.

$$\begin{array}{r} 18x - 15z = 51 \\ \underline{35x + 15z = -210} \\ 53x = -159 \\ x = -3 \end{array}$$

Replace x with −3 in equation (1).

$$6x - 5z = 17$$
$$6(-3) - 5z = 17$$
$$-18 - 5z = 17$$
$$-5z = 35$$
$$z = -7$$

Replace x with −3 in equation (3).

$$2x + y = -41$$
$$2(-3) + y = -41$$
$$-6 + y = -41$$
$$y = -35$$

The solution of the system is (−3, −35, −7).

19. $\begin{cases} x + y + z = 8 & (1) \\ 2x - y - z = 10 & (2) \\ x - 2y - 3z = 22 & (3) \end{cases}$

Add equations (1) and (2) to eliminate y and z.

$$\begin{array}{r} x + y + z = 8 \\ \underline{2x - y - z = 10} \\ 3x = 18 \\ x = 6 \end{array}$$

Multiply equation (1) by 2 and add it to equation (3) to eliminate y.

$$\begin{array}{r} 2x + 2y + 2z = 16 \\ \underline{x - 2y - 3z = 22} \\ 3x - z = 38 \quad (4) \end{array}$$

Replace x with 6 in equation (4).

$$3x - z = 38$$
$$3(6) - z = 38$$
$$18 - z = 38$$
$$z = -20$$

Replace x with 6 and z with −20 in equation (1).

$$\begin{aligned} x+y+z&=8 \\ 6+y+(-20)&=8 \\ y-14&=8 \\ y&=22 \end{aligned}$$

The solution of the system is (6, 22, −20).

21. $\begin{cases} x+2y-z=5 & (1) \\ 6x+y+z=7 & (2) \\ 2x+4y-2z=5 & (3) \end{cases}$

Add equations (1) and (2) to eliminate z.

$$\begin{array}{l} x+2y-z=5 \\ \underline{6x+y+z=7} \\ 7x+3y \quad =12 \quad (4) \end{array}$$

Multiply equation (2) by 2 and add to equation (3) to eliminate z.

$$\begin{array}{l} 12x+2y+2z=14 \\ \underline{2x+4y-2z=5} \\ 14x+6y \quad =19 \quad (5) \end{array}$$

Multiply equation (4) by −2 and add to equation (5).

$$\begin{array}{r} -14x-6y=-24 \\ \underline{14x+6y=19} \\ 0=-5 \end{array}$$

The statement 0 = −5 is false, so the system has no solution.

23. $\begin{cases} 2x-3y+z=2 & (1) \\ x-5y+5z=3 & (2) \\ 3x+y-3z=5 & (3) \end{cases}$

Multiply equation (1) by −5 and add it to equation (2) to eliminate z.

$$\begin{array}{l} -10x+15y-5z=-10 \\ \underline{x-5y+5z=3} \\ -9x+10y \quad =-7 \quad (4) \end{array}$$

Multiply equation (1) by 3 and add it to equation (3) to eliminate z.

$$\begin{array}{l} 6x-9y+3z=6 \\ \underline{3x+y-3z=5} \\ 9x-8y \quad =11 \quad (5) \end{array}$$

Add equations (4) and (5) to eliminate x.

$$\begin{array}{r} -9x+10y=-7 \\ \underline{9x-8y=11} \\ 2y=4 \\ y=2 \end{array}$$

Replace y with 2 in equation (5).

$$\begin{aligned} 9x-8y&=11 \\ 9x-8(2)&=11 \\ 9x-16&=11 \\ 9x&=27 \\ x&=3 \end{aligned}$$

Replace x with 3 and y with 2 in equation (1).

$$\begin{aligned} 2x-3y+z&=2 \\ 2(3)-3(2)+z&=2 \\ 6-6+z&=2 \\ z&=2 \end{aligned}$$

The solution of the system is (3, 2, 2).

25. $\begin{cases} -2x-4y+6z=-8 & (1) \\ x+2y-3z=4 & (2) \\ 4x+8y-12z=16 & (3) \end{cases}$

Multiply equation (2) by −2.

$-2x - 4y + 6z = -8$

This is equation (1).

Multiply equation (2) by 4.

$4x + 8y - 12z = 16$

This is equation (3).

The system is dependent and the solution set is $\{(x, y, z) | x + 2y - 3z = 4\}$.

27. $\begin{cases} 2x+2y-3z=1 & (1) \\ y+2z=-14 & (2) \\ 3x-2y=-1 & (3) \end{cases}$

Multiply equation (1) by −3 and equation (3) by 2, then add the results to eliminate x.

$$\begin{array}{l} -6x-6y+9z=-3 \\ \underline{6x-4y \quad =-2} \\ -10y+9z=-5 \quad (4) \end{array}$$

Multiply equation (2) by 10 and add the result to equation (4) to eliminate y.

$$\begin{array}{r} 10y+20z=-140 \\ \underline{-10y+9z=-5} \\ 29z=-145 \\ z=-5 \end{array}$$

Replace z with −5 in equation (2).

$$\begin{aligned} y+2z&=-14 \\ y+2(-5)&=-14 \\ y-10&=-14 \\ y&=-4 \end{aligned}$$

Replace y with –4 in equation (3).

$$\begin{aligned} 3x-2y&=-1 \\ 3x-2(-4)&=-1 \\ 3x+8&=-1 \\ 3x&=-9 \\ x&=-3 \end{aligned}$$

The solution of the system is (–3, –4, –5).

29. $\begin{cases} x+2y-z=5 & (1) \\ -3x-2y-3z=11 & (2) \\ 4x+4y+5z=-18 & (3) \end{cases}$

Add equations (1) and (2) to eliminate y.

$$\begin{array}{r} x+2y-z=5 \\ \underline{-3x-2y-3z=11} \\ -2x-4z=16 \\ x+2z=-8\ (4) \end{array}$$

Multiply equation (2) by 2 and add to equation (3) to eliminate y.

$$\begin{array}{r} -6x-4y-6z=22 \\ \underline{4x+4y+5z=-18} \\ -2x-z=4\quad(5) \end{array}$$

Multiply equation (5) by 2 and add to equation (4).

$$\begin{array}{r} x+2z=-8 \\ \underline{-4x-2z=8} \\ -3x=0 \\ x=0 \end{array}$$

Replace x with 0 in equation (4).

$$\begin{aligned} x+2z&=-8 \\ 0+2z&=-8 \\ 2z&=-8 \\ z&=-4 \end{aligned}$$

Replace x with 0 and z with –4 in equation (1).

$$\begin{aligned} x+2y-z&=5 \\ 0+2y-(-4)&=5 \\ 2y+4&=5 \\ 2y&=1 \\ y&=\frac{1}{2} \end{aligned}$$

The solution of the system is $\left(0, \frac{1}{2}, -4\right)$.

31. $\begin{cases} \frac{3}{4}x-\frac{1}{3}y+\frac{1}{2}z=9 & (1) \\ \frac{1}{6}x+\frac{1}{3}y-\frac{1}{2}z=2 & (2) \\ \frac{1}{2}x-y+\frac{1}{2}z=2 & (3) \end{cases}$

Multiply equation (1) by 12, equation (2) by 6, and equation (3) by 2 to clear fractions.

$\begin{cases} 9x-4y+6z=108 & (4) \\ x+2y-3z=12 & (5) \\ x-2y+z=4 & (6) \end{cases}$

Multiply equation (5) by 2 and add it to equation (4) to eliminate y and z.

$$\begin{array}{r} 9x-4y+6z=108 \\ \underline{2x+4y-6z=24} \\ 11x=132 \\ x=12 \end{array}$$

Add equations (5) and (6) to eliminate y.

$$\begin{array}{r} x+2y-3z=12 \\ \underline{x-2y+z=4} \\ 2x-2z=16\quad(7) \end{array}$$

Replace x with 12 in equation (7).

$$\begin{aligned} 2x-2z&=16 \\ 2(12)-2z&=16 \\ 24-2z&=16 \\ -2z&=-8 \\ z&=4 \end{aligned}$$

Replace x with 12 and z with 4 in equation (6).

$$\begin{aligned} x-2y+z&=4 \\ 12-2y+4&=4 \\ -2y+16&=4 \\ -2y&=-12 \\ y&=6 \end{aligned}$$

The solution of the system is (12, 6, 4).

33.
$$\begin{aligned} 2(x-1)-3x&=x-12 \\ 2x-2-3x&=x-12 \\ -x-2&=x-12 \\ -2&=2x-12 \\ 10&=2x \\ 5&=x \end{aligned}$$

The solution set is {5}.

35.
$$\begin{aligned} -y-5(y+5) &= 3y-10 \\ -y-5y-25 &= 3y-10 \\ -6y-25 &= 3y-10 \\ -25 &= 9y-10 \\ -15 &= 9y \\ -\frac{15}{9} &= y \\ -\frac{5}{3} &= y \end{aligned}$$

The solution set is $\left\{-\frac{5}{3}\right\}$.

37. Let x be one number. Then $2x$ is the other number.

$$\begin{aligned} x+2x &= 45 \\ 3x &= 45 \\ x &= 15 \end{aligned}$$

$2x = 2(15) = 30$

The numbers are 15 and 30.

39. answers may vary

41. answers may vary

43.
$$\begin{cases} x+y+z=1 & (1) \\ 2x-y+z=0 & (2) \\ -x+2y+2z=-1 & (3) \end{cases}$$

Add equations (1) and (2) to eliminate y.

$$\begin{aligned} x+y+z &= 1 \\ 2x-y+z &= 0 \\ \hline 3x \quad +2z &= 1 \quad (4) \end{aligned}$$

Multiply equation (2) by 2 and add it to equation (3) to eliminate y.

$$\begin{aligned} 4x-2y+2z &= 0 \\ -x+2y+2z &= -1 \\ \hline 3x \quad +4z &= -1 \quad (5) \end{aligned}$$

Multiply equation (4) by -1 and add it to equation (5) to eliminate x.

$$\begin{aligned} -3x-2z &= -1 \\ 3x+4z &= -1 \\ \hline 2z &= -2 \\ z &= -1 \end{aligned}$$

Replace z with -1 in equation (4).

$$\begin{aligned} 3x+2z &= 1 \\ 3x+2(-1) &= 1 \\ 3x-2 &= 1 \\ 3x &= 3 \\ x &= 1 \end{aligned}$$

Replace x with 1 and z with -1 in equation (1).

$$\begin{aligned} x+y+z &= 1 \\ 1+y+(-1) &= 1 \\ y &= 1 \end{aligned}$$

The solution of the system is (1, 1, −1).

$$\begin{aligned} \frac{1}{24} &= \frac{x}{8}+\frac{y}{4}+\frac{z}{3} \\ \frac{1}{24} &= \frac{1}{8}+\frac{1}{4}-\frac{1}{3} \\ \frac{1}{24} &= \frac{3}{24}+\frac{6}{24}-\frac{8}{24} \\ \frac{1}{24} &= \frac{1}{24} \end{aligned}$$

45.
$$\begin{cases} x+y \qquad -w=0 & (1) \\ \quad y+2z+w=3 & (2) \\ x \quad -z \qquad =1 & (3) \\ 2x-y \qquad -w=-1 & (4) \end{cases}$$

Add equations (1) and (2) to eliminate w.

$$\begin{aligned} x+y \qquad -w &= 0 \\ y+2z+w &= 3 \\ \hline x+2y+2z \qquad &= 3 \quad (5) \end{aligned}$$

Add equations (2) and (4) to eliminate w.

$$\begin{aligned} y+2z+w &= 3 \\ 2x-y \qquad -w &= -1 \\ \hline 2x \quad +2z \qquad &= 2 \\ x+z &= 1 \quad (6) \end{aligned}$$

Add equations (3) and (6).

$$\begin{aligned} x-z &= 1 \\ x+z &= 1 \\ \hline 2x \quad &= 2 \\ x &= 1 \end{aligned}$$

Replace x with 1 in equation (3).

$$\begin{aligned} x-z &= 1 \\ 1-z &= 1 \\ -z &= 0 \\ z &= 0 \end{aligned}$$

Replace x with 1 and z with 0 in equation (5).

$x + 2y + 2z = 3$
$1 + 2y + 2(0) = 3$
$1 + 2y = 3$
$2y = 2$
$y = 1$

Replace x with 1 and y with 1 in equation (1).

$x + y - w = 0$
$1 + 1 - w = 0$
$2 - w = 0$
$-w = -2$
$w = 2$

The solution of the system is (1, 1, 0, 2).

47. $\begin{cases} x + y + z + w = 5 & (1) \\ 2x + y + z + w = 6 & (2) \\ x + y + z \quad = 2 & (3) \\ x + y \quad = 0 & (4) \end{cases}$

Multiply equation (1) by −1 and add it to equation (2) to eliminate y, z, and w.

$-x - y - z - w = -5$
$\underline{2x + y + z + w = 6}$
$x \quad = 1$

Replace x with 1 in equation (4).

$x + y = 0$
$1 + y = 0$
$y = -1$

Replace x with 1 and y with −1 in equation (3).

$x + y + z = 2$
$1 + (-1) + z = 2$
$z = 2$

Replace x with 1, y with −1, and z with 2 in equation (1).

$x + y + z + w = 5$
$1 + (-1) + 2 + w = 5$
$2 + w = 5$
$w = 3$

The solution of the system is (1, −1, 2, 3).

49. answers may vary

Section 4.3 Practice Problems

1. a. $\begin{cases} y = -0.7x + 113.5 \\ y = 2.4x + 66.3 \end{cases}$

Use substitution.

$-0.7x + 113.5 = 2.4x + 66.3$
$47.2 = 3.1x$
$15 \approx x$

The predicted year is 15 years from 2000 or in 2015.

b. Yes; answers may vary

2. Let x be the first number and y be the second number.

$\begin{cases} y = x - 7 \\ 2x = 3y + 4 \end{cases}$

Use substitution.

$2x = 3(x - 7) + 4$
$2x = 3x - 21 + 4$
$2x = 3x - 17$
$-x = -17$
$x = 17$

Replace x with 17 in the first equation.

$y = x - 7 = 17 - 7 = 10$

The numbers are 17 and 10.

3. Let x = speed of one train and y = speed of other train.

	Rate ·	Time =	Distance
One train	x	4	$4x$
Other train	y	4	$4y$

$\begin{cases} 4x + 4y = 376 \\ x = y + 10 \end{cases}$

Use substitution.

$4(y + 10) + 4y = 376$
$4y + 40 + 4y = 376$
$8y = 336$
$y = 42$

Replace y with 42 in the second equation.

$x = y + 10 = 42 + 10 = 52$

The trains are traveling at 42 mph and 52 mph.

4. Let x = amount of 20% solutions and y = mount of 50% solution.

	oz	strength	amount of pure acid
20% solution	x	20%	$0.2x$
50% solution	y	50%	$0.5y$
30% solution	60	30%	(0.3)(60) = 1.8

$$\begin{cases} x+y=60 \\ 0.2x+0.5y=1.8 \end{cases}$$

Multiply the first equation by −2 and the second equation by 10.

$$\begin{aligned} -2x-2y&=-120 \\ 2x+5y&=180 \\ \hline 3y&=60 \\ y&=20 \end{aligned}$$

Replace y with 20 in the first equation.

$$\begin{aligned} x+y&=60 \\ x+20&=60 \\ x&=40 \end{aligned}$$

Thus, 40 ounces of 20% solution should be mixed with 20 ounces of 50% solution.

5. Let x = number of packages.
$C(x)$ = total cost for x packages.
$R(x)$ = total revenue for x packages.

$R(x) = 4x$
$C(x) = 1.5x + 2000$
Solve the equation $4x = 1.5x + 2000$.

$$\begin{aligned} 4x&=1.5x+2000 \\ 2.5x&=2000 \\ x&=800 \end{aligned}$$

The company must sell 800 packages to break even.

6. Let x = degree measure of the smallest angle, y = degree measure of the largest angle, and z = degree measure of the remaining angle.

$$\begin{cases} x+y+z=180 \\ y=x+80 \\ z=x+40 \end{cases}$$

Substitute $x + 80$ for y and $x + 40$ for z in the first equation.

$$\begin{aligned} x+y+z&=180 \\ x+x+80+x+40&=180 \\ 3x+120&=180 \\ 3x&=60 \\ x&=20 \end{aligned}$$

Replace x with 20 in the second equation.
$y = x + 80 = 20 + 80 = 100$
Replace x with 20 in the third equation.
$z = x + 40 = 20 + 40 = 60$
The angles measure 20°, 100°, and 60°.

Exercise Set 4.3

1. Let m = the first number, n = the second number.

$$\begin{cases} m=n+2 \\ 2m=3n-4 \end{cases}$$

Substitute $m = n + 2$ in the second equation.

$$\begin{aligned} 2(n+2)&=3n-4 \\ 2n+4&=3n-4 \\ n&=8 \end{aligned}$$

Replace n with 8 in the first equation.
$m = 8 + 2 = 10$
The numbers are 10 and 8.

3. a. Let e = length of the Enterprise class, n = length of the Nimitz class.

$$\begin{cases} e+n=2193 \\ e-n=9 \end{cases}$$

Add the equations.

$$\begin{aligned} 2e&=2202 \\ e&=1101 \end{aligned}$$

Replace e with 1101 in the first equation.

$$\begin{aligned} 1101+n&=2193 \\ n&=1092 \end{aligned}$$

The Enterprise class is 1101 feet and the Nimitz class is 1092 feet.

b. There are 3 feet in each yard, so there are 300 feet in 100 yards.

$\frac{1101}{300} = 3.67$

The length of the Enterprise class carrier is 3.67 football fields.

5. Let p = speed of the plane in still air, w = speed of the wind.

$\begin{cases} p + w = 560 \\ p - w = 480 \end{cases}$

Add the equations.

$2p = 1040$
$p = 520$

Replace p with 520 in the first equation.

$520 + w = 560$
$w = 40$

The speed of the plane is 520 mph and the speed of the wind is 40 mph.

7. Let x = amount of 4% butterfat milk, and y = amount of 1% butterfat milk.

qt	strength	amount of butterfat
x	4%	$0.04x$
y	1%	$0.01y$
60	2%	$0.02(60) = 1.2$

$\begin{cases} x + y = 60 \\ 0.04x + 0.01y = 1.2 \end{cases}$

Multiply the second equation by −100 and add the result to the first equation.

$x + y = 60$
$\underline{-4x - y = -120}$
$-3x = -60$
$x = 20$

Replace x with 20 in the first equation.

$20 + y = 60$
$y = 40$

Thus, mix 20 quarts of 4% butterfat milk with 40 quarts of 1% butterfat milk.

9. Let x be the number of students studying in the United Kingdom, and y be the number studying in Italy.

$\begin{cases} x + y = 50,642 \\ x = y + 12,770 \end{cases}$

Replace x with y + 12,770 in the first equation.

$x + y = 50,642$
$y + 12,770 + y = 50,642$
$2y + 12,770 = 50,642$
$2y = 37,872$
$y = 18,936$

Replace y with 18,936 in the second equation.

$x = y + 12,770 = 18,936 + 12,770 = 31,706$

31,706 students studied in the United Kingdom, and 18,936 studied in Italy.

11. Let l be the number of large frames and s be the number of small frames.

$\begin{cases} l + s = 22 \\ 15l + 8s = 239 \end{cases}$

Solve the first equation for l.

$l = 22 - s$

Replace l with $22 - s$ in the second equation.

$15(22 - s) + 8s = 239$
$330 - 15s + 8s = 239$
$-7s = -91$
$s = 13$

Replace s with 13 in the first equation.

$l + 13 = 22$
$l = 9$

She bought 9 large frames and 13 small frames.

13. Let m = the first number, n = the second number.

$\begin{cases} m = n - 2 \\ 2m = 3n + 4 \end{cases}$

Substitute $m = n - 2$ in the second equation.

$2(n - 2) = 3n + 4$
$2n - 4 = 3n + 4$
$-8 = n$

Replace n with −8 in the first equation.

$m = -8 - 2 = -10$

The numbers are −10 and −8.

15. $\begin{cases} y = 5.6x + 20.7 \\ y = 4.8x + 29.4 \end{cases}$

Replace y with $5.6x + 20.7$ in the second equation.

$$\begin{aligned} y &= 4.8x + 29.4 \\ 5.6x + 20.7 &= 4.8x + 29.4 \\ 0.8x + 20.7 &= 29.4 \\ 0.8x &= 8.7 \\ x &= 10.875 \\ x &\approx 10 \end{aligned}$$

1998 + 10 = 2008

The predicted year is 2008 where the percent of females using the Internet is equal to the percent of males.

17. Let p be the price of a pen and w be the price of a writing tablet.

$\begin{cases} 7w + 4p = 6.40 \\ 2w + 19p = 5.40 \end{cases}$

Multiply the first equation by −2 and the second equation by 7 and add the resulting equations.

$$\begin{aligned} -14w \quad - 8p &= -12.8 \\ 14w + 133p &= 37.8 \\ \hline 125p &= 25 \\ p &= 0.2 \end{aligned}$$

Replace p with 0.2 in the first equation.

$$\begin{aligned} 7w + 4p &= 6.40 \\ 7w + 4(0.2) &= 6.40 \\ 7w + 0.8 &= 6.4 \\ 7w &= 5.6 \\ w &= 0.8 \end{aligned}$$

The price of a writing tablet is \$0.80 and the price of a pen is \$0.20.

19. Let p be the speed of the plane and w be the speed of the wind.

$\begin{cases} 3p + 3w = 2160 \\ 4p - 4w = 2160 \end{cases}$

Multiply the first equation by $\frac{1}{3}$ and the second equation by $\frac{1}{4}$, and add the results.

$$\begin{aligned} p + w &= 720 \\ p - w &= 540 \\ \hline 2p \quad &= 1260 \\ p &= 630 \end{aligned}$$

Replace p with 630 in the first equation.

$$\begin{aligned} 3(630) + 3w &= 2160 \\ 1890 + 3w &= 2160 \\ 3w &= 270 \\ w &= 90 \end{aligned}$$

The plane's speed is 630 mph and the wind's speed is 90 mph.

21. a. answers may vary

b. $\begin{cases} y = -0.513x + 33.5 \\ y = 0.56x + 18.4 \end{cases}$

Replace y with $0.56x + 18.4$ in the first equation.

$$\begin{aligned} y &= -0.513x + 33.5 \\ 0.56x + 18.4 &= -0.513x + 33.5 \\ 1.073x + 18.4 &= 33.5 \\ 1.073x &= 15.1 \\ x &\approx 14.07 \end{aligned}$$

The percents of jobs in the two sectors were the same 14 years after 1970, or in 1984.

23. Let x be the length of each of the equal sides and y be the length of the third side.

$\begin{cases} 2x + y = 93 \\ y = x + 9 \end{cases}$

Replace y with $x + 9$ in the first equation.

$$\begin{aligned} 2x + y &= 93 \\ 2x + x + 9 &= 93 \\ 3x &= 84 \\ x &= 28 \end{aligned}$$

Replace x with 28 in the second equation.

$y = x + 9 = 28 + 9 = 37$

The lengths of the sides are 28 cm, 28 cm, and 37 cm.

25. Let m be the number of miles.

Hertz = $25 + 0.10m$

Budget = $20 + 0.25m$

Using Budget = 2 · Hertz gives

$$20+0.25m=2(25+0.10m)$$
$$20+0.25m=50+0.20m$$
$$0.25m=30+0.20m$$
$$0.05m=30$$
$$m=\frac{30}{0.05}=600$$

The Budget charge is twice the Hertz charge for a daily mileage of 600 miles.

27. $\begin{cases} x=y-30 \\ x+y=180 \end{cases}$

Replace x with $y - 30$ in the second equation.

$$x+y=180$$
$$y-30+y=180$$
$$2y=210$$
$$y=105$$

Replace y with 105 in the first equation.
$x = y - 30 = 105 - 30 = 75$
The values are $x = 75$ and $y = 105$.

29. The break-even point is where $C(x) = R(x)$.

$$30x+10,000=46x$$
$$10,000=16x$$
$$625=x$$

625 units must be sold to break even.

31. The break-even point is where $C(x) = R(x)$.

$$1.2x+1500=1.7x$$
$$1500=0.5x$$
$$3000=x$$

3000 units must be sold to break even.

33. The break-even point is where $C(x) = R(x)$.

$$75x+160,000=200x$$
$$160,000=125x$$
$$1280=x$$

1280 units must be sold to break even.

35. a. Let x be the number of desks. The revenue from each desk is \$450, so $R(x) = 450x$.

b. The cost is \$6000 plus \$200 for each desk, so $C(x) = 200x + 6000$.

c.

$$R(x)=C(x)$$
$$450x=200x+6000$$
$$250x=6000$$
$$x=24$$

The break-even point is 24 desks.

37. Let x = number of units of Mix A,
y = number of units of Mix B,
z = number of units of Mix C.

$$\begin{cases} 4x+6y+4z=30 & (1) \\ 6x+y+z=16 & (2) \\ 3x+2y+12z=24 & (3) \end{cases}$$

Multiply equation (2) by –6 and add to equation (1).
$-32x - 2z = -66$ or
$16x + z = 33$ (4)
Multiply equation (2) by –2 and add to equation (3).
$-9x + 10z = -8$ (5)
Multiply equation (4) by –10 and add to equation (5).

$$-169x=-338$$
$$x=2$$

Replace x with 2 in equation (4).

$$16(2)+z=33$$
$$32+z=33$$
$$z=1$$

Replace x with 2 and z with 1 in equation (2).

$$6(2)+y+1=16$$
$$12+y+1=16$$
$$y+13=16$$
$$y=3$$

Combine 2 units of Mix A, 3 units of Mix B, and 1 unit of Mix C.

39. Let x = length of shortest side,
y = length of equal sides,
z = length of longest side.

$$\begin{cases} x+2y+z=29 \\ z=2x \\ y=x+2 \end{cases}$$

Replace y with $x + 2$ and z with $2x$ in the first equation.

$$
\begin{aligned}
x+2(x+2)+2x &= 29\\
x+2x+4+2x &= 29\\
5x &= 25\\
x &= 5
\end{aligned}
$$

Replace x with 5 in the third equation.
$y = 5 + 2 = 7$
Replace x with 5 in the second equation.
$z = 2(5) = 10$
The four sides measure 5 inches, 7 inches, 7 inches, and 10 inches.

41. Let x = the first number, y = the second number, and z = the third number.

$$
\begin{cases}
x+y+z=40\\
x=y+5\\
x=2z
\end{cases}
$$

$$
\begin{cases}
x+y+z=40\\
y=x-5\\
z=\frac{1}{2}x
\end{cases}
$$

Substitute $y = x - 5$ and $z=\frac{1}{2}x$ in the first equation.

$$
\begin{aligned}
x+x-5+\frac{1}{2}x &= 40\\
\frac{5}{2}x-5 &= 40\\
\frac{5}{2}x &= 45\\
x &= \frac{2}{5}(45)=18
\end{aligned}
$$

$y = x - 5 = 18 - 5 = 13$

$z=\frac{1}{2}x=\frac{1}{2}(18)=9$

The three numbers are 18, 13, and 9.

43. Let x = number of free throws,
y = number of two-point field goals, and
z = number of three-point field goals.

$$
\begin{cases}
x+2y+3z=738\\
y=4z+12\\
x=y-78
\end{cases}
$$

$$
\begin{cases}
x+2y+3z=738\\
z=\frac{y-12}{4}\\
x=y-78
\end{cases}
$$

Substitute $z=\frac{y-12}{4}$ and $x = y - 78$ in the first equation.

$$
\begin{aligned}
y-78+2y+3\left(\frac{y-12}{4}\right) &= 738\\
4\left[y-78+2y+3\left(\frac{y-12}{4}\right)\right] &= 4(738)\\
4y-312+8y+3y-36 &= 2952\\
15y &= 3300\\
y &= 220
\end{aligned}
$$

Replace y with 220 in the second and third equations.

$z=\frac{220-12}{4}=52$

$x = 220 - 78 = 142$

There were 142 free throws, 220 two-point field goals, and 52 three-point field goals.

45.
$$
\begin{cases}
x+y+z=180\\
2x+5+y=180\\
2x-5+z=180
\end{cases}
$$

$z = 185 - 2x$
$y = 175 - 2x$
Replace y with $175 - 2x$ and z with $185 - 2x$ in the first equation.

$$
\begin{aligned}
x+(175-2x)+(185-2x) &= 180\\
360-3x &= 180\\
-3x &= -180\\
x &= 60
\end{aligned}
$$

$z = 185 - 2(60) = 185 - 120 = 65$
$y = 175 - 2(60) = 175 - 120 = 55$
The values are $x = 60$, $y = 55$, and $z = 65$.

47.
$$
\begin{array}{r}
6x-2y+2z=4\\
-x+2y+3z=6\\
\hline
5x \qquad +5z=10
\end{array}
$$

49.
$$
\begin{array}{r}
-3x-6y+3z=0\\
3x+y-z=2\\
\hline
-5y+2z=2
\end{array}
$$

51. Let x = number filed in 1993, and
y = number filed in 2003.
$y = 2x - 132{,}466$
$y = x + 764{,}765$
Use substitution.

$$2x - 132{,}466 = x + 764{,}765$$
$$x - 132{,}466 = 764{,}765$$
$$x = 897{,}231$$

$y = 897{,}231 + 764{,}765 = 1{,}661{,}996$
There were 897,231 personal bankruptcy petitions filed in 1993 and 1,661,996 filed in 2003.

53. $y = ax^2 + bx + c$
For (1, 6), use $x = 1$ and $y = 6$.
$6 = a + b + c$ (1)
For (−1, −2), use $x = -1$ and $y = -2$.
$-2 = a - b + c$ (2)
For (0, −1), use $x = 0$ and $y = -1$.
$-1 = a \cdot 0 + b \cdot 0 + c$
$-1 = c$ (3)
The system is

$$\begin{aligned} 6 &= a + b + c && (1) \\ -2 &= a - b + c && (2) \\ -1 &= c && (3) \end{aligned}$$

From equation (3), we see that $c = -1$.
Multiply equation (2) by −1 and add to equation (1).

$$8 = 2b$$
$$4 = b$$

Replace b with 4 and c with −1 in equation (1).

$$6 = a + 4 - 1$$
$$6 = a + 3$$
$$3 = a$$

The solution is $a = 3$, $b = 4$, and $c = -1$.

55. $y = ax^2 + bx + c$
For (3, 925), $x = 3$ and $y = 925$.

$$925 = a(3)^2 + b(3) + c$$
$$925 = 9a + 3b + c \quad (1)$$

For (6, 1019), $x = 6$ and $y = 1019$.

$$1019 = a(6)^2 + b(6) + c$$
$$1019 = 36a + 6b + c \quad (2)$$

For (11, 1171), $x = 11$ and $y = 1171$.

$$1171 = a(11)^2 + b(11) + c$$
$$1171 = 121a + 11b + c \quad (3)$$

The system is

$$\begin{cases} 925 = 9a + 3b + c & (1) \\ 1019 = 36a + 6b + c & (2) \\ 1171 = 121a + 11b + c & (3) \end{cases}$$

Multiply equation (1) by −1 and add to equation (2).
$94 = 27a + 3b$ (4)
Multiply equation (1) by −1 and add to equation (3).
$246 = 112a + 8b$ (5)
Multiply equation (4) by −8 and equation (5) by 3. Then add the results.

$$\begin{aligned} -752 &= -216a - 24b \\ 738 &= 336a + 24b \\ \hline -14 &= 120a \end{aligned}$$
$$-\frac{7}{60} = a$$

Replace a with $-\frac{7}{60}$ in equation (4).

$$94 = 27\left(-\frac{7}{60}\right) + 3b$$
$$94 = -\frac{189}{60} + 3b$$
$$\frac{5640}{60} + \frac{189}{60} = 3b$$
$$\frac{5829}{60} = 3b$$
$$b = \frac{1943}{60} = 32\frac{23}{60}$$

Replace a with $-\frac{7}{60}$ and b with $\frac{1943}{60}$ in equation (1).

$$925 = 9\left(-\frac{7}{60}\right) + 3\left(\frac{1943}{60}\right) + c$$
$$925 = -\frac{63}{60} + \frac{5829}{60} + c$$
$$\frac{55,500}{60} + \frac{63}{60} - \frac{5829}{60} = c$$
$$\frac{49,734}{60} = c$$
$$\frac{8289}{10} = c$$
$$c = 828\frac{9}{10}$$

The solution is $a = -\frac{7}{60} \approx -0.12,\ \ b = 32\frac{23}{60} \approx 32.38,$ and $c = 828\frac{9}{10} = 828.9.$

For 2009, $x = 16$.

$$y = -\frac{7}{60} \cdot (16)^2 + \left(32\frac{23}{60}\right)(16) + 828\frac{9}{10} \approx 1317$$

According to the model, 1317 thousand students will take the ACT in 2009.

Integrated Review

1. Graph C has intersection point (1, 2).

2. Graph D has intersection point (–2, 3).

3. Graph A has parallel lines, or no solution.

4. Graph B has coinciding lines, or infinite number of solutions.

5. $\begin{cases} x + y = 4 \\ y = 3x \end{cases}$

   Replace y with $3x$ in the first equation.
   $$x + y = 4$$
   $$x + 3x = 4$$
   $$4x = 4$$
   $$x = 1$$
   Replace x with 1 in the second equation.
   $y = 3x$
   $y = 3(1)$
   $y = 3$
   The solution is (1, 3).

6. $\begin{cases} x - y = -4 \\ y = 4x \end{cases}$

   Replace y with $4x$ in the first equation.

$$x - y = -4$$
$$x - 4x = -4$$
$$-3x = -4$$
$$x = \frac{4}{3}$$

Replace x with $\frac{4}{3}$ in the second equation.

$$y = 4x = 4\left(\frac{4}{3}\right) = \frac{16}{3}$$

The solution is $\left(\frac{4}{3}, \frac{16}{3}\right)$.

7. $\begin{cases} x + y = 1 \\ x - 2y = 4 \end{cases}$

Solve the first equation for x.
$x = 1 - y$
Replace x with $1 - y$ in the second equation.

$$x - 2y = 4$$
$$1 - y - 2y = 4$$
$$-3y = 3$$
$$y = -1$$

Replace y with -1 in the equation $x = 1 - y$.
$x = 1 - (-1) = 1 + 1 = 2$
The solution is $(2, -1)$.

8. $\begin{cases} 2x - y = 8 \\ x + 3y = 11 \end{cases}$

Solve the second equation for x, $x = 11 - 3y$.
Replace x with $11 - 3y$ in the first equation.

$$2x - y = 8$$
$$2(11 - 3y) - y = 8$$
$$22 - 6y - y = 8$$
$$-7y = -14$$
$$y = 2$$

Replace y with 2 in the equation $x = 11 - 3y$.
$x = 11 - 3(2) = 11 - 6 = 5$
The solution is $(5, 2)$.

9. $\begin{cases} 2x + 5y = 8 \\ 6x + y = 10 \end{cases}$

Solve the second equation for y, $y = 10 - 6x$.
Replace y with $10 - 6x$ in the first equation.

$$2x + 5y = 8$$
$$2x + 5(10 - 6x) = 8$$
$$2x + 50 - 30x = 8$$
$$-28x = -42$$
$$x = \frac{3}{2}$$

Replace x with $\frac{3}{2}$ in the equation $y = 10 - 6x$.

$$y = 10 - 6\left(\frac{3}{2}\right) = 10 - 9 = 1$$

The solution is $\left(\frac{3}{2}, 1\right)$.

10. $\begin{cases} \frac{1}{8}x - \frac{1}{2}y = -\frac{5}{8} \\ -3x - 8y = 0 \end{cases}$

Multiply the first equation by -16 and add to the second equation.

$$\begin{array}{r} -2x + 8y = 10 \\ -3x - 8y = 0 \\ \hline -5x \qquad = 10 \end{array}$$
$$x = -2$$

Replace x with -2 in the second equation.

$$-3(-2) - 8y = 0$$
$$6 - 8y = 0$$
$$-8y = -6$$
$$y = \frac{3}{4}$$

The solution is $\left(-2, \frac{3}{4}\right)$.

11. $\begin{cases} 4x - 7y = 7 \\ 12x - 21y = 24 \end{cases}$

Multiply the first equation by -3 and add to the second equation.

$$\begin{array}{r} -12x + 21y = -21 \\ 12x - 21y = 24 \\ \hline 0 = 3 \end{array}$$

The statement $0 = 3$ is false, so there is no solution.

12. $\begin{cases} 2x-5y=3 \\ -4x+10y=-6 \end{cases}$

Multiply the first equation by 2 and add it to the second equation.

$$\begin{array}{r} 4x-10y=6 \\ -4x+10y=-6 \\ \hline 0=0 \end{array}$$

The statement 0 = 0 is true, so the system is dependent. There are an infinite number of solutions. The solution set is $\{(x, y)|2x-5y=3\}$.

13. $\begin{cases} y=\frac{1}{3}x \\ 5x-3y=4 \end{cases}$

Replace y with $\frac{1}{3}x$ in the second equation.

$$\begin{aligned} 5x-3y&=4 \\ 5x-3\left(\frac{1}{3}x\right)&=4 \\ 5x-x&=4 \\ 4x&=4 \\ x&=1 \end{aligned}$$

Replace x with 1 in the first equation.

$$y=\frac{1}{3}(1)=\frac{1}{3}$$

The solution is $\left(1, \frac{1}{3}\right)$.

14. $\begin{cases} y=\frac{1}{4}x \\ 2x-4y=3 \end{cases}$

Replace y with $\frac{1}{4}x$ in the second equation.

$$\begin{aligned} 2x-4y&=3 \\ 2x-4\left(\frac{1}{4}x\right)&=3 \\ 2x-x&=3 \\ x&=3 \end{aligned}$$

Replace x with 3 in the first equation.

$$y=\frac{1}{4}(3)=\frac{3}{4}$$

The solution is $\left(3, \frac{3}{4}\right)$.

15. $\begin{cases} x+y \quad =2 & (1) \\ -3y+z=-7 & (2) \\ 2x+y-z=-1 & (3) \end{cases}$

Solve equation (1) for x.

$x=2-y$ (4)

Solve equation (2) for z.

$z=3y-7$ (5)

Replace x with $2-y$ and z with $3y-7$ in equation (3).

$$\begin{aligned} 2x+y-z&=-1 \\ 2(2-y)+y-(3y-7)&=-1 \\ 4-2y+y-3y+7&=-1 \\ -4y&=-12 \\ y&=3 \end{aligned}$$

Replace y with 3 in equations (4) and (5).

$x=2-y=2-3=-1$

$z=3y-7=3(3)-7=9-7=2$

The solution is (–1, 3, 2).

16. $\begin{cases} y+2z=-3 & (1) \\ x-2y \quad =7 & (2) \\ 2x-y+z=5 & (3) \end{cases}$

Solve equation (1) for z.

$z=\frac{-3-y}{2}$ (4)

Solve equation (2) for x.

$x=2y+7$ (5)

Replace z with $\frac{-3-y}{2}$ and x with $2y+7$ in equation (3).

$$\begin{aligned} 2x-y+z&=5 \\ 2(2y+7)-y+\left(\frac{-3-y}{2}\right)&=5 \\ 4y+14-y-\frac{3}{2}-\frac{y}{2}&=5 \\ \frac{5}{2}y&=-\frac{15}{2} \\ y&=-3 \end{aligned}$$

Replace y with –3 in equations (4) and (5).

$z = \dfrac{-3-(-3)}{2}$ $x = 2(-3)+7$

$z = 0$ $x = -6+7$

$x = 1$

The solution is (1, –3, 0).

17. $\begin{cases} 2x+4y-6z=3 & (1) \\ -x+y-z=6 & (2) \\ x+2y-3z=1 & (3) \end{cases}$

Multiply equation (2) by 2 and add to equation (1).

$$\begin{array}{r} 2x+4y-6z=3 \\ \underline{-2x+2y-2z=12} \\ 6y-8z=15 \quad (4) \end{array}$$

Add equations (2) and (3).

$$\begin{array}{r} -x+y-z=6 \\ \underline{x+2y-3z=1} \\ 3y-4z=7 \quad (5) \end{array}$$

Multiply equation (5) by –2 and add to equation (4).

$$\begin{array}{r} 6y-8z=15 \\ \underline{-6y+8z=-14} \\ 0=1 \end{array}$$

The statement 0 = 1 is false, so there is no solution.

18. $\begin{cases} x-y+3z=2 & (1) \\ -2x+2y-6z=-4 & (2) \\ 3x-3y+9z=6 & (3) \end{cases}$

Multiply equation (2) by $-\dfrac{1}{2}$.

$x - y + 3z = 2$

This is equivalent to equation (1).

Multiply equation (3) by $\dfrac{1}{3}$.

$x - y + 3z = 2$

This is also equivalent to equation (1). The system is dependent. The solution set is $\{(x, y, z)|x - y + 3z = 2\}$.

19. $\begin{cases} x+y-4z=5 & (1) \\ x-y+2z=-2 & (2) \\ 3x+2y+4z=18 & (3) \end{cases}$

Add equations (1) and (2).

$$\begin{array}{r} x+y-4z=5 \\ \underline{x-y+2z=-2} \\ 2x \quad -2z=3 \quad (4) \end{array}$$

Multiply equation (2) by 2 and add to equation (3).

$$\begin{array}{r} 2x-2y+4z=-4 \\ \underline{3x+2y+4z=18} \\ 5x \quad +8z=14 \quad (5) \end{array}$$

Multiply equation (4) by 4 and add to equation (5).

$$\begin{array}{r} 8x-8z=12 \\ \underline{5x+8z=14} \\ 13x \quad =26 \\ x=2 \end{array}$$

Replace x with 2 in equation (4).

$$\begin{aligned} 2x-2z&=3 \\ 2(2)-2z&=3 \\ -2z&=-1 \\ z&=\frac{1}{2} \end{aligned}$$

Replace x with 2 and z with $\dfrac{1}{2}$ in equation (1).

$$\begin{aligned} x+y-4z&=5 \\ 2+y-4\left(\frac{1}{2}\right)&=5 \\ y&=5 \end{aligned}$$

The solution is $\left(2, 5, \dfrac{1}{2}\right)$.

20. $\begin{cases} 2x-y+3z=2 & (1) \\ x+y-6z=0 & (2) \\ 3x+4y-3z=6 & (3) \end{cases}$

Add equations (1) and (2).

$$\begin{array}{r} 2x-y+3z=2 \\ \underline{x+y-6z=0} \\ 3x \quad -3z=2 \quad (4) \end{array}$$

Multiply equation (1) by 4 and add to equation (3).

$$\begin{array}{r} 8x-4y+12z=8 \\ \underline{3x+4y \quad -3z=6} \\ 11x \quad +9z=14 \quad (5) \end{array}$$

Multiply equation (4) by 3 and add to

equation (5).

$$\begin{aligned} 9x - 9y &= 6 \\ 11x + 9y &= 14 \\ \hline 20x \quad &= 20 \\ x &= 1 \end{aligned}$$

Replace x with 1 in equation (4).

$$\begin{aligned} 3(1) - 3z &= 2 \\ -3z &= -1 \\ z &= \frac{1}{3} \end{aligned}$$

Replace x with 1 and z with $\frac{1}{3}$ in equation (2).

$$\begin{aligned} x + y - 6z &= 0 \\ 1 + y - 6\left(\frac{1}{3}\right) &= 0 \\ -1 + y &= 0 \\ y &= 1 \end{aligned}$$

The solution is $\left(1, 1, \frac{1}{3}\right)$.

21. Let m be the first number and n be the second number.

$$\begin{cases} m = n - 8 \\ 2m = n + 11 \end{cases}$$

Replace m with $n - 8$ in the second equation.

$$\begin{aligned} 2(n - 8) &= n + 11 \\ 2n - 16 &= n + 11 \\ n &= 27 \end{aligned}$$

Replace n with 27 in the first equation.
$m = n - 8 = 27 - 8 = 19$
The numbers are 19 and 27.

22. Let x = the measure of each of the smallest angles, y = the measure of the third angle, and
z = the measure of the fourth angle.

$$\begin{cases} 2x + y + z = 360 \\ y = x + 30 \\ z = x + 50 \end{cases}$$

Substitute $y = x + 30$ and $z = x + 50$ in the first equation.

$$\begin{aligned} 2x + y + z &= 360 \\ 2x + x + 30 + x + 50 &= 360 \\ 4x &= 280 \\ x &= 70 \end{aligned}$$

Replace x with 70 in the second equation.
$y = x + 30 = 70 + 30 = 100$
Replace x with 70 in the third equation.
$z = x + 50 = 70 + 50 = 120$
The measures of the angles are 70°, 70°, 100°, and 120°.

Section 4.4 Practice Problems

1. $\begin{cases} x + 2y = -4 \\ 2x - 3y = 13 \end{cases}$

$$\left[\begin{array}{cc|c} 1 & 2 & -4 \\ 2 & -3 & 13 \end{array}\right]$$

Multiply row 1 by −2 and add to row 2.

$$\left[\begin{array}{cc|c} 1 & 2 & -4 \\ 0 & -7 & 21 \end{array}\right]$$

Divide row 2 by −7.

$$\left[\begin{array}{cc|c} 1 & 2 & -4 \\ 0 & 1 & -3 \end{array}\right]$$

This corresponds to $\begin{cases} x + 2y = -4 \\ y = -3 \end{cases}$.

$$\begin{aligned} x + 2(-3) &= -4 \\ x - 6 &= -4 \\ x &= 2 \end{aligned}$$

The solution is (2, −3).

2. $\begin{cases} -3x + y = 0 \\ -6x + 2y = 2 \end{cases}$

$$\left[\begin{array}{cc|c} -3 & 1 & 0 \\ -6 & 2 & 2 \end{array}\right]$$

Divide row 1 by −3.

$$\left[\begin{array}{cc|c} 1 & -\frac{1}{3} & 0 \\ -6 & 2 & 2 \end{array}\right]$$

Multiply row 1 by 6 and add to row 2.

$$\left[\begin{array}{cc|c} 1 & -\frac{1}{3} & 0 \\ 0 & 0 & 2 \end{array}\right]$$

This corresponds to $\begin{cases} x - \frac{1}{3}y = 0 \\ 0 = 2 \end{cases}$.

This is an inconsistent system. The solution is $\varnothing$.

3. $\begin{cases} x + 3y + z = 5 \\ -3x + y - 3z = 5 \\ x + 2y - 2z = 9 \end{cases}$

$\left[\begin{array}{ccc|c} 1 & 3 & 1 & 5 \\ -3 & 1 & -3 & 5 \\ 1 & 2 & -2 & 9 \end{array}\right]$

Multiply row 1 by 3 and add to row 2.
Multiply row 1 by –1 and add to row 3.

$\left[\begin{array}{ccc|c} 1 & 3 & 1 & 5 \\ 0 & 10 & 0 & 20 \\ 0 & -1 & -3 & 4 \end{array}\right]$

Divide row 2 by 10.

$\left[\begin{array}{ccc|c} 1 & 3 & 1 & 5 \\ 0 & 1 & 0 & 2 \\ 0 & -1 & -3 & 4 \end{array}\right]$

Add row 2 to row 3.

$\left[\begin{array}{ccc|c} 1 & 3 & 1 & 5 \\ 0 & 1 & 0 & 2 \\ 0 & 0 & -3 & 6 \end{array}\right]$

Divide row 3 by –3.

$\left[\begin{array}{ccc|c} 1 & 3 & 1 & 5 \\ 0 & 1 & 0 & 2 \\ 0 & 0 & 1 & -2 \end{array}\right]$

This corresponds to $\begin{cases} x + 3y + z = 5 \\ y = 2 \\ z = -2 \end{cases}$.

$$\begin{aligned} x + 3y + z &= 5 \\ x + 3(2) + (-2) &= 5 \\ x + 4 &= 5 \\ x &= 1 \end{aligned}$$

The solution is (1, 2, –2).

Exercise Set 4.4

1. $\begin{cases} x + y = 1 \\ x - 2y = 4 \end{cases}$

$\left[\begin{array}{cc|c} 1 & 1 & 1 \\ 1 & -2 & 4 \end{array}\right]$

Multiply row 1 by –1 and add to row 2.

$\left[\begin{array}{cc|c} 1 & 1 & 1 \\ 0 & -3 & 3 \end{array}\right]$

Divide row 2 by –3.

$\left[\begin{array}{cc|c} 1 & 1 & 1 \\ 0 & 1 & -1 \end{array}\right]$

This corresponds to $\begin{cases} x + y = 1 \\ y = -1 \end{cases}$.

$$\begin{aligned} x + (-1) &= 1 \\ x - 1 &= 1 \\ x &= 2 \end{aligned}$$

The solution is (2, –1).

3. $\begin{cases} x + 3y = 2 \\ x + 2y = 0 \end{cases}$

$\left[\begin{array}{cc|c} 1 & 3 & 2 \\ 1 & 2 & 0 \end{array}\right]$

Multiply row 1 by –1 and add to row 2.

$\left[\begin{array}{cc|c} 1 & 3 & 2 \\ 0 & -1 & -2 \end{array}\right]$

Multiply row 2 by –1.

$\left[\begin{array}{cc|c} 1 & 3 & 2 \\ 0 & 1 & 2 \end{array}\right]$

This corresponds to $\begin{cases} x + 3y = 2 \\ y = 2 \end{cases}$.

$$\begin{aligned} x + 3(2) &= 2 \\ x + 6 &= 2 \\ x &= -4 \end{aligned}$$

The solution is (–4, 2).

5. $\begin{cases} x - 2y = 4 \\ 2x - 4y = 4 \end{cases}$

$\left[\begin{array}{cc|c} 1 & -2 & 4 \\ 2 & -4 & 4 \end{array}\right]$

Multiply row 1 by –2 and add to row 2.

$\left[\begin{array}{cc|c} 1 & -2 & 4 \\ 0 & 0 & -4 \end{array}\right]$

This is an inconsistent system. The solution is $\varnothing$.

7. $\begin{cases} 3x-3y=9 \\ 2x-2y=6 \end{cases}$

$\left[\begin{array}{cc|c} 3 & -3 & 9 \\ 2 & -2 & 6 \end{array}\right]$

Multiply row 1 by $\frac{1}{3}$ and multiply row 2 by $\frac{1}{2}$.

$\left[\begin{array}{cc|c} 1 & -1 & 3 \\ 1 & -1 & 3 \end{array}\right]$

This corresponds to $\begin{cases} x-y=3 \\ x-y=3 \end{cases}$.

The system is dependent. The solution set is $\{(x, y)|3x-3y=9\}$.

9. $\begin{cases} x+y=3 \\ 2y=10 \\ 3x+2y-4z=12 \end{cases}$

$\left[\begin{array}{ccc|c} 1 & 1 & 0 & 3 \\ 0 & 2 & 0 & 10 \\ 3 & 2 & -4 & 12 \end{array}\right]$

Multiply row 1 by –3 and add to row 3.

$\left[\begin{array}{ccc|c} 1 & 1 & 0 & 3 \\ 0 & 2 & 0 & 10 \\ 0 & -1 & -4 & 3 \end{array}\right]$

Divide row 2 by 2.

$\left[\begin{array}{ccc|c} 1 & 1 & 0 & 3 \\ 0 & 1 & 0 & 5 \\ 0 & -1 & -4 & 3 \end{array}\right]$

Add row 2 to row 3.

$\left[\begin{array}{ccc|c} 1 & 1 & 0 & 3 \\ 0 & 1 & 0 & 5 \\ 0 & 0 & -4 & 8 \end{array}\right]$

Multiply row 3 by $-\frac{1}{4}$.

$\left[\begin{array}{ccc|c} 1 & 1 & 0 & 3 \\ 0 & 1 & 0 & 5 \\ 0 & 0 & 1 & -2 \end{array}\right]$

This corresponds to $\begin{cases} x+y=3 \\ y=5 \\ z=-2 \end{cases}$.

$x+5=3$
$x=-2$

The solution is (–2, 5, –2).

11. $\begin{cases} 2y-z=-7 \\ x+4y+z=-4 \\ 5x-y+2z=13 \end{cases}$

$\left[\begin{array}{ccc|c} 0 & 2 & -1 & -7 \\ 1 & 4 & 1 & -4 \\ 5 & -1 & 2 & 13 \end{array}\right]$

Interchange rows 1 and 2.

$\left[\begin{array}{ccc|c} 1 & 4 & 1 & -4 \\ 0 & 2 & -1 & -7 \\ 5 & -1 & 2 & 13 \end{array}\right]$

Multiply row 1 by –5 and add to row 3.

$\left[\begin{array}{ccc|c} 1 & 4 & 1 & -4 \\ 0 & 2 & -1 & -7 \\ 0 & -21 & -3 & 33 \end{array}\right]$

Multiply row 2 by $\frac{1}{2}$ and multiply row 3 by $\frac{1}{3}$.

$\left[\begin{array}{ccc|c} 1 & 4 & 1 & -4 \\ 0 & 1 & -\frac{1}{2} & -\frac{7}{2} \\ 0 & -7 & -1 & 11 \end{array}\right]$

Multiply row 2 by 7 and add to row 3.

$\left[\begin{array}{ccc|c} 1 & 4 & 1 & -4 \\ 0 & 1 & -\frac{1}{2} & -\frac{7}{2} \\ 0 & 0 & -\frac{9}{2} & -\frac{27}{2} \end{array}\right]$

Multiply row 3 by $-\frac{2}{9}$.

$$\left[\begin{array}{ccc|c} 1 & 4 & 1 & -4 \\ 0 & 1 & -\frac{1}{2} & -\frac{7}{2} \\ 0 & 0 & 1 & 3 \end{array}\right]$$

This corresponds to $\begin{cases} x+4y+z=-4 \\ y-\frac{1}{2}z=-\frac{7}{2} \\ z=3 \end{cases}$.

$$y-\frac{1}{2}(3)=-\frac{7}{2}$$
$$y-\frac{3}{2}=-\frac{7}{2}$$
$$y=-\frac{4}{2}$$
$$y=-2$$
$$x+4(-2)+3=-4$$
$$x-5=-4$$
$$x=1$$

The solution is (1, –2, 3).

13. $\begin{cases} x-4=0 \\ x+y=1 \end{cases}$ or $\begin{cases} x=4 \\ x+y=1 \end{cases}$

$$\left[\begin{array}{cc|c} 1 & 0 & 4 \\ 1 & 1 & 1 \end{array}\right]$$

Multiply row 1 by –1 and add to row 2.

$$\left[\begin{array}{cc|c} 1 & 0 & 4 \\ 0 & 1 & -3 \end{array}\right]$$

This corresponds to $\begin{cases} x=4 \\ y=-3 \end{cases}$.

The solution is (4, –3).

15. $\begin{cases} x+y+z=2 \\ 2x-z=5 \\ 3y+z=2 \end{cases}$

$$\left[\begin{array}{ccc|c} 1 & 1 & 1 & 2 \\ 2 & 0 & -1 & 5 \\ 0 & 3 & 1 & 2 \end{array}\right]$$

Multiply row 1 by –2 and add to row 2.

$$\left[\begin{array}{ccc|c} 1 & 1 & 1 & 2 \\ 0 & -2 & -3 & 1 \\ 0 & 3 & 1 & 2 \end{array}\right]$$

Multiply row 2 by $-\frac{1}{2}$.

$$\left[\begin{array}{ccc|c} 1 & 1 & 1 & 2 \\ 0 & 1 & \frac{3}{2} & -\frac{1}{2} \\ 0 & 3 & 1 & 2 \end{array}\right]$$

Multiply row 2 by –3 and add to row 3.

$$\left[\begin{array}{ccc|c} 1 & 1 & 1 & 2 \\ 0 & 1 & \frac{3}{2} & -\frac{1}{2} \\ 0 & 0 & -\frac{7}{2} & \frac{7}{2} \end{array}\right]$$

Multiply row 3 by $-\frac{2}{7}$.

$$\left[\begin{array}{ccc|c} 1 & 1 & 1 & 2 \\ 0 & 1 & \frac{3}{2} & -\frac{1}{2} \\ 0 & 0 & 1 & -1 \end{array}\right]$$

This corresponds to $\begin{cases} x+y+z=2 \\ y+\frac{3}{2}z=-\frac{1}{2} \\ z=-1 \end{cases}$.

$$y+\frac{3}{2}(-1)=-\frac{1}{2}$$
$$y=1$$
$$x+1+(-1)=2$$
$$x=2$$

The solution is (2, 1, –1).

17. $\begin{cases} 5x-2y=27 \\ -3x+5y=18 \end{cases}$

$$\left[\begin{array}{cc|c} 5 & -2 & 27 \\ -3 & 5 & 18 \end{array}\right]$$

Multiply row 1 by $\frac{1}{5}$.

$$\left[\begin{array}{cc|c} 1 & -\frac{2}{5} & \frac{27}{5} \\ -3 & 5 & 18 \end{array}\right]$$

Multiply row 1 by 3 and add to row 2.

$$\left[\begin{array}{cc|c} 1 & -\frac{2}{5} & \frac{27}{5} \\ 0 & \frac{19}{5} & \frac{171}{5} \end{array}\right]$$

Multiply row 2 by $\frac{5}{19}$.

$$\left[\begin{array}{cc|c} 1 & -\frac{2}{5} & \frac{27}{5} \\ 0 & 1 & 9 \end{array}\right]$$

This corresponds to $\begin{cases} x-\frac{2}{5}y=\frac{27}{5} \\ y=9 \end{cases}$.

$$x-\frac{2}{5}(9)=\frac{27}{5}$$
$$x-\frac{18}{5}=\frac{27}{5}$$
$$x=9$$

The solution is (9, 9).

19. $\begin{cases} 4x-7y=7 \\ 12x-21y=24 \end{cases}$

$$\left[\begin{array}{cc|c} 4 & -7 & 7 \\ 12 & -21 & 24 \end{array}\right]$$

Multiply row 1 by –3 and add to row 2.

$$\left[\begin{array}{cc|c} 4 & -7 & 7 \\ 0 & 0 & 3 \end{array}\right]$$

This corresponds to $\begin{cases} 4x-7y=7 \\ 0=3 \end{cases}$.

This is an inconsistent system. The solution is ∅.

21. $\begin{cases} 4x-y+2z=5 \\ 2y+z=4 \\ 4x+y+3z=10 \end{cases}$

$$\left[\begin{array}{ccc|c} 4 & -1 & 2 & 5 \\ 0 & 2 & 1 & 4 \\ 4 & 1 & 3 & 10 \end{array}\right]$$

Multiply row 1 by $\frac{1}{4}$.

$$\left[\begin{array}{ccc|c} 1 & -\frac{1}{4} & \frac{1}{2} & \frac{5}{4} \\ 0 & 2 & 1 & 4 \\ 4 & 1 & 3 & 10 \end{array}\right]$$

Multiply row 1 by –4 and add to row 3.

$$\left[\begin{array}{ccc|c} 1 & -\frac{1}{4} & \frac{1}{2} & \frac{5}{4} \\ 0 & 2 & 1 & 4 \\ 0 & 2 & 1 & 5 \end{array}\right]$$

Multiply row 2 by $\frac{1}{2}$.

$$\left[\begin{array}{ccc|c} 1 & -\frac{1}{4} & \frac{1}{2} & \frac{5}{4} \\ 0 & 1 & \frac{1}{2} & 2 \\ 0 & 2 & 1 & 5 \end{array}\right]$$

Multiply row 2 by –2 and add to row 3.

$$\left[\begin{array}{ccc|c} 1 & -\frac{1}{4} & \frac{1}{2} & \frac{5}{4} \\ 0 & 1 & \frac{1}{2} & 2 \\ 0 & 0 & 0 & 1 \end{array}\right]$$

Row 3 corresponds to the equation 0 = 1, which is false, so the system has no solution.

23. $\begin{cases} 4x+y+z=3 \\ -x+y-2z=-11 \\ x+2y+2z=-1 \end{cases}$

$$\left[\begin{array}{ccc|c} 4 & 1 & 1 & 3 \\ -1 & 1 & -2 & -11 \\ 1 & 2 & 2 & -1 \end{array}\right]$$

Interchange row 1 and row 3.

$$\left[\begin{array}{ccc|c} 1 & 2 & 2 & -1 \\ -1 & 1 & -2 & -11 \\ 4 & 1 & 1 & 3 \end{array}\right]$$

Add row 1 to row 2. Multiply row 1 by –4 and add to row 3.

$$\left[\begin{array}{ccc|c} 1 & 2 & 2 & -1 \\ 0 & 3 & 0 & -12 \\ 0 & -7 & -7 & 7 \end{array}\right]$$

Divide row 2 by 3.

$$\left[\begin{array}{ccc|c} 1 & 2 & 2 & -1 \\ 0 & 1 & 0 & -4 \\ 0 & -7 & -7 & 7 \end{array}\right]$$

Multiply row 2 by 7 and add to row 3.

$$\left[\begin{array}{ccc|c} 1 & 2 & 2 & -1 \\ 0 & 1 & 0 & -4 \\ 0 & 0 & -7 & -21 \end{array}\right]$$

Divide row 3 by −7.

$$\left[\begin{array}{ccc|c} 1 & 2 & 2 & -1 \\ 0 & 1 & 0 & -4 \\ 0 & 0 & 1 & 3 \end{array}\right]$$

This corresponds to $\begin{cases} x+2y+2z=-1 \\ y=-4. \\ z=3 \end{cases}$

$$\begin{aligned} x+2(-4)+2(3)&=-1 \\ x-8+6&=-1 \\ x&=1 \end{aligned}$$

The solution is (1, −4, 3).

25. No vertical line intersects the graph more than once, so the graph is the graph of a function.

27. There exists a vertical line that intersects the graph more than once, so the graph is not the graph of a function.

29. $\begin{cases} x+z=7 \\ y+2z=-6 \\ 3x-y=0 \end{cases}$ can be written as

$\begin{cases} x+0y+z=7 \\ 0x+y+2z=-6 \\ 3x-y+0z=0 \end{cases}$ which corresponds to the

matrix $\left[\begin{array}{ccc|c} 1 & 0 & 1 & 7 \\ 0 & 1 & 2 & -6 \\ 3 & -1 & 0 & 0 \end{array}\right]$; choice c.

31. a. $\begin{cases} 2.3x+y=52 \\ -5.4x+y=14 \end{cases}$

$$\left[\begin{array}{cc|c} 2.3 & 1 & 52 \\ -5.4 & 1 & 14 \end{array}\right]$$

Multiply row 1 by $\frac{1}{2.3}$.

$$\left[\begin{array}{cc|c} 1 & 0.43 & 22.61 \\ -5.4 & 1 & 14 \end{array}\right]$$

Multiply row 1 by 5.4 and add to row 2.

$$\left[\begin{array}{cc|c} 1 & 0.43 & 22.61 \\ 0 & 3.32 & 136.09 \end{array}\right]$$

Multiply row 2 by $\frac{1}{3.32}$.

$$\left[\begin{array}{cc|c} 1 & 0.43 & 22.61 \\ 0 & 1 & 40.99 \end{array}\right]$$

This corresponds to $\begin{cases} x+0.43y=22.61 \\ y=40.99 \end{cases}$

$$\begin{aligned} x+0.43(40.99)&=22.61 \\ x&\approx 4.98 \end{aligned}$$

1980 + 4.98 = 1984.98

At the end of 1984, the percents were the same.

b. More households owned black-and-white television sets in 1980. More households owned microwave ovens in 1993.

The percent of households owning black-and-white television sets is decreasing and the percent of households owning microwave ovens is increasing; answers may vary.

c. Let $y = 0$.

$$\begin{aligned} 2.3x+0&=52 \\ x&\approx 22.6 \end{aligned}$$

1980 + 22.6 = 2002.6

The percent will reach 0% in 2002.

d. answers may vary

33. answers may vary

Section 4.5 Practice Problems

1. $\begin{cases} 2x \le y \\ x + 4y \ge 4 \end{cases}$

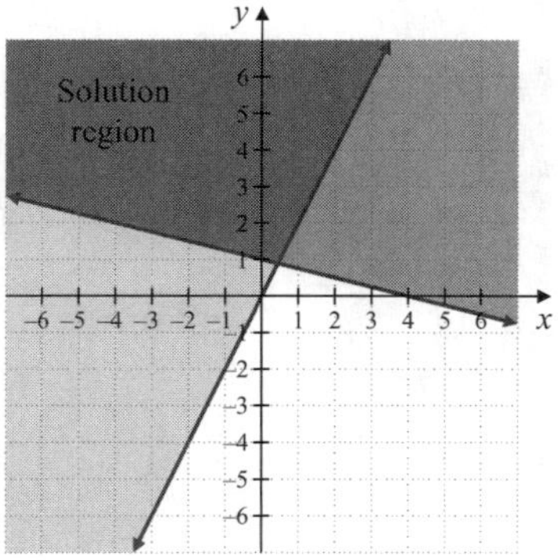

2. $\begin{cases} -x + y < 3 \\ y < 1 \\ 2x + y > -2 \end{cases}$

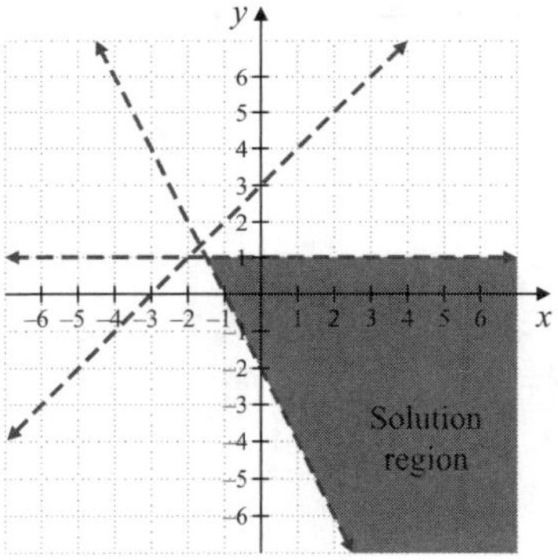

3. $\begin{cases} 2x - 3y \le 6 \\ y \ge 0 \\ y \le 4 \\ x \ge 0 \end{cases}$

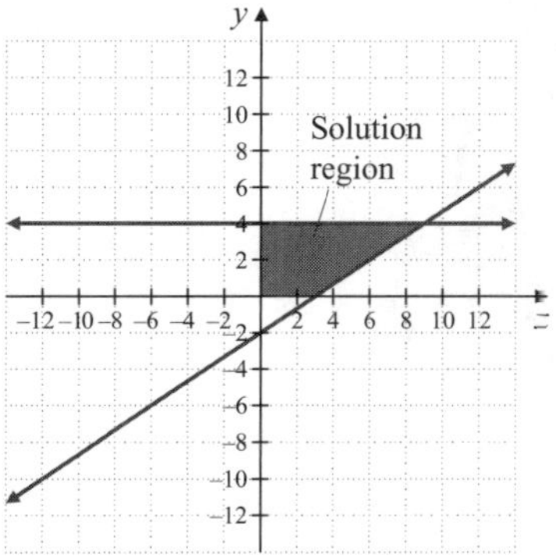

Exercise Set 4.5

1. $\begin{cases} y \ge x + 1 \\ y \ge 3 - x \end{cases}$

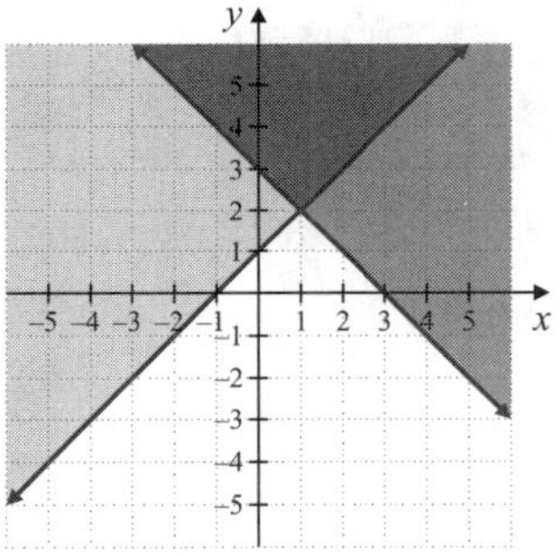

3. $\begin{cases} y < 3x - 4 \\ y \le x + 2 \end{cases}$

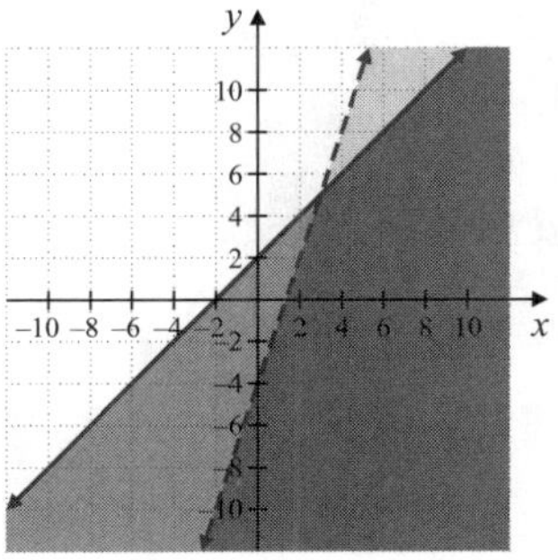

5. $\begin{cases} y < -2x - 2 \\ y > x + 4 \end{cases}$

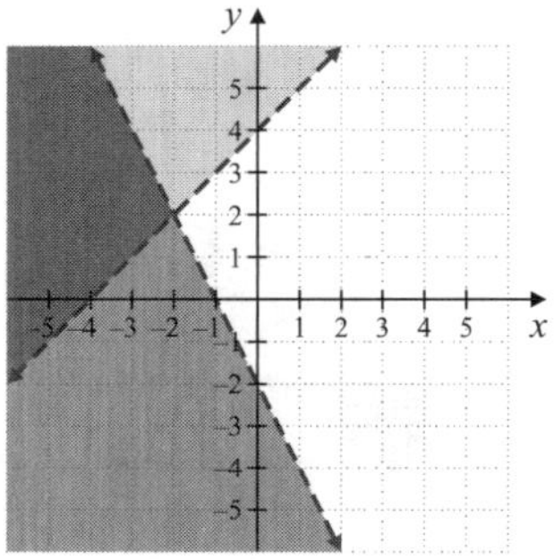

11. $\begin{cases} x \le 2 \\ y \ge -3 \end{cases}$

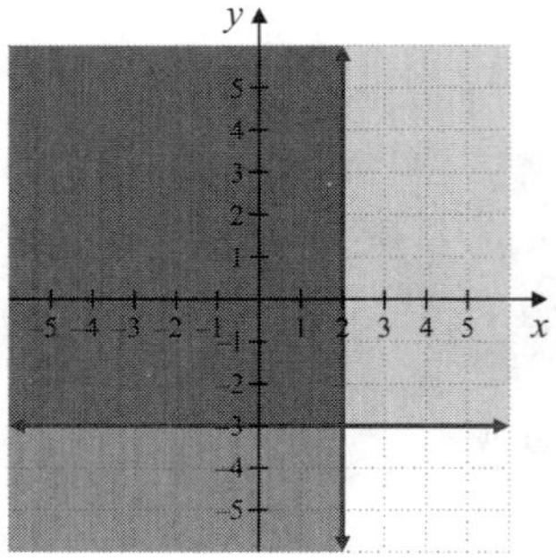

7. $\begin{cases} y \ge -x + 2 \\ y \le 2x + 5 \end{cases}$

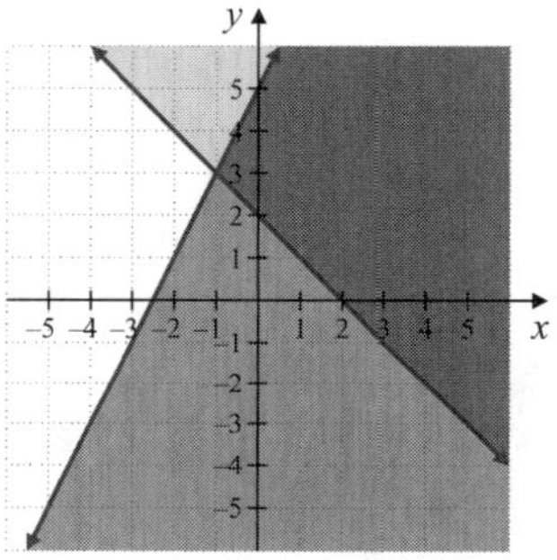

13. $\begin{cases} y \ge 1 \\ x < -3 \end{cases}$

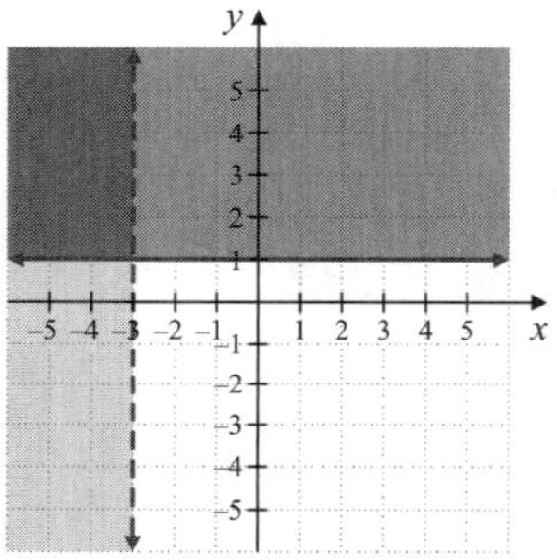

9. $\begin{cases} x \ge 3y \\ x + 3y \le 6 \end{cases}$

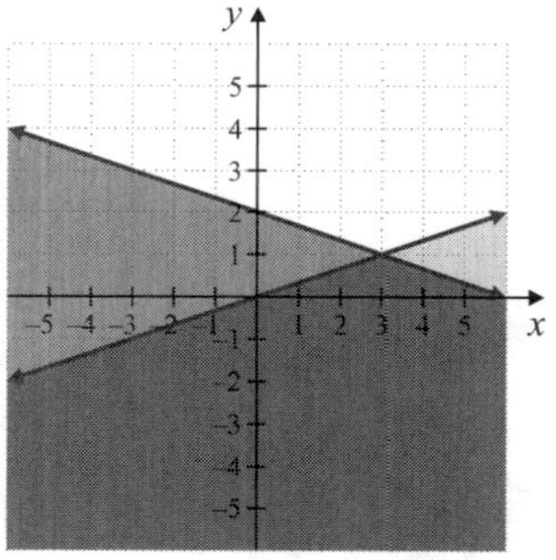

15. $\begin{cases} y + 2x \ge 0 \\ 5x - 3y \le 12 \\ y \le 2 \end{cases}$

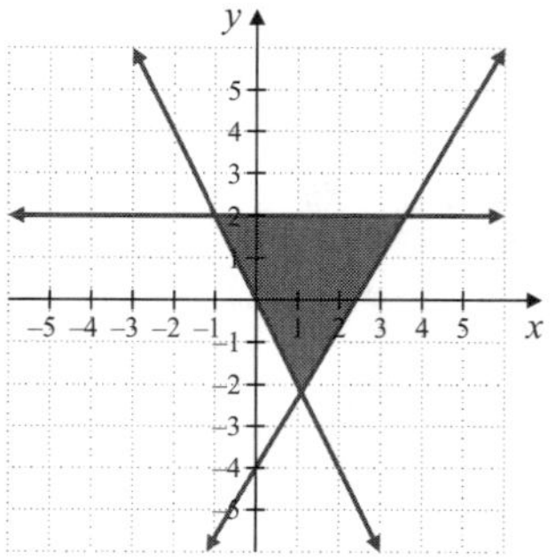

17. $\begin{cases} 3x - 4y \geq -6 \\ 2x + y \leq 7 \\ y \geq -3 \end{cases}$

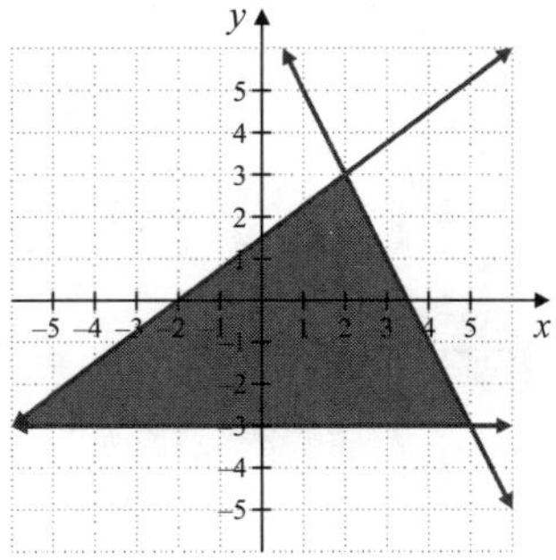

19. $\begin{cases} 2x + y \leq 5 \\ x \leq 3 \\ x \geq 0 \\ y \geq 0 \end{cases}$

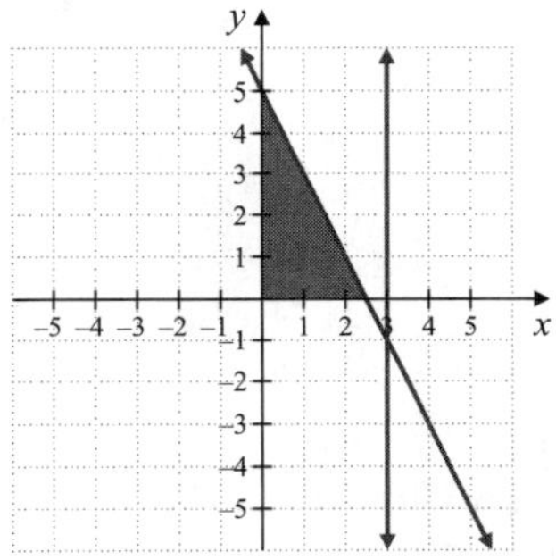

21. The graph of the system $\begin{cases} y < 5 \\ x > 3 \end{cases}$ is graph C.

23. The graph of the system $\begin{cases} y \leq 5 \\ x < 3 \end{cases}$ is graph D.

25. $(-3)^2 = (-3)(-3) = 9$

27. $\left(\frac{2}{3}\right)^2 = \left(\frac{2}{3}\right)\left(\frac{2}{3}\right) = \frac{4}{9}$

29.
$$\begin{aligned} (-2)^2 - (-3) + 2(-1) &= 4 - (-3) + 2(-1) \\ &= 4 - (-3) + (-2) \\ &= 4 + 3 + (-2) \\ &= 7 + (-2) \\ &= 5 \end{aligned}$$

31.
$$\begin{aligned} 8^2 + (-13) - 4(-2) &= 64 + (-13) - 4(-2) \\ &= 64 + (-13) + 8 \\ &= 51 + 8 \\ &= 59 \end{aligned}$$

33. The solution of the system $\begin{cases} y \leq 3 \\ y \geq 3 \end{cases}$ is the line $y = 3$.

35. answers may vary

Chapter 4 Vocabulary Check

1. Two or more linear equations in two variables form a <u>system of equations</u>.
2. A <u>solution</u> of a system of two equations in two variables is an ordered pair that makes both equations true.
3. A(n) <u>consistent</u> system of equations has at least one solution.
4. If a matrix has the same number of rows and columns, it is called a <u>square</u> matrix.
5. A(n) <u>inconsistent</u> system of equations has no solution.
6. A <u>matrix</u> is a rectangular array of numbers.

Chapter 4 Review

1. $\begin{cases} 3x+10y=1 \\ x+2y=-1 \end{cases}$

a. 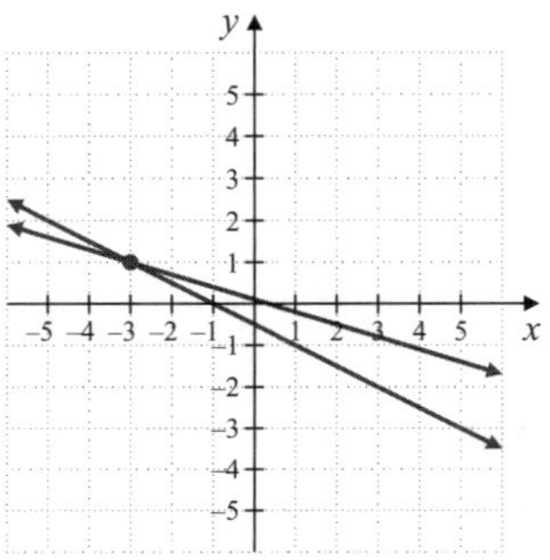

The solution is (−3, 1).

b. Solve the second equation for x.
$x = -2y - 1$
Replace x with $-2y - 1$ in the first equation.

$$\begin{aligned} 3x+10y&=1 \\ 3(-2y-1)+10y&=1 \\ -6y-3+10y&=1 \\ 4y&=4 \\ y&=1 \end{aligned}$$

Replace y with 1 in the equation $x = -2y - 1$.
$x = -2(1) - 1 = -2 - 1 = -3$
The solution is (−3, 1).

c. Multiply the second equation by −3 and add to the first equation.

$$\begin{aligned} 3x+10y&=1 \\ -3x-6y&=3 \\ \hline 4y&=4 \\ y&=1 \end{aligned}$$

Replace y with 1 in the second equation.

$$\begin{aligned} x+2y&=-1 \\ x+2(1)&=-1 \\ x&=-3 \end{aligned}$$

The solution is (−3, 1).

2. $\begin{cases} y=\frac{1}{2}x+\frac{2}{3} \\ 4x+6y=4 \end{cases}$

a. 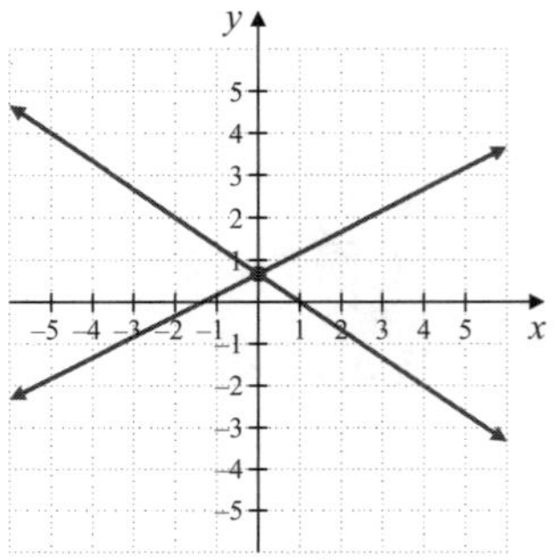

The solution is $\left(0, \frac{2}{3}\right)$.

b. Replace y with $\frac{1}{2}x+\frac{2}{3}$ in the second equation.

$$\begin{aligned} 4x+6y&=4 \\ 4x+6\left(\frac{1}{2}x+\frac{2}{3}\right)&=4 \\ 4x+3x+4&=4 \\ 7x&=0 \\ x&=0 \end{aligned}$$

Replace x with 0 in the first equation.

$$\begin{aligned} y&=\frac{1}{2}x+\frac{2}{3} \\ y&=\frac{1}{2}(0)+\frac{2}{3} \\ y&=\frac{2}{3} \end{aligned}$$

The solution is $\left(0, \frac{2}{3}\right)$.

c. Multiply the first equation by −6, rearrange, and add it to the second equation.

$$\begin{aligned} 3x-6y&=-4 \\ 4x+6y&=4 \\ \hline 7x\qquad&=0 \\ x&=0 \end{aligned}$$

Replace x with 0 in the first equation.

$y = \frac{1}{2}x + \frac{2}{3}$

$y = \frac{1}{2}(0) + \frac{2}{3}$

$y = \frac{2}{3}$

The solution is $\left(0, \frac{2}{3}\right)$.

3. $\begin{cases} 2x - 4y = 22 \\ 5x - 10y = 15 \end{cases}$

a.

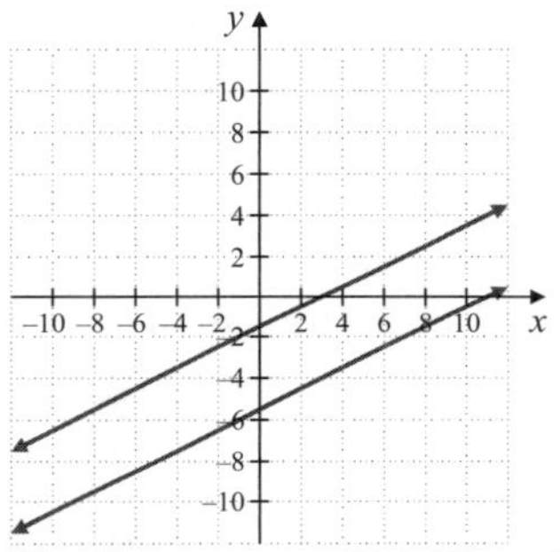

The lines are parallel. There is no solution, ∅.

b. Solve the first equation for x.

$2x - 4y = 22$

$x - 2y = 11$

$x = 2y + 11$

Replace x with $2y + 11$ in the second equation.

$5x - 10y = 15$

$5(2y + 11) - 10y = 15$

$10y + 55 - 10y = 15$

$55 = 15$

The statement 55 = 15 is false, so there is no solution.

c. Multiply the first equation by $-\frac{5}{2}$ and add it to the second equation.

$-5x + 10y = -55$

$5x - 10y = 15$

$0 = -40$

The statement 0 = –40 is false, so there is no solution.

4. $\begin{cases} 3x - 6y = 12 \\ 2y = x - 4 \end{cases}$

a.

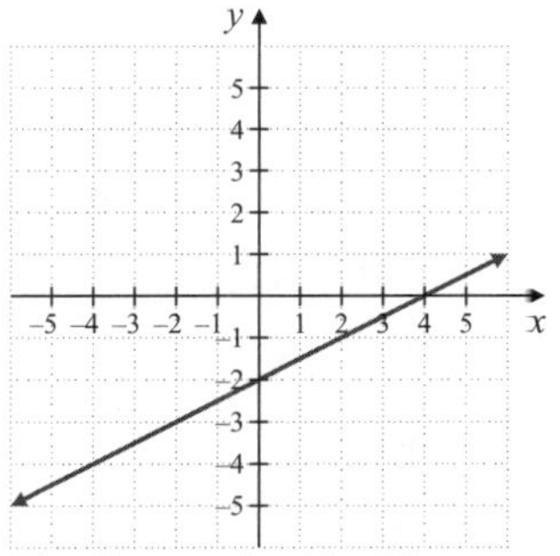

The lines coincide. The system is dependent. The solution set is $\{(x, y) | 3x - 6y = 12\}$.

b. Solve the second equation for y.

$y = \frac{x-4}{2}$

Replace y with $\frac{x-4}{2}$ in the first equation.

$3x - 6y = 12$

$3x - 6\left(\frac{x-4}{2}\right) = 12$

$3x - 3x + 12 = 12$

$12 = 12$

This statement is true, so the system is dependent. The solution set is $\{(x, y) | 3x - 6y = 12\}$.

c. Multiply the first equation by $\frac{1}{3}$ and add to the second equation.

$x - 2y = 4$

$-x + 2y = -4$

$0 = 0$

This statement is true, so the system is dependent. The solution set is $\{(x, y) | 3x - 6y = 12\}$.

5. $\begin{cases} \frac{1}{2}x - \frac{3}{4}y = -\frac{1}{2} \\ \frac{1}{8}x + \frac{3}{4}y = \frac{19}{8} \end{cases}$

a.

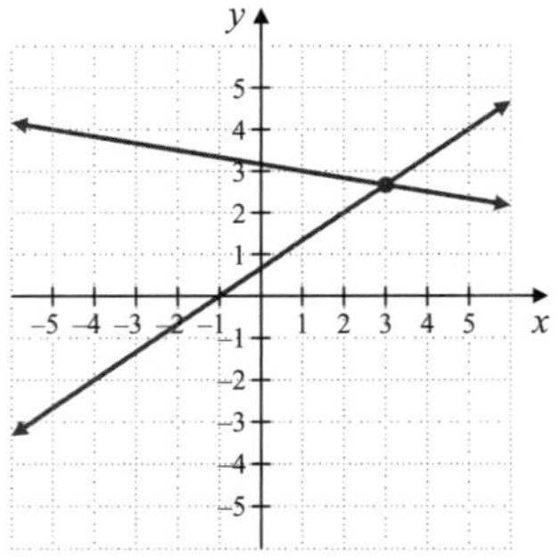

The solution is $\left(3, \frac{8}{3}\right)$.

b. Solve the first equation for y.

$$\frac{1}{2}x - \frac{3}{4}y = -\frac{1}{2}$$
$$-4\left(\frac{1}{2}x - \frac{3}{4}y\right) = -4\left(-\frac{1}{2}\right)$$
$$-2x + 3y = 2$$
$$3y = 2 + 2x$$
$$y = \frac{2+2x}{3}$$

Replace y with $\frac{2+2x}{3}$ in the second equation.

$$\frac{1}{8}x + \frac{3}{4}y = \frac{19}{8}$$
$$\frac{1}{8}x + \frac{3}{4}\left(\frac{2+2x}{3}\right) = \frac{19}{8}$$
$$x + 2(2+2x) = 19$$
$$x + 4 + 4x = 19$$
$$5x = 15$$
$$x = 3$$

Replace x with 3 in the equation $y = \frac{2+2x}{3}$.

$$y = \frac{2+2(3)}{3}$$
$$y = \frac{8}{3}$$

The solution is $\left(3, \frac{8}{3}\right)$.

c. Add the equations.

$$\frac{1}{2}x - \frac{3}{4}y = -\frac{1}{2}$$
$$\frac{1}{8}x + \frac{3}{4}y = \frac{19}{8}$$
$$\frac{1}{2}x + \frac{1}{8}x = -\frac{1}{2} + \frac{19}{8}$$
$$\frac{5}{8}x = \frac{15}{8}$$
$$x = 3$$

Replace x with 3 in the second equation.

$$\frac{1}{8}x + \frac{3}{4}y = \frac{19}{8}$$
$$\frac{1}{8}(3) + \frac{3}{4}y = \frac{19}{8}$$
$$\frac{3}{4}y = 2$$
$$y = \frac{8}{3}$$

The solution is $\left(3, \frac{8}{3}\right)$.

6. $\begin{cases} y = 32x \\ y = 15x + 25{,}500 \end{cases}$

Use substitution.

$$32x = 15x + 25{,}500$$
$$17x = 25{,}500$$
$$x = 1500$$

The company must sell 1500 backpacks to break even.

7. $\begin{cases} x \quad + z = 4 & (1) \\ 2x - y \quad = 4 & (2) \\ x + y - z = 0 & (3) \end{cases}$

Solve equation (1) for z, $z = 4 - x$.
Solve equation (2) for y, $y = 2x - 4$.

Replace z with $4 - x$ and y with $2x - 4$ in equation (3).

$$\begin{aligned} x+y-z&=0\\ x+2x-4-(4-x)&=0\\ x+2x-4-4+x&=0\\ 4x&=8\\ x&=2 \end{aligned}$$

Replace x with 2 in the equation $z = 4 - x$.
$z = 4 - 2 = 2$
Replace x with 2 in the equation $y = 2x - 4$.
$y = 2(2) - 4 = 4 - 4 = 0$
The solution is (2, 0, 2).

8. $\begin{cases} 2x+5y = 4 & (1)\\ x-5y+z=-1 & (2)\\ 4x - z = 11 & (3) \end{cases}$

Add equations (1) and (2).

$$\begin{array}{l} 2x+5y \quad = 4\\ \underline{x-5y+z=-1}\\ 3x \quad +z=3 \quad (4) \end{array}$$

Add equations (3) and (4).

$$\begin{array}{l} 4x-z=11\\ \underline{3x+z=3}\\ 7x \quad =14\\ \quad x=2 \end{array}$$

Replace x with 2 in equation (4).

$$\begin{aligned} 3x+z&=3\\ 3(2)+z&=3\\ z&=-3 \end{aligned}$$

Replace x with 2 in equation (1).

$$\begin{aligned} 2x+5y&=4\\ 2(2)+5y&=4\\ 5y&=0\\ y&=0 \end{aligned}$$

The solution is (2, 0, –3).

9. $\begin{cases} 4y+2z=5 & (1)\\ 2x+8y = 5 & (2)\\ 6x +4z=1 & (3) \end{cases}$

Multiply equation (1) by –2 and add to equation (2).

$$\begin{array}{l} -8y-4z=-10\\ \underline{2x+8y \quad =5}\\ 2x \quad -4z=-5 \quad (4) \end{array}$$

Add equations (3) and (4).

$$\begin{array}{l} 6x+4z=1\\ \underline{2x-4z=-5}\\ 8x \quad =-4\\ \quad x=-\frac{1}{2} \end{array}$$

Replace x with $-\frac{1}{2}$ in equation (2).

$$\begin{aligned} 2x+8y&=5\\ 2\left(-\frac{1}{2}\right)+8y&=5\\ 8y&=6\\ y&=\frac{3}{4} \end{aligned}$$

Replace x with $-\frac{1}{2}$ in equation (3).

$$\begin{aligned} 6x+4z&=1\\ 6\left(-\frac{1}{2}\right)+4z&=1\\ -3+4z&=1\\ 4z&=4\\ z&=1 \end{aligned}$$

The solution is $\left(-\frac{1}{2}, \frac{3}{4}, 1\right)$.

10. $\begin{cases} 5x+7y = 9 & (1)\\ 14y-z=28 & (2)\\ 4x +2z=-4 & (3) \end{cases}$

Multiply equation (1) by –2 and add to equation (2).

$$\begin{array}{l} -10x-14y \quad =-18\\ \underline{\quad 14y-z=28}\\ -10x \quad -z=10 \quad (4) \end{array}$$

Multiply equation (3) by $\frac{1}{2}$ and add to equation (4).

$$\begin{array}{l} 2x+z=-2\\ \underline{-10x-z=10}\\ -8x \quad =8\\ \quad x=-1 \end{array}$$

Replace x with –1 in equation (1).

$$\begin{aligned} 5x+7y&=9\\ 5(-1)+7y&=9\\ 7y&=14\\ y&=2 \end{aligned}$$

Replace y with 2 in equation (2).

$$\begin{aligned} 14y-z&=28\\ 14(2)-z&=28\\ -z&=0\\ z&=0 \end{aligned}$$

The solution is (–1, 2, 0).

11. $\begin{cases} 3x-2y+2z=5 & (1)\\ -x+6y+z=4 & (2)\\ 3x+14y+7z=20 & (3) \end{cases}$

Multiply equation (1) by 3 and add to equation (2).

$$\begin{array}{r} 9x-6y+6z=15\\ \underline{-x+6y+z=4}\\ 8x\qquad+7z=19 \end{array}\quad(4)$$

Multiply equation (1) by 7 and add to equation (3).

$$\begin{array}{r} 21x-14y+14z=35\\ \underline{3x+14y+7z=20}\\ 24x\qquad+21z=55 \end{array}\quad(5)$$

Multiply equation (4) by –3 and add to equation (5).

$$\begin{array}{r} -24x-21z=-57\\ \underline{24x+21z=55}\\ 0=-2 \end{array}$$

The statement 0 = –2 is false, so there is no solution, or ∅.

12. $\begin{cases} x+2y+3z=11 & (1)\\ y+2z=3 & (2)\\ 2x+2z=10 & (3) \end{cases}$

Solve equation (2) for y, $y = 3 - 2z$. Solve equation (3) for x, $x = 5 - z$. Replace y with $3 - 2z$ and x with $5 - z$ in equation (1).

$$\begin{aligned} x+2y+3z&=11\\ 5-z+2(3-2z)+3z&=11\\ 5-z+6-4z+3z&=11\\ -2z&=0\\ z&=0 \end{aligned}$$

Replace z with 0 in the equation $y = 3 - 2z$.

$y = 3 - 2(0) = 3 - 0 = 3$

Replace z with 0 in the equation $x = 5 - z$.

$x = 5 - 0 = 5$

The solution is (5, 3, 0).

13. $\begin{cases} 7x-3y+2z=0 & (1)\\ 4x-4y-z=2 & (2)\\ 5x+2y+3z=1 & (3) \end{cases}$

Multiply equation (2) by 2 and add to equation (1).

$$\begin{array}{r} 7x-3y+2z=0\\ \underline{8x-8y-2z=4}\\ 15x-11y\qquad=4 \end{array}\quad(4)$$

Multiply equation (2) by 3 and add to equation (3).

$$\begin{array}{r} 12x-12y-3z=6\\ \underline{5x+2y+3z=1}\\ 17x-10y\qquad=7 \end{array}\quad(5)$$

Multiply equation (4) by –10 and equation (5) by 11. Then add the results.

$$\begin{array}{r} -150x+110y=-40\\ \underline{187x-110y=77}\\ 37x\qquad=37\\ x=1 \end{array}$$

Replace x with 1 in equation (4).

$$\begin{aligned} 15x-11y&=4\\ 15(1)-11y&=4\\ -11y&=-11\\ y&=1 \end{aligned}$$

Replace x with 1 and y with 1 in equation (2).

$$\begin{aligned} 4x-4y-z&=2\\ 4(1)-4(1)-z&=2\\ z&=-2 \end{aligned}$$

The solution is (1, 1, –2).

14. $\begin{cases} x-3y-5z=-5 & (1)\\ 4x-2y+3z=13 & (2)\\ 5x+3y+4z=22 & (3) \end{cases}$

Multiply equation (1) by –4 and add to equation (2).

$$\begin{array}{r} -4x+12y+20z=20\\ \underline{4x-2y+3z=13}\\ 10y+23z=33 \end{array}\quad(4)$$

Multiply equation (1) by −5 and add to equation (3).

$$\begin{array}{r} -5x+15y+25z=25 \\ 5x\ +3y\ +4z=22 \\ \hline 18y+29z=47 \quad (5) \end{array}$$

Multiply equation (4) by −18 and equation (5) by 10. Then add the results.

$$\begin{array}{r} -180y-414z=-594 \\ 180y+290z=470 \\ \hline -124z=-124 \\ z=1 \end{array}$$

Replace z with 1 in equation (4).

$$\begin{aligned} 10y+23z&=33 \\ 10y+23(1)&=33 \\ 10y&=10 \\ y&=1 \end{aligned}$$

Replace y with 1 and z with 1 in equation (1).

$$\begin{aligned} x-3y-5z&=-5 \\ x-3(1)-5(1)&=-5 \\ x-8&=-5 \\ x&=3 \end{aligned}$$

The solution is (3, 1, 1).

15. Let a be the first number, b be the second number, and c be the third number.

$$\begin{cases} a+b+c=98 & (1) \\ a+b\quad =c+2 & (2) \\ \quad b\quad =4a & (3) \end{cases}$$

Rearrange equation (2) and add it to equation (1).

$$\begin{array}{r} a\ +b+c=98 \\ a\ +b-c=2 \\ \hline 2a+2b\quad =100 \quad (4) \end{array}$$

Replace b with $4a$ in equation (4).

$$\begin{aligned} 2a+2(4a)&=100 \\ 10a&=100 \\ a&=10 \end{aligned}$$

Replace a with 10 in equation (3).

$b = 4a = 4(10) = 40$

Replace a with 10 and b with 40 in equation (1).

$$\begin{aligned} 10+40+c&=98 \\ c&=48 \end{aligned}$$

The numbers are 10, 40, and 48.

16. Let a be the first number and b be the second number.

$$\begin{cases} a=3b \\ 2(a+b)=168 \end{cases} \quad \text{or} \quad \begin{cases} a\quad =3b & (1) \\ a+b=84 & (2) \end{cases}$$

Replace a with $3b$ in equation (2).

$$\begin{aligned} 3b+b&=84 \\ 4b&=84 \\ b&=21 \end{aligned}$$

Replace b with 21 in equation (1).

$a = 3(21) = 63$

The numbers are 63 and 21.

17. Let e be the rate of the east-bound car and w be the rate of the west-bound car.

rate ·	time	= distance
e	4	$4e$
w	4	$4w$

$$\begin{cases} 4e+4w=492 & (1) \\ w=e+7 & (2) \end{cases}$$

Replace w with $e + 7$ in equation (1).

$$\begin{aligned} 4e+4(e+7)&=492 \\ 4e+4e+28&=492 \\ 8e&=464 \\ e&=58 \end{aligned}$$

Replace e with 58 in equation (2).

$w = 58 + 7 = 65$

The cars are traveling at 58 mph and 65 mph.

18. Let l = length, and w = width.

$$\begin{cases} l\quad =3w & (1) \\ 2l+2w=296 & (2) \end{cases}$$

Replace l with $3w$ in equation (2).

$$\begin{aligned} 2l+2w&=296 \\ 2(3w)+2w&=296 \\ 8w&=296 \\ w&=37 \end{aligned}$$

Replace w with 37 in equation (1).

$l = 3(37) = 111$

The width is 37 feet and the length is 111 feet.

19. Let x = amount of 10% solution, and y = amount of 60% solution.

liters	strength	total amount of alcohol
x	10%	$0.1x$
y	60%	$0.6y$
50	40%	$50(0.4) = 20$

$$\begin{cases} x+y=50 & (1) \\ 0.1x+0.6y=20 & (2) \end{cases}$$

Solve equation (1) for x, $x = 50 - y$.
Replace x with $50 - y$ in equation (2).

$$\begin{aligned} 0.1(50-y)+0.6y &= 20 \\ 5-0.1y+0.6y &= 20 \\ 0.5y &= 15 \\ y &= 30 \end{aligned}$$

Replace y with 30 in equation (1).

$$\begin{aligned} x+y &= 50 \\ x+30 &= 50 \\ x &= 20 \end{aligned}$$

Mix 20 liters of 10% solution and 30 liters of 60% solution.

20. Let c = pounds of creme-filled chocolates, n = pounds of chocolate-covered nuts, and r = pounds of chocolate-covered raisins.

$$\begin{cases} c+n+r=45 & (1) \\ 3c+2.7n+2.25r=45(2.8) & (2) \\ r=2n & (3) \end{cases}$$

Multiply equation (1) by –3 and add to equation (2).

$$\begin{aligned} -3c \quad -3n \quad -3r &= -135 \\ 3c+2.7n+2.25r &= 126 \\ \hline -0.3n-0.75r &= -9 \quad (4) \end{aligned}$$

Replace r with $2n$ in equation (4).

$$\begin{aligned} -0.3n-0.75(2n) &= -9 \\ -0.3n-1.5n &= -9 \\ -1.8n &= -9 \\ n &= 5 \end{aligned}$$

Replace n with 5 in equation (3).
$r = 2(5) = 10$
Replace n with 5 and r with 10 in equation (1).

$$\begin{aligned} c+5+10 &= 45 \\ c &= 30 \end{aligned}$$

She should use 30 pounds of creme-filled chocolates, 5 pounds of chocolate-covered nuts, and 10 pounds of chocolate-covered raisins.

21. Let p = number of pennies, n = number of nickels, and d = number of dimes.

$$\begin{cases} p+n+d=53 & (1) \\ n=d+4 & (2) \\ 0.01p+0.05n+0.10d=2.77 & (3) \end{cases}$$

Multiply equation (1) by –0.01 and add to equation (3).

$$\begin{aligned} -0.01p-0.01n-0.01d &= -0.53 \\ 0.01p+0.05n+0.10d &= 2.77 \\ \hline 0.04n+0.09d &= 2.24 \quad (4) \end{aligned}$$

Replace n with $d + 4$ in equation (4).

$$\begin{aligned} 0.04(d+4)+0.09d &= 2.24 \\ 0.04d+0.16+0.09d &= 2.24 \\ 0.13d &= 2.08 \\ d &= 16 \end{aligned}$$

Replace d with 16 in equation (2).
$n = d + 4 = 16 + 4 = 20$
Replace d with 16 and n with 20 in equation (1).

$$\begin{aligned} p+20+16 &= 53 \\ p &= 17 \end{aligned}$$

She has 17 pennies, 20 nickels, and 16 dimes.

22. Let l be the rate of larger investment, and s be the rate of smaller investment.

$$\begin{cases} 10{,}000l+4000s=1250 & (1) \\ l=s+0.02 & (2) \end{cases}$$

Replace l with $s + 0.02$ in equation (1).

$$\begin{aligned} 10{,}000(s+0.02)+4000s &= 1250 \\ 10{,}000s+200+4000s &= 1250 \\ 14{,}000s &= 1050 \\ s &= 0.075 \end{aligned}$$

Replace s with 0.075 in equation (2).
$l = 0.075 + 0.02 = 0.095$
The rate of the larger investment is 9.5% and the rate of the smaller investment is 7.5%.

23. Let x be the length of each of the equal sides, and y be the unequal side.

$$\begin{cases} 2x + y = 73 & (1) \\ y = x + 7 & (2) \end{cases}$$

Replace y with $x + 7$ in the first equation.

$$2x + x + 7 = 73$$
$$3x = 66$$
$$x = 22$$

Replace x with 22 in equation (2).
$y = 22 + 7 = 29$
The two equal sides measure 22 centimeters each and the third side measures 29 centimeters.

24. Let a = first number, b = second number, and c = third number.

$$\begin{cases} a + b + c = 295 & (1) \\ a = b + 5 & (2) \\ a = 2c & (3) \end{cases}$$

Solve equation (2) for b, $b = a - 5$. Solve equation (3) for c, $c = \frac{a}{2}$.

Replace b with $a - 5$ and c with $\frac{a}{2}$ in equation (1).

$$a + a - 5 + \frac{a}{2} = 295$$
$$\frac{5a}{2} = 300$$
$$a = 120$$

Replace a with 120 in the equation $b = a - 5$.
$b = 120 - 5 = 115$

Replace a with 120 in the equation $c = \frac{a}{2}$.

$$c = \frac{120}{2} = 60$$

The numbers are 120, 115, and 60.

25. $\begin{cases} 3x + 10y = 1 \\ x + 2y = -1 \end{cases}$

$$\left[\begin{array}{cc|c} 3 & 10 & 1 \\ 1 & 2 & -1 \end{array}\right]$$

Interchange rows 1 and 2.

$$\left[\begin{array}{cc|c} 1 & 2 & -1 \\ 3 & 10 & 1 \end{array}\right]$$

Multiply row 1 by −3 and add to row 2.

$$\left[\begin{array}{cc|c} 1 & 2 & -1 \\ 0 & 4 & 4 \end{array}\right]$$

Multiply row 2 by $\frac{1}{4}$.

$$\left[\begin{array}{cc|c} 1 & 2 & -1 \\ 0 & 1 & 1 \end{array}\right]$$

This corresponds to $\begin{cases} x + 2y = -1 \\ y = 1 \end{cases}$.

$$x + 2(1) = -1$$
$$x = -3$$

The solution is (−3, 1).

26. $\begin{cases} 3x - 6y = 12 \\ 2y = x - 4 \end{cases}$

or

$$\begin{cases} 3x - 6y = 12 \\ x - 2y = 4 \end{cases}$$

$$\left[\begin{array}{cc|c} 3 & -6 & 12 \\ 1 & -2 & 4 \end{array}\right]$$

Multiply row 1 by $\frac{1}{3}$.

$$\left[\begin{array}{cc|c} 1 & -2 & 4 \\ 1 & -2 & 4 \end{array}\right]$$

This corresponds to $\begin{cases} x - 2y = 4 \\ x - 2y = 4 \end{cases}$.

The system is dependent. The solution set is $\{(x, y) | x - 2y = 4\}$.

27. $\begin{cases} 3x - 2y = -8 \\ 6x + 5y = 11 \end{cases}$

$$\left[\begin{array}{cc|c} 3 & -2 & -8 \\ 6 & 5 & 11 \end{array}\right]$$

Multiply row 1 by −2 and add to row 2.

$$\left[\begin{array}{cc|c} 3 & -2 & -8 \\ 0 & 9 & 27 \end{array}\right]$$

Multiply row 1 by $\frac{1}{3}$ and row 2 by $\frac{1}{9}$.

$$\left[\begin{array}{cc|c} 1 & -\frac{2}{3} & -\frac{8}{3} \\ 0 & 1 & 3 \end{array}\right]$$

This corresponds to $\begin{cases} x-\dfrac{2}{3}y=-\dfrac{8}{3} \\ y=3 \end{cases}$.

$$x-\frac{2}{3}(3)=-\frac{8}{3}$$
$$x-2=-\frac{8}{3}$$
$$x=-\frac{8}{3}+\frac{6}{3}$$
$$x=-\frac{2}{3}$$

The solution is $\left(-\frac{2}{3}, 3\right)$.

28. $\begin{cases} 6x-6y=-5 \\ 10x-2y=1 \end{cases}$

$$\left[\begin{array}{cc|c} 6 & -6 & -5 \\ 10 & -2 & 1 \end{array}\right]$$

Multiply row 1 by $\frac{1}{6}$.

$$\left[\begin{array}{cc|c} 1 & -1 & -\frac{5}{6} \\ 10 & -2 & 1 \end{array}\right]$$

Multiply row 1 by −10 and add to row 2.

$$\left[\begin{array}{cc|c} 1 & -1 & -\frac{5}{6} \\ 0 & 8 & \frac{28}{3} \end{array}\right]$$

Multiply row 2 by $\frac{1}{8}$.

$$\left[\begin{array}{cc|c} 1 & -1 & -\frac{5}{6} \\ 0 & 1 & \frac{7}{6} \end{array}\right]$$

This corresponds to $\begin{cases} x-y=-\dfrac{5}{6} \\ y=\dfrac{7}{6} \end{cases}$.

$$x-\frac{7}{6}=-\frac{5}{6}$$
$$x=\frac{1}{3}$$

The solution is $\left(\frac{1}{3}, \frac{7}{6}\right)$.

29. $\begin{cases} 3x-6y=0 \\ 2x+4y=5 \end{cases}$

$$\left[\begin{array}{cc|c} 3 & -6 & 0 \\ 2 & 4 & 5 \end{array}\right]$$

Multiply row 1 by $\frac{1}{3}$.

$$\left[\begin{array}{cc|c} 1 & -2 & 0 \\ 2 & 4 & 5 \end{array}\right]$$

Multiply row 1 by −2 and add to row 2.

$$\left[\begin{array}{cc|c} 1 & -2 & 0 \\ 0 & 8 & 5 \end{array}\right]$$

Multiply row 2 by $\frac{1}{8}$.

$$\left[\begin{array}{cc|c} 1 & -2 & 0 \\ 0 & 1 & \frac{5}{8} \end{array}\right]$$

This corresponds to $\begin{cases} x-2y=0 \\ y=\dfrac{5}{8} \end{cases}$.

$$x-2\left(\frac{5}{8}\right)=0$$
$$x-\frac{10}{8}=0$$
$$x=\frac{5}{4}$$

The solution is $\left(\frac{5}{4}, \frac{5}{8}\right)$.

30. $\begin{cases} 5x-3y=10 \\ -2x+y=-1 \end{cases}$

$$\left[\begin{array}{cc|c} 5 & -3 & 10 \\ -2 & 1 & -1 \end{array}\right]$$

Multiply row 1 by $\frac{1}{5}$.

$$\left[\begin{array}{cc|c} 1 & -\frac{3}{5} & 2 \\ -2 & 1 & -1 \end{array}\right]$$

Multiply row 1 by 2 and add it to row 2.

$$\left[\begin{array}{cc|c} 1 & -\frac{3}{5} & 2 \\ 0 & -\frac{1}{5} & 3 \end{array}\right]$$

Multiply row 2 by –5.

$$\left[\begin{array}{cc|c} 1 & -\frac{3}{5} & 2 \\ 0 & 1 & -15 \end{array}\right]$$

This corresponds to $\begin{cases} x - \dfrac{3}{5}y = 2 \\ y = -15 \end{cases}$

$$\begin{aligned} x - \frac{3}{5}(-15) &= 2 \\ x + 9 &= 2 \\ x &= -7 \end{aligned}$$

The solution is (–7, –15).

31. $\begin{cases} 0.2x - 0.3y = -0.7 \\ 0.5x + 0.3y = 1.4 \end{cases}$

$$\left[\begin{array}{cc|c} 0.2 & 0.3 & -0.7 \\ 0.5 & 0.3 & 1.4 \end{array}\right]$$

Multiply both rows by 10.

$$\left[\begin{array}{cc|c} 2 & -3 & -7 \\ 5 & 3 & 14 \end{array}\right]$$

Multiply row 1 by $\dfrac{1}{2}$.

$$\left[\begin{array}{cc|c} 1 & -\frac{3}{2} & -\frac{7}{2} \\ 5 & 3 & 14 \end{array}\right]$$

Multiply row 1 by –5 and add to row 2.

$$\left[\begin{array}{cc|c} 1 & -\frac{3}{2} & -\frac{7}{2} \\ 0 & \frac{21}{2} & \frac{63}{2} \end{array}\right]$$

Multiply row 2 by $\dfrac{2}{21}$.

$$\left[\begin{array}{cc|c} 1 & -\frac{3}{2} & -\frac{7}{2} \\ 0 & 1 & 3 \end{array}\right]$$

This corresponds to $\begin{cases} x - \dfrac{3}{2}y = -\dfrac{7}{2}. \\ y = 3 \end{cases}$

$$\begin{aligned} x - \frac{3}{2}(3) &= -\frac{7}{2} \\ x - \frac{9}{2} &= -\frac{7}{2} \\ x &= 1 \end{aligned}$$

The solution is (1, 3).

32. $\begin{cases} 3x + 2y = 8 \\ 3x - y = 5 \end{cases}$

$$\left[\begin{array}{cc|c} 3 & 2 & 8 \\ 3 & -1 & 5 \end{array}\right]$$

Multiply row 1 by –1 and add to row 2.

$$\left[\begin{array}{cc|c} 3 & 2 & 8 \\ 0 & -3 & -3 \end{array}\right]$$

Multiply row 1 by $\dfrac{1}{3}$ and row 2 by $-\dfrac{1}{3}$.

$$\left[\begin{array}{cc|c} 1 & \frac{2}{3} & \frac{8}{3} \\ 0 & 1 & 1 \end{array}\right]$$

This corresponds to $\begin{cases} x + \dfrac{2}{3}y = \dfrac{8}{3}. \\ y = 1 \end{cases}$

$$\begin{aligned} x + \frac{2}{3}(1) &= \frac{8}{3} \\ x &= 2 \end{aligned}$$

The solution is (2, 1).

33. $\begin{cases} x \quad\;\; + z = 4 \\ 2x - y \quad\;\; = 0 \\ x + y - z = 0 \end{cases}$

$$\left[\begin{array}{ccc|c} 1 & 0 & 1 & 4 \\ 2 & -1 & 0 & 0 \\ 1 & 1 & -1 & 0 \end{array}\right]$$

Multiply row 1 by –2 and add to row 2.
Multiply row 1 by –1 and add to row 3.

$$\left[\begin{array}{ccc|c} 1 & 0 & 1 & 4 \\ 0 & -1 & -2 & -8 \\ 0 & 1 & -2 & -4 \end{array}\right]$$

Add row 2 to row 3.

$$\left[\begin{array}{ccc|c} 1 & 0 & 1 & 4 \\ 0 & -1 & -2 & -8 \\ 0 & 0 & -4 & -12 \end{array}\right]$$

Multiply row 2 by –1 and row 3 by $-\frac{1}{4}$.

$$\left[\begin{array}{ccc|c} 1 & 0 & 1 & 4 \\ 0 & 1 & 2 & 8 \\ 0 & 0 & 1 & 3 \end{array}\right]$$

This corresponds to $\begin{cases} x+z=4 \\ y+2z=8 \\ z=3 \end{cases}$.

$y+2(3)=8$
$y=2$
$x+3=4$
$x=1$

The solution is (1, 2, 3).

34. $\begin{cases} 2x+5y \quad =4 \\ x-5y+z=-1 \\ 4x \quad -z=11 \end{cases}$

$$\left[\begin{array}{ccc|c} 2 & 5 & 0 & 4 \\ 1 & -5 & 1 & -1 \\ 4 & 0 & -1 & 11 \end{array}\right]$$

Interchange rows 1 and 2.

$$\left[\begin{array}{ccc|c} 1 & -5 & 1 & -1 \\ 2 & 5 & 0 & 4 \\ 4 & 0 & -1 & 11 \end{array}\right]$$

Multiply row 1 by –2 and add to row 2.
Multiply row 1 by –4 and add to row 3.

$$\left[\begin{array}{ccc|c} 1 & -5 & 1 & -1 \\ 0 & 15 & -2 & 6 \\ 0 & 20 & -5 & 15 \end{array}\right]$$

Multiply row 2 by $\frac{1}{15}$. Multiply row 3 by $-\frac{1}{5}$.

$$\left[\begin{array}{ccc|c} 1 & -5 & 1 & -1 \\ 0 & 1 & -\frac{2}{15} & \frac{2}{5} \\ 0 & -4 & 1 & -3 \end{array}\right]$$

Multiply row 2 by 4 and add to row 3.

$$\left[\begin{array}{ccc|c} 1 & -5 & 1 & -1 \\ 0 & 1 & -\frac{2}{15} & \frac{2}{5} \\ 0 & 0 & \frac{7}{15} & -\frac{7}{5} \end{array}\right]$$

Multiply row 3 by $\frac{15}{7}$.

$$\left[\begin{array}{ccc|c} 1 & -5 & 1 & -1 \\ 0 & 1 & -\frac{2}{15} & \frac{2}{5} \\ 0 & 0 & 1 & -3 \end{array}\right]$$

This corresponds to $\begin{cases} x-5y \quad +z=-1 \\ y-\frac{2}{15}z=\frac{2}{5} \\ z=-3 \end{cases}$.

$y-\frac{2}{15}(-3)=\frac{2}{5}$
$y=0$
$x-5(0)+(-3)=-1$
$x=2$

The solution is (2, 0, –3).

35. $\begin{cases} 3x-y \quad =11 \\ x \quad +2z=13 \\ y-z=-7 \end{cases}$

$$\left[\begin{array}{ccc|c} 3 & -1 & 0 & 11 \\ 1 & 0 & 2 & 13 \\ 0 & 1 & -1 & -7 \end{array}\right]$$

Interchange rows 1 and 2.

$$\left[\begin{array}{ccc|c} 1 & 0 & 2 & 13 \\ 3 & -1 & 0 & 11 \\ 0 & 1 & -1 & -7 \end{array}\right]$$

Multiply row 1 by –3 and add to row 2.

$$\left[\begin{array}{ccc|c} 1 & 0 & 2 & 13 \\ 0 & -1 & -6 & -28 \\ 0 & 1 & -1 & -7 \end{array}\right]$$

Add row 2 to row 3.

$$\left[\begin{array}{ccc|c} 1 & 0 & 2 & 13 \\ 0 & -1 & -6 & -28 \\ 0 & 0 & -7 & -35 \end{array}\right]$$

Multiply row 2 by –1 and row 3 by $-\frac{1}{7}$.

$$\left[\begin{array}{ccc|c} 1 & 0 & 2 & 13 \\ 0 & 1 & 6 & 28 \\ 0 & 0 & 1 & 5 \end{array}\right]$$

This corresponds to $\begin{cases} x+2z=13 \\ y+6z=28. \\ z=5 \end{cases}$

$$y+6(5)=28$$
$$y+30=28$$
$$y=-2$$
$$x+2(5)=13$$
$$x=3$$

The solution is (3, −2, 5).

36. $\begin{cases} 5x+7y+3z=9 \\ 14y-z=28 \\ 4x+2z=-4 \end{cases}$

$$\left[\begin{array}{ccc|c} 5 & 7 & 3 & 9 \\ 0 & 14 & -1 & 28 \\ 4 & 0 & 2 & -4 \end{array}\right]$$

Multiply row 1 by $\frac{1}{5}$ and row 2 by $\frac{1}{14}$.

$$\left[\begin{array}{ccc|c} 1 & \frac{7}{5} & \frac{3}{5} & \frac{9}{5} \\ 0 & 1 & -\frac{1}{14} & 2 \\ 4 & 0 & 2 & -4 \end{array}\right]$$

Multiply row 1 by −4 and add to row 3.

$$\left[\begin{array}{ccc|c} 1 & \frac{7}{5} & \frac{3}{5} & \frac{9}{5} \\ 0 & 1 & -\frac{1}{14} & 2 \\ 0 & -\frac{28}{5} & -\frac{2}{5} & -\frac{56}{5} \end{array}\right]$$

Multiply row 2 by $\frac{28}{5}$ and add to row 3.

$$\left[\begin{array}{ccc|c} 1 & \frac{7}{5} & \frac{3}{5} & \frac{9}{5} \\ 0 & 1 & -\frac{1}{14} & 2 \\ 0 & 0 & -\frac{4}{5} & 0 \end{array}\right]$$

Multiply row 3 by $-\frac{5}{4}$.

$$\left[\begin{array}{ccc|c} 1 & \frac{7}{5} & \frac{3}{5} & \frac{9}{5} \\ 0 & 1 & -\frac{1}{14} & 2 \\ 0 & 0 & 1 & 0 \end{array}\right]$$

This corresponds to $\begin{cases} x+\frac{7}{5}y+\frac{3}{5}z=\frac{9}{5} \\ y-\frac{1}{14}z=2 \\ z=0 \end{cases}$.

$$y-\frac{1}{14}(0)=2$$
$$y=2$$
$$x+\frac{7}{5}(2)+\frac{3}{5}(0)=\frac{9}{5}$$
$$x+\frac{14}{5}=\frac{9}{5}$$
$$x=-1$$

The solution is (−1, 2, 0).

37. $\begin{cases} 7x-3y+2z=0 \\ 4x-4y-z=2 \\ 5x+2y+3z=1 \end{cases}$

$$\left[\begin{array}{ccc|c} 7 & -3 & 2 & 0 \\ 4 & -4 & -1 & 2 \\ 5 & 2 & 3 & 1 \end{array}\right]$$

Multiply row 1 by $\frac{1}{7}$.

$$\left[\begin{array}{ccc|c} 1 & -\frac{3}{7} & \frac{2}{7} & 0 \\ 4 & -4 & -1 & 2 \\ 5 & 2 & 3 & 1 \end{array}\right]$$

Multiply row 1 by −4 and add to row 2.
Multiply row 1 by −5 and add to row 3.

$$\left[\begin{array}{ccc|c} 1 & -\frac{3}{7} & \frac{2}{7} & 0 \\ 0 & -\frac{16}{7} & -\frac{15}{7} & 2 \\ 0 & \frac{29}{7} & \frac{11}{7} & 1 \end{array}\right]$$

Multiply row 2 by $-\frac{7}{16}$ and row 3 by 7.

$$\left[\begin{array}{ccc|c} 1 & -\frac{3}{7} & \frac{2}{7} & 0 \\ 0 & 1 & \frac{15}{16} & -\frac{7}{8} \\ 0 & 29 & 11 & 7 \end{array}\right]$$

Multiply row 2 by −29 and add to row 3.

$$\left[\begin{array}{ccc|c} 1 & -\frac{3}{7} & \frac{2}{7} & 0 \\ 0 & 1 & \frac{15}{16} & -\frac{7}{8} \\ 0 & 0 & -\frac{259}{16} & \frac{259}{8} \end{array}\right]$$

Multiply row 3 by $-\frac{16}{259}$.

$$\left[\begin{array}{ccc|c} 1 & -\frac{3}{7} & \frac{2}{7} & 0 \\ 0 & 1 & \frac{15}{16} & -\frac{7}{8} \\ 0 & 0 & 1 & -2 \end{array}\right]$$

This corresponds to $\begin{cases} x - \frac{3}{7}y + \frac{2}{7}z = 0 \\ y + \frac{15}{16}z = -\frac{7}{8} \\ z = -2 \end{cases}$.

$$y + \frac{15}{16}(-2) = -\frac{7}{8}$$
$$y = -\frac{7}{8} + \frac{15}{8}$$
$$y = 1$$

$$x - \frac{3}{7}(1) + \frac{2}{7}(-2) = 0$$
$$x - \frac{7}{7} = 0$$
$$x = 1$$

The solution is (1, 1, –2).

38. $\begin{cases} x + 2y + 3z = 14 \\ y + 2z = 3 \\ 2x - 2z = 10 \end{cases}$

$$\left[\begin{array}{ccc|c} 1 & 2 & 3 & 14 \\ 0 & 1 & 2 & 3 \\ 2 & 0 & -2 & 10 \end{array}\right]$$

Multiply row 1 by –2 and add to row 3.

$$\left[\begin{array}{ccc|c} 1 & 2 & 3 & 14 \\ 0 & 1 & 2 & 3 \\ 0 & -4 & -8 & -18 \end{array}\right]$$

Multiply row 2 by 4 and add to row 3.

$$\left[\begin{array}{ccc|c} 1 & 2 & 3 & 14 \\ 0 & 1 & 2 & 3 \\ 0 & 0 & 0 & -6 \end{array}\right]$$

The last row corresponds to the equation 0 = –6 which is false. the system is inconsistent and has no solution, ∅.

39. $\begin{cases} y \ge 2x - 3 \\ y \le -2x + 1 \end{cases}$

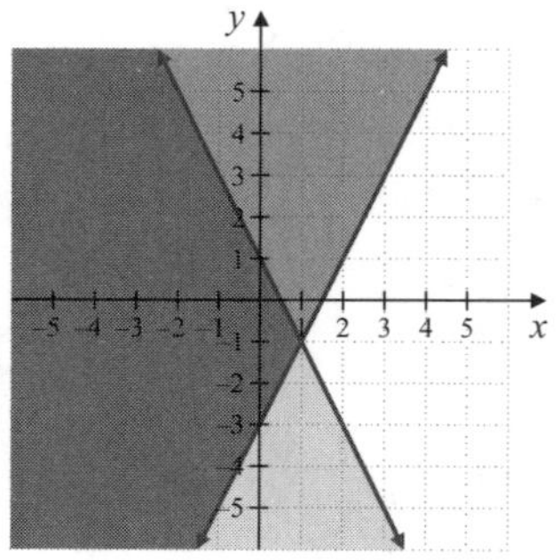

40. $\begin{cases} y \le -3x - 3 \\ y \le 2x + 7 \end{cases}$

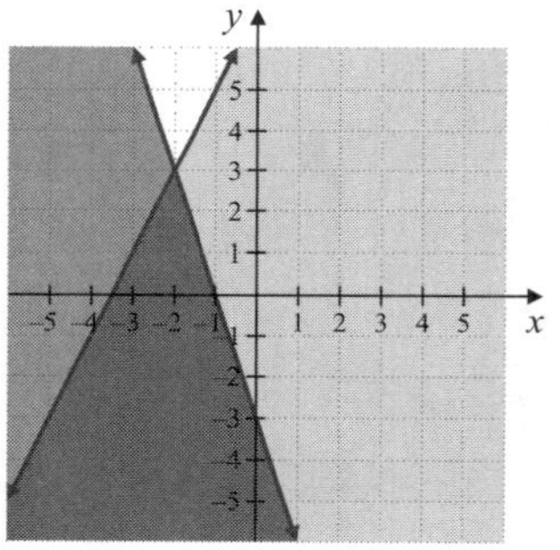

41. $\begin{cases} x + 2y > 0 \\ x - y \le 6 \end{cases}$

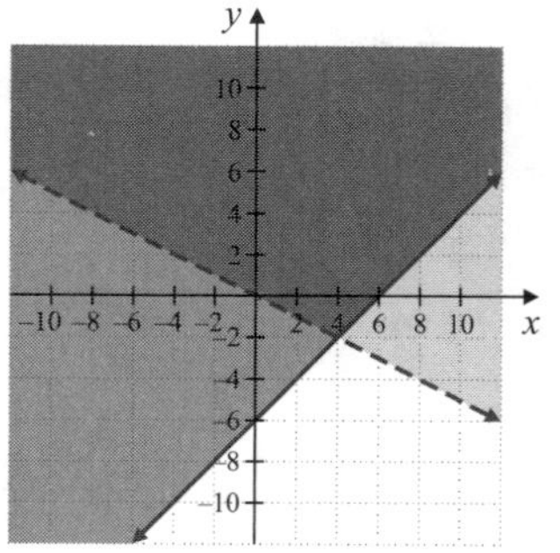

42. $\begin{cases} x-2y \ge 7 \\ x+y \le -5 \end{cases}$

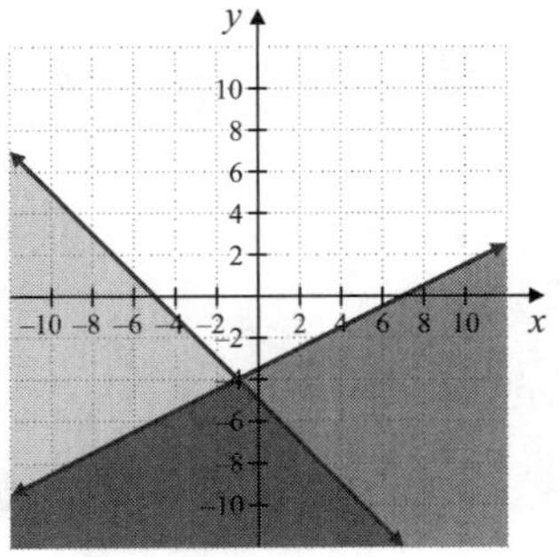

43. $\begin{cases} 3x-2y \le 4 \\ 2x+y \ge 5 \\ y \le 4 \end{cases}$

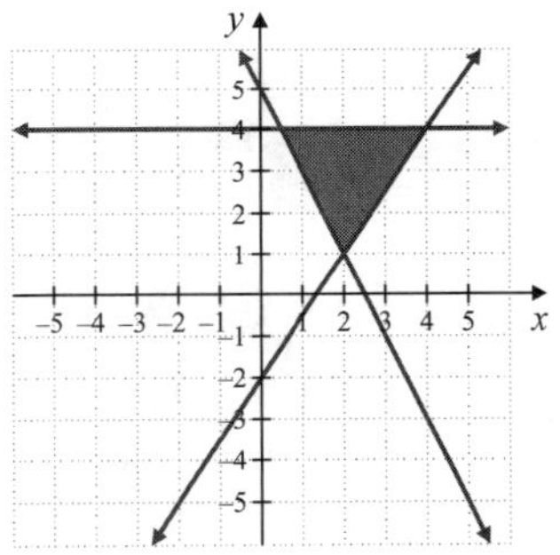

44. $\begin{cases} 4x-y \le 0 \\ 3x-2y \ge -5 \\ y \ge -4 \end{cases}$

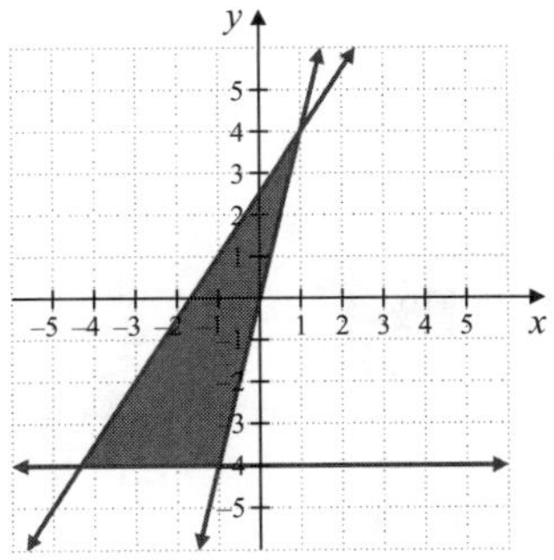

45. $\begin{cases} x+2y \le 5 \\ x \le 2 \\ x \ge 0 \\ y \ge 0 \end{cases}$

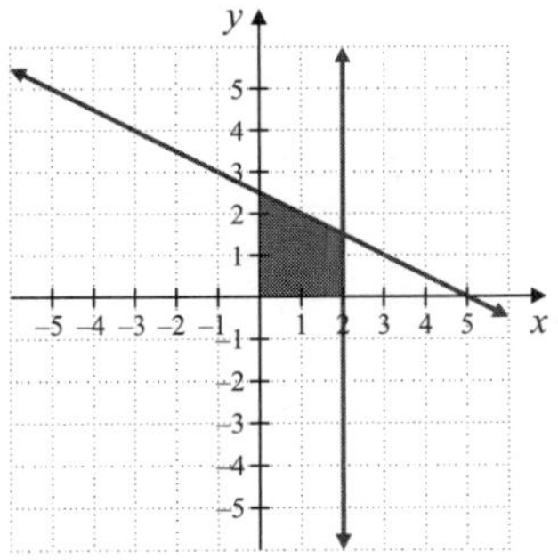

46. $\begin{cases} x+3y \le 7 \\ y \le 5 \\ x \ge 0 \\ y \ge 0 \end{cases}$

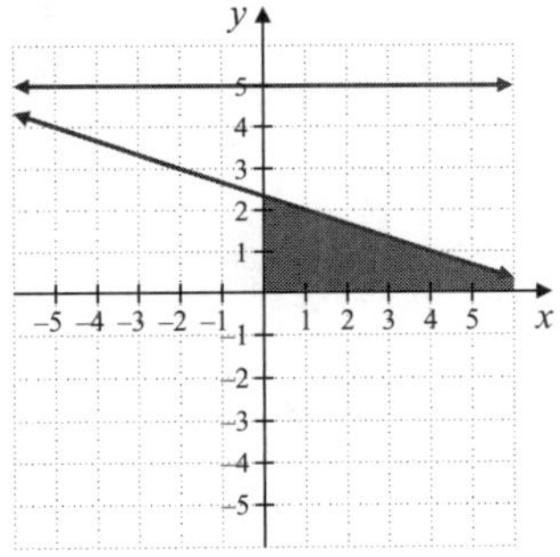

47. $\begin{cases} y = x-5 \\ y = -2x+2 \end{cases}$

Use substitution.

$$x-5 = -2x+2$$
$$3x = 7$$
$$x = \frac{7}{3}$$

Replace x with $\frac{7}{3}$ in the first equation.

$y=\frac{7}{3}-5$

$y=\frac{7}{3}-\frac{15}{3}$

$y=-\frac{8}{3}$

The solution is $\left(\frac{7}{3}, -\frac{8}{3}\right)$.

48. $\begin{cases} \frac{2}{5}x+\frac{3}{4}y=1 \\ x+3y=-2 \end{cases}$

Solve the second equation for *x*.

$x=-3y-2$

Replace *x* with $-3y-2$ in the first equation.

$\frac{2}{5}(-3y-2)+\frac{3}{4}y=1$

$-\frac{6}{5}y-\frac{4}{5}+\frac{3}{4}y=1$

$-\frac{6}{5}y+\frac{3}{4}y=\frac{4}{5}+1$

$-\frac{24}{20}y+\frac{15}{20}y=\frac{9}{5}$

$-\frac{9}{20}y=\frac{9}{5}$

$y=-4$

Replace *y* with -4 in the second equation.

$x+3(-4)=-2$

$x=10$

The solution is (10, −4).

49. $\begin{cases} 5x-2y=10 \\ x=\frac{2}{5}y+2 \end{cases}$

Replace *x* with $\frac{2}{5}y+2$ in the first equation.

$5\left(\frac{2}{5}y+2\right)-2y=10$

$2y+10-2y=10$

$10=10$

This is a true statement. The system is dependent. The solution set is $\{(x, y)|5x-2y=10\}$.

50. $\begin{cases} x-4y=4 \\ \frac{1}{8}x-\frac{1}{2}y=3 \end{cases}$

Multiply the second equation by −8 and add it to the first equation.

$$\begin{array}{r} x-4y=4 \\ -x+4y=-24 \\ \hline 0=-20 \end{array}$$

This statement is false. The system is inconsistent and has no solution, ∅.

51. $\begin{cases} x-3y+2z=0 \\ 9y-z=22 \\ 5x+3z=10 \end{cases}$

$$\left[\begin{array}{ccc|c} 1 & -3 & 2 & 0 \\ 0 & 9 & -1 & 22 \\ 5 & 0 & 3 & 10 \end{array}\right]$$

Multiply row 1 by −5 and add to row 3.

$$\left[\begin{array}{ccc|c} 1 & -3 & 2 & 0 \\ 0 & 9 & -1 & 22 \\ 0 & 15 & -7 & 10 \end{array}\right]$$

Multiply row 2 by $\frac{1}{9}$.

$$\left[\begin{array}{ccc|c} 1 & -3 & 2 & 0 \\ 0 & 1 & -\frac{1}{9} & \frac{22}{9} \\ 0 & 15 & -7 & 10 \end{array}\right]$$

Multiply row 2 by −15 and add to row 3.

$$\left[\begin{array}{ccc|c} 1 & -3 & 2 & 0 \\ 0 & 1 & -\frac{1}{9} & \frac{22}{9} \\ 0 & 0 & -\frac{16}{3} & -\frac{80}{3} \end{array}\right]$$

Multiply row 3 by $-\frac{3}{16}$.

$$\left[\begin{array}{ccc|c} 1 & -3 & 2 & 0 \\ 0 & 1 & -\frac{1}{9} & \frac{22}{9} \\ 0 & 0 & 1 & 5 \end{array}\right]$$

This corresponds to $\begin{cases} x-3y+2z=0 \\ y-\frac{1}{9}z=\frac{22}{9} \\ z=5 \end{cases}$.

$$y-\frac{1}{9}(5)=\frac{22}{9}$$
$$y=\frac{27}{9}=3$$
$$x-3(3)+2(5)=0$$
$$x=-1$$
The solution is (−1, 3, 5).

52. Let a be the first number and b be the second number.
$$\begin{cases} a=3b-5 \\ a+b=127 \end{cases}$$
Replace a with $3b-5$ in the second equation.
$$a+b=127$$
$$3b-5+b=127$$
$$4b=132$$
$$b=33$$
Replace b with 33 in the first equation.
$a=3(33)-5$
$a=99-5$
$a=94$
The numbers are 33 and 94.

53. Let x = length of shortest side, y = length of second side, and z = length of third side.
$$\begin{cases} x+y+z=126 \\ y=2x \\ z=x+14 \end{cases}$$
Replace y with $2x$ and z with $x+14$ in the first equation.
$$x+2x+x+14=126$$
$$4x=112$$
$$x=28$$
Replace x with 28 in the second and third equations.
$y=2(28)=56$
$z=28+14=42$
The sides' lengths are 28 units, 56 units, and 42 units.

54. $$\begin{cases} y\le 3x-\frac{1}{2} \\ 3x+4y\ge 6 \end{cases}$$

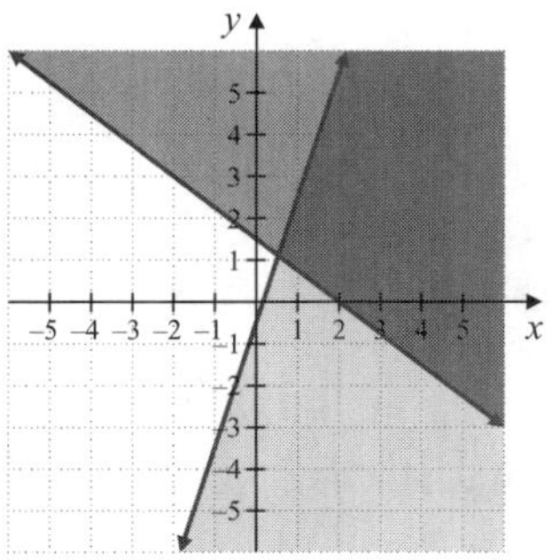

55. $$\begin{cases} y=-443x+2584 \\ y=500x+551 \end{cases}$$
Use substitution.
$$-443x+2584=500x+551$$
$$-943x=-2033$$
$$x\approx 2.2$$
1998 + 2.2 = 2000.2
The year in which the amount of money spent on VCR decks equals the amount of money spent on DVD players is 2000.

Chapter 4 Test

1. $$\begin{cases} 2x-y=-1 \\ 5x+4y=17 \end{cases}$$

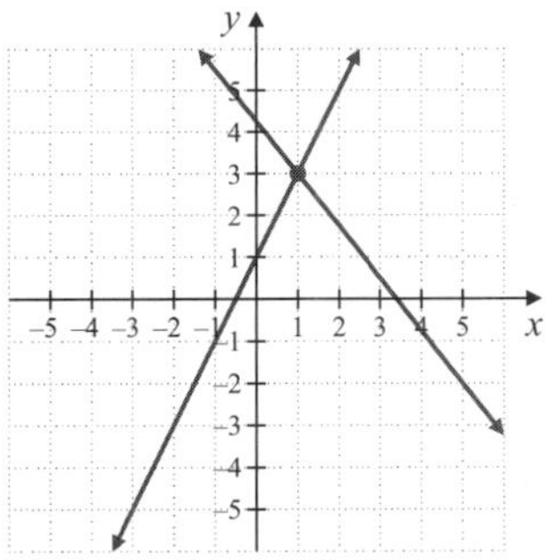

The solution is (1, 3).
Solve the first equation for y.
$$2x-y=-1$$
$$2x+1=y$$
Replace y by $2x+1$ in the second equation.

$$\begin{aligned} 5x+4y&=17\\ 5x+4(2x+1)&=17\\ 5x+8x+4&=17\\ 13x+4&=17\\ 13x&=13\\ x&=1 \end{aligned}$$

Replace x with 1 in the first equation.

$$\begin{aligned} 2x-y&=-1\\ 2(1)-y&=-1\\ -y&=-3\\ y&=3 \end{aligned}$$

The solution is (1, 3).

2. $\begin{cases} 7x-14y=5 \\ x=2y \end{cases}$

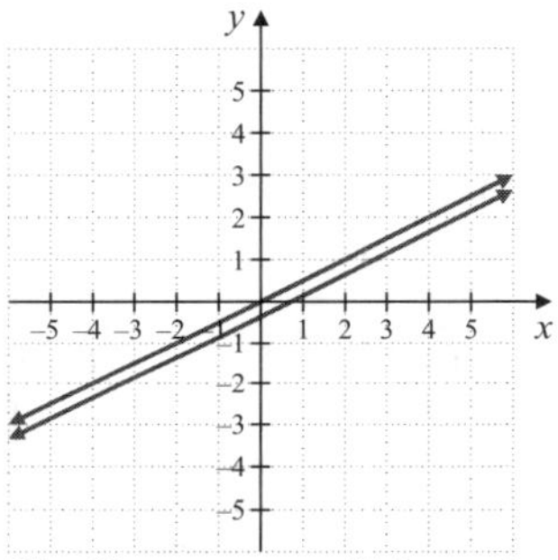

The lines are parallel, so the system is inconsistent. There is no solution, ∅.

Replace x with $2y$ in the first equation.

$$\begin{aligned} 7x-14y&=5\\ 7(2y)-14y&=5\\ 14y-14y&=5\\ 0&=5 \end{aligned}$$

The statement is false. The system is inconsistent and there is no solution, ∅.

3. $\begin{cases} 4x-7y=29 \\ 2x+5y=-11 \end{cases}$

Multiply the second equation by –2 and add to the first equation.

$$\begin{aligned} -4x-7y&=29\\ -4x-10y&=22\\ \hline -17y&=51\\ y&=-3 \end{aligned}$$

Replace y with –3 in the first equation.

$$\begin{aligned} 4x-7y&=29\\ 4x-7(-3)&=29\\ 4x+21&=29\\ 4x&=8\\ x&=2 \end{aligned}$$

The solution is (2, –3).

4. $\begin{cases} 15x+6y=15 \\ 10x+4y=10 \end{cases}$

Multiply the first equation by $-\frac{1}{3}$ and the second equation by $\frac{1}{2}$. Then add the results.

$$\begin{aligned} -5x-2y&=-5\\ 5x+2y&=5\\ \hline 0&=0 \end{aligned}$$

This is a true statement. The system is dependent. The solution set is $\{(x, y)|10x+4y=10\}$.

5. $\begin{cases} 2x-3y \quad\;\; =4 & (1) \\ \quad\;\; 3y+2z=2 & (2) \\ x \quad\quad\; -z=-5 & (3) \end{cases}$

Add equation (1) to equation (2).

$2x+2z=6$ or $x+z=3$ (5)

Add equation (3) to equation (5).

$$\begin{aligned} 2x&=-2\\ x&=-1 \end{aligned}$$

Replace x with –1 in equation (3).

$$\begin{aligned} -1-z&=-5\\ -z&=-4\\ z&=4 \end{aligned}$$

Replace x with –1 in equation (1).

$$\begin{aligned} 2(-1)-3y&=4\\ -2-3y&=4\\ -3y&=6\\ y&=-2 \end{aligned}$$

The solution is (–1, –2, 4).

6. $\begin{cases} 3x-2y-z=-1 & (1) \\ 2x-2y \quad\;\; =4 & (2) \\ 2x \quad\;\; -2z=-12 & (3) \end{cases}$

Multiply the first equation by –2 and add to

the third equation.

$$\begin{array}{r} -6x+4y+2z=2 \\ 2x \qquad -2z=-12 \\ \hline -4x+4y \qquad =-10 \end{array} \quad (4)$$

Multiply equation (2) by 2 and add to equation (4).

$$\begin{array}{r} 4x-4y=8 \\ -4x+4y=-10 \\ \hline 0=-2 \end{array}$$

This statement is false. The system is inconsistent, so there is no solution, ∅.

7. $\begin{cases} \dfrac{x}{2}+\dfrac{y}{4}=-\dfrac{3}{4} & (1) \\ x+\dfrac{3}{4}y=-4 & (2) \end{cases}$

Multiply equation (1) by −8 and equation (2) by 4. then add the results.

$$\begin{array}{r} -4x-2y=6 \\ 4x+3y=-16 \\ \hline y=-10 \end{array}$$

Replace y with −10 in equation (2).

$$x+\frac{3}{4}(-10)=-4$$
$$x-\frac{15}{2}=-4$$
$$x=\frac{15}{2}-\frac{8}{2}$$
$$x=\frac{7}{2}$$

The solution is $\left(\frac{7}{2}, -10\right)$.

8. $\begin{cases} x-y=-2 \\ 3x-3y=-6 \end{cases}$

$$\left[\begin{array}{rr|r} 1 & -1 & -2 \\ 3 & -3 & -6 \end{array}\right]$$

Multiply row 1 by −3 and add to row 2.

$$\left[\begin{array}{rr|r} 1 & -1 & -2 \\ 0 & 0 & 0 \end{array}\right]$$

This corresponds to $\begin{cases} x-y=-2 \\ 0=0 \end{cases}$.

The last equation is true. The system is dependent. The solution is $\{(x, y)|x-y=-2\}$.

9. $\begin{cases} x+2y=-1 \\ 2x+5y=-5 \end{cases}$

$$\left[\begin{array}{rr|r} 1 & 2 & -1 \\ 2 & 5 & -5 \end{array}\right]$$

Multiply row 1 by −2 and add to row 2.

$$\left[\begin{array}{rr|r} 1 & 2 & -1 \\ 0 & 1 & -3 \end{array}\right]$$

This corresponds to $\begin{cases} x+2y=-1 \\ y=-3 \end{cases}$.

$$x+2(-3)=-1$$
$$x-6=-1$$
$$x=5$$

The solution is (5, −3).

10. $\begin{cases} x \quad -y \quad -z=0 \\ 3x \quad -y-5z=-2 \\ 2x+3y \qquad =-5 \end{cases}$

$$\left[\begin{array}{rrr|r} 1 & -1 & -1 & 0 \\ 3 & -1 & -5 & -2 \\ 2 & 3 & 0 & -5 \end{array}\right]$$

Multiply row 1 by −3 and add to row 2.
Multiply row 1 by −2 and add to row 3.

$$\left[\begin{array}{rrr|r} 1 & -1 & -1 & 0 \\ 0 & 2 & -2 & -2 \\ 0 & 5 & 2 & -5 \end{array}\right]$$

Multiply row 2 by $\frac{1}{2}$.

$$\left[\begin{array}{rrr|r} 1 & -1 & -1 & 0 \\ 0 & 1 & -1 & -1 \\ 0 & 5 & 2 & -5 \end{array}\right]$$

Multiply row 2 by −5 and add to row 3.

$$\left[\begin{array}{rrr|r} 1 & -1 & -1 & 0 \\ 0 & 1 & -1 & -1 \\ 0 & 0 & 7 & 0 \end{array}\right]$$

Multiply row 3 by $\frac{1}{7}$.

$$\left[\begin{array}{ccc|c} 1 & -1 & -1 & 0 \\ 0 & 1 & -1 & -1 \\ 0 & 0 & 1 & 0 \end{array}\right]$$

This corresponds to $\begin{cases} x-y-z=0 \\ y-z=-1 \\ z=0 \end{cases}$.

$y-0=-1$
$y=-1$
$x-(-1)-0=0$
$x=-1$

The solution is (−1, −1, 0).

11. Let d = number of double rooms, and s = number of single rooms.

$$\begin{cases} 90d+80s=6930 & (1) \\ d+s=80 & (2) \end{cases}$$

Solve equation (2) for d.
$d = 80 - s$
Replace d with $80 - s$ in equation (1).

$$\begin{aligned} 90(80-s)+80s &= 6930 \\ 7200-90s+80s &= 6930 \\ -10s &= -270 \\ s &= 27 \end{aligned}$$

Replace s with 27 in the equation $d = 80 - s$.
$d = 80 - 27 = 53$
There are 53 double rooms occupied and 27 single rooms occupied.

12. Let x = amount of 10% solution, and y = amount of 20% solution.

gallons	concentration	amount of fructose solution
x	10%	$0.1x$
y	20%	$0.2y$
20	17.5%	20(0.175) = 3.5

$$\begin{cases} x+y=20 & (1) \\ 0.1x+0.2y=3.5 & (2) \end{cases}$$

Multiply equation (2) by −10 and add to equation (1).

$$\begin{aligned} x+y &= 20 \\ -x-2y &= -35 \\ \hline -y &= -15 \\ y &= 15 \end{aligned}$$

Replace y with 15 in equation (1).
$x+15=20$
$x=5$
Mix 5 gallons of 10% solution with 15 gallons of 20% solution.

13. Let x be the number of packages of gift boxes sold. Then the cost of producing x boxes is $C(x) = 2000 + 1.50x$ and the revenue from selling x boxes is $R(x) = 4x$. The break-even point is where $C(x) = R(x)$.

$$\begin{aligned} 2000+1.5x &= 4x \\ 2000 &= 2.5x \\ 800 &= x \end{aligned}$$

800 packages of gift boxes must be sold for the company to break even.

14. Let a = measure of smallest angle, b = measure of middle angle, and c = measure of largest angle.

$$\begin{cases} a+b+c=180 & (1) \\ c=5a-3 & (2) \\ b=2a-1 & (3) \end{cases}$$

Replace b with $2a - 1$ and c with $5a - 3$ in equation (1).

$$\begin{aligned} a+2a-1+5a-3 &= 180 \\ 8a-4 &= 180 \\ 8a &= 184 \\ a &= 23 \end{aligned}$$

Replace a with 23 in equations (2) and (3).
$c = 5(23) - 3 = 112$
$b = 2(23) - 1 = 45$
The angles measure 23°, 45°, and 112°.

15. $\begin{cases} 2y - x \geq 1 \\ x + y \geq -4 \\ y \leq 2 \end{cases}$

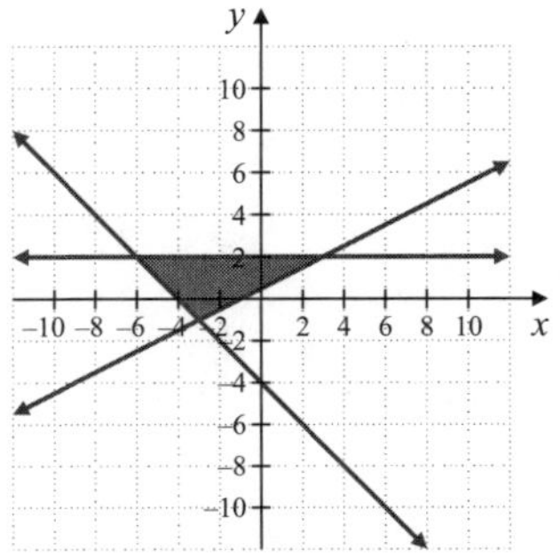

Cumulative Review Chapters 1–4

1. $7 \notin \{1, 2, 3\}$ is true, since 7 is not an element of the set $\{1, 2, 3\}$.

2. $6 \in \{1, 2, 3\}$ is false, since 6 is not an element of the set $\{1, 2, 3\}$.

3. False; the number $\frac{1}{5}$ is a rational number since it is in the form $\frac{a}{b}$ with a and b integers and $b \neq 0$.

4. True; every integer is a real number.

5. -5 is to the left of 5 on a number line, so $-5 < 5$.

6. -21 is to the right of -23 on a number line, so $-21 > -23$.

7. -2.5 is to the left of -2.1, so $-2.5 < -2.1$.

8. $-\frac{3}{3} = -1$, $-\frac{7.2}{7.2} = -1$, so $-\frac{3}{3} = -\frac{7.2}{7.2}$.

9. By dividing, we see that $\frac{3}{4} = 0.75$ and $\frac{2}{3} = 0.666....$ Thus, $\frac{2}{3} < \frac{3}{4}$ since $0.666... < 0.75$.

10. $-\frac{1}{11} < \frac{1}{11}$ since $-\frac{1}{11}$ is to the left of $\frac{1}{11}$.

11. $\left|-\frac{1}{7}\right| = -\left(-\frac{1}{7}\right) = \frac{1}{7}$

12. $|-7.8| = -(-7.8) = 7.8$

13. $-|-8| = -8$

14. $-\left|-\frac{3}{8}\right| = -\frac{3}{8}$

15. $\frac{|-2|^3 + 1}{-7 - \sqrt{4}} = \frac{2^3 + 1}{-7 - 2} = \frac{8 + 1}{-9} = \frac{9}{-9} = -1$

16. **a.** $-7 - (-2) = -7 + 2 = -5$

    b. $14 - 38 = 14 + (-38) = -24$

17. $(3x)^{-1} = \frac{1}{(3x)^1} = \frac{1}{3x}$

18. $9^{-2} = \frac{1}{9^2} = \frac{1}{81}$

19. $2^{-1} + 3^{-2} = \frac{1}{2^1} + \frac{1}{3^2} = \frac{1}{2} + \frac{1}{9} = \frac{9}{18} + \frac{2}{18} = \frac{11}{18}$

20. $(x^2)^3 \cdot x^{10} = x^6 \cdot x^{10} = x^{6+10} = x^{16}$

21. Replace x with -15.
$$x - 9 = -24$$
$$-15 - 9 \stackrel{?}{=} -24$$
$$-24 = -24$$
Since a true statement results, -15 is a solution.

22.
$$\begin{aligned} 8y-14 &= 6y-14 \\ 8y-14-6y &= 6y-14-6y \\ 2y-14 &= -14 \\ 2y-14+14 &= -14+14 \\ 2y &= 0 \\ \frac{2y}{2} &= \frac{0}{2} \\ y &= 0 \end{aligned}$$
The solution set is $\{0\}$.

23. Let x = the original price of the computer.
$$\begin{aligned} x-0.08x &= 2162 \\ 0.92x &= 2162 \\ x &= \frac{2162}{0.92} = 2350 \end{aligned}$$
The original price of the computer was $2350.

24. Let x = measure of each of the equal angles. Then $x + 10$ = measure of third angle, and $\frac{1}{2}x$ = measure of fourth angle.
$$\begin{aligned} x+x+x+10+\frac{1}{2}x &= 360 \\ \frac{7}{2}x+10 &= 360 \\ \frac{7}{2}x &= 350 \\ x &= 100 \end{aligned}$$
$x + 10 = 100 + 10 = 110$
$\frac{1}{2}x = \frac{1}{2}(100) = 50$
The angles measure 100°, 100°, 110°, and 50°.

25.
$$\begin{aligned} A &= \frac{1}{2}(B+b)h \\ 2\cdot A &= 2\cdot\frac{1}{2}(B+b)h \\ 2A &= (B+b)h \\ 2A &= Bh+bh \\ 2A-Bh &= bh \\ \frac{2A-Bh}{h} &= \frac{bh}{h} \end{aligned}$$
$\frac{2A-Bh}{h} = b$ or $b = \frac{2A-Bh}{h}$ or $b = \frac{2A}{h} - B$

26.
$$\begin{aligned} 2(m-6)-m &= 4(m-3)-3m \\ 2m-12-m &= 4m-12-3m \\ m-12 &= m-12 \\ -12 &= -12 \end{aligned}$$
This is a true statement. There are an infinite number of solutions. The solution set is $\{m|m \text{ is a real number}\}$.

27.
$$\begin{aligned} 4x-2 &< 5x \\ 4x-2-4x &< 5x-4x \\ -2 &< x \text{ or } x > -2 \end{aligned}$$
The solution set is $\{x|x > -2\}$, or in interval notation $(-2, \infty)$.

28.
$$\begin{aligned} 5(2x-1) &> -5 \\ 10x-5 &> -5 \\ 10x &> 0 \\ x &> 0 \end{aligned}$$
The solution set is $\{x|x > 0\}$, or in interval notation $(0, \infty)$.

29. $\{2, 4, 6, 8\} \cap \{3, 4, 5, 6\} = \{4, 6\}$
The numbers 4 and 6 are in both sets.

30. $\{2, 4, 6, 8\} \cup \{3, 4, 5, 6\} = \{2, 3, 4, 5, 6, 8\}$

31. $|y| = 0$
We are looking for all real numbers whose distance from 0 is zero units. The only number is 0. The solution set is $\{0\}$.

32. $|x - 5| = 4$
$x-5 = -4$ or $x-5 = 4$
$x = 1$ $\quad$ $x = 9$
The solution set is $\{1, 9\}$.

33. $y = -3$

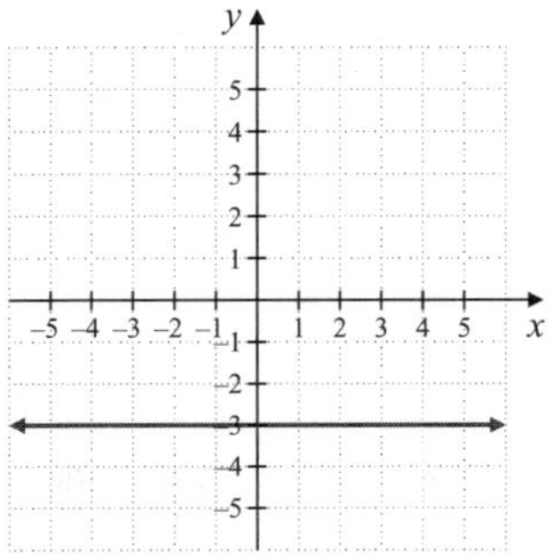

34. $|2x + 1| > 5$

$$2x+1 > 5 \quad \text{or} \quad 2x+1 < -5$$
$$2x > 4 \qquad\qquad 2x < -6$$
$$x > 2 \qquad\qquad x < -3$$

The solution is $(-\infty, -3) \cup (2, \infty)$.

35.
$$3x - 4y = 4$$
$$-4y = -3x + 4$$
$$\frac{-4y}{-4} = \frac{-3x}{-4} + \frac{4}{-4}$$
$$y = \frac{3}{4}x - 1$$

The coefficient of x, $\frac{3}{4}$, is the slope, and $(0, -1)$ is the y-intercept.

36. **a.** $(-1, -5)$ is in quadrant III.

b. $(4, -2)$ is in quadrant IV.

c. $(0, 2)$ is on the y-axis.

37. $m = -3$, $(x_1, y_1) = (1, -5)$

$$y - y_1 = m(x - x_1)$$
$$y - (-5) = -3(x - 1)$$
$$y + 5 = -3x + 3$$
$$y = -3x - 2$$

38. $f(x) = 3x^2$

a. $f(5) = 3(5)^2 = 3(25) = 75$

b. $f(-2) = 3(-2)^2 = 3(4) = 12$

39. The relation $y = 2x + 1$ is a function if each x-value corresponds to just one y-value. For each x-value substituted in the equation $y = 2x + 1$, the multiplication and addition performed gives a single result, so only one y-value will be associated with each x-value. Thus, $y = 2x + 1$ is a function.

40. $(x_1, y_1) = (-2, 6)$, $(x_2, y_2) = (0, 9)$

$$m = \frac{y_2 - y_1}{x_2 - x_1} = \frac{9-6}{0-(-2)} = \frac{3}{0+2} = \frac{3}{2}$$

41. $x \geq 1$ and $y \geq 2x - 1$

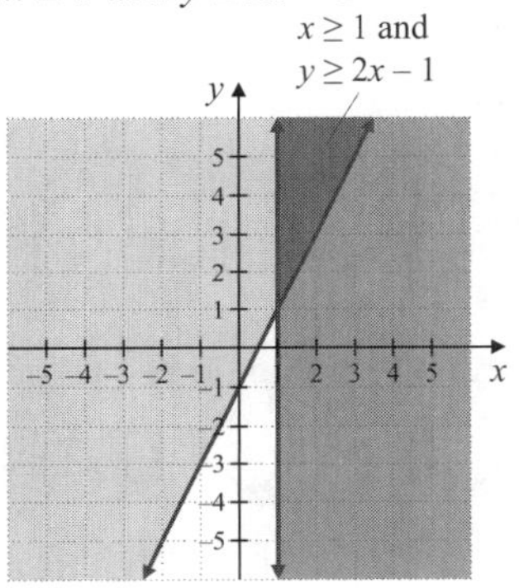

42. $f(x) = \frac{1}{2}x - \frac{1}{3}$

$y = mx + b$

$m = \frac{1}{2}$

A line perpendicular to this line has slope $m = -2$.

Let $(x_1, y_1) = (-2, 6)$.

$$y - y_1 = m(x - x_1)$$
$$y - 6 = -2[x - (-2)]$$
$$y - 6 = -2(x + 2)$$
$$y - 6 = -2x - 4$$
$$y = -2x + 2$$

43.
$$\begin{cases} 3x - 2y = 10 \\ 4x - 3y = 15 \end{cases}$$

Multiply the first equation by 3 and the second equation by -2. Then add the results.

$$\begin{array}{r} 9x-6y=30 \\ -8x+6y=-30 \\ \hline x=0 \end{array}$$

Replace x with 0 in the first equation.

$$3x-2y=10$$
$$3(0)-2y=10$$
$$-2y=10$$
$$y=-5$$

The solution is (0, –5).

44. $\begin{cases} 5x+y=-2 \\ 4x-2y=-10 \end{cases}$

Multiply the first equation by 2 and add to the second equation.

$$\begin{array}{r} 10x+2y=-4 \\ 4x-2y=-10 \\ \hline 14x=-14 \end{array}$$
$$x=-1$$

Replace x with –1 in the first equation.

$$5x+y=-2$$
$$5(-1)+y=-2$$
$$-5+y=-2$$
$$y=3$$

The solution is (–1, 3).

45. $\begin{cases} 2x-4y+8z=2 & (1) \\ -x-3y+z=11 & (2) \\ x-2y+4z=0 & (3) \end{cases}$

Add equations (2) and (3) to eliminate x.

$-5y + 5z = 11$ (4)

Multiply equation (2) by 2 and add to equation (1).

$$\begin{array}{r} 2x-4y+8z=2 \\ -2x-6y+2z=22 \\ \hline -10y+10z=24 \end{array} \quad (5)$$

Multiply equation (4) by –2 and add to equation (5).

$$\begin{array}{r} 10y-10z=-22 \\ -10y+10z=24 \\ \hline 0=2 \end{array}$$

Since the statement is false, this system is inconsistent and has no solution. The solution set is { } or ∅.

46. $\begin{cases} x-2y+z=0 & (1) \\ 3x-y-2z=-15 & (2) \\ 2x-3y+3z=7 & (3) \end{cases}$

Multiply equation (1) by –3 and add to equation (2).

$$\begin{array}{r} -3x+6y-3z=0 \\ 3x-y-2z=-15 \\ \hline 5y-5z=-15 \end{array} \quad (4)$$

Multiply equation (1) by –2 and add to equation (3).

$$\begin{array}{r} -2x+4y-2z=0 \\ 2x-3y+3z=7 \\ \hline y+z=7 \end{array} \quad (5)$$

Multiply equation (5) by 5 and add to equation (4).

$$\begin{array}{r} 5y-5z=-15 \\ 5y+5z=35 \\ \hline 10y=20 \end{array}$$
$$y=2$$

Replace y with 2 in equation (5).

$$2+z=7$$
$$z=5$$

Replace y with 2 and z with 5 in equation (1).

$$x-2(2)+5=0$$
$$x+1=0$$
$$x=-1$$

The solution is (–1, 2, 5).

47. Let x = first number and y = second number.

$\begin{cases} x=y-4 \\ 4x=2y+6 \end{cases}$

Replace x with y – 4 in the second equation.

$$4x=2y+6$$
$$4(y-4)=2y+6$$
$$4y-16=2y+6$$
$$2y=22$$
$$y=11$$

Replace y with 11 in the first equation.

$x = y - 4$

$x = 11 - 4$

$x = 7$

The numbers are 7 and 11.

48. $\begin{cases} -6x+8y=0 \\ 9x-12y=2 \end{cases}$

Multiply the first equation by 9 and the second equation by 6. Add the results.

$$\begin{array}{r} -54x+72y=0 \\ \underline{54x-72y=12} \\ 0=12 \end{array}$$

Since the statement is false, there is no solution, $\varnothing$.

Chapter 5

Section 5.1 Practice Problems

1. $2x^3$; the exponent on x is 3, so the degree of the term is 3.

2. 7^2x^4; the exponent on x is 4, so the degree of the term is 4.

3. The degree of x, or x^1, is 1.

4. $15xy^2z^4$; the degree is the sum of the exponents on the variables, or 1 + 2 + 4 = 7.

5. The degree of 9.1, which can be written as $9.1x^0$, is 0.

6. $4x^5+7x^3-1$ has 3 terms, so it is a trinomial. The term with the greatest degree is $4x^5$, which has degree 5, so the degree of the trinomial is 5.

7. $-2xy^2z$ has one term, so it is a monomial. The term has degree 1 + 2 + 1 = 4, so the degree of the monomial is 4.

8. y^3+6y has 2 terms, so it is a binomial. The term with the greatest degree is y^3, which has degree 3, so the degree of the binomial is 3.

9. $7x^2y-6x^2yz+2-4y^3$; the term with the greatest degree is $-6x^2yz$, which has degree 2 + 1 + 1 = 4, so the degree of the polynomial is 4.

10. $$\begin{aligned}10x^3-12x^3-3x &= (10-12)x^3-3x\\ &= -2x^3-3x\end{aligned}$$

11. $$\begin{aligned}&-6ab+2a+12ab-a\\ &= -6ab+12ab+2a-a\\ &= (-6+12)ab+(2-1)a\\ &= 6ab+a\end{aligned}$$

12. $$\begin{aligned}&(14x^4-6x^3+x^2-6)+(x^3-5x^2+1)\\ &= 14x^4-6x^3+x^3+x^2-5x^2-6+1\\ &= 14x^4-5x^3-4x^2-5\end{aligned}$$

13. $$\begin{array}{r}10y^3-y^2+4y-11\\ \underline{y^3-4y^2+3y}\\ 11y^3-5y^2+7y-11\end{array}$$

14. $$\begin{aligned}&(4x^2y-xy^2+5)+(-6x^2y-1)\\ &= 4x^2y-xy^2+5-6x^2y-1\\ &= 4x^2y-6x^2y-xy^2+5-1\\ &= -2x^2y-xy^2+4\end{aligned}$$

15. $$\begin{aligned}&(7x^4-8x^2+x)-(-9x^4+x^2-18)\\ &= 7x^4-8x^2+x+9x^4-x^2+18\\ &= 7x^4+9x^4-8x^2-x^2+x+18\\ &= 16x^4-9x^2+x+18\end{aligned}$$

16. $$\begin{array}{r}2y^4+4y\\ \underline{-(6y^4+7y^3-3y)}\end{array} \rightarrow \begin{array}{r}2y^4+0y^3+4y\\ \underline{-6y^4-7y^3+3y}\\ -4y^4-7y^3+7y\end{array}$$

17. $$\begin{aligned}&(7a^2b^3-ab^2)-(3a^2b^3-4ab^2+6a)\\ &= 7a^2b^3-ab^2-3a^2b^3+4ab^2-6a\\ &= 4a^2b^3+3ab^2-6a\end{aligned}$$

18. $P(x)=5x^4-3x^2+7$
$$\begin{aligned}P(2) &= 5(2)^4-3(2)^2+7\\ &= 5(16)-3(4)+7\\ &= 80-12+7\\ &= 75\end{aligned}$$

19. $P(x) = 5x^4 - 3x^2 + 7$

$P(-1) = 5(-1)^4 - 3(-1)^2 + 7$
$= 5(1) - 3(1) + 7$
$= 9$

20. $P(t) = -16t^2 + 1125$

$P(3) = -16(3)^2 + 1125$
$P(3) = -144 + 1125$
$P(3) = 981$

At 3 seconds, the height is 981 feet.

$P(7) = -16(7)^2 + 1125$
$P(7) = -16(49) + 1125$
$P(7) = -784 + 1125$
$P(7) = 341$

At 7 seconds, the height is 341 feet.

Calculator Explorations

1. $(2x^2 + 7x + 6) + (x^3 - 6x^2 - 14)$
$= 2x^2 + 7x + 6 + x^3 - 6x^2 - 14$
$= x^3 - 4x^2 + 7x - 8$

Let $Y_1 = (2x^2 + 7x + 6) + (x^3 - 6x^2 - 14)$
and $Y_2 = x^3 - 4x^2 + 7x - 8$.
The graphs coincide.

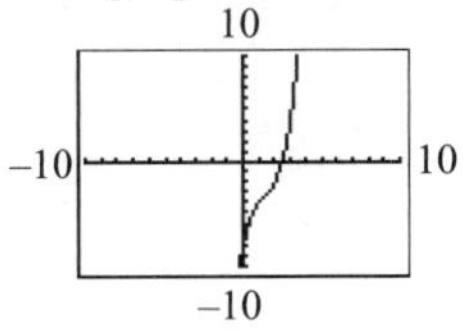

2. $(-14x^3 - x + 2) + (-x^3 + 3x^2 + 4x)$
$= -14x^3 - x + 2 - x^3 + 3x^2 + 4x$
$= -15x^3 + 3x^2 + 3x + 2$

Let $Y_1 = (-14x^3 - x + 2) + (-x^3 + 3x^2 + 4x)$
and $Y_2 = -15x^3 + 3x^2 + 3x + 2$.
The graphs coincide.

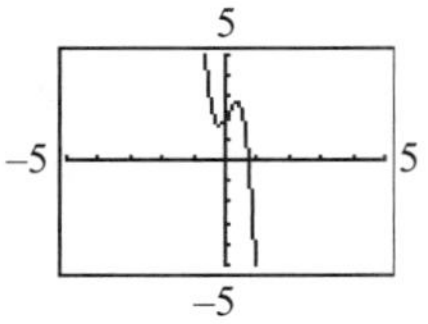

3. $(1.8x^2 - 6.8x - 1.7) - (3.9x^2 - 3.6x)$
$= 1.8x^2 - 6.8x - 1.7 - 3.9x^2 + 3.6x$
$= -2.1x^2 - 3.2x - 1.7$

Let
$Y_1 = (1.8x^2 - 6.8x - 1.7) - (3.9x^2 - 3.6x)$
and $Y_2 = -2.1x^2 - 3.2x - 1.7$.
The graphs coincide.

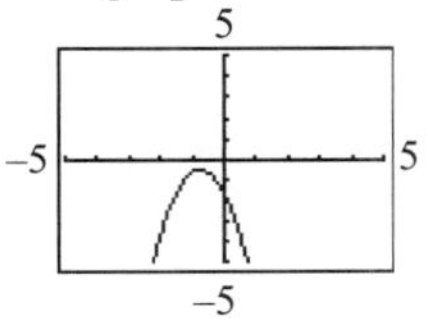

4. $(-4.8x^2 + 12.5x - 7.8) - (3.1x^2 - 7.8x)$
$= -4.8x^2 + 12.5x - 7.8 - 3.1x^2 + 7.8x$
$= -7.9x^2 + 20.3x - 7.8$

Let
$Y_1 = (-4.8x^2 + 12.5x - 7.8) - (3.1x^2 - 7.8x)$
and $Y_2 = -7.9x^2 + 20.3x - 7.8$.
The graphs coincide.

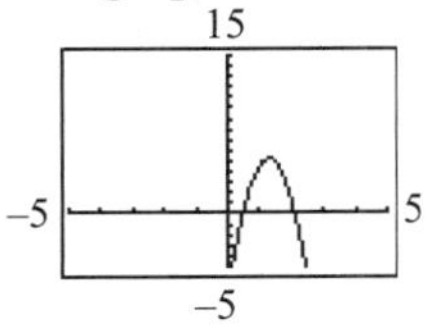

5. $(1.29x - 5.68) + (7.69x^2 - 2.55x + 10.98)$
$= 1.29x - 5.68 + 7.69x^2 - 2.55x + 10.98$
$= 7.69x^2 - 1.26x + 5.3$

Let
$Y_1 = (1.29x - 5.68) + (7.69x^2 - 2.55x + 10.98)$
and $Y_2 = 7.69x^2 - 1.26x + 5.3$.
The graphs coincide.

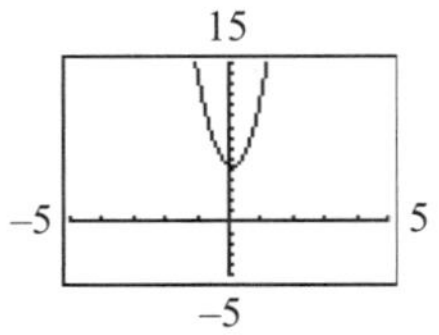

6. $(-0.98x^2 - 1.56x + 5.57) + (4.36x - 3.71)$
$= -0.98x^2 - 1.56x + 5.57 + 4.36x - 3.71$
$= -0.98x^2 + 2.8x + 1.86$
Let
$Y_1 = (-0.98x^2 - 1.56x + 5.57) + (4.36x - 3.71)$
and $Y_2 = -0.98x^2 + 2.8x + 1.86.$
The graphs coincide.

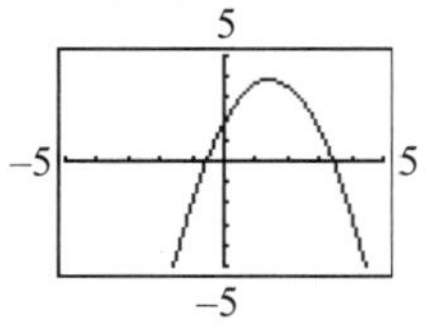

Mental Math

1. $7x + 3x = 10x$

2. $8x - 2x = 6x$

3. $14y - 9y = 5y$

4. $14y + 9y = 23y$

5. $3z - 12z = -9z$

6. $2z - 6z = -4z$

Exercise Set 5.1

1. 4 can be written as $4x^0$. The exponent on x is 0, so the degree of 4 is 0.

3. $5x^2$; the exponent on x is 2, so the degree of the term is 2.

5. The degree is the sum of the exponents on the variables, so the degree of $-3xy^2$ is $1 + 2 = 3$.

7. $-8^7 y^3$; the exponent on y is 3, so the degree of the term is 3.

9. The degree is the sum of the exponents on the variables, so the degree of $3.78ab^3c^5$ is $1 + 3 + 5 = 9$.

11. $6x + 0.3$ has 2 terms, so it is a binomial. The term with the variable has degree 1, so the degree of the binomial is 1.

13. $3x^2 - 2x + 5$ has three terms, so it is a trinomial. The term with the greatest degree is $3x^2$, which has degree 2, so the degree of the trinomial is 2.

15. $-3^4 xy^2$ is a monomial of degree $1 + 2 = 3$.

17. $x^2y - 4xy^2 + 5x + y^4$ has 4 terms so it is not a monomial, binomial, or trinomial. The term with the greatest degree is y^4, which has degree 4, so the degree of the polynomial is 4.

19. $5y + y - 6y^2 - y^2 = (5+1)y + (-6-1)y^2$
$= 6y - 7y^2$

21. $4x + 7x - 3x^4 = (4+7)x - 3x^4$
$= 11x - 3x^4$ or $-3x^4 + 11x$

23. $4x^2y + 2x - 3x^2y - \frac{1}{2} - 7x$
$= 4x^2y - 3x^2y + 2x - 7x - \frac{1}{2}$
$= x^2y - 5x - \frac{1}{2}$

25. $(9y^2 + y - 8) + (9y^2 - y - 9)$
$= 9y^2 + y - 8 + 9y^2 - y - 9$
$= 9y^2 + 9y^2 + y - y - 8 - 9$
$= 18y^2 - 17$

27. $(x^2+xy-y^2)+(2x^2-4xy+7y^2)=x^2+xy-y^2+2x^2-4xy+7y^2$
$=x^2+2x^2+xy-4xy-y^2+7y^2$
$=3x^2-3xy+6y^2$

29.
$$\begin{array}{r} x^2-6x+3 \\ +\quad (2x+5) \\ \hline x^2-4x+8 \end{array}$$

31. $(7x^3y-4xy+8)+(5x^3y+4xy+8x)=7x^3y-4xy+8+5x^3y+4xy+8x$
$=7x^3y+5x^3y-4xy+4xy+8x+8$
$=12x^3y+8x+8$

33. $(0.6x^3+1.2x^2-4.5x+9.1)+(3.9x^3-x^2+0.7x)=0.6x^3+1.2x^2-4.5x+9.1+3.9x^3-x^2+0.7x$
$=0.6x^3+3.9x^3+1.2x^2-x^2-4.5x+0.7x+9.1$
$=4.5x^3+0.2x^2-3.8x+9.1$

35. $(9y^2-7y+5)-(8y^2-7y+2)=9y^2-7y+5-8y^2+7y-2$
$=9y^2-8y^2-7y+7y+5-2$
$=y^2+3$

37. $(4x^2+2x)-(6x^2-3x)=4x^2+2x-6x^2+3x$
$=4x^2-6x^2+2x+3x$
$=-2x^2+5x$

39.
$$\begin{array}{r} 6y^2-6y+4 \\ \underline{-(-y^2+6y+7)} \end{array} \rightarrow \begin{array}{r} 6y^2-6y+4 \\ +\ y^2-6y-7 \\ \hline 7y^2-12y-3 \end{array}$$

41. $(9x^3-2x^2+4x-7)-(2x^3-6x^2-4x+3)=9x^3-2x^2+4x-7-2x^3+6x^2+4x-3$
$=9x^3-2x^3-2x^2+6x^2+4x+4x-7-3$
$=7x^3+4x^2+8x-10$

43. $\left(-19y^2+7yx+\frac{1}{7}\right)-\left(y^2+4yx+\frac{1}{7}\right)=-19y^2+7yx+\frac{1}{7}-y^2-4yx-\frac{1}{7}$
$=-19y^2-y^2+7yx-4yx+\frac{1}{7}-\frac{1}{7}$
$=-20y^2+3yx$

45. $(-3x+8)+(-3x^2+3x-5) = -3x+8-3x^2+3x-5$
$= -3x^2-3x+3x+8-5$
$= -3x^2+3$

47. $(5y^4-7y^2+x^2-3)+(-3y^4+2y^2+4) = 5y^4-7y^2+x^2-3-3y^4+2y^2+4$
$= 5y^4-3y^4-7y^2+2y^2+x^2-3+4$
$= 2y^4-5y^2+x^2+1$

49. $(4x^2-6x+2)-(-x^2+3x+5) = 4x^2-6x+2+x^2-3x-5$
$= 4x^2+x^2-6x-3x+2-5$
$= 5x^2-9x-3$

51. $(5x^2+x+9)-(2x^2-9) = 5x^2+x+9-2x^2+9$
$= 5x^2-2x^2+x+9+9$
$= 3x^2+x+18$

53. $(5x-11)+(-x-2) = 5x-11-x-2$
$= 5x-x-11-2$
$= 4x-13$

55. $(3x^3-b+2a-6)+(-4x^3+b+6a-6) = 3x^3-b+2a-6-4x^3+b+6a-6$
$= 3x^3-4x^3-b+b+2a+6a-6-6$
$= -x^3+8a-12$

57. $(14ab-10a^2b+6b^2)-(18a^2-20a^2b-6b^2) = 14ab-10a^2b+6b^2-18a^2+20a^2b+6b^2$
$= 14ab-10a^2b+20a^2b-18a^2+6b^2+6b^2$
$= 14ab+10a^2b-18a^2+12b^2$

59.
$$\begin{array}{r} 3x^2+15x+8 \\ +\ (2x^2+7x+8) \\ \hline 5x^2+22x+16 \end{array}$$

61. $(7x^2-5)+(-3x^2-2)-(4x^2-7) = 7x^2-5-3x^2-2-4x^2+7$
$= 7x^2-3x^2-4x^2-5-2+7$
$= 0$

63. $(-3+4x^2+7xy^2)+(2x^3-x^2+xy^2) = -3+4x^2+7xy^2+2x^3-x^2+xy^2$
$= 7xy^2+xy^2+2x^3+4x^2-x^2-3$
$= 8xy^2+2x^3+3x^2-3$

65.
$$\begin{array}{r} 3x^2 - 4x \ +8 \\ - \quad (5x \ -7) \\ \hline \end{array}$$
Is equivalent to:
$$\begin{array}{r} 3x^2 - 4x \ +8 \\ + \quad (-5x \ +7) \\ \hline 3x^2 - 9x + 15 \end{array}$$

67. $(7x^2 + 4x + 9) + (8x^2 + 7x - 8) - (3x + 7)$
$= 7x^2 + 4x + 9 + 8x^2 + 7x - 8 - 3x - 7$
$= 7x^2 + 8x^2 + 4x + 7x - 3x + 9 - 8 - 7$
$= 15x^2 + 8x - 6$

69. $\left(\frac{2}{3}x^2 - \frac{1}{6}x + \frac{5}{6}\right) - \left(\frac{1}{3}x^2 + \frac{5}{6}x - \frac{1}{6}\right)$
$= \frac{2}{3}x^2 - \frac{1}{6}x + \frac{5}{6} - \frac{1}{3}x^2 - \frac{5}{6}x + \frac{1}{6}$
$= \frac{2}{3}x^2 - \frac{1}{3}x^2 - \frac{1}{6}x - \frac{5}{6}x + \frac{5}{6} + \frac{1}{6}$
$= \frac{1}{3}x^2 - x + 1$

71. $P(x) = x^2 + x + 1$
$P(7) = 7^2 + 7 + 1 = 49 + 7 + 1 = 57$

73. $Q(x) = 5x^2 - 1$
$Q(-10) = 5(-10)^2 - 1$
$= 5(100) - 1$
$= 500 - 1$
$= 499$

75. $P(x) = x^2 + x + 1$
$P(0) = 0^2 + 0 + 1 = 1$

77. Use $2HL + 2LW + 2HW$ with $L = 5$, $W = 4$, and $H = 9$.
$2HL + 2LW + 2HW$
$= 2(9)(5) + 2(5)(4) + 2(9)(4)$
$= 90 + 40 + 72$
$= 202$
The surface area is 202 square inches.

79. $P(t) = -16t^2 + 300t$

a. $P(1) = -16(1)^2 + 300(1) = 284$ ft

b. $P(2) = -16(2)^2 + 300(2) = 536$ ft

c. $P(3) = -16(3)^2 + 300(3) = 756$ ft

d. $P(4) = -16(4)^2 + 300(4) = 944$ ft

e. answers may vary

f. Let $P(t) = 0$ and solve for t.
$0 = -16t^2 + 300t$
$0 = -16t(t - 18.75)$
$-16t = 0$ or $t - 18.75 = 0$
$t = 0$ or $t = 18.75$
The object hits the ground after 19 seconds.

81. $P(x) = 45x - 100{,}000$
$P(4000) = 45(4000) - 100{,}000$
$= 180{,}000 - 100{,}000$
$= 80{,}000$
The profit from selling 4000 computer briefcases is \$80,000.

83. $R(x) = 2x$
$R(20{,}000) = 2(20{,}000) = 40{,}000$
The revenue from selling 20,000 tapes per week is \$40,000.

85. $-7(2z - 6y) = -7 \cdot 2z - (-7)(6y)$
$= -14z + 42y$

87. $5(-3y^2 - 2y + 7)$
$= 5 \cdot (-3y^2) + 5 \cdot (-2y) + 5 \cdot 7$
$= -15y^2 - 10y + 35$

89. The opposite of $8x - 6$ is $-(8x - 6) = -8x + 6$, which are choices a and c.

91. $(12x-1.7)-(15x+6.2)$
$=12x-1.7-15x-6.2$
$=-3x-7.9$

93. answers may vary

95. answers may vary

97. $f(x)=0.014x^2+0.043x+0.584$

a. June 2003 is 0 months after June 2003, so $x = 0$.
$f(0)=0.014(0)^2+0.043(0)+0.584$
$=0.584$
≈ 0.6
There were approximately 0.6 million Web logs on the Internet in June 2003.

b. June 2004 is 12 months after June 2003, so $x = 12$.
$f(12)=0.014(12)^2+0.043(12)+0.584$
$=2.016+0.516+0.584$
$=3.116$
≈ 3.1
There were approximately 3.1 million Web logs on the Internet in June 2004.

c. June 2005 is 24 months after June 2003, so $x = 24$.
$f(24)$
$=0.014(24)^2+0.043(24)+0.584$
$=8.064+1.032+0.584$
$=9.68$
≈ 9.7
There were approximately 9.7 million Web logs on the Internet in June 2005.

d. answers may vary

99. $P(x)+Q(x)=(3x+3)+(4x^2-6x+3)$
$=3x+3+4x^2-6x+3$
$=4x^2+3x-6x+3+3$
$=4x^2-3x+6$

101. $P(x) = 2x - 3$
$P(a) = 2a - 3$
$P(-x) = 2(-x) - 3 = -2x - 3$
$P(x + h) = 2(x + h) - 3 = 2x + 2h - 3$

103. $(14z^{5x}+3z^{2x}+z)-(2z^{5x}-10z^{2x}+3z)$
$=14z^{5x}+3z^{2x}+z-2z^{5x}+10z^{2x}-3z$
$=14z^{5x}-2z^{5x}+3z^{2x}+10z^{2x}+z-3z$
$=12z^{5x}+13z^{2x}-2z$

105. $(2x^2-xy+7y^2)+(5xy-y^2)+(3xy+4y^2)$
$=2x^2-xy+7y^2+5xy-y^2+3xy+4y^2$
$=2x^2-xy+5xy+3xy+7y^2-y^2+4y^2$
$=2x^2+7xy+10y^2$
The perimeter is $(2x^2+7xy+10y^2)$ units.

Section 5.2 Practice Problems

1. $(7y^2)(4y^5)=7(4)(y^2)(y^5)$
$=28y^{2+5}$
$=28y^7$

2. $(-a^2b^3c)(10ab^2c^{12})$
$=-1(10)(a^2a)(b^3b^2)(cc^{12})$
$=-10a^{2+1}b^{3+2}c^{1+12}$
$=-10a^3b^5c^{13}$

3. $4x(3x-2)=4x(3x)+4x(-2)=12x^2-8x$

4. $-2y^3(5y^2-2y+6)$
$=-2y^3(5y^2)+(-2y^3)(-2y)+(-2y^3)(6)$
$=-10y^5+4y^4-12y^3$

5. $-a^2b(4a^3-2ab+b^2)$
$=-a^2b(4a^3)+(-a^2b)(-2ab)+(-a^2b)(b^2)$
$=-4a^5b+2a^3b^2-a^2b^3$

6. $(x+2)(3x+1) = x(3x+1)+2(3x+1)$
$= 3x^2 + x + 6x + 2$
$= 3x^2 + 7x + 2$

7. $(5x-1)(2x^2 - x + 4)$
$= 5x(2x^2 - x + 4) + (-1)(2x^2 - x + 4)$
$= 10x^3 - 5x^2 + 20x - 2x^2 + x - 4$
$= 10x^3 - 7x^2 + 21x - 4$

8.

$$\begin{array}{r} x^2 - 6x + 1 \\ \underline{3x^2 \qquad + 5} \\ 5x^2 - 30x + 5 \\ \underline{3x^4 - 18x^3 + 3x^2 \qquad\qquad\quad} \\ 3x^4 - 18x^3 + 8x^2 - 30x + 5 \end{array}$$

9. $(x-7)(x+5)$
$= x \cdot x + x \cdot 5 + (-7) \cdot x + (-7)(-5)$
$= x^2 + 5x - 7x - 35$
$= x^2 - 2x - 35$

10. $(4x-3)(x-6)$
$= 4x \cdot x + 4x(-6) + (-3) \cdot x + (-3)(-6)$
$= 4x^2 - 24x - 3x + 18$
$= 4x^2 - 27x + 18$

11. $(6x^2 + 5y)(2x^2 - y)$
$= 12x^4 - 6x^2y + 10x^2y - 5y^2$
$= 12x^4 + 4x^2y - 5y^2$

12. $(x+3)^2 = x^2 + 2 \cdot x \cdot 3 + 3^2 = x^2 + 6x + 9$

13. $(y-6)^2 = y^2 - 2 \cdot y \cdot 6 + 6^2 = y^2 - 12y + 36$

14. $(2x+5y)^2 = (2x)^2 + 2(2x)(5y) + (5y)^2$
$= 4x^2 + 20xy + 25y^2$

15. $(6a^2 - 2b)^2 = (6a^2)^2 - 2(6a^2)(2b) + (2b)^2$
$= 36a^4 - 24a^2b + 4b^2$

16. $(x+4)(x-4) = x^2 - 4^2 = x^2 - 16$

17. $(3m-6)(3m+6) = (3m)^2 - 6^2 = 9m^2 - 36$

18. $(a^2 + 5y)(a^2 - 5y) = (a^2)^2 - (5y)^2$
$= a^4 - 25y^2$

19. $\left(4y^2 - \frac{1}{3}\right)\left(4y^2 + \frac{1}{3}\right) = (4y^2)^2 - \left(\frac{1}{3}\right)^2$
$= 16y^4 - \frac{1}{9}$

20. $[2-(3x+y)]^2$
$= 2^2 - 2(2)(3x+y) + (3x+y)^2$
$= 4 - 4(3x+y) + (3x)^2 + 2(3x)y + y^2$
$= 4 - 12x - 4y + 9x^2 + 6xy + y^2$

21. $[(2x+3y)-2][(2x+3y)+2]$
$= (2x+3y)^2 - 2^2$
$= (2x)^2 + 2(2x)(3y) + (3y)^2 - 4$
$= 4x^2 + 12xy + 9y^2 - 4$

22. $(y-2)(y+2)(y^2-4) = (y^2-4)(y^2-4)$
$= (y^2-4)^2$
$= y^4 - 8y^2 + 16$

23. $f(x) = x^2 - 6x + 1$
$f(b-1) = (b-1)^2 - 6(b-1) + 1$
$= b^2 - 2b + 1 - 6b + 6 + 1$
$= b^2 - 8b + 8$

Calculator Explorations

1. $(x+4)(x-4) = x^2 - 16$

Let $Y_1 = (x+4)(x-4)$ and $Y_2 = x^2 - 16$. The graphs coincide.

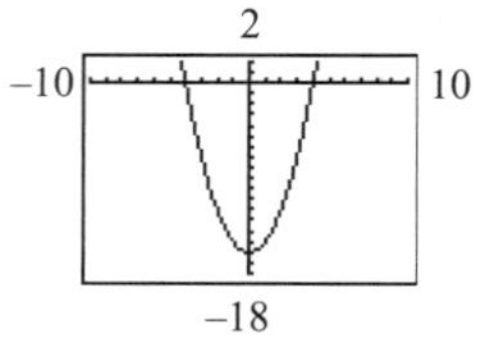

2. $(x+3)(x+3) = x^2 + 6x + 9$

Let $Y_1 = (x+3)(x+3)$ and

$Y_2 = x^2 + 6x + 9$.

The graphs coincide.

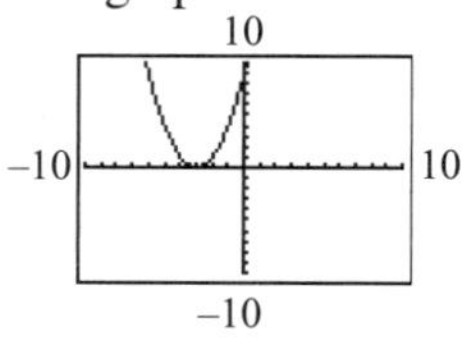

3. $(3x-7)^2 = (3x)^2 - 2(3x)(7) + 7^2$
$= 9x^2 - 42x + 49$

Let $Y_1 = (3x-7)^2$ and

$Y_2 = 9x^2 - 42x + 49$.

The graphs coincide.

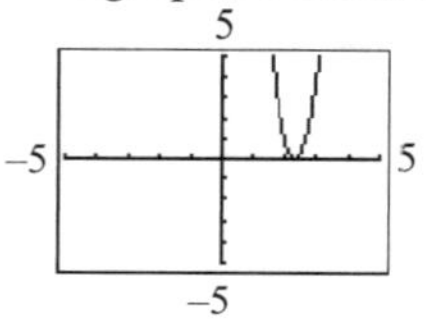

4. $(5x-2)^2 = (5x)^2 - 2(5x)(2) + 2^2$
$= 25x^2 - 20x + 4$

Let $Y_1 = (5x-2)^2$ and

$Y_2 = 25x^2 - 20x + 4$.

The graphs coincide.

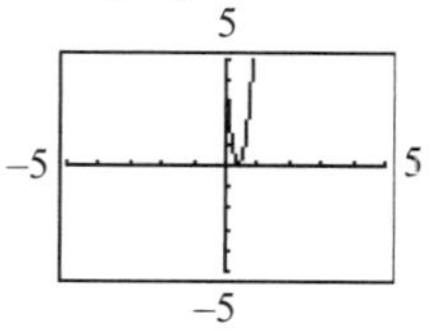

5. $(5x+1)(x^2 - 3x - 2)$
$= 5x(x^2 - 3x - 2) + 1(x^2 - 3x - 2)$
$= 5x^3 - 15x^2 - 10x + x^2 - 3x - 2$
$= 5x^3 - 14x^2 - 13x - 2$

Let $Y_1 = (5x+1)(x^2 - 3x - 2)$ and

$Y_2 = 5x^3 - 14x^2 - 13x - 2$.

The graphs coincide.

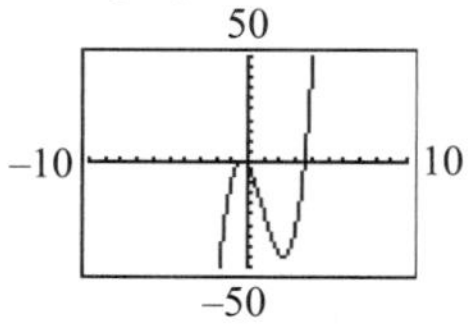

6. $(7x+4)(2x^2 + 3x - 5)$
$= 7x(2x^2 + 3x - 5) + 4(2x^2 + 3x - 5)$
$= 14x^3 + 21x^2 - 35x + 8x^2 + 12x - 20$
$= 14x^3 + 29x^2 - 23x - 20$

Let $Y_1 = (7x+4)(2x^2 + 3x - 5)$ and

$Y_2 = 14x^3 + 29x^2 - 23x - 20$.

The graphs coincide.

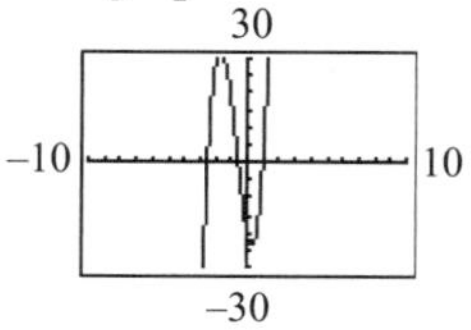

Exercise Set 5.2

1. $(-4x^3)(3x^2) = -4(3)(x^3)(x^2)$
$= -12x^{3+2}$
$= -12x^5$

3. $(8.6a^4b^5c)(10ab^3c^2)$
$= 8.6(10)(a^4a)(b^5b^3)(cc^2)$
$= 86a^5b^8c^3$

5. $3x(4x+7) = 3x(4x) + 3x(7) = 12x^2 + 21x$

7. $-6xy(4x+y) = -6xy(4x) + (-6xy)(y)$
$= -24x^2y - 6xy^2$

9. $-4ab(xa^2 + ya^2 - 3) = -4ab(xa^2) + (-4ab)(ya^2) + (-4ab)(-3)$
$= -4a^3bx - 4a^3by + 12ab$

11. $(x-3)(2x+4) = x(2x+4) - 3(2x+4)$
$= 2x^2 + 4x - 6x - 12$
$= 2x^2 - 2x - 12$

13. $(2x+3)(x^3 - x + 2) = 2x(x^3 - x + 2) + 3(x^3 - x + 2)$
$= 2x^4 - 2x^2 + 4x + 3x^3 - 3x + 6$
$= 2x^4 + 3x^3 - 2x^2 + x + 6$

15.
$$\begin{array}{r} 3x-2 \\ \times\ 5x+1 \\ \hline 3x-2 \\ 15x^2-10x \\ \hline 15x^2-7x-2 \end{array}$$

17.
$$\begin{array}{r} 3m^2+2m-1 \\ \times\qquad\qquad 5m+2 \\ \hline 6m^2+4m-2 \\ 15m^3+10m^2-5m \\ \hline 15m^3+16m^2-\ m-2 \end{array}$$

19. $-6a^2b^2(5a^2b^2 - 6a - 6b) = -6a^2b^2(5a^2b^2) + (-6a^2b^2)(-6a) + (-6a^2b^2)(-6b)$
$= -30a^4b^4 + 36a^3b^2 + 36a^2b^3$

21. $(x-3)(x+4) = x^2 + 4x - 3x - 12 = x^2 + x - 12$

23. $(5x-8y)(2x-y) = 10x^2 - 5xy - 16xy + 8y^2$
$= 10x^2 - 21xy + 8y^2$

25. $\left(4x+\frac{1}{3}\right)\left(4x-\frac{1}{2}\right) = 16x^2 - 2x + \frac{4}{3}x - \frac{1}{6}$
$= 16x^2 - \frac{2}{3}x - \frac{1}{6}$

27. $(5x^2 - 2y^2)(x^2 - 3y^2) = 5x^4 - 15x^2y^2 - 2x^2y^2 + 6y^4$
$= 5x^4 - 17x^2y^2 + 6y^4$

29. $(x+4)^2 = x^2 + 2\cdot x\cdot 4 + 4^2 = x^2 + 8x + 16$

31. $(6y-1)(6y+1) = (6y)^2 - 1^2 = 36y^2 - 1$

33. $$\begin{aligned}(3x-y)^2 &= (3x)^2 - 2\cdot 3x\cdot y + y^2\\ &= 9x^2 - 6xy + y^2\end{aligned}$$

35. $$\begin{aligned}(7ab+3c)(7ab-3c) &= (7ab)^2 - (3c)^2\\ &= 49a^2b^2 - 9c^2\end{aligned}$$

37. $\left(3x+\frac{1}{2}\right)\left(3x-\frac{1}{2}\right) = (3x)^2 - \left(\frac{1}{2}\right)^2 = 9x^2 - \frac{1}{4}$

39. $$\begin{aligned}[3+(4b+1)]^2 &= 3^2 + 2\cdot 3\cdot(4b+1) + (4b+1)^2\\ &= 9 + 6(4b+1) + (4b)^2 + 2\cdot 4b\cdot 1 + 1^2\\ &= 9 + 24b + 6 + 16b^2 + 8b + 1\\ &= 16b^2 + 32b + 16\end{aligned}$$

41. $$\begin{aligned}[(2s-3)-1][(2s-3)+1] &= (2s-3)^2 - 1^2\\ &= (2s)^2 - 2\cdot 2s\cdot 3 + 3^2 - 1\\ &= 4s^2 - 12s + 9 - 1\\ &= 4s^2 - 12s + 8\end{aligned}$$

43. $$\begin{aligned}(x+y)(2x-1)(x+1) &= (2x^2 - x + 2xy - y)(x+1)\\ &= 2x^3 - x^2 + 2x^2y - xy + 2x^2 - x + 2xy - y\\ &= 2x^3 + 2x^2y + x^2 + xy - x - y\end{aligned}$$

45. $$\begin{aligned}(x-2)^4 &= (x-2)(x-2)(x-2)(x-2)\\ &= [(x-2)^2][(x-2)^2]\\ &= [x^2 - 2\cdot x\cdot 2 + 2^2][x^2 - 2\cdot x\cdot 2 + 2^2]\\ &= (x^2 - 4x + 4)(x^2 - 4x + 4)\\ &= x^2(x^2 - 4x + 4) - 4x(x^2 - 4x + 4) + 4(x^2 - 4x + 4)\\ &= x^4 - 4x^3 + 4x^2 - 4x^3 + 16x^2 - 16x + 4x^2 - 16x + 16\\ &= x^4 - 8x^3 + 24x^2 - 32x + 16\end{aligned}$$

47. $$\begin{aligned}(x-5)(x+5)(x^2+25) &= (x^2 - 25)(x^2 + 25)\\ &= (x^2)^2 - 25^2\\ &= x^4 - 625\end{aligned}$$

49. $-8a^2b(3b^2-5b+20)=-24a^2b^3+40a^2b^2-160a^2b$

51. $(6x+1)^2=(6x)^2+2\cdot 6x\cdot 1+1^2$
$=36x^2+12x+1$

53. $(5x^3+2y)(5x^3-2y)=(5x^3)^2-(2y)^2$
$=25x^6-4y^2$

55. $(2x^3+5)(5x^2+4x+1)=2x^3(5x^2+4x+1)+5(5x^2+4x+1)$
$=10x^5+8x^4+2x^3+25x^2+20x+5$

57. $(3x^2+2x-1)^2=(3x^2+2x-1)(3x^2+2x-1)$
$=3x^2(3x^2+2x-1)+2x(3x^2+2x-1)-1(3x^2+2x-1)$
$=9x^4+6x^3-3x^2+6x^3+4x^2-2x-3x^2-2x+1$
$=9x^4+12x^3-2x^2-4x+1$

59. $(3x-1)(x+3)=3x(x+3)-1(x+3)$
$=3x^2+9x-x-3$
$=3x^2+8x-3$

61. $(3x^4+1)(3x^2+5)=9x^6+15x^4+3x^2+5$

63. $(3x+1)^2=(3x)^2+2\cdot 3x\cdot 1+1^2=9x^2+6x+1$

65. $(3b-6y)(3b+6y)=(3b)^2-(6y)^2=9b^2-36y^2$

67. $(7x-3)(7x+3)=(7x)^2-3^2=49x^2-9$

69.
$$\begin{array}{r} 3x^2+4x-4 \\ \times \qquad\qquad 3x+6 \\ \hline 18x^2+24x-24 \\ 9x^3+12x^2-12x \qquad \\ \hline 9x^3+30x^2+12x-24 \end{array}$$

71. $(4x^2-2x+5)(3x+1)=4x^2(3x+1)-2x(3x+1)+5(3x+1)$
$=12x^3+4x^2-6x^2-2x+15x+5$
$=12x^3-2x^2+13x+5$

73. $[(xy+4)-6]^2 = [xy+4-6]^2$
$= (xy-2)^2$
$= (xy)^2 - 2\cdot xy\cdot 2 + 2^2$
$= x^2y^2 - 4xy + 4$

75. $(11a^2+1)(2a+1) = 22a^3 + 11a^2 + 2a + 1$

77. $\left(\frac{2}{3}n-2\right)\left(\frac{1}{2}n-9\right) = \frac{1}{3}n^2 - 6n - n + 18$
$= \frac{1}{3}n^2 - 7n + 18$

79. $(3x+1)(3x-1)(2y+5x)$
$= [(3x)^2 - 1^2](2y+5x)$
$= (9x^2-1)(2y+5x)$
$= 18x^2y + 45x^3 - 2y - 5x$
or $45x^3 + 18x^2y - 5x - 2y$

81. $f(x) = x^2 - 3x$
$f(a) = a^2 - 3a$

83. $f(x) = x^2 - 3x$
$f(a+h) = (a+h)^2 - 3(a+h)$
$= a^2 + 2ah + h^2 - 3a - 3h$

85. $f(x) = x^2 - 3x$
$f(b-2) = (b-2)^2 - 3(b-2)$
$= b^2 - 4b + 4 - 3b + 6$
$= b^2 - 7b + 10$

87. $\frac{6x^3}{3x} = \frac{3x\cdot 2x^2}{3x} = 2x^2$

89. $\frac{20a^3b^5}{18ab^2} = \frac{20}{18}\cdot\frac{a^3}{a}\cdot\frac{b^5}{b^2}$
$= \frac{10}{9}\cdot a^{3-1}\cdot b^{5-2}$
$= \frac{10}{9}a^2b^3$
$= \frac{10a^2b^3}{9}$

91. $\frac{8m^4n}{12mn} = \frac{8}{12}\cdot\frac{m^4}{m}\cdot\frac{n}{n}$
$= \frac{2}{3}\cdot m^{4-1}\cdot n^{1-1}$
$= \frac{2}{3}m^3n^0$
$= \frac{2m^3}{3}$

93. $7y(3z-2)+1 = 7y\cdot 3z + 7y(-2) + 1$
$= 21yz - 14y + 1$

95. answers may vary

97. $5x^2y^n(6y^{n+1}-2)$
$= 5x^2y^n\cdot 6y^{n+1} + 5x^2y^n(-2)$
$= 30x^2y^{n+n+1} - 10x^2y^n$
$= 30x^2y^{2n+1} - 10x^2y^n$

99. $(x^a+5)(x^{2a}-3) = x^a(x^{2a}-3) + 5(x^{2a}-3)$
$= x^{a+2a} - 3x^a + 5x^{2a} - 15$
$= x^{3a} - 3x^a + 5x^{2a} - 15$

101. **a.** $(3x+5)+(3x+7) = 3x+5+3x+7$
$= 6x+12$

b. $(3x+5)(3x+7) = 9x^2 + 21x + 15x + 35$
$= 9x^2 + 36x + 35$

c. One operation is addition; the other is multiplication.

103. $P(x)\cdot R(x) = 5x(x+5) = 5x^2 + 25x$

105. $f(x) = x^3 - 2x^2$
$f(a) = a^3 - 2a^2$

107. $(3x-2)^2 - x^2$
$= (3x)^2 - 2\cdot(3x)\cdot(2) + 2^2 - x^2$
$= 9x^2 - 12x + 4 - x^2$
$= 8x^2 - 12x + 4$
The area of the shaded region is $(8x^2 - 12x + 4)$ square inches.

109. $A = \pi r^2$
$= \pi(5x-2)^2$
$= \pi[(5x)^2 - 2\cdot 5x \cdot 2 + 2^2]$
$= \pi(25x^2 - 20x + 4)$
The area is $\pi(25x^2 - 20x + 4)$ or $(25\pi x^2 - 20\pi x + 4\pi)$ square kilometers.

Section 5.3 Practice Problems

1. $$\frac{16y^3 - 8y^2 + 6y}{2y} = \frac{16y^3}{2y} - \frac{8y^2}{2y} + \frac{6y}{2y}$$
$$= 8y^2 - 4y + 3$$

2. $$\frac{9a^3b^3 - 6a^2b^2 + a^2b - 4a}{a^2b}$$
$$= \frac{9a^3b^3}{a^2b} - \frac{6a^2b^2}{a^2b} + \frac{a^2b}{a^2b} - \frac{4a}{a^2b}$$
$$= 9ab^2 - 6b + 1 - \frac{4}{ab}$$

3. $$\begin{array}{r} 6x-1 \\ x+2\overline{)6x^2+11x-2} \\ \underline{6x^2+12x} \\ -x-2 \\ \underline{-x-2} \\ 0 \end{array}$$
$$\frac{6x^2 + 11x - 2}{x+2} = 6x - 1$$

4. $$\begin{array}{r} 2x-3 \\ 5x-1\overline{)10x^2-17x+5} \\ \underline{10x^2-2x} \\ -15x+5 \\ \underline{-15x+3} \\ 2 \end{array}$$
$$\frac{10x^2 - 17x + 5}{5x-1} = 2x - 3 + \frac{2}{5x-1}$$

5. $$\begin{array}{r} 5x^2+6x+15 \\ x-2\overline{)5x^3-4x^2+3x-4} \\ \underline{5x^3-10x^2} \\ 6x^2+3x \\ \underline{6x^2-12x} \\ 15x-4 \\ \underline{15x-30} \\ 26 \end{array}$$
$$\frac{5x^3 - 4x^2 + 3x - 4}{x-2} = 5x^2 + 6x + 15 + \frac{26}{x-2}$$

6. $$\begin{array}{r} 3x^2+1 \\ x^2+0x+1\overline{)3x^4+0x^3+4x^2-6x+1} \\ \underline{3x^4+0x^3+3x^2} \\ x^2-6x+1 \\ \underline{x^2+0x+1} \\ -6x \end{array}$$
$$\frac{3x^4 + 4x^2 - 6x + 1}{x^2+1} = 3x^2 + 1 - \frac{6x}{x^2+1}$$

7. $$\begin{array}{r} 16x^2+12x+9 \\ 4x-3\overline{)64x^3+0x^2+0x-27} \\ \underline{64x^3-48x^2} \\ 48x^2+0x \\ \underline{48x^2-36x} \\ 36x-27 \\ \underline{36x-27} \\ 0 \end{array}$$
$$\frac{64x^3 - 27}{4x-3} = 16x^2 + 12x + 9$$

8.

2	3	−2	5	4
		6	8	26
	3	4	13	30

$$\frac{3x^3-2x^2+5x+4}{x-2}=3x^2+4x+13+\frac{30}{x-2}$$

9.

−1	1	3	0	−5	4
		−1	−2	2	3
	1	2	−2	−3	7

$$\frac{x^4+3x^3-5x+4}{x+1}=x^3+2x^2-2x-3+\frac{7}{x+1}$$

Exercise Set 5.3

1. $\dfrac{4a^2+8a}{2a}=\dfrac{4a^2}{2a}+\dfrac{8a}{2a}=2a+4$

3. $\dfrac{12a^5b^2+16a^4b}{4a^4b}=\dfrac{12a^5b^2}{4a^4b}+\dfrac{16a^4b}{4a^4b}$
$=3ab+4$

5. $\dfrac{4x^2y^2+6xy^2-4y^2}{2x^2y}$
$=\dfrac{4x^2y^2}{2x^2y}+\dfrac{6xy^2}{2x^2y}-\dfrac{4y^2}{2x^2y}$
$=2y+\dfrac{3y}{x}-\dfrac{2y}{x^2}$

7.
$$\begin{array}{r} x+1 \\ x+2\overline{)x^2+3x+2} \\ \underline{x^2+2x} \\ x+2 \\ \underline{x+2} \\ 0 \end{array}$$

$$\frac{x^2+3x+2}{x+2}=x+1$$

9.
$$\begin{array}{r} 2x-8 \\ x+1\overline{)2x^2-6x-8} \\ \underline{2x^2+2x} \\ -8x-8 \\ \underline{-8x-8} \\ 0 \end{array}$$

$$\frac{2x^2-6x-8}{x+1}=2x-8$$

11.
$$\begin{array}{r} x-\frac{1}{2} \\ 2x+4\overline{)2x^2+3x-2} \\ \underline{2x^2+4x} \\ -x-2 \\ \underline{-x-2} \\ 0 \end{array}$$

$$\frac{2x^2+3x-2}{2x+4}=x-\frac{1}{2}$$

13.
$$\begin{array}{r} 2x^2-\frac{1}{2}x+5 \\ 2x+4\overline{)4x^3+7x^2+8x+20} \\ \underline{4x^3+8x^2} \\ -x^2+8x \\ \underline{-x^2-2x} \\ 10x+20 \\ \underline{10x+20} \\ 0 \end{array}$$

$$\frac{4x^3+7x^2+8x+20}{2x+4}=2x^2-\frac{1}{2}x+5$$

15.
$$\begin{array}{r} 2x^2 \qquad -6 \\ 3x+1\overline{)6x^3+2x^2-18x-6} \\ \underline{6x^3+2x^2} \\ -18x-6 \\ \underline{-18x-6} \\ 0 \end{array}$$

$$\frac{2x^2+6x^3-18x-6}{3x+1}=2x^2-6$$

17.
$$
\begin{array}{r l}
 & 3x^3 + 5x + 4 \\
x^2 - 2 \,\big)\!\! & 3x^5 - x^3 + 4x^2 - 12x - 8 \\
 & \underline{3x^5 - 6x^3} \\
 & 5x^3 + 4x^2 - 12x \\
 & \underline{5x^3 \qquad - 10x} \\
 & 4x^2 - 2x - 8 \\
 & \underline{4x^2 \qquad - 8} \\
 & -2x
\end{array}
$$

$$\frac{3x^5 - x^3 + 4x^2 - 12x - 8}{x^2 - 2} = 3x^3 + 5x + 4 - \frac{2x}{x^2 - 2}$$

19.
$$
\begin{array}{r l}
 & 2x^3 + \frac{9}{2}x^2 + 10x + 21 \\
x - 2 \,\big)\!\! & 2x^4 + \frac{1}{2}x^3 + x^2 + x + 0 \\
 & \underline{2x^4 - 4x^3} \\
 & \frac{9}{2}x^3 + x^2 \\
 & \underline{\frac{9}{2}x^3 - 9x^2} \\
 & 10x^2 + x \\
 & \underline{10x^2 - 20x} \\
 & 21x + 0 \\
 & \underline{21x - 42} \\
 & 42
\end{array}
$$

$$\frac{2x^4 + \frac{1}{2}x^3 + x^2 + x}{x - 2} = 2x^3 + \frac{9}{2}x^2 + 10x + 21 + \frac{42}{x - 2}$$

21. $x - 5 = x - c$ where $c = 5$.

$$
\begin{array}{c|ccc}
5 & 1 & 3 & -40 \\
 & & 5 & 40 \\
\hline
 & 1 & 8 & 0
\end{array}
$$

$$\frac{x^2 + 3x - 40}{x - 5} = x + 8$$

23. $x + 6 = x - c$ where $c = -6$.

$$
\begin{array}{c|ccc}
-6 & 1 & 5 & -6 \\
 & & -6 & 6 \\
\hline
 & 1 & -1 & 0
\end{array}
$$

$$\frac{x^2 + 5x - 6}{x + 6} = x - 1$$

25. $x - 2 = x - c$ where $c = 2$.

$$
\begin{array}{c|cccc}
2 & 1 & -7 & -13 & 5 \\
 & & 2 & -10 & -46 \\
\hline
 & 1 & -5 & -23 & -41
\end{array}
$$

$$\frac{x^3 - 7x^2 - 13x + 5}{x - 2} = x^2 - 5x - 23 - \frac{41}{x - 2}$$

27. $x - 2 = x - c$ where $c = 2$.

$$
\begin{array}{c|ccc}
2 & 4 & 0 & -9 \\
 & & 8 & 16 \\
\hline
 & 4 & 8 & 7
\end{array}
$$

$$\frac{4x^2 - 9}{x - 2} = 4x + 8 + \frac{7}{x - 2}$$

29. $\dfrac{4x^7y^4 + 8xy^2 + 4xy^3}{4xy^3}$

$$= \frac{4x^7y^4}{4xy^3} + \frac{8xy^2}{4xy^3} + \frac{4xy^3}{4xy^3}$$

$$= x^6y + \frac{2}{y} + 1$$

31.
$$
\begin{array}{r l}
 & 5x^2 \qquad - 6 \\
2x - 1 \,\big)\!\! & 10x^3 - 5x^2 - 12x + 1 \\
 & \underline{10x^3 - 5x^2} \\
 & -12x + 1 \\
 & \underline{-12x + 6} \\
 & -5
\end{array}
$$

$$\frac{10x^3 - 5x^2 - 12x + 1}{2x - 1} = 5x^2 - 6 - \frac{5}{2x - 1}$$

33. $x - 4$ has the form $x - c$, where $c = 4$, so synthetic division can be used.
$2x^3 - 6x^2 - 4 = 2x^3 - 6x^2 + 0x - 4$

$$\begin{array}{r|rrrr} 4 & 2 & -6 & 0 & -4 \\ & & 8 & 8 & 32 \\ \hline & 2 & 2 & 8 & 28 \end{array}$$

$$\frac{2x^3 - 6x^2 - 4}{x-4} = 2x^2 + 2x + 8 + \frac{28}{x-4}$$

35. $x - 5$ has the form $x - c$, where $c = 5$, so synthetic division can be used.

$$\begin{array}{r|rrrrr} 5 & 2 & -13 & 16 & -9 & 20 \\ & & 10 & -15 & 5 & -20 \\ \hline & 2 & -3 & 1 & -4 & 0 \end{array}$$

$$\frac{2x^4 - 13x^3 + 16x^2 - 9x + 20}{x-5}$$
$$= 2x^3 - 3x^2 + x - 4$$

37. $x + 1$ has the form $x - c$, where $c = -1$, so synthetic division can be used.
$7x^2 - 4x + 12 + 3x^3 = 3x^3 + 7x^2 - 4x + 12$

$$\begin{array}{r|rrrr} -1 & 3 & 7 & -4 & 12 \\ & & -3 & -4 & 8 \\ \hline & 3 & 4 & -8 & 20 \end{array}$$

$$\frac{7x^2 - 4x + 12 + 3x^3}{x+1} = 3x^2 + 4x - 8 + \frac{20}{x+1}$$

39. $x - \frac{1}{3}$ has the form $x - c$, where $c = \frac{1}{3}$, so synthetic division can be used.

$$\begin{array}{r|rrrr} \frac{1}{3} & 3 & 2 & -4 & 1 \\ & & 1 & 1 & -1 \\ \hline & 3 & 3 & -3 & 0 \end{array}$$

$$\frac{3x^3 + 2x^2 - 4x + 1}{x - \frac{1}{3}} = 3x^2 + 3x - 3$$

41. $x - 1$ has the form $x - c$, where $c = 1$, so synthetic division can be used.
$x^3 - 1 = x^3 + 0x^2 + 0x - 1$

$$\begin{array}{r|rrrr} 1 & 1 & 0 & 0 & -1 \\ & & 1 & 1 & 1 \\ \hline & 1 & 1 & 1 & 0 \end{array}$$

$$\frac{x^3 - 1}{x-1} = x^2 + x + 1$$

43. $$\frac{25xy^2 + 75xyz + 125x^2yz}{-5x^2y}$$
$$= \frac{25xy^2}{-5x^2y} + \frac{75xyz}{-5x^2y} + \frac{125x^2yz}{-5x^2y}$$
$$= -\frac{5y}{x} - \frac{15z}{x} - 25z$$

45. $9x^5 + 6x^4 - 6x^2 - 4x$
$= 9x^5 + 6x^4 + 0x^3 - 6x^2 - 4x + 0$

$$\begin{array}{r} 3x^4 - 2x \\ 3x+2\overline{\smash{)}\,9x^5 + 6x^4 + 0x^3 - 6x^2 - 4x + 0} \\ \underline{9x^5 + 6x^4} \qquad\qquad\qquad\qquad \\ 0 + 0x^3 - 6x^2 - 4x \\ \underline{-6x^2 - 4x} \\ 0 \end{array}$$

$(9x^5 + 6x^4 - 6x^2 - 4x) \div (3x + 2) = 3x^4 - 2x$

47. $|x + 5| < 4$
$-4 < x + 5 < 4$
$-4 - 5 < x + 5 - 5 < 4 - 5$
$-9 < x < -1$
The solution set is (–9, 1).

49. $|2x + 7| \geq 9$
$2x + 7 \leq -9$ or $2x + 7 \geq 9$
$2x \leq -16$ $\quad$ $2x \geq 2$
$x \leq -8$ $\quad$ $x \geq 1$
The solution set is $(-\infty, -8] \cup [1, \infty)$.

51. $(5x^2 - 3x + 2) \div (x + 2)$ is a candidate for synthetic division since $x + 2$ is in the form $x - c$, where $c = -2$.

53. $(x^7 - 2) \div (x^5 + 1)$ is not a candidate for synthetic division since $x^5 + 1$ does not have the form $x - c$.

55. The degree of the remainder must be less than that of the divisor, or 3 in this case. The choices are a or d.

57. $\dfrac{3x^4 + 6x^2 - 18}{3} = \dfrac{3x^4}{3} + \dfrac{6x^2}{3} - \dfrac{18}{3}$
$= x^4 + 2x^2 - 6$

The length of each piece is $(x^4 + 2x^2 - 6)$ meters.

59.
$$\begin{array}{r} 3x - 7 \\ 5x + 2\overline{)15x^2 - 29x - 14} \\ \underline{15x^2 + 6x} \\ -35x - 14 \\ \underline{-35x - 14} \\ 0 \end{array}$$

The width is $(3x - 7)$ inches.

61. $P(x) = 3x^3 + 2x^2 - 4x + 3$

$P(1) = 3(1)^3 + 2(1)^2 - 4(1) + 3$
$= 3 + 2 - 4 + 3$
$= 4$

Use synthetic division to divide $3x^3 + 2x^2 - 4x + 3$ by $x - 1$.

1	3	2	−4	3
		3	5	1
	3	5	1	4

The remainder is 4; answers may vary.

63. $P(x) = 5x^4 - 2x^2 + 3x - 6$

$P(-3) = 5(-3)^4 - 2(-3)^2 + 3(-3) - 6$
$= 405 - 18 - 9 - 6$
$= 372$

Use synthetic division to divide $5x^4 - 2x^2 + 3x - 6$ by $x + 3$.

−3	5	0	−2	3	−6
		−15	45	−129	378
	5	−15	43	−126	372

The remainder is 372; answers may vary.

65. answers may vary

67.

1	1	$\frac{2}{3}$	0	1	0
		1	$\frac{5}{3}$	$\frac{5}{3}$	$\frac{8}{3}$
	1	$\frac{5}{3}$	$\frac{5}{3}$	$\frac{8}{3}$	$\frac{8}{3}$

$\left(x^4 + \dfrac{2}{3}x^3 + x\right) \div (x - 1)$
$= x^3 + \dfrac{5}{3}x^2 + \dfrac{5}{3}x + \dfrac{8}{3} + \dfrac{8}{3(x-1)}$

69. $\dfrac{f(x)}{g(x)} = \dfrac{25x^2 - 5x + 30}{5x}$
$= \dfrac{25x^2}{5x} - \dfrac{5x}{5x} + \dfrac{30}{5x}$
$= 5x - 1 + \dfrac{6}{x}$

Since $5x$ cannot be 0, x cannot be 0.

71. $\dfrac{f(x)}{g(x)} = \dfrac{7x^4 - 3x^2 + 2}{x - 2}$

2	7	0	−3	0	2
		14	28	50	100
	7	14	25	50	102

$\dfrac{f(x)}{g(x)} = 7x^3 + 14x^2 + 25x + 50 - \dfrac{102}{x - 2}$

Since $x - 2$ cannot be 0, $x \neq 2$.

73. answers may vary

75. answers may vary

Section 5.4 Practice Problems

1. $9x + 3 = 3 \cdot 3x + 3 \cdot 1 = 3(3x + 1)$

2. $20y^2 - 4y^3 = 4y^2 \cdot 5 - 4y^2 \cdot y = 4y^2(5 - y)$

3. $6a - 7$; there is no common factor other than 1.

4. $6a^4b^2 - 3a^2b^2 = 3a^2b^2 \cdot 2a^2 - 3a^2b^2 \cdot 1$
$= 3a^2b^2(2a^2 - 1)$

5. $-2x^2y - 4xy + 10y = 2y(-x^2 - 2x + 5)$
or
$-2x^2y - 4xy + 10y$
$= -2y(x^2) + (-2y)(2x) + (-2y)(-5)$
$= -2y(x^2 + 2x - 5)$

6. $3(x + 7) + 5y(x + 7) = (x + 7)(3 + 5y)$

7. $6a(2a + 3b) - (2a + 3b)$
$= 6a(2a + 3b) - 1(2a + 3b)$
$= (2a + 3b)(6a - 1)$

8. $xy - 5y + 3x - 15 = (xy - 5y) + (3x - 15)$
$= y(x - 5) + 3(x - 5)$
$= (x - 5)(y + 3)$

9. $y^3 + 6y^2 + 4y + 24 = (y^3 + 6y^2) + (4y + 24)$
$= y^2(y + 6) + 4(y + 6)$
$= (y + 6)(y^2 + 4)$

10. $a^2b^2 + a^2 - 3b^2 - 3$
$= (a^2b^2 + a^2) + (-3b^2 - 3)$
$= a^2(b^2 + 1) - 3(b^2 + 1)$
$= (b^2 + 1)(a^2 - 3)$

11. $ab + 5a - b - 5 = (ab + 5a) + (-b - 5)$
$= a(b + 5) - 1(b + 5)$
$= (b + 5)(a - 1)$

Mental Math

1. $6 = 6$
$12 = 2 \cdot 6$
GCF = 6

2. $9 = 9$
$27 = 9 \cdot 3$
GCF = 9

3. $15x = 3x \cdot 5$
$10 = 2 \cdot 5$
GCF = 5

4. $9x = 3x \cdot 3$
$12 = 4 \cdot 3$
GCF = 3

5. $13x = 13 \cdot x$
$2x = 2 \cdot x$
GCF = x

6. $4y = 4 \cdot y$
$5y = 5 \cdot y$
GCF = y

7. $7x = 7 \cdot x$
$14x = 7 \cdot x \cdot 2$
GCF = $7x$

8. $8z = 4z \cdot 2$
$4z = 4z$
GCF = $4z$

Exercise Set 5.4

1. $18x - 12 = 6 \cdot 3x - 6 \cdot 2 = 6(3x - 2)$

3. $4y^2 - 16xy^3 = 4y^2 \cdot 1 - 4y^2 \cdot 4xy$
$= 4y^2(1 - 4xy)$

5. $6x^5 - 8x^4 + 2x^3$
$= 2x^3 \cdot 3x^2 - 2x^3 \cdot 4x + 2x^3 \cdot 1$
$= 2x^3(3x^2 - 4x + 1)$

7. $8a^3b^3 - 4a^2b^2 + 4ab + 16ab^2$
$= 4ab \cdot 2a^2b^2 - 4ab \cdot ab + 4ab \cdot 1 + 4ab \cdot 4b$
$= 4ab(2a^2b^2 - ab + 1 + 4b)$

9. $6(x + 3) + 5a(x + 3) = (x + 3)(6 + 5a)$

11. $2x(z+7) + (z+7) = 2x(z+7) + 1(z+7)$
$= (z+7)(2x+1)$

13. $3x(6x^2 + 5) - 2(6x^2 + 5) = (6x^2 + 5)(3x - 2)$

15. $ab + 3a + 2b + 6 = (ab + 3a) + (2b + 6)$
$= a(b+3) + 2(b+3)$
$= (b+3)(a+2)$

17. $ac + 4a - 2c - 8 = (ac + 4a) + (-2c - 8)$
$= a(c+4) + (-2)(c+4)$
$= (c+4)(a-2)$

19. $2xy - 3x - 4y + 6 = (2xy - 3x) + (-4y + 6)$
$= x(2y-3) - 2(2y-3)$
$= (2y-3)(x-2)$

21. $12xy - 8x - 3y + 2$
$= (12xy - 8x) + (-3y + 2)$
$= 4x(3y-2) + (-1)(3y-2)$
$= (3y-2)(4x-1)$

23. $6x^3 + 9 = 3 \cdot 2x^3 + 3 \cdot 3 = 3(2x^3 + 3)$

25. $x^3 + 3x^2 = x^2 \cdot x + x^2 \cdot 3 = x^2(x+3)$

27. $8a^3 - 4a = 4a \cdot 2a^2 - 4a \cdot 1 = 4a(2a^2 - 1)$

29. $-20x^2y + 16xy^3 = 4xy \cdot (-5x) + 4xy \cdot 4y^2$
$= 4xy(-5x + 4y^2)$
or
$-20x^2y + 16xy^3 = -4xy \cdot 5x + (-4xy)(-4y^2)$
$= -4xy(5x - 4y^2)$

31. $10a^2b^3 + 5ab^2 - 15ab^3$
$= 5ab^2 \cdot 2ab + 5ab^2 \cdot 1 - 5ab^2 \cdot 3b$
$= 5ab^2(2ab + 1 - 3b)$

33. $9abc^2 + 6a^2bc - 6ab + 3bc$
$= 3b \cdot 3ac^2 + 3b \cdot 2a^2c + 3b \cdot (-2a) + 3b \cdot c$
$= 3b(3ac^2 + 2a^2c - 2a + c)$

35. $4x(y - 2) - 3(y - 2) = (y - 2)(4x - 3)$

37. $6xy + 10x + 9y + 15 = (6xy + 10x) + (9y + 15)$
$= 2x(3y+5) + 3(3y+5)$
$= (3y+5)(2x+3)$

39. $xy + 3y - 5x - 15 = (xy + 3y) + (-5x - 15)$
$= y(x+3) - 5(x+3)$
$= (x+3)(y-5)$

41. $6ab - 2a - 9b + 3 = (6ab - 2a) + (-9b + 3)$
$= 2a(3b-1) + (-3)(3b-1)$
$= (3b-1)(2a-3)$

43. $12xy + 18x + 2y + 3 = (12xy + 18x) + (2y + 3)$
$= 6x(2y+3) + 1(2y+3)$
$= (2y+3)(6x+1)$

45. $2m(n-8) - (n-8) = 2m(n-8) - 1(n-8)$
$= (n-8)(2m-1)$

47. $15x^3y^2 - 18x^2y^2 = 3x^2y^2 \cdot 5x - 3x^2y^2 \cdot 6$
$= 3x^2y^2(5x - 6)$

49. $2x^2+3xy+4x+6y$
$=(2x^2+3xy)+(4x+6y)$
$=x(2x+3y)+2(2x+3y)$
$=(2x+3y)(x+2)$

51. $5x^2+5xy-3x-3y=5x(x+y)-3(x+y)$
$=(x+y)(5x-3)$

53. $x^3+3x^2+4x+12=(x^3+3x^2)+(4x+12)$
$=x^2(x+3)+4(x+3)$
$=(x+3)(x^2+4)$

55. $x^3-x^2-2x+2=(x^3-x^2)+(-2x+2)$
$=x^2(x-1)-2(x-1)$
$=(x-1)(x^2-2)$

57. $(x+2)(x-5)=x^2-5x+2x-10$
$=x^2-3x-10$

59. $(x+3)(x+2)=x^2+2x+3x+6$
$=x^2+5x+6$

61. $(y-3)(y-1)=y^2-y-3y+3=y^2-4y+3$

63. $10x^2-2x-2=2\cdot 5x^2-2\cdot x-2\cdot 1$
$=2(5x^2-x-1)$
Choice d is correct.

65. $2\pi r^2+2\pi rh=2\pi r\cdot r+2\pi r\cdot h=2\pi r(r+h)$

67. $A = 5600 + 5600rt$
$A = 5600 \cdot 1 + 5600 \cdot rt$
$A = 5600(1 + rt)$

69. answers may vary

71. **a.** $(2-x)(3-y)=6-2y-3x+xy$
$=xy-3x-2y+6$

b. $(-2+x)(-3+y)=6-2y-3x+xy$
$=xy-3x-2y+6$

c. $(y-3)(x-2)=yx-2y-3x+6$
$=xy-3x-2y+6$

d. $(-x+2)(-y+3)=xy-3x-2y+6$

None; all are factored forms of $xy - 3x - 2y + 6$.

73. $12x^2+9x+3=3\cdot 4x^2+3\cdot 3x+3\cdot 1$
$=3(4x^2+3x+1)$
Choice a is correct.

75. $A=P+PRT$
$A=P\cdot 1+P\cdot RT$
$A=P(1+RT)$

77. $3y^n+3y^{2n}+5y^{8n}$
$=y^n\cdot 3+y^n\cdot 3y^n+y^n\cdot 5y^{7n}$
$=y^n(3+3y^n+5y^{7n})$

79. $3x^{5a}-6x^{3a}+9x^{2a}$
$=3x^{2a}\cdot x^{3a}-3x^{2a}\cdot 2x^a-3x^{2a}\cdot 3$
$=3x^{2a}(x^{3a}-2x^a+3)$

81. **a.** $h(t)=-16t^2+224$
$=-16\cdot t^2+(-16)(-14)$
$=-16(t^2-14)$

b. $h(t)=-16t^2+224$
$h(2)=-16(2)^2+224$
$=-64+224$
$=160$
$h(t)=-16(t^2-14)$
$h(2)=-16(2^2-14)$
$=-16(4-14)$
$=-16(-10)$
$=160$

c. answers may vary

83. $f(x) = -2x^2 + 22x + 28$

$f(x) = 2 \cdot (-x^2) + 2 \cdot 11x + 2 \cdot 14$
$= 2(-x^2 + 11x + 14)$

or

$f(x) = -2 \cdot x^2 - 2 \cdot (-11x) - 2 \cdot (-14)$
$= -2(x^2 - 11x - 14)$

Section 5.5 Practice Problems

1. $x^2 + 8x + 15$
$5 \cdot 3 = 15$ and $5 + 3 = 8$
$x^2 + 8x + 15 = (x+5)(x+3)$

2. $x^2 - 10x + 24$
$-4(-6) = 24$ and $-4 + (-6) = -10$
$x^2 - 10x + 24 = (x-4)(x-6)$

3. $6x^3 + 24x^2 - 30x = 6x(x^2 + 4x - 5)$
$-1(5) = -5$ and $-1 + 5 = 4$
$6x(x^2 + 4x - 5) = 6x(x-1)(x+5)$

4. $3y^2 + 6y + 6 = 3(y^2 + 2y + 2)$
There is no pair of numbers whose product is 2 and whose sum is 2, so $y^2 + 2y + 2$ is prime.
$3y^2 + 6y + 6 = 3(y^2 + 2y + 2)$

5. $3x^2 + 13x + 4$
Factors of $3x^2$: $3x \cdot x$
Factors of 4: $1 \cdot 4$, $2 \cdot 2$
$3x^2 + 13x + 4 = (3x+1)(x+4)$

6. $5x^2 + 13x - 6$
Factors of $5x^2$: $5x \cdot x$
Factors of –6: 6(–1), –6(1), 2(–3), –2(3)
$5x^2 + 13x - 6 = (x+3)(5x-2)$

7. $24x^2y^2 - 42xy^2 + 9y^2 = 3y^2(8x^2 - 14x + 3)$
Factors of $8x^2$: $8x \cdot x$, $4x \cdot 2x$
Negative factors of 3: $(-1) \cdot (-3)$
$3y^2(8x^2 - 14x + 3) = 3y^2(2x-3)(4x-1)$

8. $4x^2 + 28xy + 49y^2$
Factors of $4x^2$: $4x \cdot x$, $2x \cdot 2x$
Factors of $49y^2$: $49y \cdot y$, $7y \cdot 7y$
$4x^2 + 28xy + 49y^2 = (2x+7y)(2x+7y)$
$= (2x+7y)^2$

9. $12x^2 + 11x + 2 = 12x^2 + 3x + 8x + 2$
$= (12x^2 + 3x) + (8x + 2)$
$= 3x(4x+1) + 2(4x+1)$
$= (4x+1)(3x+2)$

10. $14x^2 - x - 3 = 14x^2 + 6x - 7x - 3$
$= (14x^2 + 6x) + (-7x - 3)$
$= 2x(7x+3) - 1(7x+3)$
$= (7x+3)(2x-1)$

11. Substitute x for $(z + 2)$.
$3(z+2)^2 - 19(z+2) + 6 = 3x^2 - 19x + 6$
$= (3x-1)(x-6)$
Now replace x with $(z + 2)$.
$(3x-1)(x-6) = [3(z+2)-1][(z+2)-6]$
$= (3z+6-1)(z+2-6)$
$= (3z+5)(z-4)$

12. Substitute y for x^2.
$14x^4 + 23x^2 + 3 = 14y^2 + 23y + 3$
$= (2y+3)(7y+1)$
Now replace y with x^2.
$(2y+3)(7y+1) = (2x^2+3)(7x^2+1)$

Mental Math

1. $5 \cdot 2 = 10$
$5 + 2 = 7$
5, 2

2. $2 \cdot 6 = 12$
$2 + 6 = 8$
2, 6

3. $8 \cdot 3 = 24$
$8 + 3 = 11$
8, 3

4. $10 \cdot 3 = 30$
$10 + 3 = 13$
10, 3

Exercise Set 5.5

1. $x^2 + 9x + 18$
$3 \cdot 6 = 18$ and $3 + 6 = 9$
$x^2 + 9x + 18 = (x+3)(x+6)$

3. $x^2 - 12x + 32$
$-8 \cdot (-4) = 32$ and $-8 + (-4) = -12$
$x^2 - 12x + 32 = (x-8)(x-4)$

5. $x^2 + 10x - 24$
$12 \cdot (-2) = -24$ and $12 + (-2) = 10$
$x^2 + 10x - 24 = (x+12)(x-2)$

7. $x^2 - 2x - 24$
$-6(4) = -24$ and $-6 + 4 = -2$
$x^2 - 2x - 24 = (x-6)(x+4)$

9. $3x^2 - 18x + 24 = 3(x^2 - 6x + 8)$
$-2(-4) = 8$ and $-2 + (-4) = -6$
$3x^2 - 18x + 24 = 3(x-2)(x-4)$

11. $4x^2z + 28xz + 40z = 4z(x^2 + 7x + 10)$
$2 \cdot 5 = 10$ and $2 + 5 = 7$
$4x^2z + 28xz + 40z = 4z(x+2)(x+5)$

13. $2x^2 - 24x - 64 = 2(x^2 - 12x - 32)$
There is no pair of numbers whose product is -32 and whose sum is -12, so
$x^2 - 12x - 32$ is prime.
$2x^2 - 24x - 64 = 2(x^2 - 12x - 32)$

15. $5x^2 + 16x + 3$
Factors of $5x^2$: $5x \cdot x$
Factors of 3: $3 \cdot 1$
$5x^2 + 16x + 3 = (5x+1)(x+3)$

17. $2x^2 - 11x + 12$
Factors of $2x^2$: $2x \cdot x$
Negative factors of 12: $-12(-1)$, $-6(-2)$, $-4(-3)$
$2x^2 - 11x + 12 = (2x-3)(x-4)$

19. $2x^2 + 25x - 20$
Factors of $2x^2$: $2x \cdot x$
Factors of -20: $-20 \cdot 1$, $20 \cdot (-1)$, $-4 \cdot 5$, $4 \cdot (-5)$
$2x^2 + 25x - 20$ is prime.

21. $4x^2 - 12x + 9$
Factors of $4x^2$: $4x \cdot x$, $2x \cdot 2x$
Negative factors of 9: $-9(-1)$, $-3(-3)$
$4x^2 - 12x + 9 = (2x-3)(2x-3) = (2x-3)^2$

23. $12x^2 + 10x - 50 = 2(6x^2 + 5x - 25)$
Factors of $6x^2$: $6x \cdot x$, $3x \cdot 2x$
Factors of -25: $-25 \cdot 1$, $25 \cdot (-1)$, $-5 \cdot 5$
$12x^2 + 10x - 50 = 2(3x-5)(2x+5)$

25. $3y^4 - y^3 - 10y^2 = y^2(3y^2 - y - 10)$
Factors of $3y^2$: $3y \cdot y$
Factors of -10: $10(-1)$, $-10 \cdot 1$, $5(-2)$, $-5 \cdot 2$
$3y^4 - y^3 - 10y^2 = y^2(3y+5)(y-2)$

27. $6x^3 + 8x^2 + 24x = 2x(3x^2 + 4x + 12)$
Factors of $3x^2$: $3x \cdot x$
Factors of 12: $12 \cdot 1$, $6 \cdot 2$, $4 \cdot 3$
$3x^2 + 4x + 12$ is prime.

29. $2x^2 - 5xy - 3y^2$

Factors of $2x^2$: $2x \cdot x$

Factors of $-3y^2$: $3y(-y)$, $-3y \cdot y$

$2x^2 - 5xy - 3y^2 = (2x + y)(x - 3y)$

31. $28y^2 + 22y + 4 = 2(14y^2 + 11y + 2)$

Factors of $14y^2$: $14y \cdot y$, $7y \cdot 2y$

Factors of 4: $4 \cdot 1$, $2 \cdot 2$

$28y^2 + 22y + 4 = 2(7y + 2)(2y + 1)$

33. $2x^2 + 15x - 27$

Factors of $2x^2$: $2x \cdot x$

Factors of -27: $27(-1)$, $-27 \cdot 1$, $9(-3)$, $-9 \cdot 3$

$2x^2 + 15x - 27 = (2x - 3)(x + 9)$

35. Substitute y for x^2.

$x^4 + x^2 - 6 = y^2 + y - 6$

$3 \cdot (-2) = -6$ and $3 - 2 = 1$

$y^2 + y - 6 = (y + 3)(y - 2)$

Now replace y with x^2.

$(y + 3)(y - 2) = (x^2 + 3)(x^2 - 2)$

37. Substitute y for $5x + 1$.

$(5x + 1)^2 + 8(5x + 1) + 7 = y^2 + 8y + 7$

$1 \cdot 7 = 7$ and $1 + 7 = 8$

$y^2 + 8y + 7 = (y + 7)(y + 1)$

Now replace y with $5x + 1$.

$$\begin{aligned}(y + 7)(y + 1) &= [(5x + 1) + 7][(5x + 1) + 1] \\ &= (5x + 8)(5x + 2)\end{aligned}$$

$(5x + 1)^2 + 8(5x + 1) + 7 = (5x + 8)(5x + 2)$

39. Substitute y for x^3.

$x^6 - 7x^3 + 12 = y^2 - 7y + 12$

$-4 \cdot (-3) = 12$ and $-4 + (-3) = -7$

$y^2 - 7y + 12 = (y - 4)(y - 3)$

Now replace y with x^3.

$(y - 4)(y - 3) = (x^3 - 4)(x^3 - 3)$

$x^6 - 7x^3 + 12 = (x^3 - 4)(x^3 - 3)$

41. Substitute x for $a + 5$.

$(a + 5)^2 - 5(a + 5) - 24 = x^2 - 5x - 24$

$-8 \cdot 3 = -24$ and $-8 + 3 = -5$

$x^2 - 5x - 24 = (x - 8)(x + 3)$

Now replace x with $a + 5$.

$$\begin{aligned}(x - 8)(x - 3) &= [(a + 5) - 8][(a + 5) + 3] \\ &= (a - 3)(a + 8)\end{aligned}$$

$(a + 5)^2 - 5(a + 5) - 24 = (a - 3)(a + 8)$

43. $x^2 - 24x - 81$

$-27 \cdot 3 = -81$ and $-27 + 3 = -24$

$x^2 - 24x - 81 = (x - 27)(x + 3)$

45. $x^2 - 15x - 54$

$-18 \cdot 3 = -54$ and $-18 + 3 = -15$

$x^2 - 15x - 54 = (x - 18)(x + 3)$

47. $3x^2 - 6x + 3 = 3(x^2 - 2x + 1)$

$-1 \cdot (-1) = 1$ and $-1 - 1 = -2$

$3(x^2 - 2x + 1) = 3(x - 1)(x - 1) = 3(x - 1)^2$

49. $3x^2 - 5x - 2$

Factors of $3x^2$: $3x \cdot x$

Factors of -2: $2(-1)$, $-2 \cdot 1$

$3x^2 - 5x - 2 = (3x + 1)(x - 2)$

51. $8x^2 - 26x + 15$

Factors of $8x^2$: $8x \cdot x$, $4x \cdot 2x$

Negative Factors of 15: $-15 \cdot (-1)$, $-5 \cdot (-3)$

$8x^2 - 26x + 15 = (4x - 3)(2x - 5)$

53. $18x^4 + 21x^3 + 6x^2 = 3x^2(6x^2 - 7x + 2)$

Factors of $6x^2$: $6x \cdot x$, $3x \cdot 2x$

Factors of 2: $2 \cdot 1$

$18x^4 + 21x^3 + 6x^2 = 3x^2(2x + 1)(3x + 2)$

55. $x^2 + 8xz + 7z^2$

$z \cdot 7z = 7z^2$ and $z + 7z = 8z$

$x^2 + 8xz + 7z^2 = (x + 7z)(x + z)$

57. $x^2 - x - 12$

$-4 \cdot 3 = -12$ and $-4 + 3 = -1$

$x^2 - x - 12 = (x - 4)(x + 3)$

59. $3a^2 + 12ab + 12b^2 = 3(a^2 + 4ab + 4b^2)$

$2b \cdot 2b = 4b^2$ and $2b + 2b = 4b$

$$\begin{aligned} 3a^2 + 12ab + 12b^2 &= 3(a + 2b)(a + 2b) \\ &= 3(a + 2b)^2 \end{aligned}$$

61. $x^2 + 4x + 5$

There is no pair of numbers whose product is 5 and whose sum is 4. $x^2 + 4x + 5$ is a prime polynomial.

63. Substitute y for $x + 4$.

$2(x + 4)^2 + 3(x + 4) - 5 = 2y^2 + 3y - 5$

Factors of $2y^2$: $2y \cdot y$

Factors of -5: $-5 \cdot 1$, $5 \cdot (-1)$

$2y^2 + 3y - 5 = (2y + 5)(y - 1)$

Now replace y with $x + 4$.

$$\begin{aligned} (2y + 5)(y - 1) &= [2(x + 4) + 5][(x + 4) - 1] \\ &= (2x + 8 + 5)(x + 4 - 1) \\ &= (2x + 13)(x + 3) \end{aligned}$$

$2(x + 4)^2 + 3(x + 4) - 5 = (2x + 13)(x + 3)$

65. $6x^2 - 49x + 30$

Factors of $6x^2$: $6x \cdot x$, $3x \cdot 2x$

Negative factors of 30: $-30(-1)$, $-15(-2)$, $-10(-3)$, $-6(-5)$

$6x^2 - 49x + 30 = (3x - 2)(2x - 15)$

67. Substitute y for x^2.

$x^4 - 5x^2 - 6 = y^2 - 5y - 6$

$-6 \cdot (1) = -6$ and $-6 + 1 = -5$

$y^2 - 5y - 6 = (y - 6)(y + 1)$

Now replace y with x^2.

$(y - 6)(y + 1) = (x^2 - 6)(x^2 + 1)$

$x^4 - 5x^2 - 6 = (x^2 - 6)(x^2 + 1)$

69. $6x^3 - x^2 - x = x(6x^2 - x - 1)$

Factors of $6x^2$: $6x \cdot x$, $3x \cdot 2x$

Factors of -1: $-1 \cdot 1$

$6x^3 - x^2 - x = x(3x + 1)(2x - 1)$

71. $12a^2 - 29ab + 15b^2$

Factors of $12a^2$: $12a \cdot a$, $6a \cdot 2a$, $4a \cdot 3a$

Negative factors of $15b^2$: $-15b \cdot (-b)$, $-5b \cdot (-3b)$

$12a^2 - 29ab + 15b^2 = (4a - 3b)(3a - 5b)$

73. $9x^2 + 30x + 25$

Factors of $9x^2$: $9x \cdot x$, $3x \cdot 3x$

Factors of 25: $25 \cdot 1$, $5 \cdot 5$

$9x^2 + 30x + 25 = (3x + 5)(3x + 5) = (3x + 5)^2$

75. $3x^2y - 11xy + 8y = y(3x^2 - 11x + 8)$

Factors of $3x^2$: $3x \cdot x$

Negative factors of 8: $-8(-1)$, $-4(-2)$

$3x^2y - 11xy + 8y = y(3x - 8)(x - 1)$

77. $2x^2 + 2x - 12 = 2(x^2 + x - 6)$

$3(-2) = -6$ and $3 + (-2) = 1$

$2x^2 + 2x - 12 = 2(x + 3)(x - 2)$

79. Substitute y for $x - 4$.

$(x - 4)^2 + 3(x - 4) - 18 = y^2 + 3y - 18$

$-3 \cdot 6 = -18$ and $-3 + 6 = 3$

$y^2 + 3y - 18 = (y - 3)(y + 6)$

Now replace y with $x - 4$.

$(y-3)(y+6) = [(x-4)-3][(x-4)+6]$
$= (x-7)(x+2)$

$(x-4)^2 + 3(x-4) - 18 = (x-7)(x+2)$

81. $2x^6 + 3x^3 - 9 = 2(x^3)^2 + 3x^3 - 9$

Substitute y for x^3.

$2x^6 + 3x^3 - 9 = 2y^2 + 3y - 9$

Factors of $2y^2$: $2y \cdot y$

Factors of –9: 9(–1), –9 · 1, –3 · 3

$2y^2 + 3y - 9 = (2y-3)(y+3)$

Replace y with x^3.

$(2y-3)(y+3) = (2x^3-3)(x^3+3)$

$2x^6 + 3x^3 - 9 = (2x^3-3)(x^3+3)$

83. $72xy^4 - 24xy^2z + 2xz^2$

$= 2x(36y^4 - 12y^2z + z^2)$

Factors of $36y^4$: $36y^2 \cdot y^2, 18y^2 \cdot 2y^2,$
$9y^2 \cdot 4y^2,\ 12y^2 \cdot 3y^2, 6y^2 \cdot 6y^2$

Negative factors of z^2: $-z(-z)$

$72xy^4 - 24xy^2z + 2xz^2$

$= 2x(6y^2 - z)(6y^2 - z)$

$= 2x(6y^2 - z)^2$

85. $2x^3y + 2x^2y - 12xy = 2xy(x^2 + x - 6)$

3(–2) = –6 and 3 + (–2) = 1

$2x^3y + 2x^2y - 12xy = 2xy(x+3)(x-2)$

87. $x^2 + 6xy + 5y^2$

$5y \cdot y = 5y^2$ and $5y + y = 6y$

$x^2 + 6xy + 5y^2 = (x+5y)(x+y)$

89. $(x-3)(x+3) = x^2 - 3^2 = x^2 - 9$

91. $(2x+1)^2 = (2x)^2 + 2(2x)(1) + 1^2$
$= 4x^2 + 4x + 1$

93. $(x-2)(x^2+2x+4)$

$= x(x^2+2x+4) - 2(x^2+2x+4)$

$= x^3 + 2x^2 + 4x - 2x^2 - 4x - 8$

$= x^3 - 8$

95. Factors of 6: 6 · 1, –6 · (–1), 3 · 2, –3 · (–2)

$(x+6)(x+1) = x^2 + x + 6x + 6$
$= x^2 + 7x + 6$

$(x-6)(x-1) = x^2 - x - 6x + 6$
$= x^2 - 7x + 6$

$(x+3)(x+2) = x^2 + 2x + 3x + 6$
$= x^2 + 5x + 6$

$(x-3)(x-2) = x^2 - 2x - 3x + 6$
$= x^2 - 5x + 6$

$b = \pm 5, \pm 7$

97. $x^3 + 2x^2 - 8x = x(x^2 + 2x - 8)$

4(–2) = –8 and 4 + (–2) = 2

$x^3 + 2x^2 - 8x = x(x+4)(x-2)$

99. a. $h(t) = -16t^2 + 80t + 576$

$h(0) = -16(0)^2 + 80(0) + 576 = 576$ ft

$h(2) = -16(2)^2 + 80(2) + 576 = 672$ ft

$h(4) = -16(4)^2 + 80(4) + 576 = 640$ ft

$h(6) = -16(6)^2 + 80(6) + 576 = 480$ ft

b. answers may vary

c. $-16t^2 + 80t + 576 = -16(t^2 - 5t - 36)$

4 · (–9) = –36 and 4 – 9 = –5

$-16t^2 + 80t + 576 = -16(t+4)(t-9)$

101. $x^{2n} + 10x^n + 16 = (x^n)^2 + 10x^n + 16$

2 · 8 = 16 and 2 + 8 = 10

$x^{2n} + 10x^n + 16 = (x^n + 2)(x^n + 8)$

103. $x^{2n} - 3x^n - 18 = (x^n)^2 - 3x^n - 18$

–6 · 3 = –18 and –6 + 3 = –3

$x^{2n} - 3x^n - 18 = (x^n - 6)(x^n + 3)$

105. $2x^{2n}+11x^n+5=2(x^n)^2+11x^n+5$

Factors of $2x^{2n}$: $2x^n \cdot x^n$

Factors of 5: $5 \cdot 1$

$2x^{2n}+11x^n+5=(2x^n+1)(x^n+5)$

107. $4x^{2n}-12x^n+9=4(x^n)^2-12x^n+9$

Factors of $4x^{2n}$: $4x^n \cdot x^n$, $2x^n \cdot 2x^n$

Negative factors of 9: $-9(-1)$, $-3(-3)$

$$\begin{aligned}4x^{2n}-12x^n+9&=(2x^n-3)(2x^n-3)\\&=(2x^n-3)^2\end{aligned}$$

109. $$\begin{aligned}x^4+6x^3+5x^2&=x^2(x^2+6x+5)\\&=x^2(x+5)(x+1)\end{aligned}$$

111. $$\begin{aligned}30x^3+9x^2-3x&=3x(10x^2+3x-1)\\&=3x(5x-1)(2x+1)\end{aligned}$$

Section 5.6 Practice Problems

1. $x^2+8x+16=x^2+2(x)(4)+4^2=(x+4)^2$

2. $$\begin{aligned}9x^2+6x+1&=(3x)^2+2\cdot 3x\cdot 1+1^2\\&=(3x+1)^2\end{aligned}$$

3. $$\begin{aligned}25x^2-20x+4&=(5x)^2-2(5x)(2)+2^2\\&=(5x-2)^2\end{aligned}$$

4. $$\begin{aligned}&4x^3-32x^2y+64xy^2\\&=4x(x^2-8xy+16y^2)\\&=4x[x^2-2\cdot x\cdot 4y+(4y)^2]\\&=4x(x-4y)^2\end{aligned}$$

5. $x^2-49=x^2-7^2=(x+7)(x-7)$

6. $4y^2-81=(2y)^2-9^2=(2y+9)(2y-9)$

7. $$\begin{aligned}12-3a^2&=3(4-a^2)\\&=3(2^2-a^2)\\&=3(2+a)(2-a)\end{aligned}$$

8. $y^2-\frac{1}{25}=y^2-\left(\frac{1}{5}\right)^2=\left(y+\frac{1}{5}\right)\left(y-\frac{1}{5}\right)$

9. $$\begin{aligned}a^4-81&=(a^2)^2-9^2\\&=(a^2+9)(a^2-9)\\&=(a^2+9)(a^2-3^2)\\&=(a^2+9)(a+3)(a-3)\end{aligned}$$

10. $$\begin{aligned}(x+1)^2-9&=(x+1)^2-3^2\\&=[(x+1)+3][(x+1)-3]\\&=(x+1+3)(x+1-3)\\&=(x+4)(x-2)\end{aligned}$$

11. $$\begin{aligned}a^2+2a+1-b^2&=(a^2+2a+1)-b^2\\&=(a+1)^2-b^2\\&=[(a+1)+b][(a+1)-b]\\&=(a+1+b)(a+1-b)\end{aligned}$$

12. $$\begin{aligned}z^3+27&=z^3+3^3\\&=(z+3)(z^2-z\cdot 3+3^2)\\&=(z+3)(z^2-3z+9)\end{aligned}$$

13. $$\begin{aligned}x^3+64y^3&=x^3+(4y)^3\\&=(x+4y)[x^2-x\cdot 4y+(4y)^2]\\&=(x+4y)(x^2-4xy+16y^2)\end{aligned}$$

14. $$\begin{aligned}y^3-8&=y^3-2^3\\&=(y-2)(y^2+y\cdot 2+2^2)\\&=(y-2)(y^2+2y+4)\end{aligned}$$

15. $$\begin{aligned}27a^2-b^3a^2&=a^2(27-b^3)\\&=a^2(3^3-b^3)\\&=a^2(3-b)(3^2+3\cdot b+b^2)\\&=a^2(3-b)(9+3b+b^2)\end{aligned}$$

Exercise Set 5.6

1. $x^2+6x+9=(x)^2+2\cdot x\cdot 3+(3)^2=(x+3)^2$

3. $4x^2 - 12x + 9 = (2x)^2 - 2 \cdot 2x \cdot 3 + 3^2$
$= (2x-3)^2$

5. $4a^2 + 12a + 9 = (2a)^2 + 2 \cdot 2a \cdot 3 + (3)^2$
$= (2a+3)^2$

7. $3x^2 - 24x + 48 = 3(x^2 - 8x + 16)$
$= 3(x^2 - 2 \cdot x \cdot 4 + 4^2)$
$= 3(x-4)^2$

9. $9y^2x^2 + 12yx^2 + 4x^2$
$= x^2(9y^2 + 12y + 4)$
$= x^2[(3y)^2 + 2 \cdot 3y \cdot 2 + (2)^2]$
$= x^2(3y+2)^2$

11. $16x^2 - 56xy + 49y^2$
$= (4x)^2 - 2 \cdot 4x \cdot 7y + (7y)^2$
$= (4x - 7y)^2$

13. $x^2 - 25 = (x)^2 - (5)^2 = (x+5)(x-5)$

15. $\frac{1}{9} - 4z^2 = \left(\frac{1}{3}\right)^2 - (2z)^2 = \left(\frac{1}{3} + 2z\right)\left(\frac{1}{3} - 2z\right)$

17. $(y+2)^2 - 49 = (y+2)^2 - (7)^2$
$= [(y+2)+7][(y+2)-7]$
$= (y+9)(y-5)$

19. $64x^2 - 100 = 4(16x^2 - 25)$
$= 4[(4x)^2 - 5^2]$
$= 4(4x+5)(4x-5)$

21. $(x+2y)^2 - 9 = (x+2y)^2 - (3)^2$
$= [(x+2y)+3][(x+2y)-3]$
$= (x+2y+3)(x+2y-3)$

23. $x^2 + 16x + 64 - x^4$
$= (x^2 + 16x + 64) - x^4$
$= (x^2 + 2 \cdot x \cdot 8 + 8^2) - x^4$
$= (x+8)^2 - (x^2)^2$
$= [(x+8) + x^2][(x+8) - x^2]$
$= (x+8+x^2)(x+8-x^2)$

25. $x^3 + 27 = x^3 + (3)^3 = (x+3)(x^2 - 3x + 9)$

27. $z^3 - 1 = z^3 - 1^3 = (z-1)(z^2 + z + 1)$

29. $m^3 + n^3 = (m+n)(m^2 - mn + n^2)$

31. $x^3y^2 - 27y^2 = y^2(x^3 - 27)$
$= y^2(x^3 - 3^3)$
$= y^2(x-3)(x^2 + x \cdot 3 + 3^2)$
$= y^2(x-3)(x^2 + 3x + 9)$

33. $64q^2 - q^2p^3 = q^2(64 - p^3)$
$= q^2[(4)^3 - p^3]$
$= q^2(4-p)(4^2 + 4p + p^2)$
$= q^2(4-p)(16 + 4p + p^2)$

35. $250y^3 - 16x^3$
$= 2(125y^3 - 8x^3)$
$= 2[(5y)^3 - (2x)^3]$
$= 2(5y-2x)[(5y)^2 + 5y \cdot 2x + (2x)^2]$
$= 2(5y-2x)(25y^2 + 10xy + 4x^2)$

37. $x^2 - 12x + 36 = x^2 - 2 \cdot x \cdot 6 + 6^2 = (x-6)^2$

39. $18x^2y - 2y = 2y(9x^2 - 1)$
$= 2y[(3x)^2 - 1^2]$
$= 2y(3x+1)(3x-1)$

41. $9x^2 - 49 = (3x)^2 - 7^2 = (3x+7)(3x-7)$

43. $x^4-1=(x^2)^2-1^2$
$=(x^2+1)(x^2-1)$
$=(x^2+1)(x+1)(x-1)$

45. $x^6-y^3=(x^2)^3-y^3$
$=(x^2-y)[(x^2)^2+x^2y+y^2]$
$=(x^2-y)(x^4+x^2y+y^2)$

47. $8x^3+27y^3$
$=(2x)^3+(3y)^3$
$=(2x+3y)[(2x)^2-2x\cdot 3y+(3y)^2]$
$=(2x+3y)(4x^2-6xy+9y^2)$

49. $4x^2+4x+1-z^2$
$=(4x^2+4x+1)-z^2$
$=[(2x)^2+2\cdot 2x\cdot 1+1^2]-z^2$
$=(2x+1)^2-z^2$
$=(2x+1+z)(2x+1-z)$

51. $3x^6y^2+81y^2$
$=3y^2(x^6+27)$
$=3y^2[(x^2)^3+3^3]$
$=3y^2(x^2+3)[(x^2)^2-x^2\cdot 3+3^2]$
$=3y^2(x^2+3)(x^4-3x^2+9)$

53. $n^3-\frac{1}{27}=n^3-\left(\frac{1}{3}\right)^3$
$=\left(n-\frac{1}{3}\right)\left[n^2+n\cdot\frac{1}{3}+\left(\frac{1}{3}\right)^2\right]$
$=\left(n-\frac{1}{3}\right)\left(n^2+\frac{1}{3}n+\frac{1}{9}\right)$

55. $-16y^2+64=-16(y^2-4)$
$=-16(y^2-2^2)$
$=-16(y+2)(y-2)$

57. $x^2-10x+25-y^2=(x^2-10x+25)-y^2$
$=(x^2-2\cdot x\cdot 5+5^2)-y^2$
$=(x-5)^2-y^2$
$=(x-5+y)(x-5-y)$

59. $a^3b^3+125=(ab)^3+5^3$
$=(ab+5)[(ab)^2-ab\cdot 5+5^2]$
$=(ab+5)(a^2b^2-5ab+25)$

61. $\frac{x^2}{25}-\frac{y^2}{9}=\left(\frac{x}{5}\right)^2-\left(\frac{y}{3}\right)^2=\left(\frac{x}{5}+\frac{y}{3}\right)\left(\frac{x}{5}-\frac{y}{3}\right)$

63. $(x+y)^3+125$
$=(x+y)^3+5^3$
$=[(x+y)+5][(x+y)^2-(x+y)\cdot 5+5^2]$
$=(x+y+5)(x^2+2xy+y^2-5x-5y+25)$

65. $x-5=0$
$x=5$
The solution set is $\{5\}$.

67. $3x+1=0$
$3x=-1$
$x=-\frac{1}{3}$
The solution set is $\left\{-\frac{1}{3}\right\}$.

69. $-2x=0$
$x=\frac{0}{-2}$
$x=0$
The solution set is $\{0\}$.

71. $-5x+25=0$
$-5x=-25$
$x=5$
The solution set is $\{5\}$.

73. Since $x^2-4=x^2-2^2=(x+2)(x-2)$, the polynomial $5x(x^2-4)$ is not factored completely.

75. $7y(a^2+a+1)$ is factored completely.

77. If the washer didn't have a hole, the area would be πR^2. Subtracting the hole, the area is $\pi R^2 - \pi r^2$.
$\pi R^2 - \pi r^2 = \pi(R^2 - r^2) = \pi(R+r)(R-r)$

79. $x^3 - y^2x = x(x^2 - y^2) = x(x+y)(x-y)$

81. $6x = 2 \cdot x \cdot 3$
$(x+3)^2 = x^2 + 6x + 9$
The value of c is 9.

83. $-14m = -2 \cdot m \cdot 7$
$(m-7)^2 = m^2 - 14m + 49$
The value of c is 49.

85. a.
$$\begin{aligned} &x^6 - 1 \\ &= (x^3)^2 - 1^2 \\ &= (x^3+1)(x^3-1) \\ &= (x^3+1^3)(x^3-1^3) \\ &= (x+1)(x^2-x+1)(x-1)(x^2+x+1) \end{aligned}$$

b.
$$\begin{aligned} x^6 - 1 &= (x^2)^3 - 1 \\ &= (x^2-1)[(x^2)^2 + x^2 + 1] \\ &= (x^2-1^2)(x^4+x^2+1) \\ &= (x+1)(x-1)(x^4+x^2+1) \end{aligned}$$

c. answers may vary

87. $x^{2n} - 36 = (x^n)^2 - 6^2 = (x^n+6)(x^n-6)$

89.
$$\begin{aligned} 25x^{2n} - 81 &= (5x^n)^2 - 9^2 \\ &= (5x^n+9)(5x^n-9) \end{aligned}$$

91.
$$\begin{aligned} x^{4n} - 625 &= (x^{2n})^2 - (25)^2 \\ &= (x^{2n}+25)(x^{2n}-25) \\ &= (x^{2n}+25)[(x^n)^2 - 5^2] \\ &= (x^{2n}+25)(x^n+5)(x^n-5) \end{aligned}$$

Integrated Review

1.
$$\begin{aligned} &(-y^2+6y-1)+(3y^2-4y-10) \\ &= -y^2+6y-1+3y^2-4y-10 \\ &= -y^2+3y^2+6y-4y-1-10 \\ &= 2y^2+2y-11 \end{aligned}$$

2.
$$\begin{aligned} &(5z^4-6z^2+z+1)-(7z^4-2z+1) \\ &= 5z^4-6z^2+z+1-7z^4+2z-1 \\ &= 5z^4-7z^4-6z^2+z+2z+1-1 \\ &= -2z^4-6z^2+3z \end{aligned}$$

3.
$$\begin{aligned} (x^2-6x+2)-(x-5) &= x^2-6x+2-x+5 \\ &= x^2-6x-x+2+5 \\ &= x^2-7x+7 \end{aligned}$$

4.
$$\begin{aligned} &(2x^2+6x-5)+(5x^2-10x) \\ &= 2x^2+6x-5+5x^2-10x \\ &= 2x^2+5x^2+6x-10x-5 \\ &= 7x^2-4x-5 \end{aligned}$$

5.
$$\begin{aligned} (5x-3)^2 &= (5x-3)(5x-3) \\ &= 5x(5x-3)-3(5x-3) \\ &= 25x^2-15x-15x+9 \\ &= 25x^2-30x+9 \end{aligned}$$

6.
$$\begin{array}{r} x-3 \\ 5x+1\overline{)5x^2-14x-3} \\ \underline{5x^2+x} \\ -15x-3 \\ \underline{-15x-3} \\ 0 \end{array}$$

$(5x^2-14x-3) \div (5x+1) = x-3$

7.

-2	2	0	-3	5	-2
		-4	8	-10	10
	2	-4	5	-5	8

$(2x^4-3x^2+5x-2) \div (x+2)$
$= 2x^3-4x^2+5x-5+\dfrac{8}{x+2}$

8. $(4x-1)(x^2-3x-2)$
$=4x(x^2-3x-2)-1(x^2-3x-2)$
$=4x^3-12x^2-8x-x^2+3x+2$
$=4x^3-13x^2-5x+2$

9. $x^2-8x+16-y^2=(x^2-8x+16)-y^2$
$=(x-4)^2-y^2$
$=(x-4+y)(x-4-y)$

10. $12x^2-22x-20=2(6x^2-11x-10)$

Factors of $6x^2$: $6x\cdot x$, $3x\cdot 2x$
Factors of -10: $-10\cdot 1$, $10\cdot(-1)$, $-5\cdot(2)$, $5\cdot(-2)$
$12x^2-22x-20=2(3x+2)(2x-5)$

11. $x^4-x=x(x^3-1)$
$=x(x^3-1^3)$
$=x(x-1)(x^2+x+1)$

12. Substitute y for $2x+1$.
$(2x+1)^2-3(2x+1)+2=y^2-3y+2$
$-2\cdot(-1)=2$ and $-2-1=-3$
$y^2-3y+2=(y-2)(y-1)$
Now replace y with $2x+1$.
$(y-2)(y-1)=[(2x+1)-2][(2x+1)-1]$
$=(2x-1)(2x)$
$=2x(2x-1)$

13. $14x^2y-2xy=2xy\cdot 7x+2xy\cdot(-1)$
$=2xy(7x-1)$

14. $24ab^2-6ab=6ab\cdot 4b+6ab\cdot(-1)$
$=6ab(4b-1)$

15. $4x^2-16=4(x^2-4)$
$=4(x^2-2^2)$
$=4(x+2)(x-2)$

16. $9x^2-81=9(x^2-9)$
$=9(x^2-3^2)$
$=9(x+3)(x-3)$

17. $3x^2-8x-11$
Factors of $3x^2$: $3x\cdot x$
Factors of -11: $-11\cdot 1$, $11\cdot(-1)$
$3x^2-8x-11=(3x-11)(x+1)$

18. $5x^2-2x-3$
Factors of $5x^2$: $5x\cdot x$
Factors of -3: $-3\cdot 1$, $3\cdot(-1)$
$5x^2-2x-3=(5x+3)(x-1)$

19. $4x^2+8x-12=4(x^2+2x-3)$
$3\cdot(-1)=-3$ and $3-1=2$
$4x^2+8x-12=4(x+3)(x-1)$

20. $6x^2-6x-12=6(x^2-x-2)$
$1\cdot(-2)=-2$ and $1-2=-1$
$6x^2-6x-12=6(x+1)(x-2)$

21. $4x^2+36x+81=(2x)^2+2\cdot 2x\cdot 9+9^2$
$=(2x+9)^2$

22. $25x^2+40x+16=(5x)^2+2\cdot 5x\cdot 4+4^2$
$=(5x+4)^2$

23. $8x^3+125y^3$
$=(2x)^3+(5y)^3$
$=(2x+5y)[(2x)^2-2x\cdot 5y+(5y)^2]$
$=(2x+5y)(4x^2-10xy+25y^2)$

24. $27x^3-64y^3$
$=(3x)^3-(4y)^3$
$=(3x-4y)[(3x)^2+3x\cdot 4y+(4y)^2]$
$=(3x-4y)(9x^2+12xy+16y^2)$

25. $64x^2y^3-8x^2$
$=8x^2(8y^3-1)$
$=8x^2[(2y)^3-1^3]$
$=8x^2(2y-1)[(2y)^2+2y\cdot 1+1^2]$
$=8x^2(2y-1)(4y^2+2y+1)$

26. $27x^5y^4 - 216x^2y$
$= 27x^2y(x^3y^3 - 8)$
$= 27x^2y[(xy)^3 - 2^3]$
$= 27x^2y(xy-2)[(xy)^2 + xy\cdot 2 + 2^2]$
$= 27x^2y(xy-2)(x^2y^2 + 2xy + 4)$

27. $(x+5)^3 + y^3$
$= (x+5+y)[(x+5)^2 - (x+5)\cdot y + y^2]$
$= (x+5+y)(x^2 + 10x + 25 - xy - 5y + y^2)$

28. $(y-1)^3 + 27x^3$
$= (y-1)^3 + (3x)^3$
$= (y-1+3x)[(y-1)^2 - (y-1)\cdot 3x + (3x)^2]$
$= (y-1+3x)(y^2 - 2y + 1 - 3xy + 3x + 9x^2)$

29. Substitute y for $5a - 3$.
$(5a-3)^2 - 6(5a-3) + 9 = y^2 - 6y + 9$
$= y^2 - 2\cdot y\cdot 3 + 3^2$
$= (y-3)^2$
Now replace y with $5a - 3$.
$(y-3)^2 = [(5a-3)-3]^2$
$= (5a-3-3)^2$
$= (5a-6)^2$
$(5a-3)^2 - 6(5a-3) + 9 = (5a-6)^2$

30. Substitute x for $4r + 1$.
$(4r+1)^2 + 8(4r+1) + 16 = x^2 + 8x + 16$
$= x^2 + 2\cdot x\cdot 4 + 4^2$
$= (x+4)^2$
Now replace x with $4r + 1$.
$(x+4)^2 = [(4r+1)+4]^2$
$= (4r+1+4)^2$
$= (4r+5)^2$
$(4r+1)^2 + 8(4r+1) + 16 = (4r+5)^2$

31. $7x^2 - 63x = 7x\cdot x + 7x\cdot(-9) = 7x(x-9)$

32. $20x^2 + 23x + 6$
Factors of $20x^2$: $20x \cdot x$, $10x \cdot 2x$, $5x \cdot 4x$
Factors of 6: $6 \cdot 1$, $3 \cdot 2$
$20x^2 + 23x + 6 = (4x+3)(5x+2)$

33. $ab - 6a + 7b - 42 = (ab - 6a) + (7b - 42)$
$= a(b-6) + 7(b-6)$
$= (b-6)(a+7)$

34. $20x^2 - 220x + 600 = 20(x^2 - 11x + 30)$
$-6 \cdot (-5) = 30$ and $-6 - 5 = -11$
$20x^2 - 220x + 600 = 20(x-6)(x-5)$

35. $x^4 - 1 = (x^2)^2 - 1^2$
$= (x^2+1)(x^2-1)$
$= (x^2+1)(x+1)(x-1)$

36. $15x^2 - 20x = 5x\cdot 3x + 5x\cdot(-4) = 5x(3x-4)$

37. $10x^2 - 7x - 33$
Factors of $10x^2$: $10x \cdot x$, $5x \cdot 2x$
Factors of -33: $-33 \cdot 1$, $33 \cdot (-1)$, $-11 \cdot 3$, $11 \cdot (-3)$
$10x^2 - 7x - 33 = (5x-11)(2x+3)$

38. $45m^3n^3 - 27m^2n^2$
$= 9m^2n^2\cdot 5mn + 9m^2n^2(-3)$
$= 9m^2n^2(5mn-3)$

39. $5a^3b^3 - 50a^3b = 5a^3b\cdot b^2 + 5a^3b\cdot(-10)$
$= 5a^3b(b^2-10)$

40. $x^4 + x = x(x^3+1)$
$= x(x^3 + 1^3)$
$= x(x+1)(x^2 - x + 1)$

41. $16x^2 + 25$ is prime.

42. $20x^3 + 20y^3 = 20(x^3 + y^3)$
$= 20(x+y)(x^2 - xy + y^2)$

43. $10x^3 - 210x^2 + 1100x = 10x(x^2 - 21x + 110)$
$-10(-11) = 110$ and $-10 - 11 = -21$
$10x^3 - 210x^2 + 1100x = 10x(x-10)(x-11)$

44. $9y^2 - 42y + 49 = (3y)^2 - 2\cdot 3y\cdot 7 + 7^2$
$= (3y-7)^2$

45. $64a^3b^4 - 27a^3b$
$= a^3b(64b^3 - 27)$
$= a^3b[(4b)^3 - 3^3]$
$= a^3b(4b-3)[(4b)^2 + 4b\cdot 3 + 3^2]$
$= a^3b(4b-3)(16b^2 + 12b + 9)$

46. $y^4 - 16 = (y^2)^2 - 4^2$
$= (y^2+4)(y^2-4)$
$= (y^2+4)(y+2)(y-2)$

47. $2x^3 - 54 = 2(x^3 - 27)$
$= 2(x^3 - 3^3)$
$= 2(x-3)(x^2 + x\cdot 3 + 3^2)$
$= 2(x-3)(x^2 + 3x + 9)$

48. $2sr + 10s - r - 5 = (2sr + 10s) + (-r - 5)$
$= 2s(r+5) - 1(r+5)$
$= (r+5)(2s-1)$

49. $3y^5 - 5y^4 + 6y - 10$
$= (3y^5 - 5y^4) + (6y - 10)$
$= y^4(3y-5) + 2(3y-5)$
$= (3y-5)(y^4+2)$

50. $64a^2 + b^2$ is a prime polynomial.

51. $100z^3 + 100 = 100(z^3 + 1)$
$= 100(z^3 + 1^3)$
$= 100(z+1)(z^2 - z + 1)$

52. $250x^4 - 16x$
$= 2x(125x^3 - 8)$
$= 2x[(5x)^3 - 2^3]$
$= 2x(5x-2)[(5x)^2 + 5x\cdot 2 + 2^2]$
$= 2x(5x-2)(25x^2 + 10x + 4)$

53. $4b^2 - 36b + 81 = (2b)^2 - 2\cdot 2b\cdot 9 + 9^2$
$= (2b-9)^2$

54. $2a^5 - a^4 + 6a - 3 = (2a^5 - a^4) + (6a - 3)$
$= a^4(2a-1) + 3(2a-1)$
$= (2a-1)(a^4+3)$

55. Substitute x for $y - 6$.
$(y-6)^2 + 3(y-6) + 2 = x^2 + 3x + 2$
$2 \cdot 1 = 2$ and $2 + 1 = 3$
$x^2 + 3x + 2 = (x+2)(x+1)$
Now replace x with $y - 6$.
$(x+2)(x+1) = [(y-6)+2][(y-6)+1]$
$= (y-4)(y-5)$

56. Substitute x for $c + 2$.
$(c+2)^2 - 6(c+2) + 5 = x^2 - 6x + 5$
$-5 \cdot (-1) = 5$ and $-5 - 1 = -6$
$x^2 - 6x + 5 = (x-5)(x-1)$
Now replace x with $c + 2$.
$(x-5)(x-1) = [(c+2)-5][(c+2)-1]$
$= (c-3)(c+1)$

57. $A = 3^2 - 4x^2$
$= 9 - 4x^2$
$= (3)^2 - (2x)^2$
$= (3+2x)(3-2x)$

Section 5.7 Practice Problems

1. $(x-3)(x+5) = 0$
$x - 3 = 0$ or $x + 5 = 0$
$x = 3$ or $x = -5$
The solution set is $\{3, -5\}$.

2. $3x^2 + 5x - 2 = 0$

$(3x-1)(x+2) = 0$

$3x - 1 = 0$ or $x + 2 = 0$

$3x = 1$

$x = \frac{1}{3}$ or $x = -2$

The solution set is $\left\{-2, \frac{1}{3}\right\}$.

3. $x(5x-7) = -2$

$5x^2 - 7x = -2$

$5x^2 - 7x + 2 = 0$

$(5x-2)(x-1) = 0$

$5x - 2 = 0$ or $x - 1 = 0$

$5x = 2$

$x = \frac{2}{5}$ or $x = 1$

The solution set is $\left\{\frac{2}{5}, 1\right\}$.

4. $2(x^2+5)+10 = -2(x^2+10x) - 5$

$2x^2 + 10 + 10 = -2x^2 - 20x - 5$

$4x^2 + 20x + 25 = 0$

$(2x+5)(2x+5) = 0$

$2x + 5 = 0$ or $2x + 5 = 0$

$2x = -5$ or $2x = -5$

$x = -\frac{5}{2}$ or $x = -\frac{5}{2}$

The solution set is $\left\{-\frac{5}{2}\right\}$.

5. $2x^2 + \frac{5}{2}x = 3$

$2\left(2x^2 + \frac{5}{2}x\right) = 2(3)$

$4x^2 + 5x = 6$

$4x^2 + 5x - 6 = 0$

$(4x-3)(x+2) = 0$

$4x - 3 = 0$ or $x + 2 = 0$

$4x = 3$

$x = \frac{3}{4}$ or $x = -2$

The solution set is $\left\{-2, \frac{3}{4}\right\}$.

6. $x^3 = x^2 + 6x$

$x^3 - x^2 - 6x = 0$

$x(x^2 - x - 6) = 0$

$x(x-3)(x+2) = 0$

$x = 0$ or $x - 3 = 0$ or $x + 2 = 0$

$x = 3$ or $x = -2$

The solution set is $\{-2, 0, 3\}$.

7. $x^3 + 7x^2 = 4x + 28$

$x^3 + 7x^2 - 4x - 28 = 0$

$(x^3 + 7x^2) + (-4x - 28) = 0$

$x^2(x+7) - 4(x+7) = 0$

$(x+7)(x^2 - 4) = 0$

$(x+7)(x+2)(x-2) = 0$

$x + 7 = 0$ or $x + 2 = 0$ or $x - 2 = 0$

$x = -7$ or $x = -2$ or $x = 2$

The solution set is $\{-7, -2, 2\}$.

8. $h = -16t^2 + 112t$

$0 = -16t^2 + 112t$

$0 = -16t(t-7)$

$-16t = 0$ or $t - 7 = 0$

$t = 0$ or $t = 7$

The rocket hits the ground after 7 seconds.

9. Let x = first integer and $x + 2$ = second even integer and $x + 4$ = third even integer.

$$x^2 + (x+2)^2 = (x+4)^2$$
$$x^2 + x^2 + 4x + 4 = x^2 + 8x + 16$$
$$x^2 - 4x - 12 = 0$$
$$(x-6)(x+2) = 0$$
$$x - 6 = 0 \quad \text{or} \quad x + 2 = 0$$
$$x = 6 \quad \text{or} \quad x = -2$$

Discard $x = -2$ since length cannot be negative.
$x = 6$, $x + 2 = 8$, $x + 4 = 10$
The lengths are 6, 8, and 10 units.

10. a. This graph has three x-intercepts, (5, 0), (1, 0), and (−1, 0). Therefore, this is the graph of the function $g(x) = (x - 5)(x - 1)(x + 1)$.

b. This graph has two x-intercepts, (−1, 0) and (3, 0). Therefore, this is the graph of the function $f(x) = (x + 1)(x - 3)$.

c. This graph has three x-intercepts, (0, 0), (−1, 0), and (1, 0). Therefore, this is the graph of the function $h(x) = x(x - 1)(x + 1)$.

Calculator Explorations

1. $x^2 + 3x - 2 = 0$

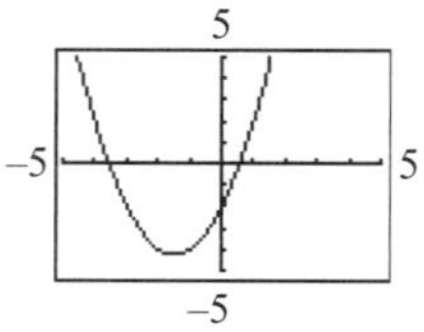

The solution set is {−3.562, 0.562}.

2. $5x^2 - 7x + 1 = 0$

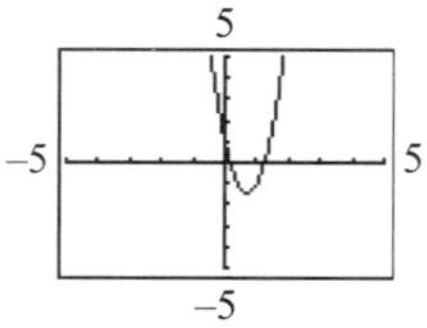

The solution set is {0.161, 1.239}.

3. $2.3x^2 - 4.4x - 5.6 = 0$

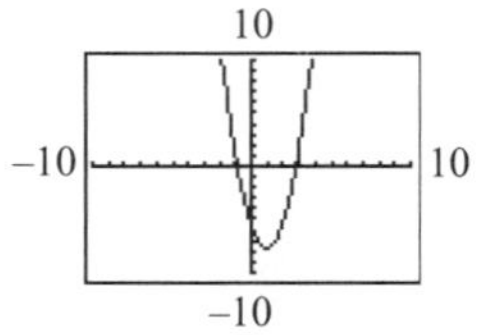

The solution set is {−0.874, 2.787}.

4. $0.2x^2 + 6.2x + 2.1 = 0$

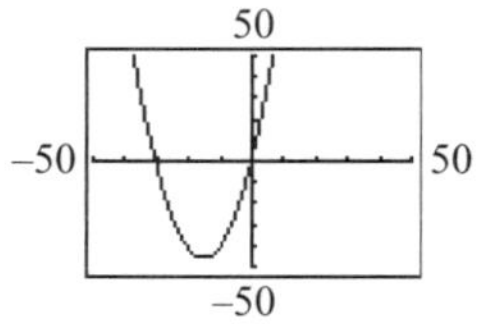

The solution set is {−30.658, −0.342}.

5. $0.09x^2 - 0.13x - 0.08 = 0$

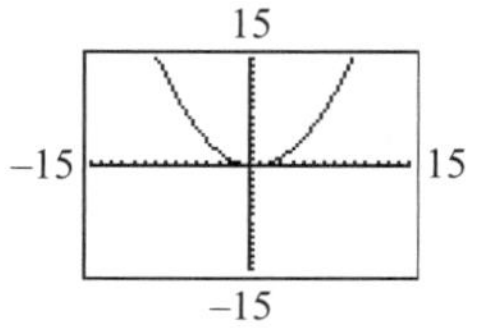

The solution set is {−0.465, 1.910}.

6. $x^2 + 0.08x - 0.01 = 0$

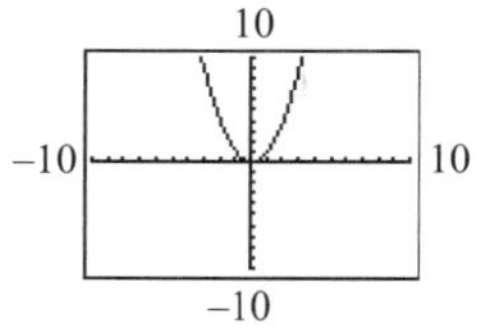

The solution set is {−0.148, 0.068}.

Mental Math

1. $(x - 3)(x + 5) = 0$
The solution set is {3, −5}.

2. $(y + 5)(y + 3) = 0$
The solution set is {−5, −3}.

3. $(z - 3)(z + 7) = 0$
The solution set is {3, −7}.

4. $(c - 2)(c - 4) = 0$
The solution set is {2, 4}.

5. $x(x-9)=0$
The solution set is {0, 9}.

6. $w(w+7)=0$
The solution set is {0, −7}.

Exercise Set 5.7

1. $(x+3)(3x-4)=0$

$$x+3=0 \quad \text{or} \quad 3x-4=0$$
$$3x=4$$
$$x=-3 \quad \text{or} \quad x=\frac{4}{3}$$

The solution set is $\left\{-3, \frac{4}{3}\right\}$.

3. $3(2x-5)(4x+3)=0$

$$2x-5=0 \quad \text{or} \quad 4x+3=0$$
$$2x=5 \quad \text{or} \quad 4x=-3$$
$$x=\frac{5}{2} \quad \text{or} \quad x=-\frac{3}{4}$$

The solution set is $\left\{-\frac{3}{4}, \frac{5}{2}\right\}$.

5. $x^2+11x+24=0$

$$(x+3)(x+8)=0$$
$$x+3=0 \quad \text{or} \quad x+8=0$$
$$x=-3 \quad \text{or} \quad x=-8$$

The solution set is {−3, −8}.

7. $12x^2+5x-2=0$

$$(4x-1)(3x+2)=0$$
$$4x-1=0 \quad \text{or} \quad 3x+2=0$$
$$4x=1 \quad \text{or} \quad 3x=-2$$
$$x=\frac{1}{4} \quad \text{or} \quad x=-\frac{2}{3}$$

The solution set is $\left\{\frac{1}{4}, -\frac{2}{3}\right\}$.

9. $z^2+9=10z$

$$z^2-10z+9=0$$
$$(z-1)(z-9)=0$$
$$z-1=0 \quad \text{or} \quad z-9=0$$
$$z=1 \quad \text{or} \quad z=9$$

The solution set is {1, 9}.

11. $x(5x+2)=3$

$$5x^2+2x=3$$
$$5x^2+2x-3=0$$
$$(5x-3)(x+1)=0$$
$$5x-3=0 \quad \text{or} \quad x+1=0$$
$$5x=3$$
$$x=\frac{3}{5} \quad \text{or} \quad x=-1$$

The solution set is $\left\{\frac{3}{5}, -1\right\}$.

13. $x^2-6x=x(8+x)$

$$x^2-6x=8x+x^2$$
$$0=14x$$
$$x=0$$

The solution set is {0}.

15. $\frac{z^2}{6}-\frac{z}{2}-3=0$

$$6\left(\frac{z^2}{6}-\frac{z}{2}-3\right)=6(0)$$
$$z^2-3z-18=0$$
$$(z-6)(z+3)=0$$
$$z-6=0 \quad \text{or} \quad z+3=0$$
$$z=6 \quad \text{or} \quad z=-3$$

The solution set is {6, −3}.

17. $\frac{x^2}{2}+\frac{x}{20}=\frac{1}{10}$

$10x^2+x=2$

$10x^2+x-2=0$

$(5x-2)(2x+1)=0$

$5x-2=0$ or $2x+1=0$

$5x=2$ or $2x=-1$

$x=\frac{2}{5}$ or $x=-\frac{1}{2}$

The solution set is $\left\{\frac{2}{5}, -\frac{1}{2}\right\}$.

19. $\frac{4t^2}{5}=\frac{t}{5}+\frac{3}{10}$

$8t^2=2t+3$

$8t^2-2t-3=0$

$(4t-3)(2t+1)=0$

$4t-3=0$ or $2t+1=0$

$4t=3$ or $2t=-1$

$t=\frac{3}{4}$ or $t=-\frac{1}{2}$

The solution set is $\left\{\frac{3}{4}, -\frac{1}{2}\right\}$.

21. $(x+2)(x-7)(3x-8)=0$

$x+2=0$ or $x-7=0$ or $3x-8=0$

$x=-2$ or $x=7$ or $x=\frac{8}{3}$

The solution set is $\left\{-2, 7, \frac{8}{3}\right\}$.

23. $y^3=9y$

$y^3-9y=0$

$y(y^2-9)=0$

$y(y-3)(y+3)=0$

$y=0$ or $y-3=0$ or $y+3=0$

$y=3$ or $y=-3$

The solution set is $\{0, 3, -3\}$.

25. $x^3-x=2x^2-2$

$x^3-2x^2-x+2=0$

$x^2(x-2)-(x-2)=0$

$(x^2-1)(x-2)=0$

$(x+1)(x-1)(x-2)=0$

$x+1=0$ or $x-1=0$ or $x-2=0$

$x=-1$ or $x=1$ or $x=2$

The solution set is $\{-1, 1, 2\}$.

27. $(2x+7)(x-10)=0$

$2x+7=0$ or $x-10=0$

$2x=-7$

$x=-\frac{7}{2}$ or $x=10$

The solution set is $\left\{-\frac{7}{2}, 10\right\}$.

29. $3x(x-5)=0$

$3x=0$ or $x-5=0$

$x=0$ or $x=5$

The solution set is $\{0, 5\}$.

31. $x^2-2x-15=0$

$(x+3)(x-5)=0$

$x+3=0$ or $x-5=0$

$x=-3$ or $x=5$

The solution set is $\{-3, 5\}$.

33. $12x^2+2x-2=0$

$2(6x^2+x-1)=0$

$2(2x+1)(3x-1)=0$

$2x+1=0$ or $3x-1=0$

$2x=-1$ or $3x=1$

$x=-\frac{1}{2}$ or $x=\frac{1}{3}$

The solution set is $\left\{-\frac{1}{2}, \frac{1}{3}\right\}$.

35. $w^2 - 5w = 36$

$w^2 - 5w - 36 = 0$

$(w+4)(w-9) = 0$

$w+4 = 0$ or $w-9 = 0$

$w = -4$ or $w = 9$

The solution set is $\{-4, 9\}$.

37. $25x^2 - 40x + 16 = 0$

$(5x-4)(5x-4) = 0$

$5x-4 = 0$ or $5x-4 = 0$

$5x = 4$ $\quad$ $5x = 4$

$x = \frac{4}{5}$ or $x = \frac{4}{5}$

The solution set is $\left\{\frac{4}{5}\right\}$.

39. $2r^3 + 6r^2 = 20r$

$2r^3 + 6r^2 - 20r = 0$

$2r(r^2 + 3r - 10) = 0$

$2r(r+5)(r-2) = 0$

$2r = 0$ or $r+5 = 0$ or $r-2 = 0$

$r = 0$ or $r = -5$ or $r = 2$

The solution set is $\{-5, 0, 2\}$.

41. $z(5z-4)(z+3) = 0$

$z = 0$ or $5z-4 = 0$ or $z+3 = 0$

$5z = 4$

$z = 0$ or $z = \frac{4}{5}$ or $z = -3$

The solution set is $\left\{-3, 0, \frac{4}{5}\right\}$.

43. $2z(z+6) = 2z^2 + 12z - 8$

$2z^2 + 12z = 2z^2 + 12z - 8$

$0 = -8$

This statement is false. There is no solution, or $\varnothing$.

45. $(x-1)(x+4) = 24$

$x^2 + 3x - 4 = 24$

$x^2 + 3x - 28 = 0$

$(x+7)(x-4) = 0$

$x+7 = 0$ or $x-4 = 0$

$x = -7$ or $x = 4$

The solution set is $\{-7, 4\}$.

47. $\frac{x^2}{4} - \frac{5}{2}x + 6 = 0$

$4\left(\frac{x^2}{4} - \frac{5}{2}x + 6\right) = 4(0)$

$x^2 - 10x + 24 = 0$

$(x-4)(x-6) = 0$

$x-4 = 0$ or $x-6 = 0$

$x = 4$ or $x = 6$

The solution set is $\{4, 6\}$.

49. $y^2 + \frac{1}{4} = -y$

$4y^2 + 1 = -4y$

$4y^2 + 4y + 1 = 0$

$(2y+1)(2y+1) = 0$

$2y+1 = 0$ or $2y+1 = 0$

$2y = -1$ or $2y = -1$

$y = -\frac{1}{2}$ or $y = -\frac{1}{2}$

The solution set is $\left\{-\frac{1}{2}\right\}$.

51. $y^3 + 4y^2 = 9y + 36$

$y^3 + 4y^2 - 9y - 36 = 0$

$y^2(y+4) - 9(y+4) = 0$

$(y+4)(y^2 - 9) = 0$

$(y+4)(y-3)(y+3) = 0$

$y+4 = 0$ or $y-3 = 0$ or $y+3 = 0$

$y = -4$ or $y = 3$ or $y = -3$

The solution set is $\{-4, -3, 3\}$.

53.
$$2x^3 = 50x$$
$$2x^3 - 50x = 0$$
$$2x(x^2 - 25) = 0$$
$$2x(x+5)(x-5) = 0$$
$$2x = 0 \quad \text{or} \quad x+5=0 \quad \text{or} \quad x-5=0$$
$$x = 0 \quad \text{or} \quad x=-5 \quad \text{or} \quad x=5$$
The solution set is $\{-5, 0, 5\}$.

55.
$$x^2 + (x+1)^2 = 61$$
$$x^2 + x^2 + 2x + 1 = 61$$
$$2x^2 + 2x - 60 = 0$$
$$2(x^2 + x - 30) = 0$$
$$2(x+6)(x-5) = 0$$
$$x+6=0 \quad \text{or} \quad x-5=0$$
$$x=-6 \quad \text{or} \quad x=5$$
The solution set is $\{-6, 5\}$.

57.
$$m^2(3m-2) = m$$
$$3m^3 - 2m^2 = m$$
$$3m^3 - 2m^2 - m = 0$$
$$m(3m^2 - 2m - 1) = 0$$
$$m(3m+1)(m-1) = 0$$
$$m=0 \quad \text{or} \quad 3m+1=0 \quad \text{or} \quad m-1=0$$
$$3m=-1$$
$$m=0 \quad \text{or} \quad m=-\frac{1}{3} \quad \text{or} \quad m=1$$
The solution set is $\left\{-\frac{1}{3}, 0, 1\right\}$.

59.
$$3x^2 = -x$$
$$3x^2 + x = 0$$
$$x(3x+1) = 0$$
$$x=0 \quad \text{or} \quad 3x+1=0$$
$$x=-\frac{1}{3}$$
The solution set is $\left\{-\frac{1}{3}, 0\right\}$.

61.
$$x(x-3) = x^2 + 5x + 7$$
$$x^2 - 3x = x^2 + 5x + 7$$
$$0 = 8x + 7$$
$$-7 = 8x$$
$$-\frac{7}{8} = x$$
The solution set is $\left\{-\frac{7}{8}\right\}$.

63.
$$3(t-8) + 2t = 7 + t$$
$$3t - 24 + 2t = 7 + t$$
$$4t = 31$$
$$t = \frac{31}{4}$$
The solution set is $\left\{\frac{31}{4}\right\}$.

65.
$$-3(x-4) + x = 5(3-x)$$
$$-3x + 12 + x = 15 - 5x$$
$$12 - 2x = 15 - 5x$$
$$12 + 3x = 15$$
$$3x = 3$$
$$x = 1$$
The solution set is $\{1\}$.

67. Let x be one number. Then $x + 5$ is the other number.
$$x(x+5) = 66$$
$$x^2 + 5x = 66$$
$$x^2 + 5x - 66 = 0$$
$$(x+11)(x-6) = 0$$
$$x+11=0 \quad \text{or} \quad x-6=0$$
$$x=-11 \quad \text{or} \quad x=6$$
$$x+5=-6 \qquad x+5=11$$
The numbers are -11 and -6 or 6 and 11.

69. Let x be the length of cable needed. By the Pythagorean theorem, $45^2 + 60^2 = x^2$.

$$x^2 = 45^2 + 60^2$$
$$x^2 = 2025 + 3600$$
$$x^2 = 5625$$
$$x^2 - 5625 = 0$$
$$x^2 - 75^2 = 0$$
$$(x+75)(x-75) = 0$$
$$x + 75 = 0 \quad \text{or} \quad x - 75 = 0$$
$$x = -75 \quad \text{or} \quad x = 75$$

Since the length of a cable cannot be negative, discard $x = -75$. The cable should be 75 feet long.

71. $C(x) = x^2 - 15x + 50$

$$9500 = x^2 - 15x + 50$$
$$0 = x^2 - 15x - 9450$$
$$0 = (x-105)(x+90)$$
$$x - 105 = 0 \quad \text{or} \quad x + 90 = 0$$
$$x = 105 \quad \text{or} \quad x = -90$$

Since the number of units cannot be negative, discard $x = -90$. The number of units is 105.

73. Let the length of the longer leg be x. Then the shorter leg has length $x - 3$.

$$x^2 + (x-3)^2 = 15^2$$
$$x^2 + x^2 - 6x + 9 = 225$$
$$2x^2 - 6x - 216 = 0$$
$$2(x^2 - 3x - 108) = 0$$
$$2(x-12)(x+9) = 0$$
$$x - 12 = 0 \quad \text{or} \quad x + 9 = 0$$
$$x = 12 \quad \text{or} \quad x = -9$$

Since a length cannot be negative, discard $x = -9$.

$x - 3 = 12 - 3 = 9$

The legs of the triangle have lengths 9 centimeters and 12 centimeters.

75. Let x be the width of the border.

$$(12+2x)(16+2x) - 12 \cdot 16 = 128$$
$$192 + 24x + 32x + 4x^2 - 192 = 128$$
$$4x^2 + 56x - 128 = 0$$
$$4(x^2 + 14x - 32) = 0$$
$$4(x-2)(x+16) = 0$$
$$x - 2 = 0 \quad \text{or} \quad x + 16 = 0$$
$$x = 2 \quad \text{or} \quad x = -16$$

Since the width cannot be negative, discard $x = -16$. The border is 2 inches wide.

77. Find t when $h(t) = 0$.

$$-16t^2 + 1600 = 0$$
$$-16(t^2 - 100) = 0$$
$$-16(t+10)(t-10) = 0$$
$$t + 10 = 0 \quad \text{or} \quad t - 10 = 0$$
$$t = -10 \quad \text{or} \quad t = 10$$

Since the time cannot be negative, discard $t = -10$. The sunglasses hit the ground after 10 seconds.

79. Let x be the width; then $2x - 3$ is the length.

$$x(2x-3) = 90$$
$$2x^2 - 3x - 90 = 0$$
$$(2x-15)(x+6) = 0$$
$$2x - 15 = 0 \quad \text{or} \quad x + 6 = 0$$
$$x = \frac{15}{2} \quad \text{or} \quad x = -6$$

Since length cannot be negative, discard $x = -6$. The width is $\frac{15}{2}$ or $7\frac{1}{2}$ feet and the length is $2x - 3 = 2\left(\frac{15}{2}\right) - 3 = 12$ feet.

81. Find x when $W(x) = 50$.

$$0.5x^2 = 50$$
$$2(0.5x^2) = 2(50)$$
$$x^2 = 100$$
$$x^2 - 100 = 0$$
$$(x+10)(x-10) = 0$$

$x+10=0$ or $x-10=0$
$x=-10$ or $x=10$

Since a length cannot be negative, discard $x = -10$. The tiers of the cake are 10-inch squares.

83. The graph of $f(x)$ has two x-intercepts, (2, 0) and (–5, 0), because the equation $0 = (x - 2)(x + 5)$ has two solutions, 2 and –5. The graph of $f(x)$ is graph E.

85. The graph of $h(x)$ has three x-intercepts, (–3, 0), (0, 0), and (3, 0), because the equation $0 = x(x + 3)(x - 3)$ has three solutions, –3, 0, and 3. The graph of $h(x)$ is graph F.

87. $G(x)=2x^2+9x+4=(2x+1)(x+4)$ has two x-intercepts, $\left(-\frac{1}{2}, 0\right)$ and (–4, 0), because $0 = (2x + 1)(x + 4)$ has two solutions, $-\frac{1}{2}$ and –4. The graph of $G(x)$ is graph B.

89. x-intercept: (–3, 0)
y-intercept: (0, 2)

91. x-intercepts: (–4, 0), (4, 0)
y-intercepts: (0, 2), (0, –2)

93. The error is in the last line.
$(x - 5)(x + 2) = 0$
$x-5=0$ or $x+2=0$
$x=5$ or $x=-2$

95.
$$y(y-5)=-6$$
$$y^2-5y+6=0$$
$$(y-2)(y-3)=0$$
$y-2=0$ or $y-3=0$
$y=2$ or $y=3$

97.
$$(x^2+x-6)(3x^2-14x-5)=0$$
$$[(x+3)(x-2)][(3x-1)(x-5)]=0$$
$$(x+3)(x-2)(3x+1)(x-5)=0$$
$x+3=0$ or $x-2=0$ or $3x+1=0$ or $x-5=0$
$3x=-1$
$x=-3$ or $x=2$ or $x=-\frac{1}{3}$ or $x=5$

The solution set is $\left\{-3, -\frac{1}{3}, 2, 5\right\}$.

99. answers may vary

101. no; answers may vary

103. answers may vary

The Bigger Picture

1. $|7x - 3| = |5x + 9|$

$$7x - 3 = 5x + 9 \quad \text{or} \quad 7x - 3 = -(5x + 9)$$
$$2x = 12 \qquad 7x - 3 = -5x - 9$$
$$x = 6 \qquad 12x = -6$$
$$x = -\frac{1}{2}$$

The solution set is $\left\{-\frac{1}{2}, 6\right\}$.

2. $\left|\frac{x+2}{5}\right| < 1$

$$-1 < \frac{x+2}{5} < 1$$
$$-5 < x + 2 < 5$$
$$-7 < x < 3$$

3. $3(x - 6) + 2 = 9 + 5(3x - 1)$

$$3x - 18 + 2 = 9 + 15x - 5$$
$$3x - 16 = 15x + 4$$
$$-12x = 20$$
$$x = -\frac{20}{12} = -\frac{5}{3}$$

The solution set is $\left\{-\frac{5}{3}\right\}$.

4. $(x - 6)(2x + 3) = 0$

$$x - 6 = 0 \quad \text{or} \quad 2x + 3 = 0$$
$$x = 6 \quad \text{or} \quad x = -\frac{3}{2}$$

The solution set is $\left\{-\frac{3}{2}, 6\right\}$.

5. $|-3x + 10| \geq -2$

The absolute value of any number is always nonnegative. Therefore, the solution set is $(-\infty, \infty)$.

6. $|-2x - 5| = 11$

$$-2x - 5 = -11 \quad \text{or} \quad -2x - 5 = 11$$
$$-2x = -6 \quad \text{or} \quad -2x = 16$$
$$x = 3 \quad \text{or} \quad x = -8$$

The solution set is $\{-8, 3\}$.

7.

$$x(x - 7) = 30$$
$$x^2 - 7x = 30$$
$$x^2 - 7x - 30 = 0$$
$$(x + 3)(x - 10) = 0$$
$$x + 3 = 0 \quad \text{or} \quad x - 10 = 0$$
$$x = -3 \quad \text{or} \quad x = 10$$

The solution set is $\{-3, 10\}$.

8.

$$8x - 4 \geq 15x - 4$$
$$8x - 4 - 15x \geq 15x - 4 - 15x$$
$$-7x - 4 \geq -4$$
$$-7x - 4 + 4 \geq -4 + 4$$
$$-7x \geq 0$$
$$\frac{-7x}{-7} \leq \frac{0}{-7}$$
$$x \leq 0$$

The solution set is $(-\infty, 0]$.

Chapter 5 Vocabulary Check

1. A polynomial is a finite sum of terms in which all variables are raised to nonnegative integer powers and no variables appear in any denominator.

2. Factoring is the process of writing a polynomial as a product.

3. The degree of a term is the sum of the exponents on the variables contained in the term.

4. A monomial is a polynomial with one term.

5. A trinomial is a polynomial with three terms.

6. A polynomial equation of degree 2 is also called a quadratic equation.

7. The <u>degree of polynomial</u> is the largest degree of all of its terms.

8. A <u>binomial</u> is a polynomial with two terms.

9. If a and b are real numbers and $a \cdot b = \underline{0}$, then $a = 0$ or $b = 0$.

10. The <u>FOIL</u> order may be used when multiplying two binomials.

11. A shortcut method called <u>synthetic division</u> may be used to divide a polynomial by a binomial of the form $x - c$.

Chapter 5 Review

1. The term with the greatest degree is $-3xy^3z$, which has degree 1 + 3 + 1 = 5, so the degree of the polynomial is 5.

2. The term with the variable is $3x$, which has degree 1, so the degree of the polynomial is 1.

3. $$\begin{aligned} 4x+8x-6x^2-6x^2y &= (4+8)x-6x^2-6x^2y \\ &= 12x-6x^2-6x^2y \end{aligned}$$

4. $$\begin{aligned} -8xy^3+4xy^3-3x^3y &= (-8+4)xy^3-3x^3y \\ &= -4xy^3-3x^3y \end{aligned}$$

5. $$\begin{aligned} (3x+7y)+(4x^2-3x+7)+(y-1) &= 3x+7y+4x^2-3x+7+y-1 \\ &= 4x^2+3x-3x+7y+y+7-1 \\ &= 4x^2+8y+6 \end{aligned}$$

6. $$\begin{aligned} (4x^2-6xy+9y^2)-(8x^2-6xy-y^2) &= 4x^2-6xy+9y^2-8x^2+6xy+y^2 \\ &= 4x^2-8x^2-6xy+6xy+9y^2+y^2 \\ &= -4x^2+10y^2 \end{aligned}$$

7. $$\begin{aligned} (3x^2-4b+28)+(9x^2-30)-(4x^2-6b+20) &= 3x^2-4b+28+9x^2-30-4x^2+6b-20 \\ &= 3x^2+9x^2-4x^2-4b+6b+28-30-20 \\ &= 8x^2+2b-22 \end{aligned}$$

8. $$\begin{aligned} (9xy+4x^2+18)+(7xy-4x^3-9x) &= 9xy+4x^2+18+7xy-4x^3-9x \\ &= -4x^3+4x^2+9xy+7xy-9x+18 \\ &= -4x^3+4x^2+16xy-9x+18 \end{aligned}$$

9. $$\begin{aligned} (3x^2y-7xy-4)+(9x^2y+x)-(x-7) &= 3x^2y-7xy-4+9x^2y+x-x+7 \\ &= 3x^2y+9x^2y-7xy+x-x-4+7 \\ &= 12x^2y-7xy+3 \end{aligned}$$

10. $\begin{array}{r} x^2-5x+7 \\ -\quad (x+4) \\ \hline \end{array} \rightarrow \begin{array}{r} x^2-5x+7 \\ +\ (-x-4) \\ \hline x^2-6x+3 \end{array}$

11. $\begin{array}{l} x^3 \qquad +2xy^2-y \\ +\quad x-4xy^2 \qquad -7 \\ \hline x^3+x-2xy^2-y-7 \end{array}$

12. $P(x)=9x^2-7x+8$

$P(6)=9(6)^2-7(6)+8=324-42+8=290$

13. $P(x)=9x^2-7x+8$

$P(-2)=9(-2)^2-7(-2)+8=36+14+8=58$

14. $P(x)=9x^2-7x+8$

$P(-3)=9(-3)^2-7(-3)+8=81+21+8=110$

15. $$\begin{aligned} P(x)+Q(x) &= (2x-1)+(x^2+2x-5) \\ &= 2x-1+x^2+2x-5 \\ &= x^2+4x-6 \end{aligned}$$

16. $$\begin{aligned} 2\cdot P(x)-Q(x) &= 2(2x-1)-(x^2+2x-5) \\ &= 4x-2-x^2-2x+5 \\ &= -x^2+2x+3 \end{aligned}$$

17. $$\begin{aligned} 2(x^2y+5)+2(2x^2y-6x+1) &= 2x^2y+10+4x^2y-12x+2 \\ &= 6x^2y-12x+12 \end{aligned}$$

The perimeter is $(6x^2y-12x+12)$ centimeters.

18. $$\begin{aligned} -6x(4x^2-6x+1) &= -6x\cdot 4x^2+(-6x)\cdot(-6x)+(-6x)\cdot 1 \\ &= -24x^3+36x^2-6x \end{aligned}$$

19. $$\begin{aligned} -4ab^2(3ab^3+7ab+1) &= -4ab^2\cdot 3ab^3+(-4ab^2)\cdot(7ab)+(-4ab^2)\cdot 1 \\ &= -12a^2b^5-28a^2b^3-4ab^2 \end{aligned}$$

20. $$\begin{aligned} (x-4)(2x+9) &= x(2x+9)-4(2x+9) \\ &= 2x^2+9x-8x-36 \\ &= 2x^2+x-36 \end{aligned}$$

21. $(-3xa+4b)^2 = (-3xa)^2 + 2(-3xa)(4b) + (4b)^2$
$= 9x^2a^2 - 24xab + 16b^2$

22. $(9x^2+4x+1)(4x-3) = 9x^2(4x-3) + 4x(4x-3) + 1(4x-3)$
$= 36x^3 - 27x^2 + 16x^2 - 12x + 4x - 3$
$= 36x^3 - 11x^2 - 8x - 3$

23. $(5x-9y)(3x+9y) = 5x(3x+9y) - 9y(3x+9y)$
$= 15x^2 + 45xy - 27xy - 81y^2$
$= 15x^2 + 18xy - 81y^2$

24. $\left(x-\frac{1}{3}\right)\left(x+\frac{2}{3}\right) = x\left(x+\frac{2}{3}\right) - \frac{1}{3}\left(x+\frac{2}{3}\right)$
$= x^2 + \frac{2}{3}x - \frac{1}{3}x - \frac{2}{9}$
$= x^2 + \frac{1}{3}x - \frac{2}{9}$

25. $(x^2+9x+1)^2 = (x^2+9x+1)(x^2+9x+1)$
$= x^2(x^2+9x+1) + 9x(x^2+9x+1) + 1(x^2+9x+1)$
$= x^4 + 9x^3 + x^2 + 9x^3 + 81x^2 + 9x + x^2 + 9x + 1$
$= x^4 + 18x^3 + 83x^2 + 18x + 1$

26. $(2x-1)(x^2+2x-5) = 2x(x^2+2x-5) - 1(x^2+2x-5)$
$= 2x^3 + 4x^2 - 10x - x^2 - 2x + 5$
$= 2x^3 + 3x^2 - 12x + 5$

27. $(3x-y)^2 = (3x)^2 - 2(3x)(y) + y^2$
$= 9x^2 - 6xy + y^2$

28. $(4x+9)^2 = (4x)^2 + 2(4x)(9) + 9^2$
$= 16x^2 + 72x + 81$

29. $(x+3y)(x-3y) = (x)^2 - (3y)^2 = x^2 - 9y^2$

30. $[4+(3a-b)][4-(3a-b)] = 4^2 - (3a-b)^2$
$= 16 - [(3a)^2 - 2(3a)(b) + b^2]$
$= 16 - (9a^2 - 6ab + b^2)$
$= 16 - 9a^2 + 6ab - b^2$

31. $(3y-7z)(3y+7z)=(3y)^2-(7z)^2$
$=9y^2-49z^2$

The area of the rectangle is $(9y^2-49z^2)$ square units.

32. $\dfrac{4xy+2x^2-9}{4xy}=\dfrac{4xy}{4xy}+\dfrac{2x^2}{4xy}-\dfrac{9}{4xy}$
$=1+\dfrac{x}{2y}-\dfrac{9}{4xy}$

33. $\dfrac{12xb^2+16xb^4}{4xb^3}=\dfrac{12xb^2}{4xb^3}+\dfrac{16xb^4}{4xb^3}=\dfrac{3}{b}+4b$

34.

$$\begin{array}{r}
3x^3+9x^2+2x+6 \\
x-3\,\overline{)\,3x^4+0x^3-25x^2+0x-20} \\
\underline{3x^4-9x^3} \\
9x^3-25x^2 \\
\underline{9x^3-27x^2} \\
2x^2+0x \\
\underline{2x^2-6x} \\
6x-20 \\
\underline{6x-18} \\
-2
\end{array}$$

$(3x^4-25x^2-20)\div(x-3)$
$=3x^3+9x^2+2x+6-\dfrac{2}{x-3}$

35.

$$\begin{array}{r}
2x^3+6x^2+17x+56 \\
x-3\,\overline{)\,2x^4+0x^3-x^2+5x-12} \\
\underline{2x^4-6x^3} \\
6x^3-x^2 \\
\underline{6x^3-18x^2} \\
17x^2+5x \\
\underline{17x^2-51x} \\
56x-12 \\
\underline{56x-168} \\
156
\end{array}$$

$(-x^2+2x^4+5x-12)\div(x-3)$
$=2x^3+6x^2+17x+56+\dfrac{156}{x-3}$

36.

$$\begin{array}{r}
2x^3+2x-2 \\
x-\tfrac{1}{2}\,\overline{)\,2x^4-x^3+2x^2-3x+1} \\
\underline{2x^4-x^3} \\
2x^2-3x \\
\underline{2x^2-x} \\
-2x+1 \\
\underline{-2x+1} \\
0
\end{array}$$

$(2x^4-x^3+2x^2-3x+1)\div\left(x-\dfrac{1}{2}\right)$
$=2x^3+2x-1$

37.

$$\begin{array}{r}
x^2+\tfrac{7}{2}x-\tfrac{1}{4} \\
x-\tfrac{1}{2}\,\overline{)\,x^3+3x^2-2x+2} \\
\underline{x^3-\tfrac{1}{2}x^2} \\
\tfrac{7}{2}x^2-2x \\
\underline{\tfrac{7}{2}x^2-\tfrac{7}{4}x} \\
-\tfrac{1}{4}x+2 \\
\underline{-\tfrac{1}{4}x+\tfrac{1}{8}} \\
\tfrac{15}{8}
\end{array}$$

$(x^3+3x^2-2x+2)\div\left(x-\dfrac{1}{2}\right)$
$=x^2+\dfrac{7}{2}x-\dfrac{1}{4}+\dfrac{15}{8\left(x-\frac{1}{2}\right)}$

38.

$$\begin{array}{r}
3x^2+2x-1 \\
x^2+x+2\,\overline{)\,3x^4+5x^3+7x^2+3x-2} \\
\underline{3x^4+3x^3+6x^2} \\
2x^3+x^2+3x \\
\underline{2x^3+2x^2+4x} \\
-x^2-x-2 \\
\underline{-x^2-x-2} \\
0
\end{array}$$

$(3x^4+5x^3+7x^2+3x-2)\div(x^2+x+2)$
$=3x^2+2x-1$

39.
$$\begin{array}{r} 3x^2 \qquad\quad +6 \\ 3x^2-2x-5\overline{)\,9x^4-6x^3+3x^2-12x-30} \\ \underline{9x^4-6x^3-15x^2} \qquad\qquad \\ 18x^2-12x-30 \\ \underline{18x^2-12x-30} \\ 0 \end{array}$$

$(9x^4-6x^3+3x^2-12x-30)\div(3x^2-2x-5)$
$=3x^2+6$

40. $x-2=x-c$ where $c=2$.

2	3	0	12	−4
		6	12	48
	3	6	24	44

$(3x^3+12x-4)\div(x-2)$
$=3x^2+6x+24+\dfrac{44}{x-2}$

41. $x+\dfrac{3}{2}=x-c$ where $c=-\dfrac{3}{2}$.

$-\frac{3}{2}$	4	2	−4	−2
		−6	6	−3
	4	−4	2	−5

$(4x^3+2x^2-4x-2)\div\left(x+\dfrac{3}{2}\right)$
$=4x^2-4x+2-\dfrac{5}{x+\frac{3}{2}}$

42. $x+1=x-c$ where $c=-1$.

−1	1	0	0	0	0	−1
		−1	1	−1	1	−1
	1	−1	1	−1	1	−2

$(x^5-1)\div(x+1)=x^4-x^3+x^2-x+1-\dfrac{2}{x+1}$

43. $x-3=x-c$ where $c=3$.

3	1	0	0	−81
		3	9	27
	1	3	9	−54

$(x^3-81)\div(x-3)=x^2+3x+9-\dfrac{54}{x-3}$

44. $x-4=x-c$ where $c=4$.

4	3	1	−1	0	−2
		12	52	204	816
	3	13	51	204	814

$(x^3-x^2+3x^4-2)\div(x-4)$
$=3x^3+13x^2+51x+204+\dfrac{814}{x-4}$

45. $x+2=x-c$ where $c=-2$.
$3x^4-2x^2+10=3x^4+0x^3-2x^2+0x+10$

−2	3	0	−2	0	10
		−6	12	−20	40
	3	−6	10	−20	50

$(3x^4-2x^2+10)\div(x+2)$
$=3x^3-6x^2+10x-20+\dfrac{50}{x+2}$

46. $16x^3-24x^2=8x^2\cdot 2x-8x^2\cdot 3$
$=8x^2(2x-3)$

47. $36y-24y^2=12y\cdot 3-12y\cdot 2y=12y(3-2y)$

48. $6ab^2+8ab-4a^2b^2$
$=2ab\cdot 3b+2ab\cdot 4-2ab\cdot 2ab$
$=2ab(3b+4-2ab)$

49. $14a^2b^2-21ab^2+7ab$
$=7ab\cdot 2ab-7ab\cdot 3b+7ab\cdot 1$
$=7ab(2ab-3b+1)$

50. $6a(a+3b)-5(a+3b)=(a+3b)(6a-5)$

51. $4x(x-2y)-5(x-2y)=(x-2y)(4x-5)$

52. $xy-6y+3x-18=(xy-6y)+(3x-18)$
$=y(x-6)+3(x-6)$
$=(x-6)(y+3)$

53. $ab-8b+4a-32=(ab-8b)+(4a-32)$
$=b(a-8)+4(a-8)$
$=(a-8)(b+4)$

54. $pq-3p-5q+15=(pq-3p)+(-5q+15)$
$=p(q-3)-5(q-3)$
$=(q-3)(p-5)$

55. $x^3-x^2-2x+2=x^2(x-1)-2(x-1)$
$=(x-1)(x^2-2)$

56. $2x\cdot y-x\cdot x=2xy-x^2=x(2y-x)$

57. $x^2-14x-72$
$-18\cdot 4=-72$ and $-18+4=-14$
$x^2-14x-72=(x-18)(x+4)$

58. $x^2+16x-80$
$-4\cdot 20=-80$ and $-4+20=16$
$x^2+16x-80=(x-4)(x+20)$

59. $2x^2-18x+28=2(x^2-9x+14)$
$-2\cdot(-7)=14$ and $-2+(-7)=-9$
$2x^2-18x+28=2(x-2)(x-7)$

60. $3x^2+33x+54=3(x^2+11x+18)$
$2\cdot 9=18$ and $2+9=11$
$3x^2+33x+54=3(x+2)(x+9)$

61. $2x^3-7x^2-9x=x(2x^2-7x-9)$
Factors of $2x^2$: $2x\cdot x$
Factors of -9: $-9\cdot 1, 9\cdot(-1), -3\cdot 3$
$2x^3-7x^2-9x=x(2x-9)(x+1)$

62. $3x^2+2x-16$
Factors of $3x^2$: $3x\cdot x$
Factors of -16: $-16\cdot 1, 16\cdot(-1), -8\cdot 2, 8\cdot(-2), -4\cdot 4$
$3x^2+2x-16=(3x+8)(x-2)$

63. $6x^2+17x+10$
Factors of $6x^2$: $6x\cdot x, 3x\cdot 2x$
Factors of 10: $10\cdot 1, 5\cdot 2$
$6x^2+17x+10=(6x+5)(x+2)$

64. $15x^2-91x+6$
Factors of $15x^2$: $15x\cdot x, 5x\cdot 3x$
Negative factors of 6: $-6\cdot(-1), -3\cdot(-2)$
$15x^2-91x+6=(15x-1)(x-6)$

65. $4x^2+2x-12=2(2x^2+x-6)$
Factors of $2x^2$: $2x\cdot x$
Factors of -6: $-6\cdot 1, 6\cdot(-1), -3\cdot 2, 3\cdot(-2)$
$4x^2+2x-12=2(2x-3)(x+2)$

66. $9x^2-12x-12=3(3x^2-4x-4)$
Factors of $3x^2$: $3x\cdot x$
Factors of -4: $-4\cdot 1, 4\cdot(-1), -2\cdot 2$
$9x^2-12x-12=3(x-2)(3x+2)$

67. $y^2(x+6)^2-2y(x+6)^2-3(x+6)^2$
$=(x+6)^2(y^2-2y-3)$
$=(x+6)^2(y-3)(y+1)$

68. Substitute y for $x+5$.
$(x+5)^2+6(x+5)+8=y^2+6y+8$
$4\cdot 2=8$ and $4+2=6$
$y^2+6y+8=(y+4)(y+2)$
Now replace y with $x+5$.
$(y+4)(y+2)=[(x+5)+4][(x+5)+2]$
$=(x+9)(x+7)$
$(x+5)^2+6(x+5)+8=(x+9)(x+7)$

69. Substitute y for x^2.

$x^4 - 6x^2 - 16 = y^2 - 6y - 16$

$-8 \cdot 2 = -16$ and $-8 + 2 = -6$

$y^2 - 6y - 16 = (y-8)(y+2)$

Replace y with x^2.

$(y-8)(y+2) = (x^2-8)(x^2+2)$

70. Substitute y for x^2.

$x^4 + 8x^2 - 20 = y^2 + 8y - 20$

$-2 \cdot 10 = -20$ and $-2 + 10 = 8$

$y^2 + 8y - 20 = (y-2)(y+10)$

Replace y with x^2.

$(y-2)(y+10) = (x^2-2)(x^2+10)$

71. $x^2 - 100 = x^2 - 10^2 = (x+10)(x-10)$

72. $x^2 - 81 = x^2 - 9^2 = (x+9)(x-9)$

73. $$\begin{aligned} 2x^2 - 32 &= 2(x^2-16) \\ &= 2(x^2-4^2) \\ &= 2(x+4)(x-4) \end{aligned}$$

74. $$\begin{aligned} 6x^2 - 54 &= 6(x^2-9) \\ &= 6(x^2-3^2) \\ &= 6(x+3)(x-3) \end{aligned}$$

75. $$\begin{aligned} 81 - x^4 &= 9^2 - (x^2)^2 \\ &= (9+x^2)(9-x^2) \\ &= (9+x^2)(3^2-x^2) \\ &= (9+x^2)(3+x)(3-x) \end{aligned}$$

76. $$\begin{aligned} 16 - y^4 &= 4^2 - (y^2)^2 \\ &= (4+y^2)(4-y^2) \\ &= (4+y^2)(2^2-y^2) \\ &= (4+y^2)(2+y)(2-y) \end{aligned}$$

77. $$\begin{aligned} (y+2)^2 - 25 &= (y+2)^2 - 5^2 \\ &= [(y+2)+5][(y+2)-5] \\ &= (y+7)(y-3) \end{aligned}$$

78. $$\begin{aligned} (x-3)^2 - 16 &= (x-3)^2 - 4^2 \\ &= [(x-3)+4][(x-3)-4] \\ &= (x+1)(x-7) \end{aligned}$$

79. $$\begin{aligned} x^3 + 216 &= x^3 + 6^3 \\ &= (x+6)(x^2 - x \cdot 6 + 6^2) \\ &= (x+6)(x^2 - 6x + 36) \end{aligned}$$

80. $$\begin{aligned} y^3 + 512 &= y^3 + 8^3 \\ &= (y+8)(y^2 - y \cdot 8 + 8^2) \\ &= (y+8)(y^2 - 8y + 64) \end{aligned}$$

81. $$\begin{aligned} 8 - 27y^3 &= 2^3 - (3y)^3 \\ &= (2-3y)[2^2 + 2 \cdot 3y + (3y)^2] \\ &= (2-3y)(4+6y+9y^2) \end{aligned}$$

82. $$\begin{aligned} 1 - 64y^3 &= 1^3 - (4y)^3 \\ &= (1-4y)[1^2 + 1 \cdot 4y + (4y)^2] \\ &= (1-4y)(1+4y+16y^2) \end{aligned}$$

83. $$\begin{aligned} 6x^4y + 48xy &= 6xy(x^3+8) \\ &= 6xy(x^3+2^3) \\ &= 6xy(x+2)(x^2 - x \cdot 2 + 2^2) \\ &= 6xy(x+2)(x^2-2x+4) \end{aligned}$$

84. $$\begin{aligned} &2x^5 + 16x^2y^3 \\ &= 2x^2(x^3+8y^3) \\ &= 2x^2[x^3+(2y)^3] \\ &= 2x^2(x+2y)[x^2 - x \cdot 2y + (2y)^2] \\ &= 2x^2(x+2y)(x^2-2xy+4y^2) \end{aligned}$$

85. $x^2-2x+1-y^2=(x^2-2x+1)-y^2$
$=(x-1)^2-y^2$
$=[(x-1)+y][(x-1)-y]$
$=(x-1+y)(x-1-y)$

86. $x^2-6x+9-4y^2$
$=(x^2-6x+9)-4y^2$
$=(x-3)^2-(2y)^2$
$=[(x-3)+2y][(x-3)-2y]$
$=(x-3+2y)(x-3-2y)$

87. $4x^2+12x+9=(2x)^2+2\cdot 2x\cdot 3+3^2$
$=(2x+3)^2$

88. $16a^2-40ab+25b^2$
$=(4a)^2-2\cdot 4a\cdot 5b+(5b)^2$
$=(4a-5b)^2$

89. $\pi R^2h-\pi r^2h=\pi h(R^2-r^2)$
$=\pi h(R+r)(R-r)$
The volume is $\pi h(R+r)(R-r)$ cubic units.

90. $(3x-1)(x+7)=0$
$3x-1=0$ or $x+7=0$
$3x=1$
$x=\frac{1}{3}$ or $x=-7$
The solution set is $\left\{\frac{1}{3}, -7\right\}$.

91. $3(x+5)(8x-3)=0$
$x+5=0$ or $8x-3=0$
$x=-5$ or $8x=3$
$x=\frac{3}{8}$
The solution set is $\left\{-5, \frac{3}{8}\right\}$.

92. $5x(x-4)(2x-9)=0$
$5x=0$ or $x-4=0$ or $2x-9=0$
$2x=9$
$x=0$ or $x=4$ or $x=\frac{9}{2}$
The solution set is $\left\{0, 4, \frac{9}{2}\right\}$.

93. $6(x+3)(x-4)(5x+1)=0$
$x+3=0$ or $x-4=0$ or $5x+1=0$
$5x=-1$
$x=-3$ or $x=4$ or $x=-\frac{1}{5}$
The solution set is $\left\{-3, 4, -\frac{1}{5}\right\}$.

94. $2x^2=12x$
$2x^2-12x=0$
$2x(x-6)=0$
$2x=0$ or $x-6=0$
$x=0$ or $x=6$
The solution set is {0, 6}.

95. $4x^3-36x=0$
$4x(x^2-9)=0$
$4x(x+3)(x-3)=0$
$4x=0$ or $x+3=0$ or $x-3=0$
$x=0$ or $x=-3$ or $x=3$
The solution set is {0, –3, 3}.

96. $(1-x)(3x+2)=-4x$
$3x+2-3x^2-2x=-4x$
$-3x^2+5x+2=0$
$(-3x-1)(x-2)=0$
$-3x-1=0$ or $x-2=0$
$-3x=1$
$x=-\frac{1}{3}$ or $x=2$
The solution set is $\left\{-\frac{1}{3}, 2\right\}$.

97.
$$2x(x-12) = -40$$
$$2x^2 - 24x = -40$$
$$2x^2 - 24x + 40 = 0$$
$$2(x^2 - 12x + 20) = 0$$
$$2(x-2)(x-10) = 0$$
$$x - 2 = 0 \quad \text{or} \quad x - 10 = 0$$
$$x = 2 \quad \text{or} \quad x = 10$$
The solution set is {2, 10}.

98.
$$3x^2 + 2x = 12 - 7x$$
$$3x^2 + 2x + 7x - 12 = 0$$
$$3x^2 + 9x - 12 = 0$$
$$3(x^2 + 3x - 4) = 0$$
$$3(x+4)(x-1) = 0$$
$$x + 4 = 0 \quad \text{or} \quad x - 1 = 0$$
$$x = -4 \quad \text{or} \quad x = 1$$
The solution set is {−4, 1}.

99.
$$2x^2 + 3x = 35$$
$$2x^2 + 3x - 35 = 0$$
$$(2x-7)(x+5) = 0$$
$$2x - 7 = 0 \quad \text{or} \quad x + 5 = 0$$
$$2x = 7$$
$$x = \frac{7}{2} \quad \text{or} \quad x = -5$$
The solution set is $\left\{\frac{7}{2}, -5\right\}$.

100.
$$x^3 - 18x = 3x^2$$
$$x^3 - 3x^2 - 18x = 0$$
$$x(x^2 - 3x - 18) = 0$$
$$x(x-6)(x+3) = 0$$
$$x = 0 \quad \text{or} \quad x - 6 = 0 \quad \text{or} \quad x + 3 = 0$$
$$x = 6 \quad \text{or} \quad x = -3$$
The solution set is {0, 6, −3}.

101.
$$19x^2 - 42x = -x^3$$
$$x^3 + 19x^2 - 42x = 0$$
$$x(x^2 + 19x - 42) = 0$$
$$x(x+21)(x-2) = 0$$
$$x = 0 \quad \text{or} \quad x + 21 = 0 \quad \text{or} \quad x - 2 = 0$$
$$x = -21 \quad \text{or} \quad x = 2$$
The solution set is {−21, 0, 2}.

102.
$$12x = 6x^3 + 6x^2$$
$$0 = 6x^3 + 6x^2 - 12x$$
$$0 = 6x(x^2 + x - 2)$$
$$0 = 6x(x+2)(x-1)$$
$$6x = 0 \quad \text{or} \quad x + 2 = 0 \quad \text{or} \quad x - 1 = 0$$
$$x = 0 \quad \text{or} \quad x = -2 \quad \text{or} \quad x = 1$$
The solution set is {0, −2, 1}.

103.
$$8x^3 + 10x^2 = 3x$$
$$8x^3 + 10x^2 - 3x = 0$$
$$x(8x^2 + 10x - 3) = 0$$
$$x(2x+3)(4x-1) = 0$$
$$x = 0 \quad \text{or} \quad 2x + 3 = 0 \quad \text{or} \quad 4x - 1 = 0$$
$$2x = -3 \quad \text{or} \quad 4x = 1$$
$$x = -\frac{3}{2} \quad \text{or} \quad x = \frac{1}{4}$$
The solution set is $\left\{0, -\frac{3}{2}, \frac{1}{4}\right\}$.

104. Let x be the number.
$$x + 2x^2 = 105$$
$$2x^2 + x - 105 = 0$$
$$(2x+15)(x-7) = 0$$
$$2x + 15 = 0 \quad \text{or} \quad x - 7 = 0$$
$$x = -\frac{15}{2} \quad \text{or} \quad x = 7$$
The number is $-\frac{15}{2}$ or 7.

105. Let x be the width; then $5x - 2$ is the length.

$$x(5x-2)=16$$
$$5x^2-2x=16$$
$$5x^2-2x-16=0$$
$$(5x+8)(x-2)=0$$
$$5x+8=0 \quad \text{or} \quad x-2=0$$
$$x=-\frac{8}{5} \quad \text{or} \quad x=2$$

Since length cannot be negative, discard $x=-\frac{8}{5}$. The width is 2 meters and the length is $5x - 2 = 5(2) - 2 = 8$ meters.

106. $h=-16t^2+400$

$$0=-16t^2+400$$
$$0=-16(t^2-25)$$
$$0=-16(t+5)(t-5)$$
$$t+5=0 \quad \text{or} \quad t-5=0$$
$$t=-5 \quad \text{or} \quad t=5$$

Since time cannot be negative, discard $t = -5$. The dummy reaches the ground in 5 seconds.

107. $(x+5)(3x^2-2x+1)$

$$=x(3x^2-2x+1)+5(3x^2-2x+1)$$
$$=3x^3-2x^2+x+15x^2-10x+5$$
$$=3x^3+13x^2-9x+5$$

108. $(3x^2+4x-1.2)-(5x^2-x+5.7)$

$$=3x^2+4x-1.2-5x^2+x-5.7$$
$$=3x^2-5x^2+4x+x-1.2-5.7$$
$$=-2x^2+5x-6.9$$

109. $(3x^2+4x-1.2)+(5x^2-x+5.7)$

$$=3x^2+4x-1.2+5x^2-x+5.7$$
$$=3x^2+5x^2+4x-x-1.2+5.7$$
$$=8x^2+3x+4.5$$

110. $\left(7ab-\frac{1}{2}\right)^2=(7ab)^2-2(7ab)\left(\frac{1}{2}\right)+\left(\frac{1}{2}\right)^2$

$$=49a^2b^2-7ab+\frac{1}{4}$$

111. $P(x)=-x^2+x-4$

$$P(5)=-5^2+5-4=-25+5-4=-24$$

112. $P(x)=-x^2+x-4$

$$P(-2)=-(-2)^2+(-2)-4$$
$$=-4-2-4$$
$$=-10$$

113. $12y^5-6y^4=6y^4\cdot 2y+6y^4\cdot(-1)$

$$=6y^4(2y-1)$$

114. $x^2y+4x^2-3y-12$

$$=(x^2y+4x^2)+(-3y-12)$$
$$=x^2(y+4)-3(y+4)$$
$$=(y+4)(x^2-3)$$

115. $6x^2-34x-12=2(3x^2-17x-6)$

Factors of $3x^2$: $3x \cdot x$

Factors of –6: $-6 \cdot 1$, $6 \cdot (-1)$, $-3 \cdot 2$, $3 \cdot (-2)$

$$6x^2-34x-12=2(3x+1)(x-6)$$

116. $y^2(4x+3)^2-19y(4x+3)^2-20(4x+3)^2$

$$=(4x+3)^2(y^2-19y-20)$$
$$=(4x+3)^2(y-20)(y+1)$$

117. $4z^7-49z^5=z^5(4z^2-49)$

$$=z^5[(2z)^2-7^2]$$
$$=z^5(2z+7)(2z-7)$$

118. Substitute y for x^2.

$5x^4+4x^2-9=5y^2+4y-9$

Factors of $5y^2$: $5y \cdot y$

Factors of –9: $-9 \cdot 1, 9 \cdot (-1), 3 \cdot (-3)$

$5y^2+4y-9=(y-1)(5y+9)$

Now replace y with x^2.

$$(y-1)(5y+9)=(x^2-1)(5x^2+9)$$
$$=(x+1)(x-1)(5x^2+9)$$
$$5x^4+4x^2-9=(x+1)(x-1)(5x^2+9)$$

119.
$$8x^2=24x$$
$$8x^2-24x=0$$
$$8x(x-3)=0$$
$8x=0$ or $x-3=0$

$x=0$ or $x=3$

The solution set is {0, 3}.

120.
$$x(x-11)=26$$
$$x^2-11x=26$$
$$x^2-11x-26=0$$
$$(x+2)(x-13)=0$$
$x+2=0$ or $x-13=0$

$x=-2$ or $x=13$

The solution set is {–2, 13}.

Chapter 5 Test

1. $(4x^3-3x-4)-(9x^3+8x+5)$
$$=4x^3-3x-4-9x^3-8x-5$$
$$=4x^3-9x^3-3x-8x-4-5$$
$$=-5x^3-11x-9$$

2. $-3xy(4x+y)=-3xy\cdot 4x-3xy\cdot y$
$$=-12x^2y-3xy^2$$

3. $(3x+4)(4x-7)$
$$=3x\cdot 4x+3x\cdot(-7)+4\cdot 4x+4\cdot(-7)$$
$$=12x^2-21x+16x-28$$
$$=12x^2-5x-28$$

4. $(5a-2b)(5a+2b)=(5a)^2-(2b)^2$
$$=25a^2-4b^2$$

5. $(6m+n)^2=(6m)^2+2\cdot 6m\cdot n+n^2$
$$=36m^2+12mn+n^2$$

6. $(2x-1)(x^2-6x+4)$
$$=2x(x^2-6x+4)-1(x^2-6x+4)$$
$$=2x^3-12x^2+8x-x^2+6x-4$$
$$=2x^3-13x^2+14x-4$$

7. $\dfrac{4x^2y+9x+z}{3xz}=\dfrac{4x^2y}{3xz}+\dfrac{9x}{3xz}+\dfrac{z}{3xz}$
$$=\frac{4xy}{3z}+\frac{3}{z}+\frac{1}{3x}$$

8.
$$\begin{array}{r} 2x^4 +2x-2 \\ 2x-1\overline{)\,4x^5-2x^4+0x^3+4x^2-6x+3} \\ \underline{4x^5-2x^4} \\ 4x^2-6x \\ \underline{4x^2-2x} \\ -4x+3 \\ \underline{-4x+2} \\ 1 \end{array}$$

$(4x^5-2x^4+4x^2-6x+3)\div(2x-1)$
$$=2x^4+2x-2+\frac{1}{2x-1}$$

9. $x+3$ is $x-c$ where $c=-3$.

–3	4	–3	2	–1	–1
		–12	45	–141	426
	4	–15	47	–142	425

$$\frac{4x^4-3x^3+2x^2-x-1}{x+3}$$
$$=4x^3-15x^2+47x-142+\frac{425}{x+3}$$

10. $16x^3y-12x^2y^4=4x^2y\cdot 4x-4x^2y\cdot 3y^3$
$$=4x^2y(4x-3y^3)$$

11. $x^2 - 13x - 30$

$-15 \cdot 2 = -30$ and $-15 + 2 = -13$

$x^2 - 13x - 30 = (x - 15)(x + 2)$

12. $$\begin{aligned} 4y^2 + 20y + 25 &= (2y)^2 + 2 \cdot 2y \cdot 5 + 5^2 \\ &= (2y + 5)^2 \end{aligned}$$

13. $$\begin{aligned} 6x^2 - 15x - 9 &= 3(2x^2 - 5x - 3) \\ &= 3(2x + 1)(x - 3) \end{aligned}$$

14. $4x^2 - 25 = (2x)^2 - 5^2 = (2x + 5)(2x - 5)$

15. $$\begin{aligned} x^3 + 64 &= x^3 + 4^3 \\ &= (x + 4)(x^2 - x \cdot 4 + 4^2) \\ &= (x + 4)(x^2 - 4x + 16) \end{aligned}$$

16. $$\begin{aligned} 3x^2y - 27y^3 &= 3y(x^2 - 9y^2) \\ &= 3y[x^2 - (3y)^2] \\ &= 3y(x + 3y)(x - 3y) \end{aligned}$$

17. $$\begin{aligned} 16y^3 - 2 &= 2(8y^3 - 1) \\ &= 2[(2y)^3 - 1^3] \\ &= 2(2y - 1)[(2y)^2 + 2y \cdot 1 + 1^2] \\ &= 2(2y - 1)(4y^2 + 2y + 1) \end{aligned}$$

18. $$\begin{aligned} x^2y - 9y - 3x^2 + 27 &= (x^2y - 9y) + (-3x^2 + 27) \\ &= y(x^2 - 9) - 3(x^2 - 9) \\ &= (x^2 - 9)(y - 3) \\ &= (x + 3)(x - 3)(y - 3) \end{aligned}$$

19.
$$3n(7n-20)=96$$
$$21n^2-60n=96$$
$$21n^2-60n-96=0$$
$$3(7n^2-20n-32)=0$$
$$3(n-4)(7n+8)=0$$
$$n-4=0 \quad \text{or} \quad 7n+8=0$$
$$n=4 \quad \text{or} \quad n=-\frac{8}{7}$$
The solution set is $\left\{4, -\frac{8}{7}\right\}$.

20. $(x+2)(x-2)=5(x+4)$
$$x^2-4=5x+20$$
$$x^2-5x-24=0$$
$$(x+3)(x-8)=0$$
$$x+3=0 \quad \text{or} \quad x-8=0$$
$$x=-3 \quad \text{or} \quad x=8$$
The solution set is $\{-3, 8\}$.

21.
$$2x^3+5x^2=8x+20$$
$$2x^3+5x^2-8x-20=0$$
$$(2x^3+5x^2)+(-8x-20)=0$$
$$x^2(2x+5)+(-4)(2x+5)=0$$
$$(2x+5)(x^2-4)=0$$
$$(2x+5)(x^2-2^2)=0$$
$$(2x+5)(x+2)(x-2)=0$$
$$2x+5=0 \quad \text{or} \quad x+2=0 \quad \text{or} \quad x-2=0$$
$$2x=-5$$
$$x=-\frac{5}{2} \quad \text{or} \quad x=-2 \quad \text{or} \quad x=2$$
The solution set is $\left\{-\frac{5}{2}, -2, 2\right\}$.

22. $x^2-(2y)^2=(x+2y)(x-2y)$

23. a. $h(t)=-16t^2+96t+880$
$$h(1)=-16(1)^2+96(1)+880=960$$
The height is 960 feet when $t = 1$ second.

b. $h(5.1) = -16(5.1)^2 + 96(5.1) + 880$
$= 953.44$
The height is 953.44 feet when $t = 5.1$ seconds.

c. $-16t^2 + 96t + 880 = -16(t^2 - 6t - 55)$
$= -16(t - 11)(t + 5)$

d. Let $h(t) = 0$ and solve for t.
$h(t) = -16t^2 + 96t + 880$
$0 = -16(t - 11)(t + 5)$
$t - 11 = 0$ or $t + 5 = 0$
$t = 11$ or $t = -5$
Since time cannot be negative, discard $t = -5$. The pebble hits the ground in 11 seconds.

Cumulative Review Chapters 1–5

1. $-2(x + 3) + 7 = -2(x) + (-2)(3) + 7$
$= -2x - 6 + 7$
$= -2x + 1$

2. $6z - 9z - \frac{1}{2} + z - \frac{3}{8}$
$= (6z - 9z + z) + \left(-\frac{1}{2} - \frac{3}{8}\right)$
$= -2z + \left(-\frac{4}{8} - \frac{3}{8}\right)$
$= -2z - \frac{7}{8}$

3. $7x + 3 - 5(x - 4) = 7x + 3 - 5x + 20 = 2x + 23$

4. $7 + 2(9x - 1) - (1.7x + 5.8)$
$= 7 + 18x - 2 - 1.7x - 5.8$
$= 18x - 1.7x + 7 - 2 - 5.8$
$= 16.3x - 0.8$

5. $\frac{2000 \times 0.000021}{700} = \frac{(2 \times 10^3)(2.1 \times 10^{-5})}{7 \times 10^2}$
$= \frac{2(2.1)}{7} \cdot \frac{10^3 \cdot 10^{-5}}{10^2}$
$= 0.6 \times 10^{-4}$
$= (6 \times 10^{-1}) \times 10^{-4}$
$= 6 \times 10^{-5}$

6. a. $(4a^{-1}b^0)^{-3} = \left(\frac{4}{a}\right)^{-3} = \frac{a^3}{4^3} = \frac{a^3}{64}$

b. $\left(\frac{a^{-6}}{a^{-8}}\right)^{-2} = (a^2)^{-2} = a^{-4} = \frac{1}{a^4}$

c. $\left(\frac{2}{3}\right)^{-3} = \left(\frac{3}{2}\right)^3 = \frac{3^3}{2^3} = \frac{27}{8}$

d. $\frac{3^{-2}a^{-2}b^{12}}{a^4b^{-5}} = \frac{b^{17}}{3^2a^6} = \frac{b^{17}}{9a^6}$

7. $3(x - 3) = 5x - 9$
$3x - 9 = 5x - 9$
$3x - 9 - 5x = 5x - 9 - 5x$
$-2x - 9 = -9$
$-2x - 9 + 9 = -9 + 9$
$-2x = 0$
$\frac{-2x}{-2} = \frac{0}{-2}$
$x = 0$
The solution set is $\{0\}$.

8. $0.3y + 2.4 = 0.1y + 4$
$0.3y - 0.1y = 4 - 2.4$
$0.2y = 1.6$
$y = 8$
The solution set is $\{8\}$.

9.
$$\begin{aligned} 3y-2x&=7\\ 3y-2x+2x&=7+2x\\ 3y&=7+2x\\ \frac{3y}{3}&=\frac{7+2x}{3}\\ y&=\frac{7+2x}{3} \text{ or } y=\frac{7}{3}+\frac{2x}{3}\end{aligned}$$

10. 2 walls × 2 coats × 14 ft × 8 ft = 448
2 walls × 2 coats × 18 ft × 8 ft = 576
$$\frac{448+576}{400}=\frac{1024}{400}=2.56$$
Buy 3 gallons of paint.

11.
$$\begin{aligned} -(x-3)+2&\le 3(2x-5)+x\\ -x+3+2&\le 6x-15+x\\ 5-x&\le 7x-15\\ 5-x+x&\le 7x-15+x\\ 5&\le 8x-15\\ 5+15&\le 8x-15+15\\ 20&\le 8x\\ \frac{20}{8}&\le\frac{8x}{8}\\ \frac{5}{2}&\le x\end{aligned}$$
The solution set is $\left[\frac{5}{2}, \infty\right)$.

12.
$$\begin{aligned} x+2&<\frac{1}{4}(x-7)\\ x+2&<\frac{1}{4}x-\frac{7}{4}\\ x-\frac{1}{4}x&<-\frac{7}{4}-2\\ \frac{3}{4}x&<-\frac{15}{4}\\ x&<-5\end{aligned}$$
The solution set is $(-\infty, -5)$.

13.
$$\begin{aligned} -1\le \frac{2}{3}x+5&<2\\ 3(-1)\le 3\left(\frac{2}{3}x+5\right)&<3(2)\\ -3\le 2x+15&<6\\ -3-15\le 2x+15-15&<6-15\\ -18\le 2x&<-9\\ -\frac{18}{2}\le\frac{2x}{2}&<-\frac{9}{2}\\ -9\le x&<-\frac{9}{2}\end{aligned}$$
The solution set is $\left[-9, -\frac{9}{2}\right)$.

14.
$$\begin{aligned} -\frac{1}{3}<\frac{3x+1}{6}&\le\frac{1}{3}\\ 6\left(-\frac{1}{3}\right)<6\left(\frac{3x+1}{6}\right)&\le 6\left(\frac{1}{3}\right)\\ -2<3x+1&\le 2\\ -2-1<3x+1-1&\le 2-1\\ -3<3x&\le 1\\ -\frac{3}{3}<\frac{3x}{3}&\le\frac{1}{3}\\ -1<x&\le\frac{1}{3}\end{aligned}$$
The solution set is $\left(-1, \frac{1}{3}\right]$.

15. $|3x+2|=|5x-8|$
$$\begin{aligned} 3x+2&=5x-8 &\text{ or }&& 3x+2&=-(5x-8)\\ -2x+2&=-8 &\text{ or }&& 3x+2&=-5x+8\\ -2x&=-10 &\text{ or }&& 8x+2&=8\\ x&=5 &\text{ or }&& 8x&=6\\ &&&& x&=\frac{3}{4}\end{aligned}$$
The solution set is $\left\{\frac{3}{4}, 5\right\}$.

16.
$$\begin{aligned} |5x-1|+9&>5\\ |5x-1|&>-4\end{aligned}$$
The absolute value of any number is nonnegative. The solution set is $(-\infty, \infty)$.

17. The graph of $x = -5$ is a vertical line. The slope is undefined.

18. $f(x) = -2x - 3$
$y = mx + b$
$m = -2$
The slope is -2.

19. $y = \frac{1}{4}x - 3$
$y = mx + b$
The slope is $m = \frac{1}{4}$ and the y-intercept is $(0, -3)$.

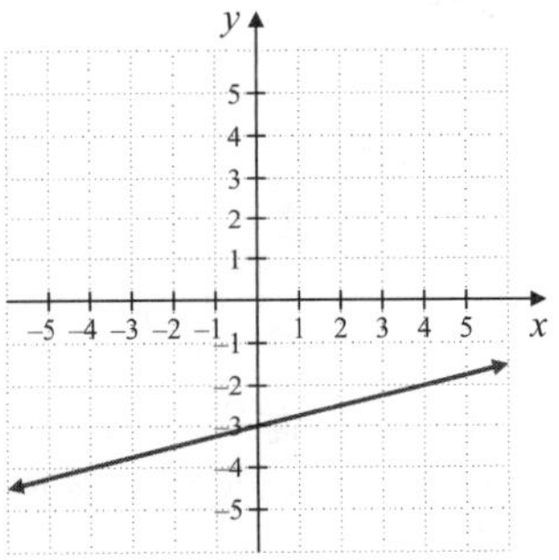

20. $y = 3x$
$y = mx + b$
The slope is $m = 3$ and the y-intercept is $(0, 0)$.

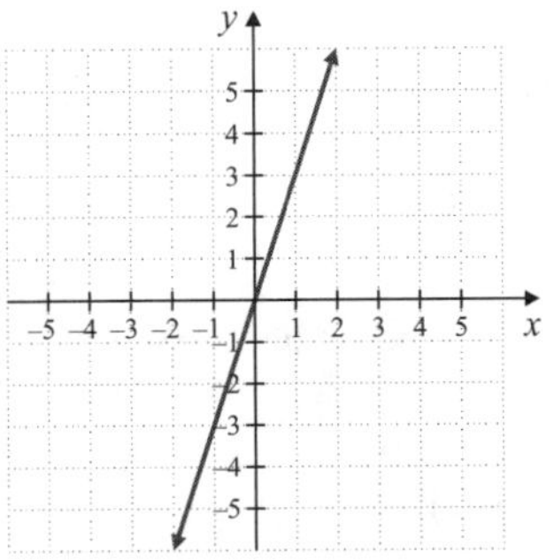

21. Find the slope of $3x + 5y = 4$.

$$3x + 5y = 4$$
$$5y = -3x + 4$$
$$y = -\frac{3}{5}x + 4$$

The slope is $-\frac{3}{5}$. A line perpendicular will have slope $\frac{5}{3}$.

$$y - 1 = \frac{5}{3}[x - (-2)]$$
$$y - 1 = \frac{5}{3}(x + 2)$$
$$y - 1 = \frac{5}{3}x + \frac{10}{3}$$
$$y = \frac{5}{3}x + \frac{13}{3}$$

22. Vertical lines are in the form $x = c$. Since the line contains the point $(-3, 2)$. The equation is $x = -3$.

23. $\{(2, 3), (2, 4), (0, -1), (3, -1)\}$
The domain is the set of all first coordinates of the ordered pairs $\{2, 0, 3\}$. The range is the set of all second coordinates $\{3, 4, -1\}$.

24. Horizontal lines have slopes of 0, and are in the form $y = c$. Since the line contains the point $(-2, 3)$, the equation is $y = 3$.

25. $g(x) = 3x - 2$
$g(0) = 3 \cdot 0 - 2 = 0 - 2 = -2$

26. $f(x) = 3x^2 + 2x + 3$

$$f(-3) = 3(-3)^2 + 2(-3) + 3$$
$$= 3(9) - 6 + 3$$
$$= 27 - 6 + 3$$
$$= 24$$

27. $f(x) = 7x^2 - 3x + 1$

$$f(1) = 7(1)^2 - 3(1) + 1 = 7 - 3 + 1 = 5$$

28. a.

$$(3y^6)(4.7y^2) = (3)(4.7)(y^6 \cdot y^2)$$
$$= 14.1y^{6+2}$$
$$= 14.1y^8$$

b. $(6a^3b^2)(-a^2bc^4)$
$= 6 \cdot (-1)(a^3 \cdot a^2)(b^2b)c^4$
$= -6a^{3+2}b^{2+1}c^4$
$= -6a^5b^3c^4$

29. $2x - y < 6$
First graph the boundary line $2x - y = 6$ as a dashed line. Test the point (0, 0).
$2(0) - 0 < 6?$
$0 < 6$ True
Shade the half-plane containing (0, 0).

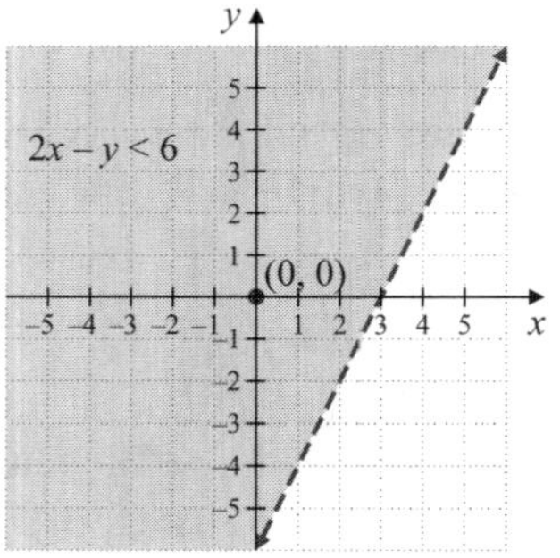

30. Let x = amount of 20% solution and y = amount of 60% solution.

amount	strength	total amount of acid
x	20%	$0.2x$
y	60%	$0.6x$
50	30%	0.3(50) = 15

$$\begin{cases} x + y = 50 \\ 0.2x + 0.6y = 15 \end{cases}$$

Multiply the first equation by −0.2 and add to the second equation.
$-0.2x - 0.2y = -10$
$0.2x + 0.6y = 15$
$0.4y = 5$
$y = 12.5$
Substitute $y = 12.5$ in the first equation.
$x + 12.5 = 50$
$x = 37.5$
Mix 37.5 ounces of 20% solution with 12.5 ounces of 60% solution.

31. $\begin{cases} 2x + 4y = -6 \\ x = 2y - 5 \end{cases}$
Substitute $2y - 5$ for x in the first equation.
$2x + 4y = -6$
$2(2y - 5) + 4y = -6$
$4y - 10 + 4y = -6$
$8y - 10 = -6$
$8y = 4$
$y = \frac{1}{2}$
Replace y with $\frac{1}{2}$ in the second equation.
$x = 2\left(\frac{1}{2}\right) - 5 = 1 - 5 = -4$
The solution is $\left(-4, \frac{1}{2}\right)$.

32. $\begin{cases} 4y = 8 \\ x + y = 7 \end{cases}$
Solve the first equation for y, $y = 2$.
Substitute 2 for y in the second equation.
$x + 2 = 7$
$x = 5$
The solution is (5, 2).

33. $\begin{cases} 3x - y + z = -15 & (1) \\ x + 2y - z = 1 & (2) \\ 2x + 3y - 2z = 0 & (3) \end{cases}$
Add equations (1) and (2).
$3x - y + z = -15$
$x + 2y - z = 1$
$4x + y = -14$ (4)
Multiply equation (1) by 2 and add to equation (3).
$6x - 2y + 2z = -30$
$2x + 3y - 2z = 0$
$8x + y = -30$ (5)
Multiply equation (4) by −1 and add to equation (5).
$-4x - y = 14$
$8x + y = -30$
$4x = -16$
$x = -4$

Replace x with –4 in equation (4).

$$4(-4)+y=-14$$
$$-16+y=-14$$
$$y=2$$

Replace x with –4 and y with 2 in equation (2).

$$-4+2(2)-z=1$$
$$-4+4-z=1$$
$$-z=1$$
$$z=-1$$

The solution is (–4, 2, –1).

34. $\begin{cases} x+y+z=0 & (1) \\ 2x-3y+z=5 & (2) \\ 2x+y+2z=2 & (3) \end{cases}$

Multiply equation (1) by –2 and add to equations (2) and (3).

$$\begin{array}{r} -2x-2y-2z=0 \\ 2x-3y+z=5 \\ \hline -5y-z=5 \end{array} \quad (4)$$

$$\begin{array}{r} -2x-2y-2z=0 \\ 2x+y+2z=2 \\ \hline -y \quad =2 \end{array}$$
$$y=-2$$

Replace y with –2 in equation (4).

$$-5y-z=5$$
$$-5(-2)-z=5$$
$$10-z=5$$
$$-z=-5$$
$$z=5$$

Replace y with –2 and z with 5 in equation (1).

$$x-2+5=0$$
$$x+3=0$$
$$x=-3$$

The solution is (–3, –2, 5).

35. $\begin{cases} x+3y=5 \\ 2x-y=-4 \end{cases}$

$$\left[\begin{array}{rr|r} 1 & 3 & 5 \\ 2 & -1 & -4 \end{array}\right]$$

Multiply row 1 by –2 and add to row 2.

$$\left[\begin{array}{rr|r} 1 & 3 & 5 \\ 0 & -7 & -14 \end{array}\right]$$

Divide row 2 by –7.

$$\left[\begin{array}{rr|r} 1 & 3 & 5 \\ 0 & 1 & 2 \end{array}\right]$$

This corresponds to the system $\begin{cases} x+3y=5 \\ y=2 \end{cases}$.

Let $y = 2$ in the first equation.

$$x+3y=5$$
$$x+3(2)=5$$
$$x+6=5$$
$$x=-1$$

The solution is (–1, 2).

36. $\begin{cases} x+y-\frac{3}{2}z=\frac{1}{2} & (1) \\ -y-2z=14 & (2) \\ x-\frac{2}{3}y=-\frac{1}{3} & (3) \end{cases}$

Multiply equation (1) by –4 and equation (2) by 3 and add the results.

$$\begin{array}{r} -4x-4y+6z=-2 \\ -3y-6z=42 \\ \hline -4x-7y \quad =40 \end{array} \quad (4)$$

Multiply equation (3) by –21 and equation (4) by 2 and add the results.

$$\begin{array}{r} -21x+14y=7 \\ -8x-14y=80 \\ \hline -29x \quad =87 \end{array}$$
$$x=-3$$

Let $x = -3$ in equation (3).

$$-3-\frac{2}{3}y=-\frac{1}{3}$$
$$-\frac{2}{3}y=\frac{8}{3}$$
$$y=-4$$

Let $y = -4$ in equation (2).

$$-(-4)-2z=14$$
$$4-2z=14$$
$$-2z=10$$
$$z=-5$$

The solution is (–3, –4, –5).

37. $\begin{cases} x - y < 2 \\ x + 2y > -1 \\ y < 2 \end{cases}$

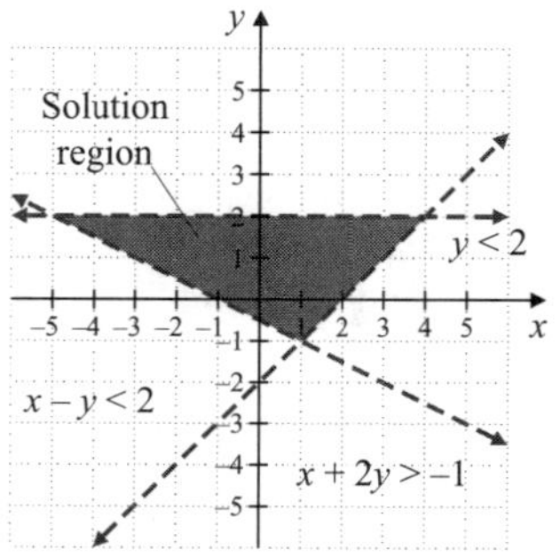

38. a. $8,250,000 = 8.25 \times 10^6$

b. $0.0000346 = 3.46 \times 10^{-5}$

39. $3xy + x^2y^2 - 5x^2 - 6.7$

The term with the greatest degree is x^2y^2 which has degree 2 + 2 = 4. Therefore, the degree of the polynomial is 4.

40. $(3x^2 - 2x) - (5x^2 + 3x)$
$= 3x^2 - 2x - 5x^2 - 3x$
$= 3x^2 - 5x^2 - 2x - 3x$
$= -2x^2 - 5x$

41. $2x(5x - 4) = 2x \cdot 5x - 2x \cdot 4 = 10x^2 - 8x$

42. $(2x - 1)(5x^2 - 6x + 2)$
$= 2x(5x^2 - 6x + 2) - 1(5x^2 - 6x + 2)$
$= 10x^3 - 12x^2 + 4x - 5x^2 + 6x - 2$
$= 10x^3 - 17x^2 + 10x - 2$

43. $-xy(7x^2y + 3xy - 11)$
$= -xy \cdot 7x^2y - xy \cdot 3xy - xy \cdot (-11)$
$= -7x^3y^2 - 3x^2y^2 + 11xy$

44. $\left(7x - \frac{1}{2}\right)^2 = (7x)^2 - 2 \cdot (7x) \cdot \left(\frac{1}{2}\right) + \left(\frac{1}{2}\right)^2$
$= 49x^2 - 7x + \frac{1}{4}$

45. $\frac{10x^3 - 5x^2 + 20x}{5x} = \frac{10x^3}{5x} - \frac{5x^2}{5x} + \frac{20x}{5x}$
$= 2x^2 - x + 4$

46. $12x^3y - 3xy^3 = 3xy \cdot 4x^2 - 3xy \cdot y^2$
$= 3xy(4x^2 - y^2)$
$= 3xy[(2x)^2 - y^2]$
$= 3xy(2x + y)(2x - y)$

47. $x^2 + 10x + 16$
$2 \cdot 8 = 16$ and $2 + 8 = 10$
$x^2 + 10x + 16 = (x + 2)(x + 8)$

48. $5a^2 + 14a - 3$
Factors of $5a^2$: $5a \cdot a$
Factors of -3: $-3 \cdot 1$, $3 \cdot (-1)$
$5a^2 + 14a - 3 = (5a - 1)(a + 3)$

49. $2x^2 + 9x - 5 = 0$
$(2x - 1)(x + 5) = 0$
$2x - 1 = 0$ or $x + 5 = 0$
$2x = 1$
$x = \frac{1}{2}$ or $x = -5$

The solution set is $\left\{-5, \frac{1}{2}\right\}$.

50. $3x^2 - 10x - 8 = 0$
$(x - 4)(3x + 2) = 0$
$x - 4 = 0$ or $3x + 2 = 0$
$3x = -2$
$x = 4$ or $x = -\frac{2}{3}$

The solution set is $\left\{4, -\frac{2}{3}\right\}$.

Chapter 6

Section 6.1 Practice Problems

1. a. $x-6=0$
$x=6$

The domain of $f(x)=\dfrac{x^2+1}{x-6}$ is $\{x|x \text{ is a real number and } x \neq 6\}$.

b. $x^2-3x-10=0$
$(x-5)(x+2)=0$
$x-5=0$ or $x+2=0$
$x=5$ or $x=-2$

The domain of $g(x)=\dfrac{5x+4}{x^2-3x-10}$ is $\{x|x \text{ is a real number and } x \neq 5 \text{ and } x \neq -2\}$.

c. 4 is never 0 so the domain of $h(x)=\dfrac{x^2-9}{4}$ is $\{x|x \text{ is a real number}\}$.

2. $\dfrac{3y^3}{6y^4-3y^3}=\dfrac{3y^3}{3y^3(2y-1)}=\dfrac{1}{2y-1}$

3. $\dfrac{5x^2+13x+6}{6x^2+7x-10}=\dfrac{(5x+3)(x+2)}{(6x-5)(x+2)}=\dfrac{5x+3}{6x-5}$

4. $\dfrac{5+x}{x+5}=\dfrac{x+5}{x+5}=1$

5. $\dfrac{5-x}{x-5}=\dfrac{-1(-5+x)}{x-5}=\dfrac{-1(x-5)}{x-5}=-1$

6. $\dfrac{3-3x^2}{x^2+x-2}=\dfrac{3(1-x^2)}{(x+2)(x-1)}$
$=\dfrac{3(1-x)(1+x)}{(x+2)(x-1)}$
$=\dfrac{(3)(-1)(-1+x)(1+x)}{(x+2)(x-1)}$
$=\dfrac{-3(x-1)(x+1)}{(x+2)(x-1)}$
$=-\dfrac{3(x+1)}{x+2}$

7. $\dfrac{x^3+27}{x+3}=\dfrac{(x+3)(x^2-3x+9)}{x+3}=x^2-3x+9$

8. $\dfrac{3x^2+6}{x^3-3x^2+2x-6}=\dfrac{3(x^2+2)}{(x^3-3x^2)+(2x-6)}$
$=\dfrac{3(x^2+2)}{x^2(x-3)+2(x-3)}$
$=\dfrac{3(x^2+2)}{(x-3)(x^2+2)}$
$=\dfrac{3}{x-3}$

9. $\dfrac{2x-3}{5x}\cdot\dfrac{5x+5}{2x^2-x-3}=\dfrac{2x-3}{5x}\cdot\dfrac{5(x+1)}{(2x-3)(x+1)}$
$=\dfrac{1}{x}$

10. $\dfrac{x^3+27}{-2x-6}\cdot\dfrac{4x^3}{x^2-3x+9}$
$=\dfrac{(x+3)(x^2-3x+9)}{-2(x+3)}\cdot\dfrac{4x^3}{x^2-3x-9}$
$=-2x^3$

11. $\frac{12y^3}{5y^2-5} \div \frac{6}{1-y} = \frac{12y^3}{5y^2-5} \cdot \frac{1-y}{6}$
$= \frac{12y^3(1-y)}{5(y^2-1)\cdot 6}$
$= \frac{12y^3(1-y)}{30(y-1)(y+1)}$
$= -\frac{2y^3}{5(y+1)}$

12. $\frac{8z^2+14z+3}{20z^2+z-1} \div \frac{2z^2+7z+6}{35z+3z-2}$
$= \frac{8z^2+14z+3}{20z^2+z-1} \cdot \frac{35z+3z-2}{2z^2+7z+6}$
$= \frac{(2z+3)(4z+1)\cdot(5z-1)(7z+2)}{(4z+1)(5z-1)\cdot(2z+3)(z+2)}$
$= \frac{7z+2}{z+2}$

13. $\frac{(x+3)^2}{x^2-9} \cdot \frac{2x-6}{5x} \div \frac{x^2+7x+12}{x}$
$= \frac{(x+3)^2}{x^2-9} \cdot \frac{2x-6}{5x} \cdot \frac{x}{x^2+7x+12}$
$= \frac{(x+3)(x+3)\cdot 2(x-3)\cdot x}{(x+3)(x-3)\cdot 5x\cdot(x+3)(x+4)}$
$= \frac{2}{5(x+4)}$

14. a. $C(100) = \frac{0.8(100)+5000}{100}$
$= \frac{5080}{100}$
$= 50.80$
The cost per book for printing 100 books is \$50.80.

b. $C(1000) = \frac{0.8(1000)+5000}{1000}$
$= \frac{5800}{1000}$
$= 5.80$
The cost per book for printing 1000 books is \$5.80.

Calculator Explorations

1. $x-6=0$
$x=6$

The domain of $f(x) = \frac{5x}{x-6}$ is $\{x|x$ is a real number and $x \neq 6\}$.

2. $x+4=0$
$x=-4$

The domain of $f(x) = \frac{x}{x+4}$ is $\{x|x$ is a real number and $x \neq -4\}$.

3. $x^2-4=0$
$(x-2)(x+2)=0$
$x-2=0$ or $x+2=0$
$x=2$ or $x=-2$

The domain of $f(x) = \frac{x+1}{x^2-4}$ is $\{x|x$ is a real number and $x \neq 2, x \neq -2\}$.

4. $x^2-9=0$
$(x-3)(x+3)=0$
$x-3=0$ or $x+3=0$
$x=3$ or $x=-3$

The domain of $g(x) = \frac{5x}{x^2-9}$ is $\{x|x$ is a real number and $x \neq 3, x \neq -3\}$.

5. $2x^2+7x-4=0$
$(2x-1)(x+4)=0$
$2x-1=0$ or $x+4=0$
$2x=1$ or $x=-4$
$x=\frac{1}{2}$ or $x=-4$

The domain of $h(x)=\frac{x^2}{2x^2+7x-4}$ is $\left\{x \middle| x \text{ is a real number and } x \neq -4,\ x \neq \frac{1}{2}\right\}$.

6. $4x^2-19x-5=0$
$(4x+1)(x-5)=0$
$4x+1=0$ or $x-5=0$
$4x=-1$ or $x=5$
$x=-\frac{1}{4}$ or $x=5$

The domain of $f(x)=\frac{3x+2}{4x^2-19x-5}$ is $\left\{x \middle| x \text{ is a real number and } x \neq 5,\ x \neq -\frac{1}{4}\right\}$.

7. 5 is never 0 so the domain of $g(x)=\frac{x^2+x+1}{5}$ is $\{x|x \text{ is a real number}\}$.

8. 2 is never 0 so the domain of $h(x)=\frac{x^2+25}{2}$ is $\{x|x \text{ is a real number}\}$.

Mental Math

1. $\frac{x}{5}\cdot\frac{y}{2}=\frac{xy}{10}$

2. $\frac{y}{6}\cdot\frac{z}{5}=\frac{yz}{30}$

3. $\frac{2}{x}\cdot\frac{y}{3}=\frac{2y}{3x}$

4. $\frac{a}{5}\cdot\frac{7}{b}=\frac{7a}{5b}$

5. $\frac{m}{6}\cdot\frac{m}{6}=\frac{m^2}{36}$

6. $\frac{9}{x}\cdot\frac{8}{x}=\frac{72}{x^2}$

Exercise Set 6.1

1. 4 is never 0, so the domain of $f(x)=\frac{5x-7}{4}$ is $\{x|x \text{ is a real number}\}$.

3. $2t=0$
$t=0$

The domain of $s(t)=\frac{t^2+1}{2t}$ is $\{t|t \text{ is a real number and } t \neq 0\}$.

5. $7-x=0$
$7=x$

The domain of $f(x)=\frac{3x}{7-x}$ is $\{x|x \text{ is a real number and } x \neq 7\}$

7. $3x-1=0$
$3x=1$
$x=\frac{1}{3}$

The domain of $f(x)=\frac{x}{3x-1}$ is $\left\{x \middle| x \text{ is a real number and } x \neq \frac{1}{3}\right\}$.

9. $x^3+x^2-2x=0$
$x(x^2+x-2)=0$
$x(x+2)(x-1)=0$
$x=0$ or $x+2=0$ or $x-1=0$
$x=0$ or $x=-2$ or $x=1$

The domain of $R(x)=\dfrac{3+2x}{x^3+x^2-2x}$ is $\{x|x \text{ is a real number and } x\neq -2, x\neq 0, x\neq 1\}$.

11. $x^2-4=0$
$(x+2)(x-2)=0$
$x+2=0$ or $x-2=0$
$x=-2$ or $x=2$

The domain of $C(x)=\dfrac{x+3}{x^2-4}$ is $\{x|x \text{ is a real number and } x\neq 2, x\neq -2\}$.

13. $\dfrac{8x-16x^2}{8x}=\dfrac{8x(1-2x)}{8x}=1-2x$

15. $\dfrac{x^2-9}{3+x}=\dfrac{(x+3)(x-3)}{(3+x)}=\dfrac{(x+3)(x-3)}{(x+3)}=x-3$

17. $\dfrac{9y-18}{7y-14}=\dfrac{9(y-2)}{7(y-2)}=\dfrac{9}{7}$

19. $\dfrac{x^2+6x-40}{x+10}=\dfrac{(x+10)(x-4)}{(x+10)}=x-4$

21. $\dfrac{x-9}{9-x}=\dfrac{-1(9-x)}{9-x}=-1$

23. $\dfrac{x^2-49}{7-x}=\dfrac{(x+7)(x-7)}{7-x}=\dfrac{(x+7)(x-7)}{-1(-7+x)}=\dfrac{(x+7)(x-7)}{-1(x-7)}=-(x+7)$

25. $\dfrac{2x^2-7x-4}{x^2-5x+4}=\dfrac{(2x+1)(x-4)}{(x-1)(x-4)}=\dfrac{2x+1}{x-1}$

27. $\dfrac{x^3-125}{2x-10}=\dfrac{(x-5)(x^2+5x+25)}{2(x-5)}=\dfrac{x^2+5x+25}{2}$

29. $\dfrac{3x^2-5x-2}{6x^3+2x^2+3x+1}=\dfrac{(3x+1)(x-2)}{2x^2(3x+1)+1(3x+1)}=\dfrac{(3x+1)(x-2)}{(3x+1)(2x^2+1)}=\dfrac{x-2}{2x^2+1}$

31. $\dfrac{9x^2-15x+25}{27x^3+125}=\dfrac{9x^2-15x+25}{(3x+5)(9x^2-15x+25)}=\dfrac{1}{3x+5}$

33. $\dfrac{2x-4}{15}\cdot\dfrac{6}{2-x}=\dfrac{2(x-2)}{3\cdot 5}\cdot\dfrac{2\cdot 3}{-(x-2)}=\dfrac{2\cdot 2}{5(-1)}=-\dfrac{4}{5}$

35. $\dfrac{18a-12a^2}{4a^2+4a+1}\cdot\dfrac{4a^2+8a+3}{4a^2-9}$
$=\dfrac{6a(3-2a)}{(2a+1)(2a+1)}\cdot\dfrac{(2a+3)(2a+1)}{(2a+3)(2a-3)}$
$=\dfrac{-6a(2a-3)}{(2a+1)(2a+1)}\cdot\dfrac{(2a+3)(2a+1)}{(2a+3)(2a-3)}$
$=-\dfrac{6a}{2a+1}$

37. $\dfrac{9x+9}{4x+8}\cdot\dfrac{2x+4}{3x^2-3}=\dfrac{9(x+1)}{4(x+2)}\cdot\dfrac{2(x+2)}{3(x^2-1)}=\dfrac{3\cdot 3(x+1)}{2\cdot 2(x+2)}\cdot\dfrac{2(x+2)}{3(x+1)(x-1)}=\dfrac{3}{2(x-1)}$

39. $$\frac{2x^3-16}{6x^2+6x-36}\cdot\frac{9x+18}{3x^2+6x+12}$$
$$=\frac{2(x^3-8)}{6(x^2+x-6)}\cdot\frac{9(x+2)}{3(x^2+2x+4)}$$
$$=\frac{2(x-2)(x^2+2x+4)}{2\cdot 3(x-2)(x+3)}\cdot\frac{3\cdot 3(x+2)}{3(x^2+2x+4)}$$
$$=\frac{x+2}{x+3}$$

41. $$\frac{a^3+a^2b+a+b}{5a^3+5a}\cdot\frac{6a^2}{2a^2-2b^2}$$
$$=\frac{a^2(a+b)+1(a+b)}{5a(a^2+1)}\cdot\frac{6a^2}{2(a^2-b^2)}$$
$$=\frac{(a+b)(a^2+1)}{5a(a^2+1)}\cdot\frac{2\cdot 3\cdot a\cdot a}{2(a+b)(a-b)}$$
$$=\frac{3a}{5(a-b)}$$

43. $$\frac{x^2-6x-16}{2x^2-128}\cdot\frac{x^2+16x+64}{3x^2+30x+48}$$
$$=\frac{(x-8)(x+2)}{2(x^2-64)}\cdot\frac{(x+8)(x+8)}{3(x^2+10x+16)}$$
$$=\frac{(x-8)(x+2)}{2(x+8)(x-8)}\cdot\frac{(x+8)(x+8)}{3(x+2)(x+8)}$$
$$=\frac{1}{2\cdot 3}$$
$$=\frac{1}{6}$$

45. $$\frac{2x}{5}\div\frac{6x+12}{5x+10}=\frac{2x}{5}\cdot\frac{5x+10}{6x+12}$$
$$=\frac{2x}{5}\cdot\frac{5(x+2)}{6(x+2)}$$
$$=\frac{2x}{5}\cdot\frac{5(x+2)}{2\cdot 3(x+2)}$$
$$=\frac{x}{3}$$

47. $$\frac{a+b}{ab}\div\frac{a^2-b^2}{4a^3b}=\frac{a+b}{ab}\cdot\frac{4a^3b}{a^2-b^2}$$
$$=\frac{a+b}{ab}\cdot\frac{4a^3b}{(a+b)(a-b)}$$
$$=\frac{4a^2}{a-b}$$

49. $$\frac{x^2-6x+9}{x^2-x-6}\div\frac{x^2-9}{4}$$
$$=\frac{x^2-6x+9}{x^2-x-6}\cdot\frac{4}{x^2-9}$$
$$=\frac{(x-3)^2}{(x-3)(x+2)}\cdot\frac{4}{(x+3)(x-3)}$$
$$=\frac{4}{(x+2)(x+3)}$$

51. $$\frac{x^2-6x-16}{2x^2-128}\div\frac{x^2+10x+16}{x^2+16x+64}$$
$$=\frac{x^2-6x-16}{2x^2-128}\cdot\frac{x^2+16x+64}{x^2+10x+16}$$
$$=\frac{(x-8)(x+2)}{2(x^2-64)}\cdot\frac{(x+8)(x+8)}{(x+2)(x+8)}$$
$$=\frac{(x-8)(x+2)}{2(x-8)(x+8)}\cdot\frac{(x+8)(x+8)}{(x+2)(x+8)}$$
$$=\frac{1}{2}$$

53. $$\frac{3x-x^2}{x^3-27}\div\frac{x}{x^2+3x+9}$$
$$=\frac{3x-x^2}{x^3-27}\cdot\frac{x^2+3x+9}{x}$$
$$=\frac{x(3-x)}{(x-3)(x^2+3x+9)}\cdot\frac{x^2+3x-9}{x}$$
$$=\frac{-x(x-3)}{(x-3)(x^2+3x+9)}\cdot\frac{x^2+3x+9}{x}$$
$$=-1$$

55. $\frac{8b+24}{3a+6} \div \frac{ab-2b+3a-6}{a^2-4a+4}$
$= \frac{8b+24}{3a+6} \cdot \frac{a^2-4a+4}{ab-2b+3a-6}$
$= \frac{8(b+3)}{3(a+2)} \cdot \frac{(a-2)(a-2)}{(a-2)(b+3)}$
$= \frac{8(a-2)}{3(a+2)}$

57. $\frac{x^2-9}{4} \cdot \frac{x^2-x-6}{x^2-6x+9}$
$= \frac{(x+3)(x-3)}{4} \cdot \frac{(x+2)(x-3)}{(x-3)^2}$
$= \frac{(x+3)(x+2)}{4}$

59. $\frac{2x^2-4x-30}{5x^2-40x-75} \div \frac{x^2-8x+15}{x^2-6x+9}$
$= \frac{2x^2-4x-30}{5x^2-40x-75} \cdot \frac{x^2-6x+9}{x^2-8x+15}$
$= \frac{2(x^2-2x-15)}{5(x^2-8x-15)} \cdot \frac{(x-3)(x-3)}{(x-3)(x-5)}$
$= \frac{2(x-5)(x+3)}{5(x^2-8x-15)} \cdot \frac{(x-3)(x-3)}{(x-3)(x-5)}$
$= \frac{2(x+3)(x-3)}{5(x^2-8x-15)}$

61. $\frac{r^3+s^3}{r+s} = \frac{(r+s)(r^2-rs+s^2)}{r+s} = r^2-rs+s^2$

63. $\frac{4}{x} \div \frac{3xy}{x^2} \cdot \frac{6x^2}{x^4} = \frac{4}{x} \cdot \frac{x^2}{3xy} \cdot \frac{6x^2}{x^4} = \frac{8}{x^2y}$

65. $\frac{3x^2-5x-2}{y^2+y-2} \cdot \frac{y^2+4y-5}{12x^2+7x+1} \div \frac{5x^2-9x-2}{8x^2-2x-1}$
$= \frac{3x^2-5x-2}{y^2+y-2} \cdot \frac{y^2+4y-5}{12x^2+7x+1} \cdot \frac{8x^2-2x-1}{5x^2-9x-2}$
$= \frac{(3x+1)(x-2)(y+5)(y-1)(4x+1)(2x-1)}{(y+2)(y-1)(4x+1)(3x+1)(5x+1)(x-2)}$
$= \frac{(y+5)(2x-1)}{(y+2)(5x+1)}$

67. $f(x) = \frac{x+8}{2x-1}$
$f(2) = \frac{2+8}{2(2)-1} = \frac{10}{4-1} = \frac{10}{3}$
$f(0) = \frac{0+8}{2(0)-1} = \frac{8}{0-1} = \frac{8}{-1} = -8$
$f(-1) = \frac{-1+8}{2(-1)-1} = \frac{7}{-2-1} = \frac{7}{-3} = -\frac{7}{3}$

69. $g(x) = \frac{x^2+8}{x^3-25x}$
$g(3) = \frac{3^2+8}{3^3-25(3)} = \frac{9+8}{27-75} = \frac{17}{-48} = -\frac{17}{48}$
$g(-2) = \frac{(-2)^2+8}{(-2)^3-25(-2)} = \frac{4+8}{-8+50} = \frac{12}{42} = \frac{2}{7}$
$g(1) = \frac{1^2+8}{1^3-25(1)} = \frac{1+8}{1-25} = \frac{9}{-24} = -\frac{3}{8}$

71. $R(x) = \frac{1000x^2}{x^2+4}$

a. $R(1) = \frac{1000 \cdot 1^2}{1^2+4} = \frac{1000}{5} = 200$
The revenue at the end of the first year is \$200 million.

b. $R(2) = \frac{1000 \cdot 2^2}{2^2 + 4}$
$= \frac{1000 \cdot 4}{4+4}$
$= \frac{4000}{8}$
$= 500$
The revenue at the end of the second year is \$500 million.

c. The revenue during the second year is equal to the revenue at the end of the second year minus the revenue at the end of the first year.
$500 - 200 = 300$
The revenue during the second year is \$300 million.

d. $x^2 + 4 = 0$
$x^2 = -4$
$\{\ \}$
Thus there exists no values of x that would make the denominator equal to zero and the function undefined. The domain of $R(x)$ is $\{x | x \text{ is a real number}\}$.

73. $\frac{4}{5} + \frac{3}{5} = \frac{7}{5}$

75. $\frac{5}{28} - \frac{2}{21} = \frac{15}{84} - \frac{8}{84} = \frac{7}{84} = \frac{1}{12}$

77. $\frac{3}{8} + \frac{1}{2} - \frac{3}{16} = \frac{6}{16} + \frac{8}{16} - \frac{3}{16} = \frac{11}{16}$

79. a. $\frac{-x}{5-x} \neq \frac{x}{5-x}$

b. $\frac{-x}{-5+x} = \frac{x}{-(-5+x)} = \frac{x}{5-x}$

c. $\frac{x}{x-5} = \frac{x}{-5+x} \neq \frac{x}{5-x}$

d. $\frac{-x}{x-5} = \frac{x}{-(x-5)} = \frac{x}{-x+5} = \frac{x}{5-x}$

81. no; answers may vary

83. $A = l \cdot w$
$= \left(\frac{x+2}{x}\right)\left(\frac{5x}{x^2-4}\right)$
$= \frac{x+2}{x} \cdot \frac{5x}{(x+2)(x-2)}$
$= \frac{5}{x-2}$

The area is $\frac{5}{x-2}$ square meters

85. Since $A = b \cdot h$, $b = \frac{A}{h}$.

$b = \frac{\frac{x^2+x-2}{x^3}}{\frac{x^2}{x-1}}$

$b = \frac{(x+2)(x-1)}{x^3} \cdot \frac{(x-1)}{x^2}$

$b = \frac{(x+2)(x-1)^2}{x^5}$

The base is $\frac{(x+2)(x-1)^2}{x^5}$ feet

87. answers may vary

89. a. $\frac{x+5}{5+x} = \frac{x+5}{x+5} = 1$

b. $\frac{x-5}{5-x} = \frac{x-5}{-(x-5)} = -1$

c. $\frac{x+5}{x-5}$ neither

d. $\frac{-x-5}{x+5} = \frac{-(x+5)}{x+5} = -1$

e. $\frac{x-5}{-x+5}=\frac{x-5}{-(x-5)}=-1$

f. $\frac{-5+x}{x-5}=\frac{x-5}{x-5}=1$

91. $\frac{x^2-4}{x^2-7x+10}\cdot\frac{?}{2x^2+11x+14}=1$

$\frac{(x+2)(x-2)}{(x-2)(x-5)}\cdot\frac{?}{(2x+7)(x+2)}=1$

$\frac{?}{(x-5)(2x+7)}=1$

For the fraction to equal 1, the numerator needs to be $(x-5)(2x+7)$.

93. $\frac{p^x-4}{4-p^x}=\frac{p^x-4}{-(-4+p^x)}=\frac{p^x-4}{-(p^x-4)}=\frac{1}{-1}=-1$

95. $\frac{x^n+4}{x^{2n}-16}=\frac{x^n+4}{(x^n+4)(x^n-4)}=\frac{1}{x^n-4}$

Section 6.2 Practice Problems

1. $\frac{9}{11x^4}+\frac{y}{11x^4}=\frac{9+y}{11x^4}$

2. $\frac{x}{6}+\frac{7x}{6}=\frac{x+7x}{6}=\frac{8x}{6}=\frac{4x}{3}$

3. $\frac{x^2}{x+3}-\frac{9}{x+3}=\frac{x^2-9}{x+3}=\frac{(x+3)(x-3)}{x+3}=x-3$

4. $\frac{a}{5b^3}-\frac{a+2}{5b^3}=\frac{a-(a+2)}{5b^3}$

$=\frac{a-a-2}{5b^3}$

$=-\frac{2}{5b^3}$

5. a. $\frac{7}{20a^2b^3}, \frac{9}{15ab^4}$

We factor each denominator.

$20a^2b^3=4\cdot5\cdot a^2\cdot b^3$

$15ab^4=3\cdot5\cdot a\cdot b^4$

$\text{LCD}=3\cdot4\cdot5\cdot a^2\cdot b^4=60a^2b^4$

b. $\frac{6x}{x-2}, \frac{5}{x+2}$

The denominators $x-2$ and $x+2$ do not factor further.

LCD = $(x-2)(x+2)$

c. $\frac{x+4}{x^2-36}, \frac{x}{x^2+12x+36}, \frac{x^3}{3x^2+19x+6}$

We factor each denominator.

$x^2-36=(x+6)(x-6)$

$x^2+12x+36=(x+6)(x+6)$

$3x^2+19x+6=(3x+1)(x+6)$

$\text{LCD}=(x-6)(x+6)^2(3x+1)$

d. $\frac{6}{x^2-1}, \frac{7}{2-2x}$

We factor each denominator.

$x^2-1=(x+1)(x-1)$

$2-2x=2(1-x)=-2(x-1)$

LCD = $-2(x-1)(x+1)$

6. $\frac{7}{a^3}+\frac{9}{2a^4}=\frac{7\cdot2a}{a^3\cdot2a}+\frac{9}{2a^4}$

$=\frac{14a}{2a^4}+\frac{9}{2a^4}$

$=\frac{14a+9}{2a^4}$

7. $\frac{1}{x+5}+\frac{6x}{x-5}=\frac{1\cdot(x-5)}{(x+5)(x-5)}+\frac{6x(x+5)}{(x-5)(x+5)}$
$=\frac{x-5}{(x+5)(x-5)}+\frac{6x^2+30x}{(x-5)(x+5)}$
$=\frac{x-5+6x^2+30x}{(x-5)(x+5)}$
$=\frac{6x^2+31x-5}{(x+5)(x-5)}$

8. $\frac{3m-26}{m-6}-\frac{8}{6-m}=\frac{3m-26}{m-6}-\frac{8}{-1(m-6)}$
$=\frac{3m-26}{m-6}-\frac{-8}{m-6}$
$=\frac{3m-26+8}{m-6}$
$=\frac{3m-18}{m-6}$
$=\frac{3(m-6)}{m-6}$
$=3$

9. $\frac{2x}{x^2-9}-\frac{3}{x^2-4x+3}$
$=\frac{2x}{(x+3)(x-3)}-\frac{3}{(x-3)(x-1)}$
$=\frac{2x(x-1)}{(x+3)(x-3)(x-1)}-\frac{3(x+3)}{(x-3)(x-1)(x+3)}$
$=\frac{2x^2-2x}{(x+3)(x-3)(x-1)}-\frac{3x+9}{(x+3)(x-3)(x-1)}$
$=\frac{2x^2-2x-3x-9}{(x+3)(x-3)(x-1)}$
$=\frac{2x^2-5x-9}{(x+3)(x-3)(x-1)}$

10. $\frac{x+1}{x^2+x-12}+\frac{2x-1}{x^2+6x+8}$
$=\frac{x+1}{(x+4)(x-3)}+\frac{2x-1}{(x+4)(x+2)}$
$=\frac{(x+1)(x+2)}{(x+4)(x-3)(x+2)}+\frac{(2x-1)(x-3)}{(x+4)(x+2)(x-3)}$
$=\frac{x^2+3x+2}{(x+4)(x-3)(x+2)}+\frac{2x^2-7x+3}{(x+4)(x-3)(x+2)}$
$=\frac{x^2+3x+2+2x^2-7x+3}{(x+4)(x-3)(x+2)}$
$=\frac{3x^2-4x+5}{(x+4)(x-3)(x+2)}$

11. $\frac{6}{x-5}+\frac{x-35}{x^2-5x}-\frac{2}{x}$
$=\frac{6}{x-5}+\frac{x-35}{x(x-5)}-\frac{2}{x}$
$=\frac{6x}{(x-5)(x)}+\frac{x-35}{x(x-5)}-\frac{2(x-5)}{x(x-5)}$
$=\frac{6x}{(x-5)(x)}+\frac{x-35}{(x-5)(x)}-\frac{2x-10}{(x-5)(x)}$
$=\frac{6x+x-35-2x+10}{(x-5)(x)}$
$=\frac{5x-25}{(x-5)(x)}$
$=\frac{5(x-5)}{(x-5)(x)}$
$=\frac{5}{x}$

Exercise Set 6.2

1. $\frac{2}{xz^2}-\frac{5}{xz^2}=\frac{2-5}{xz^2}=\frac{-3}{xz^2}=-\frac{3}{xz^2}$

3. $\frac{2}{x-2}+\frac{x}{x-2}=\frac{2+x}{x-2}=\frac{x+2}{x-2}$

5. $\frac{x^2}{x+2}-\frac{4}{x+2}=\frac{x^2-4}{x+2}$
$=\frac{(x+2)(x-2)}{x+2}$
$=x-2$

7. $\frac{2x-6}{x^2+x-6}+\frac{3-3x}{x^2+x-6}=\frac{2x-6+3-3x}{x^2+x-6}$
$=\frac{-x-3}{x^2+x-6}$
$=\frac{-1(x+3)}{(x+3)(x-2)}$
$=\frac{-1}{x-2}$ or $\frac{1}{-(x-2)}$
$=\frac{-1}{x-2}$ or $\frac{1}{2-x}$

9. $\frac{x-5}{2x}-\frac{x+5}{2x}=\frac{x-5-x-5}{2x}=\frac{-10}{2x}=-\frac{5}{x}$

11. $7=7$
$5x=5x$
$\text{LCD}=7\cdot 5x=35x$

13. $x=x$
$x+1=x+1$
$\text{LCD}=x(x+1)$

15. $x+7=x+7$
$x-7=x-7$
$\text{LCD}=(x+7)(x-7)$

17. $3x+6=3(x+2)$
$2x-4=2(x-2)$
$\text{LCD}=3\cdot 2(x+2)(x-2)=6(x+2)(x-2)$

19. $a^2-b^2=(a+b)(a-b)$
$a^2-2ab+b^2=(a-b)^2$
$\text{LCD}=(a+b)(a-b)^2$

21. $x^2-9=(x+3)(x-3)$
$x=x$
$12-4x=-4(x-3)$
$\text{LCD}=-4x(x+3)(x-3)$

23. $\frac{4}{3x}+\frac{3}{2x}=\frac{4\cdot 2}{3x(2)}+\frac{3\cdot 3}{2x(3)}=\frac{8}{6x}+\frac{9}{6x}=\frac{17}{6x}$

25. $\frac{5}{2y^2}-\frac{2}{7y}=\frac{5\cdot 7}{2y^2\cdot 7}-\frac{2\cdot 2y}{7y\cdot 2y}$
$=\frac{35}{14y^2}-\frac{4y}{14y^2}$
$=\frac{35-4y}{14y^2}$

27. $\frac{x-3}{x+4}-\frac{x+2}{x-4}$
$=\frac{(x-3)(x-4)}{(x+4)(x-4)}-\frac{(x+2)(x+4)}{(x-4)(x+4)}$
$=\frac{x^2-7x+12}{(x+4)(x-4)}-\frac{x^2+6x+8}{(x+4)(x-4)}$
$=\frac{x^2-7x+12-x^2-6x-8}{(x+4)(x-4)}$
$=\frac{-13x+4}{(x+4)(x-4)}$

29. $\frac{1}{x-5}+\frac{2x-19}{(x-5)(x+4)}$
$=\frac{1\cdot(x+4)}{(x-5)\cdot(x+4)}+\frac{2x-19}{(x-5)(x+4)}$
$=\frac{x+4+2x-19}{(x-5)(x+4)}$
$=\frac{3x-15}{(x-5)(x+4)}$
$=\frac{3(x-5)}{(x-5)(x+4)}$
$=\frac{3}{x+4}$

31. $\frac{1}{a-b}+\frac{1}{b-a}=\frac{1}{a-b}+\frac{1}{-1(-b+a)}$
$=\frac{1}{a-b}+\frac{-1}{a-b}$
$=\frac{0}{a-b}$
$=0$

33. $\frac{x+1}{1-x}+\frac{1}{x-1}=\frac{x+1}{-(x-1)}+\frac{1}{x-1}$

$$=\frac{-(x+1)}{x-1}+\frac{1}{x-1}$$
$$=\frac{-x-1+1}{x-1}$$
$$=\frac{-x}{x-1}$$
$$=-\frac{x}{x-1}$$

35. $\frac{5}{x-2}+\frac{x+4}{2-x}=\frac{5}{x-2}+\frac{x+4}{-(-2+x)}$

$$=\frac{5}{x-2}+\frac{-(x+4)}{x-2}$$
$$=\frac{5-x-4}{x-2}$$
$$=\frac{-x+1}{x-2}$$

37. $\frac{y+1}{y^2-6y+8}-\frac{3}{y^2-16}=\frac{y+1}{(y-2)(y-4)}-\frac{3}{(y+4)(y-4)}$

$$=\frac{(y+1)(y+4)}{(y-2)(y-4)(y+4)}-\frac{3(y-2)}{(y-2)(y+4)(y-4)}$$
$$=\frac{(y+1)(y+4)-3(y-2)}{(y-2)(y-4)(y+4)}$$
$$=\frac{y^2+5y+4-3y+6}{(y-2)(y-4)(y+4)}$$
$$=\frac{y^2+2y+10}{(y-2)(y-4)(y+4)}$$

39. $\frac{x+4}{3x^2+11x+6}+\frac{x}{2x^2+x-15}=\frac{x+4}{(3x+2)(x+3)}+\frac{x}{(2x-5)(x+3)}$

$=\frac{(x+4)(2x-5)}{(3x+2)(x+3)(2x-5)}+\frac{x(3x+2)}{(2x-5)(x+3)(3x+2)}$

$=\frac{2x^2+3x-20}{(3x+2)(x+3)(2x-5)}+\frac{3x^2+2x}{(3x+2)(x+3)(2x-5)}$

$=\frac{2x^2+3x-20+3x^2+2x}{(3x+2)(x+3)(2x-5)}$

$=\frac{5x^2+5x-20}{(3x+2)(x+3)(2x-5)}$

$=\frac{5(x^2+x-4)}{(3x+2)(x+3)(2x-5)}$

41. $\frac{7}{x^2-x-2}+\frac{x}{x^2+4x+3}=\frac{7}{(x-2)(x+1)}+\frac{x}{(x+3)(x+1)}$

$=\frac{7(x+3)}{(x-2)(x+1)(x+3)}+\frac{x(x-2)}{(x+3)(x+1)(x-2)}$

$=\frac{7(x+3)+x(x-2)}{(x-2)(x+1)(x+3)}$

$=\frac{7x+21+x^2-2x}{(x-2)(x+1)(x+3)}$

$=\frac{x^2+5x+21}{(x-2)(x+1)(x+3)}$

43. $\frac{x+4}{3x^2+11x+6}+\frac{x}{2x^2+x-15}=\frac{x+4}{(3x+2)(x+3)}+\frac{x}{(2x-5)(x+3)}$

$=\frac{(x+4)(2x-5)}{(3x+2)(x+3)(2x-5)}+\frac{x(3x+2)}{(2x-5)(x+3)(3x+2)}$

$=\frac{2x^2+3x-20}{(3x+2)(x+3)(2x-5)}+\frac{3x^2+2x}{(3x+2)(x+3)(2x-5)}$

$=\frac{2x^2+3x-20+3x^2+2x}{(3x+2)(x+3)(2x-5)}$

$=\frac{5x^2+5x-20}{(3x+2)(x+3)(2x-5)}$

$=\frac{5(x^2+x-4)}{(3x+2)(x+3)(2x-5)}$

45. $\frac{2}{a^2+2a+1}+\frac{3}{a^2-1}$
$=\frac{2}{(a+1)^2}+\frac{3}{(a+1)(a-1)}$
$=\frac{2(a-1)}{(a+1)^2(a-1)}+\frac{3(a+1)}{(a+1)(a-1)(a+1)}$
$=\frac{2(a-1)+3(a+1)}{(a+1)^2(a-1)}$
$=\frac{2a-2+3a+3}{(a+1)^2(a-1)}$
$=\frac{5a+1}{(a+1)^2(a-1)}$

47. $\frac{4}{3x^2y^3}+\frac{5}{3x^2y^3}=\frac{9}{3x^2y^3}=\frac{3}{x^2y^3}$

49. $\frac{13x-5}{2x}-\frac{13x+5}{2x}=\frac{13x-5-13x-5}{2x}$
$=\frac{-10}{2x}$
$=-\frac{5}{x}$

51. $\frac{3}{2x+10}+\frac{8}{3x+15}=\frac{3}{2(x+5)}+\frac{8}{3(x+5)}$
$=\frac{3\cdot 3}{2(x+5)\cdot 3}+\frac{8\cdot 2}{3(x+5)\cdot 2}$
$=\frac{9}{6(x+5)}+\frac{16}{6(x+5)}$
$=\frac{25}{6(x+5)}$

53. $\frac{-2}{x^2-3x}-\frac{1}{x^3-3x^2}=\frac{-2}{x(x-3)}-\frac{1}{x^2(x-3)}$
$=\frac{-2(x)}{x^2(x-3)}-\frac{1}{x^2(x-3)}$
$=\frac{-2x-1}{x^2(x-3)}$

55. $\frac{ab}{a^2-b^2}+\frac{b}{a+b}$
$=\frac{ab}{(a+b)(a-b)}+\frac{b}{a+b}$
$=\frac{ab}{(a+b)(a-b)}+\frac{b(a-b)}{(a+b)(a-b)}$
$=\frac{ab}{(a+b)(a-b)}+\frac{ab-b^2}{(a+b)(a-b)}$
$=\frac{ab+ab-b^2}{(a+b)(a-b)}$
$=\frac{2ab-b^2}{(a+b)(a-b)}$
$=\frac{b(2a-b)}{(a+b)(a-b)}$

57. $\frac{5}{x^2-4}-\frac{3}{x^2+4x+4}$
$=\frac{5}{(x+2)(x-2)}-\frac{3}{(x+2)^2}$
$=\frac{5(x+2)}{(x+2)(x-2)(x+2)}-\frac{3(x-2)}{(x+2)^2(x-2)}$
$=\frac{5(x+2)-3(x-2)}{(x+2)^2(x-2)}$
$=\frac{5x+10-3x+6}{(x+2)^2(x-2)}$
$=\frac{2x+16}{(x+2)^2(x-2)}$
$=\frac{2(x+8)}{(x+2)^2(x-2)}$

59. $\frac{3x}{2x^2-11x+5}+\frac{7}{x^2-2x-15}=\frac{3x}{(2x-1)(x-5)}+\frac{7}{(x-5)(x+3)}$

$=\frac{3x(x+3)}{(2x-1)(x-5)(x+3)}+\frac{7(2x-1)}{(x-5)(x+3)(2x-1)}$

$=\frac{3x^2+9x}{(2x-1)(x-5)(x+3)}+\frac{14x-7}{(2x-1)(x-5)(x+3)}$

$=\frac{3x^2+9x+14x-7}{(2x-1)(x-5)(x+3)}$

$=\frac{3x^2+23x-7}{(2x-1)(x-5)(x+3)}$

61. $\frac{2}{x+1}-\frac{3x}{3x+3}+\frac{1}{2x+2}=\frac{2}{x+1}-\frac{3x}{3(x+1)}+\frac{1}{2(x+1)}$

$=\frac{2}{x+1}-\frac{x}{x+1}+\frac{1}{2(x+1)}$

$=\frac{2\cdot 2}{2(x+1)}-\frac{2\cdot x}{2(x+1)}+\frac{1}{2(x+1)}$

$=\frac{4-2x+1}{2(x+1)}$

$=\frac{5-2x}{2(x+1)}$

63. $\frac{3}{x+3}+\frac{5}{x^2+6x+9}-\frac{x}{x^2-9}=\frac{3}{x+3}+\frac{5}{(x+3)^2}-\frac{x}{(x+3)(x-3)}$

$=\frac{3(x+3)(x-3)}{(x+3)(x+3)(x-3)}+\frac{5(x-3)}{(x+3)^2(x-3)}-\frac{x(x+3)}{(x+3)(x-3)(x+3)}$

$=\frac{3x^2-27}{(x+3)^2(x-3)}+\frac{5x-15}{(x+3)^2(x-3)}-\frac{x^2+3x}{(x+3)^2(x-3)}$

$=\frac{3x^2-27+5x-15-x^2-3x}{(x+3)^2(x-3)}$

$=\frac{2x^2+2x-42}{(x+3)^2(x-3)}$

$=\frac{2(x^2+x-21)}{(x+3)^2(x-3)}$

65. $\frac{x}{x^2-9}+\frac{3}{x^2-6x+9}-\frac{1}{x+3}=\frac{x}{(x+3)(x-3)}+\frac{3}{(x-3)^2}-\frac{1}{x+3}$

$$=\frac{x(x-3)}{(x+3)(x-3)^2}+\frac{3(x+3)}{(x-3)^2(x+3)}-\frac{1(x-3)^2}{(x+3)(x-3)^2}$$

$$=\frac{x(x-3)+3(x+3)-(x-3)^2}{(x+3)(x-3)^2}$$

$$=\frac{x^2-3x+3x+9-(x^2-6x+9)}{(x+3)(x-3)^2}$$

$$=\frac{x^2+9-x^2+6x-9}{(x+3)(x-3)^2}$$

$$=\frac{6x}{(x+3)(x-3)^2}$$

67. $\left(\frac{1}{x}+\frac{2}{3}\right)-\left(\frac{1}{x}-\frac{2}{3}\right)=\left(\frac{3}{3x}+\frac{2x}{3x}\right)-\left(\frac{3}{3x}-\frac{2x}{3x}\right)$

$$=\left(\frac{3+2x}{3x}\right)-\left(\frac{3-2x}{3x}\right)$$

$$=\frac{3+2x-3+2x}{3x}$$

$$=\frac{4}{3}$$

69. $12\left(\frac{2}{3}+\frac{1}{6}\right)=12\left(\frac{2}{3}\right)+12\left(\frac{1}{6}\right)$

$$=4(2)+2(1)$$

$$=8+2$$

$$=10$$

71. $x^2\left(\frac{4}{x^2}+1\right)=x^2\cdot\frac{4}{x^2}+x^2\cdot 1=4+x^2$

73. $\frac{2x-3}{x^2+1}-\frac{x-6}{x^2+1}=\frac{2x-3-(x-6)}{x^2+1}$

$$=\frac{2x-3-x+6}{x^2+1}$$

$$=\frac{x+3}{x^2+1}$$

75. $P = 4s$

$P = 4\cdot\left(\frac{x}{x+5}\right) = \frac{4x}{x+5}$

$A = s^2$

$A = \left(\frac{x}{x+5}\right)^2$

$A = \left(\frac{x}{x+5}\right)\left(\frac{x}{x+5}\right) = \frac{x^2}{(x+5)^2}$

The perimeter is $\frac{4x}{x+5}$ feet and the area $\frac{x^2}{(x+5)^2}$ square feet.

77. answers may vary

79. answers may vary

81. answers may vary

83. $\left(\frac{2}{3}-\frac{1}{x}\right)\cdot\left(\frac{3}{x}+\frac{1}{2}\right)$

$= \left(\frac{2x}{3x}-\frac{3}{3x}\right)\cdot\left(\frac{3\cdot 2}{2x}+\frac{x}{2x}\right)$

$= \left(\frac{2x-3}{3x}\right)\cdot\left(\frac{6+x}{2x}\right)$

$= \frac{2x^2+9x-18}{6x^2}$ or $\frac{(2x-3)(x+6)}{6x^2}$

85. $\left(\frac{2a}{3}\right)^2 \div \left(\frac{a^2}{a+1}-\frac{1}{a+1}\right)$

$= \frac{4a^2}{9} \div \frac{a^2-1}{a+1}$

$= \frac{4a^2}{9} \div \frac{(a+1)(a-1)}{a+1}$

$= \frac{4a^2}{9} \div (a-1)$

$= \frac{4a^2}{9}\cdot\frac{1}{a-1}$

$= \frac{4a^2}{9(a-1)}$

87. $\left(\frac{2x}{3}\right)^2 \div \left(\frac{x}{3}\right)^2 = \left(\frac{2^2x^2}{3^2}\right) \div \left(\frac{x^2}{3^2}\right)$

$= \frac{2^2x^2}{3^2}\cdot\frac{3^2}{x^2}$

$= 2^2$

$= 4$

89. $\left(\frac{x}{x+1}-\frac{x}{x-1}\right) \div \frac{x}{2x+2}$

$= \left(\frac{x\cdot(x-1)}{(x+1)(x-1)}-\frac{x\cdot(x+1)}{(x-1)(x+1)}\right) \div \frac{x}{2(x+1)}$

$= \frac{x(x-1)-x(x+1)}{(x+1)(x-1)} \div \frac{x}{2(x+1)}$

$= \frac{x^2-x-x^2-x}{(x+1)(x-1)} \div \frac{x}{2(x+1)}$

$= \frac{-2x}{(x+1)(x-1)}\cdot\frac{2(x+1)}{x}$

$= -\frac{4}{x-1}$

91. $\frac{4}{x}\cdot\left(\frac{2}{x+2}-\frac{2}{x-2}\right)$

$= \frac{4}{x}\cdot\left(\frac{2(x-2)}{(x+2)(x-2)}-\frac{2(x+2)}{(x-2)(x+2)}\right)$

$= \frac{4}{x}\cdot\left(\frac{2x-4}{(x+2)(x-2)}-\frac{2x+4}{(x+2)(x-2)}\right)$

$= \frac{4}{x}\cdot\left(\frac{2x-4-2x-4}{(x+2)(x-2)}\right)$

$= \frac{4}{x}\left(\frac{-8}{(x+2)(x-2)}\right)$

$= -\frac{32}{x(x+2)(x-2)}$

93. $x^{-1}+(2x)^{-1}=\frac{1}{x}+\frac{1}{2x}$
$=\frac{1\cdot 2}{x\cdot 2}+\frac{1}{2x}$
$=\frac{2+1}{2x}$
$=\frac{3}{2x}$

95. $4x^{-2}-3x^{-1}=\frac{4}{x^2}-\frac{3}{x}=\frac{4}{x^2}-\frac{3x}{x^2}=\frac{4-3x}{x^2}$

97.

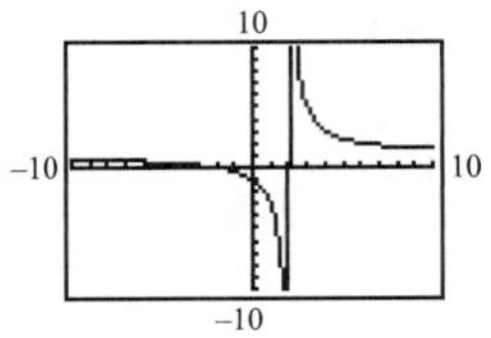

Section 6.3 Practice Problems

1. $\frac{\frac{6x}{x-5}}{\frac{12}{x+5}}=\frac{6x}{x-5}\cdot\frac{x+5}{12}=\frac{x(x+5)}{2(x-5)}$

2. $\frac{\frac{x}{y^2}-\frac{1}{y}}{\frac{y}{x^2}-\frac{1}{x}}=\frac{\frac{x}{y^2}-\frac{y}{y\cdot y}}{\frac{y}{x^2}-\frac{x}{x\cdot x}}$
$=\frac{\frac{x-y}{y^2}}{\frac{y-x}{x^2}}$
$=\frac{x-y}{y^2}\cdot\frac{x^2}{y-x}$
$=\frac{x-y}{y^2}\cdot\frac{x^2}{-(x-y)}$
$=-\frac{x^2}{y^2}$

3. $\frac{\frac{6x}{x-5}}{\frac{12}{x+5}}=\frac{\frac{6x}{x-5}(x-5)(x+5)}{\frac{12}{x+5}(x-5)(x+5)}$
$=\frac{6x(x+5)}{12(x-5)}$
$=\frac{x(x+5)}{2(x-5)}$

4. $\frac{\frac{x}{y^2}-\frac{1}{y}}{\frac{y}{x^2}-\frac{1}{x}}=\frac{\left(\frac{x}{y^2}-\frac{1}{y}\right)\cdot x^2y^2}{\left(\frac{y}{x^2}-\frac{1}{x}\right)\cdot x^2y^2}$
$=\frac{\frac{x}{y^2}\cdot x^2y^2-\frac{1}{y}\cdot x^2y^2}{\frac{y}{x^2}\cdot x^2y^2-\frac{1}{x}\cdot x^2y^2}$
$=\frac{x^3-x^2y}{y^3-xy^2}$
$=\frac{x^2(x-y)}{y^2(y-x)}$
$=\frac{x^2(x-y)}{-y^2(x-y)}$
$=-\frac{x^2}{y^2}$

5. $\frac{2x^{-1}+3y^{-1}}{x^{-1}-2y^{-1}}=\frac{\frac{2}{x}+\frac{3}{y}}{\frac{1}{x}-\frac{2}{y}}$
$=\frac{\left(\frac{2}{x}+\frac{3}{y}\right)xy}{\left(\frac{1}{x}-\frac{2}{y}\right)xy}$
$=\frac{\frac{2}{x}\cdot xy+\frac{3}{y}\cdot xy}{\frac{1}{x}\cdot xy-\frac{2}{y}\cdot xy}$
$=\frac{2y+3x}{y-2x}$

6. $\dfrac{5-3x^{-1}}{2+(3x)^{-1}} = \dfrac{5-\frac{3}{x}}{2+\frac{1}{3x}}$
$= \dfrac{\left(5-\frac{3}{x}\right)\cdot 3x}{\left(2+\frac{1}{3x}\right)\cdot 3x}$
$= \dfrac{5\cdot 3x-\frac{3}{x}\cdot 3x}{2\cdot 3x+\frac{1}{3x}\cdot 3x}$
$= \dfrac{15x-9}{6x+1}$
$= \dfrac{3(5x-3)}{6x+1}$

Exercise Set 6.3

1. $\dfrac{1+\frac{2}{5}}{2+\frac{3}{5}} = \dfrac{5\left(1+\frac{2}{5}\right)}{5\left(2+\frac{3}{5}\right)} = \dfrac{5+2}{10+3} = \dfrac{7}{13}$

3. $\dfrac{\frac{4}{x-1}}{\frac{x}{x-1}} = \dfrac{4}{x-1}\cdot\dfrac{x-1}{x} = \dfrac{4}{x}$

5. $\dfrac{1-\frac{2}{x}}{x+\frac{4}{9x}} = \dfrac{9x\left(1-\frac{2}{x}\right)}{9x\left(x+\frac{4}{9x}\right)} = \dfrac{9x-18}{9x^2+4} = \dfrac{9(x-2)}{9x^2+4}$

7. $\dfrac{\frac{10}{3x}}{\frac{5}{6x}} = \dfrac{10}{3x}\cdot\dfrac{6x}{5} = \dfrac{60x}{15x} = 4$

9. $\dfrac{\frac{4x^2-y^2}{xy}}{\frac{2}{y}-\frac{1}{x}} = \dfrac{\left(\frac{4x^2-y^2}{xy}\right)\cdot xy}{\left(\frac{2}{y}-\frac{1}{x}\right)\cdot xy}$
$= \dfrac{4x^2-y^2}{2x-y}$
$= \dfrac{(2x-y)(2x+y)}{2x-y}$
$= 2x+y$

11. $\dfrac{\frac{x+1}{3}}{\frac{2x-1}{6}} = \dfrac{x+1}{3}\cdot\dfrac{6}{2x-1} = \dfrac{2(x+1)}{2x-1}$

13. $\dfrac{\frac{2}{x}+\frac{3}{x^2}}{\frac{4}{x^2}-\frac{9}{x}} = \dfrac{\left(\frac{2}{x}+\frac{3}{x^2}\right)x^2}{\left(\frac{4}{x^2}-\frac{9}{x}\right)x^2} = \dfrac{2x+3}{4-9x}$

15. $\dfrac{\frac{1}{x}+\frac{2}{x^2}}{x+\frac{8}{x^2}} = \dfrac{x^2\left(\frac{1}{x}+\frac{2}{x^2}\right)}{x^2\left(x+\frac{8}{x^2}\right)}$
$= \dfrac{x+2}{x^3+8}$
$= \dfrac{x+2}{(x+2)(x^2-2x+4)}$
$= \dfrac{1}{x^2-2x+4}$

17. $\dfrac{\frac{4}{5-x}+\frac{5}{x-5}}{\frac{2}{x}+\frac{3}{x-5}} = \dfrac{-\frac{4}{x-5}+\frac{5}{x-5}}{\frac{2(x-5)+3x}{x(x-5)}}$
$= \dfrac{\frac{1}{x-5}}{\frac{2x-10+3x}{x(x-5)}}$
$= \dfrac{1}{x-5}\cdot\dfrac{x(x-5)}{5x-10}$
$= \dfrac{x}{5x-10}$

19. $\dfrac{\frac{x+2}{x}-\frac{2}{x-1}}{\frac{x+1}{x}+\frac{x+1}{x-1}} = \dfrac{\frac{(x+2)(x-1)-2x}{x(x-1)}}{\frac{(x+1)(x-1)+(x+1)(x)}{x(x-1)}}$
$= \dfrac{\frac{x^2+x-2-2x}{x(x-1)}}{\frac{x^2-1+x^2+x}{x(x-1)}}$
$= \dfrac{x^2-x-2}{x(x-1)}\cdot\dfrac{x(x-1)}{2x^2+x-1}$
$= \dfrac{(x-2)(x+1)}{x(x-1)}\cdot\dfrac{x(x-1)}{(2x-1)(x+1)}$
$= \dfrac{x-2}{2x-1}$

21. $\dfrac{\frac{2}{x}+3}{\frac{4}{x^2}-9}=\dfrac{\left(\frac{2}{x}+3\right)\cdot x^2}{\left(\frac{4}{x^2}-9\right)\cdot x^2}$
$=\dfrac{2x+3x^2}{4-9x^2}$
$=\dfrac{x(2+3x)}{(2+3x)(2-3x)}$
$=\dfrac{x}{2-3x}$

23. $\dfrac{1-\frac{x}{y}}{\frac{x^2}{y^2}-1}=\dfrac{\left(1-\frac{x}{y}\right)\cdot y^2}{\left(\frac{x^2}{y^2}-1\right)\cdot y^2}$
$=\dfrac{y^2-xy}{x^2-y^2}$
$=\dfrac{y(y-x)}{(x+y)(x-y)}$
$=\dfrac{-y(x-y)}{(x+y)(x-y)}$
$=-\dfrac{y}{x+y}$

25. $\dfrac{\frac{-2x}{x^2-xy}}{\frac{y}{x^2}}=\dfrac{-2x}{x^2-xy}\div\dfrac{y}{x^2}$
$=\dfrac{-2x}{x^2-xy}\cdot\dfrac{x^2}{y}$
$=\dfrac{-2x}{x(x-y)}\cdot\dfrac{x^2}{y}$
$=\dfrac{-2x\cdot x}{(x-y)y}$
$=-\dfrac{2x^2}{y(x-y)}$

27. $\dfrac{\frac{2}{x}+\frac{1}{x^2}}{\frac{y}{x^2}+1}=\dfrac{\left(\frac{2}{x}+\frac{1}{x^2}\right)x^2}{\left(\frac{y}{x^2}+1\right)x^2}=\dfrac{2x+1}{y+x^2}$

29. $\dfrac{\frac{x}{9}-\frac{1}{x}}{1+\frac{3}{x}}=\dfrac{\left(\frac{x}{9}-\frac{1}{x}\right)\cdot 9x}{\left(1+\frac{3}{x}\right)\cdot 9x}$
$=\dfrac{x^2-9}{9x+27}$
$=\dfrac{(x+3)(x-3)}{9(x+3)}$
$=\dfrac{x-3}{9}$

31. $\dfrac{\frac{x-1}{x^2-4}}{1+\frac{1}{x-2}}=\dfrac{\frac{x-1}{x^2-4}}{\frac{x-2+1}{x-2}}=\dfrac{\frac{x-1}{x^2-4}}{\frac{x-1}{x-2}}$
$=\dfrac{x-1}{x^2-4}\cdot\dfrac{x-2}{x-1}$
$=\dfrac{x-1}{(x+2)(x-2)}\cdot\dfrac{x-2}{x-1}$
$=\dfrac{1}{x+2}$

33. $\dfrac{\frac{2}{x+5}+\frac{4}{x+3}}{\frac{3x+13}{x^2+8x+15}}=\dfrac{\frac{2}{x+5}+\frac{4}{x+3}}{\frac{3x+13}{(x+5)(x+3)}}$
$=\dfrac{\left(\frac{2}{x+5}+\frac{4}{x+3}\right)(x+5)(x+3)}{\frac{3x+13}{(x+5)(x+3)}(x+5)(x+3)}$
$=\dfrac{2(x+3)+4(x+5)}{3x+13}$
$=\dfrac{2x+6+4x+20}{3x+13}$
$=\dfrac{6x+26}{3x+13}$
$=\dfrac{2(3x+13)}{3x+13}$
$=2$

35. $$\frac{x^{-1}}{x^{-2}+y^{-2}}=\frac{\frac{1}{x}}{\frac{1}{x^2}+\frac{1}{y^2}}=\frac{x^2y^2\left(\frac{1}{x}\right)}{x^2y^2\left(\frac{1}{x^2}+\frac{1}{y^2}\right)}=\frac{xy^2}{y^2+x^2}=\frac{xy^2}{x^2+y^2}$$

37. $$\frac{2a^{-1}+3b^{-2}}{a^{-1}-b^{-1}}=\frac{\frac{2}{a}+\frac{3}{b^2}}{\frac{1}{a}-\frac{1}{b}}=\frac{ab^2\left(\frac{2}{a}+\frac{3}{b^2}\right)}{ab^2\left(\frac{1}{a}-\frac{1}{b}\right)}=\frac{2b^2+3a}{b^2-ab}=\frac{2b^2+3a}{b(b-a)}$$

39. $$\frac{1}{x-x^{-1}}=\frac{1}{x-\frac{1}{x}}=\frac{x(1)}{x\left(x-\frac{1}{x}\right)}=\frac{x}{x^2-1}=\frac{x}{(x+1)(x-1)}$$

41. $$\frac{a^{-1}+1}{a^{-1}-1}=\frac{\frac{1}{a}+1}{\frac{1}{a}-1}=\frac{a\left(\frac{1}{a}+1\right)}{a\left(\frac{1}{a}-1\right)}=\frac{1+a}{1-a}$$

43. $$\frac{3x^{-1}+(2y)^{-1}}{x^{-2}}=\frac{\frac{3}{x}+\frac{1}{2y}}{\frac{1}{x^2}}=\frac{2x^2y\left(\frac{3}{x}+\frac{1}{2y}\right)}{2x^2y\left(\frac{1}{x^2}\right)}=\frac{6xy+x^2}{2y}=\frac{x(x+6y)}{2y}$$

45. $$\frac{2a^{-1}+(2a)^{-1}}{a^{-1}+2a^{-2}}=\frac{\frac{2}{a}+\frac{1}{2a}}{\frac{1}{a}+\frac{2}{a^2}}=\frac{2a^2\left(\frac{2}{a}+\frac{1}{2a}\right)}{2a^2\left(\frac{1}{a}+\frac{2}{a^2}\right)}=\frac{4a+a}{2a+4}=\frac{5a}{2(a+2)}$$

47. $$\frac{5x^{-1}+2y^{-1}}{x^{-2}y^{-2}}=\frac{\frac{5}{x}+\frac{2}{y}}{\frac{1}{x^2y^2}}=\frac{x^2y^2\left(\frac{5}{x}+\frac{2}{y}\right)}{x^2y^2\left(\frac{1}{x^2y^2}\right)}=5xy^2+2x^2y$$

49. $\dfrac{5x^{-1}-2y^{-1}}{25x^{-2}-4y^{-2}}=\dfrac{\frac{5}{x}-\frac{2}{y}}{\frac{25}{x^2}-\frac{4}{y^2}}$

$=\dfrac{x^2y^2\left(\frac{5}{x}-\frac{2}{y}\right)}{x^2y^2\left(\frac{25}{x^2}-\frac{4}{y^2}\right)}$

$=\dfrac{5xy^2-2x^2y}{25y^2-4x^2}$

$=\dfrac{xy(5y-2x)}{(5y+2x)(5y-2x)}$

$=\dfrac{xy}{5y+2x}$

51. $7x+2=x-3$

$7x=x-5$

$6x=-5$

$x=-\dfrac{5}{6}$

The solution set is $\left\{-\dfrac{5}{6}\right\}$.

53. $x^2=4x-4$

$x^2-4x+4=0$

$(x-2)(x-2)=0$

$x-2=0$ or $x-2=0$

$x=2$ or $x=2$

The solution set is $\{2\}$.

55. $\dfrac{x}{3}-5=13$

$\dfrac{x}{3}=18$

$x=54$

The solution set is $\{54\}$.

57. $\dfrac{\frac{x+1}{9}}{\frac{y-2}{5}}=\dfrac{x+1}{9}\div\dfrac{y-2}{5}=\dfrac{x+1}{9}\cdot\dfrac{5}{y-2}$

Both a and c are equivalent to the original expression.

59. $\dfrac{a}{1-\frac{s}{770}}=\dfrac{770(a)}{770\left(1-\frac{s}{770}\right)}=\dfrac{770a}{770-s}$

61. $\dfrac{\frac{1}{x}}{\frac{3}{y}}=\dfrac{1}{x}\div\dfrac{3}{y}=\dfrac{1}{x}\cdot\dfrac{y}{3}$

Both a and b are equivalent to the original expression.

63. $\dfrac{\frac{2}{y^2}-\frac{5}{xy}-\frac{3}{x^2}}{\frac{2}{y^2}+\frac{7}{xy}+\frac{3}{x^2}}=\dfrac{x^2y^2\left(\frac{2}{y^2}-\frac{5}{xy}-\frac{3}{x^2}\right)}{x^2y^2\left(\frac{2}{y^2}+\frac{7}{xy}+\frac{3}{x^2}\right)}$

$=\dfrac{2x^2-5xy-3y^2}{2x^2+7xy+3y^2}$

$=\dfrac{(2x+1y)(x-3y)}{(2x+1y)(x+3y)}$

$=\dfrac{x-3y}{x+3y}$

65. $\dfrac{1}{1+(1+x)^{-1}}=\dfrac{1}{1+\frac{1}{1+x}}$

$=\dfrac{(1+x)\cdot 1}{(1+x)\left(1+\frac{1}{1+x}\right)}$

$=\dfrac{1+x}{1+x+1}$

$=\dfrac{1+x}{2+x}$

67. $\dfrac{x}{1-\frac{1}{1+\frac{1}{x}}}=\dfrac{x}{1-\frac{x}{x+1}}$

$=\dfrac{(x+1)(x)}{(x+1)\left(1-\frac{x}{x+1}\right)}$

$=\dfrac{x(x+1)}{x+1-x}$

$=\dfrac{x(x+1)}{1}$

$=x(x+1)$

69. $f(x)=\frac{1}{x}$

a. $f(a+h)=\frac{1}{a+h}$

b. $f(a)=\frac{1}{a}$

c. $\frac{f(a+h)-f(a)}{h}=\frac{\frac{1}{a+h}-\frac{1}{a}}{h}$

d. $$\frac{\frac{1}{a+h}-\frac{1}{a}}{h}=\frac{a(a+h)\left(\frac{1}{a+h}-\frac{1}{a}\right)}{a(a+h)\cdot h}$$
$$=\frac{a-(a+h)}{ah(a+h)}$$
$$=\frac{-h}{ah(a+h)}$$
$$=\frac{-1}{a(a+h)}$$

71. $f(x)=\frac{3}{x+1}$

a. $f(a+h)=\frac{3}{a+h+1}$

b. $f(a)=\frac{3}{a+1}$

c. $\frac{f(a+h)-f(a)}{h}=\frac{\frac{3}{a+h+1}-\frac{3}{a+1}}{h}$

d. $$\frac{\frac{3}{a+h+1}-\frac{3}{a+1}}{h}$$
$$=\frac{(a+h+1)(a+1)\left(\frac{3}{a+h+1}-\frac{3}{a+1}\right)}{(a+h+1)(a+1)h}$$
$$=\frac{3(a+1)-3(a+h+1)}{(a+h+1)(a+1)h}$$
$$=\frac{3a+3-3a-3h-3}{(a+h+1)(a+1)h}$$
$$=\frac{-3h}{(a+h+1)(a+1)h}$$
$$=\frac{-3}{(a+h+1)(a+1)}$$

Section 6.4 Practice Problems

1. The LCD of 6, 2, and 3 is 6.

$$\frac{5x}{6}+\frac{1}{2}=\frac{x}{3}$$
$$6\left(\frac{5x}{6}+\frac{1}{2}\right)=6\left(\frac{x}{3}\right)$$
$$6\cdot\frac{5x}{6}+6\cdot\frac{1}{2}=6\cdot\frac{x}{3}$$
$$5x+3=2x$$
$$5x=2x-3$$
$$3x=-3$$
$$x=-1$$

The solution set is $\{-1\}$.

2. The LCD of x, $2x$, and 2 is $2x$.

$$\frac{5}{x}-\frac{3x+6}{2x}=\frac{7}{2}$$
$$2x\left(\frac{5}{x}-\frac{3x+6}{2x}\right)=2x\left(\frac{7}{2}\right)$$
$$2x\cdot\frac{5}{x}-2x\cdot\frac{3x+6}{2x}=2x\cdot\frac{7}{2}$$
$$10-(3x+6)=7x$$
$$10-3x-6=7x$$
$$4-3x=7x$$
$$4=10x$$
$$\frac{4}{10}=x$$
$$x=\frac{2}{5}$$

The solution set is $\left\{\frac{2}{5}\right\}$.

3. The LCD is $x-3$.

$$\frac{x+5}{x-3}=\frac{2(x+1)}{x-3}$$
$$(x-3)\left(\frac{x+5}{x-3}\right)=(x-3)\left(\frac{2(x+1)}{x-3}\right)$$
$$x+5=2(x+1)$$
$$x+5=2x+2$$
$$x+3=2x$$
$$3=x$$

The number 3 makes the denominator $x-3$ equal to 0, so it is not a solution. The solution set is $\varnothing$.

4. The LCD is $x(3x-1)$.

$$\frac{3x}{3x-1}+\frac{1}{x}=\frac{1}{3x-1}$$
$$x(3x-1)\left(\frac{3x}{3x-1}+\frac{1}{x}\right)=x(3x-1)\left(\frac{1}{3x-1}\right)$$
$$x(3x-1)\cdot\frac{3x}{3x-1}+x(3x-1)\cdot\frac{1}{x}=x$$
$$3x^2+3x-1=x$$
$$3x^2+2x-1=0$$
$$(3x-1)(x+1)=0$$

$$\begin{aligned} 3x-1=0 \quad &\text{or} \quad x+1=0 \\ 3x=1 \quad &\text{or} \quad x=-1 \\ x=\frac{1}{3} \quad &\text{or} \quad x=-1 \end{aligned}$$

The number $\frac{1}{3}$ makes the denominator $3x - 1$ equal to 0, so it is not a solution. The solution set is $\{-1\}$.

5. $x^2-16=(x+4)(x-4)$

The LCD is $x^2-16=(x+4)(x-4)$.

$$\frac{2x}{x-4}+\frac{10-5x}{x^2-16}=\frac{x}{x+4}$$

$$(x+4)(x-4)\cdot\frac{2x}{x-4}+(x+4)(x-4)\cdot\frac{10-5x}{x^2-16}=(x+4)(x-4)\cdot\frac{x}{x+4}$$

$$\begin{aligned} 2x(x+4)+10-5x&=x(x-4) \\ 2x^2+8x+10-5x&=x^2-4x \\ 2x^2+3x+10&=x^2-4x \\ x^2+7x+10&=0 \\ (x+5)(x+2)&=0 \end{aligned}$$

$$\begin{aligned} x+5=0 \quad &\text{or} \quad x+2=0 \\ x=-5 \quad &\text{or} \quad x=-2 \end{aligned}$$

The solution set is $\{-5, -2\}$.

6. $3z^2+7z-6=(3z-2)(z+3)$

$3z^2-2z=z(3z-2)$

The LCD is $z(3z-2)(z+3)$.

$$\frac{3z}{3z^2+7z-6}-\frac{1}{z}=\frac{4}{3z^2-2z}$$

$$z(3z-2)(z+3)\cdot\frac{3z}{3z^2+7z-6}-z(3z-2)(z+3)\cdot\frac{1}{z}=z(3z-2)(z+3)\cdot\frac{4}{3z^2-2z}$$

$$\begin{aligned} 3z^2-(3z-2)(z+3)&=4(z+3) \\ 3z^2-(3z^2+7z-6)&=4z+12 \\ 3z^2-3z^2-7z+6&=4z+12 \\ -7z+6&=4z+12 \\ -11z&=6 \\ z&=-\frac{6}{11} \end{aligned}$$

The solution set is $\left\{-\frac{6}{11}\right\}$.

Mental Math

1. $\frac{x}{2}=\frac{3x}{5}+\frac{x}{6}$ is an equation.

2. $\frac{3x}{5}+\frac{x}{6}$ is an expression.

3. $\frac{x}{x-1}+\frac{2x}{x+1}$ is an expression.

4. $\frac{x}{x-1}+\frac{2x}{x+1}=5$ is an equation.

5. $\frac{y+7}{2}=\frac{y+1}{6}+\frac{1}{y}$ is an equation.

6. $\frac{y+1}{6}+\frac{1}{y}$ is an expression.

Exercise Set 6.4

1. The LCD of 2 and 3 is 6.

$$\frac{x}{2}-\frac{x}{3}=12$$
$$6\left(\frac{x}{2}-\frac{x}{3}\right)=6(12)$$
$$6\cdot\frac{x}{2}-6\cdot\frac{x}{3}=72$$
$$3x-2x=72$$
$$x=72$$

The solution set is $\{72\}$.

3. The LCD of 3, 6, and 4 is 12.

$$\frac{x}{3}=\frac{1}{6}+\frac{x}{4}$$
$$12\left(\frac{x}{3}\right)=12\left(\frac{1}{6}+\frac{x}{4}\right)$$
$$12\cdot\frac{x}{3}=12\cdot\frac{1}{6}+12\cdot\frac{x}{4}$$
$$4x=2+3x$$
$$x=2$$

The solution set is $\{2\}$.

5. The LCD of x and 2 is $2x$.

$$\frac{2}{x}+\frac{1}{2}=\frac{5}{x}$$
$$2x\left(\frac{2}{x}+\frac{1}{2}\right)=2x\left(\frac{5}{x}\right)$$
$$2x\cdot\frac{2}{x}+2x\cdot\frac{1}{2}=10$$
$$4+x=10$$
$$x=6$$

The solution set is $\{6\}$.

7. The LCD is x.

$$\frac{x^2+1}{x}=\frac{5}{x}$$
$$x\cdot\frac{x^2+1}{x}=x\cdot\frac{5}{x}$$
$$x^2+1=5$$
$$x^2-4=0$$
$$(x+2)(x-2)=0y$$
$$x+2=0 \quad \text{or} \quad x-2=0$$
$$x=-2 \quad \text{or} \quad x=2$$

The solution set is $\{-2, 2\}$.

9. The LCD is $x + 3$.

$$\frac{x+5}{x+3}=\frac{2}{x+3}$$
$$(x+3)\left(\frac{x+5}{x+3}\right)=(x+3)\left(\frac{2}{x+3}\right)$$
$$x+5=2$$
$$x=-3$$

The number -3 makes the denominator $x + 3$ equal 0, so it is not a solution. The solution set is $\varnothing$.

11. $x^2+2x-8=(x+4)(x-2)$

The LCD is $(x + 4)(x - 2)$.

$$\frac{5}{x-2}-\frac{2}{x+4}=\frac{-4}{x^2+2x-8}$$
$$(x+4)(x-2)\cdot\frac{5}{x-2}-(x+4)(x-2)\cdot\frac{2}{x+4}=(x+4)(x-2)\cdot\frac{-4}{x^2+2x-8}$$
$$5(x+4)-2(x-2)=-4$$
$$5x+20-2x+4=-4$$
$$3x+24=-4$$
$$3x=-28$$
$$x=-\frac{28}{3}$$

The solution set is $\left\{-\frac{28}{3}\right\}$.

13. The LCD is $(x - 1)(x + 1)$.

$$\frac{1}{x-1}=\frac{2}{x+1}$$
$$(x-1)(x+1)\left(\frac{1}{x-1}\right)=(x-1)(x+1)\left(\frac{2}{x+1}\right)$$
$$x+1=2(x-1)$$
$$x+1=2x-2$$
$$1=x-2$$
$$3=x$$

The solution set is $\{3\}$.

15. $2x^2-5x-3=(2x+1)(x-3)$

The LCD is $(2x + 1)(x - 3)$.

$$\frac{y^2-23}{2x^2-5x-3}+\frac{2}{x-3}=\frac{-1}{2x+1}$$
$$(2x+1)(x-3)\cdot\frac{x^2-23}{2x^2-5x-3}+(2x+1)(x-3)\cdot\frac{2}{x-3}=(2x+1)(x-3)\cdot\frac{-1}{2x+1}$$
$$x^2-23+(2x+1)2=-1(x-3)$$
$$x^2-23+4x+2=-x+3$$
$$x^2+4x-21=-x+3$$
$$x^2+5x-24=0$$
$$(x+8)(x-3)=0$$

$x+8=0$ or $x-3=0$

$x=-8$ or $x=3$

The number 3 makes the denominator $x - 3$ equal to 0, so it is not a solution. The solution set is $\{-8\}$.

17. $x^2-16=(x+4)(x-4)$

The LCD is $x^2-16=(x+4)(x-4)$.

$$\frac{1}{x-4}-\frac{3x}{x^2-16}=\frac{2}{x+4}$$
$$(x+4)(x-4)\left(\frac{1}{x-4}\right)-(x^2-16)\left(\frac{3x}{x^2-16}\right)=(x+4)(x-4)\left(\frac{2}{x+4}\right)$$
$$x+4-3x=2(x-4)$$
$$4-2x=2x-8$$
$$4=4x-8$$
$$12=4x$$
$$3=x$$

The solution set is $\{3\}$.

19. $x^2-16=(x-4)(x+4)$

The LCD is $(x-4)(x+4)$.

$$\frac{1}{x-4}=\frac{8}{x^2-16}$$
$$(x-4)(x+4)\cdot\frac{1}{x-4}=(x-4)(x+4)\cdot\frac{8}{x^2-16}$$
$$x+4=8$$
$$x=4$$

The number 4 makes the denominators $x-4$ and x^2-16 equal to 0, so it is not a solution. The solution set is $\varnothing$.

21. $x^2-2x=x(x-2)$

The LCD is $x^2-2x=x(x-2)$.

$$\frac{1}{x-2}-\frac{2}{x^2-2x}=1$$
$$x(x-2)\left(\frac{1}{x-2}\right)-(x^2-2x)\left(\frac{2}{x^2-2x}\right)=(x^2-2x)1$$
$$x-2=x^2-2x$$
$$0=x^2-3x+2$$
$$0=(x-2)(x-1)$$

$x-2=0$ or $x-1=0$
$x=2$ or $x=1$

The number 2 makes the denominators $x-2$ and x^2-2x equal 0, so it is not a solution. The solution set is $\{1\}$.

23. The LCD of 12 and x is $12x$.

$$\frac{5}{x}=\frac{20}{12}$$
$$12x\cdot\frac{5}{x}=12x\cdot\frac{20}{12}$$
$$60=20x$$
$$3=x$$

The solution set is $\{3\}$.

25. The LCD is a.

$$1-\frac{4}{a}=5$$
$$a\cdot 1-a\cdot\frac{4}{a}=a\cdot 5$$
$$a-4=5a$$
$$-4=4a$$
$$-1=a$$

The solution set is $\{-1\}$.

27. The LCD is x.

$$\frac{x^2+5}{x}-1=\frac{5(x+1)}{x}$$
$$x\cdot\frac{x^2+5}{x}-x=x\cdot\frac{5(x+1)}{x}$$
$$x^2+5-x=5(x+1)$$
$$x^2-x+5=5x+5$$
$$x^2-6x=0$$
$$x(x-6)=0$$

$x=0$ or $x-6=0$

$x=0$ or $x=6$

The number 0 makes the denominator x equal to 0, so it is not a solution. The solution set is $\{6\}$.

29. $3x^2+3x=3x(x+1)$

The LCD is $2\cdot 3x(x+1)=6x(x+1)$.

$$\frac{1}{2x}-\frac{1}{x+1}=\frac{1}{3x^2+3x}$$

$$2\cdot 3x(x+1)\cdot\frac{1}{2x}-6x(x+1)\cdot\frac{1}{x+1}=2\cdot 3x(x+1)\cdot\frac{1}{3x(x+1)}$$

$$3(x+1)-6x=2$$

$$3x+3-6x=2$$

$$3-3x=2$$

$$-3x=-1$$

$$x=\frac{1}{3}$$

The solution set is $\left\{\frac{1}{3}\right\}$.

31. The LCD of x and 25 is $25x$.

$$\frac{1}{x}-\frac{x}{25}=0$$

$$25x\cdot\frac{1}{x}-25x\cdot\frac{x}{25}=25x\cdot 0$$

$$25-x^2=0$$

$$(5-x)(5+x)=0$$

$5-x=0$ or $5+x=0$

$-x=-5$ or $x=-5$

$x=5$ or $x=-5$

The solution set is $\{5, -5\}$.

33. The LCD is $2y-5$.

$$5-\frac{2}{2y-5}=\frac{3}{2y-5}$$

$$(2y-5)\cdot 5-(2y-5)\left(\frac{2}{2y-5}\right)=(2y-5)\left(\frac{3}{2y-5}\right)$$

$$5(2y-5)-2=3$$

$$10y-25-2=3$$

$$10y-27=3$$

$$10y=30$$

$$y=3$$

The solution set is $\{3\}$.

35. The LCD is $3(x + 2)$.

$$\frac{x-1}{x+2}=\frac{2}{3}$$
$$3(x+2)\cdot\frac{x-1}{x+2}=3(x+2)\cdot\frac{2}{3}$$
$$3(x-1)=2(x+2)$$
$$3x-3=2x+4$$
$$x-3=4$$
$$x=7$$

The solution set is $\{7\}$.

37. The LCD is $x + 2$.

$$\frac{x+3}{x+2}=\frac{1}{x+2}$$
$$(x+2)\left(\frac{x+3}{x+2}\right)=(x+2)\left(\frac{1}{x+2}\right)$$
$$x+3=1$$
$$x=-2$$

The number -2 makes the denominator $x + 2$ equal 0, so it is not a solution. The solution set is $\varnothing$.

39. $a^2-9=(a+3)(a-3)$

The LCD is $(a + 3)(a - 3)$.

$$\frac{1}{a-3}+\frac{2}{a+3}=\frac{1}{a^2-9}$$
$$(a+3)(a-3)\cdot\frac{1}{a-3}+(a+3)(a-3)\cdot\frac{2}{a+3}=(a+3)(a-3)\cdot\frac{1}{a^2-9}$$
$$a+3+2(a-3)=1$$
$$a+3+2a-6=1$$
$$3a-3=1$$
$$3a=4$$
$$a=\frac{4}{3}$$

The solution set is $\left\{\frac{4}{3}\right\}$.

41. $x^2 - 16 = (x+4)(x-4)$

The LCD is $x^2 - 16 = (x+4)(x-4)$.

$$\frac{64}{x^2-16} + 1 = \frac{2x}{x-4}$$

$$(x^2-16)\left(\frac{64}{x^2-16}\right) + (x^2-16)(1) = (x+4)(x-4)\left(\frac{2x}{x-4}\right)$$

$$64 + x^2 - 16 = 2x(x+4)$$

$$x^2 + 48 = 2x^2 + 8x$$

$$0 = x^2 + 8x - 48$$

$$0 = (x-4)(x+12)$$

$x - 4 = 0$ or $x + 12 = 0$

$x = 4$ $\quad$ $x = -12$

The number 4 makes the denominators $x^2 - 16$ and $x - 4$ equal 0, so it is not a solution. The solution set is $\{-12\}$.

43. The LCD is $4y + 1$.

$$\frac{-15}{4y+1} + 4 = y$$

$$(4y+1)\cdot\frac{-15}{4y+1} + (4y+1)\cdot 4 = (4y+1)y$$

$$-15 + 16y + 4 = 4y^2 + y$$

$$16y - 11 = 4y^2 + y$$

$$0 = 4y^2 - 15y + 11$$

$$0 = (4y-11)(y-1)$$

$4y - 11 = 0$ or $y - 1 = 0$

$4y = 11$ or $y = 1$

$y = \frac{11}{4}$ or $y = 1$

The solution set is $\left\{1, \frac{11}{4}\right\}$.

45. $x^2 - 9 = (x+3)(x-3)$

The LCD is $x^2 - 9 = (x+3)(x-3)$.

$$\frac{28}{x^2-9} + \frac{2x}{x-3} + \frac{6}{x+3} = 0$$

$$(x^2-9)\left(\frac{28}{x^2-9}\right) + (x+3)(x-3)\left(\frac{2x}{x-3}\right) + (x+3)(x-3)\left(\frac{6}{x+3}\right) = (x^2-9)(0)$$

$$28 + 2x(x+3) + 6(x-3) = 0$$

$$28 + 2x^2 + 6x + 6x - 18 = 0$$

$$2x^2 + 12x + 10 = 0$$

$$2(x^2 + 6x + 5) = 0$$

$$2(x+5)(x+1) = 0$$

$x + 5 = 0$ or $x + 1 = 0$

$x = -5$ $\quad x = -1$

The solution set is $\{-5, -1\}$.

47. $3x + 6 = 3(x+2)$

$x^2 + 7x + 10 = (x+2)(x+5)$

The LCD is $3(x + 2)(x + 5)$.

$$\frac{x+2}{x^2+7x+10} = \frac{1}{3x+6} - \frac{1}{x+5}$$

$$3(x+2)(x+5)\cdot\frac{x+2}{x^2+7x+10} = 3(x+2)(x+5)\cdot\frac{1}{3x+6} - 3(x+2)(x+5)\cdot\frac{1}{x+5}$$

$$3(x+2) = x + 5 - 3(x+2)$$

$$3x + 6 = x + 5 - 3x - 6$$

$$3x + 6 = -2x - 1$$

$$5x + 6 = -1$$

$$5x = -7$$

$$x = -\frac{7}{5}$$

The solution set is $\left\{-\frac{7}{5}\right\}$.

49. Let x be the number.

$4 + 3x = 19$

$3x = 15$

$x = 5$

The number is 5.

51. Let w be the width of a rectangle. Let $w + 5$ be the length of a rectangle.

$$P = 2w + 2l$$
$$50 = 2w + 2(w+5)$$
$$50 = 2w + 2w + 10$$
$$40 = 4w$$
$$10 = w$$
$$l = w + 5 = 10 + 5 = 15$$

The width is 10 in. and the length is 15 in.

53. From the graph, 13.5% of state and federal prison inmates are ages 45 to 54.

55. From the graph, the category that shows the highest percent of prison inmates is the 25–29 year olds.

57. 17.4% of $36{,}612 = 0.174(36{,}612) = 6370.488$

Approximately 6370 25- to 29-year-old inmates would be expected in Louisiana at the end of 2003.

59. answers may vary

61. Find x when $C(x) = 5.1$.

$$5.1 = 3.3 + \frac{5400}{x}$$
$$x(5.1) = x(3.3) + x\left(\frac{5400}{x}\right)$$
$$5.1x = 3.3x + 5400$$
$$1.8x = 5400$$
$$x = 3000$$

3000 game disks must be produced for the average cost to be \$5.10.

63.
$$x^{-2} - 19x^{-1} + 48 = 0$$
$$\frac{1}{x^2} - \frac{19}{x} + 48 = 0$$
$$x^2\left(\frac{1}{x^2}\right) - x^2\left(\frac{19}{x}\right) + 48x^2 = x^2(0)$$
$$1 - 19x + 48x^2 = 0$$
$$(1-16x)(1-3x) = 0$$

$1 - 16x = 0$ or $1 - 3x = 0$

$-16x = -1$ or $-3x = -1$

$x = \frac{1}{16}$ or $x = \frac{1}{3}$

The solution set is $\left\{\frac{1}{16}, \frac{1}{3}\right\}$.

65.

$$p^{-2}+4p^{-1}-5=0$$
$$\frac{1}{p^2}+\frac{4}{p}-5=0$$
$$p^2\left(\frac{1}{p^2}\right)+p^2\left(\frac{4}{p}\right)-p^2(5)=p^2(0)$$
$$1+4p-5p^2=0$$
$$(1+5p)(1-p)=0$$
$$1+5p=0 \quad \text{or} \quad 1-p=0$$
$$p=-\frac{1}{5} \qquad 1=p$$

The solution set is $\left\{-\frac{1}{5}, 1\right\}$.

67. $\frac{1.4}{x-2.6}=\frac{-3.5}{x+7.1}$

The LCD is $(x-2.6)(x+7.1)$.

$$(x-2.6)(x+7.1)\cdot\frac{1.4}{x-2.6}=(x-2.6)(x+7.1)\cdot\frac{-3.5}{x+7.1}$$
$$(x+7.1)1.4=(x-2.6)\cdot-3.5$$
$$1.4x+9.94=-3.5x+9.1$$
$$1.4x=-3.5x-0.84$$
$$4.9x=-0.84$$
$$x=\frac{-0.84}{4.9}$$
$$x\approx-0.17$$

The solution set is $\{-0.17\}$.

69. $\frac{10.6}{y} - 14.7 = \frac{9.92}{3.2} + 7.6$

$\frac{10.6}{y} - 14.7 = 3.1 + 7.6$

$\frac{10.6}{y} - 14.7 = 10.7$

$\frac{10.6}{y} = 25.4$

$y\left(\frac{10.6}{y}\right) = y(25.4)$

$10.6 = 25.4y$

$\frac{10.6}{25.4} = y$

$0.42 \approx y$

The solution set is {0.42}.

71. Let $u = (x - 1)$.

$(x-1)^2 + 3(x-1) + 2 = 0$

$u^2 + 3u + 2 = 0$

$(u+2)(u+1) = 0$

$u + 2 = 0$ or $u + 1 = 0$

$u = -2$ or $u = -1$

Now solve for x.

$-2 = x - 1$ $-1 = x - 1$

$-1 = x$ $0 = x$

The solution set is {−1, 0}.

73. Let $u = \frac{3}{x-1}$.

$\left(\frac{3}{x-1}\right)^2 + 2\left(\frac{3}{x-1}\right) + 1 = 0$

$u^2 + 2u + 1 = 0$

$(u+1)^2 = 0$

$u + 1 = 0$

$u = -1$

Now solve for x.

$-1 = \frac{3}{x-1}$

$(x-1)(-1) = (x-1)\left(\frac{3}{x-1}\right)$

$-x + 1 = 3$

$-x = 2$

$x = -2$

The solution set is {−2}.

75.

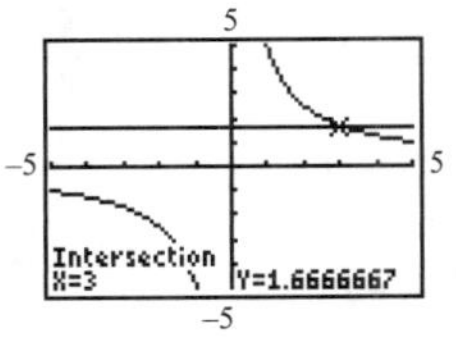

77.

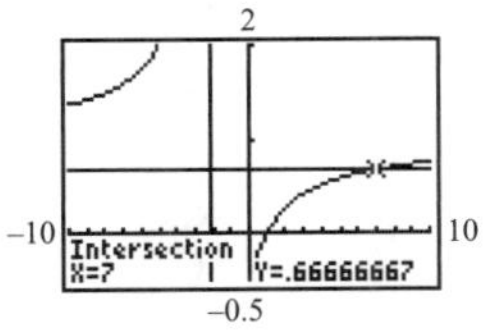

The Bigger Picture

1. $|-7x + 1| < 15$

$-15 < -7x + 1 < 15$

$-16 < -7x < 14$

$\frac{16}{7} > x > -2$

The solution set is $\left(-2, \frac{16}{7}\right)$.

2. $|-7x + 1| = 15$

$-7x + 1 = 15$ or $-7x + 1 = -15$

$-7x = 14$ $-7x = -16$

$x = -2$ $x = \frac{16}{7}$

The solution set is $\left\{-2, \frac{16}{7}\right\}$.

3. $x^2 - 121 = 0$

$(x-11)(x+11) = 0$

$x - 11 = 0$ or $x + 11 = 0$

$x = 11$ or $x = -11$

The solution set is {11, −11}.

4. $x^2 + x - 2 = (x+2)(x-1)$

The LCD is $(x + 2)(x - 1)$.

$$\frac{8}{x+2} - \frac{3}{x-1} = \frac{x+6}{x^2+x-2}$$
$$(x+2)(x-1)\cdot\frac{8}{x+2} - (x+2)(x-1)\cdot\frac{3}{x-1} = (x+2)(x-1)\cdot\frac{x+6}{x^2+x-2}$$
$$8(x-1) - 3(x+2) = x+6$$
$$8x - 8 - 3x - 6 = x+6$$
$$5x - 14 = x+6$$
$$4x - 14 = 6$$
$$4x = 20$$
$$x = 5$$

The solution set is $\{5\}$.

5. $9x + 6 = 4x - 2$

$9x = 4x - 8$

$5x = -8$

$x = -\frac{8}{5}$

The solution set is $\left\{-\frac{8}{5}\right\}$.

6. $3x \le 6$ or $-x \ge 5$

$x \le 2$ or $x \le -5$

$(-\infty, 2] \cup (-\infty, -5] = (-\infty, 2]$

The solution set is $(-\infty, 2]$.

7. $3x \le 6$ and $-x \ge 5$

$x \le 2$ and $x \le -5$

$(-\infty, 2] \cap (-\infty, -5] = (-\infty, -5]$

The solution set is $(-\infty, -5]$.

8. $-9 \le -3x + 21 < 0$

$-30 \le -3x < -21$

$10 \ge x > 7$

The solution set is $(7, 10]$.

9. $\left|\frac{2x-1}{5}\right| > 7$

$$\frac{2x-1}{5} > 7 \quad \text{or} \quad \frac{2x-1}{5} < -7$$
$$2x-1 > 35 \quad \text{or} \quad 2x-1 < -35$$
$$2x > 36 \quad \text{or} \quad 2x < -34$$
$$x > 18 \quad \text{or} \quad x < -17$$

$(-18, \infty) \cup (-\infty, -17)$

The solution set is $(-\infty, -17) \cup (18, \infty)$.

10. $15x^3 - 16x^2 = 7x$

$$15x^3 - 16x^2 - 7x = 0$$
$$x(15x^2 - 16x - 7) = 0$$
$$x(3x+1)(5x-7) = 0$$
$$x = 0 \quad \text{or} \quad 3x+1 = 0 \quad \text{or} \quad 5x-7 = 0$$
$$x = 0 \quad \text{or} \quad 3x = -1 \quad \text{or} \quad 5x = 7$$
$$x = 0 \quad \text{or} \quad x = -\frac{1}{3} \quad \text{or} \quad x = \frac{7}{5}$$

The solution set is $\left\{0, -\frac{1}{3}, \frac{7}{5}\right\}$.

Integrated Review

1. $\frac{x}{2} = \frac{1}{8} + \frac{x}{4}$

The LCD is 8.

$$8 \cdot \frac{x}{2} = 8 \cdot \frac{1}{8} + 8 \cdot \frac{x}{4}$$
$$4x = 1 + 2x$$
$$2x = 1$$
$$x = \frac{1}{2}$$

The solution set is $\left\{\frac{1}{2}\right\}$.

2. $\frac{x}{4}=\frac{3}{2}+\frac{x}{10}$

The LCD is 20.

$$20\cdot\frac{x}{4}=20\cdot\frac{3}{2}+20\cdot\frac{x}{10}$$
$$5x=30+2x$$
$$3x=30$$
$$x=10$$

The solution set is $\{10\}$.

3. $\frac{1}{8}+\frac{x}{4}=\frac{1}{8}+\frac{x}{4}\cdot\frac{2}{2}=\frac{1}{8}+\frac{2x}{8}=\frac{1+2x}{8}$

4. $\frac{3}{2}+\frac{x}{10}=\frac{5}{5}\cdot\frac{3}{2}+\frac{x}{10}=\frac{15}{10}+\frac{x}{10}=\frac{15+x}{10}$

5.
$$\frac{4}{x+2}-\frac{2}{x-1}=\frac{4}{x+2}\cdot\frac{x-1}{x-1}-\frac{2}{x-1}\cdot\frac{x+2}{x+2}$$
$$=\frac{4(x-1)}{(x+2)(x-1)}-\frac{2(x+2)}{(x+2)(x+1)}$$
$$=\frac{4x-4-2x-4}{(x+2)(x-1)}$$
$$=\frac{2x-8}{(x+2)(x-1)}$$
$$=\frac{2(x-4)}{(x+2)(x-1)}$$

6.
$$\frac{5}{x-2}-\frac{10}{x+4}=\frac{5}{x-2}\cdot\frac{x+4}{x+4}-\frac{10}{x+4}\cdot\frac{x-2}{x-2}$$
$$=\frac{5(x+4)}{(x-2)(x+4)}-\frac{10(x-2)}{(x-2)(x+4)}$$
$$=\frac{5x+20-10x+20}{(x-2)(x+4)}$$
$$=\frac{-5x+40}{(x-2)(x+4)}$$
$$=\frac{-5(x-8)}{(x-2)(x+4)}$$

7. $\dfrac{4}{x+2}=\dfrac{2}{x-1}$

The LCD is $(x + 2)(x - 1)$.

$$(x+2)(x-1)\cdot\frac{4}{x+2}=(x+2)(x-1)\cdot\frac{2}{x-1}$$
$$4(x-1)=2(x+2)$$
$$4x-4=2x+4$$
$$4x=2x+8$$
$$2x=8$$
$$x=4$$

The solution set is $\{4\}$.

8. $\dfrac{5}{x-2}=\dfrac{10}{x+4}$

The LCD is $(x - 2)(x + 4)$.

$$(x-2)(x+4)\cdot\frac{5}{x-2}=(x-2)(x+4)\cdot\frac{10}{x+4}$$
$$5(x+4)=10(x-2)$$
$$5x+20=10x-20$$
$$5x=10x-40$$
$$-5x=-40$$
$$x=8$$

The solution set is $\{8\}$.

9. $x^2-4=(x+2)(x-2)$

The LCD is $(x + 2)(x - 2)$.

$$\frac{2}{x^2-4}=\frac{1}{x+2}-\frac{3}{x-2}$$
$$(x+2)(x-2)\cdot\frac{2}{x^2-4}=(x+2)(x-2)\cdot\frac{1}{x+2}-(x+2)(x-2)\cdot\frac{3}{x-2}$$
$$2=x-2-3(x+2)$$
$$2=x-2-3x-6$$
$$2x=-10$$
$$x=-5$$

The solution set is$\{-5\}$.

10. $x^2 - 25 = (x-5)(x+5)$

The LCD is $(x - 5)(x + 5)$.

$$\frac{3}{x^2-25} = \frac{1}{x+5} + \frac{2}{x-5}$$

$$(x+5)(x-5)\cdot\frac{3}{x^2-25} = (x+5)(x-5)\cdot\frac{1}{x+5} + (x+5)(x-5)\cdot\frac{2}{x-5}$$

$$\begin{aligned} 3 &= x-5+2(x+5) \\ 3 &= x-5+2x+10 \\ 3 &= 3x+5 \\ -2 &= 3x \\ -\frac{2}{3} &= x \end{aligned}$$

The solution set is $\left\{-\frac{2}{3}\right\}$.

11. $$\begin{aligned} \frac{5}{x^2-3x} + \frac{4}{2x-6} &= \frac{5}{x(x-3)} + \frac{4}{2(x-3)} \\ &= \frac{5}{x(x-3)}\cdot\frac{2}{2} + \frac{4}{2(x-3)}\cdot\frac{x}{x} \\ &= \frac{10}{2x(x-3)} + \frac{4x}{2x(x-3)} \\ &= \frac{4x+10}{2x(x-3)} \\ &= \frac{2(2x+5)}{2x(x-3)} \\ &= \frac{2x+5}{x(x-3)} \end{aligned}$$

12. $$\begin{aligned} \frac{5}{x^2-3x} \div \frac{4}{2x-6} &= \frac{5}{x^2-3x}\cdot\frac{2x-6}{4} \\ &= \frac{5}{x(x-3)}\cdot\frac{2(x-3)}{4} \\ &= \frac{5}{2x} \end{aligned}$$

13. $x^2 - 1 = (x-1)(x+1)$

The LCD is $(x - 1)(x + 1)$.

$$\frac{x-1}{x+1} + \frac{x+7}{x-1} = \frac{4}{x^2-1}$$

$$(x-1)(x+1)\cdot\frac{x-1}{x+1} + (x-1)(x+1)\cdot\frac{x+7}{x-1} = (x-1)(x+1)\cdot\frac{4}{x^2-1}$$

$$(x-1)(x-1) + (x+1)(x+7) = 4$$

$$x^2 - 2x + 1 + x^2 + 8x + 7 = 4$$

$$2x^2 + 6x + 8 = 4$$

$$2x^2 + 6x + 4 = 0$$

$$2(x^2 + 3x + 2) = 0$$

$$2(x+1)(x+2) = 0$$

$x + 1 = 0$ or $x + 2 = 0$

$x = -1$ or $x = -2$

The number -1 makes the denominator $x + 1$ equal to 0, so it is not a solution. The solution set is $\{-2\}$.

14.
$$\left(1-\frac{y}{x}\right)\div\left(1-\frac{x}{y}\right) = \left(\frac{x}{x}-\frac{y}{x}\right)\div\left(\frac{y}{y}-\frac{x}{y}\right)$$

$$= \left(\frac{x-y}{x}\right)\div\left(\frac{y-x}{y}\right)$$

$$= \frac{x-y}{x}\cdot\frac{y}{y-x}$$

$$= \frac{x-y}{x}\cdot\frac{y}{-(x-y)}$$

$$= -\frac{y}{x}$$

15.
$$\frac{a^2-9}{a-6}\cdot\frac{a^2-5a-6}{a^2-a-6} = \frac{(a+3)(a-3)}{a-6}\cdot\frac{(a-6)(a+1)}{(a-3)(a+2)}$$

$$= \frac{(a+3)(a+1)}{(a+2)}$$

16. $\frac{2}{a-6}\cdot\frac{3a}{a^2-5a-6}-\frac{a}{5a+5}=\frac{2}{a-6}+\frac{3a}{(a-6)(a+1)}-\frac{a}{5(a+1)}$

$$=\frac{2}{a-6}\cdot\frac{5(a+1)}{5(a+1)}+\frac{3a}{(a-6)(a+1)}\cdot\frac{5}{5}-\frac{a}{5(a+1)}\cdot\frac{a-6}{a-6}$$

$$=\frac{10a+10}{5(a+1)(a-6)}+\frac{15a}{5(a+1)(a-6)}-\frac{a^2-6a}{5(a+1)(a-6)}$$

$$=\frac{10a+10+15a-a^2+6a}{5(a+1)(a-6)}$$

$$=\frac{-a^2+31a+10}{5(a+1)(a-6)}$$

17. $\frac{2x+3}{3x-2}=\frac{4x+1}{6x+1}$

The LCD is $(3x-2)(6x+1)$.

$$(3x-2)(6x+1)\cdot\frac{2x+3}{3x-2}=(3x-2)(6x+1)\cdot\frac{4x+1}{6x+1}$$

$$(6x+1)(2x+3)=(3x-2)(4x+1)$$

$$12x^2+18x+2x+3=12x^2+3x-8x-2$$

$$12x^2+20x+3=12x^2-5x-2$$

$$20x+3=-15x-2$$

$$25x+3=-2$$

$$25x=-5$$

$$x=-\frac{5}{25}$$

$$x=-\frac{1}{5}$$

The solution set is $\left\{-\frac{1}{5}\right\}$.

18. The LCD is $(2x)(4x + 1)$.

$$\frac{5x-3}{2x} = \frac{10x+3}{4x+1}$$
$$(2x)(4x+1) \cdot \frac{5x-3}{2x} = (2x)(4x+1) \cdot \frac{10x+3}{4x+1}$$
$$(4x+1)(5x-3) = (2x)(10x+3)$$
$$20x^2 - 12x + 5x - 3 = 20x^2 + 6x$$
$$20x^2 - 7x - 3 = 20x^2 + 6x$$
$$-7x - 3 = 6x$$
$$-3 = 13x$$
$$-\frac{3}{13} = x$$

The solution set is $\left\{-\frac{3}{13}\right\}$.

19. $\frac{a}{9a^2-1} + \frac{2}{6a-2}$

$$= \frac{a}{(3a-1)(3a+1)} + \frac{2}{2(3a-1)}$$
$$= \frac{a}{(3a-1)(3a+1)} \cdot \frac{2}{2} + \frac{2}{2(3a-1)} \cdot \frac{(3a+1)}{(3a+1)}$$
$$= \frac{2a}{2(3a-1)(3a+1)} + \frac{6a+2}{2(3a-1)(3a+1)}$$
$$= \frac{8a+2}{2(3a-1)(3a+1)}$$
$$= \frac{2(4a+1)}{2(3a-1)(3a+1)}$$
$$= \frac{4a+1}{(3a-1)(3a+1)}$$

20. $\frac{3}{4a-8} - \frac{a+2}{a^2-2a} = \frac{3}{4(a-2)} - \frac{a+2}{a(a-2)}$

$$= \frac{3}{4(a-2)} \cdot \frac{a}{a} - \frac{a+2}{a(a-2)} \cdot \frac{4}{4}$$
$$= \frac{3a}{4a(a-2)} - \frac{4(a+2)}{4a(a-2)}$$
$$= \frac{3a}{4a(a-2)} - \frac{4a+8}{4a(a-2)}$$
$$= \frac{3a-4a-8}{4a(a-2)}$$
$$= \frac{-a-8}{4a(a-2)}$$
$$= -\frac{(a+8)}{4a(a-2)}$$

21. The LCD is x^2.

$$-\frac{3}{x^2} - \frac{1}{x} + 2 = 0$$
$$x^2 \cdot -\frac{3}{x^2} - x^2 \cdot \frac{1}{x} + x^2 \cdot 2 = 0$$
$$-3 - x + 2x^2 = 0$$
$$2x^2 - x - 3 = 0$$
$$(2x-3)(x+1) = 0$$

$2x - 3 = 0$ or $x + 1 = 0$

$2x = 3$ or $x = -1$

$x = \frac{3}{2}$ or $x = -1$

The solution set is $\left\{-1, \frac{3}{2}\right\}$.

22. $\frac{x}{2x+6}+\frac{5}{x^2-9}=\frac{x}{2(x+3)}+\frac{5}{(x-3)(x+3)}$

$=\frac{x}{2(x+3)}\cdot\frac{x-3}{x-3}+\frac{5}{(x-3)(x+3)}\cdot\frac{2}{2}$

$=\frac{x(x-3)}{2(x+3)(x-3)}+\frac{10}{2(x+3)(x-3)}$

$=\frac{x^2-3x}{2(x+3)(x-3)}+\frac{10}{2(x+3)(x-3)}$

$=\frac{x^2-3x+10}{2(x+3)(x-3)}$

23. $\frac{x-8}{x^2-x-2}+\frac{2}{x-2}=\frac{x-8}{(x-2)(x+1)}+\frac{2}{x-2}$

$=\frac{x-8}{(x-2)(x+1)}+\frac{2}{x-2}\cdot\frac{(x+1)}{x+1}$

$=\frac{x-8}{(x-2)(x+1)}+\frac{2x+2}{(x-2)(x+1)}$

$=\frac{x-8+2x+2}{(x-2)(x+1)}$

$=\frac{3x-6}{(x-2)(x+1)}$

$=\frac{3(x-2)}{(x-2)(x+1)}$

$=\frac{3}{x+1}$

24. $x^2-x-2=(x-2)(x+1)$

The LCD is $(x-2)(x+1)$.

$$\frac{x-8}{x^2-x-2}+\frac{2}{x-2}=\frac{3}{x+1}$$

$$(x-2)(x+1)\cdot\frac{x-8}{x^2-x-2}+(x-2)(x+1)\cdot\frac{2}{x-2}=(x-2)(x+1)\cdot\frac{3}{x+1}$$

$$x-8+2(x+1)=3(x-2)$$

$$x-8+2x+2=3x-6$$

$$3x-6=3x-6$$

$$-6=-6 \quad \text{True}$$

The solution set is
$\{x|x \text{ is a real number and } x \neq 2, x \neq -1\}$.

25. The LCD is a.

$$\frac{3}{a}-5=\frac{7}{a}-1$$
$$a\cdot\frac{3}{a}-a\cdot 5=a\cdot\frac{7}{a}-a\cdot 1$$
$$3-5a=7-a$$
$$3=7+4a$$
$$-4=4a$$
$$-1=a$$

The solution set is $\{-1\}$.

26.
$$\frac{7}{3z-9}+\frac{5}{z}=\frac{7}{3z-9}\cdot\frac{z}{z}+\frac{5}{z}\cdot\frac{3z-9}{3z-9}$$
$$=\frac{7z}{z(3z-9)}+\frac{5(3z-9)}{z(3z-9)}$$
$$=\frac{7z}{z(3z-9)}+\frac{15z-45}{z(3z-9)}$$
$$=\frac{7z+15z-45}{z(3z-9)}$$
$$=\frac{22z-45}{z(3z-9)}$$
$$=\frac{22z-45}{3z(z-3)}$$

27. a. $\frac{x}{5}-\frac{x}{4}+\frac{1}{10}$ is an expression.

b. The first step to simplify this expression is to write each rational expression term so that the denominator is the LCD, 20.

c.
$$\frac{x}{5}-\frac{x}{4}+\frac{1}{10}=\frac{x}{5}\cdot\frac{4}{4}-\frac{x}{4}\cdot\frac{5}{5}+\frac{1}{10}\cdot\frac{2}{2}$$
$$=\frac{4x}{20}-\frac{5x}{20}+\frac{2}{20}$$
$$=\frac{4x-5x+2}{20}$$
$$=\frac{-x+2}{20}$$

28. a. $\frac{x}{5}-\frac{x}{4}=\frac{1}{10}$ is an equation.

b. The first step to solve this equation is to clear the equation of fractions by multiplying each term by the LCD, 20.

c.
$$\frac{x}{5}-\frac{x}{4}=\frac{1}{10}$$
$$20\cdot\frac{x}{5}-20\cdot\frac{x}{4}=20\cdot\frac{1}{10}$$
$$4x-5x=2$$
$$-x=2$$
$$x=-2$$

The solution set is $\{-2\}$.

29. $\frac{\triangle+\square}{\triangle}=\frac{\triangle}{\triangle}+\frac{\square}{\triangle}=1+\frac{\square}{\triangle}$

b is the correct answer.

30.
$$\frac{\triangle}{\square}+\frac{\square}{\triangle}=\frac{\triangle}{\square}\cdot\frac{\triangle}{\triangle}+\frac{\square}{\triangle}\cdot\frac{\square}{\square}$$
$$=\frac{\triangle\triangle}{\square\triangle}+\frac{\square\square}{\square\triangle}$$
$$=\frac{\triangle\triangle+\square\square}{\square\triangle}$$

d is the correct answer.

31. $\frac{\triangle}{\square}\cdot\frac{\bigcirc}{\triangle}=\frac{\triangle}{\square}\cdot\frac{\triangle}{\bigcirc}=\frac{\triangle\triangle}{\square\bigcirc}$

d is the correct answer.

32. $\frac{\triangle}{\square}\div\frac{\bigcirc}{\triangle}=\frac{\triangle}{\square}\cdot\frac{\triangle}{\bigcirc}=\frac{\triangle\triangle}{\square\bigcirc}$

a is the correct answer.

33. $\frac{\frac{\triangle+\square}{\bigcirc}}{\frac{\triangle}{\bigcirc}}=\frac{\triangle+\square}{\bigcirc}\div\frac{\triangle}{\bigcirc}=\frac{\triangle+\square}{\bigcirc}\cdot\frac{\bigcirc}{\triangle}=\frac{\triangle+\square}{\triangle}$

d is the correct answer.

Section 6.5 Practice Problems

1. $\frac{1}{x}+\frac{1}{y}=\frac{1}{z}$

xyz is the LCD.

$$xyz\cdot\frac{1}{x}+xyz\cdot\frac{1}{y}=xyz\cdot\frac{1}{z}$$
$$yz+xz=xy$$
$$xz=xy-yz$$
$$xz=y(x-z)$$
$$\frac{xz}{x-z}=y \text{ or } y=\frac{xz}{x-z}$$

2. Let x = the number.

$$\frac{1+x}{20-x}=\frac{2}{5}$$
$$5(20-x)\left(\frac{1+x}{20-x}\right)=5(20-x)\cdot\frac{2}{5}$$
$$5(1+x)=2(20-x)$$
$$5+5x=40-2x$$
$$5x=35-2x$$
$$7x=35$$
$$x=5$$

The number is 5.

3. Let x = number of homes in the community heated by wood.

$$\frac{1}{50}=\frac{x}{36,000}$$
$$36,000=50x$$
$$720=x$$

720 homes in the community are heated by wood.

4.

	Time	Part done in one hour
Greg	3	$\frac{1}{3}$
Phillip	5	$\frac{1}{5}$
Together	x	$\frac{1}{x}$

$$\frac{1}{3}+\frac{1}{5}=\frac{1}{x}$$
$$15x\cdot\frac{1}{3}+15x\cdot\frac{1}{5}=15x\cdot\frac{1}{x}$$
$$5x+3x=15$$
$$8x=15$$
$$x=\frac{15}{8}=1\frac{7}{8}$$

It takes $1\frac{7}{8}$ hours when they work together.

5.

	Distance	Rate	Time
Upstream	60	$25-x$	$\frac{60}{25-x}$
Downstream	60	$25+x$	$\frac{60}{25+x}$

Let x = speed of the current.

$$\frac{2}{3}\cdot\frac{60}{25-x}=\frac{60}{25+x}$$
$$3(25-x)(25+x)\cdot\frac{2}{3}\cdot\frac{60}{25-x}=3(25-x)(25+x)\cdot\frac{60}{25+x}$$
$$(25+x)\cdot 2\cdot 60=3(25-x)\cdot 60$$
$$120(25+x)=180(25-x)$$
$$3000+120x=4500-180x$$
$$120x=1500-180x$$
$$300x=1500$$
$$x=5$$

The speed of the current is 5 mph.

Exercise Set 6.5

1.
$$F=\frac{9}{5}C+32$$
$$F-32=\frac{9}{5}C$$
$$C=\frac{5}{9}(F-32)$$

3.
$$Q = \frac{A-I}{L}$$
$$QL = A - I$$
$$QL - A = -I$$
$$-QL + A = I$$
$$A - QL = I \text{ or } I = A - QL$$

5.
$$\frac{1}{R} = \frac{1}{R_1} + \frac{1}{R_2}$$
$$RR_1R_2\left(\frac{1}{R}\right) = RR_1R_2\left(\frac{1}{R_1} + \frac{1}{R_2}\right)$$
$$R_1R_2 = RR_2 + RR_1$$
$$R_1R_2 = R(R_2 + R_1)$$
$$R = \frac{R_1R_2}{R_1 + R_2}$$

7.
$$S = \frac{n(a+L)}{2}$$
$$2S = n(a+L)$$
$$\frac{2S}{a+L} = n \text{ or } n = \frac{2S}{a+L}$$

9.
$$A = \frac{h(a+b)}{2}$$
$$2A = ah + bh$$
$$2A - ah = bh$$
$$\frac{2A - ah}{h} = b \text{ or } b = \frac{2A - ah}{h}$$

11.
$$\frac{P_1V_1}{T_1} = \frac{P_2V_2}{T_2}$$
$$T_2P_1V_1 = T_1P_2V_2$$
$$T_2 = \frac{T_1P_2V_2}{P_1V_1}$$

13.
$$f = \frac{f_1f_2}{f_1 + f_2}$$
$$(f_1 + f_2)f = \left(\frac{f_1f_2}{f_1 + f_2}\right)(f_1 + f_2)$$
$$f_1f + f_2f = f_1f_2$$
$$f_1f = f_1f_2 - ff_2$$
$$f_1f = f_2(f_1 - f)$$
$$\frac{f_1f}{f_1 - f} = f_2 \text{ or } f_2 = \frac{f_1f}{f_1 - f}$$

15.
$$\lambda = \frac{2L}{n}$$
$$\lambda n = 2L$$
$$\frac{\lambda n}{2} = L \text{ or } L = \frac{\lambda n}{2}$$

17.
$$\frac{\theta}{\omega} = \frac{2L}{c}$$
$$c\omega\left(\frac{\theta}{\omega}\right) = \frac{2L}{c}(c\omega)$$
$$c\theta = 2L\omega$$
$$c = \frac{2L\omega}{\theta}$$

19. Let x = the number.
$$x + 5 \cdot \frac{1}{x} = 6$$
$$x^2 + 5 = 6x$$
$$x^2 - 6x + 5 = 0$$
$$(x-5)(x-1) = 0$$
$$x - 5 = 0 \quad \text{or} \quad x - 1 = 0$$
$$x = 5 \quad \text{or} \quad x = 1$$
The numbers are 1 and 5.

21. Let x = the number.
$$\frac{12 + x}{41 + 2x} = \frac{1}{3}$$
$$3(12 + x) = (41 + 2x) \cdot 1$$
$$36 + 3x = 41 + 2x$$
$$x = 5$$
The number is 5.

23. Let x = amount of water the camel can drink in 3 minutes. Solve the proportion:

$$\frac{x}{3}=\frac{15}{10}$$
$$10x=3\cdot 15$$
$$10x=45$$
$$x=4.5$$

An Arabian camel can drink 4.5 gallons of water in 3 minutes.

25. Let x be the number of women. Solve the proportion $\frac{x}{40,151}=\frac{10.7}{100}$.

$$\frac{x}{40,151}=\frac{10.7}{100}$$
$$100x=10.7\cdot 40,151$$
$$x=\frac{429,615.7}{100}$$
$$x=4296.157\approx 4296$$

There were approximately 4296 women in the Coast Guard in 2004.

27. Convert each time to a rate.

$$\frac{1\text{ task}}{26\text{ hours}}=\frac{1}{26}\frac{\text{task}}{\text{hour}}$$
$$\frac{1\text{ task}}{39\text{ hours}}=\frac{1}{39}\frac{\text{task}}{\text{hour}}$$

Let x = amount of time required for both roofers to complete the task together.

So $\frac{1\text{ task}}{x\text{ hours}}=\frac{1}{x}\frac{\text{task}}{\text{hour}}$. Adding we get

$$\frac{1}{26}+\frac{1}{39}=\frac{1}{x}$$
$$\frac{3}{78}+\frac{2}{78}=\frac{1}{x}$$
$$\frac{5}{78}=\frac{1}{x}$$
$$5x=78$$
$$x=15.6$$

It takes 15.6 hours for the roofers to complete the task together.

29. Convert each time to a rate.

$$\frac{1\text{ stack}}{20\text{ minutes}}=\frac{1}{20}\frac{\text{stack}}{\text{minute}}$$
$$\frac{1\text{ stack}}{30\text{ minutes}}=\frac{1}{30}\frac{\text{stack}}{\text{minute}}$$
$$\frac{1\text{ stack}}{60\text{ minutes}}=\frac{1}{60}\frac{\text{stack}}{\text{minute}}$$

Let x = amount of time required for all three workers to sort the stack.

So $\frac{1\text{ stack}}{x\text{ minutes}}=\frac{1}{x}\frac{\text{stack}}{\text{minute}}$. Adding we get,

$$\frac{1}{20}+\frac{1}{30}+\frac{1}{60}=\frac{1}{x}$$
$$60x\left(\frac{1}{20}+\frac{1}{30}+\frac{1}{60}\right)=60x\left(\frac{1}{x}\right)$$
$$3x+2x+x=60$$
$$6x=60$$
$$x=10$$

It takes 10 minutes for the workers to sort the stack.

31. Let r = the speed of the car and $r + 150$ = the speed of the plane.

Recall that $d = rt$ or $t=\frac{d}{r}$. Using the latter equation we get $\frac{150}{r}=\frac{600}{r+150}$.

$$(r+150)(150)=r\cdot 600$$
$$150r+22,500=600r$$
$$22,500=450r$$
$$50=r$$

$r + 150 = 50 + 150 = 200$

The speed of the plane was 200 mph.

33. Let r = the speed of the boat in still water.

Recall that $d = rt$ or $t=\frac{d}{r}$. Using the latter equation we get $\frac{20}{r+5}=\frac{10}{r-5}$, where $r + 5$ is the rate of the boat traveling downstream and $r - 5$ is the rate of the boat traveling upstream.

$$\frac{20}{r+5}=\frac{10}{r-5}$$
$$20(r-5)=10(r+5)$$
$$20r-100=10r+50$$
$$10r=150$$
$$r=15$$

The speed of the boat in still water is 15 mph.

35. Let x = the first integer and $x + 1$ = the next consecutive integer.

$$\frac{1}{x}+\frac{1}{x+1}=-\frac{15}{56}$$
$$56x(x+1)\left(\frac{1}{x}+\frac{1}{x+1}\right)=56x(x+1)\left(-\frac{15}{56}\right)$$
$$56(x+1)+56x=-15x(x+1)$$
$$56x+56+56x=-15x^2-15x$$
$$112x+56=-15x^2-15x$$
$$15x^2+127x+56=0$$
$$(15x+7)(x+8)=0$$
$$15x+7=0 \quad \text{or} \quad x+8=0$$
$$15x=-7 \quad \text{or} \quad x=-8$$
$$x=-\frac{7}{15} \quad \text{or} \quad x=-8$$

$x + 1 = -8 + 1 = -7$

The two integers are –8 and –7.

37. Convert times to rates.

$$\frac{1 \text{ pond}}{45 \text{ minutes}}=\frac{1}{45}\frac{\text{pond}}{\text{minute}}$$
$$\frac{1 \text{ pond}}{20 \text{ minutes}}=\frac{1}{20}\frac{\text{pond}}{\text{minute}}$$

Let x = the number of minutes required for the second hose to fill the pond. Then,

$$\frac{1 \text{ pond}}{x \text{ minute}}=\frac{1}{x}\frac{\text{pond}}{\text{minute}}.$$

$$\frac{1}{20}=\frac{1}{45}+\frac{1}{x}$$
$$\frac{1}{20}=\frac{x+45}{45x}$$
$$45x=20(x+45)$$
$$45x=20x+900$$
$$25x=900$$
$$x=\frac{900}{25}=36$$

Thus, the second hose alone can fill the pond in 36 minutes.

39. Let r = the rate of one train and $r + 15$ = the rate of the other train. Recall that $d = rt$. One train travels a total distance of $6r$ and the other travels $6(r + 15)$ miles.

$$6r+6(r+15)=630$$
$$6r+6r+90=630$$
$$12r+90=630$$
$$12r=540$$
$$r=45$$

$r + 15 = (45) + 15 = 60$

The speeds of the trains are 45 mph and 60 mph.

41. Let x = time.

$$\frac{x \text{ hour}}{1 \text{ mile}}=\frac{1 \text{ hour}}{0.17 \text{ mile}} \quad \text{or}$$
$$\frac{x}{1}=\frac{1}{0.17}$$
$$0.17x=1$$
$$x=\frac{1}{0.17}\approx 5.9$$

It takes approximately 5.9 hours to travel 1 mile.

43. Let x = time it takes to fill the quota if all 3 machines are running. Convert each time to a rate.

$\frac{1 \text{ task}}{5 \text{ hours}} = \frac{1}{5} \frac{\text{task}}{\text{hour}}$

$\frac{1 \text{ task}}{6 \text{ hours}} = \frac{1}{6} \frac{\text{task}}{\text{hour}}$

$\frac{1 \text{ task}}{7.5 \text{ hours}} = \frac{1}{7.5} \frac{\text{task}}{\text{hour}}$

$\frac{1 \text{ task}}{x \text{ hours}} = \frac{1}{x} \frac{\text{task}}{\text{hour}}$

$$\frac{1}{5}+\frac{1}{6}+\frac{1}{7.5}=\frac{1}{x}$$
$$225x\left(\frac{1}{5}+\frac{1}{6}+\frac{1}{7.5}\right)=225x\left(\frac{1}{x}\right)$$
$$45x+37.5x+30x=225$$
$$112.5x=225$$
$$x=2$$

If all 3 machines are running, they can fill the quota in 2 hours.

45.

	distance	= rate	· time
with wind	465	$x + 20$	$\frac{465}{x+20}$
against wind	345	$x - 20$	$\frac{345}{x-20}$

$$\frac{465}{x+20}=\frac{345}{x-20}$$
$$465(x-20)=345(x+20)$$
$$465x-9300=345x+6900$$
$$120x=16,200$$
$$x=135 \text{ mph}$$

The speed of the plane in still air is 135 miles per hour.

47. Let t = time of faster jogger.

time	distance	= rate	·
Faster jogger	$8t$	8	t
Slower jogger	$6\left(t+\frac{1}{2}\right)$	6	$t+\frac{1}{2}$

$$8t=6\left(t+\frac{1}{2}\right)$$
$$8t=6t+3$$
$$2t=3$$
$$t=\frac{3}{2}$$
$$8t=8\left(\frac{3}{2}\right)=12$$

They ran 12 miles.

49. Let x be he numerator of the fraction. Then the denominator is $x + 1$.

$$\frac{x-3}{(x+1)-3}=\frac{4}{5}$$
$$\frac{x-3}{x-2}=\frac{4}{5}$$
$$5(x-3)=4(x-2)$$
$$5x-15=4x-8$$
$$x-15=-8$$
$$x=7$$
$$x+1=7+1=8$$

The original fraction is $\frac{7}{8}$.

51. Convert times to rates.

$\frac{1 \text{ task}}{2 \text{ mins}} = \frac{1}{2} \frac{\text{task}}{\text{min}}$

$\frac{1 \text{ task}}{6 \text{ mins}} = \frac{1}{6} \frac{\text{task}}{\text{min}}$

Let x = the amount of time it takes to move the cans to the storage area if both belts are used.

$\frac{1 \text{ task}}{x \text{ min}} = \frac{1}{x} \frac{\text{task}}{\text{min}}$

$$\frac{1}{2}+\frac{1}{6}=\frac{1}{x}$$
$$6x\cdot\frac{1}{2}+6x\cdot\frac{1}{6}=6x\cdot\frac{1}{x}$$
$$3x+x=6$$
$$4x=6$$
$$x=\frac{6}{4}=\frac{3}{2}$$

If both belts are used, it takes $1\frac{1}{2}$ minutes.

53.

	distance =	rate	· time
upstream	3	$x-6$	$\frac{3}{x-6}$
downstream	9	$x+6$	$\frac{9}{x+6}$

$$\frac{3}{x-6}=\frac{9}{x+6}$$
$$3(x+6)=9(x-6)$$
$$3x+18=9x-54$$
$$3x+72=9x$$
$$72=6x$$
$$12=x$$

Time upstream $=\frac{3}{12-6}=\frac{3}{6}=\frac{1}{2}$ hour.

Total time is $2\left(\frac{1}{2}\right)=1$ hour.

55. Convert times to rates.

$$\frac{1 \text{ task}}{4 \text{ hours}}=\frac{1}{4}\ \frac{\text{task}}{\text{hour}}$$
$$\frac{1 \text{ task}}{5 \text{ hours}}=\frac{1}{5}\ \frac{\text{task}}{\text{hour}}$$

Let x = the amount of time it takes them to survey a stretch of road together.

$$\frac{1 \text{ task}}{x \text{ hours}}=\frac{1}{x}\ \frac{\text{task}}{\text{hour}}$$
$$\frac{1}{4}+\frac{1}{5}=\frac{1}{x}$$
$$20x\cdot\frac{1}{4}+20x\cdot\frac{1}{5}=20x\cdot\frac{1}{x}$$
$$5x+4x=20$$
$$9x=20$$
$$x=\frac{20}{9}$$

It takes them $2\frac{2}{9}$ hours to survey a stretch of road together.

57.

	Time	Part done in one hour
Experienced	3	$\frac{1}{3}$
Apprentice	6	$\frac{1}{6}$
Together	x	$\frac{1}{x}$

$$\frac{1}{3}+\frac{1}{6}=\frac{1}{x}$$
$$6x\left(\frac{1}{3}\right)+6x\left(\frac{1}{6}\right)=6x\left(\frac{1}{x}\right)$$
$$2x+x=6$$
$$3x=6$$
$$x=2$$

It takes them 2 hours if they work together.

59. $$\frac{x}{5}=\frac{x+2}{3}$$
$$3x=5(x+2)$$
$$3x=5x+10$$
$$-2x=10$$
$$x=-5$$

The solution set is $\{-5\}$.

61. The LCD is 6.

$$\frac{x-3}{2}=\frac{x-5}{6}$$
$$6\left(\frac{x-3}{2}\right)=6\left(\frac{x-5}{6}\right)$$
$$3(x-3)=x-5$$
$$3x-9=x-5$$
$$2x-9=-5$$
$$2x=4$$
$$x=2$$

The solution set is $\{2\}$.

63. 5 ft 8 in. = 68 in.

$B = \frac{705w}{h^2}$

Replace B with 25 and h with 68.

$$25 = \frac{705w}{68^2}$$
$$25 = \frac{705w}{4624}$$
$$115,600 = 705w$$
$$w = \frac{115,600}{705} \approx 164$$

His or her weight would be 164 pounds.

65. Use $h = \frac{a}{1-\frac{s}{770}}$ with $a = 329.63$ and $s = 50$.

$$h = \frac{a}{1-\frac{s}{770}} = \frac{329.63}{1-\frac{50}{770}} \approx 352.52$$

A pitch of 352.52 is closest to the pitch of the musical note F. The observer hears a higher pitch than the actual siren.

67. $\frac{1}{R} = \frac{1}{R_1} + \frac{1}{R_2}$

$$\frac{1}{R} = \frac{1}{12} + \frac{1}{12}$$
$$\frac{1}{R} = \frac{2}{12}$$
$$2R = 12$$
$$R = 6$$

The combined resistance is 6 ohms.

Section 6.6 Practice Problems

1. $y = kx$

$$24 = k(8)$$
$$3 = k$$
$$y = 3x$$

2. $d = kw$

$$8 = k(56)$$
$$\frac{8}{56} = k$$
$$\frac{1}{7} = k$$
$$d = \frac{1}{7}w$$
$$d = \frac{1}{7}(85)$$
$$d = 12\frac{1}{7}$$

An 85-pound weight stretches the spring $12\frac{1}{7}$ inches.

3. $y = \frac{k}{x}$

$$6 = \frac{k}{3}$$
$$18 = k$$
$$y = \frac{18}{x}$$

4. $r = \frac{k}{t}$

$$24 = \frac{k}{5}$$
$$120 = k$$
$$r = \frac{120}{t}$$
$$r = \frac{120}{4}$$
$$r = 30$$

The rate needed to drive the same distance in
4 hours is 30 mph.

5. $A = kbh$

6. $y = \frac{k}{x^2}$

$24 = \frac{k}{(2)^2}$

$24 = \frac{k}{4}$

$96 = k$

$y = \frac{96}{x^2}$

7. $\text{Weight} = \frac{kwh^2}{l}$

$3 = \frac{k \cdot \frac{1}{3} \cdot (1)^2}{10}$

$30 = \frac{1}{3}k$

$90 = k$

$\text{Weight} = \frac{90wh^2}{l}$

$\text{Weight} = \frac{90(1)\left(\frac{1}{3}\right)^2}{9}$

$\text{Weight} = \frac{90 \cdot 1 \cdot \frac{1}{9}}{9}$

$\text{Weight} = \frac{10}{9} = 1\frac{1}{9}$

A similar beam can support $1\frac{1}{9}$ tons.

Exercise Set 6.6

1. $y = kx$

$4 = k(20)$

$k = \frac{1}{5}$

$y = \frac{1}{5}x$

3. $y = kx$

$6 = k(4)$

$k = \frac{3}{2}$

$y = \frac{3}{2}x$

5. $y = kx$

$7 = k\left(\frac{1}{2}\right)$

$k = 14$

$y = 14x$

7. $y = kx$

$0.2 = k(0.8)$

$k = 0.25$

$y = 0.25x$

9. $W = kr^3$

$1.2 = k \cdot 2^3$

$k = \frac{1.2}{8} = 0.15$

$W = 0.15r^3$

$= 0.15(3)^3$

$= 0.15(27)$

$= 4.05$

The ball weighs 4.05 pounds.

11. $P = kN$

$260{,}000 = k(442{,}000)$

$k = \frac{10}{17}$

$P = \frac{10}{17}N$

$P = \frac{10}{17}(348{,}000) \approx 204{,}706$

St. Louis produces 204,706 tons.

13. $y = \frac{k}{x}$
$6 = \frac{k}{5}$
$k = 30$
$y = \frac{30}{x}$

15. $y = \frac{k}{x}$
$100 = \frac{k}{7}$
$k = 700$
$y = \frac{700}{x}$

17. $y = \frac{k}{x}$
$\frac{1}{8} = \frac{k}{16}$
$k = 2$
$y = \frac{2}{x}$

19. $y = \frac{k}{x}$
$0.2 = \frac{k}{0.7}$
$k = 0.14$
$y = \frac{0.14}{x}$

21. $R = \frac{k}{T}$
$45 = \frac{k}{6}$
$k = 270$
$R = \frac{270}{5} = 54$
The car's speed is 54 mph.

23. $I = \frac{k}{R}$
$40 = \frac{k}{270}$
$k = 10,800$
$I = \frac{10,800}{R} = \frac{10,800}{150} = 72$
The current is 72 amps.

25. $I_1 = \frac{k}{d^2}$
Replace d by $2d$.
$I_2 = \frac{k}{(2d)^2} = \frac{k}{4d^2} = \frac{1}{4}I_1$
Thus, the intensity is divided by 4.

27. $x = kyz$

29. $r = kst^3$

31. $y = kx^3$
$9 = k(3)^3$
$9 = 27k$
$k = \frac{1}{3}$
$y = \frac{1}{3}x^3$

33. $y = k\sqrt{x}$
$0.4 = k\sqrt{4}$
$0.4 = 2k$
$\frac{0.4}{2} = k$
$0.2 = k$
$y = 0.2\sqrt{x}$

35. $y = \frac{k}{x^2}$

$0.052 = \frac{k}{5^2}$

$k = 1.3$

$y = \frac{1.3}{x^2}$

37. $y = kxz^3$

$120 = k(5)(2^3)$

$120 = k(5)(8)$

$120 = 40k$

$3 = k$

$y = 3xz^3$

39. $\text{Weight} = \frac{kwh^2}{l}$

$12 = \frac{k\left(\frac{1}{2}\right)\left(\frac{1}{3}\right)^2}{10}$

$120 = k \cdot \frac{1}{2} \cdot \frac{1}{9}$

$k = 2160$

$\text{Weight} = \frac{2160wh^2}{l} = \frac{2160\left(\frac{2}{3}\right)\left(\frac{1}{2}\right)^2}{16} = 22.5$

The beam can support 22.5 tons.

41. $V = kr^2h$

$32\pi = k(4)^2(6)$

$32\pi = k(16)(6)$

$32\pi = 96k$

$\frac{32\pi}{96} = k$

$\frac{\pi}{3} = k$

$V = \frac{\pi}{3}r^2h$

$V = \frac{\pi}{3}(3)^2(5)$

$V = 15\pi$

The volume is 15π cubic inches.

43. $I = \frac{k}{x^2}$

$80 = \frac{k}{2^2}$

$k = 320$

$I = \frac{320}{x^2}$

$5 = \frac{320}{x^2}$

$5x^2 = 320$

$x^2 = 64$

$x = 8$

The source is 8 feet from the light source.

45. y varies directly as x is written as $y = kx$.

47. a varies inversely as b is written as $a = \frac{k}{b}$.

49. y varies jointly as x and z is written as $y = kxz$.

51. y varies inversely as x^3 is written as $y = \frac{k}{x^3}$.

53. y varies directly as x and inversely as p^2 is written as $y = \frac{kx}{p^2}$.

55. Circumference:

$C = 2\pi \cdot \text{radius} = 2\pi(6\text{ cm}) = 12\pi\text{ cm}$

Area:

$A = \pi(\text{radius})^2 = \pi(6\text{ cm})^2 = 36\pi\text{ square cm}$

57. Circumference:

$C = 2\pi \cdot \text{radius} = 2\pi(7\text{ meters}) = 14\pi\text{ meters}$

Area:

$A = \pi(\text{radius})^2$

$= \pi(7\text{ meters})^2$

$= 49\pi\text{ square meters}$

59. $m = \frac{y_2 - y_1}{x_2 - x_1} = \frac{6-6}{-2-3} = \frac{0}{-5} = 0$

61. $m = \frac{y_2 - y_1}{x_2 - x_1} = \frac{-2-(-1)}{5-4} = \frac{-2+1}{1} = -1$

63. $y = \frac{2}{3}x$ is an example of direct variation; a.

65. $y = 9ab$ is an example of joint variation; c.

67. Since the volume varies jointly as the height and the square of the radius, $V = khr^2$.

Replace h by $\frac{1}{2}h$ and r by $2r$.

$$V = k\left(\frac{1}{2}h\right)(2r)^2$$
$$= k\left(\frac{1}{2}h\right)(4r^2)$$
$$= 2khr^2$$

If the height is halved and the radius is doubled, the volume is multiplied by 2.

69. Since y varies directly as x^2, $y = kx^2$.

Replace x by $2x$.

$$y = k(2x)^2$$
$$y = 4kx^2$$

If x is doubled, y is multiplied by 4.

Chapter 6 Vocabulary Check

1. A rational expression whose numerator, denominator, or both contain one or more rational expressions is called a complex fraction.
2. In the equation $y = kx$, y varies directly as x.
3. In the equation $y = \frac{k}{x}$, y varies inversely as x.
4. The least common denominator of a list of rational expressions is a polynomial of least degree whose factors include the denominator factors in the list.
5. In the equation $y = kxz$, y varies jointly as x and z.
6. The expressions $(x - 5)$ and $(5 - x)$ are called opposites.
7. A rational expression is an expression that can be written as the quotient $\frac{P}{Q}$ of two polynomials P and Q as long as Q is not 0.
8. Which is an expression and which is an equation? An example of an equation is $\frac{2}{x} + \frac{2}{x^2} = 7$ and an example of an expression is $\frac{2}{x} + \frac{5}{x^2}$.

Chapter 6 Review

1. 7 is never 0 so the domain of $f(x) = \frac{3-5x}{7}$ is $\{x|x \text{ is a real number}\}$.

2. 11 is never 0 so the domain of $g(x) = \frac{2x+4}{11}$ is $\{x|x \text{ is a real number}\}$.

3. $x - 5 = 0$
   $x = 5$

   The domain of $F(x) = \frac{-3x^2}{x-5}$ is $\{x|x \text{ is a real number and } x \neq 5\}$.

4. $3x - 12 = 0$
   $3x = 12$
   $x = 4$

   The domain of $h(x) = \frac{4x}{3x-12}$ is $\{x|x \text{ is a real number and } x \neq 4\}$

5. $x^2+8x=0$
$x(x+8)=0$
$x=0$ or $x+8=0$
$x=0$ or $x=-8$

The domain of $f(x)=\dfrac{x^3+2}{x^2+8x}$ is $\{x|x \text{ is a real number and } x \neq 0, x \neq -8\}$.

6. $3x^2-48=0$
$3(x^2-16)=0$
$3(x+4)(x-4)=0$
$x+4=0$ or $x-4=0$
$x=-4$ or $x=4$

The domain of $G(x)=\dfrac{20}{3x^2-48}$ is $\{x|x \text{ is a real number and } x \neq -4, x \neq 4\}$.

7. $$\frac{x-12}{12-x}=\frac{x-12}{-(x-12)}=-1$$

8. $$\frac{5x-15}{25x-75}=\frac{5(x-3)}{25(x-3)}=\frac{5}{25}=\frac{1}{5}$$

9. $$\frac{2x}{2x^2-2x}=\frac{2x}{2x(x-1)}=\frac{1}{x-1}$$

10. $$\frac{x+7}{x^2-49}=\frac{x+7}{(x-7)(x+7)}=\frac{1}{x-7}$$

11. $$\frac{2x^2+4x-30}{x^2+x-20}=\frac{(2x-6)(x+5)}{(x+5)(x-4)}=\frac{2x-6}{x-4}=\frac{2(x-3)}{x-4}$$

12. $C(x)=\dfrac{35x+4200}{x}$

a. $$C(50)=\frac{35(50)+4200}{50}=\frac{1750+4200}{50}=\frac{5950}{50}=119$$
The average cost is \$119.

b. $$C(100)=\frac{35(100)+4200}{100}=\frac{3500+4200}{100}=\frac{7700}{100}=77$$
The average cost is \$77.

c. It will decrease.

13. $$\frac{4-x}{5}\cdot\frac{15}{2x-8}=\frac{4-x}{5}\cdot\frac{5\cdot 3}{2(x-4)}=\frac{-1(x-4)}{5}\cdot\frac{5\cdot 3}{2(x-4)}=-\frac{3}{2}$$

14. $$\frac{x^2-6x+9}{2x^2-18}\cdot\frac{4x+12}{5x-15}=\frac{(x-3)(x-3)}{2(x^2-9)}\cdot\frac{4(x+3)}{5(x-3)}=\frac{(x-3)(x-3)}{2(x-3)(x+3)}\cdot\frac{4(x+3)}{5(x-3)}=\frac{4}{10}=\frac{2}{5}$$

15. $\frac{a-4b}{a^2+ab}\cdot\frac{b^2-a^2}{8b-2a}$
$=\frac{a-4b}{a(a+b)}\cdot\frac{(b-a)(b+a)}{2(4b-a)}$
$=\frac{-(4b-a)}{a(a+b)}\cdot-\frac{(a-b)(a+b)}{2(4b-a)}$
$=\frac{a-b}{2a}$

16. $\frac{x^2-x-12}{2x^2-32}\cdot\frac{x^2+8x+16}{3x^2+21x+36}$
$=\frac{(x-4)(x+3)}{2(x^2-16)}\cdot\frac{(x+4)(x+4)}{3(x^2+7x+12)}$
$=\frac{(x-4)(x+3)}{2(x-4)(x+4)}\cdot\frac{(x+4)(x+4)}{3(x+4)(x+3)}$
$=\frac{1}{2\cdot 3}$
$=\frac{1}{6}$

17. $\frac{4x+8y}{3}\div\frac{5x+10y}{9}=\frac{4x+8y}{3}\cdot\frac{9}{5x+10y}$
$=\frac{4(x+2y)}{3}\cdot\frac{3\cdot 3}{5(x+2y)}$
$=\frac{12}{5}$

18. $\frac{x^2-25}{3}\div\frac{x^2-10x+25}{x^2-x-20}$
$=\frac{x^2-25}{3}\cdot\frac{x^2-x-20}{x^2-10x+25}$
$=\frac{(x+5)(x-5)}{3}\cdot\frac{(x-5)(x+4)}{(x-5)(x-5)}$
$=\frac{(x+5)(x+4)}{3}$

19. $\frac{a-4b}{a^2+ab}\div\frac{20b-5a}{b^2-a^2}$
$=\frac{a-4b}{a^2+ab}\cdot\frac{b^2-a^2}{20b-5a}$
$=\frac{a-4b}{a(a+b)}\cdot\frac{(b+a)(b-a)}{5(4b-a)}$
$=\frac{a-4b}{a(a+b)}\cdot\frac{-(a+b)(a-b)}{-5(a-4b)}$
$=\frac{a-b}{5a}$

20. $\frac{3x+3}{x-1}\div\frac{x^2-6x-7}{x^2-1}=\frac{3x+3}{x-1}\cdot\frac{x^2-1}{x^2-6x-7}$
$=\frac{3(x+1)}{x-1}\cdot\frac{(x+1)(x-1)}{(x-7)(x+1)}$
$=\frac{3(x+1)}{x-7}$

21. $\frac{2x-x^2}{x^3-8}\div\frac{x^2}{x^2+2x+4}$
$=\frac{2x-x^2}{x^3-8}\cdot\frac{x^2+2x+4}{x^2}$
$=\frac{x(2-x)}{(x-2)(x^2+2x+4)}\cdot\frac{x^2+2x+4}{x^2}$
$=\frac{-x(x-2)}{(x-2)(x^2+2x+4)}\cdot\frac{x^2+2x+4}{x^2}$
$=-\frac{1}{x}$

22. $\frac{5x-15}{3-x}\cdot\frac{x+2}{10x+20}\cdot\frac{x^2-9}{x^2-x-6}$
$=\frac{5(x-3)}{-(x-3)}\cdot\frac{x+2}{10(x+2)}\cdot\frac{(x+3)(x-3)}{(x-3)(x+2)}$
$=-\frac{x+3}{2(x+2)}$

23. $4x^2y^5 = 2\cdot 2x^2y^5$
$10x^2y^4 = 2\cdot 5x^2y^4$
$6y^4 = 2\cdot 3y^4$
The LCD of $4x^2y^5$, $10x^2y^4$, and $6y^4$ is $2\cdot 2\cdot 5\cdot 3x^2y^5 = 60x^2y^5$.

24. The LCD of $2x$ and $x-2$ is $2x(x-2)$.

25. The LCD of $5x$ and $x-5$ is $5x(x-5)$.

26. $5x^3 = 5\cdot x^3$
$x^2+3x-28 = (x+7)(x-4)$
$10x^2-30x = 10x(x-3) = 2\cdot 5\cdot x\cdot(x-3)$
The LCD of $5x^3$, $x^2+3x-28$ and $10x^2-30x$ is $2\cdot 5\cdot x^3\cdot(x-4)(x+7)(x-3)$
$= 10x^3(x-4)(x+7)(x-3)$.

27. $\frac{4}{x-4}+\frac{x}{x-4}=\frac{4+x}{x-4}$

28. $\frac{4}{3x^2}+\frac{2}{3x^2}=\frac{4+2}{3x^2}=\frac{6}{3x^2}=\frac{2}{x^2}$

29.
$$\begin{aligned}\frac{1}{x-2}-\frac{1}{4-2x} &= \frac{1}{x-2}-\frac{1}{2(2-x)}\\ &= \frac{1}{x-2}-\frac{-1}{2(x-2)}\\ &= \frac{1}{x-2}\cdot\frac{2}{2}-\frac{-1}{2(x-2)}\\ &= \frac{2}{2(x-2)}-\frac{-1}{2(x-2)}\\ &= \frac{2+1}{2(x-2)}\\ &= \frac{3}{2(x-2)}\end{aligned}$$

30.
$$\begin{aligned}\frac{1}{10-x}+\frac{x-1}{x-10} &= \frac{-1}{x-10}+\frac{x-1}{x-10}\\ &= \frac{-1+x-1}{x-10}\\ &= \frac{x-2}{x-10}\end{aligned}$$

31.
$$\begin{aligned}&\frac{x}{9-x^2}-\frac{2}{5x-15}\\ &= \frac{-x}{x^2-9}-\frac{2}{5x-15}\\ &= \frac{-x}{(x-3)(x+3)}-\frac{2}{5(x-3)}\\ &= \frac{-x}{(x-3)(x+3)}\cdot\frac{5}{5}-\frac{2}{5(x-3)}\cdot\frac{x+3}{x+3}\\ &= \frac{-5x}{5(x-3)(x+3)}-\frac{2(x+3)}{5(x-3)(x+3)}\\ &= \frac{-5x-2(x+3)}{5(x-3)(x+3)}\\ &= \frac{-5x-2x-6}{5(x-3)(x+3)}\\ &= \frac{-7x-6}{5(x-3)(x+3)}\end{aligned}$$

32.
$$\begin{aligned}2x+1-\frac{1}{x-3} &= 2x\cdot\frac{x-3}{x-3}+1\cdot\frac{x-3}{x-3}-\frac{1}{x-3}\\ &= \frac{2x^2-6x}{x-3}+\frac{x-3}{x-3}-\frac{1}{x-3}\\ &= \frac{2x^2-6x+x-3-1}{x-3}\\ &= \frac{2x^2-5x-4}{x-3}\end{aligned}$$

33. $\dfrac{2}{a^2-2a+1}+\dfrac{3}{a^2-1}$

$=\dfrac{2}{(a-1)(a-1)}+\dfrac{3}{(a+1)(a-1)}$

$=\dfrac{2}{(a-1)(a-1)}\cdot\dfrac{a+1}{a+1}+\dfrac{3}{(a+1)(a-1)}\cdot\dfrac{(a-1)}{(a-1)}$

$=\dfrac{2(a+1)}{(a-1)^2(a+1)}+\dfrac{3(a-1)}{(a-1)^2(a+1)}$

$=\dfrac{2a+2}{(a-1)^2(a+1)}+\dfrac{3a-3}{(a-1)^2(a+1)}$

$=\dfrac{2a+2+3a-3}{(a-1)^2(a+1)}$

$=\dfrac{5a-1}{(a-1)^2(a+1)}$

34. $\dfrac{x}{9x^2+12x+16}-\dfrac{3x+4}{27x^3-64}=\dfrac{x}{9x^2+12x+16}-\dfrac{3x+4}{(3x-4)(9x^2+12x+16)}$

$=\dfrac{x}{9x^2+12x+16}\cdot\dfrac{3x-4}{3x-4}-\dfrac{3x+4}{(3x-4)(9x^2+12x+16)}$

$=\dfrac{3x^2-4x-3x-4}{(3x-4)(9x^2+12x+16)}$

$=\dfrac{3x^2-7x-4}{(3x-4)(9x^2+12x+16)}$

35. $\dfrac{2}{x-1}-\dfrac{3x}{3x-3}+\dfrac{1}{2x-2}=\dfrac{2}{x-1}-\dfrac{3x}{3(x-1)}+\dfrac{1}{2(x-1)}$

$=\dfrac{2}{x-1}\cdot\dfrac{6}{6}-\dfrac{3x}{3(x-1)}\cdot\dfrac{2}{2}+\dfrac{1}{2(x-1)}\cdot\dfrac{3}{3}$

$=\dfrac{12}{6(x-1)}-\dfrac{6x}{6(x-1)}+\dfrac{3}{6(x-1)}$

$=\dfrac{12-6x+3}{6(x-1)}$

$=\dfrac{15-6x}{6(x-1)}$

$=\dfrac{3(5-2x)}{6(x-1)}$

$=\dfrac{5-2x}{2(x-1)}$

36. $\text{Perimeter} = \frac{1}{x}+\frac{1}{x}+\frac{1}{x}+\frac{2}{x}+\frac{2}{x}+\frac{3}{2x}+\frac{5}{2x}$

$= \frac{7}{x}+\frac{8}{2x}$

$= \frac{7}{x}\cdot\frac{2}{2}+\frac{8}{2x}$

$= \frac{14}{2x}+\frac{8}{2x}$

$= \frac{14+8}{2x}$

$= \frac{22}{2x}$

$= \frac{11}{x}$

37. $\frac{1-\frac{3x}{4}}{2+\frac{x}{4}} = \frac{4\left(1-\frac{3x}{4}\right)}{4\left(2+\frac{x}{4}\right)} = \frac{4-3x}{8+x}$

38. $\frac{\frac{x^2}{15}}{\frac{x+1}{5x}} = \frac{x^2}{15} \div \frac{x+1}{5x} = \frac{x^2}{15}\cdot\frac{5x}{x+1} = \frac{x^3}{3(x+1)}$

39. $\frac{2-\frac{3}{2x}}{x-\frac{2}{5x}} = \frac{10x\left(2-\frac{3}{2x}\right)}{10x\left(x-\frac{2}{5x}\right)}$

$= \frac{20x-15}{10x^2-4}$

$= \frac{5(4x-3)}{2(5x^2-2)}$

40. $\frac{1+\frac{x}{y}}{\frac{x^2}{y^2}-1} = \frac{y^2\left(1+\frac{x}{y}\right)}{y^2\left(\frac{x^2}{y^2}-1\right)}$

$= \frac{y^2+xy}{x^2-y^2}$

$= \frac{y(y+x)}{(x+y)(x-y)}$

$= \frac{y}{x-y}$

41. $\frac{\frac{5}{x}+\frac{1}{xy}}{\frac{3}{x^2}} = \frac{x^2y\left(\frac{5}{x}+\frac{1}{xy}\right)}{x^2y\left(\frac{3}{x^2}\right)} = \frac{5xy+x}{3y} = \frac{x(5y+1)}{3y}$

42. $\frac{\frac{x}{3}-\frac{3}{x}}{1+\frac{3}{x}} = \frac{3x\left(\frac{x}{3}-\frac{3}{x}\right)}{3x\left(1+\frac{3}{x}\right)}$

$= \frac{x^2-9}{3x+9}$

$= \frac{(x+3)(x-3)}{3(x+3)}$

$= \frac{x-3}{3}$

43. $\frac{\frac{1}{x-1}+1}{\frac{1}{x+1}-1} = \frac{(x+1)(x-1)\left(\frac{1}{x-1}+1\right)}{(x+1)(x-1)\left(\frac{1}{x+1}-1\right)}$

$= \frac{x+1+(x+1)(x-1)}{x-1-(x+1)(x-1)}$

$= \frac{x+1+x^2-1}{x-1-x^2+1}$

$= \frac{x+x^2}{x-x^2}$

$= \frac{x(1+x)}{x(1-x)}$

$= \frac{1+x}{1-x}$

44. $\frac{\frac{x-3}{x+3}+\frac{x+3}{x-3}}{\frac{x-3}{x+3}-\frac{x+3}{x-3}} = \frac{(x-3)(x+3)\left(\frac{x-3}{x+3}+\frac{x+3}{x-3}\right)}{(x-3)(x+3)\left(\frac{x-3}{x+3}-\frac{x+3}{x-3}\right)}$

$= \frac{(x-3)^2+(x+3)^2}{(x-3)^2-(x+3)^2}$

$= \frac{x^2-6x+9+x^2+6x+9}{x^2-6x+9-(x^2+6x+9)}$

$= \frac{2x^2+18}{x^2-6x+9-x^2-6x-9}$

$= \frac{2(x^2+9)}{-12x}$

$= -\frac{x^2+9}{6x}$

45. The LCD of x and 3 is $3x$.

$$\begin{aligned} \frac{3}{x}+\frac{1}{3}&=\frac{5}{x} \\ 3x\left(\frac{3}{x}+\frac{1}{3}\right)&=3x\left(\frac{5}{x}\right) \\ 3x\left(\frac{3}{x}\right)+3x\left(\frac{1}{3}\right)&=3x\left(\frac{5}{x}\right) \\ 9+x&=3\cdot 5 \\ 9+x&=15 \\ x&=6 \end{aligned}$$

The solution set is $\{6\}$.

46. The LCD is $2(5x-9)$.

$$\begin{aligned} \frac{2x+3}{5x-9}&=\frac{3}{2} \\ 2(5x-9)\cdot\frac{2x+3}{5x-9}&=2(5x-9)\cdot\frac{3}{2} \\ 2(2x+3)&=(5x-9)\cdot 3 \\ 4x+6&=15x-27 \\ 6&=11x-27 \\ 33&=11x \\ 3&=x \end{aligned}$$

47. The LCD is $(x-2)(x+2)$.

$$\begin{aligned} \frac{1}{x-2}-\frac{3x}{x^2-4}&=\frac{2}{x+2} \\ (x-2)(x+2)\left(\frac{1}{x-2}-\frac{3x}{x^2-4}\right)&=(x-2)(x+2)\left(\frac{2}{x+2}\right) \\ (x-2)(x+2)\cdot\frac{1}{x-2}-(x-2)(x+2)\cdot\frac{3x}{x^2-4}&=(x-2)(x+2)\cdot\frac{2}{x+2} \\ x+2-3x&=2(x-2) \\ -2x+2&=2x-4 \\ -4x+2&=-4 \\ -4x&=-6 \\ x&=\frac{-6}{-4}=\frac{3}{2} \end{aligned}$$

The solution set is $\left\{\frac{3}{2}\right\}$.

48. The LCD is $7x$.

$$\frac{7}{x}-\frac{x}{7}=0$$
$$7x\left(\frac{7}{x}-\frac{x}{7}\right)=7x(0)$$
$$7x\left(\frac{7}{x}\right)-7x\left(\frac{x}{7}\right)=0$$
$$49-x^2=0$$
$$(7-x)(7+x)=0$$
$$7-x=0 \quad \text{or} \quad 7+x=0$$
$$7=x \quad \text{or} \quad x=-7$$

The solution set is $\{-7, 7\}$.

49.
$$\frac{5}{x^2-7x}+\frac{4}{2x-14}$$
$$=\frac{5}{x(x-7)}+\frac{4}{2(x-7)}$$
$$=\frac{5}{x(x-7)}\cdot\frac{2}{2}+\frac{4}{2(x-7)}\cdot\frac{x}{x}$$
$$=\frac{10}{2x(x-7)}+\frac{4x}{2x(x-7)}$$
$$=\frac{10+4x}{2x(x-7)}$$
$$=\frac{2(5+2x)}{2x(x-7)}$$
$$=\frac{5+2x}{x(x-7)}$$

50. The LCD is x^2.

$$3-\frac{5}{x}-\frac{2}{x^2}=0$$
$$x^2\left(3-\frac{5}{x}-\frac{2}{x^2}\right)=x^2(0)$$
$$x^2\cdot 3-x^2\left(\frac{5}{x}\right)-x^2\left(\frac{2}{x^2}\right)=0$$
$$3x^2-5x-2=0$$
$$(3x+1)(x-2)=0$$
$$3x+1=0 \quad \text{or} \quad x-2=0$$
$$3x=-1 \quad \text{or} \quad x=2$$
$$x=-\frac{1}{3} \quad \text{or} \quad x=2$$

The solution set is $\left\{-\frac{1}{3}, 2\right\}$.

51.
$$\frac{4}{3-x}-\frac{7}{2x-6}+\frac{5}{x}$$
$$=\frac{-4}{x-3}-\frac{7}{2(x-3)}+\frac{5}{x}$$
$$=\frac{-4}{x-3}\cdot\frac{2x}{2x}-\frac{7}{2(x-3)}\cdot\frac{x}{x}+\frac{5}{x}\cdot\frac{2(x-3)}{2(x-3)}$$
$$=\frac{-8x}{2x(x-3)}-\frac{7x}{2x(x-3)}+\frac{10x-30}{2x(x-3)}$$
$$=\frac{-8x-7x+10x-30}{2x(x-3)}$$
$$=\frac{-5x-30}{2x(x-3)}$$
$$=\frac{-5(x+6)}{2x(x-3)}$$

52.
$$A=\frac{h(a+b)}{2}$$
$$2A=h(a+b)$$
$$\frac{2A}{h}=a+b$$
$$a=\frac{2A}{h}-b$$

53.
$$\frac{1}{R}=\frac{1}{R_1}+\frac{1}{R_2}$$
$$RR_1R_2\left(\frac{1}{R}\right)=RR_1R_2\left(\frac{1}{R_1}\right)+RR_1R_2\left(\frac{1}{R_2}\right)$$
$$R_1R_2=RR_2+RR_1$$
$$R_1R_2-RR_2=RR_1$$
$$R_2(R_1-R)=RR_1$$
$$R_2=\frac{RR_1}{R_1-R}$$

54.
$$I = \frac{E}{R+r}$$
$$I(R+r) = E$$
$$R+r = \frac{E}{I}$$
$$R = \frac{E}{I} - r$$

55.
$$A = P + Prt$$
$$A - P = Prt$$
$$\frac{A-P}{Pt} = r \text{ or } r = \frac{A-P}{Pt}$$

56.
$$H = \frac{kA(T_1 - T_2)}{L}$$
$$HL = kA(T_1 - T_2)$$
$$\frac{HL}{k(T_1 - T_2)} = A \text{ or } A = \frac{HL}{k(T_1 - T_2)}$$

57. Let x = the number.
$$x + 2\left(\frac{1}{x}\right) = 3$$
$$x\left[x + 2\left(\frac{1}{x}\right)\right] = x \cdot 3$$
$$x^2 + 2 = 3x$$
$$x^2 - 3x + 2 = 0$$
$$(x-1)(x-2) = 0$$
$$x - 1 = 0 \quad \text{or} \quad x - 2 = 0$$
$$x = 1 \quad \text{or} \quad x = 2$$
The numbers are 1 and 2.

58. Let x = the number.
$$\frac{3+x}{7+2x} = \frac{10}{21}$$
$$21(3+x) = 10(7+2x)$$
$$63 + 21x = 70 + 20x$$
$$63 + 1x = 70$$
$$x = 7$$
The number is 7.

59. Convert each time to a rate.
$$\frac{1 \text{ task}}{4 \text{ hours}} = \frac{1}{4} \cdot \frac{\text{task}}{\text{hour}}$$
$$\frac{1 \text{ task}}{5 \text{ hours}} = \frac{1}{5} \cdot \frac{\text{task}}{\text{hour}}$$
$$\frac{1 \text{ task}}{6 \text{ hours}} = \frac{1}{6} \cdot \frac{\text{task}}{\text{hour}}$$
Let x = amount of time required for all three boys to paint the fence together.
$$\frac{1 \text{ task}}{x \text{ hours}} = \frac{1}{x} \frac{\text{task}}{\text{hour}}$$
Adding we get
$$\frac{1}{4} + \frac{1}{5} + \frac{1}{6} = \frac{1}{x}$$
$$60x \cdot \frac{1}{4} + 60x \cdot \frac{1}{5} + 60x \cdot \frac{1}{6} = 60x \cdot \frac{1}{x}$$
$$15x + 12x + 10x = 60$$
$$37x = 60$$
$$x = \frac{60}{37} = 1\frac{23}{37}$$
It will take $1\frac{23}{37}$ hours for all three boys to paint the fence together.

60. Convert times to rates.
$$\frac{1 \text{ task}}{6 \text{ hours}} = \frac{1}{6} \cdot \frac{\text{task}}{\text{hour}}$$
$$\frac{1 \text{ task}}{4 \text{ hours}} = \frac{1}{4} \cdot \frac{\text{task}}{\text{hour}}$$
Let x = amount of time it takes Tom to type the mailing labels when working alone.
$$\frac{1 \text{ task}}{x \text{ hours}} = \frac{1}{x} \cdot \frac{\text{task}}{\text{hour}}$$
$$\frac{1}{6} + \frac{1}{x} = \frac{1}{4}$$
$$12x \cdot \frac{1}{6} + 12x \cdot \frac{1}{x} = 12x \cdot \frac{1}{4}$$
$$2x + 12 = 3x$$
$$12 = x$$
It takes Tom 12 hours to complete the task alone.

61. Let x = the speed of the current.

	distance =	rate ·	time
Upstream	72	$32-x$	$\frac{72}{32-x}$
Downstream	120	$32+x$	$\frac{120}{32+x}$

$$\frac{72}{32-x}=\frac{120}{32+x}$$
$$72(32+x)=120(32-x)$$
$$2304+72x=3840-120x$$
$$72x=1536-120x$$
$$192x=1536$$
$$x=8$$

The speed of the current is 8 mph.

62. Let x = the speed of the walker.

	distance =	rate ·	time
Jogger	14	$x+3$	$\frac{14}{x+3}$
Walker	8	x	$\frac{8}{x}$

$$\frac{14}{x+3}=\frac{8}{x}$$
$$14x=8(x+3)$$
$$14x=8x+24$$
$$6x=24$$
$$x=4$$

The speed of the walker is 4 mph.

63. $A=kB$

$$6=k14$$
$$k=\frac{6}{14}$$
$$A=\frac{6}{14}B=\frac{6}{14}(21)=9$$

64. $P=\frac{K}{V}$

$$1250=\frac{K}{2}$$
$$K=2500$$
$$P=\frac{2500}{V}$$
$$800=\frac{2500}{V}$$
$$800V=2500$$
$$V=3.125$$

When the pressure is 800 kilopascals, the volume is 3.125 cubic meters.

65. $\frac{22x+8}{11x+4}=\frac{2(11x+4)}{11x+4}=2$

66.
$$\frac{xy-3x+2y-6}{x^2+4x+4}=\frac{x(y-3)+2(y-3)}{(x+2)(x+2)}$$
$$=\frac{(x+2)(y-3)}{(x+2)(x+2)}$$
$$=\frac{y-3}{x+2}$$

67.
$$\frac{2}{5x}\div\frac{4-18x}{6-27x}=\frac{2}{5x}\cdot\frac{6-27x}{4-18x}$$
$$=\frac{2}{5x}\cdot\frac{3(2-9x)}{2(2-9x)}$$
$$=\frac{3}{5x}$$

68.
$$\frac{7x+28}{2x+4}\div\frac{x^2+2x-8}{x^2-2x-8}$$
$$=\frac{7x+28}{2x+4}\cdot\frac{x^2-2x-8}{x^2+2x-8}$$
$$=\frac{7(x+4)}{2(x+2)}\cdot\frac{(x-4)(x+2)}{(x+4)(x-2)}$$
$$=\frac{7(x-4)}{2(x-2)}$$

69. $$\frac{5a^2-20}{a^2+2a^2+a+2} \div \frac{7a}{a^3+a} = \frac{5a^2-20}{a^3+2a^2+a+2} \cdot \frac{a^3+a}{7a}$$
$$= \frac{5(a^2-4)}{a^2(a+2)+1(a+2)} \cdot \frac{a(a^2+1)}{7a}$$
$$= \frac{5(a+2)(a-2)}{(a+2)(a^2+1)} \cdot \frac{a(a^2+1)}{7a}$$
$$= \frac{5(a-2)}{7}$$

70. $$\frac{4a+8}{5a^2-20} \cdot \frac{3a^2-6a}{a+3} \div \frac{2a^2}{5a+15} = \frac{4a+8}{5a^2-20} \cdot \frac{3a^2-6a}{a+3} \cdot \frac{5a+15}{2a^2}$$
$$= \frac{4(a+2)}{5(a^2-4)} \cdot \frac{3a(a-2)}{a+3} \cdot \frac{5(a+3)}{2a^2}$$
$$= \frac{4(a+2)}{5(a-2)(a+2)} \cdot \frac{3a(a-2)}{a+3} \cdot \frac{5(a+3)}{2a^2}$$
$$= \frac{6}{a}$$

71. $$\frac{7}{2x} + \frac{5}{6x} = \frac{7}{2x} \cdot \frac{3}{3} + \frac{5}{6x}$$
$$= \frac{21}{6x} + \frac{5}{6x}$$
$$= \frac{21+5}{6x}$$
$$= \frac{26}{6x}$$
$$= \frac{13}{3x}$$

72. $$\frac{x-2}{x+1} - \frac{x-3}{x-1} = \frac{x-2}{x+1} \cdot \frac{x-1}{x-1} - \frac{x-3}{x-1} \cdot \frac{x+1}{x+1}$$
$$= \frac{(x-2)(x-1)}{(x+1)(x-1)} - \frac{(x-3)(x+1)}{(x+1)(x-1)}$$
$$= \frac{x^2-3x+2}{(x+1)(x-1)} - \frac{x^2-2x-3}{(x+1)(x-1)}$$
$$= \frac{x^2-3x+2-x^2+2x+3}{(x+1)(x-1)}$$
$$= \frac{-x+5}{(x+1)(x-1)}$$

73. $$\frac{2x+1}{x^2+x-6}+\frac{2-x}{x^2+x-6}=\frac{2x+1+2-x}{x^2+x-6}$$
$$=\frac{x+3}{x^2+x-6}$$
$$=\frac{x+3}{(x+3)(x-2)}$$
$$=\frac{1}{x-2}$$

74. $$\frac{2}{x^2-16}-\frac{3x}{x^2+8x+16}+\frac{3}{x+4}=\frac{2}{(x+4)(x-4)}-\frac{3x}{(x+4)^2}+\frac{3}{x+4}$$
$$=\frac{2}{(x+4)(x-4)}\cdot\frac{x+4}{x+4}-\frac{3x}{(x+4)^2}\cdot\frac{x-4}{x-4}+\frac{3}{x+4}\cdot\frac{(x+4)(x-4)}{(x+4)(x-4)}$$
$$=\frac{2x+8}{(x+4)^2(x-4)}-\frac{3x^2-12x}{(x+4)^2(x-4)}+\frac{3x^2-48}{(x+4)^2(x-4)}$$
$$=\frac{2x+8-3x^2+12x+3x^2-48}{(x+4)^2(x-4)}$$
$$=\frac{14x-40}{(x+4)^2(x-4)}$$
$$=\frac{2(7x-20)}{(x+4)^2(x-4)}$$

75. $$\frac{\frac{1}{x}-\frac{2}{3x}}{\frac{5}{2x}-\frac{1}{3}}=\frac{6x\left(\frac{1}{x}-\frac{2}{3x}\right)}{6x\left(\frac{5}{2x}-\frac{1}{3}\right)}=\frac{6-4}{15-2x}=\frac{2}{15-2x}$$

76. $$\frac{2}{1-\frac{2}{x}}=\frac{x(2)}{x\left(1-\frac{2}{x}\right)}=\frac{2x}{x-2}$$

77. $$\frac{\frac{x^2+5x-6}{4x+3}}{\frac{(x+6)^2}{8x+6}}=\frac{x^2+5x-6}{4x+3}\div\frac{(x+6)^2}{8x+6}$$
$$=\frac{x^2+5x-6}{4x+3}\cdot\frac{8x+6}{(x+6)^2}$$
$$=\frac{(x+6)(x-1)}{4x+3}\cdot\frac{2(4x+3)}{(x+6)^2}$$
$$=\frac{2(x-1)}{x+6}$$

78. $\dfrac{\frac{3}{x-1}-\frac{2}{1-x}}{\frac{2}{x-1}-\frac{2}{x}} = \dfrac{\frac{3}{x-1}-\frac{-2}{x-1}}{\frac{2}{x-1}-\frac{2}{x}}$

$$= \frac{(x)(x-1)\left(\frac{3}{x-1}-\frac{-2}{x-1}\right)}{(x)(x-1)\left(\frac{2}{x-1}-\frac{2}{x}\right)}$$

$$= \frac{3x+2x}{2x-2(x-1)}$$

$$= \frac{5x}{2x-2x+2}$$

$$= \frac{5x}{2}$$

79. $4+\dfrac{8}{x}=8$

The LCD is x.

$$x(4)+x\left(\frac{8}{x}\right)=x(8)$$

$$4x+8=8x$$

$$8=4x$$

$$x=2$$

The solution set is $\{2\}$.

80. $\dfrac{x-2}{x^2-7x+10}=\dfrac{1}{5x-10}-\dfrac{1}{x-5}$

The LCD is $5(x-2)(x-5)$.

$$5(x-2)(x-5)\cdot\frac{x-2}{x^2-7x+10}=5(x-2)(x-5)\cdot\frac{1}{5x-10}-5(x-2)(x-5)\cdot\frac{1}{x-5}$$

$$5(x-2)=x-5-5(x-2)$$

$$5x-10=x-5-5x+10$$

$$5x-10=-4x+5$$

$$9x-10=5$$

$$9x=15$$

$$x=\frac{15}{9}=\frac{5}{3}$$

The solution set is $\left\{\frac{5}{3}\right\}$.

81. Let x be the numerator of a fraction. Then $x + 2$ is the denominator.

$$\frac{x-3}{x+2+5}=\frac{2}{3}$$
$$\frac{x-3}{x+7}=\frac{2}{3}$$
$$3(x-3)=2(x+7)$$
$$3x-9=2x+14$$
$$x-9=14$$
$$x=23$$
$$x+2=23+2=25$$

The fraction is $\frac{23}{25}$.

82. Let x be the first even integer and $x + 2$ be the next consecutive even integer.

$$\frac{1}{x}+\frac{1}{x+2}=-\frac{9}{40}$$
$$40x(x+2)\cdot\frac{1}{x}+40x(x+2)\cdot\frac{1}{x+2}=40x(x+2)\cdot\frac{-9}{40}$$
$$40(x+2)+40x=-9x(x+2)$$
$$40x+80+40x=-9x^2-18x$$
$$80x+80=-9x^2-18x$$
$$9x^2+98x+80=0$$
$$(9x+8)(x+10)=0$$

$9x+8=0$ or $x+10=0$

$9x=-8$ or $x=-10$

$x=-\frac{8}{9}$ or $x=-10$

$x+2=-10+2=-8$

The two integers are –10 and –8.

83. Convert the time to rates.

$$\frac{1\text{ task}}{2.5\text{ hours}}=\frac{1}{2.5}\ \frac{\text{task}}{\text{hour}}$$
$$\frac{1\text{ task}}{2\text{ hours}}=\frac{1}{2}\ \frac{\text{task}}{\text{hour}}$$

Let x = time it takes to empty a full tank if both pipes are open.

$$-\frac{1}{2.5}+\frac{1}{2}=\frac{1}{x}$$
$$(2.5)(2x)\cdot-\frac{1}{2.5}+(2.5)(2x)\cdot\frac{1}{2}=(2.5)(2x)\cdot\frac{1}{x}$$
$$-2x+2.5x=5$$
$$0.5x=5$$
$$x=10$$

It takes 10 hours to empty a full tank if both pipes are open.

84. Let x = the speed of the car and $x + 430$ = the speed of the jet.

	distance =	rate	· time
Car	210	x	$\frac{210}{x}$
Jet	1715	$x + 430$	$\frac{1715}{x+430}$

$$\frac{210}{x} = \frac{1715}{x+430}$$
$$210(x+430) = 1715x$$
$$210x + 70,300 = 1715x$$
$$90,300 = 1505x$$
$$60 = x$$

$x + 430 = 60 + 430 = 490$

The speed of the jet is 490 mph.

85. One train traveled 382 miles in 6 hours and the second train traveled $382 - 112 = 270$ miles in 6 hours.

Recall $D = rt$ or $r = \frac{D}{t}$.

$r = \frac{82}{6} = 63\frac{2}{3}$ $\qquad$ $r = \frac{270}{6} = 45$

The speeds of the trains are $63\frac{2}{3}$ mph and 45 mph.

86.
$$C = \frac{k}{D}$$
$$12 = \frac{k}{8}$$
$$96 = k$$
$$C = \frac{96}{D} = \frac{96}{24} = 4$$

87.
$$A = kr^2$$
$$36\pi = k(3)^2$$
$$36\pi = 9k$$
$$k = 4\pi$$
$$A = 4\pi r^2$$
$$A = 4\pi(4)^2 = 4\pi \cdot 16 = 64\pi$$

The surface area is 64π square inches.

Chapter 6 Test

1.
$$1 - x = 0$$
$$1 = x$$

The domain of $f(x) = \frac{5x^2}{1-x}$ is $\{x|x \text{ is a real number and } x \neq 1\}$.

2.
$$x^2 + 4x + 3 = 0$$
$$(x+3)(x+1) = 0$$
$$x + 3 = 0 \quad \text{or} \quad x + 1 = 0$$
$$x = -3 \quad \text{or} \quad x = -1$$

The domain of $g(x) = \frac{9x^2 - 9}{x^2 + 4x + 3}$ is $\{x|x \text{ is a real number and } x \neq -3 \text{ and } x \neq -1\}$.

3. $\frac{7x-21}{24-8x} = \frac{7(x-3)}{8(3-x)} = \frac{7(x-3)}{-8(x-3)} = -\frac{7}{8}$

4. $\frac{x^2-4x}{x^2+5x-36} = \frac{x(x-4)}{(x+9)(x-4)} = \frac{x}{x+9}$

5.
$$\frac{x^3-8}{x-2} = \frac{x^3-2^3}{x-2}$$
$$= \frac{(x-2)(x^2+2x+4)}{x-2}$$
$$= x^2 + 2x + 4$$

6. $\frac{2x^3+16}{6x^2+12x}\cdot\frac{5}{x^2-2x+4}$
$=\frac{2(x^3+8)}{6x(x+2)}\cdot\frac{5}{x^2-2x+4}$
$=\frac{2(x+2)(x^2-2x+4)}{6x(x+2)}\cdot\frac{5}{x^2-2x+4}$
$=\frac{5}{3x}$

7. $\frac{5}{4x^3}+\frac{7}{4x^3}=\frac{5+7}{4x^3}=\frac{12}{4x^3}=\frac{3}{x^3}$

8. $\frac{3x^2-12}{x^2+2x-8}\div\frac{6x+18}{x+4}$
$=\frac{3x^2-12}{x^2+2x-8}\cdot\frac{x+4}{6x+18}$
$=\frac{3(x^2-4)}{(x+4)(x-2)}\cdot\frac{x+4}{6(x+3)}$
$=\frac{3(x+2)(x-2)}{(x+4)(x-2)}\cdot\frac{(x+4)}{6(x+3)}$
$=\frac{x+2}{2(x+3)}$

9. $\frac{4x-12}{2x-9}\div\frac{3-x}{4x^2-81}\cdot\frac{x+3}{5x+15}$
$=\frac{4x-12}{2x-9}\cdot\frac{4x^2-81}{3-x}\cdot\frac{x+3}{5x+15}$
$=\frac{4(x-3)}{2x-9}\cdot\frac{(2x+9)(2x-9)}{-(x-3)}\cdot\frac{x+3}{5(x+3)}$
$=\frac{4(2x+9)}{-5}$
$=-\frac{4(2x+9)}{5}$

10. $\frac{3+2x}{10-x}+\frac{13+x}{x-10}=\frac{-(3+2x)}{x-10}+\frac{13+x}{x-10}$
$=\frac{-3-2x+13+x}{x-10}$
$=\frac{-x+10}{x-10}$
$=-\frac{(x-10)}{x-10}$
$=-1$

11. $\frac{2x^2+7}{2x^4-18x^2}-\frac{6x+7}{2x^4-18x^2}$
$=\frac{2x^2+7-6x-7}{2x^4-18x^2}$
$=\frac{2x^2-6x}{2x^4-18x^2}$
$=\frac{2x(x-3)}{2x^2(x^2-9)}$
$=\frac{2x(x-3)}{2x^2(x-3)(x+3)}$
$=\frac{1}{x(x+3)}$

12. $\frac{3}{x^2-x-6}+\frac{2}{x^2-5x+6}$
$=\frac{3}{(x-3)(x+2)}+\frac{2}{(x-3)(x-2)}$
$=\frac{3}{(x-3)(x+2)}\cdot\frac{x-2}{x-2}+\frac{2}{(x-3)(x-2)}\cdot\frac{x+2}{x+2}$
$=\frac{3(x-2)}{(x-3)(x+2)(x-2)}+\frac{2(x+2)}{(x-3)(x+2)(x-2)}$
$=\frac{3x-6}{(x-3)(x+2)(x-2)}+\frac{2x+4}{(x-3)(x+2)(x-2)}$
$=\frac{3x-6+2x+4}{(x-3)(x+2)(x-2)}$
$=\frac{5x-2}{(x-3)(x+2)(x-2)}$

13. $3x - 21 = 3(x - 7)$
$2x - 14 = 2(x - 7)$
The LCD is $3 \cdot 2(x - 7) = 6(x - 7)$.

$$\frac{5}{x-7} - \frac{2x}{3x-21} + \frac{x}{2x-14} = \frac{6(5)}{6(x-7)} - \frac{2(2x)}{2(3x-21)} + \frac{3(x)}{3(2x-14)}$$
$$= \frac{30-4x+3x}{6(x-7)}$$
$$= \frac{30-x}{6(x-7)}$$

14. $$\frac{3x}{5} \cdot \left(\frac{5}{x} - \frac{5}{2x}\right) = \frac{3x}{5} \cdot \left(\frac{5}{x} \cdot \frac{2}{2} - \frac{5}{2x}\right)$$
$$= \frac{3x}{5}\left(\frac{10}{2x} - \frac{5}{2x}\right)$$
$$= \frac{3x}{5}\left(\frac{10-5}{2x}\right)$$
$$= \frac{3x}{5}\left(\frac{5}{2x}\right)$$
$$= \frac{3}{2}$$

15. $$\frac{\frac{5}{x} - \frac{7}{3x}}{\frac{9}{8x} - \frac{1}{x}} = \frac{24x\left(\frac{5}{x} - \frac{7}{3x}\right)}{24x\left(\frac{9}{8x} - \frac{1}{x}\right)} = \frac{120-56}{27-24} = \frac{64}{3}$$

16. $$\frac{\frac{x^2-5x+6}{x+3}}{\frac{x^2-4x+4}{x^2-9}} = \frac{x^2-5x+6}{x+3} \div \frac{x^2-4x+4}{x^2-9}$$
$$= \frac{x^2-5x+6}{x+3} \cdot \frac{x^2-9}{x^2-4x+4}$$
$$= \frac{(x-3)(x-2)}{x+3} \cdot \frac{(x+3)(x-3)}{(x-2)(x-2)}$$
$$= \frac{(x-3)^2}{x-2}$$

17. The LCD is $7(3x-7)$.

$$\frac{5x+3}{3x-7}=\frac{19}{7}$$
$$7(3x-7)\left(\frac{5x+3}{3x-7}\right)=7(3x-7)\left(\frac{19}{7}\right)$$
$$7(5x+3)=19(3x-7)$$
$$35x+21=57x-133$$
$$21=22x-133$$
$$154=22x$$
$$7=x$$

The solution set is $\{7\}$.

18. The LCD is $x-4$.

$$\frac{x}{x-4}=3-\frac{4}{x-4}$$
$$(x-4)\frac{x}{x-4}=(x-4)\cdot 3-(x-4)\cdot\frac{4}{x-4}$$
$$x=3(x-4)-4$$
$$x=3x-12-4$$
$$x=3x-16$$
$$-2x=-16$$
$$x=8$$

The solution set is $\{8\}$.

19. The LCD is $5x(x+2)$.

$$\frac{3}{x+2}-\frac{1}{5x}=\frac{2}{5x^2+10x}$$
$$5x(x+2)\cdot\frac{3}{x+2}-5x(x+2)\cdot\frac{1}{5x}=5x(x+2)\cdot\frac{2}{5x^2+10x}$$
$$5x\cdot 3-(x+2)=2$$
$$15x-x-2=2$$
$$14x-2=2$$
$$14x=4$$
$$x=\frac{4}{14}=\frac{2}{7}$$

The solution set is $\left\{\frac{2}{7}\right\}$.

20. The LCD is x.

$$\frac{x^2+8}{x}-1=\frac{2(x+4)}{x}$$
$$x\left(\frac{x^2+8}{x}\right)-x(1)=x\cdot\frac{2(x+4)}{x}$$
$$x^2+8-x=2(x+4)$$
$$x^2+8-x=2x+8$$
$$x^2-3x+8=8$$
$$x^2-3x=0$$
$$x(x-3)=0$$
$$x=0 \quad \text{or} \quad x-3=0$$
$$x=0 \quad \text{or} \quad x=3$$

The number 0 makes the denominator x equal to zero, so it is not a solution. The solution set is $\{3\}$.

21. The LCD is ab.

$$\frac{x+b}{a}=\frac{4x-7a}{b}$$
$$ab\left(\frac{x+b}{a}\right)=ab\left(\frac{4x-7a}{b}\right)$$
$$b(x+b)=a(4x-7a)$$
$$bx+b^2=4ax-7a^2$$
$$b^2=4ax-7a^2-bx$$
$$7a^2+b^2=4ax-bx$$
$$7a^2+b^2=x(4a-b)$$
$$\frac{7a^2+b^2}{4a-b}=x \text{ or } x=\frac{7a^2+b^2}{4a-b}$$

22. Let x = the number.

$$(x+1)\cdot\frac{2}{x}=\frac{12}{5}$$
$$12x=5\cdot 2(x+1)$$
$$12x=10x+10$$
$$2x=10$$
$$x=5$$

The number is 5.

23. Convert time to rates.

$$\frac{1 \text{ task}}{2 \text{ hours}}=\frac{1}{2}\ \frac{\text{task}}{\text{hour}}$$
$$\frac{1 \text{ task}}{1.5 \text{ hours}}=\frac{1}{1.5}\ \frac{\text{task}}{\text{hour}}$$

Let x = amount of time it takes them to week the garden together.

$$\frac{1 \text{ task}}{x \text{ hours}}=\frac{1}{x}\ \frac{\text{task}}{\text{hours}}$$
$$\frac{1}{2}+\frac{1}{1.5}=\frac{1}{x}$$
$$2\cdot 1.5x\cdot\frac{1}{2}+2\cdot 1.5x\cdot\frac{1}{1.5}=2\cdot 1.5x\cdot\frac{1}{x}$$
$$1.5x+2x=3$$
$$3.5x=3$$
$$35x=30$$
$$x=\frac{30}{35}=\frac{6}{7}$$

It takes them $\frac{6}{7}$ hour to weed the garden together.

24.
$$W=\frac{k}{V}$$
$$20=\frac{k}{12}$$
$$240=k$$
$$W=\frac{240}{V}=\frac{240}{15}=16$$

25. $Q=kRS^2$

Use $Q = 24$, $R = 3$, and $S = 4$.

$$24=k\cdot 3\cdot 4^2$$
$$24=48k$$
$$\frac{1}{2}=k$$
$$Q=\frac{1}{2}RS^2$$

$R = 2$ and $S = 3$

$$Q=\frac{1}{2}\cdot 2\cdot 3^2=9$$

26. $S = k\sqrt{d}$
$160 = k\sqrt{400}$
$160 = k(20)$
$k = 8$
$S = 8\sqrt{d}$
$128 = 8\sqrt{d}$
$16 = \sqrt{d}$
$16^2 = d$
$256 = d$
The height of the cliff is 256 feet.

27. $A = \dfrac{E}{\frac{I}{9}}$
Replace A with 2.27 and I with 166.1.
$2.27 = \dfrac{E}{\frac{166.1}{9}}$
$E = 2.27 \cdot \left(\dfrac{166.1}{9}\right) \approx 42$
There were 42 earned runs scored on Jake Peavy.

Cumulative Review Chapters 1–6

1. 3 is a natural number so the statement $3 \in \{x | x \text{ is a natural number}\}$ is true.

2. a. One third subtracted from a number is written as $x - \dfrac{1}{3}$.

b. Six less than 5 times a number is written as $5x - 6$.

c. Three more than eight times a number is written as $8x + 3$.

d. The quotient of seven and the difference of two and a number is written as $\dfrac{7}{2-x}$.

3. $-7^0 = -(7^0) = -1$

4. $(-5xy^7)(-0.2xy^{-5}) = (-5)(-0.2)xxy^7y^{-5}$
$= (-5)(-0.2)x^{1+1}y^{7+(-5)}$
$= 1x^2y^2$
$= x^2y^2$

5. $(2x+5)^0 = 1$

6. $\dfrac{(3a^{-2}b)^{-2}}{(2ab^{-3})^{-3}} = \dfrac{3^{-2}a^4b^{-2}}{2^{-3}a^{-3}b^9} = \dfrac{2^3}{3^2} \cdot \dfrac{a^7}{b^{11}} = \dfrac{8a^7}{9b^{11}}$

7. $2x \ge 0$ and $4x - 1 \le -9$
$x \ge 0$ and $4x \le -8$
$x \ge 0$ and $x \le -2$
The solution set is $[0, \infty) \cap (-\infty, -2] = \varnothing$.

8. $\left|\dfrac{2x-1}{3}\right| + 6 = 3$
$\left|\dfrac{2x-1}{3}\right| = -3$
An absolute value can never equal a negative value so this equation has no solutions. The solution set is $\varnothing$.

9. $|w + 3| = 7$
$w + 3 = 7$ or $w + 3 = -7$
$w = 4$ or $w = -10$
The solution set is $\{4, -10\}$.

10. $\left|\dfrac{3(x-1)}{4}\right| \ge 2$
$\dfrac{3(x-1)}{4} \ge 2$ or $\dfrac{3(x-1)}{4} \le -2$
$3(x-1) \ge 8$ or $3(x-1) \le -8$
$3x - 3 \ge 8$ or $3x - 3 \le -8$
$3x \ge 11$ or $3x \le -5$
$x \ge \dfrac{11}{3}$ or $x \le -\dfrac{5}{3}$
The solution set is $\left(-\infty, -\dfrac{5}{3}\right] \cup \left[\dfrac{11}{3}, \infty\right)$.

11. $3x - y = 12$
$3(0) - (-12) \stackrel{?}{=} 12$
$0 + 12 = 12$
Yes, $(0, -12)$ is a solution.

$$3(1)-(9) \stackrel{?}{=} 12$$
$$3-9 \neq 12$$
No, (1, 9) is not a solution.
$$3(2)-(-6) \stackrel{?}{=} 12$$
$$6+6=12$$
Yes, (2, −6) is a solution.

12. $f(x)=-x^2+3x-2$

a. $f(0)=-0^2+3(0)-2=0+0-2=-2$
$f(0)=-2$

b. $f(-3)=-(-3)^2+3(-3)-2$
$$=-9-9-2$$
$$=-20$$
$f(-3)=-20$

c. $f\left(\frac{1}{3}\right)=-\left(\frac{1}{3}\right)^2+3\left(\frac{1}{3}\right)-2$
$$=-\frac{1}{9}+1-2$$
$$=-\frac{1}{9}-1$$
$$=-\frac{10}{9}$$
$f\left(\frac{1}{3}\right)=-\frac{10}{9}$

13. $3x+7y=4$
$$7y=-3x+4$$
$$y=-\frac{3}{7}x+\frac{4}{7}$$

$6x+14y=7$
$$14y=-6x+7$$
$$y=-\frac{6}{14}x+\frac{7}{14}$$
$$y=-\frac{3}{7}x+\frac{1}{2}$$
Since both lines have the same slope, $-\frac{3}{7}$, then the lines are parallel.

14. $\frac{x}{7}+\frac{x}{5}=\frac{12}{5}$
$$35\left(\frac{x}{7}+\frac{x}{5}\right)=35\left(\frac{12}{5}\right)$$
$$35\cdot\frac{x}{7}+35\cdot\frac{x}{5}=35\cdot\frac{12}{5}$$
$$5x+7x=84$$
$$12x=84$$
$$x=7$$

15. $m=\frac{y_2-y_1}{x_2-x_1}=\frac{-5-0}{-4-4}=\frac{-5}{-8}=\frac{5}{8}$
$$y-y_1=m(x-x_1)$$
$$y-0=\frac{5}{8}(x-4)$$
$$y=\frac{5}{8}x-\frac{5}{2}$$
$$8y=5x-20$$
$$-5x+8y=-20$$
$$5x-8y=20$$

16. $y-y_1=m(x-x_1)$
$$y-3=\frac{1}{2}[x-(-1)]$$
$$y-3=\frac{1}{2}(x+1)$$
$$y-3=\frac{1}{2}x+\frac{1}{2}$$
$$y=\frac{1}{2}x+\frac{7}{2}$$
$$f(x)=\frac{1}{2}x+\frac{7}{2}$$

17. domain: $(-\infty, \infty)$
range: $[0, \infty)$

18. domain: $(-\infty, \infty)$
range: $(-\infty, \infty)$

19. domain: $[-4, 4]$
range: $[-2, 2]$

20. domain: $(-\infty, \infty)$
range: $(-\infty, -1]$

21. $\begin{cases} -\frac{x}{6}+\frac{y}{2}=\frac{1}{2} \\ \frac{x}{3}-\frac{y}{6}=-\frac{3}{4} \end{cases}$

$$\frac{y}{2}=\frac{1}{2}+\frac{x}{6}$$
$$y=2\left(\frac{1}{2}+\frac{x}{6}\right)=1+\frac{x}{3}$$
$$\frac{x}{3}-\frac{1+\frac{x}{3}}{6}=-\frac{3}{4}$$
$$12\left(\frac{x}{3}-\frac{1+\frac{x}{3}}{6}\right)=12\left(-\frac{3}{4}\right)$$
$$12\cdot\frac{x}{3}-12\cdot\frac{1+\frac{x}{3}}{6}=12\cdot-\frac{3}{4}$$
$$4x-2\left(1+\frac{x}{3}\right)=-9$$
$$4x-2-\frac{2}{3}x=-9$$
$$4x-\frac{2}{3}x=-7$$
$$\frac{10}{3}x=-7$$
$$10x=-21$$
$$x=-\frac{21}{10}$$
$$y=1+\frac{x}{3}=1+\frac{-\frac{21}{10}}{3}=1-\frac{21}{30}=\frac{30}{30}-\frac{21}{30}=\frac{9}{30}=\frac{3}{10}$$

The solution is $\left(-\frac{21}{10}, \frac{3}{10}\right)$.

22.
$$-2x+3y=6$$
$$3x-y=5$$
$$-y=5-3x$$
$$y=-5+3x$$
$$-2x+3(-5+3x)=6$$
$$-2x-15+9x=6$$
$$7x-15=6$$
$$7x=21$$
$$x=3$$
$y = -5 + 3(3) = -5 + 9 = 4$
The solution is (3, 4).

23. $\begin{cases} 2x+4y =1 & (1) \\ 4x -4z=-1 & (2) \\ y-4z=-3 & (3) \end{cases}$

Multiply equation (1) by −2 and add to equation (2).
$$\begin{array}{r} -4x-8y =-2 \\ 4x -4z=-1 \\ \hline -8y-4z=-3 \end{array} \quad (4)$$

Multiply equation (4) by −1 and add to equation (3).
$$\begin{array}{r} y-4z=-3 \\ 8y+4z=3 \\ \hline 9y =0 \end{array}$$
$$y=0$$

Substitute $y = 0$ in equations (1) and (3).

$2x+4y=1$, $2x+4(0)=1$, $2x=1$, $x=\frac{1}{2}$

$y-4z=-3$, $0-4z=-3$, $z=\frac{3}{4}$

The solution is $\left(\frac{1}{2}, 0, \frac{3}{4}\right)$.

24. $\begin{cases} 2x-2y+4z=6 & (1) \\ -4x-y+z=-8 & (2) \\ 3x-y+z=6 & (3) \end{cases}$

Multiply equation (2) by −2 and add to equation (1).
$$\begin{array}{r} 2x-2y+4z=6 \\ 8x+2y-2z=16 \\ \hline 10x +2z=22 \end{array} \quad (4)$$

Multiply equation (3) by −2 and add to equation (1).
$$\begin{array}{r} 2x-2y+4z=6 \\ -6x+2y-2z=-12 \\ \hline -4x +2z=-6 \end{array} \quad (5)$$

Multiply equation (4) by −1 and add to equation (5).
$$\begin{array}{r} -10x-2z=-22 \\ -4x+2z=-6 \\ \hline -14x =-28 \end{array}$$
$$x=2$$

Substitute $x = 2$ into equation (4).

$$10x + 2z = 22$$
$$10(2) + 2z = 22$$
$$2z = 2$$
$$z = 1$$

Substitute $x = 2$ and $z = 1$ in equation (2).

$$-4x - y + z = -8$$
$$-4(2) - y + 1 = -8$$
$$-y = -1$$
$$y = 1$$

The solution is (2, 1, 1).

25. Let x = the speed of one car and $x + 5$ is the speed of the faster car.
Recall that distance = rate · time.

$$3x + 3(x+5) = 297$$
$$3x + 3x + 15 = 297$$
$$6x + 15 = 297$$
$$6x = 282$$
$$x = 47$$

$x + 5 = 47 + 5 = 52$
The speeds of the cars are 47 mph and 52 mph.

26. Let x = cost of a ream of paper and y = cost of a box of manila folders.

$$3x + 2y = 21.90$$
$$5x + y = 24.25$$
$$y = 24.25 - 5x$$

$$3x + 2(24.25 - 5x) = 21.90$$
$$3x + 48.50 - 10x = 21.90$$
$$-7x + 48.50 = 21.90$$
$$-7x = -26.60$$
$$x = 3.8$$

$y = 24.25 - 5(3.8) = 24.25 - 19 = 5.25$
A ream of paper costs \$3.80 and a box of manila folders costs \$5.25.

27. $\begin{cases} x + 2y + z = 2 \\ -2x - y + 2z = 5 \\ x + 3y - 2z = -8 \end{cases}$

$$\left[\begin{array}{ccc|c} 1 & 2 & 1 & 2 \\ -2 & -1 & 2 & 5 \\ 1 & 3 & 2 & -8 \end{array}\right] \to \left[\begin{array}{ccc|c} 1 & 2 & 1 & 2 \\ 0 & 3 & 4 & 9 \\ 0 & 1 & -3 & -10 \end{array}\right]$$
$$\to \left[\begin{array}{ccc|c} 1 & 2 & 1 & 2 \\ 0 & 1 & -3 & -10 \\ 0 & 3 & 4 & 9 \end{array}\right] \to \left[\begin{array}{ccc|c} 1 & 2 & 1 & 2 \\ 0 & 1 & -3 & -10 \\ 0 & 0 & 13 & 39 \end{array}\right]$$
$$\to \left[\begin{array}{ccc|c} 1 & 2 & 1 & 2 \\ 0 & 1 & -3 & -10 \\ 0 & 0 & 1 & 3 \end{array}\right]$$

This matrix corresponds to the following system:

$$x + 2y + z = 2$$
$$y - 3z = -10$$
$$z = 3$$

$$y - 3z = -10$$
$$y - 3(3) = -10$$
$$y - 9 = -10$$
$$y = -1$$

$$x + 2y + z = 2$$
$$x + 2(-1) + (3) = 2$$
$$x - 2 + 3 = 2$$
$$x + 1 = 2$$
$$x = 1$$

The solution set is (1, −1, 3).

28. $\begin{cases} x + y + z = 9 \\ 2x - 2y + 3z = 2 \\ -3x + y - z = 1 \end{cases}$

$$\left[\begin{array}{ccc|c} 1 & 1 & 1 & 9 \\ 2 & -2 & 3 & 2 \\ -3 & 1 & -1 & 1 \end{array}\right] \to \left[\begin{array}{ccc|c} 1 & 1 & 1 & 9 \\ 0 & -4 & 1 & -16 \\ 0 & 4 & 2 & 28 \end{array}\right]$$
$$\to \left[\begin{array}{ccc|c} 1 & 1 & 1 & 9 \\ 0 & -4 & 1 & -16 \\ 0 & 0 & 3 & 12 \end{array}\right] \to \left[\begin{array}{ccc|c} 1 & 1 & 1 & 9 \\ 0 & -4 & 1 & -16 \\ 0 & 0 & 1 & 4 \end{array}\right]$$

This matrix corresponds to the following system:

$$\begin{aligned} x+y+z&=9 \\ -4y+z&=-16 \\ z&=4 \end{aligned}$$

$$\begin{aligned} -4y+z&=-16 \\ -4y+(4)&=-16 \\ -4y&=-20 \\ y&=5 \end{aligned}$$

$$\begin{aligned} x+y+z&=9 \\ x+(5)+(4)&=9 \\ x+9&=9 \\ x&=0 \end{aligned}$$

The solution set is (0, 5, 4).

29. $\left(\frac{3x^2y}{y^{-9}z}\right)^{-2} = \left(\frac{3x^2y^{10}}{z}\right)^{-2}$
$= \frac{3^{-2}x^{-4}y^{-20}}{z^{-2}}$
$= \frac{z^2}{3^2x^4y^{20}}$
$= \frac{z^2}{9x^4y^{20}}$

30. a. $2^{-2}+3^{-1} = \frac{1}{2^2}+\frac{1}{3}$
$= \frac{1}{4}+\frac{1}{3}$
$= \frac{3}{12}+\frac{4}{12}$
$= \frac{7}{12}$

b. $-6a^0 = -6(1) = -6$

c. $\frac{x^{-5}}{x^{-2}} = x^{-5-(-2)} = x^{-5+2} = x^{-3} = \frac{1}{x^3}$

31. $\begin{cases} 3x \geq y \\ x+2y \leq 8 \end{cases}$

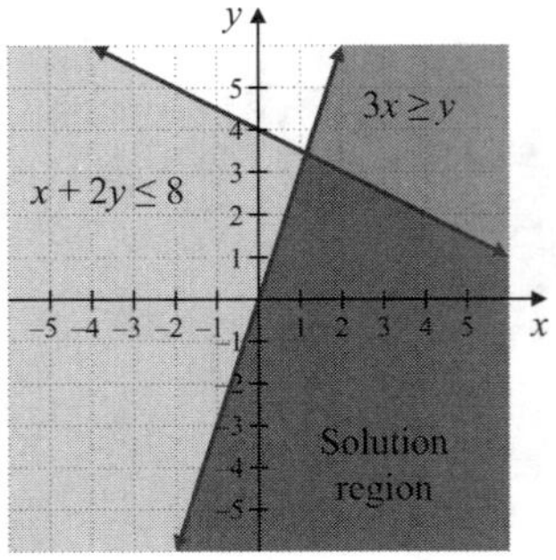

32. a. $3x^{4a}(4x^{-a})^2 = 3x^{4a}\cdot 4^2\cdot x^{-2a}$
$= 3\cdot 16\cdot x^{4a}x^{-2a}$
$= 48x^{2a}$

b. $\frac{(y^{4b})^3}{y^{2b-3}} = \frac{y^{12b}}{y^{2b-3}}$
$= y^{12b-(2b-3)}$
$= y^{12b-2b+3}$
$= y^{10b+3}$

33. $-12x^2+7x^2-6x = -5x^2-6x$

34. $2x^2+8x-3-(2x-7)$
$= 2x^2+8x-3-2x+7$
$= 2x^2+6x+4$

35. $3xy - 2x + 5xy - x = 8xy - 3x$

36. $7+2(x-7y)+\frac{1}{2}-\frac{1}{4}(8x-16y)$
$= 7+2x-14y+\frac{1}{2}-2x+4y$
$= \frac{14}{2}+\frac{1}{2}-14y+4y$
$= -10y+\frac{15}{2}$

37. $(4x^2+7)(x^2+2x+8)$

$$\begin{array}{r} x^2+2x+8 \\ \underline{4x^2+7} \\ 7x^2+14x+56 \\ \underline{4x^4+8x^3+32x^2 \qquad\qquad} \\ 4x^4+8x^3+39x^2+14x+56 \end{array}$$

38. $[4+(3x-y)]^2 = [4+(3x-y)][4+(3x-y)]$
$= 16+4(3x-y)+4(3x-y)+(3x-y)(3x-y)$
$= 16+12x-4y+12x-4y+9x^2-6xy+y^2$
$= 16+24x-8y+9x^2-6xy+y^2$

39.
$$\begin{array}{r} 2x \quad -5 \\ x+2\overline{)\ 2x^2 \quad -x-10} \\ \underline{-(2x^2+4x)} \\ -5x-10 \\ \underline{-(-5x-10)} \\ 0 \end{array}$$

40. $xy+2x-5y-10 = x(y+2)-5(y+2)$
$= (y+2)(x-5)$

41. $-3x^3y+2x^2y-5xy = xy(-3x^2+2x-5)$ or $-xy(3x^2-2x+5)$

42. $6x^2-x-35 = (2x-5)(3x+7)$

43. $x^2+10x+16 = (x+2)(x+8)$

44. $4x^2-4x+1-9y^2 = (2x-1)^2-9y^2$
$= [(2x-1)-3y][(2x-1)+3y]$
$= (2x-1-3y)(2x-1+3y)$

45. $p^4-16 = (p^2-4)(p^2+4)$
$= (p-2)(p+2)(p^2+4)$

46. **a.** domain: $(-\infty, \infty)$
range: $[-4, \infty)$

b. x-intercepts: $(-2, 0)$, $(2, 0)$
y-intercepts: $(0, -4)$

c. There is no such point with the greatest y-value.

d. (0, –4) is the point with the least y-value.

e. The x-values whose y-values are zero are 2 and –2.

f. The x-values whose y-values are less than zero are between $x = -2$ and $x = 2$.

g. The solutions of $f(x) = 0$ are $\{2, -2\}$.

47. $(x + 2)(x - 6) = 0$

$$x+2=0 \quad \text{or} \quad x-6=0$$
$$x=-2 \quad \text{or} \quad x=6$$

The solution set is $\{-2, 6\}$.

48. $2x(3x + 1)(x - 3) = 0$

$$2x=0 \quad \text{or} \quad 3x+1=0 \quad \text{or} \quad x-3=0$$
$$x=0 \quad \text{or} \quad 3x=-1 \quad \text{or} \quad x=3$$
$$x=0 \quad \text{or} \quad x=-\frac{1}{3} \quad \text{or} \quad x=3$$

The solution set is $\left\{0, -\frac{1}{3}, 3\right\}$.

49.
$$\frac{\frac{5x}{x+2}}{\frac{10}{x-2}} = \frac{5x}{x+2} \div \frac{10}{x-2}$$
$$= \frac{5x}{x+2} \cdot \frac{x-2}{10}$$
$$= \frac{x(x-2)}{2(x+2)}$$

50.
$$\frac{3x-4}{2x} = -\frac{8}{x}$$
$$x(3x-4) = -8(2x)$$
$$3x^2 - 4x = -16x$$
$$3x^2 + 12x = 0$$
$$3x(x+4) = 0$$
$$3x=0 \quad \text{or} \quad x+4=0$$
$$x=0 \quad \text{or} \quad x=-4$$

The number 0 makes the denominators $2x$ and x equal to 0 so it is not a solution. The solution set is $\{-4\}$.

Chapter 7

Section 7.1 Practice Problems

1. Since $6^2 = 36$ and $(-6)^2 = 36$, the square roots of 36 are 6 and −6.

2. Since $9^2 = 81$ and $(-9)^2 = 81$, the square roots of 81 are 9 and −9.

3. There are no real numbers whose square is −16. The number −16 has no real number square roots.

4. $\sqrt{25} = 5$ because $(5)^2 = 25$.

5. $\sqrt{0} = 0$ because $0^2 = 0$.

6. $\sqrt{\frac{9}{25}} = \frac{3}{5}$ because $\left(\frac{3}{5}\right)^2 = \frac{9}{25}$.

7. $\sqrt{0.36} = 0.6$ because $(0.6)^2 = 0.36$.

8. $\sqrt{x^{10}} = x^5$ because $(x^5)^2 = x^{10}$.

9. $\sqrt{36x^6} = 6x^3$ because $(6x^3)^2 = 36x^6$.

10. $-\sqrt{25} = -5$ because $(5)^2 = 25$.

11. $\sqrt{-25}$ is not a real number.

12. $\sqrt{30} \approx 5.477$

13. $\sqrt[3]{0} = 0$ because $0^3 = 0$.

14. $\sqrt[3]{-8} = -2$ because $(-2)^3 = -8$.

15. $\sqrt[3]{\frac{1}{64}} = \frac{1}{4}$ because $\left(\frac{1}{4}\right)^3 = \frac{1}{64}$.

16. $\sqrt[3]{x^9} = x^3$ because $(x^3)^3 = x^9$.

17. $\sqrt[3]{-64x^6} = -4x^2$ because $(-4x^2)^3 = -64x^6$.

18. $\sqrt[4]{16} = 2$ because $2^4 = 16$.

19. $\sqrt[5]{-32} = -2$ because $(-2)^5 = -32$.

20. $-\sqrt{36} = -6$ because $6^2 = 36$.

21. $\sqrt[4]{-16}$ is not a real number.

22. $\sqrt[3]{8x^6} = 2x^2$ because $(2x^2)^3 = 8x^6$.

23. $\sqrt{(-5)^2} = |-5| = 5$

24. $\sqrt{x^6} = |x^3|$

25. $\sqrt[4]{(x+6)^4} = |x+6|$

26. $\sqrt[3]{(-3)^3} = -3$

27. $\sqrt[5]{(7x-1)^5} = 7x-1$

28. $\sqrt{36x^2} = 6|x|$

29. $\sqrt{x^2+6x+9} = \sqrt{(x+3)^2} = |x+3|$

30. $f(x) = \sqrt{x+2}$
    $f(7) = \sqrt{7+2} = \sqrt{9} = 3$

31. $g(x) = \sqrt[3]{x-1}$
    $g(9) = \sqrt[3]{9-1} = \sqrt[3]{8} = 2$

32. $f(x) = \sqrt{x+2}$
    $f(0) = \sqrt{0+2} = \sqrt{2}$

33. $g(x)=\sqrt[3]{x-1}$
$g(10)=\sqrt[3]{10-1}=\sqrt[3]{9}$

34.

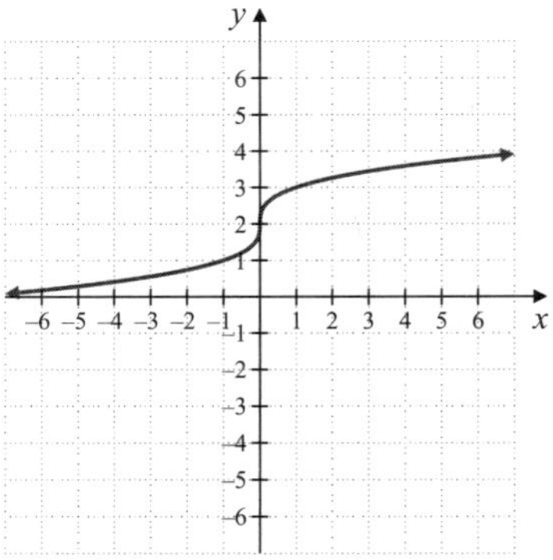

x	$h(x)=\sqrt[3]{x}+2$
0	2
1	3
−1	1
8	4
−8	0

Exercise Set 7.1

1. Since $2^2=4$ and $(-2)^2=4$, the square roots of 4 are 2 and −2.

3. There are no real number square roots for −25.

5. Since $10^2=100$ and $(-10)^2=100$, the square roots of 100 are 10 and −10.

7. $\sqrt{100}=10$ because $10^2=100$.

9. $\sqrt{\frac{1}{4}}=\frac{1}{2}$ because $\left(\frac{1}{2}\right)^2=\frac{1}{4}$.

11. $\sqrt{0.0001}=0.01$ because $(0.01)^2=0.0001$.

13. $-\sqrt{36}=-6$ because $(6)^2=36$.

15. $\sqrt{x^{10}}=x^5$ because $(x^5)^2=x^{10}$.

17. $\sqrt{16y^6}=4y^3$ because $(4y^3)^2=16y^6$.

19. $\sqrt{7}\approx 2.646$
Since $4<7<9$, then $\sqrt{4}<\sqrt{7}<\sqrt{9}$ or $2<\sqrt{7}<3$. The approximation is between 2 and 3 and is thus reasonable.

21. $\sqrt{38}\approx 6.164$
Since $36<38<49$, then $\sqrt{36}<\sqrt{38}<\sqrt{49}$ or $6<\sqrt{38}<7$. The approximation is between 6 and 7 and is thus reasonable.

23. $\sqrt{200}\approx 14.142$
Since $196<200<225$, then $\sqrt{196}<\sqrt{200}<\sqrt{225}$ or $14<\sqrt{200}<15$. The approximation is between 14 and 15 and is thus reasonable.

25. $\sqrt[3]{64}=4$ because $(4)^3=64$.

27. $\sqrt[3]{\frac{1}{8}}=\frac{1}{2}$ because $\left(\frac{1}{2}\right)^3=\frac{1}{8}$.

29. $\sqrt[3]{-1}=-1$ because $(-1)^3=-1$.

31. $\sqrt[3]{x^{12}}=x^4$ because $(x^4)^3=x^{12}$.

33. $\sqrt[3]{-27x^9}=-3x^3$ because $(-3x^3)^3=-27x^9$.

35. $-\sqrt[4]{16}=-2$ because $2^4=16$.

37. $\sqrt[4]{-16}$ is not a real number.

39. $\sqrt[5]{-32}=-2$ because $(-2)^5=-32$.

41. $\sqrt[5]{x^{20}}=x^4$ because $(x^4)^5=x^{20}$.

43. $\sqrt[6]{64x^{12}} = 2x^2$ because $(2x^2)^6 = 64x^{12}$.

45. $\sqrt{81x^4} = 9x^2$ because $(9x^2)^2 = 81x^4$

47. $\sqrt[4]{256x^8} = 4x^2$ because $(4x^2)^4 = 256x^8$.

49. $\sqrt{(-8)^2} = |-8| = 8$

51. $\sqrt[3]{(-8)^3} = -8$

53. $\sqrt{4x^2} = |2x| = 2|x|$

55. $\sqrt[3]{x^3} = x$

57. $\sqrt[4]{(x-2)^4} = |x-2|$

59. $\sqrt{x^2+4x+4} = \sqrt{(x+2)^2} = |x+2|$

61. $-\sqrt{121} = -11$ because -11 is the opposite of $\sqrt{121}$.

63. $\sqrt[3]{8x^3} = 2x$ because $(2x)^3 = 8x^3$.

65. $\sqrt{y^{12}} = y^6$ because $(y^6)^2 = y^{12}$.

67. $\sqrt{25a^2b^{20}} = 5ab^{10}$ because $(5ab^{10})^2 = 25a^2b^{20}$.

69. $\sqrt[3]{-27x^{12}y^9} = -3x^4y^3$ because $(-3x^4y^3)^3 = -27x^{12}y^9$.

71. $\sqrt[4]{a^{16}b^4} = a^4b$ because $(a^4b)^4 = a^{16}b^4$.

73. $\sqrt[5]{-32x^{10}y^5} = -2x^2y$ because $(-2x^2y)^5 = -32x^{10}y^5$.

75. $\sqrt{\frac{25}{49}} = \frac{5}{7}$ because $\left(\frac{5}{7}\right)^2 = \frac{25}{49}$.

77. $\sqrt{\frac{x^2}{4y^2}} = \frac{x}{2y}$ because $\left(\frac{x}{2y}\right)^2 = \frac{x^2}{4y^2}$.

79. $-\sqrt[3]{\frac{z^{21}}{27x^3}} = -\frac{z^7}{3x}$ because $\left(\frac{z^7}{3x}\right)^3 = \frac{z^{21}}{27x^3}$.

81. $\sqrt[4]{\frac{x^4}{16}} = \frac{x}{2}$ because $\left(\frac{x}{2}\right)^4 = \frac{x^4}{16}$.

83. $f(x) = \sqrt{2x+3}$
$f(0) = \sqrt{2(0)+3} = \sqrt{3}$

85. $g(x) = \sqrt[3]{x-8}$
$g(7) = \sqrt[3]{7-8} = \sqrt[3]{-1} = -1$

87. $g(x) = \sqrt[3]{x-8}$
$g(-19) = \sqrt[3]{-19-8} = \sqrt[3]{-27} = -3$

89. $f(x) = \sqrt{2x+3}$
$f(2) = \sqrt{2(2)+3} = \sqrt{7}$

91. The domain of $f(x) = \sqrt[3]{x}+1$ is all real numbers.

x	$f(x) = \sqrt[3]{x}+1$
0	$\sqrt[3]{0}+1=1$
1	$\sqrt[3]{1}+1=2$
8	$\sqrt[3]{8}+1=2+1=3$
−8	$\sqrt[3]{-8}+1=-2+1=-1$
−1	$\sqrt[3]{-1}+1=-1+1=0$

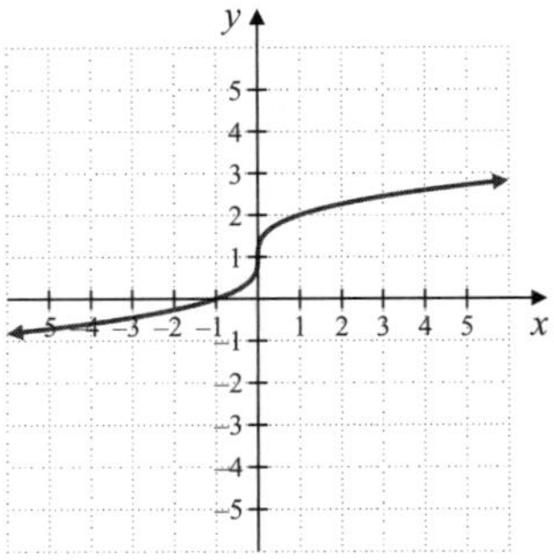

93. The domain of $g(x) = \sqrt[3]{x-1}$ is all real numbers.

x	$g(x) = \sqrt[3]{x-1}$
0	$\sqrt[3]{0-1} = \sqrt[3]{-1} = -1$
1	$\sqrt[3]{1-1} = \sqrt[3]{0} = 0$
2	$\sqrt[3]{2-1} = \sqrt[3]{1} = 1$
7	$\sqrt[3]{7-1} = \sqrt[3]{6} \approx 1.8$
−5	$\sqrt[3]{-5-1} = \sqrt[3]{-6} \approx -1.8$
9	$\sqrt[3]{9-1} = \sqrt[3]{8} = 2$
−7	$\sqrt[3]{-7-1} = \sqrt[3]{-8} = -2$

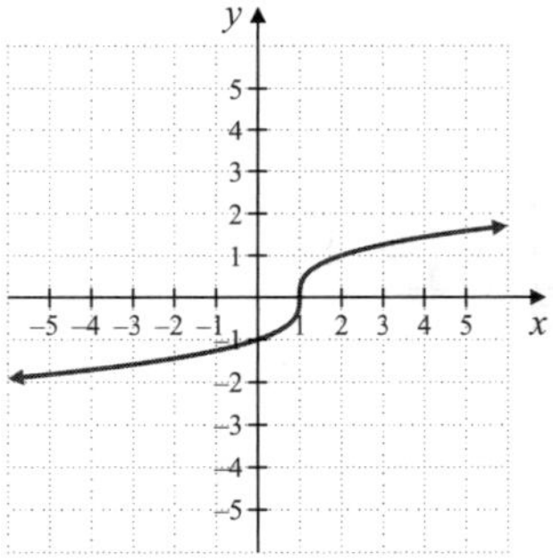

95. $(-2x^3y^2)^5 = -32x^{15}y^{10}$

97. $(-3x^2y^3z^5)(20x^5y^7) = -3 \cdot 20x^2x^5y^3y^7z^5$
$= -60x^7y^{10}z^5$

99. $\dfrac{7x^{-1}y}{14(x^5y^2)^{-2}} = \dfrac{7y(x^5y^2)^2}{14x}$
$= \dfrac{7yx^{10}y^4}{14x}$
$= \dfrac{x^9y^5}{2}$

101. $\sqrt{-17}$ is not a real number.

103. $\sqrt[10]{-17}$ is not a real number.

105. answers may vary

107. $144 < 160 < 169$ so $\sqrt{144} < \sqrt{160} < \sqrt{169}$
$12 < \sqrt{160} < 13$
$\sqrt{160}$ is closest to 13; b.

109. $25 < 30 < 36$ so $5 = \sqrt{25} < \sqrt{30} < \sqrt{36} = 6$
$9 < 10 < 16$ so $3 = \sqrt{9} < \sqrt{10} < \sqrt{16} = 4$
$81 < 90 < 100$ so
$9 = \sqrt{81} < \sqrt{90} < \sqrt{100} = 10$
The perimeter is $\sqrt{30} + \sqrt{10} + \sqrt{90}$ is between 5 + 3 + 9 = 17 and 6 + 4 + 10 = 20.
The perimeter is closest to 18; b.

111. $B = \sqrt{\dfrac{hw}{3131}}$
$= \sqrt{\dfrac{66 \cdot 135}{3131}}$
$= \sqrt{\dfrac{8910}{3131}}$
$= \sqrt{2.84}$
$= 1.69$
The body surface area is 1.69 square meters.

113. answers may vary

Section 7.2 Practice Problems

1. $25^{1/2} = \sqrt{25} = 5$

2. $27^{1/3} = \sqrt[3]{27} = 3$

3. $x^{1/5} = \sqrt[5]{x}$

4. $-25^{1/2} = -\sqrt{25} = -5$

5. $(-27y^6)^{1/3} = \sqrt[3]{-27y^6} = -3y^2$

6. $7x^{1/5} = 7\sqrt[5]{x}$

7. $9^{3/2} = \left(\sqrt{9}\right)^3 = 3^3 = 27$

8. $-256^{3/4} = -\left(\sqrt[4]{256}\right)^3 = -(4)^3 = -64$

9. $(-32)^{2/5} = \left(\sqrt[5]{-32}\right)^2 = (-2)^2 = 4$

10. $\left(\frac{1}{4}\right)^{3/2} = \left(\sqrt{\frac{1}{4}}\right)^3 = \left(\frac{1}{2}\right)^3 = \frac{1}{8}$

11. $(2x+1)^{2/7} = \sqrt[7]{(2x+1)^2}$

12. $27^{-2/3} = \frac{1}{27^{2/3}} = \frac{1}{\left(\sqrt[3]{27}\right)^2} = \frac{1}{(3)^2} = \frac{1}{9}$

13.
$$\begin{aligned} -256^{-3/4} &= -\frac{1}{256^{3/4}} \\ &= -\frac{1}{\left(\sqrt[4]{256}\right)^3} \\ &= -\frac{1}{4^3} \\ &= -\frac{1}{64} \end{aligned}$$

14. $x^{1/3}x^{1/4} = x^{\frac{1}{3}+\frac{1}{4}} = x^{\frac{4}{12}+\frac{3}{12}} = x^{7/12}$

15. $\frac{9^{2/5}}{9^{12/5}} = 9^{\frac{2}{5}-\frac{12}{5}} = 9^{\frac{-10}{5}} = 9^{-2} = \frac{1}{9^2} = \frac{1}{81}$

16. $y^{-3/10} \cdot y^{6/10} = y^{-\frac{3}{10}+\frac{6}{10}} = y^{3/10}$

17. $(11^{2/9})^3 = 11^{\frac{2}{9}\cdot 3} = 11^{6/9} = 11^{2/3}$

18.
$$\begin{aligned} \frac{(3x^{2/3})^3}{x^2} &= \frac{3^3(x^{2/3})^3}{x^2} \\ &= \frac{27x^2}{x^2} \\ &= 27x^{2-2} \\ &= 27x^0 \\ &= 27\cdot 1 \\ &= 27 \end{aligned}$$

19. $\sqrt[10]{y^5} = y^{5/10} = y^{1/2} = \sqrt{y}$

20. $\sqrt[4]{9} = 9^{1/4} = (3^2)^{1/4} = 3^{2/4} = 3^{1/2} = \sqrt{3}$

21.
$$\begin{aligned} \sqrt[9]{a^6b^3} &= (a^6b^3)^{1/9} \\ &= a^{6/9}b^{3/9} \\ &= a^{2/3}b^{1/3} \\ &= \sqrt[3]{a^2b} \end{aligned}$$

22.
$$\begin{aligned} \sqrt{y}\cdot\sqrt[3]{y} &= y^{1/2}\cdot y^{1/3} \\ &= y^{3/6}\cdot y^{2/6} \\ &= y^{5/6} \\ &= \sqrt[6]{y^5} \end{aligned}$$

23. $\frac{\sqrt[3]{x}}{\sqrt[4]{x}} = \frac{x^{1/3}}{x^{1/4}} = x^{\frac{1}{3}-\frac{1}{4}} = x^{\frac{4}{12}-\frac{3}{12}} = x^{1/12} = \sqrt[12]{x}$

24.
$$\begin{aligned} \sqrt{5}\cdot\sqrt[3]{2} &= 5^{1/2}\cdot 2^{1/3} \\ &= 5^{3/6}\cdot 2^{2/6} \\ &= (5^3\cdot 2^2)^{1/6} \\ &= \sqrt[6]{5^3\cdot 2^2} \\ &= \sqrt[6]{500} \end{aligned}$$

Exercise Set 7.2

1. $49^{1/2} = \sqrt{49} = 7$

3. $27^{1/3} = \sqrt[3]{27} = 3$

5. $\left(\frac{1}{16}\right)^{1/4} = \sqrt[4]{\frac{1}{16}} = \frac{1}{2}$

7. $169^{1/2} = \sqrt{169} = 13$

9. $2m^{1/3} = 2\sqrt[3]{m}$

11. $(9x^4)^{1/2} = \sqrt{9x^4} = 3x^2$

13. $(-27)^{1/3} = \sqrt[3]{-27} = -3$

15. $-16^{1/4} = -\sqrt[4]{16} = -2$

17. $16^{3/4} = \left(\sqrt[4]{16}\right)^3 = 2^3 = 8$

19. $(-64)^{2/3} = \left(\sqrt[3]{-64}\right)^2 = (-4)^2 = 16$

21. $(-16)^{3/4} = \left(\sqrt[4]{-16}\right)^3$ is not a real number.

23. $(2x)^{3/5} = \sqrt[5]{(2x)^3}$

25. $(7x+2)^{2/3} = \sqrt[3]{(7x+2)^2}$ or $\left(\sqrt[3]{7x+2}\right)^2$

27. $\left(\frac{16}{9}\right)^{3/2} = \left(\sqrt{\frac{16}{9}}\right)^3 = \left(\frac{4}{3}\right)^3 = \frac{64}{27}$

29. $8^{-4/3} = \frac{1}{8^{4/3}} = \frac{1}{(8^{1/3})^4} = \frac{1}{2^4} = \frac{1}{16}$

31. $(-64)^{-2/3} = \frac{1}{\left(\sqrt[3]{-64}\right)^2} = \frac{1}{(-4)^2} = \frac{1}{16}$

33. $(-4)^{-3/2} = \frac{1}{(-4)^{3/2}} = \frac{1}{\left(\sqrt{-4}\right)^3}$ is not a real number.

35. $x^{-1/4} = \frac{1}{x^{1/4}}$

37. $\frac{1}{a^{-2/3}} = a^{2/3}$

39. $\frac{5}{7x^{-3/4}} = \frac{5x^{3/4}}{7}$

41. $a^{2/3}a^{5/3} = a^{2/3+5/3} = a^{7/3}$

43. $x^{-2/5} \cdot x^{7/5} = x^{-\frac{2}{5}+\frac{7}{5}} = x^{5/5} = x$

45. $3^{1/4} \cdot 3^{3/8} = 3^{1/4+3/8} = 3^{2/8+3/8} = 3^{5/8}$

47. $\frac{y^{1/3}}{y^{1/6}} = y^{\frac{1}{3}-\frac{1}{6}} = y^{\frac{2}{6}-\frac{1}{6}} = y^{1/6}$

49.
$$\begin{aligned}(4u^2)^{3/2} &= 4^{3/2}(u^2)^{3/2}\\ &= (4^{1/2})^3 u^{2\cdot 3/2}\\ &= 2^3 \cdot u^3\\ &= 8u^3\end{aligned}$$

51.
$$\begin{aligned}\frac{b^{1/2}b^{3/4}}{-b^{1/4}} &= -b^{\frac{1}{2}+\frac{3}{4}-\frac{1}{4}}\\ &= -b^{\frac{2}{4}+\frac{3}{4}-\frac{1}{4}}\\ &= -b^{4/4}\\ &= -b\end{aligned}$$

53.
$$\begin{aligned}\frac{(x^3)^{1/2}}{x^{7/2}} &= \frac{x^{3\cdot 1/2}}{x^{7/2}}\\ &= \frac{x^{3/2}}{x^{7/2}}\\ &= x^{3/2-7/2}\\ &= x^{-2}\\ &= \frac{1}{x^2}\end{aligned}$$

55. $\frac{(3x^{1/4})^3}{x^{1/12}} = \frac{3^3x^{3/4}}{x^{1/12}}$
$= 27x^{\frac{9}{12}-\frac{1}{12}}$
$= 27x^{8/12}$
$= 27x^{2/3}$

57. $\frac{(y^3z)^{1/6}}{y^{-1/2}z^{1/3}} = \frac{y^{3/6}z^{1/6}}{y^{-1/2}z^{1/3}}$
$= y^{3/6-(-1/2)}z^{1/6-1/3}$
$= y^{1/2+1/2}z^{1/6-2/6}$
$= y^1z^{-1/6}$
$= \frac{y}{z^{1/6}}$

59. $\frac{(x^3y^2)^{1/4}}{(x^{-5}y^{-1})^{-1/2}} = \frac{x^{3/4}y^{2/4}}{x^{5/2}y^{1/2}}$
$= x^{\frac{3}{4}-\frac{5}{2}}y^{\frac{2}{4}-\frac{1}{2}}$
$= x^{\frac{3}{4}-\frac{10}{4}}y^{\frac{1}{2}-\frac{1}{2}}$
$= x^{-7/4}$
$= \frac{1}{x^{7/4}}$

61. $\sqrt[6]{x^3} = x^{3/6} = x^{1/2} = \sqrt{x}$

63. $\sqrt[6]{4} = 4^{1/6} = 2^{2/6} = 2^{1/3} = \sqrt[3]{2}$

65. $\sqrt[4]{16x^2} = 16^{1/4}x^{2/4} = 2x^{1/2} = 2\sqrt{x}$

67. $\sqrt[4]{(x+3)^2} = (x+3)^{2/4} = (x+3)^{1/2} = \sqrt{x+3}$

69. $\sqrt[8]{x^4y^4} = x^{4/8}y^{4/8} = x^{1/2}y^{1/2} = \sqrt{xy}$

71. $\sqrt[12]{a^8b^4} = a^{8/12}b^{4/12} = a^{2/3}b^{1/3} = \sqrt[3]{a^2b}$

73. $\sqrt[3]{y}\cdot\sqrt[5]{y^2} = y^{1/3}\cdot y^{2/5}$
$= y^{5/15}\cdot y^{6/15}$
$= y^{11/15}$
$= \sqrt[15]{y^{11}}$

75. $\frac{\sqrt[3]{b^2}}{\sqrt[4]{b}} = b^{2/3}\cdot b^{-1/4}$
$= b^{8/12}b^{-3/12}$
$= b^{5/12}$
$= \sqrt[12]{b^5}$

77. $\sqrt[3]{x}\cdot\sqrt[4]{x}\cdot\sqrt[8]{x^3} = x^{1/3}\cdot x^{1/4}\cdot x^{3/8}$
$= x^{8/24}\cdot x^{6/24}\cdot x^{9/24}$
$= x^{23/24}$
$= \sqrt[24]{x^{23}}$

79. $\frac{\sqrt[3]{a^2}}{\sqrt[6]{a}} = a^{2/3}a^{-1/6}$
$= a^{4/6}a^{-1/6}$
$= a^{3/6}$
$= a^{1/2}$
$= \sqrt{a}$

81. $\sqrt{3}\cdot\sqrt[3]{4} = 3^{1/2}\cdot 4^{1/3}$
$= 3^{3/6}\cdot 4^{2/6}$
$= (3^3\cdot 4^2)^{1/6}$
$= (432)^{1/6}$
$= \sqrt[6]{432}$

83. $\sqrt[5]{7}\cdot\sqrt[3]{y} = 7^{1/5}\cdot y^{1/3}$
$= 7^{3/15}y^{5/15}$
$= (7^3y^5)^{1/15}$
$= (343y^5)^{1/15}$
$= \sqrt[15]{343y^5}$

85. $\sqrt{5r}\cdot\sqrt[3]{s}=(5r)^{1/2}\cdot s^{1/3}$
$=(5r)^{3/6}\cdot s^{2/6}$
$=[(5r)^3\cdot s^2]^{1/6}$
$=(125r^3s^2)^{1/6}$
$=\sqrt[6]{125r^3s^2}$

87. $75=25\cdot 3$

89. $48=16\cdot 3$ or $4\cdot 12$

91. $16=8\cdot 2$

93. $54=27\cdot 2$

95. $B(w)=70w^{3/4}$
$B(60)=70(60)^{3/4}=70(21.558)=1509$
The BMR is 1509 calories.

97. $f(x)=12x^{7/6}$
2004 is 10 years after 1994, so $x=10$.
$f(10)=12(10)^{7/6}\approx 176.1$
There were approximately 176.1 million cellular telephone subscriptions in 2004.

99. answers may vary

101. Since $\frac{1}{3}+\frac{2}{3}=\frac{3}{3}$, then
$a^{1/3}\cdot a^{2/3}=a^{1/3+2/3}=a^{3/3}=a^1=a.$

103. Since $\frac{1}{5}-\left(-\frac{2}{5}\right)=\frac{3}{5}$ then
$\frac{x^{1/5}}{x^{-2/5}}=x^{\frac{1}{5}-\left(-\frac{2}{5}\right)}=x^{\frac{1}{5}+\frac{2}{5}}=x^{3/5}.$

105. $8^{1/4}\approx 1.6818$

107. $\frac{\sqrt{t}}{\sqrt{u}}=\frac{t^{1/2}}{u^{1/2}}$

Section 7.3 Practice Problems

1. $\sqrt{2}\cdot\sqrt{13}=\sqrt{2\cdot 13}=\sqrt{26}$

2. $\sqrt{17}\cdot\sqrt{y}=\sqrt{17y}$

3. $\sqrt[3]{2}\cdot\sqrt[3]{32}=\sqrt[3]{2\cdot 32}=\sqrt[3]{64}=4$

4. $\sqrt[4]{6}\cdot\sqrt[4]{3x^2}=\sqrt[4]{6\cdot 3x^2}=\sqrt[4]{18x^2}$

5. $\sqrt{\frac{3}{x}}\cdot\sqrt{\frac{y}{2}}=\sqrt{\frac{3}{x}\cdot\frac{y}{2}}=\sqrt{\frac{3y}{2x}}$

6. $\sqrt{\frac{9}{25}}=\frac{\sqrt{9}}{\sqrt{25}}=\frac{3}{5}$

7. $\sqrt{\frac{y}{36}}=\frac{\sqrt{y}}{\sqrt{36}}=\frac{\sqrt{y}}{6}$

8. $\sqrt[3]{\frac{27}{64}}=\frac{\sqrt[3]{27}}{\sqrt[3]{64}}=\frac{3}{4}$

9. $\sqrt[5]{\frac{7}{32x^5}}=\frac{\sqrt[5]{7}}{\sqrt[5]{32x^5}}=\frac{\sqrt[5]{7}}{2x}$

10. $\sqrt{18}=\sqrt{9\cdot 2}=\sqrt{9}\cdot\sqrt{2}=3\sqrt{2}$

11. $\sqrt[3]{40}=\sqrt[3]{8\cdot 5}=\sqrt[3]{8}\sqrt[3]{5}=2\sqrt[3]{5}$

12. $\sqrt{14}$ cannot be simplified.

13. $\sqrt[4]{162}=\sqrt[4]{81\cdot 2}=\sqrt[4]{81}\sqrt[4]{2}=3\sqrt[4]{2}$

14. $\sqrt{49a^5}=\sqrt{49a^4}\sqrt{a}=7a^2\sqrt{a}$

15. $\sqrt[3]{24x^9y^7}=\sqrt[3]{8x^9y^6}\sqrt[3]{3y}=2x^3y^2\sqrt[3]{3y}$

16. $\sqrt[4]{16z^9}=\sqrt[4]{16z^8}\sqrt[4]{z}=2z^2\sqrt[4]{z}$

17. $\frac{\sqrt{75}}{\sqrt{3}}=\sqrt{\frac{75}{3}}=\sqrt{25}=5$

18. $\frac{\sqrt{80y}}{3\sqrt{5}} = \frac{1}{3}\sqrt{\frac{80y}{5}} = \frac{\sqrt{16y}}{3} = \frac{4\sqrt{y}}{3}$

19. $\frac{5\sqrt[3]{162x^8}}{\sqrt[3]{3x^2}} = 5\sqrt[3]{\frac{162x^8}{3x^2}}$
$= 5\sqrt[3]{54x^6}$
$= 5\sqrt[3]{27 \cdot 2 \cdot x^6}$
$= 5\sqrt[3]{27x^6} \cdot \sqrt[3]{2}$
$= 5 \cdot 3 \cdot x^2\sqrt[3]{2}$
$= 15x^2\sqrt[3]{2}$

20. $\frac{3\sqrt[4]{243x^9y^6}}{\sqrt[4]{x^{-3}y}} = 3\sqrt[4]{\frac{243x^9y^6}{x^{-3}y}}$
$= 3\sqrt[4]{243x^{12}y^5}$
$= 3\sqrt[4]{81 \cdot 3x^{12}y^5}$
$= 9x^3y\sqrt[4]{3y}$

21. $d = \sqrt{(x_2 - x_1)^2 + (y_2 - y_1)^2}$
$= \sqrt{(-1-(-2))^2 + (3-6)^2}$
$= \sqrt{1^2 + (-3)^2}$
$= \sqrt{1+9}$
$= \sqrt{10}$
≈ 3.162

The distance is exactly $\sqrt{10}$ units, or approximately 3.162 units.

22. midpoint $= \left(\frac{x_1 + x_2}{2}, \frac{y_1 + y_2}{2}\right)$
$= \left(\frac{-2+4}{2}, \frac{5+(-6)}{2}\right)$
$= \left(\frac{2}{2}, -\frac{1}{2}\right)$
$= \left(1, -\frac{1}{2}\right)$

The midpoint of the segment is $\left(1, -\frac{1}{2}\right)$.

Exercise Set 7.3

1. $\sqrt{7} \cdot \sqrt{2} = \sqrt{7 \cdot 2} = \sqrt{14}$

3. $\sqrt[4]{8} \cdot \sqrt[4]{2} = \sqrt[4]{8 \cdot 2} = \sqrt[4]{16} = 2$

5. $\sqrt[3]{4} \cdot \sqrt[3]{9} = \sqrt[3]{4 \cdot 9} = \sqrt[3]{36}$

7. $\sqrt{2} \cdot \sqrt{3x} = \sqrt{2 \cdot 3x} = \sqrt{6x}$

9. $\sqrt{\frac{7}{x}} \cdot \sqrt{\frac{2}{y}} = \sqrt{\frac{7 \cdot 2}{x \cdot y}} = \sqrt{\frac{14}{xy}}$

11. $\sqrt[4]{4x^3} \cdot \sqrt[4]{5} = \sqrt[4]{4x^3 \cdot 5} = \sqrt[4]{20x^3}$

13. $\sqrt{\frac{6}{49}} = \frac{\sqrt{6}}{\sqrt{49}} = \frac{\sqrt{6}}{7}$

15. $\sqrt{\frac{2}{49}} = \frac{\sqrt{2}}{\sqrt{49}} = \frac{\sqrt{2}}{7}$

17. $\sqrt[4]{\frac{x^3}{16}} = \frac{\sqrt[4]{x^3}}{\sqrt[4]{16}} = \frac{\sqrt[4]{x^3}}{2}$

19. $\sqrt[3]{\frac{4}{27}} = \frac{\sqrt[3]{4}}{\sqrt[3]{27}} = \frac{\sqrt[3]{4}}{3}$

21. $\sqrt[4]{\frac{8}{x^8}} = \frac{\sqrt[4]{8}}{\sqrt[4]{x^8}} = \frac{\sqrt[4]{8}}{x^2}$

23. $\sqrt[3]{\frac{2x}{81y^{12}}} = \frac{\sqrt[3]{2x}}{\sqrt[3]{81y^{12}}} = \frac{\sqrt[3]{2x}}{3y^4\sqrt[3]{3}}$

25. $\sqrt{\frac{x^2y}{100}} = \frac{\sqrt{x^2} \cdot \sqrt{y}}{\sqrt{100}} = \frac{x\sqrt{y}}{10}$

27. $\sqrt{\frac{5x^2}{169y^2}} = \frac{\sqrt{5x^2}}{\sqrt{169y^2}} = \frac{x\sqrt{5}}{13y}$

29. $-\sqrt[3]{\dfrac{z^7}{125x^3}} = \dfrac{-\sqrt[3]{z^7}}{\sqrt[3]{125x^3}}$
$= \dfrac{-\sqrt[3]{z^6 z}}{\sqrt[3]{125}\cdot\sqrt[3]{x^3}}$
$= \dfrac{-\sqrt[3]{z^6}\cdot\sqrt[3]{z}}{5x}$
$= -\dfrac{z^2\sqrt[3]{z}}{5x}$

31. $\sqrt{32} = \sqrt{16(2)} = \sqrt{16}\sqrt{2} = 4\sqrt{2}$

33. $\sqrt[3]{192} = \sqrt[3]{64(3)} = \sqrt[3]{64}\cdot\sqrt[3]{3} = 4\sqrt[3]{3}$

35. $5\sqrt{75} = 5\sqrt{25(3)} = 5\sqrt{25}\sqrt{3} = 25\sqrt{3}$

37. $\sqrt{24} = \sqrt{4\cdot 6} = \sqrt{4}\cdot\sqrt{6} = 2\sqrt{6}$

39. $\sqrt{100x^5} = \sqrt{100}\sqrt{x^5} = 10x^2\sqrt{x}$

41. $\sqrt[3]{16y^7} = \sqrt[3]{(8y^6)(2y)}$
$= \sqrt[3]{8}\cdot\sqrt[3]{y^6}\cdot\sqrt[3]{2y}$
$= 2y^2\sqrt[3]{2y}$

43. $\sqrt[4]{a^8b^7} = \sqrt[4]{a^8b^4}\sqrt[4]{b^3} = a^2b\sqrt[4]{b^3}$

45. $\sqrt{y^5} = \sqrt{y^4y} = \sqrt{y^4}\cdot\sqrt{y} = y^2\sqrt{y}$

47. $\sqrt{25a^2b^3} = \sqrt{25}\sqrt{a^2}\sqrt{b^3} = 5ab\sqrt{b}$

49. $\sqrt[5]{-32x^{10}y} = \sqrt[5]{-32}\cdot\sqrt[5]{x^{10}}\cdot\sqrt[5]{y} = -2x^2\sqrt[5]{y}$

51. $\sqrt[3]{50x^{14}} = \sqrt[3]{50}\sqrt[3]{x^{12}}\sqrt[3]{x^2} = x^4\sqrt[3]{50x^2}$

53. $-\sqrt{32a^8b^7} = -\sqrt{16a^8b^6(2b)}$
$= -\sqrt{16}\cdot\sqrt{a^8}\cdot\sqrt{b^6}\cdot\sqrt{2b}$
$= -4a^4b^3\sqrt{2b}$

55. $\sqrt{9x^7y^9} = \sqrt{9}\sqrt{x^6}\sqrt{x}\sqrt{y^8}\sqrt{y} = 3x^3y^4\sqrt{xy}$

57. $\sqrt[3]{125r^9s^{12}} = 5r^3s^4$

59. $\dfrac{\sqrt{14}}{\sqrt{7}} = \sqrt{\dfrac{14}{7}} = \sqrt{2}$

61. $\dfrac{\sqrt[3]{24}}{\sqrt[3]{3}} = \sqrt[3]{\dfrac{24}{3}} = \sqrt[3]{8} = 2$

63. $\dfrac{5\sqrt[4]{48}}{\sqrt[4]{3}} = 5\sqrt[4]{\dfrac{48}{3}} = 5\sqrt[4]{16} = 5\cdot 2 = 10$

65. $\dfrac{\sqrt{x^5y^3}}{\sqrt{xy}} = \sqrt{\dfrac{x^5y^3}{xy}} = \sqrt{x^4y^2} = x^2y$

67. $\dfrac{8\sqrt[3]{54m^7}}{\sqrt[3]{2m}} = 8\sqrt[3]{\dfrac{54m^7}{2m}}$
$= 8\sqrt[3]{27m^6}$
$= 8\cdot 3m^2$
$= 24m^2$

69. $\dfrac{3\sqrt{100x^2}}{2\sqrt{2x^{-1}}} = \dfrac{3}{2}\sqrt{\dfrac{100x^2}{2x^{-1}}}$
$= \dfrac{3}{2}\sqrt{50x^3}$
$= \dfrac{3}{2}\sqrt{25x^2\cdot 2x}$
$= \dfrac{3}{2}\cdot 5x\sqrt{2x}$
$= \dfrac{15x}{2}\sqrt{2x}$

71. $\dfrac{\sqrt[4]{96a^{10}b^3}}{\sqrt[4]{3a^2b^3}} = \sqrt[4]{\dfrac{96a^{10}b^3}{3a^2b^3}} = \sqrt[4]{32a^8} = 2a^2\sqrt[4]{2}$

73. $(x_1, y_1) = (5, 1), (x_2, y_2) = (8, 5)$

$$\begin{aligned} d &= \sqrt{(x_2 - x_1)^2 + (y_2 - y_1)^2} \\ &= \sqrt{(5-1)^2 + (8-5)^2} \\ &= \sqrt{4^2 + 3^2} \\ &= \sqrt{16+9} \\ &= \sqrt{25} \\ &= 5 \end{aligned}$$

The distance is 5 units.

75. $(x_1, y_1) = (-3, 2), (x_2, y_2) = (1, -3)$

$$\begin{aligned} d &= \sqrt{(x_2 - x_1)^2 + (y_2 - y_1)^2} \\ &= \sqrt{(1-(-3))^2 + (-3-2)^2} \\ &= \sqrt{4^2 + (-5)^2} \\ &= \sqrt{16+25} \\ &= \sqrt{41} \end{aligned}$$

The distance is $\sqrt{41}$ units approximately 6.403 units.

77. $(x_1, y_1) = \left(0, -\sqrt{2}\right), (x_2, y_2) = \left(\sqrt{3}, 0\right)$

$$\begin{aligned} d &= \sqrt{(x_2 - x_1)^2 + (y_2 - y_1)^2} \\ &= \sqrt{\left(\sqrt{3}-0\right)^2 + \left[0-\left(-\sqrt{2}\right)\right]^2} \\ &= \sqrt{\left(\sqrt{3}\right)^2 + \left(\sqrt{2}\right)^2} \\ &= \sqrt{3+2} \\ &= \sqrt{5} \end{aligned}$$

The distance is $\sqrt{5}$ units, approximately 2.236 units.

79. $(x_1, y_1) = (1.7, -3.6), (x_2, y_2) = (-8.6, 5.7)$

$$\begin{aligned} d &= \sqrt{(x_2 - x_1)^2 + (y_2 - y_1)^2} \\ &= \sqrt{(-8.6-1.7)^2 + (5.7-(-3.6))^2} \\ &= \sqrt{(-10.3)^2 + (9.3)^2} \\ &= \sqrt{106.09+86.49} \\ &= \sqrt{192.58} \end{aligned}$$

The distance is $\sqrt{192.58}$ units, approximately 13.877 units.

81. $(x_1, y_1) = (6, -8), (x_2, y_2) = (2, 4)$

$$\begin{aligned} \text{midpoint} &= \left(\frac{x_1 + x_2}{2}, \frac{y_1 + y_2}{2}\right) \\ &= \left(\frac{6+2}{2}, \frac{-8+4}{2}\right) \\ &= \left(\frac{8}{2}, \frac{-4}{2}\right) \\ &= (4, -2) \end{aligned}$$

83. $(x_1, y_1) = (-2, -1), (x_2, y_2) = (-8, 6)$

$$\begin{aligned} \text{midpoint} &= \left(\frac{x_1 + x_2}{2}, \frac{y_1 + y_2}{2}\right) \\ &= \left(\frac{-2+(-8)}{2}, \frac{-1+6}{2}\right) \\ &= \left(\frac{-10}{2}, \frac{5}{2}\right) \\ &= \left(-5, \frac{5}{2}\right) \end{aligned}$$

85. $(x_1, y_1) = \left(\frac{1}{2}, \frac{3}{8}\right), (x_2, y_2) = \left(-\frac{3}{2}, \frac{5}{8}\right)$

$$\begin{aligned} \text{midpoint} &= \left(\frac{x_1 + x_2}{2}, \frac{y_1 + y_2}{2}\right) \\ &= \left(\frac{\frac{1}{2}+\left(-\frac{3}{2}\right)}{2}, \frac{\frac{3}{8}+\frac{5}{8}}{2}\right) \\ &= \left(\frac{-1}{2}, \frac{1}{2}\right) \\ &= \left(-\frac{1}{2}, \frac{1}{2}\right) \end{aligned}$$

87. $(x_1, y_1) = (\sqrt{2}, 3\sqrt{5}), (x_2, y_2) = (\sqrt{2}, -2\sqrt{5})$

$$\text{midpoint} = \left(\frac{x_1+x_2}{2}, \frac{y_1+y_2}{2}\right) = \left(\frac{\sqrt{2}+\sqrt{2}}{2}, \frac{3\sqrt{5}-2\sqrt{5}}{2}\right) = \left(\frac{2\sqrt{2}}{2}, \frac{\sqrt{5}}{2}\right) = \left(\sqrt{2}, \frac{\sqrt{5}}{2}\right)$$

89. $6x + 8x = (6 + 8)x = 14x$

91. $(2x+3)(x-5) = 2x^2 - 10x + 3x - 15 = 2x^2 - 7x - 15$

93. $9y^2 - 8y^2 = (9-8)y^2 = y^2$

95. $(x-4)^2 = (x-4)(x-4) = x^2 - 4x - 4x + 16 = x^2 - 8x + 16$

97. $\frac{\sqrt[3]{64}}{\sqrt{64}} = \frac{4}{8} = \frac{1}{2}$

99. $r = \sqrt{\frac{A}{4\pi}} = \sqrt{\frac{32.17}{4\pi}} = \sqrt{2.56} = 1.6$

The radius of a standard zorb is 1.6 meters.

101. $F(x) = 0.6\sqrt{49 - x^2}$

a. $F(3) = 0.6\sqrt{49-3^2} = 0.6\sqrt{49-9} = 0.6\sqrt{40} \approx 3.8$

The demand is 3.8 times per week.

b. $F(5) = 0.6\sqrt{49-5^2} = 0.6\sqrt{49-25} = 0.6\sqrt{24} \approx 2.9$

The demand is 2.9 times per week.

c. answers may vary

Section 7.4 Practice Problems

1. $5\sqrt{15} + 2\sqrt{15} = (5+2)\sqrt{15} = 7\sqrt{15}$

2. $9\sqrt[3]{2y} - 15\sqrt[3]{2y} = (9-15)\sqrt[3]{2y} = -6\sqrt[3]{2y}$

3. $6\sqrt{10} - 3\sqrt[3]{10}$ cannot be simplified.

4. $\sqrt{50} + 5\sqrt{18} = \sqrt{25\cdot 2} + 5\sqrt{9\cdot 2} = 5\sqrt{2} + 5\cdot 3\sqrt{2} = 20\sqrt{2}$

5. $\sqrt[3]{24} - 4\sqrt[3]{192} + \sqrt[3]{3} = \sqrt[3]{8\cdot 3} - 4\sqrt[3]{64\cdot 3} + \sqrt[3]{3} = 2\sqrt[3]{3} - 4\cdot 4\sqrt[3]{3} + \sqrt[3]{3} = -13\sqrt[3]{3}$

6. $\sqrt{20x} - 6\sqrt{16x} + \sqrt{45x} = \sqrt{4\cdot 5x} - 6\sqrt{16x} + \sqrt{9\cdot 5x} = 2\sqrt{5x} - 6\cdot 4\sqrt{x} + 3\sqrt{5x} = 5\sqrt{5x} - 24\sqrt{x}$

7. $\sqrt[4]{32} + \sqrt{32} = \sqrt[4]{32} + \sqrt{16\cdot 2} = \sqrt[4]{16\cdot 2} + \sqrt{16\cdot 2} = 2\sqrt[4]{2} + 4\sqrt{2}$

8. $\sqrt[3]{8y^5} + \sqrt[3]{27y^5} = \sqrt[3]{8y^3\cdot y^2} + \sqrt[3]{27y^3\cdot y^2} = 2y\sqrt[3]{y^2} + 3y\sqrt[3]{y^2} = 5y\sqrt[3]{y^2}$

9. $\frac{\sqrt{75}}{9} - \frac{\sqrt{3}}{2} = \frac{5\sqrt{3}}{9} - \frac{\sqrt{3}}{2} = \frac{10\sqrt{3}}{18} - \frac{9\sqrt{3}}{18} = \frac{\sqrt{3}}{18}$

10. $\sqrt[3]{\frac{5x}{27}}+4\sqrt[3]{5x}=\frac{\sqrt[3]{5x}}{\sqrt[3]{27}}+4\sqrt[3]{5x}$
$=\frac{\sqrt[3]{5x}}{3}+4\sqrt[3]{5x}$
$=\frac{\sqrt[3]{5x}}{3}+\frac{12\sqrt[3]{5x}}{3}$
$=\frac{13\sqrt[3]{5x}}{3}$

11. $\sqrt{2}\left(6+\sqrt{10}\right)=6\sqrt{2}+\sqrt{20}=6\sqrt{2}+2\sqrt{5}$

12. $\left(\sqrt{3}-\sqrt{5}\right)\left(\sqrt{2}+7\right)=\sqrt{6}+7\sqrt{3}-\sqrt{10}-7\sqrt{5}$

13. $\left(\sqrt{5y}+2\right)\left(\sqrt{5y}-2\right)$
$=\left(\sqrt{5y}\right)^2-2\sqrt{5y}+2\sqrt{5y}-4$
$=5y-4$

14. $\left(\sqrt{3}-7\right)^2=\left(\sqrt{3}-7\right)\left(\sqrt{3}-7\right)$
$=3-7\sqrt{3}-7\sqrt{3}+49$
$=52-14\sqrt{3}$

15. $\left(\sqrt{x+1}+2\right)^2=\left(\sqrt{x+1}+2\right)\left(\sqrt{x+1}+2\right)$
$=x+1+2\sqrt{x+1}+2\sqrt{x+1}+4$
$=x+5+4\sqrt{x+1}$

Mental Math

1. $2\sqrt{3}+4\sqrt{3}=6\sqrt{3}$

2. $5\sqrt{7}+3\sqrt{7}=8\sqrt{7}$

3. $8\sqrt{x}-5\sqrt{x}=3\sqrt{x}$

4. $3\sqrt{y}+10\sqrt{y}=13\sqrt{y}$

5. $7\sqrt[3]{x}+5\sqrt[3]{x}=12\sqrt[3]{x}$

6. $8\sqrt[3]{z}-2\sqrt[3]{z}=6\sqrt[3]{z}$

7. $\left(\sqrt{3}\right)^2=\sqrt{3}\cdot\sqrt{3}=3$

8. $\left(\sqrt{4x+1}\right)^2=\left(\sqrt{4x+1}\right)\left(\sqrt{4x+1}\right)=4x+1$

Exercise Set 7.4

1. $\sqrt{8}-\sqrt{32}=\sqrt{4(2)}-\sqrt{16(2)}$
$=\sqrt{4}\sqrt{2}-\sqrt{16}\sqrt{2}$
$=2\sqrt{2}-4\sqrt{2}$
$=-2\sqrt{2}$

3. $2\sqrt{2x^3}+4x\sqrt{8x}=2\sqrt{2}\sqrt{x^2}\sqrt{x}+4x\sqrt{4}\sqrt{2x}$
$=2x\sqrt{2x}+8x\sqrt{2x}$
$=10x\sqrt{2x}$

5. $2\sqrt{50}-3\sqrt{125}+\sqrt{98}$
$=2\sqrt{25}\sqrt{2}-3\sqrt{25}\sqrt{5}+\sqrt{49}\sqrt{2}$
$=2(5)\sqrt{2}-3(5)\sqrt{5}+7\sqrt{2}$
$=10\sqrt{2}-15\sqrt{5}+7\sqrt{2}$
$=17\sqrt{2}-15\sqrt{5}$

7. $\sqrt[3]{16x}-\sqrt[3]{54x}=\sqrt[3]{8}\sqrt[3]{2}-\sqrt[3]{27}\sqrt[3]{2}$
$=2\sqrt[3]{2}-3\sqrt[3]{2}$
$=-\sqrt[3]{2}$

9. $\sqrt{9b^3}-\sqrt{25b^3}+\sqrt{49b^3}$
$=\sqrt{9b^2}\sqrt{b}-\sqrt{25b^2}\sqrt{b}+\sqrt{49b^2}\sqrt{b}$
$=3b\sqrt{b}-5b\sqrt{b}+7b\sqrt{b}$
$=5b\sqrt{b}$

11. $\frac{5\sqrt{2}}{3}+\frac{2\sqrt{2}}{5}=\frac{25\sqrt{2}}{15}+\frac{6\sqrt{2}}{15}=\frac{31\sqrt{2}}{15}$

13. $\sqrt[3]{\frac{11}{8}}-\frac{\sqrt[3]{11}}{6}=\frac{\sqrt[3]{11}}{\sqrt[3]{8}}-\frac{\sqrt[3]{11}}{6}$
$=\frac{\sqrt[3]{11}}{2}-\frac{\sqrt[3]{11}}{6}$
$=\frac{3\sqrt[3]{11}-\sqrt[3]{11}}{6}$
$=\frac{2\sqrt[3]{11}}{6}$
$=\frac{\sqrt[3]{11}}{3}$

15. $\frac{\sqrt{20x}}{9}+\sqrt{\frac{5x}{9}}=\frac{\sqrt{20x}}{9}+\frac{\sqrt{5x}}{3}$
$=\frac{2\sqrt{5x}}{9}+\frac{3\sqrt{5x}}{9}$
$=\frac{5\sqrt{5x}}{9}$

17. $7\sqrt{9}-7+\sqrt{3}=7(3)-7+\sqrt{3}$
$=21-7+\sqrt{3}$
$=14+\sqrt{3}$

19. $2+3\sqrt{y^2}-6\sqrt{y^2}+5=2+3y-6y+5$
$=7-3y$

21. $3\sqrt{108}-2\sqrt{18}-3\sqrt{48}$
$=3\sqrt{36}\sqrt{3}-2\sqrt{9}\sqrt{2}-3\sqrt{16}\sqrt{3}$
$=3(6)\sqrt{3}-2(3)\sqrt{2}-3(4)\sqrt{3}$
$=18\sqrt{3}-6\sqrt{2}-12\sqrt{3}$
$=6\sqrt{3}-6\sqrt{2}$

23. $-5\sqrt[3]{625}+\sqrt[3]{40}=-5\sqrt[3]{125}\sqrt[3]{5}+\sqrt[3]{8}\sqrt[3]{5}$
$=-25\sqrt[3]{5}+2\sqrt[3]{5}$
$=-23\sqrt[3]{5}$

25. $\sqrt{9b^3}-\sqrt{25b^3}+\sqrt{16b^3}$
$=\sqrt{9b^2}\sqrt{b}-\sqrt{25b^2}\sqrt{b}+\sqrt{16b^2}\sqrt{b}$
$=3b\sqrt{b}-5b\sqrt{b}+4b\sqrt{b}$
$=(3-5+4)b\sqrt{b}$
$=2b\sqrt{b}$

27. $5y\sqrt{8y}+2\sqrt{50y^3}$
$=5y\sqrt{4}\sqrt{2y}+2\sqrt{25y^2}\sqrt{2y}$
$=10y\sqrt{2y}+10y\sqrt{2y}$
$=20y\sqrt{2y}$

29. $\sqrt[3]{54xy^3}-5\sqrt[3]{2xy^3}+y\sqrt[3]{128x}$
$=\sqrt[3]{27y^3}\sqrt[3]{2x}-5\sqrt[3]{y^3}\sqrt[3]{2x}+y\sqrt[3]{64}\sqrt[3]{2x}$
$=3y\sqrt[3]{2x}-5y\sqrt[3]{2x}+y(4)\sqrt[3]{2x}$
$=-2y\sqrt[3]{2x}+4y\sqrt[3]{2x}$
$=2y\sqrt[3]{2x}$

31. $6\sqrt[3]{11}+8\sqrt{11}-12\sqrt{11}=6\sqrt[3]{11}-4\sqrt{11}$

33. $-2\sqrt[4]{x^7}+3\sqrt[4]{16x^7}$
$=-2\sqrt[4]{x^4}\sqrt[4]{x^3}+3\sqrt[4]{16x^4}\sqrt[4]{x^3}$
$=-2x\sqrt[4]{x^3}+3(2x)\sqrt[4]{x^3}$
$=-2x\sqrt[4]{x^3}+6x\sqrt[4]{x^3}$
$=4x\sqrt[4]{x^3}$

35. $\frac{4\sqrt{3}}{3}-\frac{\sqrt{12}}{3}=\frac{4\sqrt{3}}{3}-\frac{\sqrt{4}\sqrt{3}}{3}$
$=\frac{4\sqrt{3}}{3}-\frac{2\sqrt{3}}{3}$
$=\frac{2\sqrt{3}}{3}$

37. $\frac{\sqrt[3]{8x^4}}{7}+\frac{3x\sqrt[3]{x}}{7}=\frac{\sqrt[3]{8x^3}\sqrt[3]{x}+3x\sqrt[3]{x}}{7}$
$=\frac{2x\sqrt[3]{x}+3x\sqrt[3]{x}}{7}$
$=\frac{5x\sqrt[3]{x}}{7}$

39. $\sqrt{\frac{28}{x^2}}+\sqrt{\frac{7}{4x^2}}=\frac{\sqrt{4}\sqrt{7}}{\sqrt{x^2}}+\frac{\sqrt{7}}{\sqrt{4x^2}}$
$=\frac{2\sqrt{7}}{x}+\frac{\sqrt{7}}{2x}$
$=\frac{4\sqrt{7}}{2x}+\frac{\sqrt{7}}{2x}$
$=\frac{5\sqrt{7}}{2x}$

41. $\sqrt[3]{\frac{16}{27}}-\frac{\sqrt[3]{54}}{6}=\frac{\sqrt[3]{16}}{\sqrt[3]{27}}-\frac{\sqrt[3]{27}\sqrt[3]{2}}{6}$
$=\frac{\sqrt[3]{8}\sqrt[3]{2}}{3}-\frac{3\sqrt[3]{2}}{6}$
$=\frac{2(2)\sqrt[3]{2}}{6}-\frac{3\sqrt[3]{2}}{6}$
$=\frac{4\sqrt[3]{2}-3\sqrt[3]{2}}{6}$
$=\frac{\sqrt[3]{2}}{6}$

43. $-\frac{\sqrt[3]{2x^4}}{9}+\sqrt[3]{\frac{250x^4}{27}}$
$=\frac{-\sqrt[3]{x^3}\sqrt[3]{2x}}{9}+\frac{\sqrt[3]{125x^3}\sqrt[3]{2x}}{\sqrt[3]{27}}$
$=\frac{-x\sqrt[3]{2x}}{9}+\frac{5x\sqrt[3]{2x}}{3}$
$=\frac{-x\sqrt[3]{2x}}{9}+\frac{15x\sqrt[3]{2x}}{9}$
$=\frac{14x\sqrt[3]{2x}}{9}$

45. $P=2\sqrt{12}+\sqrt{12}+2\sqrt{27}+3\sqrt{3}$
$=2\sqrt{4}\sqrt{3}+\sqrt{4}\sqrt{3}+2\sqrt{9}\sqrt{3}+3\sqrt{3}$
$=2\cdot 2\sqrt{3}+2\sqrt{3}+2\cdot 3\sqrt{3}+3\sqrt{3}$
$=(4+2+6+3)\sqrt{3}$
$=15\sqrt{3}$

The perimeter of the trapezoid is $15\sqrt{3}$ inches.

47. $\sqrt{7}\left(\sqrt{5}+\sqrt{3}\right)=\sqrt{7}\sqrt{5}+\sqrt{7}\sqrt{3}$
$=\sqrt{35}+\sqrt{21}$

49. $\left(\sqrt{5}-\sqrt{2}\right)^2=\left(\sqrt{5}\right)^2-2\sqrt{5}\sqrt{2}+\left(\sqrt{2}\right)^2$
$=5-2\sqrt{10}+2$
$=7-2\sqrt{10}$

51. $\sqrt{3x}\left(\sqrt{3}-\sqrt{x}\right)=\sqrt{3x}\sqrt{3}-\sqrt{3x}\sqrt{x}$
$=3\sqrt{x}-x\sqrt{3}$

53. $\left(2\sqrt{x}-5\right)\left(3\sqrt{x}+1\right)$
$=\left(2\sqrt{x}\right)\left(3\sqrt{x}\right)+\left(2\sqrt{x}\right)1-5\left(3\sqrt{x}\right)-5\cdot 1$
$=6x+2\sqrt{x}-15\sqrt{x}-5$
$=6x-13\sqrt{x}-5$

55. $\left(\sqrt[3]{a}-4\right)\left(\sqrt[3]{a}+5\right)$
$=\sqrt[3]{a}\sqrt[3]{a}+5\sqrt[3]{a}-4\sqrt[3]{a}-20$
$=\sqrt[3]{a^2}+\sqrt[3]{a}-20$

57. $6\left(\sqrt{2}-2\right)=6\sqrt{2}-6\cdot 2=6\sqrt{2}-12$

59. $\sqrt{2}\left(\sqrt{2}+x\sqrt{6}\right)=\sqrt{2}\sqrt{2}+\sqrt{2}x\sqrt{6}$
$=2+x\sqrt{12}$
$=2+2x\sqrt{3}$

61. $\left(2\sqrt{7}+3\sqrt{5}\right)\left(\sqrt{7}-2\sqrt{5}\right)=2\left(\sqrt{7}\right)^2-\left(2\sqrt{7}\right)\left(2\sqrt{5}\right)+\left(3\sqrt{5}\right)\sqrt{7}-3\cdot 2\left(\sqrt{5}\right)^2$
$=2\cdot 7-4\sqrt{35}+3\sqrt{35}-6\cdot 5$
$=14-\sqrt{35}-30$
$=-16-\sqrt{35}$

63. $\left(\sqrt{x}-y\right)\left(\sqrt{x}+y\right)=\sqrt{x}\sqrt{x}+y\sqrt{x}-y\sqrt{x}-y^2$
$=x-y^2$

65. $\left(\sqrt{3}+x\right)^2=\left(\sqrt{3}\right)^2+2\sqrt{3}x+x^2$
$=3+2x\sqrt{3}+x^2$

67. $\left(\sqrt{5x}-3\sqrt{2}\right)\left(\sqrt{5x}-3\sqrt{3}\right)=\sqrt{5x}\sqrt{5x}-3\sqrt{3}\sqrt{5x}-3\sqrt{2}\sqrt{5x}-3\sqrt{2}\left(3\sqrt{3}\right)$
$=5x-3\sqrt{15x}-3\sqrt{10x}-9\sqrt{6}$

69. $\left(\sqrt[3]{4}+2\right)\left(\sqrt[3]{2}-1\right)=\sqrt[3]{4}\sqrt[3]{2}-\sqrt[3]{4}\cdot 1+2\sqrt[3]{2}-2\cdot 1$
$=\sqrt[3]{8}-\sqrt[3]{4}+2\sqrt[3]{2}-2$
$=2-\sqrt[3]{4}+2\sqrt[3]{2}-2$
$=-\sqrt[3]{4}+2\sqrt[3]{2}$

71. $\left(\sqrt[3]{x}+1\right)\left(\sqrt[3]{x}-4\sqrt{x}+7\right)=\sqrt[3]{x}\sqrt[3]{x}-4\sqrt[3]{x}\sqrt{x}+7\sqrt[3]{x}+\sqrt[3]{x}-4\sqrt{x}+7$
$=\sqrt[3]{x^2}-4\sqrt[6]{x^5}+8\sqrt[3]{x}-4\sqrt{x}+7$

73. $\left(\sqrt{x-1}+5\right)^2=\left(\sqrt{x-1}\right)^2+2\sqrt{x-1}(5)+5^2$
$=x-1+10\sqrt{x-1}+25$
$=x+24+10\sqrt{x-1}$

75. $\left(\sqrt{2x+5}-1\right)^2=\left(\sqrt{2x+5}\right)^2+2(-1)\left(\sqrt{2x+5}\right)+1^2$
$=2x+5-2\sqrt{2x+5}+1$
$=2x+6-2\sqrt{2x+5}$

77. $\dfrac{2x-14}{2}=\dfrac{2(x-7)}{2}=x-7$

79. $\dfrac{7x-7y}{x^2-y^2}=\dfrac{7(x-y)}{(x-y)(x+y)}=\dfrac{7}{x+y}$

81. $\frac{6a^2b-9ab}{3ab}=\frac{3ab(2a-3)}{3ab}=2a-3$

83. $\frac{-4+2\sqrt{3}}{6}=\frac{2\left(-2+\sqrt{3}\right)}{6}=\frac{-2+\sqrt{3}}{3}$

85. $P=2\left(3\sqrt{20}\right)+2\sqrt{125}$
$=6\sqrt{4}\sqrt{5}+2\sqrt{25}\sqrt{5}$
$=6(2)\sqrt{5}+2(5)\sqrt{5}$
$=12\sqrt{5}+10\sqrt{5}$
$=22\sqrt{5}$

The perimeter is $22\sqrt{5}$ feet.

$A=3\sqrt{20}\cdot\sqrt{125}$
$=3\cdot 2\sqrt{5}\cdot 5\sqrt{5}$
$=30\left(\sqrt{5}\right)^2$
$=30\cdot 5$
$=150$

The area is 150 square feet.

87. a. $\sqrt{3}+\sqrt{3}=2\sqrt{3}$

b. $\sqrt{3}\sqrt{3}=\sqrt{9}=3$

c. answers may vary

Section 7.5 Practice Problems

1. $\frac{7}{\sqrt{2}}=\frac{7}{\sqrt{2}}\cdot\frac{\sqrt{2}}{\sqrt{2}}=\frac{7\sqrt{2}}{2}$

2. $\frac{2\sqrt{9}}{\sqrt{16y}}=\frac{2\sqrt{9}}{\sqrt{16y}}\cdot\frac{\sqrt{16y}}{\sqrt{16y}}=\frac{24\sqrt{y}}{16y}=\frac{3\sqrt{y}}{2y}$

3. $\sqrt[3]{\frac{2}{25}}=\frac{\sqrt[3]{2}}{\sqrt[3]{5^2}}\cdot\frac{\sqrt[3]{5}}{\sqrt[3]{5}}=\frac{\sqrt[3]{10}}{5}$

4. $\sqrt{\frac{5m}{11n}}=\frac{\sqrt{5m}}{\sqrt{11n}}\cdot\frac{\sqrt{11n}}{\sqrt{11n}}=\frac{\sqrt{55mn}}{11n}$

5. $\frac{\sqrt[5]{a^2}}{\sqrt[5]{32b^{12}}}=\frac{\sqrt[5]{a^2}}{\sqrt[5]{32b^{10}}\sqrt[5]{b^2}}$
$=\frac{\sqrt[5]{a^2}}{2b^2\sqrt[5]{b^2}}\cdot\frac{\sqrt[5]{b^3}}{\sqrt[5]{b^3}}$
$=\frac{\sqrt[5]{a^2b^3}}{2b^3}$

6. $\frac{3}{2\sqrt{5}+1}=\frac{3}{2\sqrt{5}+1}\cdot\frac{\left(2\sqrt{5}-1\right)}{\left(2\sqrt{5}-1\right)}$
$=\frac{6\sqrt{5}-3}{20-1}$
$=\frac{3\left(2\sqrt{5}-1\right)}{19}$

7. $\frac{\sqrt{5}+3}{\sqrt{3}-\sqrt{2}}=\frac{\sqrt{5}+3}{\sqrt{3}-\sqrt{2}}\cdot\frac{\left(\sqrt{3}+\sqrt{2}\right)}{\left(\sqrt{3}+\sqrt{2}\right)}$
$=\frac{\sqrt{5}\sqrt{3}+\sqrt{5}\sqrt{2}+3\sqrt{3}+3\sqrt{2}}{\left(\sqrt{3}\right)^2-\left(\sqrt{2}\right)^2}$
$=\frac{\sqrt{15}+\sqrt{10}+3\sqrt{3}+3\sqrt{2}}{3-2}$
$=\sqrt{15}+\sqrt{10}+3\sqrt{3}+3\sqrt{2}$

8. $\frac{3}{2-\sqrt{x}}=\frac{3}{2-\sqrt{x}}\cdot\frac{\left(2+\sqrt{x}\right)}{\left(2+\sqrt{x}\right)}$
$=\frac{6+3\sqrt{x}}{2^2-\left(\sqrt{x}\right)^2}$
$=\frac{6+3\sqrt{x}}{4-x}$

9. $\frac{\sqrt{18}}{\sqrt{75}}=\frac{3\sqrt{2}}{5\sqrt{3}}\cdot\frac{\sqrt{2}}{\sqrt{2}}=\frac{6}{5\sqrt{6}}$

10. $\frac{\sqrt[3]{3a}}{\sqrt[3]{7b}}=\frac{\sqrt[3]{3a}}{\sqrt[3]{7b}}\cdot\frac{\sqrt[3]{3^2a^2}}{\sqrt[3]{3^2a^2}}=\frac{3a}{\sqrt[3]{63a^2b}}$

11. $\frac{\sqrt{x}+5}{3} = \frac{\sqrt{x}+5}{3}\cdot\frac{\sqrt{x}-5}{\sqrt{x}-5}$
$= \frac{\left(\sqrt{x}\right)^2-5^2}{3\sqrt{x}-15}$
$= \frac{x-25}{3\left(\sqrt{x}-5\right)}$

Mental Math

1. $\sqrt{2}+x$
conjugate $= \sqrt{2}-x$

2. $\sqrt{3}+y$
conjugate $= \sqrt{3}-y$

3. $5-\sqrt{a}$
conjugate $= 5+\sqrt{a}$

4. $6-\sqrt{b}$
conjugate $= 6+\sqrt{b}$

5. $7\sqrt{4}+8\sqrt{x}$
conjugate $= 7\sqrt{4}-8\sqrt{x}$

6. $9\sqrt{2}-6\sqrt{y}$
conjugate $= 9\sqrt{2}+6\sqrt{y}$

Exercise Set 7.5

1. $\frac{\sqrt{2}}{\sqrt{7}} = \frac{\sqrt{2}\cdot\sqrt{7}}{\sqrt{7}\cdot\sqrt{7}} = \frac{\sqrt{14}}{7}$

3. $\sqrt{\frac{1}{5}} = \frac{\sqrt{1}}{\sqrt{5}}\cdot\frac{\sqrt{5}}{\sqrt{5}} = \frac{\sqrt{5}}{5}$

5. $\frac{4}{\sqrt[3]{3}} = \frac{4}{\sqrt[3]{3}}\cdot\frac{\sqrt[3]{9}}{\sqrt[3]{9}} = \frac{4\sqrt[3]{9}}{\sqrt[3]{27}} = \frac{4\sqrt[3]{9}}{3}$

7. $\frac{3}{\sqrt{8x}} = \frac{3}{\sqrt{8x}}\cdot\frac{\sqrt{8x}}{\sqrt{8x}} = \frac{6\sqrt{2x}}{8x} = \frac{3\sqrt{2x}}{4x}$

9. $\frac{3}{\sqrt[3]{4x^2}} = \frac{3}{\sqrt[3]{4x^2}}\cdot\frac{\sqrt[3]{2x}}{\sqrt[3]{2x}} = \frac{3\sqrt[3]{2x}}{\sqrt[3]{8x^3}} = \frac{3\sqrt[3]{2x}}{2x}$

11. $\frac{9}{\sqrt{3a}} = \frac{9}{\sqrt{3a}}\cdot\frac{\sqrt{3a}}{\sqrt{3a}} = \frac{9\sqrt{3a}}{3a} = \frac{3\sqrt{3a}}{a}$

13. $\frac{3}{\sqrt[3]{2}} = \frac{3}{\sqrt[3]{2}}\cdot\frac{\sqrt[3]{4}}{\sqrt[3]{4}} = \frac{3\sqrt[3]{4}}{\sqrt[3]{8}} = \frac{3\sqrt[3]{4}}{2}$

15. $\frac{2\sqrt{3}}{\sqrt{7}} = \frac{2\sqrt{3}}{\sqrt{7}}\cdot\frac{\sqrt{7}}{\sqrt{7}} = \frac{2\sqrt{21}}{7}$

17. $\sqrt{\frac{2x}{5y}} = \frac{\sqrt{2x}}{\sqrt{5y}} = \frac{\sqrt{2x}\cdot\sqrt{5y}}{\sqrt{5y}\cdot\sqrt{5y}} = \frac{\sqrt{10xy}}{5y}$

19. $\sqrt[3]{\frac{3}{5}} = \frac{\sqrt[3]{3}}{\sqrt[3]{5}}\cdot\frac{\sqrt[3]{25}}{\sqrt[3]{25}} = \frac{\sqrt[3]{75}}{5}$

21. $\sqrt{\frac{3x}{50}} = \frac{\sqrt{3x}}{\sqrt{50}}$
$= \frac{\sqrt{3x}}{5\sqrt{2}}$
$= \frac{\sqrt{3x}\cdot\sqrt{2}}{5\sqrt{2}\cdot\sqrt{2}}$
$= \frac{\sqrt{6x}}{5\cdot 2}$
$= \frac{\sqrt{6x}}{10}$

23. $\frac{1}{\sqrt{12z}} = \frac{1}{2\sqrt{3z}}\cdot\frac{\sqrt{3z}}{\sqrt{3z}} = \frac{\sqrt{3z}}{6z}$

25. $\frac{\sqrt[3]{2y^2}}{\sqrt[3]{9x^2}} = \frac{\sqrt[3]{2y^2}\cdot\sqrt[3]{3x}}{\sqrt[3]{9x^2}\cdot\sqrt[3]{3x}} = \frac{\sqrt[3]{6xy^2}}{3x}$

27. $\sqrt[4]{\frac{16}{9x^7}} = \frac{\sqrt[4]{16}}{\sqrt[4]{9x^7}} = \frac{2}{x\sqrt[4]{9x^3}} \cdot \frac{\sqrt[4]{9x}}{\sqrt[4]{9x}} = \frac{2\sqrt[4]{9x}}{3x^2}$

29. $\frac{5a}{\sqrt[5]{8a^9b^{11}}} = \frac{5a}{ab^2\sqrt[5]{8a^4b}}$

$= \frac{5a \cdot \sqrt[5]{4ab^4}}{ab^2\sqrt[5]{8a^4b} \cdot \sqrt[5]{4ab^4}}$

$= \frac{5a\sqrt[5]{4ab^4}}{2a^2b^3}$

$= \frac{5\sqrt[5]{4ab^4}}{2ab^3}$

31. $\frac{6}{2-\sqrt{7}} = \frac{6}{2-\sqrt{7}} \cdot \frac{(2+\sqrt{7})}{(2+\sqrt{7})}$

$= \frac{6(2+\sqrt{7})}{4-7}$

$= \frac{6(2+\sqrt{7})}{-3}$

$= -2(2+\sqrt{7})$

33. $\frac{-7}{\sqrt{x}-3} = \frac{(-7)(\sqrt{x}+3)}{(\sqrt{x}-3)(\sqrt{x}+3)}$

$= \frac{-7(\sqrt{x}+3)}{(\sqrt{x})^2 - 3^2}$

$= \frac{-7(\sqrt{x}+3)}{x-9}$

$= \frac{7(\sqrt{x}+3)}{9-x}$

35. $\frac{\sqrt{2}-\sqrt{3}}{\sqrt{2}+\sqrt{3}} = \frac{\sqrt{2}-\sqrt{3}}{\sqrt{2}+\sqrt{3}} \cdot \frac{\sqrt{2}-\sqrt{3}}{\sqrt{2}-\sqrt{3}}$

$= \frac{\sqrt{2}^2 - 2\sqrt{3}\sqrt{2} + \sqrt{3}^2}{\sqrt{2}^2 - \sqrt{3}^2}$

$= \frac{2-2\sqrt{6}+3}{2-3}$

$= -5 + 2\sqrt{6}$

37. $\frac{\sqrt{a}+1}{2\sqrt{a}-\sqrt{b}}$

$= \frac{(\sqrt{a}+1)(2\sqrt{a}+\sqrt{b})}{(2\sqrt{a}-\sqrt{b})(2\sqrt{a}+\sqrt{b})}$

$= \frac{\sqrt{a}(2\sqrt{a}) + \sqrt{a}\sqrt{b} + 1(2\sqrt{a}) - 1\sqrt{b}}{(2\sqrt{a})^2 - (\sqrt{b})^2}$

$= \frac{2a + \sqrt{ab} + 2\sqrt{a} + \sqrt{b}}{4a-b}$

39. $\frac{8}{1+\sqrt{10}} = \frac{8}{1+\sqrt{10}} \cdot \frac{1-\sqrt{10}}{1-\sqrt{10}}$

$= \frac{8-8\sqrt{10}}{1^2 - (\sqrt{10})^2}$

$= -\frac{8(1-\sqrt{10})}{9}$

41. $\frac{\sqrt{x}}{\sqrt{x}+\sqrt{y}} = \frac{(\sqrt{x})(\sqrt{x}-\sqrt{y})}{(\sqrt{x}+\sqrt{y})(\sqrt{x}-\sqrt{y})}$

$= \frac{\sqrt{x}\sqrt{x} - \sqrt{x}\sqrt{y}}{(\sqrt{x})^2 - (\sqrt{y})^2}$

$= \frac{x - \sqrt{xy}}{x-y}$

43. $\dfrac{2\sqrt{3}+\sqrt{6}}{4\sqrt{3}-\sqrt{6}}$

$= \dfrac{2\sqrt{3}+\sqrt{6}}{4\sqrt{3}-\sqrt{6}} \cdot \dfrac{4\sqrt{3}+\sqrt{6}}{4\sqrt{3}+\sqrt{6}}$

$= \dfrac{2\sqrt{3}\left(4\sqrt{3}\right)+2\sqrt{3}\sqrt{6}+4\sqrt{3}\sqrt{6}+\sqrt{6}\sqrt{6}}{\left(4\sqrt{3}\right)^2-\left(\sqrt{6}\right)^2}$

$= \dfrac{24+6\sqrt{2}+12\sqrt{2}+6}{48-6}$

$= \dfrac{30+18\sqrt{2}}{42}$

$= \dfrac{5+3\sqrt{2}}{7}$

45. $\sqrt{\dfrac{5}{3}} = \dfrac{\sqrt{5}}{\sqrt{3}} = \dfrac{\sqrt{5}\cdot\sqrt{5}}{\sqrt{3}\cdot\sqrt{5}} = \dfrac{5}{\sqrt{15}}$

47. $\sqrt{\dfrac{18}{5}} = \dfrac{\sqrt{18}}{\sqrt{5}} = \dfrac{\sqrt{18}}{\sqrt{5}}\cdot\dfrac{\sqrt{18}}{\sqrt{18}} = \dfrac{18}{3\sqrt{10}} = \dfrac{6}{\sqrt{10}}$

49. $\dfrac{\sqrt{4x}}{7} = \dfrac{2\sqrt{x}}{7} = \dfrac{2\sqrt{x}\cdot\sqrt{x}}{7\cdot\sqrt{x}} = \dfrac{2x}{7\sqrt{x}}$

51. $\dfrac{\sqrt[3]{5y^2}}{\sqrt[3]{4x}} = \dfrac{\sqrt[3]{5y^2}}{\sqrt[3]{4x}}\cdot\dfrac{\sqrt[3]{5^2y}}{\sqrt[3]{5^2y}} = \dfrac{5y}{\sqrt[3]{100xy}}$

53. $\sqrt{\dfrac{2}{5}} = \dfrac{\sqrt{2}}{\sqrt{5}} = \dfrac{\sqrt{2}\cdot\sqrt{2}}{\sqrt{5}\cdot\sqrt{2}} = \dfrac{2}{\sqrt{10}}$

55. $\dfrac{\sqrt{2x}}{11} = \dfrac{\sqrt{2x}}{11}\cdot\dfrac{\sqrt{2x}}{\sqrt{2x}} = \dfrac{2x}{11\sqrt{2x}}$

57. $\sqrt[3]{\dfrac{7}{8}} = \dfrac{\sqrt[3]{7}}{\sqrt[3]{8}}$

$= \dfrac{\sqrt[3]{7}}{2}$

$= \dfrac{\sqrt[3]{7}\cdot\sqrt[3]{7^2}}{2\cdot\sqrt[3]{7^2}}$

$= \dfrac{\sqrt[3]{7^3}}{2\sqrt[3]{7^2}}$

$= \dfrac{7}{2\sqrt[3]{49}}$

59. $\dfrac{\sqrt[3]{3x^5}}{10} = \dfrac{x\sqrt[3]{3x^2}}{10}\cdot\dfrac{\sqrt[3]{9x}}{\sqrt[3]{9x}} = \dfrac{3x^2}{10\sqrt[3]{9x}}$

61. $\sqrt{\dfrac{18x^4y^6}{3z}} = \dfrac{\sqrt{18x^4y^6}}{\sqrt{3z}}$

$= \dfrac{3x^2y^3\sqrt{2}}{\sqrt{3z}}$

$= \dfrac{3x^2y^3\sqrt{2}\cdot\sqrt{2}}{\sqrt{3z}\cdot\sqrt{2}}$

$= \dfrac{6x^2y^3}{\sqrt{6z}}$

63. answers may vary

65. $\dfrac{2-\sqrt{11}}{6} = \dfrac{\left(2-\sqrt{11}\right)\left(2+\sqrt{11}\right)}{6\left(2+\sqrt{11}\right)}$

$= \dfrac{4-11}{12+6\sqrt{11}}$

$= \dfrac{-7}{12+6\sqrt{11}}$

67. $\dfrac{2-\sqrt{7}}{-5}=\dfrac{2-\sqrt{7}}{-5}\cdot\dfrac{2+\sqrt{7}}{2+\sqrt{7}}$
$=\dfrac{4-\sqrt{49}}{-5(2+\sqrt{7})}$
$=\dfrac{-3}{-5(2+\sqrt{7})}$
$=\dfrac{3}{10+5\sqrt{7}}$

69. $\dfrac{\sqrt{x}+3}{\sqrt{x}}=\dfrac{(\sqrt{x}+3)(\sqrt{x}-3)}{\sqrt{x}(\sqrt{x}-3)}$
$=\dfrac{(\sqrt{x})^2-3^2}{(\sqrt{x})^2-3\sqrt{x}}$
$=\dfrac{x-9}{x-3\sqrt{x}}$

71. $\dfrac{\sqrt{x}+1}{\sqrt{x}-1}=\dfrac{\sqrt{x}+1}{\sqrt{x}-1}\cdot\dfrac{\sqrt{x}-1}{\sqrt{x}-1}$
$=\dfrac{\sqrt{x^2}-1}{\sqrt{x^2}-2\sqrt{x}+1}$
$=\dfrac{x-1}{x-2\sqrt{x}+1}$

73. $2x-7=3(x-4)$
$2x-7=3x-12$
$2x+5=3x$
$5=x$
The solution set is {5}.

75. $(x-6)(2x+1)=0$
$x-6=0$ or $2x+1=0$
$x=6$ $\quad 2x=-1$
$x=-\dfrac{1}{2}$
The solution set is $\left\{-\dfrac{1}{2}, 6\right\}$.

77. $x^2-8x=-12$
$x^2-8x+12=0$
$(x-2)(x-6)=0$
$x-2=0$ or $x-6=0$
$x=2$ $\quad x=6$
The solution set is {2, 6}.

79. $r=\sqrt{\dfrac{A}{4\pi}}=\dfrac{\sqrt{A}}{\sqrt{4\pi}}=\dfrac{\sqrt{A}}{2\sqrt{\pi}}\cdot\dfrac{\sqrt{\pi}}{\sqrt{\pi}}=\dfrac{\sqrt{A\pi}}{2\pi}$

81. answers may vary

83. $\dfrac{9}{\sqrt[3]{5}}=\dfrac{9}{\sqrt[3]{5}}\cdot\dfrac{\sqrt[3]{25}}{\sqrt[3]{25}}=\dfrac{9\sqrt[3]{25}}{\sqrt[3]{125}}=\dfrac{9\sqrt[3]{25}}{5}$
The smaller number is $\sqrt[3]{25}$.

Integrated Review

1. $\sqrt{81}=9$ because $9^2=81$.
2. $\sqrt[3]{-8}=-2$ because $(-2)^3=-8$.
3. $\sqrt[4]{\dfrac{1}{16}}=\dfrac{1}{2}$ because $\left(\dfrac{1}{2}\right)^4=\dfrac{1}{16}$.
4. $\sqrt{x^6}=x^3$ because $(x^3)^2=x^6$.
5. $\sqrt[3]{y^9}=y^3$ because $(y^3)^3=y^9$
6. $\sqrt{4y^{10}}=2y^5$ because $(2y^5)^2=4y^{10}$.
7. $\sqrt[5]{-32y^5}=-2y$ because $(-2y)^5=-32y^5$.
8. $\sqrt[4]{81b^{12}}=3b^3$ because $(3b^3)^4=81b^{12}$.
9. $36^{1/2}=\sqrt{36}=6$
10. $(3y)^{1/4}=\sqrt[4]{3y}$

11. $64^{-2/3} = \dfrac{1}{\left(\sqrt[3]{64}\right)^2} = \dfrac{1}{(4)^2} = \dfrac{1}{16}$

12. $(x+1)^{3/5} = \sqrt[5]{(x+1)^3}$

13. $y^{-1/6} \cdot y^{7/6} = y^{6/6} = y$

14. $\dfrac{(2x^{1/3})^4}{x^{5/6}} = 16x^{4/3}x^{-5/6}$
$= 16x^{\frac{8}{6}-\frac{5}{6}}$
$= 16x^{3/6}$
$= 16x^{1/2}$

15. $\dfrac{x^{1/4}x^{3/4}}{x^{-1/4}} = x^{\frac{1}{4}+\frac{3}{4}+\frac{1}{4}} = x^{5/4}$

16. $4^{1/3} \cdot 4^{2/5} = 4^{\frac{1}{3}+\frac{2}{5}} = 4^{\frac{5}{15}+\frac{6}{15}} = 4^{11/15}$

17. $\sqrt[3]{8x^6} = 2x^2$

18. $\sqrt[12]{a^9b^6} = (a^9b^6)^{1/12}$
$= a^{9/12}b^{6/12}$
$= a^{3/4}b^{1/2}$
$= a^{3/4}b^{2/4}$
$= \sqrt[4]{a^3b^2}$

19. $\sqrt[4]{x} \cdot \sqrt{x} = x^{1/4} \cdot x^{1/2} = x^{\frac{1}{4}+\frac{2}{4}} = x^{3/4} = \sqrt[4]{x^3}$

20. $\sqrt{5} \cdot \sqrt[3]{2} = 5^{1/2} \cdot 2^{1/3}$
$= 5^{3/6} \cdot 2^{2/6}$
$= \sqrt[6]{5^3 \cdot 2^2}$
$= \sqrt[6]{500}$

21. $\sqrt{40} = \sqrt{4}\sqrt{10} = 2\sqrt{10}$

22. $\sqrt[4]{16x^7y^{10}} = \sqrt[4]{16x^4y^8}\sqrt[4]{x^3y^2} = 2xy^2\sqrt[4]{x^3y^2}$

23. $\sqrt[3]{54x^4} = \sqrt[3]{27x^3}\sqrt[3]{2x} = 3x\sqrt[3]{2x}$

24. $\sqrt[5]{-64b^{10}} = \sqrt[5]{-32b^{10}}\sqrt[5]{2} = -2b^2\sqrt[5]{2}$

25. $\sqrt{5} \cdot \sqrt{x} = \sqrt{5x}$

26. $\sqrt[3]{8x} \cdot \sqrt[3]{8x^2} = \sqrt[3]{64x^3} = 4x$

27. $\dfrac{\sqrt{98y^6}}{\sqrt{2y}} = \sqrt{\dfrac{98y^6}{2y}}$
$= \sqrt{49y^5}$
$= \sqrt{49y^4} \cdot \sqrt{y}$
$= 7y^2\sqrt{y}$

28. $\dfrac{\sqrt[4]{48a^9b^3}}{\sqrt[4]{ab^3}} = \sqrt[4]{\dfrac{48a^9b^3}{ab^3}}$
$= \sqrt[4]{48a^8}$
$= \sqrt[4]{16a^8} \cdot \sqrt[4]{3}$
$= 2a^2\sqrt[4]{3}$

29. $\sqrt{20} - \sqrt{75} + 5\sqrt{7} = \sqrt{4}\sqrt{5} - \sqrt{25}\sqrt{3} + 5\sqrt{7}$
$= 2\sqrt{5} - 5\sqrt{3} + 5\sqrt{7}$

30. $\sqrt[3]{54y^4} - y\sqrt[3]{16y} = \sqrt[3]{27y^3}\sqrt[3]{2y} - y\sqrt[3]{8}\sqrt[3]{2y}$
$= 3y\sqrt[3]{2y} - 2y\sqrt[3]{2y}$
$= y\sqrt[3]{2y}$

31. $\sqrt{3}\left(\sqrt{5} - \sqrt{2}\right) = \sqrt{3}\sqrt{5} - \sqrt{3}\sqrt{2} = \sqrt{15} - \sqrt{6}$

32. $\left(\sqrt{7} + \sqrt{3}\right)^2 = \left(\sqrt{7}\right)^2 + 2\sqrt{7}\sqrt{3} + \left(\sqrt{3}\right)^2$
$= 7 + 2\sqrt{21} + 3$
$= 10 + 2\sqrt{21}$

33. $\left(2x - \sqrt{5}\right)\left(2x + \sqrt{5}\right) = (2x)^2 - \left(\sqrt{5}\right)^2$
$= 4x^2 - 5$

34. $\left(\sqrt{x+1}-1\right)^2 = \left(\sqrt{x+1}\right)^2 - 2\left(\sqrt{x+1}\right) + 1^2$
$= x+1-2\sqrt{x+1}+1$
$= x+2-2\sqrt{x+1}$

35. $\sqrt{\frac{7}{3}} = \frac{\sqrt{7}}{\sqrt{3}} = \frac{\sqrt{7}}{\sqrt{3}} \cdot \frac{\sqrt{3}}{\sqrt{3}} = \frac{\sqrt{21}}{3}$

36. $\frac{5}{\sqrt[3]{2x^2}} = \frac{5}{\sqrt[3]{2x^2}} \cdot \frac{\sqrt[3]{4x}}{\sqrt[3]{4x}} = \frac{5\sqrt[3]{4x}}{2x}$

37. $\frac{\sqrt{3}-\sqrt{7}}{2\sqrt{3}+\sqrt{7}}$
$= \frac{\sqrt{3}-\sqrt{7}}{2\sqrt{3}+\sqrt{7}} \cdot \frac{\left(2\sqrt{3}-\sqrt{7}\right)}{\left(2\sqrt{3}-\sqrt{7}\right)}$
$= \frac{\sqrt{3}\left(2\sqrt{3}\right)-\sqrt{3}\sqrt{7}-\sqrt{7}\left(2\sqrt{3}\right)+\sqrt{7}\sqrt{7}}{\left(2\sqrt{3}\right)^2-\left(\sqrt{7}\right)^2}$
$= \frac{6-\sqrt{21}-2\sqrt{21}+7}{12-7}$
$= \frac{13-3\sqrt{21}}{5}$

38. $\sqrt{\frac{7}{3}} = \frac{\sqrt{7}}{\sqrt{3}} = \frac{\sqrt{7}}{\sqrt{3}} \cdot \frac{\sqrt{7}}{\sqrt{7}} = \frac{7}{\sqrt{21}}$

39. $\sqrt[3]{\frac{9y}{11}} = \frac{\sqrt[3]{9y}}{\sqrt[3]{11}} = \frac{\sqrt[3]{9y}}{\sqrt[3]{11}} \cdot \frac{\sqrt[3]{3y^2}}{\sqrt[3]{3y^2}} = \frac{3y}{\sqrt[3]{33y^2}}$

40. $\frac{\sqrt{x}-2}{\sqrt{x}} = \frac{\sqrt{x}-2}{\sqrt{x}} \cdot \frac{\sqrt{x}+2}{\sqrt{x}+2}$
$= \frac{\left(\sqrt{x}\right)^2-2^2}{\sqrt{x}\sqrt{x}+2\sqrt{x}}$
$= \frac{x-4}{x+2\sqrt{x}}$

Section 7.6 Practice Problems

1. $\sqrt{3x-2} = 5$
$\left(\sqrt{3x-2}\right)^2 = 5^2$
$3x-2 = 25$
$3x = 27$
$x = 9$
The solution set is $\{9\}$.

2. $\sqrt{9x-2} - 2x = 0$
$\sqrt{9x-2} = 2x$
$\left(\sqrt{9x-2}\right)^2 = (2x)^2$
$9x-2 = 4x^2$
$0 = 4x^2 - 9x + 2$
$(4x-2)(x-2) = 0$
$4x-1=0$ or $x-2=0$
$4x = 1$ $\quad x = 2$
$x = \frac{1}{4}$
The solution set is $\left\{\frac{1}{4}, 2\right\}$.

3. $\sqrt[3]{x-5} + 2 = 1$
$\sqrt[3]{x-5} = -1$
$\left(\sqrt[3]{x-5}\right)^3 = (-1)^3$
$x-5 = -1$
$x = 4$
The solution set is $\{4\}$.

4. $\sqrt{9+x} = x+3$
$\left(\sqrt{9+x}\right)^2 = (x+3)^2$
$9+x = x^2+6x+9$
$0 = x^2+5x$
$0 = x(x+5)$
$x = 0$ or $x+5 = 0$
$x = -5$

Check:

$$\sqrt{9+x}=x+3$$
$$\sqrt{9+0}=0+3$$
$$\sqrt{9}=3$$
$$3=3 \quad \text{True}$$
$$\sqrt{9+x}=x+3$$
$$\sqrt{9+(-5)}=-5+3$$
$$\sqrt{4}=-2$$
$$2=-2 \quad \text{False}$$

The solution set is $\{0\}$.

5. $\sqrt{3x+1}+\sqrt{3x}=2$

$$\sqrt{3x+1}=2-\sqrt{3x}$$
$$\left(\sqrt{3x+1}\right)^2=\left(2-\sqrt{3x}\right)^2$$
$$3x+1=4-2(2)\sqrt{3x}+3x$$
$$-3=-4\sqrt{3x}$$
$$(-3)^2=\left(-4\sqrt{3x}\right)^2$$
$$9=16(3x)$$
$$9=48x$$
$$\frac{9}{48}=x$$
$$\frac{3}{16}=x$$

The solution set is $\left\{\frac{3}{16}\right\}$.

6. $a^2+b^2=c^2$

$$3^2+b^2=6^2$$
$$9+b^2=36$$
$$b^2=27$$
$$b=\sqrt{27}=3\sqrt{3}$$

The unknown length is $3\sqrt{3}$ cm.

7. $a^2+b^2=c^2$

$$45^2+45^2=c^2$$
$$2025+2025=c^2$$
$$4050=c^2$$
$$45\sqrt{2} \text{ or } 63.64 \approx c$$

The longest strip is $45\sqrt{2}$ or 63.64 inches.

Calculator Explorations

1. $\sqrt{x+7}=x$

$Y=\sqrt{x+7}$ and $Y_2=x$

The solution set is $\{3.19\}$.

2. $\sqrt{3x+5}=2x$

$Y_1=\sqrt{3x+5}$ and $Y_2=2x$

The solution set is $\{1.55\}$.

3. $\sqrt{2x+1}=\sqrt{2x+2}$

$Y_1=\sqrt{2x+1}$ and $Y_2=\sqrt{2x+2}$

The solution set is $\varnothing$.

4. $\sqrt{10x-1}=\sqrt{-10x+10}-1$

$Y_1=\sqrt{10x-1}$ and $Y_2=\sqrt{-10x+10}-1$

The solution set is $\{0.34\}$.

5. $1.2x=\sqrt{3.1x+5}$

$Y_1=1.2x$ and $Y_2=\sqrt{3.1x+5}$

The solution set is $\{3.23\}$.

6. $\sqrt{1.9x^2-2.2}=-0.8x+2$

$Y_1=\sqrt{1.9x^2-2.2}$ and $Y_2=-0.8x+2$

The solution set is $\{-5.44, 1.63\}$.

Exercise Set 7.6

1. $\sqrt{2x}=4$

$$\left(\sqrt{2x}\right)^2=4^2$$
$$2x=16$$
$$x=8$$

The solution set is $\{8\}$.

3. $\sqrt{x-3}=2$

$\left(\sqrt{x-3}\right)^2=2^2$

$x-3=4$

$x=7$

The solution set is $\{7\}$.

5. $\sqrt{2x}=-4$

There is no solution since a principle square root does not yield a negative number. The solution set is $\varnothing$.

7. $\sqrt{4x-3}-5=0$

$\sqrt{4x-3}=5$

$\left(\sqrt{4x-3}\right)^2=25$

$4x=28$

$x=7$

The solution set is $\{7\}$.

9. $\sqrt{2x-3}-2=1$

$\sqrt{2x-3}=3$

$\left(\sqrt{2x-3}\right)^2=3^2$

$2x-3=9$

$2x=12$

$x=6$

The solution set is $\{6\}$.

11. $\sqrt[3]{6x}=-3$

$\left(\sqrt[3]{6x}\right)^3=(-3)^3$

$6x=-27$

$x=\frac{-27}{6}=-\frac{9}{2}$

The solution set is $\left\{-\frac{9}{2}\right\}$.

13. $\sqrt[3]{x-2}-3=0$

$\sqrt[3]{x-2}=3$

$\left(\sqrt[3]{x-2}\right)^3=3^3$

$x-2=27$

$x=29$

The solution set is $\{29\}$.

15. $\sqrt{13-x}=x-1$

$\left(\sqrt{13-x}\right)^2=(x-1)^2$

$13-x=x^2-2x+1$

$0=x^2-x-12$

$(x-4)(x+3)=0$

$x-4=0$ or $x+3=0$

$x=4$ $\quad$ $x=-3$

We discard the -3 as extraneous leaving $x=4$ as the only solution. The solution set is $\{4\}$.

17. $x-\sqrt{4-3x}=-8$

$x+8=\sqrt{4-3x}$

$(x+8)^2=\left(\sqrt{4-3x}\right)^2$

$x^2+16x+64=4-3x$

$x^2+16x+64=4-3x$

$x^2+19x+60=0$

$(x+4)(x+15)=0$

$x+4=0$ or $x+15=0$

$x=-4$ $\quad$ $x=-15$

We discard $x=-15$ as extraneous, leaving $x=-4$ as the only solution. The solution set is $\{-4\}$.

19. $\sqrt{y+5}=2-\sqrt{y-4}$

$y+5=4-4\sqrt{y-4}+y-4$

$5=-4\sqrt{y-4}$

$25=16y-64$

$89=16y$

$\frac{89}{16}=y$

We discard the $\frac{89}{16}$ as extraneous. There is no solution. The solution set is $\varnothing$.

21. $\sqrt{x-3}+\sqrt{x+2}=5$

$$\sqrt{x-3}=5-\sqrt{x+2}$$
$$\left(\sqrt{x-3}\right)^2=\left(5-\sqrt{x+2}\right)^2$$
$$x-3=25-10\sqrt{x+2}+x+2$$
$$-3=27-10\sqrt{x+2}$$
$$-30=-10\sqrt{x+2}$$
$$3=\sqrt{x+2}$$
$$3^2=\left(\sqrt{x+2}\right)^2$$
$$9=x+2$$
$$7=x$$

The solution set is $\{7\}$.

23. $\sqrt{3x-2}=5$

$$3x-2=25$$
$$3x=27$$
$$x=9$$

The solution set is $\{9\}$.

25. $-\sqrt{2x}+4=-6$

$$10=\sqrt{2x}$$
$$10^2=\left(\sqrt{2x}\right)^2$$
$$100=2x$$
$$x=50$$

The solution set is $\{50\}$.

27. $\sqrt{3x+1}+2=0$

$$\sqrt{3x+1}=-2$$

There are no real solutions to this equation. The solution set is $\varnothing$.

29. $\sqrt[4]{4x+1}-2=0$

$$\sqrt[4]{4x+1}=2$$
$$\left(\sqrt[4]{4x+1}\right)^4=2^4$$
$$4x+1=16$$
$$4x=15$$
$$x=\frac{15}{4}$$

The solution set is $\left\{\frac{15}{4}\right\}$.

31. $\sqrt{3x+4}=5$

$$3x+4=25$$
$$3x=21$$
$$x=7$$

The solution set is $\{7\}$.

33. $\sqrt[3]{6x-3}-3=0$

$$\sqrt[3]{6x-3}=3$$
$$\left(\sqrt[3]{6x-3}\right)^3=3^3$$
$$6x-3=27$$
$$6x=30$$
$$x=5$$

The solution set is $\{5\}$.

35. $\sqrt[3]{2x-3}-2=-5$

$$\sqrt[3]{2x-3}=-3$$
$$2x-3=-27$$
$$2x=-24$$
$$x=-12$$

The solution set is $\{-12\}$.

37. $\sqrt{x+4}=\sqrt{2x-5}$

$$x+4=2x-5$$
$$9=x$$
$$x=9$$

The solution set is $\{9\}$.

39. $x-\sqrt{1-x}=-5$

$$x+5=\sqrt{1-x}$$
$$x^2+10x+25=1-x$$
$$x^2+11x+24=0$$
$$(x+8)(x+3)=0$$
$$x+8=0 \quad \text{or} \quad x+3=0$$
$$x=-8 \qquad\qquad x=-3$$

We discard the -8 or extraneous, leaving $x=-3$ as the only solution. The solution set is $\{-3\}$.

41. $\sqrt[3]{-6x-1} = \sqrt[3]{-2x-5}$
$-6x-1 = -2x-5$
$4 = 4x$
$x = 1$
The solution set is $\{1\}$.

43. $\sqrt{5x-1} - \sqrt{x} + 2 = 3$
$\sqrt{5x-1} = \sqrt{x} + 1$
$5x-1 = x + 2\sqrt{x} + 1$
$4x-2 = 2\sqrt{x}$
$2x-1 = \sqrt{x}$
$4x^2 - 4x + 1 = x$
$4x^2 - 5x + 1 = 0$
$(4x-1)(x-1) = 0$
$4x-1 = 0$ or $x-1 = 0$
$4x = 1$ $\quad x = 1$
$x = \frac{1}{4}$

We discard the $\frac{1}{4}$ or extraneous, leaving $x = 1$ as the only solution. The solution set is $\{1\}$.

45. $\sqrt{2x-1} = \sqrt{1-2x}$
$2x-1 = 1-2x$
$4x = 2$
$x = \frac{1}{2}$

The solution set is $\left\{\frac{1}{2}\right\}$.

47. $\sqrt{3x+4} - 1 = \sqrt{2x+1}$
$3x+4 - 2\sqrt{3x+4} + 1 = 2x+1$
$x+4 = 2\sqrt{3x+4}$
$x^2 + 8x + 16 = 4(3x+4)$
$x^2 - 4x = 0$
$x(x-4) = 0$
$x = 0$ or $x-4 = 0$
$x = 4$
The solution set is $\{0, 4\}$.

49. $\sqrt{y+3} - \sqrt{y-3} = 1$
$\sqrt{y+3} = 1 + \sqrt{y-3}$
$y+3 = \left(1+\sqrt{y-3}\right)^2$
$y+3 = 1 + 2\sqrt{y-3} + y - 3$
$5 = 2\sqrt{y-3}$
$25 = 4(y-3)$
$\frac{25}{4} = y-3$
$\frac{25}{4} + \frac{12}{4} = y$
$\frac{37}{4} = y$

The solution is $\left\{\frac{37}{4}\right\}$.

51. Let c = the length of the unknown side of the right triangle. By the Pythagorean Theorem,
$6^2 + 3^2 = c^2$
$36 + 9 = c^2$
$45 = c^2$
$3\sqrt{5} = c$

The length of the unknown side is $3\sqrt{5}$ feet.

53. Let b = the length of the unknown leg of the right triangle. Use the Pythagorean theorem.
$7^2 = 3^2 + b^2$
$49 = 9 + b^2$
$b^2 = 40$
$b = \sqrt{40}$
$= \sqrt{4}\sqrt{10}$
$= 2\sqrt{10}$
The length of the unknown side is $2\sqrt{10}$ meters.

55. Let b = the length of the leg of the right triangle. By the Pythagorean Theorem,

$$9^2 + b^2 = \left(11\sqrt{5}\right)^2$$
$$81 + b^2 = 605$$
$$b^2 = 524$$
$$b = 2\sqrt{131}$$

The length of the unknown side is $2\sqrt{131}$ meters ≈ 22.9 meters.

57. Let c = the length of the hypotenuse of the right triangle. Use the Pythagorean theorem.

$$c^2 = 7^2 + (7.2)^2$$
$$c^2 = 49 + 51.84$$
$$c^2 = 100.84$$
$$c = \sqrt{100.84}$$

The length of the unknown side is $\sqrt{100.84}$ millimeters ≈ 10.0 millimeters.

59. $15^2 + 8^2 = c^2$

$$225 + 64 = c^2$$
$$289 = c^2$$
$$17 = c$$

The cable should be 17 feet.

61. $x^2 = 5^2 + 12^2$

$$x^2 = 25 + 144$$
$$x^2 = 169$$
$$x = \sqrt{169}$$
$$x = 13$$

A 13 foot ladder is needed.

63.

$$r = \sqrt{\frac{A}{4\pi}}$$
$$1080 = \sqrt{\frac{A}{4\pi}}$$
$$1{,}166{,}400 = \frac{A}{4\pi}$$
$$14{,}657{,}415 = A$$

The surface area is 14,657,415 square miles.

65.

$$v = \sqrt{2gh}$$
$$80 = \sqrt{2(32)h}$$
$$(80)^2 = \left(\sqrt{2(32)h}\right)^2$$
$$6400 = 2(32)\cdot h$$
$$100 = h$$

The object fell 100 feet.

67.

$$S = 2\sqrt{I} - 9$$
$$11 = 2\sqrt{I} - 9$$
$$20 = 2\sqrt{I}$$
$$10 = \sqrt{I}$$
$$100 = I$$

The IQ is 100.

69. $P = 2\pi\sqrt{\frac{l}{32}}$

$$= 2\pi\sqrt{\frac{2}{32}}$$
$$= 2\pi\sqrt{\frac{1}{16}}$$
$$= 2\pi\left(\frac{1}{4}\right)$$
$$= \frac{2\pi}{4}$$
$$= \frac{\pi}{2}$$

The period is $\frac{\pi}{2}$ seconds ≈ 1.57 seconds.

71.

$$P = 2\pi\sqrt{\frac{l}{32}}$$
$$4 = 2\pi\sqrt{\frac{l}{32}}$$
$$0.6366 = \sqrt{\frac{l}{32}}$$
$$0.4053 = \frac{l}{32}$$
$$12.97 = l$$

The length is 12.97 feet.

73. answers may vary

75. $s = \frac{1}{2}(a+b+c)$

$s = \frac{1}{2}(6+14+10)$

$s = \frac{30}{2} = 15$

$A = \sqrt{s(s-a)(s-b)(s-c)}$

$= \sqrt{15(15-6)(15-14)(15-10)}$

$= \sqrt{15(9)(1)(5)}$

$= \sqrt{15(15)(3)}$

$= 15\sqrt{3}$

The area is
$15\sqrt{3}$ square miles ≈ 25.98 square miles.

77. answers may vary

79. $D(h) = 111.7\sqrt{h}$

$80 = 111.7\sqrt{h}$

$0.716 = \sqrt{h}$

$0.51 = h$

The height would be 0.51 kilometer.

81. $\frac{\frac{x}{6}}{\frac{2x}{3}+\frac{1}{2}} = \frac{\left(\frac{x}{6}\right)6}{\left(\frac{2x}{3}+\frac{1}{2}\right)6}$

$= \frac{x}{\left(\frac{2x}{3}\right)6+\left(\frac{1}{2}\right)6}$

$= \frac{x}{4x+3}$

83. $\frac{\frac{z}{5}+\frac{1}{10}}{\frac{z}{20}-\frac{z}{5}} = \frac{\left(\frac{z}{5}+\frac{1}{10}\right)20}{\left(\frac{z}{20}-\frac{z}{5}\right)20} = \frac{4z+2}{z-4z} = -\frac{4z+2}{3z}$

85. $\sqrt{\sqrt{x+3}+\sqrt{x}} = \sqrt{3}$

$\left(\sqrt{\sqrt{x+3}+\sqrt{x}}\right)^2 = \left(\sqrt{3}\right)^2$

$\sqrt{x+3}+\sqrt{x} = 3$

$\sqrt{x+3} = 3-\sqrt{x}$

$\left(\sqrt{x+3}\right)^2 = \left(3-\sqrt{x}\right)^2$

$x+3 = 9-6\sqrt{x}+x$

$-6 = -6\sqrt{x}$

$1 = \sqrt{x}$

$1^2 = \left(\sqrt{x}\right)^2$

$1 = x$

The solution set is $\{1\}$.

87. $\sqrt{5x-1}+4 = 7$

$\sqrt{5x-1} = 3$

$\left(\sqrt{5x-1}\right)^2 = 3^2$

$5x-1 = 9$

$5x = 10$

$x = 2$

89. $80\sqrt[3]{x}+500 < 1620$

$80\sqrt[3]{x} < 1120$

$\sqrt[3]{x} < 14$

$\left(\sqrt[3]{x}\right)^3 < 14^3$

$x < 2744$

The company can make at most 2743 deliveries.

The Bigger Picture

1. $\frac{x}{4}+\frac{x+18}{20} = \frac{x-5}{5}$

$20\left(\frac{x}{4}\right)+20\left(\frac{x+18}{20}\right) = 20\left(\frac{x-5}{5}\right)$

$5x+(x+18) = 4(x-5)$

$6x+18 = 4x-20$

$2x = -38$

$x = -19$

The solution set is $\{-19\}$.

2. $|3x-5|=10$

$3x-5=-10$ or $3x-5=10$

$3x=-5$ $\quad$ $3x=15$

$x=-\frac{5}{3}$ $\quad$ $x=5$

The solution set is $\left\{-\frac{5}{3}, 5\right\}$.

3. $2x^2-x=45$

$2x^2-x-45=0$

$(2x+9)(x-5)=0$

$2x+9=0$ or $x-5=0$

$x=-\frac{9}{2}$ $\quad$ $x=5$

The solution set is $\left\{-\frac{9}{2}, 5\right\}$.

4. $-6\le -5x-1\le 10$

$-5\le -5x\le 11$

$1\ge x\ge -\frac{11}{5}$

$-\frac{11}{5}\le x\le 1$

The solution is $\left[-\frac{11}{5}, 1\right]$.

5. $4(x-1)+3x>1+2(x-6)$

$4x-4+3x>1+2x-12$

$7x-4>2x-11$

$5x>-7$

$x>-\frac{7}{5}$

The solution is $\left(-\frac{7}{5}, \infty\right)$.

6. $\sqrt{x}+14=x-6$

$\sqrt{x}=x-20$

$\left(\sqrt{x}\right)^2=(x-20)^2$

$x=x^2-40x+400$

$0=x^2-41x+400$

$0=(x-25)(x-16)$

$x-25=0$ or $x-16=0$

$x=25$ $\quad$ $x=16$

Discard $x = 16$ as an extraneous solution. The solution set is $\{25\}$.

7. $x\ge 10$ or $-x<5$

$x\ge 10$ or $x>-5$

The solution is $(-5, \infty)$.

8. $\sqrt{3x-1}+4=1$

$\sqrt{3x-1}=-3$

There is no real number whose square root is negative. The solution set is $\varnothing$.

9. $|x-2|>15$

$x-2<-15$ or $x-2>15$

$x<-13$ or $x>17$

The solution is $(-\infty, -13)\cup(17, \infty)$.

10. $5x-4[x-2(3x+1)]=25$

$5x-4(x-6x-2)=25$

$5x-4(-5x-2)=25$

$5x+20x+8=25$

$25x=17$

$x=\frac{17}{25}$

The solution set is $\left\{\frac{17}{25}\right\}$.

Section 7.7 Practice Problems

1. $\sqrt{-25}=\sqrt{-1}\cdot\sqrt{25}=5i$

2. $\sqrt{-17}=\sqrt{-1}\cdot\sqrt{17}=i\sqrt{17}$

3. $-\sqrt{-50}=-\sqrt{-1}\cdot\sqrt{25}\sqrt{2}=-5i\sqrt{2}$

4. $\sqrt{-3}\cdot\sqrt{-7}=i\sqrt{3}\cdot\left(i\sqrt{7}\right)=i^2\sqrt{21}=-\sqrt{21}$

5. $\sqrt{-25}\cdot\sqrt{-1}=\sqrt{-1}\cdot\sqrt{25}\sqrt{-1}=i^2(5)=-5$

6. $\sqrt{27}\cdot\sqrt{-3}=\sqrt{27}\cdot\sqrt{3}\cdot\sqrt{-1}=i\sqrt{81}=9i$

7. $\dfrac{\sqrt{-8}}{\sqrt{2}}=\dfrac{i\sqrt{8}}{\sqrt{2}}=i\sqrt{4}=2i$

8. $(5+2i)+(4-3i)=(5+4)+(2-3)i=9-i$

9. $$\begin{aligned}6i-(2-i)&=6i-2+i\\&=2+(6+1)i\\&=-2+7i\end{aligned}$$

10. $(-2-4i)-(-3)=(-2+3)-4i=1-4i$

11. $-5i\cdot 3i=-15i^2=(-15)(-1)=15$

12. $$\begin{aligned}-2i(6-2i)&=-12i+4i^2\\&=(-1)(4)-12i\\&=-4-12i\end{aligned}$$

13. $$\begin{aligned}(3-4i)(6+i)&=18+3i-24i-4i^2\\&=18-21i-4(-1)\\&=22-21i\end{aligned}$$

14. $$\begin{aligned}(1-2i)^2&=1^2-2(2i)+(2i)^2\\&=1-4i+4(-1)\\&=-3-4i\end{aligned}$$

15. $$\begin{aligned}(6+5i)(6-5i)&=6^2-(5i)^2\\&=36-25(-1)\\&=61\end{aligned}$$

16. $$\begin{aligned}\frac{3+i}{2-3i}&=\frac{(3+i)(2+3i)}{(2-3i)(2+3i)}\\&=\frac{6+9i+2i+3i^2}{2^2-(3i)^2}\\&=\frac{6+11i-3}{4+9}\\&=\frac{3+11i}{13}\\&=\frac{3}{13}+\frac{11}{13}i\end{aligned}$$

17. $$\begin{aligned}\frac{6}{5i}&=\frac{6(-5i)}{(5i)(-5i)}\\&=\frac{-30i}{-25i^2}\\&=\frac{-30i}{-25(-1)}\\&=\frac{-30i}{25}\\&=-\frac{6}{5}i\end{aligned}$$

18. $i^{11}=(i^4)^2\cdot i^3=1^2(-i)=-i$

19. $i^{40}=(i^4)^{10}=1^{10}=1$

20. $i^{50}=(i^4)^{12}i^2=1^{12}\cdot(-1)=-1$

21. $i^{-10}=\dfrac{1}{i^{10}}=\dfrac{1}{(i^4)^2i^2}=\dfrac{1}{1^2\cdot(-1)}=-1$

Mental Math

1. $\sqrt{-81}=\sqrt{-1}\sqrt{81}=9i$

2. $\sqrt{-49}=\sqrt{-1}\sqrt{49}=7i$

3. $\sqrt{-7}=\sqrt{-1}\sqrt{7}=i\sqrt{7}$

4. $\sqrt{-3}=\sqrt{-1}\sqrt{3}=i\sqrt{3}$

5. $-\sqrt{16}=-4$

6. $-\sqrt{4} = -2$

7. $\sqrt{-64} = \sqrt{-1}\sqrt{64} = 8i$

8. $\sqrt{-100} = \sqrt{-1}\sqrt{100} = 10i$

Exercise Set 7.7

1. $\sqrt{-24} = \sqrt{4}\sqrt{6}\sqrt{-1} = 2i\sqrt{6}$

3. $-\sqrt{-36} = -\sqrt{36}\sqrt{-1} = -6i$

5. $8\sqrt{-63} = 8\sqrt{9}\sqrt{7}\sqrt{-1} = 8 \cdot 3i\sqrt{7} = 24i\sqrt{7}$

7. $-\sqrt{54} = -\sqrt{9}\sqrt{6} = -3\sqrt{6}$

9. $\sqrt{-2} \cdot \sqrt{-7} = \left(i\sqrt{2}\right)\left(i\sqrt{7}\right)$
$= i^2\sqrt{14}$
$= -\sqrt{14}$

11. $\sqrt{-5} \cdot \sqrt{-10} = \left(\sqrt{5}i\right)\left(\sqrt{10}i\right)$
$= \sqrt{25}\sqrt{2}i^2$
$= -5\sqrt{2}$

13. $\sqrt{16} \cdot \sqrt{-1} = 4i$

15. $\dfrac{\sqrt{-9}}{\sqrt{3}} = \dfrac{\sqrt{9}i}{\sqrt{3}} = i\sqrt{3}$

17. $\dfrac{\sqrt{-80}}{\sqrt{-10}} = \dfrac{i\sqrt{80}}{i\sqrt{10}} = \sqrt{\dfrac{80}{10}} = \sqrt{8} = \sqrt{4}\sqrt{2} = 2\sqrt{2}$

19. $(4-7i)+(2+3i) = (4+2)+(-7+3)i$
$= 6-4i$

21. $(6+5i)-(8-i) = (6-8)+[5-(-1)]i$
$= -2+6i$

23. $6-(8+4i) = (6-8)-4i = -2-4i$

25. $(6-3i)-(4-2i) = 6-3i-4+2i$
$= (6-4)+(-3+2)i$
$= 2-i$

27. $(5-6i)-4i = 5+(-6-4)i = 5-10i$

29. $(2+4i)+(6-5i) = (2+6)+(4-5)i = 8-i$

31. $6i \cdot 2i = 12i^2 = 12(-1) = -12$

33. $-9i \cdot 7i = -63i^2 = -63(-1) = 63$

35. $-10i \cdot -4i = 40i^2 = 40(-1) = -40$

37. $6i(2-3i) = 6i(2)-6i(3i)$
$= 12i-18i^2$
$= 12i-18(-1)$
$= 18+12i$

39. $-3i(-1+9i) = 3i-27i^2$
$= 3i-27(-1)$
$= 27+3i$

41. $(4+i)(5+2i) = 20+8i+5i+2i^2$
$= 20+13i+2(-1)$
$= 20+13i-2$
$= 18+13i$

43. $\left(\sqrt{3}+2i\right)\left(\sqrt{3}-2i\right) = \left(\sqrt{3}\right)^2-(2i)^2$
$= 3-4i^2$
$= 3-4(-1)$
$= 7$

45. $(4-2i)^2 = 16-2 \cdot 4 \cdot 2i+4i^2$
$= 16-16i+4(-1)$
$= 16-4-16i$
$= 12-16i$

47. $(6-2i)(3+i) = 18+6i-6i-2i^2$
$= 18-2(-1)$
$= 20$

49. $(1-i)(1+i) = 1^2-i^2 = 1^2-(-1) = 1+1 = 2$

51. $(9+8i)^2 = 81+144i+64i^2$
$= 81+144i+64(-1)$
$= 17+144i$

53. $(1-i)^2 = 1^2 - 2(1)(i) + i^2 = 1-2i-1 = -2i$

55. $\frac{4}{i} = \frac{4}{i}\cdot\frac{i}{i} = \frac{4i}{i^2} = -4i$

57. $\frac{7}{4+3i} = \frac{7}{4+3i}\cdot\frac{4-3i}{4-3i}$
$= \frac{28-21i}{4^2+3^2}$
$= \frac{28-21i}{16+9}$
$= \frac{28-21i}{25}$
$= \frac{28}{25} - \frac{21}{25}i$

59. $\frac{6i}{1-2i} = \frac{6i}{1-2i}\cdot\frac{1+2i}{1+2i}$
$= \frac{6i+12i^2}{1+4}$
$= \frac{-12+6i}{5}$
$= -\frac{12}{5} + \frac{6i}{5}$

61. $\frac{3+5i}{1+i} = \frac{3+5i}{1+i}\cdot\frac{1-i}{1-i}$
$= \frac{3-3i+5i-5i^2}{1^2+1^2}$
$= \frac{3+2i-5(-1)}{1+1}$
$= \frac{3+5+2i}{2}$
$= \frac{8+2i}{2}$
$= 4+i$

63. $\frac{4-5i}{2i} = \frac{4-5i}{2i}\cdot\frac{i}{i} = \frac{4i+5}{-2} = -\frac{5}{2} - 2i$

65. $\frac{16+15i}{-3i} = \frac{(16+15i)i}{-3i^2}$
$= \frac{16i+15i^2}{-3(-1)}$
$= \frac{16i+15(-1)}{3}$
$= \frac{-15}{3} + \frac{16}{3}i$
$= -5 + \frac{16}{3}i$

67. $\frac{2}{3+i} = \frac{2}{3+i}\cdot\frac{3-i}{3-i} = \frac{6-2i}{9+1} = \frac{6-2i}{10} = \frac{3}{5} - \frac{1}{5}i$

69. $\frac{2-3i}{2+i} = \frac{(2-3i)(2-i)}{(2+i)(2-i)}$
$= \frac{4-2i-6i+3i^2}{2^2+1^2}$
$= \frac{4-8i+3(-1)}{4+1}$
$= \frac{4-3-8i}{5}$
$= \frac{1}{5} - \frac{8}{5}i$

71. $i^8 = (i^4)^2 = 1^2 = 1$

73. $i^{21} = i^{20}i = (i^4)^5 i = 1^5 i = 1i = i$

75. $i^{11} = (i^4)^2 i^3 = 1^2(-i) = -i$

77. $i^{-6} = (i^2)^{-3} = (-1)^{-3} = \frac{1}{(-1)^3} = \frac{1}{-1} = -1$

79. $(2i)^6 = 2^6 i^6 = 2^6 i^4 i^2 = 64(1)(-1) = -64$

81. $(-3i)^5 = (-3)^5 i^5$
$= -243 \cdot i^4 \cdot i$
$= -243(1)(i)$
$= -243i$

83. 5 people reported an average checking balance of \$201 to \$300.

85. $5 + 9 = 14$
14 people reported an average balance of \$200 or less.

87. $\frac{5}{30} = 0.167$ or 16.7%
16.7% of people reported an average balance of \$201 to \$300.

89. $i^3 + i^4 = -i + 1 = 1 - i$

91. $i^6 + i^8 = (-1) + 1 = 0$

93. $2 + \sqrt{-9} = 2 + 3i$

95. $\frac{6+\sqrt{-18}}{3} = \frac{6+3i\sqrt{2}}{3} = 2 + i\sqrt{2}$

97.
$$\frac{5-\sqrt{-75}}{10} = \frac{5-5i\sqrt{3}}{10} = \frac{5}{10} - \frac{5i\sqrt{3}}{10} = \frac{1}{2} - \frac{\sqrt{3}}{2}i$$

99. answers may vary

101.
$$\left(8-\sqrt{-4}\right)-\left(2+\sqrt{-16}\right) = (8-2i)-(2+4i) = (8-2)+(-2-4)i = 6-6i$$

103.
$$x^2 + 2x = -2$$
$$(-1+i)^2 + 2(-1+i) \stackrel{?}{=} -2$$
$$1 - 2i - 1 - 2 + 2i \stackrel{?}{=} -2$$
$$-2 = -2 \quad \text{True}$$
$-1 + i$ is a solution.

Chapter 7 Vocabulary Check

1. The <u>conjugate</u> of $\sqrt{3}+2$ is $\sqrt{3}-2$.
2. The <u>principal square root</u> of a nonnegative number a is written as $\sqrt{a}$.
3. The process of writing a radical expression as an equivalent expression but without a radical in the denominator is called <u>rationalizing</u> the denominator.
4. The <u>imaginary unit</u>, written i, is the number whose square is -1.
5. The <u>cube root</u> of a number is written as $\sqrt[3]{a}$.
6. In the notation $\sqrt[n]{a}$, n is called the <u>index</u> and a is called the <u>radicand</u>.
7. Radicals with the same index and the same radicand are called <u>like radicals</u>.
8. A <u>complex number</u> is a number that can be written in the form $a + bi$, where a and b are real numbers.
9. The <u>distance</u> formula is $d = \sqrt{(x_2 - x_1)^2 + (y_2 - y_1)^2}$.
10. The <u>midpoint</u> formula is $\left(\frac{x_1 + x_2}{2}, \frac{y_1 + y_2}{2}\right)$.

Chapter 7 Review

1. $\sqrt{81} = 9$ because $9^2 = 81$.
2. $\sqrt[4]{81} = 3$ because $3^4 = 81$.
3. $\sqrt[3]{-8} = -2$ because $(-2)^3 = -8$.
4. $\sqrt[4]{-16}$ is not a real number.
5. $-\sqrt{\frac{1}{49}} = -\frac{\sqrt{1}}{\sqrt{49}} = -\frac{1}{7}$ because $\left(\frac{1}{7}\right)^2 = \frac{1}{49}$.
6. $\sqrt{x^{64}} = x^{32}$ because $(x^{32})^2 = x^{64}$.

7. $-\sqrt{36} = -6$ because $6^2 = 36$.

8. $\sqrt[3]{64} = 4$ because $4^3 = 64$.

9. $\sqrt[3]{-a^6b^9} = -a^2b^3$ because $(-a^2b^3)^3 = -a^6b^9$.

10. $\sqrt{16a^4b^{12}} = 4a^2b^6$ because $(4a^2b^6)^2 = 16a^4b^{12}$.

11. $\sqrt[5]{32a^5b^{10}} = 2ab^2$ because $(2ab^2)^5 = 32a^5b^{10}$.

12. $\sqrt[5]{-32x^{15}y^{20}} = -2x^3y^4$ because $(-2x^3y^4)^5 = -32x^{15}y^{20}$.

13. $\sqrt{\dfrac{x^{12}}{36y^2}} = \dfrac{\sqrt{x^{12}}}{\sqrt{36y^2}} = \dfrac{x^6}{6y}$ because $\left(\dfrac{x^6}{6y}\right)^2 = \dfrac{x^{12}}{36y^2}$.

14. $\sqrt[3]{\dfrac{27y^3}{z^{12}}} = \dfrac{\sqrt[3]{27y^3}}{\sqrt[3]{z^{12}}} = \dfrac{3y}{z^4}$ because $\left(\dfrac{3y}{z^4}\right)^3 = \dfrac{27y^3}{z^{12}}$.

15. $\sqrt{x^2} = |x|$

16. $\sqrt[4]{(x^2-4)^4} = |x^2-4|$

17. $\sqrt[3]{(-27)^3} = -27$

18. $\sqrt[5]{(-5)^5} = -5$

19. $-\sqrt[5]{x^5} = -x$

20. $\sqrt[4]{16(2y+z)^4} = 2|2y+z|$

21. $\sqrt{25(x-y)^2} = 5|x-y|$

22. $\sqrt[5]{y^5} = y$

23. $\sqrt[6]{x^6} = |x|$

24. $f(x) = \sqrt{x}+3$
$f(0) = \sqrt{0}+3 = 0+3 = 3$
$f(9) = \sqrt{9}+3 = 3+3 = 6$

25. $g(x) = \sqrt[3]{x-3}$
$g(11) = \sqrt[3]{11-3} = \sqrt[3]{8} = 2$
$g(20) = \sqrt[3]{20-3} = \sqrt[3]{17}$

26. $\left(\dfrac{1}{81}\right)^{1/4} = \dfrac{\sqrt[4]{1}}{\sqrt[4]{81}} = \dfrac{1}{3}$

27. $\left(-\dfrac{1}{27}\right)^{1/3} = \dfrac{\sqrt[3]{-1}}{\sqrt[3]{27}} = -\dfrac{1}{3}$

28. $(-27)^{-1/3} = \dfrac{1}{\sqrt[3]{-27}} = -\dfrac{1}{3}$

29. $(-64)^{-1/3} = \dfrac{1}{\sqrt[3]{-64}} = -\dfrac{1}{4}$

30. $-9^{3/2} = -\left(\sqrt{9}\right)^3 = -(3)^3 = -27$

31. $64^{-1/3} = \dfrac{1}{\sqrt[3]{64}} = \dfrac{1}{4}$

32. $(-25)^{5/2} = \left(\sqrt{-25}\right)^5$ is not a real number.

33. $\left(\frac{25}{49}\right)^{-3/2} = \frac{1}{\left(\frac{25}{49}\right)^{3/2}}$
$= \left(\frac{49}{25}\right)^{3/2}$
$= \left(\frac{\sqrt{49}}{\sqrt{25}}\right)^3$
$= \left(\frac{7}{5}\right)^3$
$= \frac{343}{125}$

34. $\left(\frac{8}{27}\right)^{-2/3} = \frac{1}{\left(\frac{8}{27}\right)^{2/3}}$
$= \left(\frac{27}{8}\right)^{2/3}$
$= \left(\frac{\sqrt[3]{27}}{\sqrt[3]{8}}\right)^2$
$= \left(\frac{3}{2}\right)^2$
$= \frac{9}{4}$

35. $\left(-\frac{1}{36}\right)^{-1/4}$ is not a real number.

36. $\sqrt[3]{x^2} = x^{2/3}$

37. $\sqrt[5]{5x^2y^3} = 5^{1/5}x^{2/5}y^{3/5}$

38. $y^{4/5} = \sqrt[5]{y^4}$

39. $5(xy^2z^5)^{1/3} = 5\sqrt[3]{xy^2z^5}$

40. $(x+2y)^{-1/2} = \frac{1}{\sqrt{x+2y}}$

41. $a^{1/3}a^{4/3}a^{1/2} = a^{2/6}a^{8/6}a^{3/6} = a^{13/6}$

42. $\frac{b^{1/3}}{b^{4/3}} = b^{\frac{1}{3}-\frac{4}{3}} = b^{-3/3} = b^{-1} = \frac{1}{b}$

43. $(a^{1/2}a^{-2})^3 = a^{3/2}a^{-6}$
$= a^{\frac{3}{2}-\frac{12}{2}}$
$= a^{-9/2}$
$= \frac{1}{a^{9/2}}$

44. $(x^{-3}y^6)^{1/3} = x^{-3/3}y^{6/3} = x^{-1}y^2 = \frac{y^2}{x}$

45. $\left(\frac{b^{3/4}}{a^{-1/2}}\right)^8 = \frac{b^{24/4}}{a^{-8/2}} = \frac{b^6}{a^{-4}} = a^4b^6$

46. $\frac{x^{1/4}x^{-1/2}}{x^{2/3}} = x^{\frac{1}{4}-\frac{1}{2}-\frac{2}{3}}$
$= x^{\frac{3}{12}-\frac{6}{12}-\frac{8}{12}}$
$= x^{-\frac{11}{12}}$
$= \frac{1}{x^{11/12}}$

47. $\left(\frac{49c^{5/3}}{a^{-1/4}b^{5/6}}\right)^{-1} = \left(\frac{a^{-1/4}b^{5/6}}{49c^{5/3}}\right)$
$= \frac{b^{5/6}}{49a^{1/4}c^{5/3}}$

48. $a^{-1/4}(a^{5/4} - a^{9/4}) = a^{-\frac{1}{4}+\frac{5}{4}} - a^{-\frac{1}{4}+\frac{9}{4}}$
$= a^{\frac{4}{4}} - a^{\frac{8}{4}}$
$= a - a^2$

49. $\sqrt{20} \approx 4.472$

50. $\sqrt[3]{-39} \approx -3.391$

51. $\sqrt[4]{726} \approx 5.191$

52. $56^{1/3} \approx 3.826$

53. $-78^{3/4} \approx -26.246$

54. $105^{-2/3} \approx 0.045$

55. $\sqrt[3]{2} \cdot \sqrt{7} = 2^{1/3} \cdot 7^{1/2}$
$= 2^{2/6} \cdot 7^{3/6}$
$= \sqrt[6]{2^2 \cdot 7^3}$
$= \sqrt[6]{1372}$

56. $\sqrt[3]{3} \cdot \sqrt[4]{x} = 3^{1/3} x^{1/4}$
$= 3^{4/12} \cdot x^{3/12}$
$= \sqrt[12]{3^4 x^3}$
$= \sqrt[12]{81x^3}$

57. $\sqrt{3} \cdot \sqrt{8} = \sqrt{24} = \sqrt{4}\sqrt{6} = 2\sqrt{6}$

58. $\sqrt[3]{7y} \cdot \sqrt[3]{x^2 z} = \sqrt[3]{7x^2 yz}$

59. $\dfrac{\sqrt{44x^3}}{\sqrt{11x}} = \sqrt{\dfrac{44x^3}{11x}} = \sqrt{4x^2} = 2x$

60. $\dfrac{\sqrt[4]{a^6 b^{13}}}{\sqrt[4]{a^2 b}} = \sqrt[4]{\dfrac{a^6 b^{13}}{a^2 b}} = \sqrt[4]{a^4 b^{12}} = ab^3$

61. $\sqrt{60} = \sqrt{4}\sqrt{15} = 2\sqrt{15}$

62. $-\sqrt{75} = -\sqrt{25}\sqrt{3} = -5\sqrt{3}$

63. $\sqrt[3]{162} = \sqrt[3]{27}\sqrt[3]{6} = 3\sqrt[3]{6}$

64. $\sqrt[3]{-32} = \sqrt[3]{-8}\sqrt[3]{4} = -2\sqrt[3]{4}$

65. $\sqrt{36x^7} = \sqrt{36x^6}\sqrt{x} = 6x^3\sqrt{x}$

66. $\sqrt[3]{24a^5b^7} = \sqrt[3]{8a^3b^6}\sqrt[3]{3a^2b^1} = 2ab^2\sqrt[3]{3a^2b}$

67. $\sqrt{\dfrac{p^{17}}{121}} = \dfrac{\sqrt{p^{16}}\sqrt{p}}{\sqrt{121}} = \dfrac{p^8\sqrt{p}}{11}$

68. $\sqrt[3]{\dfrac{y^5}{27x^6}} = \dfrac{\sqrt[3]{y^3}\sqrt[3]{y^2}}{\sqrt[3]{27x^6}} = \dfrac{y\sqrt[3]{y^2}}{3x^2}$

69. $\sqrt[4]{\dfrac{xy^6}{81}} = \dfrac{\sqrt[4]{y^4}\sqrt[4]{xy^2}}{\sqrt[4]{81}} = \dfrac{y\sqrt[4]{xy^2}}{3}$

70. $\sqrt{\dfrac{2x^3}{49y^4}} = \dfrac{\sqrt{x^2}\sqrt{2x}}{\sqrt{49y^4}} = \dfrac{x\sqrt{2x}}{7y^2}$

71. a. $r = \sqrt{\dfrac{A}{\pi}}$
$= \sqrt{\dfrac{25}{\pi}}$
$= \dfrac{\sqrt{25}}{\sqrt{\pi}}$
$= \dfrac{5}{\sqrt{\pi}}$ meters or $\dfrac{5\sqrt{\pi}}{\pi}$ meters

b. $r = \sqrt{\dfrac{A}{\pi}} = \sqrt{\dfrac{104}{\pi}} = 5.75$ inches

72. $(x_1, y_1) = (-6, 3), (x_2, y_2) = (8, 4)$
$d = \sqrt{(x_2 - x_1)^2 + (y_2 - y_1)^2}$
$= \sqrt{(8+6)^2 + (4-3)^2}$
$= \sqrt{196+1}$
$= \sqrt{197}$ units ≈ 14.036

73. $(x_1, y_1) = (-4, -6), (x_2, y_2) = (-1, 5)$
$d = \sqrt{(x_2 - x_1)^2 + (y_2 - y_1)^2}$
$= \sqrt{(-1+4)^2 + (5+6)^2}$
$= \sqrt{9+121}$
$= \sqrt{130}$ units ≈ 11.402

74. $(x_1, y_1) = (-1, 5), (x_2, y_2) = (2, -3)$

$$\begin{aligned} d &= \sqrt{(x_2 - x_1)^2 + (y_2 - y_1)^2} \\ &= \sqrt{(2+1)^2 + (-3-5)^2} \\ &= \sqrt{9+64} \\ &= \sqrt{73} \text{ units} \approx 8.544 \end{aligned}$$

75. $(x_1, y_1) = \left(-\sqrt{2}, 0\right), (x_2, y_2) = \left(0, -4\sqrt{6}\right)$

$$\begin{aligned} d &= \sqrt{(x_2 - x_1)^2 + (y_2 - y_1)^2} \\ &= \sqrt{\left(0+\sqrt{2}\right)^2 + \left(-4\sqrt{6}-0\right)^2} \\ &= \sqrt{2+96} \\ &= \sqrt{98} \\ &= 7\sqrt{2} \text{ units} \approx 9.899 \end{aligned}$$

76. $(x_1, y_1) = \left(-\sqrt{5}, -\sqrt{11}\right)$,
$(x_2, y_2) = \left(-\sqrt{5}, -3\sqrt{11}\right)$

$$\begin{aligned} d &= \sqrt{(x_2 - x_1)^2 + (y_2 - y_1)^2} \\ &= \sqrt{\left(-\sqrt{5}+\sqrt{5}\right)^2 + \left(-3\sqrt{11}+\sqrt{11}\right)^2} \\ &= \sqrt{0+44} \\ &= \sqrt{44} \\ &= 2\sqrt{11} \text{ units} \approx 6.633 \end{aligned}$$

77. $(x_1, y_1) = (7.4, -8.6), (x_2, y_2) = (-1.2, 5.6)$

$$\begin{aligned} d &= \sqrt{(-1.2-7.4)^2 + (5.6+8.6)^2} \\ &= \sqrt{(-8.6)^2 + (14.2)^2} \\ &= \sqrt{73.96+201.64} \\ &= \sqrt{275.6} \text{ units} \approx 16.601 \end{aligned}$$

78. $(x_1, y_1) = (2, 6), (x_2, y_2) = (-12, 4)$

$$\begin{aligned} \text{midpoint} &= \left(\frac{x_1+x_2}{2}, \frac{y_1+y_2}{2}\right) \\ &= \left(\frac{2-12}{2}, \frac{6+4}{2}\right) \\ &= \left(\frac{-10}{2}, \frac{10}{2}\right) \\ &= (-5, 5) \end{aligned}$$

79. $(x_1, y_1) = (-6, -5), (x_2, y_2) = (-9, 7)$

$$\begin{aligned} \text{midpoint} &= \left(\frac{x_1+x_2}{2}, \frac{y_1+y_2}{2}\right) \\ &= \left(\frac{-6-9}{2}, \frac{-5+7}{2}\right) \\ &= \left(-\frac{15}{2}, 1\right) \end{aligned}$$

80. $(x_1, y_1) = (4, -6), (x_2, y_2) = (-15, 2)$

$$\begin{aligned} \text{midpoint} &= \left(\frac{x_1+x_2}{2}, \frac{y_1+y_2}{2}\right) \\ &= \left(\frac{4-15}{2}, \frac{-6+2}{2}\right) \\ &= \left(-\frac{11}{2}, -2\right) \end{aligned}$$

81. $(x_1, y_1) = \left(0, -\frac{3}{8}\right), (x_2, y_2) = \left(\frac{1}{10}, 0\right)$

$$\begin{aligned} \text{midpoint} &= \left(\frac{x_1+x_2}{2}, \frac{y_1+y_2}{2}\right) \\ &= \left(\frac{0+\frac{1}{10}}{2}, \frac{-\frac{3}{8}+0}{2}\right) \\ &= \left(\frac{1}{20}, -\frac{3}{16}\right) \end{aligned}$$

82. $(x_1, y_1) = \left(\frac{3}{4}, -\frac{1}{7}\right), (x_2, y_2) = \left(-\frac{1}{4}, -\frac{3}{7}\right)$

$$\begin{aligned} \text{midpoint} &= \left(\frac{\frac{3}{4}-\frac{1}{4}}{2}, \frac{-\frac{1}{7}-\frac{3}{7}}{2}\right) \\ &= \left(\frac{\frac{1}{2}}{2}, \frac{-\frac{11}{7}}{2}\right) \\ &= \left(\frac{1}{4}, -\frac{2}{7}\right) \end{aligned}$$

83. $(x_1, y_1) = \left(\sqrt{3}, -2\sqrt{6}\right), (x_2, y_2) = \left(\sqrt{3}, -4\sqrt{6}\right)$

$$\begin{aligned}\text{midpoint} &= \left(\frac{x_1 + x_2}{2}, \frac{y_1 + y_2}{2}\right)\\ &= \left(\frac{\sqrt{3}+\sqrt{3}}{2}, \frac{-2\sqrt{6}-4\sqrt{6}}{2}\right)\\ &= \left(\frac{2\sqrt{3}}{2}, \frac{-6\sqrt{6}}{2}\right)\\ &= \left(\sqrt{3}, -3\sqrt{6}\right)\end{aligned}$$

84. $\sqrt{20} + \sqrt{45} - 7\sqrt{5} = 2\sqrt{5} + 3\sqrt{5} - 7\sqrt{5} = -2\sqrt{5}$

85. $x\sqrt{75x} - \sqrt{27x^3} = 5x\sqrt{3x} - 3x\sqrt{3x} = 2x\sqrt{3x}$

86. $\sqrt[3]{128} + \sqrt[3]{250} = 4\sqrt[3]{2} + 5\sqrt[3]{2} = 9\sqrt[3]{2}$

87. $3\sqrt[4]{32a^5} - a\sqrt[4]{162a} = 6a\sqrt[4]{2a} - 3a\sqrt[4]{2a} = 3a\sqrt[4]{2a}$

88. $\frac{5}{\sqrt{4}} + \frac{\sqrt{3}}{3} = \frac{5}{2} + \frac{\sqrt{3}}{3} = \frac{15}{6} + \frac{2\sqrt{3}}{6} = \frac{15 + 2\sqrt{3}}{6}$

89.
$$\begin{aligned}\sqrt{\frac{8}{x^2}} - \sqrt{\frac{50}{16x^2}} &= \frac{2\sqrt{2}}{x} - \frac{5\sqrt{2}}{4x}\\ &= \frac{8\sqrt{2} - 5\sqrt{2}}{4x}\\ &= \frac{3\sqrt{2}}{4x}\end{aligned}$$

90.
$$\begin{aligned}2\sqrt{50} - 3\sqrt{125} + \sqrt{98} &= 10\sqrt{2} - 15\sqrt{5} + 7\sqrt{2}\\ &= 17\sqrt{2} - 15\sqrt{5}\end{aligned}$$

91.
$$\begin{aligned}2a\sqrt[4]{32b^5} - 3b\sqrt[4]{162a^4b} + \sqrt[4]{2a^4b^5} &= 4ab\sqrt[4]{2b} - 9ab\sqrt[4]{2b} + ab\sqrt[4]{2b}\\ &= -4ab\sqrt[4]{2b}\end{aligned}$$

92.
$$\begin{aligned}\sqrt{3}\left(\sqrt{27} - \sqrt{3}\right) &= \sqrt{3}\sqrt{27} - \sqrt{3}\sqrt{3}\\ &= \sqrt{81} - \sqrt{9}\\ &= 9 - 3\\ &= 6\end{aligned}$$

93. $\left(\sqrt{x}-3\right)^2=\left(\sqrt{x}\right)^2-(2)(3)\sqrt{x}+9=x-6\sqrt{x}+9$

94. $\left(\sqrt{5}-5\right)\left(2\sqrt{5}+2\right)=\sqrt{5}\left(2\sqrt{5}\right)+2\sqrt{5}-10\sqrt{5}-10$
$=10-8\sqrt{5}-10$
$=-8\sqrt{5}$

95. $\left(2\sqrt{x}-3\sqrt{y}\right)\left(2\sqrt{x}+3\sqrt{y}\right)=\left(2\sqrt{x}\right)^2-\left(3\sqrt{y}\right)^2$
$=4x-9y$

96. $\left(\sqrt{a}+3\right)\left(\sqrt{a}-3\right)=\left(\sqrt{a}\right)^2-(3)^2=a-9$

97. $\left(\sqrt[3]{a}+2\right)^2=\left(\sqrt[3]{a}\right)^2+2(2)\left(\sqrt[3]{a}\right)+2^2$
$=\sqrt[3]{a^2}+4\sqrt[3]{a}+4$

98. $\left(\sqrt[3]{5x}+9\right)\left(\sqrt[3]{5x}-9\right)=\left(\sqrt[3]{5x}\right)^2-9^2$
$=\sqrt[3]{25x^2}-81$

99. $\left(\sqrt[3]{a}+4\right)\left(\sqrt[3]{a^2}-4\sqrt[3]{a}+16\right)=\sqrt[3]{a}\sqrt[3]{a^2}-\sqrt[3]{a}\left(4\sqrt[3]{a}\right)+16\sqrt[3]{a}+4\sqrt[3]{a^2}-16\sqrt[3]{a}+64$
$=a-4\sqrt[3]{a^2}+4\sqrt[3]{a^2}+64$
$=a+64$

100. $\dfrac{3}{\sqrt{7}}=\dfrac{3}{\sqrt{7}}\cdot\dfrac{\sqrt{7}}{\sqrt{7}}=\dfrac{3\sqrt{7}}{7}$

101. $\sqrt{\dfrac{x}{12}}=\dfrac{\sqrt{x}}{\sqrt{12}}=\dfrac{\sqrt{x}}{2\sqrt{3}}\cdot\dfrac{\sqrt{3}}{\sqrt{3}}=\dfrac{\sqrt{3x}}{6}$

102. $\dfrac{5}{\sqrt[3]{4}}=\dfrac{5}{\sqrt[3]{4}}\cdot\dfrac{\sqrt[3]{2}}{\sqrt[3]{2}}=\dfrac{5\sqrt[3]{2}}{2}$

103. $\sqrt{\dfrac{24x^5}{3y^2}} = \dfrac{\sqrt{24x^5}}{\sqrt{3y^2}}$

$= \dfrac{2x^2\sqrt{6x}}{y\sqrt{3}} \cdot \dfrac{\sqrt{3}}{\sqrt{3}}$

$= \dfrac{2x^2\sqrt{9x}}{3y}$

$= \dfrac{6x^2\sqrt{2x}}{3y}$

$= \dfrac{2x^2\sqrt{2x}}{y}$

104. $\sqrt[3]{\dfrac{15x^6y^7}{z^2}} = \dfrac{\sqrt[3]{15x^6y^7}}{\sqrt[3]{z^2}}$

$= \dfrac{x^2y^2\sqrt[3]{15y}}{\sqrt[3]{z^2}} \cdot \dfrac{\sqrt[3]{z}}{\sqrt[3]{z}}$

$= \dfrac{x^2y^2\sqrt[3]{15yz}}{z}$

105. $\sqrt[4]{\dfrac{81}{8x^{10}}} = \dfrac{\sqrt[4]{81}}{\sqrt[4]{8x^{10}}}$

$= \dfrac{3}{x^2\sqrt[4]{8x^2}} \cdot \dfrac{\sqrt[4]{2x^2}}{\sqrt[4]{2x^2}}$

$= \dfrac{3\sqrt[4]{2x^2}}{2x^3}$

106. $\dfrac{3}{\sqrt{y}-2} = \dfrac{3}{\sqrt{y}-2} \cdot \dfrac{\sqrt{y}+2}{\sqrt{y}+2} = \dfrac{3\sqrt{y}+6}{y-4}$

107. $\dfrac{\sqrt{2}-\sqrt{3}}{\sqrt{2}+\sqrt{3}} = \dfrac{\sqrt{2}-\sqrt{3}}{\sqrt{2}+\sqrt{3}} \cdot \dfrac{\sqrt{2}-\sqrt{3}}{\sqrt{2}-\sqrt{3}}$

$= \dfrac{2-2\sqrt{6}+3}{2-3}$

$= -5+2\sqrt{6}$

108. $\dfrac{\sqrt{11}}{3} = \dfrac{\sqrt{11}}{3} \cdot \dfrac{\sqrt{11}}{\sqrt{11}} = \dfrac{11}{3\sqrt{11}}$

109. $\sqrt{\dfrac{18}{y}} = \dfrac{\sqrt{18}}{\sqrt{y}} = \dfrac{3\sqrt{2}}{\sqrt{y}} \cdot \dfrac{\sqrt{2}}{\sqrt{2}} = \dfrac{6}{\sqrt{2y}}$

110. $\dfrac{\sqrt[3]{9}}{7} = \dfrac{\sqrt[3]{9}}{7} \cdot \dfrac{\sqrt[3]{3}}{\sqrt[3]{3}} = \dfrac{3}{7\sqrt[3]{3}}$

111. $\sqrt{\dfrac{24x^5}{3y^2}} = \dfrac{\sqrt{24x^5}}{\sqrt{3y^2}}$

$= \dfrac{2x^2\sqrt{6x}}{y\sqrt{3}} \cdot \dfrac{\sqrt{6x}}{\sqrt{6x}}$

$= \dfrac{12x^3}{3y\sqrt{2x}}$

$= \dfrac{4x^3}{y\sqrt{2x}}$

112. $\sqrt[3]{\dfrac{xy^2}{10z}} = \dfrac{\sqrt[3]{xy^2}}{\sqrt[3]{10z}} = \dfrac{\sqrt[3]{xy^2}}{\sqrt[3]{10z}} \cdot \dfrac{\sqrt[3]{x^2y}}{\sqrt[3]{x^2y}} = \dfrac{xy}{\sqrt[3]{10x^2yz}}$

113. $\dfrac{\sqrt{x}+5}{-3} = \dfrac{\sqrt{x}+5}{-3} \cdot \dfrac{\sqrt{x}-5}{\sqrt{x}-5} = \dfrac{x-25}{-3\sqrt{x}+15}$

114. $\sqrt{y-7} = 5$

$y-7 = 25$

$y = 32$

The solution set is $\{32\}$.

115. $\sqrt{2x}+10 = 4$

$\sqrt{2x} = -6$

No solution since a principal square root does not yield a negative number. The solution set is $\varnothing$.

116. $\sqrt[3]{2x-6} = 4$

$\left(\sqrt[3]{2x-6}\right)^3 = 4^3$

$2x-6 = 64$

$2x = 70$

$x = 35$

The solution set is $\{35\}$.

117. $\sqrt{x+6}=\sqrt{x+2}$

$\left(\sqrt{x+6}\right)^2=\left(\sqrt{x+2}\right)^2$

$x+6=x+2$

$6=2$ False

The solution set is $\varnothing$.

118. $2x-5\sqrt{x}=3$

$2x-3=5\sqrt{x}$

$4x^2-12x+9=25x$

$4x^2-37x+9=0$

$(4x-1)(x-9)=0$

$4x-1=0$ or $x-9=0$

$4x=1$ $\quad x=9$

$x=\frac{1}{4}$

We discard the $\frac{1}{4}$ as extraneous, leaving $x=9$ as the only solution. The solution set is $\{9\}$.

119. $\sqrt{x+9}=2+\sqrt{x-7}$

$x+9=4+(2)2\sqrt{x-7}+x-7$

$12=4\sqrt{x-7}$

$3=\sqrt{x-7}$

$9=x-7$

$16=x$

The solution set is $\{16\}$.

120. $a^2+b^2=c^2$

$3^2+3^2=c^2$

$9+9=c^2$

$18=c^2$

$\sqrt{18}=c$

$3\sqrt{2}=c$

The unknown length is $3\sqrt{2}$ centimeters.

121. $a^2+b^2=c^2$

$7^2+\left(8\sqrt{3}\right)^2=c^2$

$49+192=c^2$

$241=c^2$

$\sqrt{241}=c$

The unknown length is $\sqrt{241}$ feet.

122. $a^2+b^2=c^2$

$a^2+40^2=65^2$

$a^2+1600=4225$

$a^2=2625$

$a=51.2$

The width of the lake is 51.2 feet.

123. $a^2+b^2=c^2$

$3^2+3^2=c^2$

$9+9=c^2$

$18=c^2$

$\sqrt{18}=c$

$3\sqrt{2}$ or $4.24=c$

The length is 4.24 feet.

124. $\sqrt{-8}=\sqrt{-1}\cdot\sqrt{8}=2i\sqrt{2}$

125. $-\sqrt{-6}=-\sqrt{-1}\cdot\sqrt{6}=-i\sqrt{6}$

126. $\sqrt{-4}+\sqrt{-16}=\sqrt{-1}\sqrt{4}+\sqrt{-1}\sqrt{16}$

$=2i+4i$

$=6i$

127. $\sqrt{-2}\cdot\sqrt{-5}=\sqrt{-1}\cdot\sqrt{2}\cdot\sqrt{-1}\cdot\sqrt{5}$

$=i\sqrt{2}\cdot i\sqrt{5}$

$=i^2\sqrt{10}$

$=-\sqrt{10}$

128. $(12-6i)+(3+2i)=(12+3)+(-6+2)i$

$=15-4i$

129. $(-8-7i)-(5-4i)=(-8-5)+[-7-(-4)]i$

$=-13-3i$

130. $(2i)^6 = 2^6 \cdot i^6 = 64 \cdot (-1) = -64$

131. $-3i(6 - 4i) = -18i + 12(-1) = -12 - 18i$

132. $(3 + 2i)(1 + i) = 3 + 3i + 2i - 2 = 1 + 5i$

133. $(2-3i)^2 = 2^2 - 2(2)(3i) + (3i)^2$
$= 4 - 12i - 9$
$= -5 - 12i$

134. $\left(\sqrt{6} - 9i\right)\left(\sqrt{6} + 9i\right) = \left(\sqrt{6}\right)^2 - (9i)^2$
$= 6 + 81$
$= 87$

135. $\dfrac{2+3i}{2i} = \dfrac{2+3i}{2i} \cdot \dfrac{-2i}{-2i} = \dfrac{-4i+6}{4} = \dfrac{3}{2} - i$

136. $\dfrac{1+i}{-3i} = \dfrac{1+i}{-3i} \cdot \dfrac{3i}{3i} = \dfrac{3i-3}{9} = \dfrac{-1+i}{3} = -\dfrac{1}{3} + \dfrac{1}{3}i$

137. $\sqrt[3]{x^3} = x$

138. $\sqrt{(x+2)^2} = |x+2|$

139. $-\sqrt{100} = -10$

140. $\sqrt[3]{-x^{12}y^3} = -x^4 y$

141. $\sqrt[4]{\dfrac{y^{20}}{16x^{12}}} = \dfrac{\sqrt[4]{y^{20}}}{\sqrt[4]{16x^{12}}} = \dfrac{y^5}{2x^3}$

142. $9^{1/2} = \sqrt{9} = 3$

143. $64^{-1/2} = \dfrac{1}{64^{1/2}} = \dfrac{1}{\sqrt{64}} = \dfrac{1}{8}$

144. $\left(\dfrac{27}{64}\right)^{-2/3} = \left(\dfrac{64}{27}\right)^{2/3}$
$= \left(\sqrt[3]{\dfrac{64}{27}}\right)^2$
$= \left(\dfrac{4}{3}\right)^2$
$= \dfrac{16}{9}$

145. $\dfrac{(x^{2/3}x^{-3})^3}{x^{-1/2}} = \dfrac{x^{6/3}x^{-9}}{x^{-1/2}}$
$= x^{2-9+\frac{1}{2}}$
$= x^{-13/2}$
$= \dfrac{1}{x^{13/2}}$

146. $\sqrt{200x^9} = 10x^4\sqrt{2x}$

147. $\sqrt{\dfrac{3n^3}{121m^{10}}} = \dfrac{\sqrt{3n^3}}{\sqrt{121m^{10}}} = \dfrac{n\sqrt{3n}}{11m^5}$

148. $3\sqrt{20} - 7x\sqrt[3]{40} + 3\sqrt[3]{5x^3}$
$= 3\sqrt{4}\sqrt{5} - 7x\sqrt[3]{8}\sqrt[3]{5} + 3\sqrt[3]{x^3}\sqrt[3]{5}$
$= 6\sqrt{5} - 14x\sqrt[3]{5} + 3x\sqrt[3]{5}$
$= 6\sqrt{5} - 11x\sqrt[3]{5}$

149. $\left(2\sqrt{x} - 5\right)^2 = \left(2\sqrt{x}\right)^2 - 2(5)\left(2\sqrt{x}\right) + 5^2$
$= 4x - 20\sqrt{x} + 25$

150. $(x_1, y_1) = (-3, 5), (x_2, y_2) = (-3, 9)$
$d = \sqrt{(x_2 - x_1)^2 + (y_2 - y_1)^2}$
$= \sqrt{(-8+3)^2 + (9-5)^2}$
$= \sqrt{(-5)^2 + (4)^2}$
$= \sqrt{25+16}$
$= \sqrt{41}$
The distance is $\sqrt{41}$ units.

151. $(x_1, y_1) = (-3, 8), (x_2, y_2) = (11, 24)$

$$\text{midpoint} = \left(\frac{x_1 + x_2}{2}, \frac{y_1 + y_2}{2}\right) = \left(\frac{-3+11}{2}, \frac{8+24}{2}\right) = \left(\frac{8}{2}, \frac{32}{2}\right) = (4, 16)$$

152. $\frac{7}{\sqrt{13}} = \frac{7}{\sqrt{13}} \cdot \frac{\sqrt{13}}{\sqrt{13}} = \frac{7\sqrt{13}}{13}$

153. $\frac{2}{\sqrt{x}+3} = \frac{2}{\sqrt{x}+3} \cdot \frac{\sqrt{x}-3}{\sqrt{x}-3} = \frac{2\sqrt{x}-6}{x-9}$

154. $\sqrt{x} + 2 = x$

$$\sqrt{x} = x - 2$$
$$\left(\sqrt{x}\right)^2 = (x-2)^2$$
$$x = x^2 - 4x + 4$$
$$0 = x^2 - 5x + 4$$
$$0 = (x-4)(x-1)$$

$x - 4 = 0$ or $x - 1 = 0$

$x = 4$ $\quad x = 1$

Discard the extraneous solution $x = 1$. The solution set is $\{4\}$.

Chapter 7 Test

1. $\sqrt{216} = \sqrt{36 \cdot 6} = 6\sqrt{6}$

2. $-\sqrt[4]{x^{64}} = -x^{16}$

3. $\left(\frac{1}{125}\right)^{1/3} = \sqrt[3]{\frac{1}{125}} = \frac{1}{5}$

4. $\left(\frac{1}{125}\right)^{-1/3} = (125)^{1/3} = \sqrt[3]{125} = 5$

5. $\left(\frac{8x^3}{27}\right)^{2/3} = \frac{8^{2/3}(x^3)^{2/3}}{27^{2/3}} = \frac{(8^{1/3})^2 x^2}{(27^{1/3})^2} = \frac{2^2 x^2}{3^2} = \frac{4x^2}{9}$

6. $\sqrt[3]{-a^{18}b^9} = -a^6b^3$

7. $\left(\frac{64c^{4/3}}{a^{-2/3}b^{5/6}}\right)^{1/2} = \frac{(64)^{1/2}c^{4/6}}{a^{-2/6}b^{5/12}} = \frac{8a^{1/3}c^{2/3}}{b^{5/12}}$

8. $a^{-2/3}(a^{5/4} - a^3) = a^{-\frac{2}{3}+\frac{5}{4}} - a^{-\frac{2}{3}+3} = a^{-\frac{8}{12}+\frac{15}{12}} - a^{-\frac{2}{3}+\frac{9}{3}} = a^{7/12} - a^{7/3}$

9. $\sqrt[4]{(4xy)^4} = |4xy|$ or $4|xy|$

10. $\sqrt[3]{(-27)^3} = -27$

11. $\sqrt{\frac{9}{y}} = \frac{\sqrt{9}}{\sqrt{y}} = \frac{3}{\sqrt{y}} \cdot \frac{\sqrt{y}}{\sqrt{y}} = \frac{3\sqrt{y}}{y}$

12. $\frac{4-\sqrt{x}}{4+2\sqrt{x}} = \frac{4-\sqrt{x}}{4+2\sqrt{x}} \cdot \frac{4-2\sqrt{x}}{4-2\sqrt{x}}$

$= \frac{4^2 - 4(2\sqrt{x}) - 4\sqrt{x} + \sqrt{x}(2\sqrt{x})}{4^2 - (2\sqrt{x})^2}$

$= \frac{16 - 8\sqrt{x} - 4\sqrt{x} + 2x}{16-4x}$

$= \frac{16 - 12\sqrt{x} + 2x}{16-4x}$

$= \frac{8 - 6\sqrt{x} + x}{8-2x}$

13. $\sqrt[3]{\frac{8}{9x}} = \frac{\sqrt[3]{8}}{\sqrt[3]{9x}} = \frac{2 \cdot \sqrt[3]{3x^2}}{\sqrt[3]{9x} \cdot \sqrt[3]{3x^2}} = \frac{2\sqrt[3]{3x^2}}{3x}$

14. $\frac{\sqrt{6}+x}{8} = \frac{\sqrt{6}+x}{8} \cdot \frac{\sqrt{6}-x}{\sqrt{6}-x}$

$= \frac{\sqrt{6}^2 - x^2}{8(\sqrt{6}-x)}$

$= \frac{6-x^2}{8(\sqrt{6}-x)}$

15. $\sqrt{125x^3} - 3\sqrt{20x^3} = 5x\sqrt{5x} - 6x\sqrt{5x}$

$= -x\sqrt{5x}$

16. $\sqrt{3}(\sqrt{16} - \sqrt{2}) = \sqrt{3}(4) - \sqrt{3}\sqrt{2} = 4\sqrt{3} - \sqrt{6}$

17. $(\sqrt{x}+1)^2 = (\sqrt{x})^2 + 2\sqrt{x} + 1 = x + 2\sqrt{x} + 1$

18. $(\sqrt{2}-4)(\sqrt{3}+1) = \sqrt{6} + \sqrt{2} - 4\sqrt{3} - 4$

19. $(\sqrt{5}+5)(\sqrt{5}-5) = (\sqrt{5})^2 - 5^2$

$= 5 - 25$

$= -20$

20. $\sqrt{561} \approx 23.685$

21. $386^{-2/3} \approx 0.019$

22.

$x = \sqrt{x-2} + 2$

$x - 2 = \sqrt{x-2}$

$(x-2)^2 = (\sqrt{x-2})^2$

$x^2 - 4x + 4 = x - 2$

$x^2 - 5x + 6 = 0$

$(x-3)(x-2) = 0$

$x = 3$ or $x = 2$

The solution set is {2, 3}.

23. $\sqrt{x^2-7} + 3 = 0$

$\sqrt{x^2-7} = -3$

No solution since a principal square root does not yield a negative number. The solution set is ∅.

24. $\sqrt{x+5} = \sqrt{2x-1}$

$x + 5 = 2x - 1$

$6 = x$

The solution set is {6}.

25. $\sqrt{-2} = \sqrt{-1} \cdot \sqrt{2} = i\sqrt{2}$

26. $-\sqrt{-8} = -\sqrt{-1} \cdot \sqrt{4} \cdot \sqrt{2} = -2i\sqrt{2}$

27. $(12-6i) - (12-3i) = 12 - 6i - 12 + 3i$

$= (12-12) + (-6+3)i$

$= 0 - 3i$

$= -3i$

28. $(6-2i)(6+2i) = 6^2 - (2i)^2 = 36 + 4 = 40$

29. $(4+3i)^2 = 16 + 24i + 9i^2$

$= 16 + 24i + 9(-1)$

$= (16-9) + 24i$

$= 7 + 24i$

30. $\frac{1+4i}{1-i} = \frac{(1+4i)}{(1-i)} \cdot \frac{(1+i)}{(1+i)}$
$= \frac{1+4i+i-4}{1+1}$
$= \frac{-3+5i}{2}$
$= -\frac{3}{2} + \frac{5i}{2}$

31. $a^2 + b^2 = c^2$
$x^2 + x^2 = 5^2$
$2x^2 = 25$
$x^2 = \frac{25}{2}$
$x = \frac{\sqrt{25}}{\sqrt{2}} = \frac{5}{\sqrt{2}} \cdot \frac{\sqrt{2}}{\sqrt{2}} = \frac{5\sqrt{2}}{2}$ inches

32. $g(x) = \sqrt{x+2}$
$g(0) = \sqrt{0+2} = \sqrt{2}$
$g(23) = \sqrt{23+2} = \sqrt{25} = 5$

33. $(x_1, y_1) = (-6, 3), (x_2, y_2) = (-8, -7)$
$d = \sqrt{(-8-(-6))^2 + (-7-3)^2}$
$= \sqrt{(-2)^2 + (-10)^2}$
$= \sqrt{4+100}$
$= \sqrt{104}$
$= \sqrt{4 \cdot 26}$
$= 2\sqrt{26}$
The distance is $2\sqrt{26}$ units.

34. $(x_1, y_1) = \left(-2\sqrt{5}, \sqrt{10}\right)$,
$(x_2, y_2) = \left(-\sqrt{5}, 4\sqrt{10}\right)$
$d = \sqrt{(x_2 - x_1)^2 + (y_2 - y_1)^2}$
$= \sqrt{\left(-\sqrt{5} + 2\sqrt{5}\right)^2 + \left(4\sqrt{10} - \sqrt{10}\right)^2}$
$= \sqrt{\left(\sqrt{5}\right)^2 + \left(3\sqrt{10}\right)^2}$
$= \sqrt{5+90}$
$= \sqrt{95}$
The distance is $\sqrt{95}$ units.

35. $(x_1, y_1) = (-2, -5), (x_2, y_2) = (-6, 12)$
$\text{midpoint} = \left(\frac{x_1 + x_2}{2}, \frac{y_1 + y_2}{2}\right)$
$= \left(\frac{-2-6}{2}, \frac{-5+12}{2}\right)$
$= \left(-\frac{8}{2}, \frac{7}{2}\right)$
$= \left(-4, \frac{7}{2}\right)$

36. $(x_1, y_1) = \left(-\frac{2}{3}, -\frac{1}{5}\right), (x_2, y_2) = \left(-\frac{1}{3}, \frac{4}{5}\right)$
$\text{midpoint} = \left(\frac{x_1 + x_2}{2}, \frac{y_1 + y_2}{2}\right)$
$= \left(\frac{-\frac{2}{3} - \frac{1}{3}}{2}, \frac{-\frac{1}{5} + \frac{4}{5}}{2}\right)$
$= \left(\frac{-\frac{3}{3}}{2}, \frac{\frac{3}{5}}{2}\right)$
$= \left(-\frac{1}{2}, \frac{3}{10}\right)$

37. $V = \sqrt{2.5(300)} \approx 27$
The maximum safe speed is 27 mph.

38. $V = \sqrt{2.5r}$
$30 = \sqrt{2.5r}$
$900 = 2.5r$
$360 = r$
The radius is 360 feet.

Cumulative Review Chapters 1–7

1. $x^{-b}(2x^b)^2 = x^{-b}4x^{2b} = 4x^b$

2. $\left(-\frac{1}{3}x^a\right)\left(-\frac{2}{5}x^a\right) = \frac{2}{15}x^{2a}$

3. $\frac{(y^{3a})^2}{y^{a-6}} = \frac{y^{6a}}{y^{a-6}} = y^{6a-(a-6)} = y^{5a+6}$

4. $\frac{2.8y^{7a}}{-2y^{-2a}} = \frac{2.8}{-2}y^{7a-(-2a)} = -1.4y^{9a}$

5. $|x+1| = 6$
$x+1=-6$ or $x+1=6$
$x=-7$ $\quad x=5$
The solution set is $\{-7, 5\}$.

6. $|3x-2|+5=5$
$|3x-2|=0$
$3x=2$
$x=\frac{2}{3}$
The solution set is $\left\{\frac{2}{3}\right\}$.

7. $2x+y=-6$
$y=-2x-6$
$m=-2$
A parallel line has slope of –2.
$y-y_1=m(x-x_1)$
$y-4=-2(x-4)$
$y-4=-2x+8$
$y=-2x+12$

8. $\frac{a-1}{2}+a=2-\frac{2a+7}{8}$
$8\left(\frac{a-1}{2}+a\right)=8\left(2-\frac{2a+7}{8}\right)$
$4a-4+8a=16-2a-7$
$12a-4=-2a+9$
$14a=13$
$a=\frac{13}{14}$

9. $x-5y=-12$
$\underline{-x+y=4}$
$-4y=-8$
$y=2$
Substitute $y = 2$ in the second equation and solve for x.
$-x+y=4$
$-x+2=4$
$-x=2$
$x=-2$
The solution to the system is (–2, 2).

10. $y=-3$
$y=0x-3$
$m=0$

11. $\begin{cases} x-5y-2z=6 \\ -2x+10y+4z=-12 \\ \frac{1}{2}x-\frac{5}{2}y-z=3 \end{cases}$

$\rightarrow \begin{cases} x-5y-2z=6 \\ -\frac{1}{2}(2x+10y+4z)=-\frac{1}{2}(-12) \\ 2\left(\frac{1}{2}x-\frac{5}{2}y-z\right)=2(3) \end{cases}$

$\rightarrow \begin{cases} x-5y-2z=6 \\ x-5y-2z=6 \\ x-5y-2z=6 \end{cases}$

Since all three equations are identical, there are infinitely many solutions of this system.
$\{(x, y, z) | x-5y-2z=6\}$

12. $\begin{cases} \frac{x}{6}-\frac{y}{2}=1 \\ \frac{x}{3}-\frac{y}{4}=2 \end{cases} \rightarrow \begin{cases} x-3y=6 \\ 4x-3y=24 \end{cases}$

Solve the first equation for x.

$x-3y=6$

$x=3y+6$

Substitute 3*y* + 6 for *x* in the second equation and solve for *y*.

$4x-3y=24$

$4(3y+6)-3y=24$

$12y+24-3y=24$

$9y=0$

$y=0$

Now solve for *x*.

$x = 3y + 6 = 3(0) + 6 = 6$

The solution to the system is {6, 0}.

13. $\begin{cases} 2x-y=3 \\ 4x-2y=5 \end{cases}$

$\left[\begin{array}{cc|c} 2 & -1 & 3 \\ 4 & -2 & 5 \end{array}\right]$

$\left[\begin{array}{cc|c} 1 & -\frac{1}{2} & \frac{3}{2} \\ 4 & -2 & 5 \end{array}\right]$

$\rightarrow \left[\begin{array}{cc|c} 1 & -\frac{1}{2} & \frac{3}{2} \\ -4(1)+4 & -4\left(-\frac{1}{2}\right)-2 & -4\left(\frac{3}{2}\right)+5 \end{array}\right]$

$\left[\begin{array}{cc|c} 1 & -\frac{1}{2} & \frac{3}{2} \\ 0 & 0 & -1 \end{array}\right]$

$x-\frac{1}{2}y=\frac{3}{2}$

$0=-1$

Since the equation 0 = −1 is false for all *y* or *x* values, the system is inconsistent and has no solution. The solution set is ∅.

14. a. $2(x-3)+(5x+3)=2x-6+5x+3$
$=7x-3$

b. $4(3x+2)-3(5x-1)=12x+8-15x+3$
$=-3x+11$

c. $7x+2(x-7)-3x=7x+2x-14-3x$
$=6x-14$

15. $(7x^3y-xy^3+11)+(6x^3y-4)$
$=13x^3y-xy^3+7$

16. $\frac{0.0000035\times 4000}{0.28}=\frac{(3.5\times 10^{-6})(4\times 10^3)}{2.8\times 10^{-1}}$
$=5\times 10^{-2}$

17. $P(x)=3x^4-2x^2-5$

$P(1)=3(1)^4-2(1)^2-5=3-2-5=-4$

18. $P(x)=3x^4-2x^2-5$

$P(0)=3(0)^4-2(0)^2-5=0-0-5=-5$

19. $P(x)=3x^4-2x^2-5$

$P(-2)=3(-2)^4-2(-2)^2-5$
$=48-8-5$
$=35$

20. $P(x)=3x^4-2x^2-5$

$P\left(\frac{1}{2}\right)=3\left(\frac{1}{2}\right)^4-2\left(\frac{1}{2}\right)^2-5$
$=\frac{3}{16}-\frac{1}{2}-5$
$=\frac{3}{16}-\frac{8}{16}-\frac{80}{16}$
$=-\frac{85}{16}$

21. $(x+5)^2=x^2+2(5x)+5^2=x^2+10x+25$

22. $(y-2)(3y+4)=3y^2+4y-6y-8$
$=3y^2-2y-8$

23. $(4m^2-3n)^2=(4m^2)^2-2(3n)(4m)+(3n)^2$
$=16m^4-24mn+9n^2$

24. $(3y-1)(2y^2+3y-1)$
$=6y^3+9y^2-3y-2y^2-3y+1$
$=6y^3+7y^2-6y+1$

25.

3	2	−1	−13	1
		6	15	6
	2	5	2	7

$$\frac{2x^3-x^2-13x+1}{x-3}=2x^2+5x+2+\frac{7}{x-3}$$

26.

3	4	−12	−1	12
		12	0	−3
	4	0	−1	9

$$\frac{4y^3-12y^2-y+12}{y-3}=4y^2-1+\frac{9}{y-3}$$

27. $ab-6a+2b-12=a(b-6)+2(b-6)$
$=(b-6)(a+2)$

28. $x^3-x^2+4x-4=x^2(x-1)+4(x-1)$
$=(x-1)(x^2+4)$

29. $2n^2-38n+80=2(n^2-19n+40)$

30. $2x^3-16=2(x^3-8)=2(x-2)(x^2+2x+4)$

31. $16x^2+24xy+9y^2$
$=(4x)^2+2(4x)(3y)+(3y)^2$
$=(4x+3y)^2$

32. $x^2+2x+1-y^2=(x+1)^2-y^2$
$=[(x+1)+y][(x+1)-y]$
$=(x+1+y)(x+1-y)$

33. $50-8y^2=2(25-4y^2)=2(5-2y)(5+2y)$

34. $x^4-1=(x^2-1)(x^2+1)$
$=(x+1)(x-1)(x^2+1)$

35. $x(2x-7)=4$
$2x^2-7x-4=0$
$(2x+1)(x-4)=0$
$2x+1=0$ or $x-4=0$
$2x=-1$ $x=4$
$x=-\frac{1}{2}$

The solution set is $\left\{-\frac{1}{2}, 4\right\}$.

36. $\frac{x^3-2x^2+3x-6}{x-2}=\frac{x^2(x-2)+3(x-2)}{x-2}$
$=\frac{(x-2)(x^2+3)}{x-2}$
$=x^2+3$

37. $x^3=4x$
$x^2=4$
$\sqrt{x^2}=\pm\sqrt{4}$
$x=\pm 2$

The solution set is $\{-2, 0, 2\}$.

38. a. $\frac{\frac{y-2}{16}}{\frac{2y+3}{12}}=\frac{y-2}{16}\cdot\frac{12}{2y+3}=\frac{3(y-2)}{4(2y+3)}$

b. $\frac{\frac{x}{16}-\frac{1}{x}}{1-\frac{4}{x}}=\frac{\frac{x^2-16}{16x}}{\frac{x-4}{x}}=\frac{x^2-16}{16x}\cdot\frac{x}{x-4}=\frac{x+4}{16}$

39. $\frac{2+x}{x+2}=\frac{x+2}{x+2}=1$

40. $\frac{a^3-8}{2-a}=\frac{(a-2)(a^2+2a+4)}{-(a-2)}=-a^2-2a-4$

41. $\frac{2x^2}{10x^3-2x^2}=\frac{2x^2}{2x^2(5x-1)}=\frac{1}{5x-1}$

42. $$\frac{3a^2-3}{a^3+5a^2-a-5} = \frac{3(a^2-1)}{a^2(a+5)-1(a+5)} = \frac{3(a^2-1)}{(a+5)(a^2-1)} = \frac{3}{a+5}$$

43. $$\frac{5}{7z^2}+\frac{x}{7z^2}=\frac{5+x}{7z^2}$$

44. a. $$\frac{3}{xy^2}-\frac{2}{3x^2y}=\frac{9x}{3x^2y^2}-\frac{2y}{3x^2y^2}=\frac{9x-2y}{3x^2y^2}$$

b. $$\frac{5x}{x+3}-\frac{2x}{x-3} = \frac{5x(x-3)}{(x+3)(x-3)}-\frac{2x(x+3)}{(x-3)(x+3)} = \frac{5x^2-15x}{(x+3)(x-3)}-\frac{2x^2+6x}{(x+3)(x-3)} = \frac{3x^2-21x}{(x+3)(x-3)} = \frac{3x(x-7)}{(x+3)(x-3)}$$

c. $$\frac{x}{x-2}-\frac{5}{2-x}=\frac{x}{x-2}-\frac{5}{-(x-2)}=\frac{x+5}{x-2}$$

45. $$\frac{3}{x}-\frac{x+21}{3x}=\frac{5}{3}$$
$$3x\left[\frac{3}{x}-\frac{x+21}{3x}\right]=3x\left(\frac{5}{3}\right)$$
$$9-x-21=5x$$
$$-12=6x$$
$$-2=x$$
The solution set is $\{-2\}$.

46.
$$\frac{28}{9-a^2}=\frac{2a}{a-3}+\frac{6}{a+3}$$
$$(a-3)(a+3)\left[-\frac{28}{a^2-9}\right]=(a-3)(a+3)\left(\frac{2a}{a-3}+\frac{6}{a+3}\right)$$
$$-28=2a(a+3)+6(a-3)$$
$$-28=2a^2+6a+6a-18$$
$$0=2a^2+12a+10$$
$$0=2(a^2+6a+5)$$
$$0=2(a+5)(a+1)$$

$a+5=0$ or $a+1=0$

$a=-5$ $\quad a=-1$

The solution set is $\{-1, -5\}$.

47. $16^{-3/4}=\frac{1}{\left(\sqrt[4]{16}\right)^3}=\frac{1}{2^3}=\frac{1}{8}$

48. $81^{-3/4}=\frac{1}{\left(\sqrt[4]{81}\right)^3}=\frac{1}{3^3}=\frac{1}{27}$

49. $(-27)^{-2/3}=\frac{1}{\left(\sqrt[3]{-27}\right)^2}=\frac{1}{(-3)^2}=\frac{1}{9}$

50. $(-125)^{-2/3}=\frac{1}{\left(\sqrt[3]{-125}\right)^2}=\frac{1}{(-5)^2}=\frac{1}{25}$

Chapter 8

Section 8.1 Practice Problems

1. $x^2 = 45$

$x = \pm\sqrt{45}$

$x = \pm 3\sqrt{5}$

The solution set is $\{-3\sqrt{5}, 3\sqrt{5}\}$.

2. $5x^2 = 55$

$x^2 = 11$

$x = \pm\sqrt{11}$

The solution set is $\{-\sqrt{11}, \sqrt{11}\}$.

3. $(x+2)^2 = 18$

$x+2 = \pm\sqrt{18}$

$x+2 = \pm 3\sqrt{2}$

$x = -2 \pm 3\sqrt{2}$

The solution set is $\{-2-3\sqrt{2}, -2+3\sqrt{2}\}$.

4. $(3x-1)^2 = -4$

$3x-1 = \pm\sqrt{-4}$

$3x-1 = \pm 2i$

$3x = 1 \pm 2i$

$x = \dfrac{1 \pm 2i}{3}$

The solution set is $\left\{\dfrac{1+2i}{3}, \dfrac{1-2i}{3}\right\}$.

5. $x^2 + 12x$

$\frac{1}{2}(12) = 6$ and $6^2 = 36$

$x^2 + 12x + 36 = (x+6)^2$

6. $y^2 - 5y$

$\frac{1}{2}(-5) = -\frac{5}{2}$ and $\left(-\frac{5}{2}\right)^2 = \frac{25}{4}$

$y^2 - 5y + \frac{25}{4} = \left(y - \frac{5}{2}\right)^2$

7. $x^2 + 8x = 1$

$x^2 + 8x + \left(\frac{8}{2}\right)^2 = 1 + \left(\frac{8}{2}\right)^2$

$x^2 + 8x + 16 = 1 + 16$

$(x+4)^2 = 17$

$x + 4 = \pm\sqrt{17}$

$x = -4 \pm \sqrt{17}$

The solution set is $\{-4-\sqrt{17}, -4+\sqrt{17}\}$.

8. $y^2 - 5y + 2 = 0$

$y^2 - 5y = -2$

$y^2 - 5y + \left(\frac{-5}{2}\right)^2 = -2 + \left(\frac{-5}{2}\right)^2$

$y^2 - 5y + \frac{25}{4} = -\frac{8}{4} + \frac{25}{4}$

$\left(y - \frac{5}{2}\right)^2 = \frac{17}{4}$

$y - \frac{5}{2} = \pm\sqrt{\frac{17}{4}}$

$y - \frac{5}{2} = \pm\frac{\sqrt{17}}{2}$

$y = \frac{5}{2} \pm \frac{\sqrt{17}}{2}$

The solution set is $\left\{\dfrac{5-\sqrt{17}}{2}, \dfrac{5+\sqrt{17}}{2}\right\}$.

9.
$$2x^2 - 2x + 7 = 0$$
$$x^2 - x + \frac{7}{2} = 0$$
$$x^2 - x = -\frac{7}{2}$$
$$x^2 - x + \left(-\frac{1}{2}\right)^2 = -\frac{7}{2} + \left(-\frac{1}{2}\right)^2$$
$$x^2 - x + \frac{1}{4} = -\frac{14}{4} + \frac{1}{4}$$
$$\left(x - \frac{1}{2}\right)^2 = -\frac{13}{4}$$
$$x - \frac{1}{2} = \pm\sqrt{-\frac{13}{4}}$$
$$x - \frac{1}{2} = \pm\frac{i\sqrt{13}}{2}$$
$$x = \frac{1}{2} \pm \frac{i\sqrt{13}}{2}$$

The solution set is $\left\{\frac{1+i\sqrt{13}}{2}, \frac{1-i\sqrt{13}}{2}\right\}$.

10. Let $P = 1600$, $A = 1764$, and $t = 2$.
$$A = P(1+r)^t$$
$$1764 = 1600(1+r)^2$$
$$\frac{1764}{1600} = (1+r)^2$$
$$\frac{441}{400} = (1+r)^2$$
$$\pm\sqrt{\frac{441}{400}} = 1+r$$
$$\pm\frac{21}{20} = 1+r$$
$$-\frac{20}{20} \pm \frac{21}{20} = r$$
$$r = \frac{1}{20} \text{ or } r = -\frac{41}{20}$$

The rate cannot be negative, so reject $-\frac{41}{20}$.

The interest rate is $\frac{1}{20} = 0.05 = 5\%$.

Exercise Set 8.1

1.
$$x^2 = 16$$
$$x = \pm\sqrt{16}$$
$$x = \pm 4$$

The solution set is $\{-4, 4\}$.

3.
$$x^2 - 7 = 0$$
$$x^2 = 7$$
$$x = \pm\sqrt{7}$$

The solution set is $\{-\sqrt{7}, \sqrt{7}\}$.

5.
$$x^2 = 18$$
$$x = \pm\sqrt{18}$$
$$x = \pm\sqrt{9}\sqrt{2}$$
$$x = \pm 3\sqrt{2}$$

The solution set is $\{-3\sqrt{2}, 3\sqrt{2}\}$.

7.
$$3z^2 - 30 = 0$$
$$3z^2 = 30$$
$$z^2 = 10$$
$$z = \pm\sqrt{10}$$

The solution set is $\{-\sqrt{10}, \sqrt{10}\}$.

9.
$$(x+5)^2 = 9$$
$$x + 5 = \pm\sqrt{9}$$
$$x + 5 = \pm 3$$
$$x = -5 \pm 3$$
$$x = -8 \text{ or } x = -2$$

The solution set is $\{-8, -2\}$.

11.
$$(z-6)^2 = 18$$
$$z - 6 = \pm\sqrt{18}$$
$$z - 6 = \pm\sqrt{9}\sqrt{2}$$
$$z - 6 = \pm 3\sqrt{2}$$
$$z = 6 \pm 3\sqrt{2}$$

The solution set is $\{6 - 3\sqrt{2}, 6 + 3\sqrt{2}\}$.

13. $(2x-3)^2 = 8$
$$2x-3 = \pm\sqrt{8}$$
$$2x-3 = \pm\sqrt{4}\sqrt{2}$$
$$2x-3 = \pm 2\sqrt{2}$$
$$2x = 3 \pm 2\sqrt{2}$$
$$x = \frac{3 \pm 2\sqrt{2}}{2}$$
The solution set is $\left\{\frac{3-2\sqrt{2}}{2}, \frac{3+2\sqrt{2}}{2}\right\}$.

15. $x^2 + 9 = 0$
$$x^2 = -9$$
$$x = \pm\sqrt{-9}$$
$$x = \pm 3i$$
The solution set is $\{-3i, 3i\}$.

17. $x^2 - 6 = 0$
$$x^2 = 6$$
$$x = \pm\sqrt{6}$$
The solution set is $\{-\sqrt{6}, \sqrt{6}\}$.

19. $2z^2 + 16 = 0$
$$2z^2 = -16$$
$$z^2 = -8$$
$$z = \pm\sqrt{-8}$$
$$z = \pm\sqrt{-4}\sqrt{2}$$
$$z = \pm 2i\sqrt{2}$$
The solution set is $\{-2i\sqrt{2}, 2i\sqrt{2}\}$.

21. $(3x-1)^2 = -16$
$$3x-1 = \pm\sqrt{-16}$$
$$3x-1 = \pm 4i$$
$$3x = 1 \pm 4i$$
$$x = \frac{1 \pm 4i}{3}$$
The solution set is $\left\{\frac{1-4i}{3}, \frac{1+4i}{3}\right\}$.

23. $(z+7)^2 = 5$
$$z+7 = \pm\sqrt{5}$$
$$z = -7 \pm \sqrt{5}$$
The solution set is $\{-7-\sqrt{5}, -7+\sqrt{5}\}$.

25. $(x+3)^2 + 8 = 0$
$$(x+3)^2 = -8$$
$$x+3 = \pm\sqrt{-8}$$
$$x+3 = \pm\sqrt{4}\sqrt{2}\sqrt{-1}$$
$$x+3 = \pm 2i\sqrt{2}$$
$$x = -3 \pm 2i\sqrt{2}$$
The solution set is $\{-3-2i\sqrt{2}, -3+2i\sqrt{2}\}$.

27. $x^2 + 16x + \left(\frac{16}{2}\right)^2 = x^2 + 16x + 64 = (x+8)^2$

29. $z^2 - 12z + \left(-\frac{12}{2}\right)^2 = z^2 - 12z + 36$
$$= (z-6)^2$$

31. $p^2 + 9p + \left(\frac{9}{2}\right)^2 = p^2 + 9p + \frac{81}{4} = \left(p + \frac{9}{2}\right)^2$

33. $r^2 - r + \left(-\frac{1}{2}\right)^2 = r^2 - r + \frac{1}{4} = \left(r - \frac{1}{2}\right)^2$

35. $x^2 + 8x = -15$
$$x^2 + 8x + \left(\frac{8}{2}\right)^2 = -15 + 16$$
$$(x+4)^2 = 1$$
$$x+4 = \pm\sqrt{1}$$
$$x+4 = \pm 1$$
$$x = -4 \pm 1$$
$x = -5$ or $x = -3$
The solution set is $\{-5, -3\}$.

37.

$$x^2+6x+2=0$$
$$x^2+6x=-2$$
$$x^2+6x+\left(\frac{6}{2}\right)^2=-2+9$$
$$(x+3)^2=7$$
$$x+3=\pm\sqrt{7}$$
$$x=-3\pm\sqrt{7}$$

The solution set is $\left\{-3-\sqrt{7}, -3+\sqrt{7}\right\}$.

39.

$$x^2+x-1=0$$
$$x^2+x=1$$
$$x^2+x+\left(\frac{1}{2}\right)^2=1+\frac{1}{4}$$
$$\left(x+\frac{1}{2}\right)^2=\frac{5}{4}$$
$$x+\frac{1}{2}=\pm\sqrt{\frac{5}{4}}$$
$$x+\frac{1}{2}=\pm\frac{\sqrt{5}}{2}$$
$$x=-\frac{1}{2}\pm\frac{\sqrt{5}}{2}$$

The solution set is $\left\{\frac{-1-\sqrt{5}}{2}, \frac{-1+\sqrt{5}}{2}\right\}$.

41.

$$x^2+2x-5=0$$
$$x^2+2x=5$$
$$x^2+2x+\left(\frac{2}{2}\right)^2=5+1$$
$$(x+1)^2=6$$
$$x+1=\pm\sqrt{6}$$
$$x=-1\pm\sqrt{6}$$

The solution set is $\left\{-1-\sqrt{6}, -1+\sqrt{6}\right\}$.

43.

$$y^2+y-7=0$$
$$y^2+y=7$$
$$y^2+y+\left(\frac{1}{2}\right)^2=7+\frac{1}{4}$$
$$\left(y+\frac{1}{2}\right)^2=\frac{29}{4}$$
$$y+\frac{1}{2}=\pm\sqrt{\frac{29}{4}}$$
$$y+\frac{1}{2}=\pm\frac{\sqrt{29}}{2}$$
$$y=-\frac{1}{2}\pm\frac{\sqrt{29}}{2}$$

The solution set is $\left\{\frac{-1-\sqrt{29}}{2}, \frac{-1+\sqrt{29}}{2}\right\}$.

45.

$$x^2+8x+1=0$$
$$x^2+8x=-1$$
$$x^2+8x+\left(\frac{8}{2}\right)^2=-1+16$$
$$(x+4)^2=15$$
$$x+4=\pm\sqrt{15}$$
$$x=-4\pm\sqrt{15}$$

The solution set is $\left\{-4-\sqrt{15}, -4+\sqrt{15}\right\}$.

47.
$$3p^2 - 12p + 2 = 0$$
$$3p^2 - 12p = -2$$
$$p^2 - 4p = -\frac{2}{3}$$
$$p^2 - 4p + \left(-\frac{4}{2}\right)^2 = -\frac{2}{3} + 4$$
$$(p-2)^2 = \frac{10}{3}$$
$$p - 2 = \pm\sqrt{\frac{10}{3}}$$
$$p - 2 = \pm\frac{\sqrt{30}}{3}$$
$$p = 2 \pm \frac{\sqrt{30}}{3}$$
$$p = \frac{6 \pm \sqrt{30}}{3}$$
The solution set is $\left\{\frac{6-\sqrt{30}}{3}, \frac{6+\sqrt{30}}{3}\right\}$.

49.
$$2x^2 + 7x = 4$$
$$x^2 + \frac{7}{2}x = 2$$
$$x^2 + \frac{7}{2}x + \left(\frac{\frac{7}{2}}{2}\right)^2 = 2 + \frac{49}{16}$$
$$\left(x + \frac{7}{4}\right)^2 = \frac{81}{16}$$
$$x + \frac{7}{4} = \pm\sqrt{\frac{81}{16}}$$
$$x = -\frac{7}{4} \pm \frac{9}{4}$$
$x = -4$ or $x = \frac{1}{2}$

The solution set is $\left\{-4, \frac{1}{2}\right\}$.

51.
$$3y^2 + 6y - 4 = 0$$
$$3y^2 + 6y = 4$$
$$y^2 + 2y = \frac{4}{3}$$
$$y^2 + 2y + \left(\frac{2}{2}\right)^2 = \frac{4}{3} + 1$$
$$(y+1)^2 = \frac{7}{3}$$
$$y + 1 = \pm\sqrt{\frac{7}{3}}$$
$$y + 1 = \pm\frac{\sqrt{21}}{3}$$
$$y = -1 \pm \frac{\sqrt{21}}{3}$$
$$y = \frac{-3 \pm \sqrt{21}}{3}$$
The solutions set is $\left\{\frac{-3-\sqrt{21}}{3}, \frac{-3+\sqrt{21}}{3}\right\}$.

53.
$$y^2 + 2y + 2 = 0$$
$$y^2 + 2y = -2$$
$$y^2 + 2y + \left(\frac{2}{2}\right)^2 = -2 + 1$$
$$(y+1)^2 = -1$$
$$y + 1 = \pm\sqrt{-1}$$
$$y + 1 = \pm i$$
$$y = -1 \pm i$$
The solution set is $\{-1 - i, -1 + i\}$.

55.
$$2a^2 + 8a = -12$$
$$a^2 + 4a = -6$$
$$a^2 + 4a + \left(\frac{4}{2}\right)^2 = -6 + 4$$
$$(a+2)^2 = -2$$
$$a + 2 = \pm\sqrt{-2}$$
$$a + 2 = \pm i\sqrt{2}$$
$$a = -2 \pm i\sqrt{2}$$
The solution set is $\left\{-2 - i\sqrt{2}, -2 + i\sqrt{2}\right\}$.

57.
$$\begin{aligned} 2x^2 - x + 6 &= 0 \\ 2x^2 - x &= -6 \\ x^2 - \frac{1}{2}x &= -3 \\ x^2 - \frac{1}{2}x + \left(\frac{-\frac{1}{2}}{2}\right)^2 &= -3 + \frac{1}{16} \\ \left(x - \frac{1}{4}\right)^2 &= -\frac{47}{16} \\ x - \frac{1}{4} &= \pm\sqrt{-\frac{47}{16}} \\ x - \frac{1}{4} &= \pm\frac{\sqrt{47}\sqrt{-1}}{\sqrt{16}} \\ x - \frac{1}{4} &= \pm\frac{i\sqrt{47}}{4} \\ x &= \frac{1 \pm i\sqrt{47}}{4} \end{aligned}$$

The solution set is $\left\{\frac{1 - i\sqrt{47}}{4}, \frac{1 + i\sqrt{47}}{4}\right\}$.

59.
$$\begin{aligned} x^2 + 10x + 28 &= 0 \\ x^2 + 10x &= -28 \\ x^2 + 10x + \left(\frac{10}{2}\right)^2 &= -28 + 25 \\ (x+5)^2 &= -3 \\ x + 5 &= \pm\sqrt{-3} \\ x + 5 &= \pm i\sqrt{3} \\ x &= -5 \pm i\sqrt{3} \end{aligned}$$

The solution set is $\left\{-5 - i\sqrt{3}, -5 + i\sqrt{3}\right\}$.

61.
$$\begin{aligned} z^2 + 3z - 4 &= 0 \\ z^2 + 3z &= 4 \\ z^2 + 3z + \left(\frac{3}{2}\right)^2 &= 4 + \frac{9}{4} \\ \left(z + \frac{3}{2}\right)^2 &= \frac{25}{4} \\ z + \frac{3}{2} &= \pm\sqrt{\frac{25}{4}} \\ z + \frac{3}{2} &= \pm\frac{5}{2} \\ z &= -\frac{3}{2} \pm \frac{5}{2} \end{aligned}$$

$z = -4$ or $z = 1$

The solution set is $\{-4, 1\}$.

63.
$$\begin{aligned} 2x^2 - 4x + 3 &= 0 \\ 2x^2 - 4x &= -3 \\ x^2 - 2x &= -\frac{3}{2} \\ x^2 - 2x + \left(\frac{-2}{2}\right)^2 &= -\frac{3}{2} + 1 \\ (x-1)^2 &= -\frac{1}{2} \\ x - 1 &= \pm\sqrt{-\frac{1}{2}} \\ x - 1 &= \pm\frac{i\sqrt{2}}{2} \\ x &= 1 \pm \frac{i\sqrt{2}}{2} \\ x &= \frac{2 \pm i\sqrt{2}}{2} \end{aligned}$$

The solution set is $\left\{\frac{2 - i\sqrt{2}}{2}, \frac{2 + i\sqrt{2}}{2}\right\}$.

65.
$$3x^2+3x=5$$
$$x^2+x=\frac{5}{3}$$
$$x^2+x+\left(\frac{1}{2}\right)^2=\frac{5}{3}+\frac{1}{4}$$
$$\left(x+\frac{1}{2}\right)^2=\frac{23}{12}$$
$$x+\frac{1}{2}=\pm\sqrt{\frac{23}{12}}$$
$$x+\frac{1}{2}=\pm\frac{\sqrt{23}}{2\sqrt{3}}\cdot\frac{\sqrt{3}}{\sqrt{3}}$$
$$x+\frac{1}{2}=\pm\frac{\sqrt{69}}{6}$$
$$x=-\frac{1}{2}\pm\frac{\sqrt{69}}{6}$$
$$x=-\frac{3}{6}\pm\frac{\sqrt{69}}{6}$$
$$x=\frac{-3\pm\sqrt{69}}{6}$$

The solution set is $\left\{\frac{-3-\sqrt{69}}{6}, \frac{-3+\sqrt{69}}{6}\right\}$.

67. Use $A=P(1+r)^t$ with $A = 4320$, $P = 3000$, and
$t = 2$.
$$A=P(1+r)^t$$
$$4320=3000(1+r)^2$$
$$\frac{4320}{3000}=(1+r)^2$$
$$\frac{36}{25}=(1+r)^2$$
$$\pm\sqrt{\frac{36}{25}}=1+r$$
$$\pm\frac{6}{5}=1+r$$
$$-1\pm\frac{6}{5}=r$$
$$-\frac{5}{5}\pm\frac{6}{5}=r$$
$$r=\frac{1}{5} \text{ or } r=-\frac{11}{5}$$

The rate cannot be negative, so $-\frac{11}{5}$ is rejected. $\frac{1}{5}=0.2=20\%$, so \$3000 grows to \$4320 in
2 years with an interest rate of 20%.

69. Use $A=P(1+r)^t$ with $A = 1000$, $P = 810$, and
$t = 2$.
$$A=P(1+r)^t$$
$$1000=810(1+r)^2$$
$$\frac{1000}{810}=(1+r)^2$$
$$\frac{100}{81}=(1+r)^2$$
$$\pm\sqrt{\frac{100}{81}}=1+r$$
$$\pm\frac{10}{9}=1+r$$
$$-1\pm\frac{10}{9}=r$$
$$-\frac{9}{9}\pm\frac{10}{9}=r$$
$$r=\frac{1}{9} \text{ or } r=-\frac{19}{9}$$

The rate cannot be negative, so $-\frac{19}{9}$ is rejected. $\frac{1}{9}=0.1\overline{1}\approx 11\%$, so \$810 grows to approximately \$1000 in 2 years with an interest rate of 11%.

71.
$$\begin{aligned} s(t) &= 16t^2 \\ 1483 &= 16t^2 \\ \frac{1483}{16} &= t^2 \\ \pm\sqrt{\frac{1483}{16}} &= t \\ \pm\frac{\sqrt{1483}}{4} &= t \end{aligned}$$

The time cannot be negative, so $-\frac{\sqrt{1483}}{4}$ is rejected. It takes the object $\frac{\sqrt{1483}}{4} \approx 9.63$ seconds to fall from the top of the tower.

73.
$$\begin{aligned} s(t) &= 16t^2 \\ 1100 &= 16t^2 \\ \frac{1100}{16} &= t^2 \\ \frac{275}{4} &= t^2 \\ \pm\sqrt{\frac{275}{4}} &= t^2 \\ \pm\frac{5\sqrt{11}}{2} &= t \end{aligned}$$

The time cannot be negative, so $-\frac{5\sqrt{11}}{2}$ is rejected.

It takes the object $\frac{5\sqrt{11}}{2} \approx 8.29$ seconds to fall from the top to the base of the dam.

75. Let x be the length of one side.
$$\begin{aligned} x^2 &= 225 \\ x &= \pm\sqrt{225} \\ x &= \pm 15 \end{aligned}$$
The length cannot be negative, so -15 is rejected. The dimensions are 15 feet by 15 feet.

77.
$$\begin{aligned} a^2 + b^2 &= c^2 \\ x^2 + x^2 &= 20^2 \\ 2x^2 &= 400 \\ x^2 &= 200 \\ x &= \pm\sqrt{200} \\ x &= \pm 10\sqrt{2} \end{aligned}$$
The length cannot be negative, so $-10\sqrt{2}$ is rejected. The length of each leg is $10\sqrt{2}$ centimeters.

79. $\frac{10-20\sqrt{3}}{2} = \frac{2\left(5-10\sqrt{3}\right)}{2} = 5-10\sqrt{3}$

81. $\frac{12-8\sqrt{7}}{16} = \frac{4\left(3-2\sqrt{7}\right)}{4\cdot 4} = \frac{3-2\sqrt{7}}{4}$

83.
$$\begin{aligned} \sqrt{b^2-4ac} &= \sqrt{6^2-4(1)(2)} \\ &= \sqrt{36-8} \\ &= \sqrt{28} \\ &= \sqrt{4}\sqrt{7} \\ &= 2\sqrt{7} \end{aligned}$$

85.
$$\begin{aligned} \sqrt{b^2-4ac} &= \sqrt{(-3)^2-4(1)(-1)} \\ &= \sqrt{9+4} \\ &= \sqrt{13} \end{aligned}$$

87. $(y-5)^2 = -9$

Since the value that the squared term equals is negative, the equation has complex, but not real numbers as solutions.

89.
$$\begin{aligned} 4x^2 &= 17 \\ x^2 &= \frac{17}{4} \end{aligned}$$
Since the value that the squared term equals is positive, the equation has real solutions.

91.
$$\begin{aligned} (3m+2)^2 + 4 &= 1 \\ (3m+2)^2 &= -3 \end{aligned}$$
Since the value that the squared term equals is negative, the equation has complex, but not real numbers as solutions.

93. Compound interest is preferable; answers may vary.

95. $16 = 4^2$

$4 \cdot 2 = 8$

$x^2 + 8x + 16 = (x+4)^2$

$x^2 - 8x + 16 = (x-4)^2$

The missing term can be $-8x$ or $8x$.

97.
$$\begin{aligned} p &= -x^2 + 47 \\ 11 &= -x^2 + 47 \\ -36 &= -x^2 \\ 36 &= x^2 \\ \pm\sqrt{36} &= x \\ \pm 6 &= x \end{aligned}$$

Since the quantity demanded cannot be negative, $x = 6$ and the demand is 6 thousand scissors.

Section 8.2 Practice Problems

1. $2x^2 + 9x + 10 = 0$

$a = 2, b = 9, c = 10$

$$\begin{aligned} x &= \frac{-b \pm \sqrt{b^2 - 4ac}}{2a} \\ &= \frac{-9 \pm \sqrt{9^2 - 4(2)(10)}}{2(2)} \\ &= \frac{-9 \pm \sqrt{81 - 80}}{4} \\ &= \frac{-9 \pm \sqrt{1}}{4} \\ &= \frac{-9 \pm 1}{4} \end{aligned}$$

$x = \frac{-9-1}{4} = -\frac{5}{2}$ or $x = \frac{-9+1}{4} = -2$

The solution set is $\left\{-\frac{5}{2}, -2\right\}$.

2. $2x^2 - 6x - 1 = 0$

$a = 1, b = -6, c = -1$

$$\begin{aligned} x &= \frac{-b \pm \sqrt{b^2 - 4ac}}{2a} \\ &= \frac{-(-6) \pm \sqrt{(-6)^2 - 4(2)(-1)}}{2(2)} \\ &= \frac{6 \pm \sqrt{36 + 8}}{4} \\ &= \frac{6 \pm \sqrt{44}}{4} \\ &= \frac{6 \pm 2\sqrt{11}}{4} \\ &= \frac{2\left(3 \pm \sqrt{11}\right)}{2 \cdot 2} \\ &= \frac{3 \pm \sqrt{11}}{2} \end{aligned}$$

The solution set is $\left\{\frac{3-\sqrt{11}}{2}, \frac{3+\sqrt{11}}{2}\right\}$.

3.
$$\begin{aligned} \frac{1}{6}x^2 - \frac{1}{3}x - 1 &= 0 \\ 6\left(\frac{1}{6}x^2 - \frac{1}{3}x - 1\right) &= 6(0) \\ x^2 - 2x - 6 &= 0 \end{aligned}$$

$a = 1, b = -2, c = -6$

$$\begin{aligned} x &= \frac{-b \pm \sqrt{b^2 - 4ac}}{2a} \\ &= \frac{-(-2) \pm \sqrt{(-2)^2 - 4(1)(-6)}}{2(1)} \\ &= \frac{2 \pm \sqrt{4 + 24}}{2} \\ &= \frac{2 \pm \sqrt{28}}{2} \\ &= \frac{2 \pm 2\sqrt{7}}{2} \\ &= \frac{2\left(1 \pm \sqrt{7}\right)}{2} \\ &= 1 \pm \sqrt{7} \end{aligned}$$

The solution set is $\left\{1 - \sqrt{7}, 1 + \sqrt{7}\right\}$.

4. $x = -4x^2 - 4$

$4x^2 + x + 4 = 0$

$a = 4,\ b = 1,\ c = 4$

$$x = \frac{-b \pm \sqrt{b^2 - 4ac}}{2a} = \frac{-1 \pm \sqrt{1^2 - 4(4)(4)}}{2(4)} = \frac{-1 \pm \sqrt{1-64}}{8} = \frac{-1 \pm \sqrt{-63}}{8} = \frac{-1 \pm 3i\sqrt{7}}{8}$$

The solution set is $\left\{\frac{-1-3i\sqrt{7}}{8}, \frac{-1+3i\sqrt{7}}{8}\right\}$.

5. $x^2 + 4x + 4 = 0$

$a = 1,\ b = 4,\ c = 4$

$b^2 - 4ac = 4^2 - 4(1)(4) = 16 - 16 = 0$

Since $b^2 - 4ac = 0$, this quadratic equation has one real solution.

6. $5x^2 + 7 = 0$

$a = 5,\ b = 0,\ c = 7$

$b^2 - 4ac = 0^2 - 4(5)(7) = 0 - 140 = -140$

Since $b^2 - 4ac$ is negative, this quadratic equation has two complex but not real solutions.

7. $3x^2 - 2x - 2 = 0$

$a = 3,\ b = -2,\ c = -2$

$b^2 - 4ac = (-2)^2 - 4(3)(-2) = 4 + 24 = 28$

Since $b^2 - 4ac$ is positive, this quadratic equation has two real solutions.

8. $x^2 + (x+8)^2 = 20^2$

$x^2 + x^2 + 16x + 64 = 400$

$2x^2 + 16x - 336 = 0$

$x^2 + 8x - 168 = 0$

$a = 1,\ b = 8,\ c = -168$

$$x = \frac{-b \pm \sqrt{b^2 - 4ac}}{2a} = \frac{-8 \pm \sqrt{8^2 - 4(1)(-168)}}{2(1)} = \frac{-8 \pm \sqrt{64 + 672}}{2} = \frac{-8 \pm \sqrt{736}}{2} = \frac{-8 \pm 4\sqrt{46}}{2} = -4 \pm 2\sqrt{46}$$

The length of a side cannot be negative, so reject $-4 - 2\sqrt{46}$. Since $-4 + 2\sqrt{46}$ the walking distance on the sidewalk is $x + x + 8 = 10 + 10 + 8 = 28$ feet. The person saves $28 - 20 = 8$ feet.

9. $h = -16t^2 + 12t + 200$

$100 = -16t^2 + 12t + 200$

$0 = -16t^2 + 12t + 100$

$0 = 4t^2 - 3t - 25$

$a = 4,\ b = -3,\ c = -25$

$$t = \frac{-b \pm \sqrt{b^2 - 4ac}}{2a} = \frac{-(-3) \pm \sqrt{(-3)^2 - 4(4)(-25)}}{2(4)} = \frac{3 \pm \sqrt{9 + 400}}{8} = \frac{3 \pm \sqrt{409}}{8}$$

Since time won't be negative, reject $\frac{3 - \sqrt{409}}{8}$. The time it takes for the object to be 100 feet from the ground is $\frac{3+\sqrt{409}}{8} \approx 2.9$ seconds.

Calculator Explorations

1. $x(x-5)=8$
   $Y_1 = x(x-5),\ Y_2 = 8$

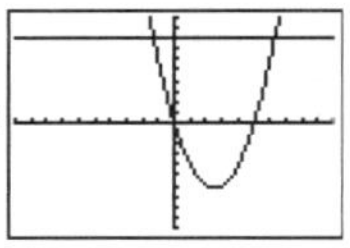

The solution set is $\{-1.27, 6.27\}$.

2. $x(x+2)=5$
   $Y_1 = x(x+2),\ Y_2 = 5$

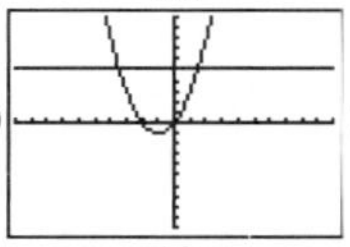

The solution set is $\{-3.45, 1.45\}$.

3. $x^2+0.5x = 0.3x+1$
   $Y_1 = x^2+0.5x,\ Y_2 = 0.3x+1$

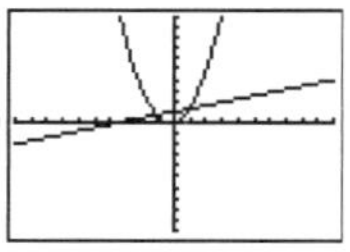

The solution set is $\{-1.10, 0.90\}$.

4. $x^2-2.6x = -2.2x+3$

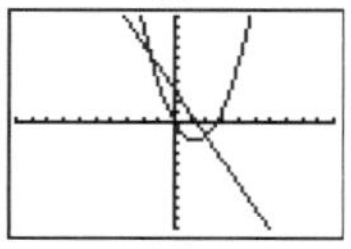

The solution set is $\{-1.54, 1.94\}$.

5. $(2x-5)^2 = -16$
   $Y_1 = (2x-5)^2,\ Y_2 = -16$

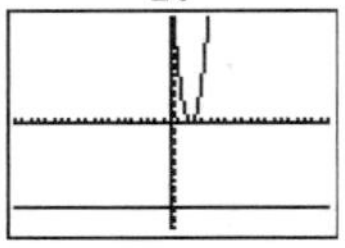

The solution set is $\varnothing$.

6. answers may vary

Mental Math

1. $x^2+3x+1=0$
   $a=1, b=3, c=1$

2. $2x^2-5x-7=0$
   $a=2, b=-5, c=-7$

3. $7x^2-4=0$
   $a=7, b=0, c=-4$

4. $x^2+9=0$
   $a=1, b=0, c=9$

5. $6x^2-x=0$
   $a=6, b=-1, c=0$

6. $5x^2+3x=0$
   $a=5, b=3, c=0$

Exercise Set 8.2

1. $m^2+5m-6=0$
   $a=1, b=5, c=-6$
   $$\begin{aligned} m &= \frac{-b\pm\sqrt{b^2-4ac}}{2a} \\ &= \frac{-5\pm\sqrt{5^2-4(1)(-6)}}{2(1)} \\ &= \frac{-5\pm\sqrt{25+24}}{2} \\ &= \frac{-5\pm\sqrt{49}}{2} \\ &= \frac{-5\pm 7}{2} \end{aligned}$$

$m = \frac{-5-7}{2} = -6$ or $m = \frac{-5+7}{2} = 1$
The solution set is $\{-6, 1\}$.

3. $2y = 5y^2 - 3$
$0 = 5y^2 - 2y - 3$
$a = 5, b = -2, c = -3$

$$y = \frac{-b \pm \sqrt{b^2 - 4ac}}{2a}$$
$$= \frac{-(-2) \pm \sqrt{(-2)^2 - 4(5)(-3)}}{2(5)}$$
$$= \frac{2 \pm \sqrt{4+60}}{10}$$
$$= \frac{2 \pm \sqrt{64}}{10}$$
$$= \frac{2 \pm 8}{10}$$

$y = \frac{2-8}{10} = -\frac{3}{5}$ or $y = \frac{2+8}{10} = 1$
The solution set is $\left\{-\frac{3}{5}, 1\right\}$.

5. $x^2 - 6x + 9 = 0$
$a = 1, b = -6, c = 9$

$$x = \frac{-b \pm \sqrt{b^2 - 4ac}}{2a}$$
$$= \frac{-(-6) \pm \sqrt{(-6)^2 - 4(1)(9)}}{2(1)}$$
$$= \frac{6 \pm \sqrt{36-36}}{2}$$
$$= \frac{6 \pm \sqrt{0}}{2}$$
$$= 3$$

The solution set is $\{3\}$.

7. $x^2 + 7x + 4 = 0$
$a = 1, b = 7, c = 4$

$$x = \frac{-b \pm \sqrt{b^2 - 4ac}}{2a}$$
$$= \frac{-7 \pm \sqrt{7^2 - 4(1)(4)}}{2(1)}$$
$$= \frac{-7 \pm \sqrt{49-16}}{2}$$
$$= \frac{-7 \pm \sqrt{33}}{2}$$

The solution set is $\left\{\frac{-7-\sqrt{33}}{2}, \frac{-7+\sqrt{33}}{2}\right\}$.

9. $8m^2 - 2m = 7$
$8m^2 - 2m - 7 = 0$
$a = 8, b = -2, c = -7$

$$m = \frac{-b \pm \sqrt{b^2 - 4ac}}{2a}$$
$$= \frac{-(-2) \pm \sqrt{(-2)^2 - 4(8)(-7)}}{2(8)}$$
$$= \frac{2 \pm \sqrt{4+224}}{16}$$
$$= \frac{2 \pm \sqrt{228}}{16}$$
$$= \frac{2 \pm 2\sqrt{57}}{16}$$
$$= \frac{2\left(1 \pm \sqrt{57}\right)}{16}$$
$$= \frac{1 \pm \sqrt{57}}{8}$$

The solution set is $\left\{\frac{1-\sqrt{57}}{8}, \frac{1+\sqrt{57}}{8}\right\}$.

11. $3m^2 - 7m = 3$

$3m^2 - 7m - 3 = 0$

$a = 3, b = -7, c = -3$

$$m = \frac{-b \pm \sqrt{b^2 - 4ac}}{2a}$$
$$= \frac{-(-7) \pm \sqrt{(-7)^2 - 4(3)(-3)}}{2(3)}$$
$$= \frac{7 \pm \sqrt{49+36}}{6}$$
$$= \frac{7 \pm \sqrt{85}}{6}$$

The solution set is $\left\{\frac{7-\sqrt{85}}{6}, \frac{7+\sqrt{85}}{6}\right\}$.

13. $\frac{1}{2}x^2 - x - 1 = 0$

$x^2 - 2x - 2 = 0$

$a = 1, b = -2, c = -2$

$$x = \frac{-b \pm \sqrt{b^2 - 4ac}}{2a}$$
$$= \frac{-(-2) \pm \sqrt{(-2)^2 - 4(1)(-2)}}{2(1)}$$
$$= \frac{2 \pm \sqrt{4+8}}{2}$$
$$= \frac{2 \pm \sqrt{12}}{2}$$
$$= \frac{2 \pm 2\sqrt{3}}{2}$$
$$= \frac{2\left(1 \pm \sqrt{3}\right)}{2}$$
$$= 1 \pm \sqrt{3}$$

The solution set is $\left\{1-\sqrt{3}, 1+\sqrt{3}\right\}$.

15. $\frac{2}{5}y^2 + \frac{1}{5}y = \frac{3}{5}$

$2y^2 + y = 3$

$2y^2 + y - 3 = 0$

$a = 2, b = 1, c = -3$

$$y = \frac{-b \pm \sqrt{b^2 - 4ac}}{2a}$$
$$= \frac{-1 \pm \sqrt{1^2 - 4(2)(-3)}}{2(2)}$$
$$= \frac{-1 \pm \sqrt{1+24}}{4}$$
$$= \frac{-1 \pm \sqrt{25}}{4}$$
$$= \frac{-1 \pm 5}{4}$$

$y = \frac{-1-5}{4} = -\frac{3}{2}$ or $y = \frac{-1+5}{4} = 1$

The solution set is $\left\{-\frac{3}{2}, 1\right\}$.

17. $\frac{1}{3}y^2 - y - \frac{1}{6} = 0$

$2y^2 - 6y - 1 = 0$

$a = 2, b = -6, c = -1$

$$y = \frac{-b \pm \sqrt{b^2 - 4ac}}{2a}$$
$$= \frac{-(-6) \pm \sqrt{(-6)^2 - 4(2)(-1)}}{2(2)}$$
$$= \frac{6 \pm \sqrt{36+8}}{4}$$
$$= \frac{6 \pm \sqrt{44}}{4}$$
$$= \frac{6 \pm 2\sqrt{11}}{4}$$
$$= \frac{2\left(3 + \sqrt{11}\right)}{4}$$
$$= \frac{3 \pm \sqrt{11}}{2}$$

The solution set is $\left\{\frac{3-\sqrt{11}}{2}, \frac{3+\sqrt{11}}{2}\right\}$.

19. $x^2 + 5x = -2$

$$x^2 + 5x + 2 = 0$$

$a = 1$, $b = 5$, $c = 2$

$$\begin{aligned} x &= \frac{-b \pm \sqrt{b^2 - 4ac}}{2a} \\ &= \frac{-5 \pm \sqrt{5^2 - 4(1)(2)}}{2(1)} \\ &= \frac{-5 \pm \sqrt{25 - 8}}{2} \\ &= \frac{-5 \pm \sqrt{17}}{2} \end{aligned}$$

The solution set is $\left\{\frac{-5 - \sqrt{17}}{2}, \frac{-5 + \sqrt{17}}{2}\right\}$.

21. $(m + 2)(2m - 6) = 5(m - 1) - 12$

$$2m^2 - 2m - 12 = 5m - 5 - 12$$

$$2m^2 - 7m + 5 = 0$$

$a = 2$, $b = -7$, $c = 5$

$$\begin{aligned} m &= \frac{-b \pm \sqrt{b^2 - 4ac}}{2a} \\ &= \frac{-(-7) \pm \sqrt{(-7)^2 - 4(2)(5)}}{2(2)} \\ &= \frac{7 \pm \sqrt{49 - 40}}{4} \\ &= \frac{7 \pm \sqrt{9}}{4} \\ &= \frac{7 \pm 3}{4} \end{aligned}$$

$m = \frac{7 - 3}{4} = 1$ or $m = \frac{7 + 3}{4} = \frac{10}{4} = \frac{5}{2}$

The solution set is $\left\{1, \frac{5}{2}\right\}$.

23. $x^2 + 6x + 13 = 0$

$a = 1$, $b = 6$, $c = 13$

$$\begin{aligned} x &= \frac{-b \pm \sqrt{b^2 - 4ac}}{2a} \\ &= \frac{-6 \pm \sqrt{6^2 - 4(1)(13)}}{2(1)} \\ &= \frac{-6 \pm \sqrt{36 - 52}}{2} \\ &= \frac{-6 \pm \sqrt{-16}}{2} \\ &= \frac{-6 \pm 4i}{2} \\ &= \frac{2(-3 \pm 2i)}{2} \\ &= -3 \pm 2i \end{aligned}$$

The solution set is $\{-3 - 2i, -3 - 2i\}$.

25. $(x + 5)(x - 1) = 2$

$$x^2 + 4x - 5 = 2$$

$$x^2 + 4x - 7 = 0$$

$a = 1$, $b = 4$, $c = -7$

$$\begin{aligned} x &= \frac{-b \pm \sqrt{b^2 - 4ac}}{2a} \\ &= \frac{-4 \pm \sqrt{4^2 - 4(1)(-7)}}{2(1)} \\ &= \frac{-4 \pm \sqrt{16 + 28}}{2} \\ &= \frac{-4 \pm \sqrt{44}}{2} \\ &= \frac{-4 \pm 2\sqrt{11}}{2} \\ &= \frac{2\left(-2 \pm \sqrt{11}\right)}{2} \\ &= -2 \pm \sqrt{11} \end{aligned}$$

The solution set is $\left\{-2 - \sqrt{11}, -2 + \sqrt{11}\right\}$.

27. $6 = -4x^2 + 3x$

$$4x^2 - 3x + 6 = 0$$

$a = 4$, $b = -3$, $c = 6$

$$\begin{aligned} x &= \frac{-b \pm \sqrt{b^2 - 4ac}}{2a} \\ &= \frac{-(-3) \pm \sqrt{(-3)^2 - 4(4)(6)}}{2(4)} \\ &= \frac{3 \pm \sqrt{9-96}}{8} \\ &= \frac{3 \pm \sqrt{-87}}{8} \\ &= \frac{3 \pm i\sqrt{87}}{8} \end{aligned}$$

The solution set is $\left\{\frac{3 - i\sqrt{87}}{8}, \frac{3 + i\sqrt{87}}{8}\right\}$.

29. $\frac{x^2}{3} - x = \frac{5}{3}$

$$\frac{x^2}{3} - x - \frac{5}{3} = 0$$

$$x^2 - 3x - 5 = 0$$

$a = 1$, $b = -3$, $c = -5$

$$\begin{aligned} x &= \frac{-b \pm \sqrt{b^2 - 4ac}}{2a} \\ &= \frac{-(-3) \pm \sqrt{(-3)^2 - 4(1)(-5)}}{2(1)} \\ &= \frac{3 \pm \sqrt{9+20}}{2} \\ &= \frac{3 \pm \sqrt{29}}{2} \end{aligned}$$

The solution set is $\left\{\frac{3 - \sqrt{29}}{2}, \frac{3 + \sqrt{29}}{2}\right\}$.

31. $10y^2 + 10y + 3 = 0$

$a = 10$, $b = 10$, $c = 3$

$$\begin{aligned} y &= \frac{-b \pm \sqrt{b^2 - 4ac}}{2a} \\ &= \frac{-10 \pm \sqrt{10^2 - 4(10)(3)}}{2(10)} \\ &= \frac{-10 \pm \sqrt{100-120}}{20} \\ &= \frac{-10 \pm \sqrt{-20}}{20} \\ &= \frac{-10 \pm 2i\sqrt{5}}{20} \\ &= \frac{2\left(-5 \pm i\sqrt{5}\right)}{2 \cdot 10} \\ &= \frac{-5 \pm i\sqrt{5}}{10} \end{aligned}$$

The solution set is $\left\{\frac{-5 - i\sqrt{5}}{10}, \frac{-5 + i\sqrt{5}}{10}\right\}$.

33. $x(6x + 2) = 3$

$$6x^2 + 2x = 3$$

$$6x^2 + 2x - 3 = 0$$

$a = 6$, $b = 2$, $c = -3$

$$\begin{aligned} x &= \frac{-b \pm \sqrt{b^2 - 4ac}}{2a} \\ &= \frac{-2 \pm \sqrt{2^2 - 4(6)(-3)}}{2(6)} \\ &= \frac{-2 \pm \sqrt{4+72}}{12} \\ &= \frac{-2 \pm \sqrt{76}}{12} \\ &= \frac{-2 \pm 2\sqrt{19}}{12} \\ &= \frac{2\left(-1 \pm \sqrt{19}\right)}{12} \\ &= \frac{-1 \pm \sqrt{19}}{6} \end{aligned}$$

The solution set is $\left\{\frac{-1 - \sqrt{19}}{6}, \frac{-1 + \sqrt{19}}{6}\right\}$.

35. $\frac{2}{5}y^2+\frac{1}{5}y+\frac{3}{5}=0$

$2y^2+y+3=0$

$a=2,\ b=1,\ c=3$

$$y=\frac{-b\pm\sqrt{b^2-4ac}}{2a}=\frac{-1\pm\sqrt{1^2-4(2)(3)}}{2(2)}=\frac{-1\pm\sqrt{1-24}}{4}=\frac{-1\pm\sqrt{-23}}{4}=\frac{-1\pm i\sqrt{23}}{4}$$

The solution set is $\left\{\frac{-1-i\sqrt{23}}{4}, \frac{-1+i\sqrt{23}}{4}\right\}$.

37. $\frac{1}{2}y^2=y-\frac{1}{2}$

$\frac{1}{2}y^2-y+\frac{1}{2}=0$

$y^2-2y+1=0$

$a=1,\ b=-2,\ c=1$

$$y=\frac{-b\pm\sqrt{b^2-4ac}}{2a}=\frac{-(-2)\pm\sqrt{(-2)^2-4(1)(1)}}{2(1)}=\frac{2\pm\sqrt{4-4}}{2}=\frac{2\pm\sqrt{0}}{2}=1$$

The solution set is $\{1\}$.

39. $(n-2)^2=2n$

$n^2-4n+4=2n$

$n^2-6n+4=0$

$a=1,\ b=-6,\ c=4$

$$n=\frac{-b\pm\sqrt{b^2-4ac}}{2a}=\frac{-(-6)\pm\sqrt{(-6)^2-4(1)(4)}}{2(1)}=\frac{6\pm\sqrt{36-16}}{2}=\frac{6\pm\sqrt{20}}{2}=\frac{6\pm2\sqrt{5}}{2}=\frac{2\left(3\pm\sqrt{5}\right)}{2}=3\pm\sqrt{5}$$

The solution set is $\left\{3-\sqrt{5}, 3+\sqrt{5}\right\}$.

41. $x^2-5=0$

$x^2+0x-5=0$

$a=1,\ b=0,\ c=-5$

$b^2-4ac=0^2-4(1)(-5)=0+20=20$

Since b^2-4ac is positive, the equation has two real solutions.

43. $4x^2+12x=-9$

$4x^2+12x+9=0$

$a=4,\ b=12,\ c=9$

$b^2-4ac=12^2-4(4)(9)=144-144=0$

Since $b^2-4ac=0$, the equation has one real solution.

45. $3x=-2x^2+7$

$0=-2x^2-3x+7$

$a=-2,\ b=-3,\ c=7$

$b^2-4ac=(-3)^2-4(-2)(7)=9+56=65$

Since b^2-4ac is positive, the equation has two real solutions.

47. $6 = 4x - 5x^2$

$5x^2 - 4x + 6 = 0$

$a = 5, b = -4, c = 6$

$b^2 - 4ac = (-4)^2 - 4(5)(6) = 16 - 120 = -104$

Since $b^2 - 4ac$ is negative, the equation has two complex but not real solutions.

49. $9x - 2x^2 + 5 = 0$

$-2x^2 + 9x + 5 = 0$

$a = -2, b = 9, c = 5$

$b^2 - 4ac = 9^2 - 4(-2)(5) = 81 + 40 = 121$

Since $b^2 - 4ac$ is positive, the equation has two real solutions.

51. $x^2 + (x+8)^2 = (36)^2$

$x^2 + x^2 + 16x + 64 = 1296$

$2x^2 + 16x - 1232 = 0$

$x^2 + 8x - 616 = 0$

$a = 1, b = 8, c = -616$

$$x = \frac{-b \pm \sqrt{b^2 - 4ac}}{2a} = \frac{-8 \pm \sqrt{8^2 - 4(1)(-616)}}{2(1)} = \frac{-8 \pm \sqrt{64 + 2464}}{2} = \frac{-8 \pm \sqrt{2528}}{2}$$

Since the length cannot be negative, reject $\frac{-8 - \sqrt{2528}}{2}$. Therefore

$$x = \frac{-8 + \sqrt{2528}}{2} \approx 21 \text{ feet.}$$

The legs have a total length of $x + (x + 8) = 21 + (21 + 8) = 50$ feet.

They save a distance of $50 - 36 = 14$ feet.

53. Let x be the length of the legs, then the length of the hypotenuse is $x + 2$.

$x^2 + x^2 = (x+2)^2$

$2x^2 = x^2 + 4x + 4$

$x^2 - 4x - 4 = 0$

$a = 1, b = -4, c = -4$

$$x = \frac{-b \pm \sqrt{b^2 - 4ac}}{2a} = \frac{-(-4) \pm \sqrt{(-4)^2 - 4(1)(-4)}}{2(1)} = \frac{4 \pm \sqrt{16 + 16}}{2} = \frac{4 \pm \sqrt{32}}{2} = \frac{4 \pm 4\sqrt{2}}{2} = \frac{2\left(2 \pm 2\sqrt{2}\right)}{2} = 2 \pm 2\sqrt{2}$$

Since the lengths cannot be negative, reject $2 - 2\sqrt{2}$.

$x + 2 = \left(2 + 2\sqrt{2}\right) + 2 = 4 + 2\sqrt{2}$

The legs both have length $\left(2 + 2\sqrt{2}\right)$ centimeters, and the hypotenuse has length $\left(4 + 2\sqrt{2}\right)$ centimeters.

55. Let x be the width. Then $x + 10$ is the length.

$x(x+10) = 400$

$x^2 + 10x - 400 = 0$

$a = 1, b = 10, c = -400$

$$x = \frac{-b \pm \sqrt{b^2 - 4ac}}{2a}$$
$$= \frac{-10 \pm \sqrt{10^2 - 4(1)(-400)}}{2(1)}$$
$$= \frac{-10 \pm \sqrt{100 + 1600}}{2}$$
$$= \frac{-10 \pm \sqrt{1700}}{2}$$
$$= \frac{-10 \pm 10\sqrt{17}}{2}$$
$$= -5 \pm 5\sqrt{17}$$

Since the length cannot be negative, reject $-5 - 5\sqrt{17}$.

$$x + 10 = \left(-5 + 5\sqrt{17}\right) + 10 = 5 + 5\sqrt{17}$$

The width is $\left(-5 + 5\sqrt{17}\right)$ feet and the length is $\left(5 + 5\sqrt{17}\right)$ feet.

57. a. Let x be the length of the sides.

$$x^2 + x^2 = 100^2$$
$$2x^2 = 10{,}000$$
$$x^2 = 5000$$
$$x^2 - 5000 = 0$$

$a = 1$, $b = 0$, $c = -5000$

$$x = \frac{-b \pm \sqrt{b^2 - 4ac}}{2a}$$
$$= \frac{-0 \pm \sqrt{0^2 - 4(1)(-5000)}}{2(1)}$$
$$= \frac{\pm\sqrt{20{,}000}}{2}$$
$$= \frac{\pm 100\sqrt{2}}{2}$$
$$= \pm 50\sqrt{2}$$

Since the length cannot be negative, reject $-50\sqrt{2}$. The sides of the pen are $50\sqrt{2}$ meters long.

b. $A = s^2$

$$= \left(50\sqrt{2}\right)^2$$
$$= 50^2\left(\sqrt{2}\right)^2$$
$$= 2500(2)$$
$$= 5000$$

The area of the pen is 5000 square meters.

59. Let x be the width. Then $x + 1.1$ is the height.

$$x(x + 1.1) = 1439.9$$
$$x^2 + 1.1x - 1439.9 = 0$$

$a = 1$, $b = 1.1$, $c = -1439.9$

$$x = \frac{-b \pm \sqrt{b^2 - 4ac}}{2a}$$
$$= \frac{-1.1 \pm \sqrt{(1.1)^2 - 4(1)(-1439.9)}}{2(1)}$$
$$= \frac{-1.1 \pm \sqrt{5760.81}}{2}$$

Since the length cannot be negative, reject $\frac{-1.1 - \sqrt{5760.81}}{2}$. The width is $\frac{-1.1 + \sqrt{5760.81}}{2} = 37.4$ feet and the height is $37.4 + 1.1 = 38.5$ feet.

61. Let h be the height of the triangle. Then the base is $2h + 4$.

$$A = \frac{1}{2}bh$$
$$42 = \frac{1}{2}(2h + 4) \cdot h$$
$$42 = h^2 + 2h$$
$$0 = h^2 + 2h - 42$$

$a = 1$, $b = 2$, $c = -42$

$$h = \frac{-b \pm \sqrt{b^2 - 4ac}}{2a}$$
$$= \frac{-2 \pm \sqrt{2^2 - 4(1)(-42)}}{2(1)}$$
$$= \frac{-2 \pm \sqrt{4+168}}{2}$$
$$= \frac{-2 \pm \sqrt{172}}{2}$$
$$= \frac{-2 \pm 2\sqrt{43}}{2}$$
$$= \frac{2\left(-1 \pm \sqrt{43}\right)}{2}$$
$$= -1 \pm \sqrt{43}$$

Since the height cannot be negative, reject $-1-\sqrt{43}$.

$$2h + 4 = 2\left(-1+\sqrt{43}\right) + 4$$
$$= -2 + 2\sqrt{43} + 4$$
$$= 2 + 2\sqrt{43}$$

The height of the triangle is $\left(-1+\sqrt{43}\right)$ centimeters and the base is $\left(2+2\sqrt{43}\right)$ centimeters.

63. Find t when $h = 0$.

$$h = -16t^2 + 20t + 1100$$
$$0 = -16t^2 + 20t + 1100$$
$$0 = 4t^2 - 5t - 275$$
$$a = 4,\ b = -5,\ c = -275$$
$$t = \frac{-b \pm \sqrt{b^2 - 4ac}}{2a}$$
$$= \frac{-(-5) \pm \sqrt{(-5)^2 - 4(4)(-275)}}{2(4)}$$
$$= \frac{5 \pm \sqrt{25+4400}}{8}$$
$$= \frac{5 \pm \sqrt{4425}}{8}$$

Since the time cannot be negative, reject $\frac{5-\sqrt{4425}}{8}$. The pebble will hit the ground $\frac{5+\sqrt{4425}}{8} \approx 8.9$ seconds after it is thrown.

65. Find t when $h = 0$.

$$h = -16t^2 - 20t + 180$$
$$0 = -16t^2 - 20t + 180$$
$$0 = -4t^2 - 5t + 45$$
$$a = -4,\ b = -5,\ c = 45$$
$$t = \frac{-b \pm \sqrt{b^2 - 4ac}}{2a}$$
$$= \frac{-(-5) \pm \sqrt{(-5)^2 - 4(-4)(45)}}{2(-4)}$$
$$= \frac{5 \pm \sqrt{25+720}}{-8}$$
$$= \frac{5 \pm \sqrt{745}}{-8}$$
$$= \frac{-5 \pm \sqrt{745}}{8}$$

Since the time cannot be negative, reject $\frac{-5-\sqrt{745}}{8}$. The ball strikes the ground $\frac{-5+\sqrt{745}}{8} \approx 2.8$ seconds after it is thrown.

67.
$$\sqrt{5x-2} = 3$$
$$\left(\sqrt{5x-2}\right)^2 = 3^2$$
$$5x - 2 = 9$$
$$5x = 11$$
$$x = \frac{11}{5}$$

The solution set is $\left\{\frac{11}{5}\right\}$.

69. The LCD is $5x$.

$$\frac{1}{x}+\frac{2}{5}=\frac{7}{x}$$
$$5x\left(\frac{1}{x}\right)+5x\left(\frac{2}{5}\right)=5x\left(\frac{7}{x}\right)$$
$$5+2x=35$$
$$2x=30$$
$$x=15$$

The solution set is {15}.

71. Let $y=x^2$.

$x^4+x^2-20=y^2+y-20$

$5(-4)=-20$ and $5-4=1$

$y^2+y-20=(y+5)(y-4)$

Replace y with x^2.

$$(y+5)(y-4)=(x^2+5)(x^2-4)$$
$$=(x^2+5)(x+2)(x-2)$$

$x^4+x^2-20=(x^2+5)(x+2)(x-2)$

73.
$$z^4-13z^2+36=(z^2-9)(z^2-4)$$
$$=(z^2-3^2)(z^2-2^2)$$
$$=(z+3)(z-3)(z+2)(z-2)$$

75.
$$x^2=-10$$
$$x^2+10=0$$
$$x^2+0x+10=0$$

$a=1$, $b=0$, $c=10$

The correct choice is b.

77. answers may vary

79. $2x^2-6x+3=0$

$a=2$, $b=-6$, $c=3$

$$x=\frac{-b\pm\sqrt{b^2-4ac}}{2a}$$
$$=\frac{-(-6)\pm\sqrt{(-6)^2-4(2)(3)}}{2(2)}$$
$$=\frac{6\pm\sqrt{36-24}}{4}$$
$$=\frac{6\pm\sqrt{12}}{4}$$
$$=\frac{6\pm2\sqrt{3}}{4}$$
$$=\frac{3\pm\sqrt{3}}{2}$$

$x=\frac{3-\sqrt{3}}{2}\approx0.6$ or $x=\frac{3+\sqrt{3}}{2}\approx2.4$

The solution set is {0.6, 2.4}.

81. The greatest decrease occurs from Sunday to Monday.

83. Wednesday has the lowest temperature.

85. $f(x)=3x^2-18x+57$

Thursday is 4 days from Sunday.

$$f(4)=3(4)^2-18(4)+57$$
$$=48-72+57$$
$$=33$$

The temperature was approximately 33°F on Thursday; answers may vary.

87. **a.** 2004 is 3 years after 2001, so $x=3$.

$f(x)=213x^2-66.4x+4689$

$f(3)=213(3)^2-66.4(3)+4689=6406.8$

The net earnings in 2004 were \$6406.8 million.

b. Find x when $f(x) = 9682$.

$$f(x) = 213x^2 - 66.4x + 4689$$
$$9682 = 213x^2 - 66.4x + 4689$$
$$0 = 213x^2 - 66.4x - 4993$$

$a = 213$, $b = -66.4$, $c = -4993$

$$x = \frac{-b \pm \sqrt{b^2 - 4ac}}{2a}$$
$$= \frac{-(-66.4) \pm \sqrt{(-66.4)^2 - 4(213)(-4993)}}{2(213)}$$
$$= \frac{66.4 \pm \sqrt{4,258,444.96}}{426}$$
$$= \frac{66.4 \pm 2063.6}{426}$$

$x \approx -4.7$ or $x = 5$

Reject −4.7 since time cannot be negative.

2001 + 5 = 2006

The net earnings will be \$9682 million in 2006.

89. a. 2001 is 1 year after 2000, so $x = 1$.

$$y = -13x^2 + 221x + 8476$$
$$y = -13(1)^2 + 221(1) + 8476 = 8684$$

The average total daily supply of motor gasoline in 2001 was 8684 thousand barrels per day.

b. Find x when $y = 9334$.

$$y = -13x^2 + 221x + 8476$$
$$9334 = -13x^2 + 221x + 8476$$
$$0 = -13x^2 + 221x - 858$$
$$0 = x^2 - 17x + 66$$

$a = 1$, $b = -17$, $c = 66$

$$x = \frac{-b \pm \sqrt{b^2 - 4ac}}{2a}$$
$$= \frac{-(-17) \pm \sqrt{(-17)^2 - 4(1)(66)}}{2(1)}$$
$$= \frac{17 \pm \sqrt{289 - 264}}{2}$$
$$= \frac{17 \pm \sqrt{25}}{2}$$
$$= \frac{17 \pm 5}{2}$$

$x = \dfrac{17-5}{2} = 6$ or $x = \dfrac{17+5}{6} = 11$

The average total daily supply of motor gasoline will be 9334 thousand barrels per day 6 or 11 years after 2000, so in 2006 or 2011.

Section 8.3 Practice Problems

1. $$\begin{aligned} x - \sqrt{x-1} - 3 &= 0 \\ x - 3 &= \sqrt{x-1} \\ x^2 - 6x + 9 &= x - 1 \\ x^2 - 7x + 10 &= 0 \\ (x-5)(x-2) &= 0 \end{aligned}$$

$x - 5 = 0$ or $x - 2 = 0$

$x = 5$ $\quad$ $x = 2$

The value 2 does not check, so the solution set is {5}.

2. $$\begin{aligned} \frac{2x}{x-1} - \frac{x+2}{x} &= \frac{5}{x(x-1)} \\ x(x-1)\left(\frac{2x}{x-1}\right) - x(x-1)\left(\frac{x+2}{x}\right) &= x(x-1)\left[\frac{5}{x(x-1)}\right] \\ 2x^2 - (x-1)(x+2) &= 5 \\ 2x^2 - (x^2 + x - 2) &= 5 \\ 2x^2 - x^2 - x + 2 &= 5 \\ x^2 - x - 3 &= 0 \end{aligned}$$

$a = 1$, $b = -1$, $c = -3$

$$\begin{aligned} x &= \frac{-b \pm \sqrt{b^2 - 4ac}}{2a} \\ &= \frac{-(-1) \pm \sqrt{(-1)^2 - 4(1)(-3)}}{2(1)} \\ &= \frac{1 \pm \sqrt{13}}{2} \end{aligned}$$

The solution set is $\left\{\dfrac{1-\sqrt{13}}{2}, \dfrac{1+\sqrt{13}}{2}\right\}$.

3.
$$x^4 - 5x^2 - 36 = 0$$
$$(x^2 - 9)(x^2 + 4) = 0$$
$$(x + 3)(x - 3)(x^2 + 4) = 0$$
$x + 3 = 0$ or $x - 3 = 0$ or $x^2 + 4 = 0$
$x = -3$ $\quad x = 3$ $\quad x^2 = -4$
$x = \pm\sqrt{-4} = \pm 2i$

The solution set is $\{-3, 3, -2i, 2i\}$.

4. Let $u = x + 4$.
$$(x + 4)^2 - (x + 4) - 6 = 0$$
$$u^2 - u - 6 = 0$$
$$(u - 3)(u + 2) = 0$$
$u - 3 = 0$ or $u + 2 = 0$
$u = 3$ $\quad u = -2$

Replace u with $x + 4$
$x + 4 = 3$ or $x + 4 = -2$
$x = -1$ $\quad x = -6$

The solution set is $\{-1, -6\}$.

5. Let $m = x^{1/3}$.
$$x^{2/3} - 7x^{1/3} + 10 = 0$$
$$m^2 - 7m + 10 = 0$$
$$(m - 5)(m - 2) = 0$$
$m - 5 = 0$ or $m - 2 = 0$
$m = 5$ $\quad m = 2$

Replace m with $x^{1/3}$.
$x^{1/3} = 5$ or $x^{1/3} = 2$
$x = 5^3 = 125$ $\quad x = 2^3 = 8$

The solution set is $\{8, 125\}$.

6. Let x be the amount of time it takes Doug to complete the job alone. Then $x - 1$ is the amount of time it takes Karen to complete the job alone.

	Total hours	Part of job in 1 hour
Doug	x	$\frac{1}{x}$
Karen	$x - 1$	$\frac{1}{x-1}$
Together	5	$\frac{1}{5}$

$$\frac{1}{x}+\frac{1}{x-1}=\frac{1}{5}$$
$$5x(x-1)\left(\frac{1}{x}+\frac{1}{x-1}\right)=5x(x-1)\left(\frac{1}{5}\right)$$
$$5(x-1)+5x=x(x-1)$$
$$5x-5+5x=x^2-x$$
$$0=x^2-11x+5$$

$a=1, b=-11, c=5$

$$x=\frac{-b\pm\sqrt{b^2-4ac}}{2a}$$
$$=\frac{-(-11)\pm\sqrt{(-11)^2-4(1)(5)}}{2(1)}$$
$$=\frac{11\pm\sqrt{101}}{2}$$

$x\approx 0.5$ or $x\approx 10.5$

Reject 0.5, since $0.5-1=-0.5$. It takes Doug 10.5 hours and Karen 9.5 hours for each to complete the job alone.

7. Let x be the time to the beach.

	Distance =	Rate	· Time
To	500	x	$\frac{500}{x}$
From	500	$x+10$	$\frac{500}{x+10}$

$$\frac{500}{x}+\frac{500}{x+10}=18\frac{1}{3}$$
$$\frac{500}{x}+\frac{500}{x+10}=\frac{55}{3}$$
$$3x(x+10)\left(\frac{500}{x}+\frac{500}{x+10}\right)=3x(x+10)\left(\frac{55}{3}\right)$$
$$3x(x+10)\frac{500}{x}+3x(x+10)\frac{500}{x+10}=3x(x+10)\left(\frac{55}{3}\right)$$
$$3(x+10)500+3x(500)=x(x+10)(55)$$
$$1500x+15,000+1500x=55x^2+550x$$
$$0=55x^2-2450x-15,000$$

$a=55, b=-2450, c=-15,000$

$$x = \frac{-b \pm \sqrt{b^2 - 4ac}}{2a}$$
$$= \frac{-(-2450) \pm \sqrt{(-2450)^2 - 4(55)(-15,000)}}{2(55)}$$
$$= \frac{2450 \pm \sqrt{9,302,500}}{110}$$
$$= \frac{2450 \pm 3050}{110}$$
$x \approx -5.5$ or $x = 50$

Reject –5.5 since speed cannot be negative. They drove 50 mph to the beach and 50 + 10 = 60 mph on their return.

Exercise Set 8.3

1.
$$2x = \sqrt{10+3x}$$
$$(2x)^2 = \left(\sqrt{10+3x}\right)^2$$
$$4x^2 = 10+3x$$
$$4x^2 - 3x - 10 = 0$$
$$(4x+5)(x-2) = 0$$
$4x+5=0$ or $x-2=0$

$4x = -5$

$x = -\frac{5}{4}$ $\quad x = 2$

The value $-\frac{5}{4}$ does not check, so the solution set is $\{2\}$.

3.
$$x - 2\sqrt{x} = 8$$
$$x - 8 = 2\sqrt{x}$$
$$(x-8)^2 = \left(2\sqrt{x}\right)^2$$
$$x^2 - 16x + 64 = 4x$$
$$x^2 - 20x + 64 = 0$$
$$(x-16)(x-4) = 0$$
$x-16=0$ or $x-4=0$

$x = 16$ $\quad x = 4$

The value 4 does not check, so the solution set is $\{16\}$.

5.
$$\sqrt{9x} = x + 2$$
$$\left(\sqrt{9x}\right)^2 = (x+2)^2$$
$$9x = x^2 + 4x + 4$$
$$0 = x^2 - 5x + 4$$
$$0 = (x-1)(x-4)$$
$x - 1 = 0$ or $x - 4 = 0$

$x = 1$ $\quad$ $x = 4$

The solution set is $\{1, 4\}$.

7. The LCD is $x(x - 1)$.
$$\frac{2}{x} + \frac{3}{x-1} = 1$$
$$x(x-1)\left(\frac{2}{x}\right) + x(x-1)\left(\frac{3}{x-1}\right) = x(x-1)(1)$$
$$2(x-1) + 3x = x(x-1)$$
$$2x - 2 + 3x = x^2 - x$$
$$0 = x^2 - 6x + 2$$
$a = 1$, $b = -6$, $c = 2$
$$x = \frac{-b \pm \sqrt{b^2 - 4ac}}{2a}$$
$$= \frac{-(-6) \pm \sqrt{(-6)^2 - 4(1)(2)}}{2(1)}$$
$$= \frac{6 \pm \sqrt{36 - 8}}{2}$$
$$= \frac{6 \pm \sqrt{28}}{2}$$
$$= \frac{6 \pm 2\sqrt{7}}{2}$$
$$= 3 \pm \sqrt{7}$$
Neither value will make a denominator 0, so the solution set is $\left\{3 - \sqrt{7}, 3 + \sqrt{7}\right\}$.

9. The LCD is $x(x + 2)$.
$$\frac{3}{x} + \frac{4}{x+2} = 2$$
$$x(x+2)\left(\frac{3}{x}\right) + x(x+2)\left(\frac{4}{x+2}\right) = x(x+2)(2)$$
$$3(x+2) + 4x = 2x(x+2)$$
$$3x + 6 + 4x = 2x^2 + 4x$$
$$0 = 2x^2 - 3x - 6$$
$a = 2$, $b = -3$, $c = -6$

$$x=\frac{-b\pm\sqrt{b^2-4ac}}{2a}$$
$$=\frac{-(-3)\pm\sqrt{(-3)^2-4(2)(-6)}}{2(2)}$$
$$=\frac{3\pm\sqrt{9+48}}{4}$$
$$=\frac{3\pm\sqrt{57}}{4}$$

Neither value will make a denominator 0, so the solution set is $\left\{\frac{3-\sqrt{57}}{4}, \frac{3+\sqrt{57}}{4}\right\}$.

11. $x^2-5x+6=(x-3)(x-2)$

The LCD is $(x-3)(x-2)$.

$$\frac{7}{x^2-5x+6}=\frac{2x}{x-3}-\frac{x}{x-2}$$
$$(x-3)(x-2)\left[\frac{7}{(x-3)(x-2)}\right]=(x-3)(x-2)\left(\frac{2x}{x-3}\right)-(x-3)(x-2)\left(\frac{x}{x-2}\right)$$
$$7=2x(x-2)-x(x-3)$$
$$7=2x^2-4x-x^2+3x$$
$$0=x^2-x-7$$

$a=1, b=-1, c=-7$

$$x=\frac{-b\pm\sqrt{b^2-4ac}}{2a}$$
$$=\frac{-(-1)\pm\sqrt{(-1)^2-4(1)(-7)}}{2(1)}$$
$$=\frac{1\pm\sqrt{1+28}}{2}$$
$$=\frac{1\pm\sqrt{29}}{2}$$

Neither value will make a denominator 0, so the solution set is $\left\{\frac{1-\sqrt{29}}{2}, \frac{1+\sqrt{29}}{2}\right\}$.

13.

$$p^4 - 16 = 0$$
$$(p^2 + 4)(p^2 - 4) = 0$$
$$(p^2 + 4)(p + 2)(p - 2) = 0$$
$$p^2 + 4 = 0 \quad \text{or} \quad p + 2 = 0 \quad \text{or} \quad p - 2 = 0$$
$$p^2 = -4$$
$$p = \pm\sqrt{-4} = \pm 2i \qquad p = -2 \qquad p = 2$$

The solution set is $\{-2, 2, -2i, 2i\}$.

15.

$$z^4 - 13z^2 + 36 = 0$$
$$(z^2 - 4)(z^2 - 9) = 0$$
$$(z + 2)(z - 2)(z + 3)(z - 3) = 0$$
$$z + 2 = 0 \quad \text{or} \quad z - 2 = 0 \quad \text{or} \quad z + 3 = 0 \quad \text{or} \quad z - 3 = 0$$
$$z = -2 \qquad z = 2 \qquad z = -3 \qquad z = 3$$

The solution set is $\{-3, 3, -2, 2\}$.

17.

$$4x^4 + 11x^2 = 3$$
$$4x^4 + 11x^2 - 3 = 0$$
$$(4x^2 - 1)(x^2 + 3) = 0$$
$$(2x + 1)(2x - 1)(x^2 + 3) = 0$$
$$2x + 1 = 0 \quad \text{or} \quad 2x - 1 = 0 \quad \text{or} \quad x^2 + 3 = 0$$
$$2x = -1 \qquad 2x = 1 \qquad x^2 = -3$$
$$x = -\frac{1}{2} \qquad x = \frac{1}{2} \qquad x = \pm\sqrt{-3} = \pm i\sqrt{3}$$

The solution set is $\left\{-\frac{1}{2}, \frac{1}{2}, -i\sqrt{3}, i\sqrt{3}\right\}$.

19. Let $y = x^{1/3}$.

$$x^{2/3} - 3x^{1/3} - 10 = 0$$
$$y^2 - 3y - 10 = 0$$
$$(y - 5)(y + 2) = 0$$
$$y - 5 = 0 \quad \text{or} \quad y + 2 = 0$$
$$y = 5 \qquad y = -2$$
$$x^{1/3} = 5 \qquad x^{1/3} = -2$$
$$x = 5^3 \qquad x = (-2)^3$$
$$x = 125 \qquad x = -8$$

The solution set is $\{125, -8\}$.

21. Let $y = 5n + 1$.

$$(5n+1)^2 + 2(5n+1) - 3 = 0$$
$$y^2 + 2y - 3 = 0$$
$$(y+3)(y-1) = 0$$

$y+3=0$ or $y-1=0$

$y=-3$ $\quad$ $y=1$

$5n+1=-3$ $\quad$ $5n+1=1$

$5n=-4$ $\quad$ $5n=0$

$n=-\frac{4}{5}$ $\quad$ $n=0$

The solution set is $\left\{-\frac{4}{5}, 0\right\}$.

23. Let $y = x^{1/3}$.

$$2x^{2/3} - 5x^{1/3} = 3$$
$$2x^{2/3} - 5x^{1/3} - 3 = 0$$
$$2y^2 - 5y - 3 = 0$$
$$(2y+1)(y-3) = 0$$

$2y+1=0$ or $y-3=0$

$2y=-1$ $\quad$ $y=3$

$y=-\frac{1}{2}$ $\quad$ $x^{1/3}=3$

$x^{1/3}=-\frac{1}{2}$ $\quad$ $x=3^3$

$x=\left(-\frac{1}{2}\right)^3$ $\quad$ $x=27$

$x=-\frac{1}{8}$

The solution set is $\left\{-\frac{1}{8}, 27\right\}$.

25. Let $x = \frac{1}{3t-2}$.

$$1 + \frac{2}{3t-2} = \frac{8}{(3t-2)^2}$$
$$1 + 2x = 8x^2$$
$$0 = 8x^2 - 2x - 1$$
$$0 = (2x-1)(4x+1)$$

$2x-1=0$ or $4x+1=0$

$2x=1$ $\quad$ $4x=-1$

$x=\frac{1}{2}$ $\quad$ $x=-\frac{1}{4}$

$\frac{1}{3t-2}=\frac{1}{2}$ $\quad$ $\frac{1}{3t-2}=-\frac{1}{4}$

$2=3t-2$ $\quad$ $-4=3t-2$

$4=3t$ $\quad$ $-2=3t$

$\frac{4}{3}=t$ $\quad$ $-\frac{2}{3}=t$

The solution set is $\left\{-\frac{2}{3}, \frac{4}{3}\right\}$.

27. Let $y = x^{1/3}$.

$$20x^{2/3} - 6x^{1/3} - 2 = 0$$
$$20y^2 - 6y - 2 = 0$$
$$2(10y^2 - 3y - 1) = 0$$
$$2(5y+1)(2y-1) = 0$$

$5y+1=0$ or $2y-1=0$

$5y=-1$ $\quad$ $2y=1$

$y=-\frac{1}{5}$ $\quad$ $y=\frac{1}{2}$

$x^{1/3}=-\frac{1}{5}$ $\quad$ $x^{1/3}=\frac{1}{2}$

$x=\left(-\frac{1}{5}\right)^3$ $\quad$ $x=\left(\frac{1}{2}\right)^3$

$x=-\frac{1}{125}$ $\quad$ $x=\frac{1}{8}$

The solution set is $\left\{-\frac{1}{125}, \frac{1}{8}\right\}$.

29. $a^4 - 5a^2 + 6 = 0$

$$(a^2-2)(a^2-3) = 0$$

$a^2-2=0$ or $a^2-3=0$

$a^2=2$ $\quad$ $a^2=3$

$a=\pm\sqrt{2}$ $\quad$ $a=\pm\sqrt{3}$

The solution set is $\left\{-\sqrt{2}, \sqrt{2}, -\sqrt{3}, \sqrt{3}\right\}$.

31. The LCD is $(x-2)(x+3)$.

$$\frac{2x}{x-2}+\frac{x}{x+3}=\frac{-5}{x+3}$$

$$(x-2)(x+3)\left(\frac{2x}{x-2}\right)+(x-2)(x+3)\left(\frac{x}{x+3}\right)=(x-2)(x+3)\left(\frac{-5}{x+3}\right)$$

$$2x(x+3)+x(x-2)=-5(x-2)$$

$$2x^2+6x+x^2-2x=-5x+10$$

$$3x^2+9x-10=0$$

$a=3$, $b=9$, $c=-10$

$$x=\frac{-b\pm\sqrt{b^2-4ac}}{2a}=\frac{-9\pm\sqrt{9^2-4(3)(-10)}}{2(3)}=\frac{-9\pm\sqrt{201}}{6}$$

The solution set is $\left\{\frac{-9-\sqrt{201}}{6}, \frac{-9+\sqrt{201}}{6}\right\}$.

33.

$$(p+2)^2=9(p+2)-20$$

$$(p+2)^2-9(p+2)+20=0$$

Let $x=p+2$.

$$x^2-9x+20=0$$

$$(x-4)(x-5)=0$$

$x-4=0$ or $x-5=0$

$x=4$ $x=5$

$p+2=4$ $p+2=5$

$p=2$ $p=3$

The solution set is $\{2, 3\}$.

35.

$$2x=\sqrt{11x+3}$$

$$(2x)^2=\left(\sqrt{11x+3}\right)^2$$

$$4x^2=11x+3$$

$$4x^2-11x-3=0$$

$$(4x+1)(x-3)=0$$

$4x+1=0$ or $x-3=0$

$x=-\frac{1}{4}$ $x=3$

The value $-\frac{1}{4}$ does not check, so the solution set is $\{3\}$.

37. Let $y = x^{1/3}$.

$$x^{2/3} - 8x^{1/3} + 15 = 0$$
$$(x^{1/3})^2 - 8x^{1/3} + 15 = 0$$
$$y^2 - 8y + 15 = 0$$
$$(y-3)(y-5) = 0$$
$$y - 3 = 0 \quad \text{or} \quad y - 5 = 0$$
$$y = 3 \qquad y = 5$$
$$x^{1/3} = 3 \qquad x^{1/3} = 5$$
$$(x^{1/3})^3 = 3^3 \qquad (x^{1/3})^3 = 5^3$$
$$x = 27 \qquad x = 125$$

The solution set is {27, 125}.

39. $x - \sqrt{19-2x} - 2 = 0$

$$x - 2 = \sqrt{19-2x}$$
$$(x-2)^2 = \left(\sqrt{19-2x}\right)^2$$
$$x^2 - 4x + 4 = 19 - 2x$$
$$x^2 - 2x - 15 = 0$$
$$(x-5)(x+3) = 0$$
$$x - 5 = 0 \quad \text{or} \quad x + 3 = 0$$
$$x = 5 \qquad x = -3$$

The value –3 does not check, so the solution set is {5}.

41. Let $y = x^{1/3}$.

$$2x^{2/3} + 3x^{1/3} - 2 = 0$$
$$2(x^{1/3})^2 + 3x^{1/3} - 2 = 0$$
$$2y^2 + 3y - 2 = 0$$
$$(y+2)(2y-1) = 0$$
$$y + 2 = 0 \quad \text{or} \quad 2y - 1 = 0$$
$$y = -2 \qquad 2y = 1$$
$$y = \frac{1}{2}$$
$$x^{1/3} = -2 \qquad x^{1/3} = \frac{1}{2}$$
$$(x^{1/3})^3 = (-2)^3 \qquad (x^{1/3})^3 = \left(\frac{1}{2}\right)^3$$
$$x = -8 \qquad x = \frac{1}{8}$$

The solution set is $\left\{-8, \frac{1}{8}\right\}$.

43. Let $x = t + 3$.

$$(t+3)^2 - 2(t+3) - 8 = 0$$
$$x^2 - 2x - 8 = 0$$
$$(x-4)(x+2) = 0$$
$$x - 4 = 0 \quad \text{or} \quad x + 2 = 0$$
$$x = 4 \qquad x = -2$$
$$t + 3 = 4 \qquad t + 3 = -2$$
$$t = 1 \qquad t = -5$$

The solution set is {–5, 1}.

45.

$$x - \sqrt{x} = 2$$
$$x - 2 = \sqrt{x}$$
$$(x-2)^2 = \left(\sqrt{x}\right)^2$$
$$x^2 - 4x + 4 = x$$
$$x^2 - 5x + 4 = 0$$
$$(x-1)(x-4) = 0$$
$$x - 1 = 0 \quad \text{or} \quad x - 4 = 0$$
$$x = 1 \quad \text{or} \quad x = 4$$

The value $x = 1$ does not check, so the solution set is {4}.

47. The LCD is $(x-1)(x+1)$.

$$\frac{x}{x-1}+\frac{1}{x+1}=\frac{2}{x^2-1}$$

$$(x-1)(x+1)\left(\frac{x}{x-1}\right)+(x-1)(x+1)\left(\frac{1}{x+1}\right)=(x-1)(x+1)\left[\frac{2}{(x-1)(x+1)}\right]$$

$$x(x+1)+1(x-1)=2$$

$$x^2+x+x-1=2$$

$$x^2+2x-3=0$$

$$(x+3)(x-1)=0$$

$x+3=0$ or $x-1=0$

$x=-3$ $\quad x=1$

The value 1 does not check, so the solution set is $\{-3\}$.

49. $p^4-p^2-20=0$

$(p^2-5)(p^2+4)=0$

$p^2-5=0$ or $p^2+4=0$

$p^2=5$ $\quad p^2=-4$

$p=\pm\sqrt{5}$ $\quad p=\pm\sqrt{-4}=\pm 2i$

The solution set is $\left\{-\sqrt{5}, \sqrt{5}, -2i, 2i\right\}$.

51. Let $y=\dfrac{1}{x-7}$.

$1=\dfrac{4}{x-7}+\dfrac{5}{(x-7)^2}$

$1=4y+5y^2$

$0=5y^2+4y-1$

$0=(5y-1)(y+1)$

$5y-1=0$ or $y+1=0$

$y=\dfrac{1}{5}$ $\quad y=-1$

$\dfrac{1}{x-7}=\dfrac{1}{5}$ $\quad \dfrac{1}{x-7}=-1$

$5=x-7$ $\quad 1=-x+7$

$12=x$ $\quad x=6$

The solution set is $\{6, 12\}$.

53.

$$27y^4 + 15y^2 = 2$$
$$27y^4 + 15y^2 - 2 = 0$$
$$(3y^2 + 2)(9y^2 - 1) = 0$$

$3y^2 + 2 = 0$ or $9y^2 - 1 = 0$

$$3y^2 = -2 \qquad 9y^2 = 1$$
$$y^2 = -\frac{2}{3} \qquad y^2 = \frac{1}{9}$$
$$y = \pm\sqrt{-\frac{2}{3}} \qquad y = \pm\sqrt{\frac{1}{9}}$$
$$y = \pm\frac{i\sqrt{2}}{\sqrt{3}} \cdot \frac{\sqrt{3}}{\sqrt{3}} = \pm\frac{i\sqrt{6}}{3} \qquad y = \pm\frac{1}{3}$$

The solution set is $\left\{-\frac{1}{3}, \frac{1}{3}, -\frac{i\sqrt{6}}{3}, \frac{i\sqrt{6}}{3}\right\}$.

55. Let x be the jogger's speed on the first part of the run. Then $x - 1$ is her speed on the second part.

	rate, mph	distance, miles	time = $\frac{\text{distance}}{\text{rate}}$
1st part	x	3	$\frac{3}{x}$
2nd part	$x - 1$	4	$\frac{4}{x-1}$

$$\frac{3}{x} + \frac{4}{x-1} = 1\frac{3}{5}$$
$$\frac{3}{x} + \frac{4}{x-1} = \frac{8}{5}$$

The LCD is $5x(x - 1)$.

$$5x(x-1)\left(\frac{3}{x}\right) + 5x(x-1)\left(\frac{4}{x-1}\right) = 5x(x-1)\left(\frac{8}{5}\right)$$
$$15(x-1) + 20x = 8x(x-1)$$
$$15x - 15 + 20x = 8x^2 - 8x$$
$$0 = 8x^2 - 43x + 15$$
$$0 = (8x-3)(x-5)$$

$8x - 3 = 0$ or $x - 5 = 0$

$$x = \frac{3}{8} \qquad x = 5$$

Since $x - 1 = \frac{3}{8} - 1$ is negative, reject $\frac{3}{8}$ since rate cannot be negative. The jogger traveled at 5 mph for the first part of her run and $x - 1 = 5 - 1 = 4$ mph for the second part of her run.

57. Let x be the amount of time it takes the hose alone to fill the pond. Then the inlet pipe can fill the pond in $(x - 1)$ hours.

	Time to fill	Part filled in 1 hour
Hose	x hours	$\frac{1}{x}$
Inlet pipe	$(x - 1)$ hours	$\frac{1}{x-1}$
Together	8 hours	$\frac{1}{8}$

The LCD is $8x(x - 1)$.

$$\frac{1}{x}+\frac{1}{x-1}=\frac{1}{8}$$
$$8x(x-1)\left(\frac{1}{x}\right)+8x(x-1)\left(\frac{1}{x-1}\right)=8x(x-1)\left(\frac{1}{8}\right)$$
$$8(x-1)+8x=x(x-1)$$
$$8x-8+8x=x^2-x$$
$$0=x^2-17x+8$$

$a = 1, b = -17, c = 8$

$$\begin{aligned}x&=\frac{-b\pm\sqrt{b^2-4ac}}{2a}\\&=\frac{-(-17)\pm\sqrt{(-17)^2-4(1)(8)}}{2(1)}\\&=\frac{17\pm\sqrt{289-32}}{2}\\&=\frac{17\pm\sqrt{257}}{2}\end{aligned}$$

Since $x - 1$ would be negative and the time cannot be negative, reject $x=\frac{17-\sqrt{257}}{2}\approx 0.5$.

The hose fills the pond in $\frac{17+\sqrt{257}}{2}\approx 16.5$ hours. The inlet pipe takes about 15.5 hours.

59. Let x be the rate to Tucson. Then $x + 11$ is the rate returning home.

	rate, mph	distance, miles	time = $\frac{\text{distance}}{\text{rate}}$
To	x	330	$\frac{330}{x}$
From	$x + 11$	330	$\frac{330}{x+11}$

The LCD is $x(x + 11)$.

$$\frac{330}{x} - \frac{330}{x+11} = 1$$

$$x(x+11)\left(\frac{330}{x}\right) - x(x+11)\left(\frac{330}{x+11}\right) = x(x+11)(1)$$

$$330(x+11) - 330x = x(x+11)$$

$$330x + 3630 - 330x = x^2 + 11x$$

$$0 = x^2 + 11x - 3630$$

$$0 = (x+66)(x-55)$$

$x + 66 = 0$ or $x - 55 = 0$

$x = -66$ $\quad$ $x = 55$

Reject –66 since rate cannot be negative. The speed to Tucson is 55 mph and the speed returning is $x + 11 = 55 + 11 = 66$ mph.

61. Let x be the amount of time it takes Billy to clean the house. Then Billy's dad takes $(x - 1)$ hours.

	Time	Part in 1 hour
Dad	$x - 1$	$\frac{1}{x-1}$
Billy	x	$\frac{1}{x}$
Together	4	$\frac{1}{4}$

The LCD is $4x(x - 1)$.

$$\frac{1}{x} + \frac{1}{x-1} = \frac{1}{4}$$

$$4x(x-1)\left(\frac{1}{x}\right) + 4x(x-1)\left(\frac{1}{x-1}\right) = 4x(x-1)\left(\frac{1}{4}\right)$$

$$4(x-1) + 4x = x(x-1)$$

$$4x - 4 + 4x = x^2 - x$$

$$0 = x^2 - 9x + 4$$

$a = 1$, $b = -9$, $c = 4$

$$x = \frac{-b \pm \sqrt{b^2 - 4ac}}{2a}$$

$$= \frac{-(-9) \pm \sqrt{(-9)^2 - 4(1)(4)}}{2(1)}$$

$$= \frac{9 \pm \sqrt{81 - 16}}{2}$$

$$= \frac{9 \pm \sqrt{65}}{2}$$

$$x = \frac{9-\sqrt{65}}{2} \approx 0.5 \text{ or } x = \frac{9+\sqrt{65}}{2} \approx 8.5$$

Since Billy's dad takes one hour less to clean the house, reject 0.5.
It takes Billy 8.5 hours to clean the house.

63. Let x be the number. Then $x - 4$ is 4 less than the number.

$$x(x-4) = 96$$
$$x^2 - 4x = 96$$
$$x^2 - 4x - 96 = 0$$
$$(x-12)(x+8) = 0$$
$$x - 12 = 0 \quad \text{or} \quad x + 8 = 0$$
$$x = 12 \qquad\qquad x = -8$$

The number is 12 or –8.

65. a. The ? in the drawing is the original length, x, less the two 3-inch squares, or $(x - 6)$ inches.

b. The height of the box is 3 inches, while both the length and the width are $(x - 6)$ inches, and the volume is 300 cubic inches.

$$V = l \cdot w \cdot h$$
$$300 = (x-6)\cdot(x-6)\cdot 3$$

c.

$$300 = 3(x^2 - 12x + 36)$$
$$100 = x^2 - 12x + 36$$
$$0 = x^2 - 12x - 64$$
$$0 = (x+4)(x-16)$$
$$x + 4 = 0 \quad \text{or} \quad x - 16 = 0$$
$$x = -4 \qquad\qquad x = 16$$

Since the length cannot be negative, reject –4.
The piece of cardboard is 16 inches by 16 inches. The sides of the box have length 16 – 6 = 10 inches each and the volume is
$V = lwh = 10 \cdot 10 \cdot 3 = 300$ cubic inches.

67. Let r be the radius. Each side of the square has length $\sqrt{920}$ feet, since $\text{Area} = (\text{side})^2$. Using the Pythagorean Theorem,

$$c^2 = a^2 + b^2$$
$$(2r)^2 = \left(\sqrt{920}\right)^2 + \left(\sqrt{920}\right)^2$$
$$4r^2 = 920 + 920$$
$$4r^2 = 1840$$
$$r^2 = 460$$
$$r = \pm\sqrt{460}$$
$$r \approx -21.4 \quad \text{or} \quad r = 21.4$$

Reject –21.4 since length cannot be negative. In order for the entire garden to be watered, the radius must be 22 feet.

69.

$$\frac{5x}{3} + 2 \le 7$$
$$\frac{5x}{3} \le 5$$
$$5x \le 15$$
$$x \le 3$$

The solution set is $(-\infty, 3]$.

71.

$$\frac{y-1}{15} > -\frac{2}{5}$$
$$15\left(\frac{y-1}{15}\right) > 15\left(-\frac{2}{5}\right)$$
$$y - 1 > -6$$
$$y - 1 + 1 > -6 + 1$$
$$y > -5$$

The solution set is $(-5, \infty)$.

73.

$$y^3 + 9y - y^2 - 9 = 0$$
$$y(y^2 + 9) - (y^2 + 9) = 0$$
$$(y^2 + 9)(y - 1) = 0$$
$$y^2 + 9 = 0 \quad \text{or} \quad y - 1 = 0$$
$$y^2 = -9 \qquad\qquad y = 1$$
$$y = \pm\sqrt{-9} = \pm 3i$$

The solution set is $\{1, -3i, 3i\}$.

75. $x^{-2} - x^{-1} - 6 = 0$

$$\frac{1}{x^2} - \frac{1}{x} - 6 = 0$$

Let $y = \frac{1}{x}$.

$$y^2 - y - 6 = 0$$
$$(y-3)(y+2) = 0$$
$$y - 3 = 0 \quad \text{or} \quad y + 2 = 0$$
$$y = 3 \qquad y = -2$$
$$\frac{1}{x} = 3 \qquad \frac{1}{x} = -2$$
$$x = \frac{1}{3} \qquad x = -\frac{1}{2}$$

The solution set is $\left\{-\frac{1}{2}, \frac{1}{3}\right\}$.

77.

$$2x^3 = -54$$
$$2x^3 + 54 = 0$$
$$x^3 + 27 = 0$$
$$x^3 + 3^3 = 0$$
$$(x+3)(x^2 - 3x + 9) = 0$$
$$x + 3 = 0 \quad \text{or} \quad x^2 - 3x + 9 = 0$$
$$x = -3$$

$$x^2 - 3x + 9 = 0$$
$$a = 1, b = -3, c = 9$$
$$x = \frac{-b \pm \sqrt{b^2 - 4ac}}{2a}$$
$$= \frac{-(-3) \pm \sqrt{(-3)^2 - 4(1)(9)}}{2(1)}$$
$$= \frac{3 \pm \sqrt{9 - 36}}{2}$$
$$= \frac{3 \pm \sqrt{-27}}{2}$$
$$= \frac{3 \pm 3i\sqrt{3}}{2}$$

The solution set is $\left\{-3, \frac{3 - 3i\sqrt{3}}{2}, \frac{3 + 3i\sqrt{3}}{2}\right\}$.

79. answers may vary

81. a. Let x be Junqueira's fastest speed. Using $d = r \cdot t$ or $t = \frac{d}{r}$, Junqueira's time for that lap was $\frac{11,119.68}{x}$ seconds. Bourdais's fastest speed is then (x – 0.1159) feet per second and the time for Bourdais's fastest lap was $\frac{11,119.68}{x-0.1159}$ seconds, which is also $\frac{11,119.68}{x}+0.036$.
The LCD is $x(x - 0.1159)$.

$$\frac{11,119.68}{x}+0.036=\frac{11,119.68}{x-0.1159}$$
$$x(x-0.1159)\left(\frac{11,119.68}{x}\right)+x(x-0.1159)(0.036)=x(x-0.1159)\left(\frac{11,119.68}{x-0.1159}\right)$$
$$11,119.68(x-0.1159)+0.036x(x-0.1159)=11,119.68x$$
$$11,119.68x-1288.770912+0.036x^2-0.0041724x=11,119.68x$$
$$0.036x^2-0.0041724x-1288.770912=0$$

$a = 0.036$, $b = -0.0041724$, $c = -1288.770912$

$$x=\frac{-b\pm\sqrt{b^2-4ac}}{2a}$$
$$=\frac{-(-0.0041724)\pm\sqrt{(-0.0041724)^2-4(0.036)(-1288.770912)}}{2(0.036)}$$

$x \approx -189.1488$ or $x \approx 189.2647$

Since the speed cannot be negative, reject –189.1488.
Junqueira's fastest speed was approximately 189.265 feet per second.

b. Bourdais's fastest speed was 189.2647 – 0.1159 ≈ 189.149 feet per second.

c. $$189.149 \text{ feet/sec} = \frac{189.149 \text{ feet}}{1 \text{ sec}}\cdot\frac{3600 \text{ sec}}{1 \text{ hour}}\cdot\frac{1 \text{ mile}}{5280 \text{ feet}}$$
$$\approx 128.965 \text{ miles/hour}$$

Bourdais's fastest speed was about 128.965 miles per hour.

Integrated Review

1. $x^2 - 10 = 0$
$x^2 = 10$
$x = \pm\sqrt{10}$

The solution set is $\left\{-\sqrt{10}, \sqrt{10}\right\}$.

2. $3x^2 + 24 = 0$

$$3x^2 = -24$$
$$x^2 = -8$$
$$x = \pm\sqrt{-8}$$
$$x = \pm\sqrt{-4}\sqrt{2}$$
$$x = \pm 2i\sqrt{2}$$

The solution set is $\left\{-2i\sqrt{2}, 2i\sqrt{2}\right\}$.

3. $(x-1)^2 = 8$

$$x-1 = \pm\sqrt{8}$$
$$x = 1 \pm \sqrt{8}$$
$$x = 1 \pm 2\sqrt{2}$$

The solution set is $\left\{1-2\sqrt{2}, 1+2\sqrt{2}\right\}$.

4. $(2x+5)^2 = 12$

$$2x+5 = \pm\sqrt{12}$$
$$2x+5 = \pm 2\sqrt{3}$$
$$2x = -5 \pm 2\sqrt{3}$$
$$x = \frac{-5 \pm 2\sqrt{3}}{2}$$

The solution set is $\left\{\frac{-5-2\sqrt{3}}{2}, \frac{-5+2\sqrt{3}}{2}\right\}$.

5. $x^2 + 2x - 12 = 0$

$$x^2 + 2x = 12$$
$$x^2 = 2x + \left(\frac{2}{2}\right)^2 = 12 + 1$$
$$(x+1)^2 = 13$$
$$x+1 = \pm\sqrt{13}$$
$$x = -1 \pm \sqrt{13}$$

The solution set is $\left\{-1-\sqrt{13}, -1+\sqrt{13}\right\}$.

6. $x^2 - 12x + 11 = 0$

$$x^2 - 12x = -11$$
$$x^2 - 12x + \left(-\frac{12}{2}\right)^2 = -11 + 36$$
$$(x-6)^2 = 25$$
$$x-6 = \pm\sqrt{25}$$
$$x-6 = \pm 5$$
$$x = 6 \pm 5$$

$x = 6 - 5 = 1$ or $x = 6 + 5 = 11$

The solution set is $\{1, 11\}$.

7. $4x^2 + 12x = 8$

$$x^2 + 3x = 2$$
$$x^2 + 3x + \left(\frac{3}{2}\right)^2 = 2 + \frac{9}{4}$$
$$\left(x + \frac{3}{2}\right)^2 = \frac{17}{4}$$
$$x + \frac{3}{2} = \pm\sqrt{\frac{17}{4}}$$
$$x + \frac{3}{2} = \pm\frac{\sqrt{17}}{2}$$
$$x = -\frac{3}{2} \pm \frac{\sqrt{17}}{2}$$

The solution set is $\left\{\frac{-3-\sqrt{17}}{2}, \frac{-3+\sqrt{17}}{2}\right\}$.

8. $16y^2+16y=1$

$$y^2+y=\frac{1}{16}$$

$$y^2+y+\left(\frac{1}{2}\right)^2=\frac{1}{16}+\frac{1}{4}$$

$$\left(y+\frac{1}{2}\right)^2=\frac{5}{16}$$

$$y+\frac{1}{2}=\pm\sqrt{\frac{5}{16}}$$

$$y+\frac{1}{2}=\pm\frac{\sqrt{5}}{4}$$

$$y=-\frac{1}{2}\pm\frac{\sqrt{5}}{4}$$

$$y=-\frac{2}{4}\pm\frac{\sqrt{5}}{4}$$

The solution set is $\left\{\frac{-2-\sqrt{5}}{4}, \frac{-2+\sqrt{5}}{4}\right\}$.

9. $2x^2-4x+1=0$

$a=2, b=-4, c=1$

$$x=\frac{-b\pm\sqrt{b^2-4ac}}{2a}$$

$$=\frac{-(-4)\pm\sqrt{(-4)^2-4(2)(1)}}{2(2)}$$

$$=\frac{4\pm\sqrt{16-8}}{4}$$

$$=\frac{4\pm\sqrt{8}}{4}$$

$$=\frac{4\pm2\sqrt{2}}{4}$$

$$=\frac{2\left(2\pm\sqrt{2}\right)}{2\cdot2}$$

$$=\frac{2\pm\sqrt{2}}{2}$$

The solution set is $\left\{\frac{2-\sqrt{2}}{2}, \frac{2+\sqrt{2}}{2}\right\}$.

10. $\frac{1}{2}x^2+3x+2=0$

$x^2+6x+4=0$

$a=1, b=6, c=4$

$$x=\frac{-b\pm\sqrt{b^2-4ac}}{2a}$$

$$=\frac{-6\pm\sqrt{6^2-4(1)(4)}}{2(1)}$$

$$=\frac{-6\pm\sqrt{36-16}}{2}$$

$$=\frac{-6\pm\sqrt{20}}{2}$$

$$=\frac{-6\pm2\sqrt{5}}{2}$$

$$=\frac{2\left(-3\pm\sqrt{5}\right)}{2}$$

$$=-3\pm\sqrt{5}$$

The solution set is $\left\{-3-\sqrt{5}, -3+\sqrt{5}\right\}$.

11. $x^2+4x=-7$

$x^2+4x+7=0$

$a=1, b=4, c=7$

$$x=\frac{-b\pm\sqrt{b^2-4ac}}{2a}$$

$$=\frac{-4\pm\sqrt{4^2-4(1)(7)}}{2(1)}$$

$$=\frac{-4\pm\sqrt{16-28}}{2}$$

$$=\frac{-4\pm\sqrt{-12}}{2}$$

$$=\frac{-4\pm\sqrt{-4}\sqrt{3}}{2}$$

$$=\frac{-4\pm2i\sqrt{3}}{2}$$

$$=\frac{2\left(-2\pm i\sqrt{3}\right)}{2}$$

$$=-2\pm i\sqrt{3}$$

The solution set is $\left\{-2-i\sqrt{3}, -2+i\sqrt{3}\right\}$.

12.
$$5x^2+6x=-3$$
$$5x^2+6x+3=0$$
$$a=5,\ b=6,\ c=3$$
$$\begin{aligned}x&=\frac{-b\pm\sqrt{b^2-4ac}}{2a}\\&=\frac{-6\pm\sqrt{6^2-4(5)(3)}}{2(5)}\\&=\frac{-6\pm\sqrt{36-60}}{10}\\&=\frac{-6\pm\sqrt{-24}}{10}\\&=\frac{-6\pm\sqrt{-4}\sqrt{6}}{10}\\&=\frac{-6\pm 2i\sqrt{6}}{10}\\&=\frac{2\left(-3\pm i\sqrt{6}\right)}{2\cdot 5}\\&=\frac{-3\pm i\sqrt{6}}{5}\end{aligned}$$
The solution set is $\left\{\frac{-3-i\sqrt{6}}{5}, \frac{-3+i\sqrt{6}}{5}\right\}$.

13. $x^2+3x+6=0$
$$a=1,\ b=3,\ c=6$$
$$\begin{aligned}x&=\frac{-b\pm\sqrt{b^2-4ac}}{2a}\\&=\frac{-3\pm\sqrt{3^2-4(1)(6)}}{2(1)}\\&=\frac{-3\pm\sqrt{9-24}}{2}\\&=\frac{-3\pm\sqrt{-15}}{2}\\&=\frac{-3\pm i\sqrt{15}}{2}\end{aligned}$$
The solution set is $\left\{\frac{-3-i\sqrt{15}}{2}, \frac{-3+i\sqrt{15}}{2}\right\}$.

14. $2x^2+18=0$
$$\begin{aligned}2x^2&=-18\\x^2&=-9\\x&=\pm\sqrt{-9}\\x&=\pm 3i\end{aligned}$$
The solution set is $\{-3i, 3i\}$.

15. $x^2+17x=0$
$$x(x+17)=0$$
$$x=0 \quad\text{or}\quad x+17=0$$
$$x=-17$$
The solution set is $\{0, -17\}$.

16. $4x^2-2x-3=0$
$$a=4,\ b=-2,\ c=-3$$
$$\begin{aligned}x&=\frac{-b\pm\sqrt{b^2-4ac}}{2a}\\&=\frac{-(-2)\pm\sqrt{(-2)^2-4(4)(-3)}}{2(4)}\\&=\frac{2\pm\sqrt{4+48}}{8}\\&=\frac{2\pm\sqrt{52}}{8}\\&=\frac{2\pm 2\sqrt{13}}{8}\\&=\frac{2\left(1\pm\sqrt{13}\right)}{2\cdot 4}\\&=\frac{1\pm\sqrt{13}}{4}\end{aligned}$$
The solution set is $\left\{\frac{1-\sqrt{13}}{4}, \frac{1+\sqrt{13}}{4}\right\}$.

17.
$$\begin{aligned}(x-2)^2&=27\\\sqrt{(x-2)^2}&=\pm\sqrt{27}\\x-2&=\pm 3\sqrt{3}\\x&=2\pm 3\sqrt{3}\end{aligned}$$
The solution set is $\left\{2+3\sqrt{3}, 2-3\sqrt{3}\right\}$.

18. $\frac{1}{2}x^2 - 2x + \frac{1}{2} = 0$

$$x^2 - 4x + 1 = 0$$
$$a = 1, b = -4, c = 1$$
$$x = \frac{-b \pm \sqrt{b^2 - 4ac}}{2a} = \frac{-(-4) \pm \sqrt{(-4)^2 - 4(1)(1)}}{2(1)} = \frac{4 \pm \sqrt{16 - 4}}{2} = \frac{4 \pm \sqrt{12}}{2} = \frac{4 \pm 2\sqrt{3}}{2} = 2 \pm \sqrt{3}$$

The solution set is $\{2 - \sqrt{3}, 2 + \sqrt{3}\}$.

19. $3x^2 + 2x = 8$

$$3x^2 + 2x - 8 = 0$$
$$(3x - 4)(x + 2) = 0$$
$$3x - 4 = 0 \quad \text{or} \quad x + 2 = 0$$
$$x = \frac{4}{3} \qquad x = -2$$

The solution set is $\left\{-2, \frac{4}{3}\right\}$.

20. $2x^2 = -5x - 1$

$$2x^2 + 5x + 1 = 0$$
$$a = 2, b = 5, c = 1$$
$$x = \frac{-b \pm \sqrt{b^2 - 4ac}}{2a} = \frac{-5 \pm \sqrt{5^2 - 4(2)(1)}}{2(2)} = \frac{-5 \pm \sqrt{25 - 8}}{4} = \frac{-5 \pm \sqrt{17}}{4}$$

The solution set is $\left\{\frac{-5 - \sqrt{17}}{4}, \frac{-5 + \sqrt{17}}{4}\right\}$.

21. $x(x - 2) = 5$

$$x^2 - 2x = 5$$
$$x^2 - 2x - 5 = 0$$
$$a = 1, b = -2, c = -5$$
$$x = \frac{-b \pm \sqrt{b^2 - 4ac}}{2a} = \frac{-(-2) \pm \sqrt{(-2)^2 - 4(1)(-5)}}{2(1)} = \frac{2 \pm \sqrt{4 + 20}}{2} = \frac{2 \pm \sqrt{24}}{2} = \frac{2 \pm 2\sqrt{6}}{2} = 1 \pm \sqrt{6}$$

The solution set is $\{1 - \sqrt{6}, 1 + \sqrt{6}\}$.

22. $x^2 - 31 = 0$

$$x^2 = 31$$
$$x = \pm\sqrt{31}$$

The solution set is $\{-\sqrt{31}, \sqrt{31}\}$.

23. $4x^2 - 48 = 0$

$$4x^2 = 48$$
$$x^2 = 12$$
$$x = \pm\sqrt{12}$$
$$x = \pm 2\sqrt{3}$$

The solution set is $\{-2\sqrt{3}, 2\sqrt{3}\}$.

24. $5x^2 + 55 = 0$

$$5x^2 = -55$$
$$x^2 = -11$$
$$x = \pm\sqrt{-11}$$
$$x = \pm i\sqrt{11}$$

The solution set is $\{-i\sqrt{11}, i\sqrt{11}\}$.

25.
$$x(x+5)=66$$
$$x^2+5x=66$$
$$x^2+5x-66=0$$
$$(x+11)(x-6)=0$$
$$x+11=0 \quad \text{or} \quad x-6=0$$
$$x=-11 \qquad\qquad x=6$$
The solution set is $\{-11, 6\}$.

26. $5x^2+6x-2=0$

$a=5,\ b=6,\ c=-2$
$$x=\frac{-b\pm\sqrt{b^2-4ac}}{2a}$$
$$=\frac{-6\pm\sqrt{6^2-4(5)(-2)}}{2(5)}$$
$$=\frac{-6\pm\sqrt{36+40}}{10}$$
$$=\frac{-6\pm\sqrt{76}}{10}$$
$$=\frac{-6\pm2\sqrt{19}}{10}$$
$$=\frac{2\left(-3\pm\sqrt{19}\right)}{2\cdot5}$$
$$=\frac{-3\pm\sqrt{19}}{5}$$
The solution set is $\left\{\frac{-3-\sqrt{19}}{5}, \frac{-3+\sqrt{19}}{5}\right\}$.

27.
$$2x^2+3x=1$$
$$2x^2+3x-1=0$$
$a=2,\ b=3,\ c=-1$
$$x=\frac{-b\pm\sqrt{b^2-4ac}}{2a}$$
$$=\frac{-3\pm\sqrt{3^2-4(2)(-1)}}{2(2)}$$
$$=\frac{-3\pm\sqrt{9+8}}{4}$$
$$=\frac{-3\pm\sqrt{17}}{4}$$
The solution set is $\left\{\frac{-3-\sqrt{17}}{4}, \frac{-3+\sqrt{17}}{4}\right\}$.

28. $x-\sqrt{13-3x}-3=0$

$$x-3=\sqrt{13-3x}$$
$$(x-3)^2=\left(\sqrt{13-3x}\right)^2$$
$$x^2-6x+9=13-3x$$
$$x^2-3x-4=0$$
$$(x-4)(x+1)=0$$
$$x-4=0 \quad \text{or} \quad x+1=0$$
$$x=4 \qquad\qquad x=-1$$

The value -1 does not check, so the solution set is $\{4\}$.

29. The LCD is $x(x-2)$.

$$\frac{5x}{x-2}-\frac{x+1}{x}=\frac{3}{x(x-2)}$$
$$x(x-2)\left(\frac{5x}{x-2}\right)-x(x-2)\left(\frac{x+1}{x}\right)=x(x-2)\left[\frac{3}{x(x-2)}\right]$$
$$x(5x)-(x-2)(x+1)=3$$
$$5x^2-(x^2-x-2)=3$$
$$5x^2-x^2+x+2=3$$
$$4x^2+x-1=0$$

$a=4$, $b=1$, $c=-1$

$$x=\frac{-b\pm\sqrt{b^2-4ac}}{2a}$$
$$=\frac{-1\pm\sqrt{1^2-4(4)(-1)}}{2(4)}$$
$$=\frac{-1\pm\sqrt{1+16}}{8}$$
$$=\frac{-1\pm\sqrt{17}}{8}$$

The solution set is $\left\{\frac{-1-\sqrt{17}}{8}, \frac{-1+\sqrt{17}}{8}\right\}$.

30. $x^2+x^2=20^2$

$$2x^2=400$$
$$x^2=200$$
$$x=\pm\sqrt{200}$$
$$x=\pm 10\sqrt{2}$$

Reject $-10\sqrt{2}$ since length cannot be negative. The length of the room is $10\sqrt{2}\approx 14.1$ feet.

31. Let x be the time that Diane can complete the job alone. Then Lucy can complete it in $x - 2$.

	Time	Part in 1 hour
Diane	x	$\frac{1}{x}$
Lucy	$x - 2$	$\frac{1}{x-2}$
Together	4	$\frac{1}{4}$

The LCD is $4x(x - 2)$.

$$\frac{1}{x}+\frac{1}{x-2}=\frac{1}{4}$$

$$4x(x-2)\left(\frac{1}{x}\right)+4x(x-2)\left(\frac{1}{x-2}\right)=4x(x-2)\left(\frac{1}{4}\right)$$

$$4(x-2)+4x(1)=x(x-2)$$

$$4x-8+4x=x^2-2x$$

$$0=x^2-10x+8$$

$a = 1, b = -10, c = 8$

$$x=\frac{-b\pm\sqrt{b^2-4ac}}{2a}$$

$$=\frac{-(-10)\pm\sqrt{(-10)^2-4(1)(8)}}{2(1)}$$

$$=\frac{10\pm\sqrt{100-32}}{2}$$

$$=\frac{10\pm\sqrt{68}}{2}$$

$$=\frac{10\pm2\sqrt{17}}{2}$$

$$=5\pm\sqrt{17}$$

$x \approx 0.9$ or $x \approx 9.1$

Reject 0.9, since $0.9 - 2$ is negative. Diane can complete the meal in 9.1 hours and Lucy can complete it in $9.1 - 2 = 7.1$ hours.

32. Let x be the rate during the first part. Then $x + 1$ is the rate for the second part.

	rate, mph	distance, miles	time $= \frac{\text{distance}}{\text{rate}}$
1st part	x	5	$\frac{5}{x}$
2nd part	$x + 1$	2	$\frac{2}{x+1}$

$$\frac{5}{x}+\frac{2}{x+1}=1\frac{1}{3}$$
$$\frac{5}{x}+\frac{2}{x+1}=\frac{4}{3}$$

The LCD is $3x(x + 1)$.

$$3x(x+1)\left(\frac{5}{x}\right)+3x(x+1)\left(\frac{2}{x+1}\right)=3x(x+1)\left(\frac{4}{3}\right)$$
$$3(x+1)(5)+3x(2)=x(x+1)(4)$$
$$15x+15+6x=4x^2+4x$$
$$0=4x^2-17x-15$$
$$0=(4x+3)(x-5)$$

$4x+3=0$ or $x-5=0$

$x=-\frac{3}{4}$ $\qquad x=5$

Reject $-\frac{3}{4}$ since rate cannot be negative. He runs 5 mph during the first part and 5 + 1 = 6 mph during the second part.

Section 8.4 Practice Problems

1. $(x-2)(x+4)>0$

$(x-2)(x+4)=0$

$x-2=0$ or $x+4=0$

$x=2$ $\qquad x=-4$

Region	Test Point Value	$(x-2)(x+4)>0$	Result
$(-\infty, -4)$	-5	$(-7)(-1)>0$	True
$(-4, 2)$	0	$(-2)(4)>0$	False
$(2, \infty)$	3	$(1)(7)>0$	True

The solution set is $(-\infty, -4) \cup (2, \infty)$.

2. $x^2-6x\le 0$

$x(x-6)\le 0$

$x(x-6)=0$

$x=0$ or $x=6$

Region	Test Point Value	$x(x-6) \le 0$	Result
$(-\infty, 0)$	-1	$-1(-7) \le 0$	False
$(0, 6)$	1	$1(-5) \le 0$	True
$(6, \infty)$	7	$7(1) \le 0$	False

The solution set is $[0, 6]$.

3. $(x-2)(x+1)(x+5) \le 0$

$(x-2)(x+1)(x+5) = 0$

$x-2=0$ or $x+1=0$ or $x+5=0$

$x=2$ $\quad x=-1$ $\quad x=-5$

Region	Test Point Value	$(x-2)(x+1)(x+5) \le 0$	Result
$(-\infty, -5)$	-6	$(-8)(-5)(-1) \le 0$	True
$(-5, -1)$	-2	$(-4)(-1)(3) \le 0$	False
$(-1, 2)$	0	$(-2)(1)(5) \le 0$	True
$(2, \infty)$	3	$(1)(4)(8) \le 0$	False

The solution set is $(-\infty, -5] \cup [-1, 2]$.

4. $\dfrac{x-3}{x+5} \le 0$

$x+5=0$

$x=-5$ makes the denominator 0.

$\dfrac{x-3}{x+5} = 0$

$x-3=0$

$x=3$

Region	Test Point Value	$\frac{x-3}{x+5} \le 0$	Result
$(-\infty, -5)$	-6	$\frac{-9}{-1} \le 0$	False
$(-5, 3)$	0	$\frac{-3}{5} \le 0$	True
$(3, \infty)$	4	$\frac{1}{9} \le 0$	False

The solution set is $(-5, 3]$.

5. $\frac{3}{x-2} < 2$

$x-2=0$

$x=2$ makes the denominator 0.

$\frac{3}{x-2}=2$

$3=2x-4$

$7=2x$

$\frac{7}{2}=x$

Region	Test Point Value	$\frac{3}{x-2}<2$	Result
$(-\infty, 2)$	0	$\frac{3}{-2}<2$	True
$\left(2, \frac{7}{2}\right)$	3	$\frac{3}{1}<2$	False
$\left(\frac{7}{2}, \infty\right)$	5	$\frac{3}{3}<2$	True

The solution set is $(-\infty, 2) \cup \left(\frac{7}{2}, \infty\right)$.

Exercise Set 8.4

1. $(x+1)(x+5)>0$

$(x+1)(x+5)=0$

$x+1=0$ or $x+5=0$

$x=-1$ $\quad$ $x=-5$

Region	Test Point Value	$(x+1)(x+5)>0$	Result
$(-\infty, -5)$	-6	$(-5)(-1)>0$	True
$(-5, -1)$	-2	$(-1)(3)>0$	False
$(-1, \infty)$	0	$(1)(5)>0$	True

The solution set is $(-\infty, -5) \cup (-1, \infty)$.

3. $(x-3)(x+4) \le 0$

$(x-3)(x+4)=0$

$x-3=0$ or $x+4=0$

$x=3$ $\quad$ $x=-4$

Region	Test Point Value	$(x-3)(x+4) \le 0$	Result
$(-\infty, -4)$	-5	$(-8)(-1) \le 0$	False
$(-4, 3)$	0	$(-3)(4) \le 0$	True
$(3, \infty)$	4	$(1)(8) \le 0$	False

The solution set is $[-4, 3]$.

5. $x^2 + 8x + 15 \ge 0$

$(x+5)(x+3) \ge 0$

$(x+5)(x+3) = 0$

$x+5=0$ or $x+3=0$

$x=-5$ $\quad x=-3$

Region	Test Point Value	$(x+5)(x+3) \ge 0$	Result
$(-\infty, -5)$	-6	$(-1)(-3) \ge 0$	True
$(-5, -3)$	-4	$(1)(-1) \ge 0$	False
$(-3, \infty)$	0	$(5)(3) \ge 0$	True

The solution set is $(-\infty, -5] \cup [-3, \infty)$.

7. $3x^2 + 16x < -5$

$3x^2 + 16x + 5 < 0$

$(3x+1)(x+5) < 0$

$(3x+1)(x+5) = 0$

$3x+1=0$ or $x+5=0$

$x = -\frac{1}{3}$ $\quad x=-5$

Region	Test Point Value	$(3x+1)(x+5) < 0$	Result
$(-\infty, -5)$	-6	$(-17)(-1) < 0$	False
$\left(-5, -\frac{1}{3}\right)$	-3	$(-8)(2) < 0$	True
$\left(-\frac{1}{3}, \infty\right)$	0	$(1)(5) < 0$	False

The solution set is $\left(-5, -\frac{1}{3}\right)$.

9. $(x-6)(x-4)(x-2)>0$
$(x-6)(x-4)(x-2)=0$
$x-6=0$ or $x-4=0$ or $x-2=0$
$x=6$ $x=4$ $x=2$

Region	Test Point Value	$(x-6)(x-4)(x-2)>0$	Result
$(-\infty, 2)$	0	$(-6)(-4)(-2)>0$	False
$(2, 4)$	3	$(-3)(-1)(1)>0$	True
$(4, 6)$	5	$(-1)(1)(3)>0$	False
$(6, \infty)$	7	$(1)(3)(5)>0$	True

The solution set is $(2, 4) \cup (6, \infty)$.

11. $x(x-1)(x+4)\le 0$
$x(x-1)(x+4)=0$
$x=0$ or $x-1=0$ or $x+4=0$
$x=1$ $x=-4$

Region	Test Point Value	$x(x-1)(x+4)\le 0$	Result
$(-\infty, -4)$	-5	$-5(-6)(-1)\le 0$	True
$(-4, 0)$	-2	$-2(-3)(2)\le 0$	False
$(0, 1)$	$\frac{1}{2}$	$\frac{1}{2}\left(-\frac{1}{2}\right)\left(4\frac{1}{2}\right)\le 0$	True
$(1, \infty)$	2	$2(1)(6)\le 0$	False

The solution set is $(-\infty, -4] \cup [0, 1]$.

13. $(x^2-9)(x^2-4)>0$
$(x+3)(x-3)(x+2)(x-2)>0$
$(x+3)(x-3)(x+2)(x-2)=0$
$x+3=0$ or $x-3=0$ or $x+2=0$ or $x-2=0$
$x=-3$ $x=3$ $x=-2$ $x=2$

Region	Test Point Value	$(x+3)(x-3)(x+2)(x-2) > 0$	Result
$(-\infty, -3)$	-4	$(-1)(-7)(-2)(-6) > 0$	True
$(-3, -2)$	$-\frac{5}{2}$	$\left(\frac{1}{2}\right)\left(-\frac{11}{2}\right)\left(-\frac{1}{2}\right)\left(-\frac{9}{2}\right) > 0$	False
$(-2, 2)$	0	$(3)(-3)(2)(-2) > 0$	True
$(2, 3)$	$\frac{5}{2}$	$\left(\frac{11}{2}\right)\left(-\frac{1}{2}\right)\left(\frac{9}{2}\right)\left(\frac{1}{2}\right) > 0$	False
$(3, \infty)$	4	$(7)(1)(6)(2) > 0$	True

The solution set is $(-\infty, -3) \cup (-2, 2) \cup (3, \infty)$.

15. $\frac{x+7}{x-2} < 0$

$x - 2 = 0$

$x = 2$ makes the denominator 0

$$\frac{x+7}{x-2} = 0$$

$$(x-2)\left(\frac{x+7}{x-2}\right) = (x-2)(0)$$

$$x + 7 = 0$$

$$x = -7$$

Region	Test Point Value	$\frac{x+7}{x-2} < 0$	Result
$(-\infty, -7)$	-8	$\frac{-1}{-10} < 0$	False
$(-7, 2)$	0	$\frac{7}{-2} < 0$	True
$(2, \infty)$	3	$\frac{10}{1} < 0$	False

The solution set is $(-7, 2)$.

17. $\frac{5}{x+1} > 0$

$x + 1 = 0$

$x = -1$ makes the denominator 0.

Region	Test Point Value	$\frac{5}{x+1} > 0$	Result
$(-\infty, -1)$	-2	$\frac{5}{-1} > 0$	False
$(-1, \infty)$	0	$\frac{5}{1} > 0$	True

The solution set is $(-1, \infty)$.

19. $\dfrac{x+1}{x-4} \geq 0$

$x-4=0$

$x=4$ makes the denominator 0.

$$\frac{x+1}{x-4}=0$$
$$(x-4)\left(\frac{x+1}{x-4}\right)=(x-4)(0)$$
$$x+1=0$$
$$x=-1$$

Region	Test Point Value	$\frac{x+1}{x-4} \geq 0$	Result
$(-\infty, -1)$	-2	$\frac{-1}{-6} \geq 0$	True
$(-1, 4)$	0	$\frac{1}{-4} \geq 0$	False
$(4, \infty)$	5	$\frac{6}{1} \geq 0$	True

The solution set is $(-\infty, -1] \cup (4, \infty)$.

21. $\dfrac{3}{x-2} < 4$

$x-2=0$

$x=2$ makes the denominator 0.

$$\frac{3}{x-2}=4$$
$$(x-2)\left(\frac{3}{x-2}\right)=(x-2)(4)$$
$$3=4x-8$$
$$11=4x$$
$$\frac{11}{4}=x$$

Region	Test Point Value	$\frac{3}{x-2} < 4$	Result
$(-\infty, 2)$	0	$\frac{3}{-2} < 4$	True
$\left(2, \frac{11}{4}\right)$	$\frac{5}{2}$	$\frac{3}{\frac{1}{2}} < 4$	False
$\left(\frac{11}{4}, \infty\right)$	3	$\frac{3}{1} < 4$	True

The solution set is $(-\infty, 2) \cup \left(\frac{11}{4}, \infty\right)$.

23. $\frac{x^2+6}{5x} \geq 1$

$5x = 0$

$x = 0$ makes the denominator 0.

$$\frac{x^2+6}{5x} = 1$$

$$5x\left(\frac{x^2+6}{5x}\right) = 5x(1)$$

$$x^2 + 6 = 5x$$

$$x^2 - 5x + 6 = 0$$

$$(x-3)(x-2) = 0$$

$x - 3 = 0$ or $x - 2 = 0$

$x = 3$ $\quad\quad$ $x = 2$

Region	Test Point Value	$\frac{x^2+6}{5x} \geq 1$	Result
$(-\infty, 0)$	-1	$\frac{7}{-5} \geq 1$	False
$(0, 2)$	1	$\frac{7}{5} \geq 1$	True
$(2, 3)$	2.5	$\frac{12.25}{12.5} \geq 1$	False
$(3, \infty)$	4	$\frac{22}{20} \geq 1$	True

The solution set is $(0, 2] \cup [3, \infty)$.

25. $\frac{x+2}{x-3} < 1$

$x - 3 = 0$

$x = 3$ makes the denominator 0.

$$\frac{x+2}{x-3} = 1$$

$$(x-3)\left(\frac{x+2}{x-3}\right) = (x-3)(1)$$

$$x + 2 = x - 3$$

$$2 = -3 \quad \text{False}$$

The related equation has no solution.

Region	Test Point Value	$\frac{x+2}{x-3} < 1$	Result
$(-\infty, 3)$	0	$\frac{2}{-3} < 1$	True
$(3, \infty)$	4	$\frac{6}{1} < 1$	False

The solution set is $(-\infty, 3)$.

27. $(2x - 3)(4x + 5) \le 0$

$(2x - 3)(4x + 5) = 0$

$2x - 3 = 0$ or $4x + 5 = 0$

$2x = 3 \qquad 4x = -5$

$x = \frac{3}{2} \qquad x = -\frac{5}{4}$

Region	Test Point Value	$(2x - 3)(4x + 5) \le 0$	Result
$\left(-\infty, -\frac{5}{4}\right)$	−2	$(-7)(-3) \le 0$	False
$\left(-\frac{5}{4}, \frac{3}{2}\right)$	0	$(-3)(5) \le 0$	True
$\left(\frac{3}{2}, \infty\right)$	2	$(1)(13) \le 0$	False

The solution set is $\left[-\frac{5}{4}, \frac{3}{2}\right]$.

29. $x^2 > x$

$x^2 = x$

$x^2 - x = 0$

$x(x-1) = 0$

$x = 0$ or $x - 1 = 0$

$x = 1$

Region	Test Point Value	$x^2 > x$	Result
$(-\infty, 0)$	-1	$(-1)^2 > -1$	True
$(0, 1)$	$\frac{1}{2}$	$\left(\frac{1}{2}\right)^2 > \frac{1}{2}$	False
$(1, \infty)$	2	$2^2 > 2$	True

The solution set is $(-\infty, 0) \cup (1, \infty)$.

31. $\frac{x}{x-10} < 0$

$x - 10 = 0$

$x = 10$ makes the denominator 0.

$\frac{x}{x-10} = 0$

$(x-10)\left(\frac{x}{x-10}\right) = (x-10)(0)$

$x = 0$

Region	Test Point Value	$\frac{x}{x-10} < 0$	Result
$(-\infty, 0)$	-1	$\frac{-1}{-11} < 0$	False
$(0, 10)$	1	$\frac{1}{-9} < 0$	True
$(10, \infty)$	11	$\frac{11}{1} < 0$	False

The solution set is $(0, 10)$.

33. $(2x-8)(x+4)(x-6) \le 0$

$(2x-8)(x+4)(x-6) = 0$

$2x - 8 = 0$ or $x + 4 = 0$ or $x - 6 = 0$

$2x = 8$

$x = 4$ $\quad x = -4$ $\quad x = 6$

Region	Test Point Value	$(2x-8)(x+4)(x-6) \le 0$	Result
$(-\infty, -4)$	-5	$(-18)(-1)(-11) \le 0$	True
$(-4, 4)$	0	$(-8)(4)(-6) \le 0$	False
$(4, 6)$	5	$(2)(9)(-1) \le 0$	True
$(6, \infty)$	7	$(6)(11)(1) \le 0$	False

The solution set is $(-\infty, -4] \cup [4, 6]$.

35.
$$6x^2 - 5x \ge 6$$
$$6x^2 - 5x = 6$$
$$6x^2 - 5x - 6 = 0$$
$$(3x+2)(2x-3) = 0$$
$$3x+2=0 \quad \text{or} \quad 2x-3=0$$
$$3x=-2 \qquad 2x=3$$
$$x=-\frac{2}{3} \qquad x=\frac{3}{2}$$

Region	Test Point Value	$6x^2 - 5x \ge 6$	Result
$\left(-\infty, -\frac{2}{3}\right)$	-1	$6+5 \ge 6$	True
$\left(-\frac{2}{3}, \frac{3}{2}\right)$	0	$0-0 \ge 6$	False
$\left(\frac{3}{2}, \infty\right)$	2	$24-10 \ge 6$	True

The solution set is $\left(-\infty, -\frac{2}{3}\right] \cup \left[\frac{3}{2}, \infty\right)$.

37. $\frac{x-5}{x+4} \ge 0$

$x+4=0$

$x=-4$ makes the denominator 0.

$$\frac{x-5}{x+4} = 0$$
$$(x+4)\left(\frac{x-5}{x+4}\right) = (x+4)(0)$$
$$x-5=0$$
$$x=5$$

Region	Test Point Value	$\frac{x-5}{x+4} \geq 0$	Result
$(-\infty, -4)$	-5	$\frac{-10}{-1} \geq 0$	True
$(-4, 5)$	0	$\frac{-5}{4} \geq 0$	False
$(5, \infty)$	6	$\frac{1}{10} \geq 0$	True

The solution set is $(-\infty, -4) \cup [5, \infty)$.

39. $\frac{-1}{x-1} > -1$

$x - 1 = 0$

$x = 1$ makes the denominator 0.

$$\frac{-1}{x-1} = -1$$

$$(x-1)\left(\frac{-1}{x-1}\right) = (x-1)(-1)$$

$$-1 = -x + 1$$

$$x = 2$$

Region	Test Point Value	$\frac{-1}{x-1} > -1$	Result
$(-\infty, 1)$	0	$\frac{-1}{-1} > -1$	True
$(1, 2)$	1.5	$\frac{-1}{0.5} > -1$	False
$(2, \infty)$	3	$\frac{-1}{2} > -1$	True

The solution set is $(-\infty, 1) \cup (2, \infty)$.

41. $4x^3 + 16x^2 - 9x - 36 > 0$

$$4x^2(x+4) - 9(x+4) > 0$$

$$(x+4)(4x^2 - 9) > 0$$

$$(x+4)(2x+3)(2x-3) > 0$$

$$(x+4)(2x+3)(2x-3) = 0$$

$x + 4 = 0$ or $2x + 3 = 0$ or $2x - 3 = 0$

$2x = -3$ $\quad$ $2x = 3$

$x = -4$ $\quad$ $x = -\frac{3}{2}$ $\quad$ $x = \frac{3}{2}$

Region	Test Point Value	$(x+4)(2x+3)(2x-3)>0$	Result
$(-\infty, -4)$	-5	$(-1)(-7)(-13)>0$	False
$\left(-4, -\frac{3}{2}\right)$	-2	$(2)(-1)(-7)>0$	True
$\left(-\frac{3}{2}, \frac{3}{2}\right)$	0	$(4)(3)(-3)>0$	False
$\left(\frac{3}{2}, \infty\right)$	2	$(6)(7)(1)>0$	True

The solution set is $\left(-4, -\dfrac{3}{2}\right) \cup \left(\dfrac{3}{2}, \infty\right)$.

43.

$$x^4 - 26x^2 + 25 \geq 0$$
$$(x^2 - 25)(x^2 - 1) \geq 0$$
$$(x+5)(x-5)(x+1)(x-1) \geq 0$$
$$(x+5)(x-5)(x+1)(x-1) = 0$$
$$x+5=0 \quad \text{or} \quad x-5=0 \quad \text{or} \quad x+1=0 \quad \text{or} \quad x-1=0$$
$$x=-5 \qquad\qquad x=5 \qquad\qquad x=-1 \qquad\qquad x=1$$

Region	Test Point Value	$(x+5)(x-5)(x+1)(x-1) \geq 0$	Result
$(-\infty, -5)$	-6	$(-1)(-11)(-5)(-7) \geq 0$	True
$(-5, -1)$	-2	$(3)(-7)(-1)(-3) \geq 0$	False
$(-1, 1)$	0	$(5)(-5)(1)(-1) \geq 0$	True
$(1, 5)$	2	$(7)(-3)(3)(1) \geq 0$	False
$(5, \infty)$	6	$(11)(1)(7)(5) \geq 0$	True

The solution set is $(-\infty, -5] \cup [-1, 1] \cup [5, \infty)$.

45. $\dfrac{x(x+6)}{(x-7)(x+1)} \geq 0$

$(x-7)(x+1) = 0$

$x - 7 = 0$ or $x + 1 = 0$

$x = 7$ and $x = -1$ make the denominator 0.

$$\frac{x(x+6)}{(x-7)(x+1)} = 0$$

$$(x-7)(x+1)\left(\frac{x(x+6)}{(x-7)(x+1)}\right) = (x-7)(x+1)(0)$$

$$x(x+6) = 0$$

$x = 0$ or $x+6=0$

$x = -6$

Region	Test Point Value	$\frac{x(x+6)}{(x-7)(x+1)} \geq 0$	Result
$(-\infty, -6)$	-7	$\frac{(-7)(-1)}{(-14)(-6)} \geq 0$	True
$(-6, -1)$	-2	$\frac{(-2)(4)}{(-9)(-1)} \geq 0$	False
$(-1, 0)$	$-\frac{1}{2}$	$\frac{\left(-\frac{1}{2}\right)\left(\frac{11}{2}\right)}{\left(-\frac{15}{2}\right)\left(\frac{1}{2}\right)} \geq 0$	True
$(0, 7)$	1	$\frac{(1)(7)}{(-6)(2)} \geq 0$	False
$(7, \infty)$	8	$\frac{8(14)}{(1)(9)} \geq 0$	True

The solution set is $(-\infty, -6] \cup (-1, 0] \cup (7, \infty)$.

47. $\dfrac{x}{x+4} \leq 2$

$x + 4 = 0$

$x = -4$ makes the denominator 0.

$$\frac{x}{x+4} = 2$$

$$(x+4)\left(\frac{x}{x+4}\right) = (x+4)(2)$$

$$x = 2x + 8$$

$$-8 = x$$

Region	Test Point Value	$\frac{x}{x+4} \le 2$	Result
$(-\infty, -8)$	-9	$\frac{-9}{-5} \le 2$	True
$(-8, -4)$	-5	$\frac{-5}{-1} \le 2$	False
$(-4, \infty)$	0	$\frac{0}{4} \le 2$	True

The solution set is $(-\infty, -8] \cup (-4, \infty)$.

49. $(2x-7)(3x+5) > 0$
$(2x-7)(3x+5) = 0$
$2x-7=0$ or $3x+5=0$
$2x=7$ $\quad$ $3x=-5$
$x=\frac{7}{2}$ $\quad$ $x=-\frac{5}{3}$

Region	Test Point Value	$(2x-7)(3x+5) > 0$	Result
$\left(-\infty, -\frac{5}{3}\right)$	-2	$(-11)(-1) > 0$	True
$\left(-\frac{5}{3}, \frac{7}{2}\right)$	0	$(-7)(5) > 0$	False
$\left(\frac{7}{2}, \infty\right)$	4	$(1)(17) > 0$	True

The solution set is $\left(-\infty, -\frac{5}{3}\right) \cup \left(\frac{7}{2}, \infty\right)$.

51. $\frac{z}{z-5} \ge 2z$
$z-5=0$
$z=5$ makes the denominator 0.

$$\frac{z}{z-5} = 2z$$

$$(z-5)\left(\frac{z}{z-5}\right) = (z-5)(2z)$$

$$z = 2z^2 - 10z$$

$$0 = 2z^2 - 11z$$

$$0 = z(2z-11)$$

$z = 0$ or $2z - 11 = 0$

$$2z = 11$$

$$z = \frac{11}{2}$$

Region	Test Point Value	$\frac{z}{z-5} \geq 2z$	Result
$(-\infty, 0)$	-1	$\frac{-1}{-6} \geq -2$	True
$(0, 5)$	1	$\frac{1}{-4} \geq 2$	False
$\left(5, \frac{11}{2}\right)$	5.25	$\frac{5.25}{0.25} \geq 10.5$	True
$\left(\frac{11}{2}, \infty\right)$	6	$\frac{6}{1} \geq 12$	False

The solution set is $(-\infty, 0] \cup \left(5, \frac{11}{2}\right]$.

53. $\frac{(x+1)^2}{5x} > 0$

$5x = 0$

$x = 0$ makes the denominator 0.

$$\frac{(x+1)^2}{5x} = 0$$

$$(5x)\left(\frac{(x+1)^2}{5x}\right) = (5x)(0)$$

$$(x+1)^2 = 0$$

$x + 1 = 0$

$x = -1$

Region	Test Point Value	$\frac{(x+1)^2}{5x} > 0$	Result
$(-\infty, -1)$	-2	$\frac{(-1)^2}{-10} > 0$	False
$(-1, 0)$	$-\frac{1}{2}$	$\frac{\left(\frac{1}{2}\right)^2}{-\frac{5}{2}} > 0$	False
$(0, \infty)$	1	$\frac{2^2}{5} > 0$	True

The solution set is $(0, \infty)$.

55.

x	$y = f(x) = x^2$
0	$0^2 = 0$
1	$1^2 = 1$
-1	$(-1)^2 = 1$
2	$2^2 = 4$
-2	$(-2)^2 = 4$

57.

x	$y = f(x) = -x^2$
0	$-0^2 = 0$
1	$-1^2 = -1$
-1	$-(-1)^2 = -1$
2	$-2^2 = -4$
-2	$-(-2)^2 = -4$

59. answers may vary

61. A number minus its reciprocal is written as $x-\frac{1}{x}$.

$x-\frac{1}{x}<0$

$x=0$ makes the denominator 0.

$$x-\frac{1}{x}=0$$
$$x(x)-x\left(\frac{1}{x}\right)=x(0)$$
$$x^2-1=0$$
$$(x+1)(x-1)=0$$
$$x+1=0 \quad \text{or} \quad x-1=0$$
$$x=-1 \qquad\qquad x=1$$

Region	Test Point Value	$x-\frac{1}{x}<0$	Result
$(-\infty, -1)$	-2	$-2-\frac{1}{-2}=-2+\frac{1}{2}<0$	True
$(-1, 0)$	$-\frac{1}{2}$	$-\frac{1}{2}-\frac{1}{-\frac{1}{2}}=-\frac{1}{2}+2<0$	False
$(0, 1)$	$\frac{1}{2}$	$\frac{1}{2}-\frac{1}{\frac{1}{2}}=\frac{1}{2}-2<0$	True
$(1, \infty)$	2	$2-\frac{1}{2}<0$	False

The solution set is $(-\infty, -1) \cup (0, 1)$.

63.
$$P(x)>0$$
$$-2x^2+26x-44>0$$
$$-2(x^2-13x+22)>0$$
$$-2(x-11)(x-2)>0$$
$$-2(x-11)(x-2)=0$$
$$x-11=0 \quad \text{or} \quad x-2=0$$
$$x=11 \qquad\qquad x=2$$

Region	Test Point Value	$-2(x-11)(x-2)>0$	Result
$(-\infty, 2)$	0	$-2(-11)(-2)>0$	False
$(2, 11)$	3	$-2(-8)(1)>0$	True
$(11, \infty)$	12	$-2(1)(10)>0$	False

The company makes a profit when x is between 2 and 11.

65. $x^2 - x - 56 > 0$

$(x+7)(x-8) > 0$

$(x+7)(x-8) = 0$

$x+7=0$ or $x-8=0$

$x=-7$ $\quad x=8$

Region	Test Point Value	$(x + 7)(x - 8) > 0$	Result
$(-\infty, -7)$	-8	$(-1)(-16) > 0$	True
$(-7, 8)$	0	$(7)(-8) > 0$	False
$(8, \infty)$	9	$(16)(1) > 0$	True

The solution set is $(-\infty, -7) \cup (8, \infty)$.

The Bigger Picture

1. $|x - 8| = |2x + 1|$

$x-8=2x+1$ or $x-8=-(2x+1)$

$-x-8=1$ $\quad x-8=-2x-1$

$-x=9$ $\quad 3x-8=-1$

$x=-9$ $\quad 3x=7$

$x=\frac{7}{3}$

The solution set is $\left\{-9, \frac{7}{3}\right\}$.

2.
$$0 < -x+7 < 3$$
$$0-7 < -x+7-7 < 3-7$$
$$-7 < -x < -4$$
$$7 > x > 4$$
$$\text{or } 4 < x < 7$$

The solution set is (4, 7).

3. $\sqrt{3x-11}+3=x$

$\sqrt{3x-11}=x-3$

$\left(\sqrt{3x-11}\right)^2=(x-3)^2$

$3x-11=x^2-6x+9$

$0=x^2-9x+20$

$0=(x-4)(x-5)$

$x-4=0$ or $x-5=0$

$x=4$ $\quad x=5$

The solution set is $\{4, 5\}$.

4. $x(3x+1)=1$

$3x^2+x=1$

$3x^2+x-1=0$

$a=3,\ b=1,\ c=-1$

$$x=\frac{-b\pm\sqrt{b^2-4ac}}{2a}$$

$$=\frac{-1\pm\sqrt{1^2-4(3)(-1)}}{2(3)}$$

$$=\frac{-1\pm\sqrt{1+12}}{6}$$

$$=\frac{-1\pm\sqrt{13}}{6}$$

The solution set is $\left\{\frac{-1-\sqrt{13}}{6}, \frac{-1+\sqrt{13}}{6}\right\}$.

5. $\frac{x+2}{x-7}\le 0$

$x-7=0$

$x=7$ makes the denominator 0.

$$\frac{x+2}{x-7}=0$$

$$(x-7)\left(\frac{x+2}{x-7}\right)=(x-7)(0)$$

$$x+2=0$$

$$x=-2$$

Region	Test Point Value	$\frac{x+2}{x-7}\le 0$	Result
$(-\infty, -2)$	-3	$\frac{-1}{-10}\le 0$	False
$(-2, 7)$	0	$\frac{2}{-7}\le 0$	True
$(7, \infty)$	8	$\frac{10}{1}\le 0$	False

The solution set is $[-2, 7)$.

6. $x(x-6)+4=x^2-2(3-x)$

$$x^2-6x+4=x^2-6+2x$$
$$-6x+4=-6+2x$$
$$-8x=-10$$
$$x=\frac{10}{8}$$
$$x=\frac{5}{4}$$

The solution set is $\left\{\frac{5}{4}\right\}$.

7. $x(5x-36)=-7$

$$5x^2-36x=-7$$
$$5x^2-36x+7=0$$
$$(5x-1)(x-7)=0$$
$$5x-1=0 \quad \text{or} \quad x-7=0$$
$$5x=1 \qquad\qquad x=7$$
$$x=\frac{1}{5}$$

The solution set is $\left\{\frac{1}{5}, 7\right\}$.

8. $2x^2-4\geq 7x$

$$2x^2-7x-4\geq 0$$
$$(2x+1)(x-4)\geq 0$$
$$(2x+1)(x-4)=0$$
$$2x+1=0 \quad \text{or} \quad x-4=0$$
$$2x=-1 \qquad\qquad x=4$$
$$x=-\frac{1}{2}$$

Region	Test Point Value	$(2x + 1)(x - 4) \geq 0$	Result
$\left(-\infty, -\frac{1}{2}\right)$	-1	$(-1)(-5)\geq 0$	True
$\left(-\frac{1}{2}, 4\right)$	0	$(1)(-4)\geq 0$	False
$(4, \infty)$	5	$(11)(1)\geq 0$	True

The solution set is $\left(-\infty, -\frac{1}{2}\right]\cup[4, \infty)$.

9. $\left|\dfrac{x-7}{3}\right| > 5$

$\dfrac{x-7}{3} > 5$ or $\dfrac{x-7}{3} < -5$

$x-7 > 15 \qquad x-7 < -15$

$x > 22 \qquad x < -8$

The solution set is $(-\infty, -8) \cup (22, \infty)$.

10. $2(x-5)+4 < 1+7(x-5)-x$

$2x-10+4 < 1+7x-35-x$

$2x-6 < 6x-34$

$2x-6x < -34+6$

$-4x < -28$

$\dfrac{-4x}{-4} > \dfrac{-28}{-4}$

$x > 7$

The solution set is $(7, \infty)$.

Section 8.5 Practice Problems

1. $f(x) = x^2$, $g(x) = x^2 + 4$

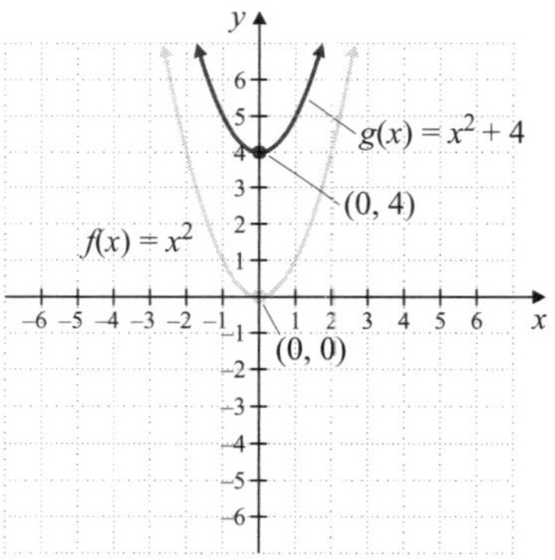

2. $F(x) = x^2 + 1$

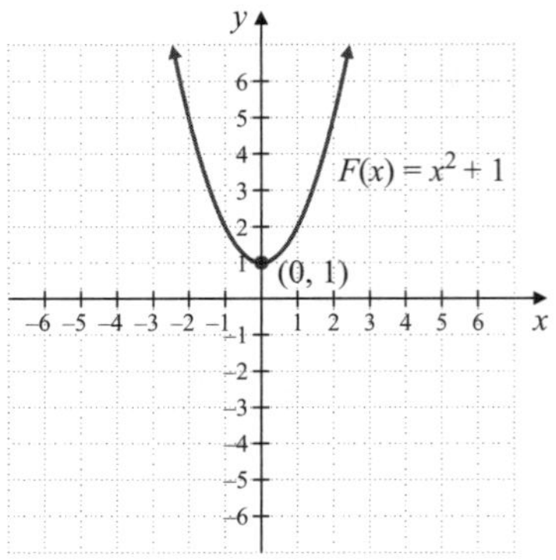

3. $g(x) = x^2 - 2$

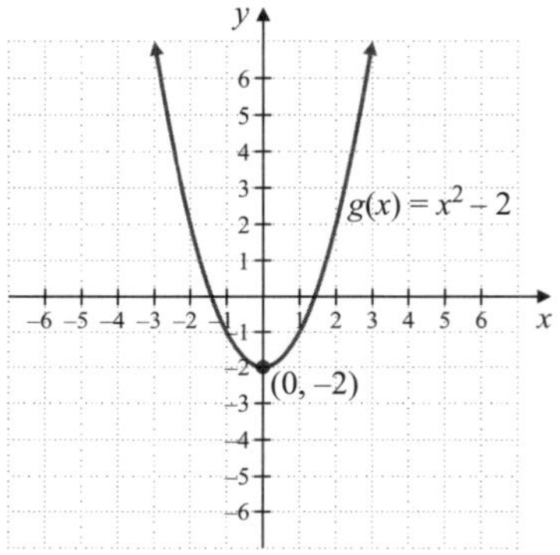

4. $f(x) = x^2$, $g(x) = (x-1)^2$

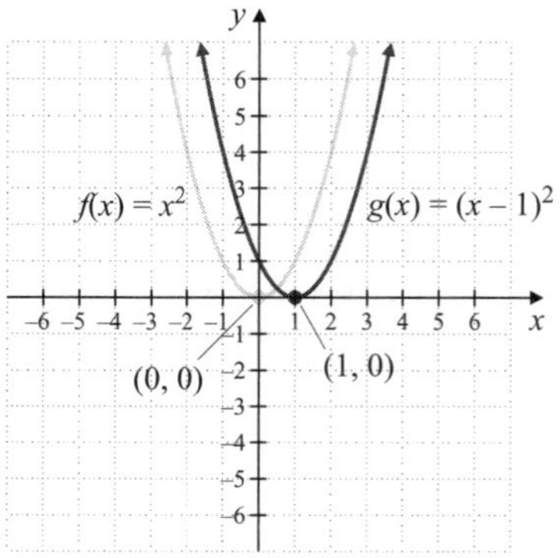

5. $G(x) = (x-4)^2$

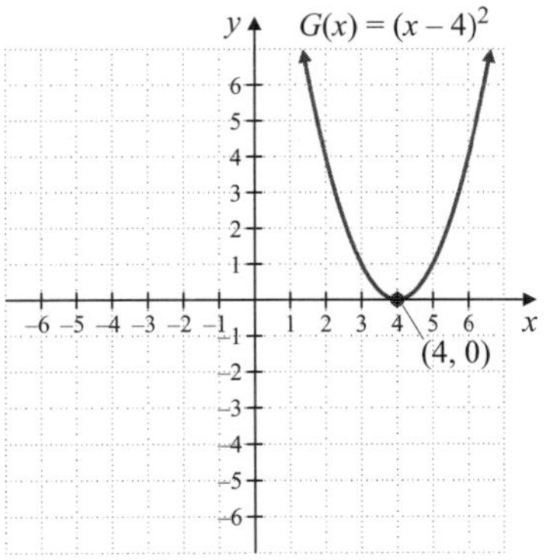

6. $F(x) = (x+2)^2$

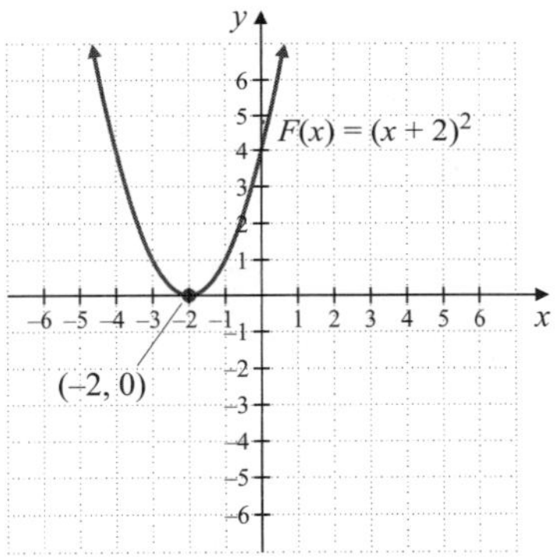

7. $F(x) = (x-2)^2 + 3$

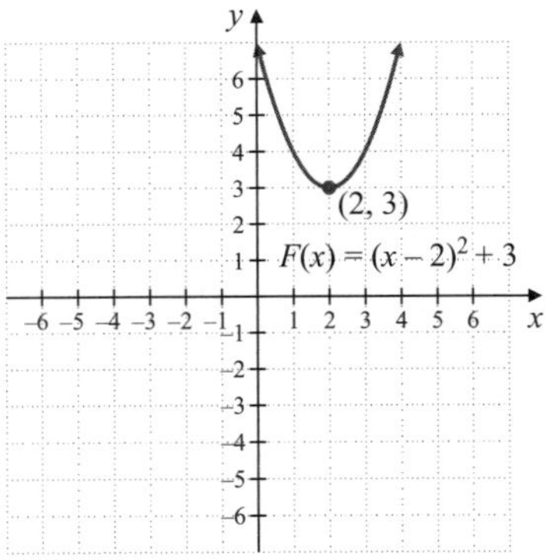

8. $f(x) = x^2,\ g(x) = 2x^2,\ h(x) = \frac{1}{3}x^2$

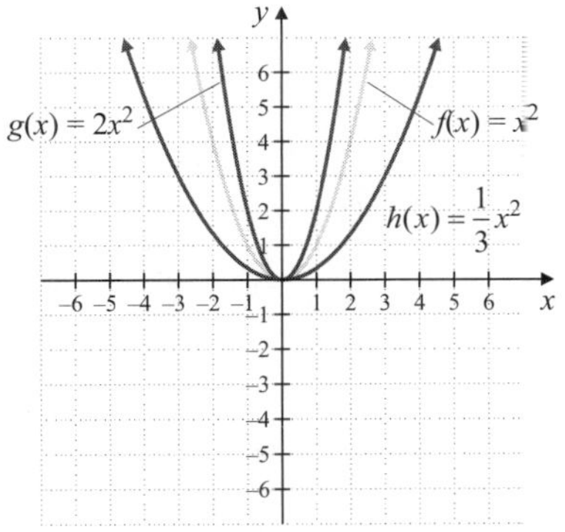

9. $f(x) = -3x^2$

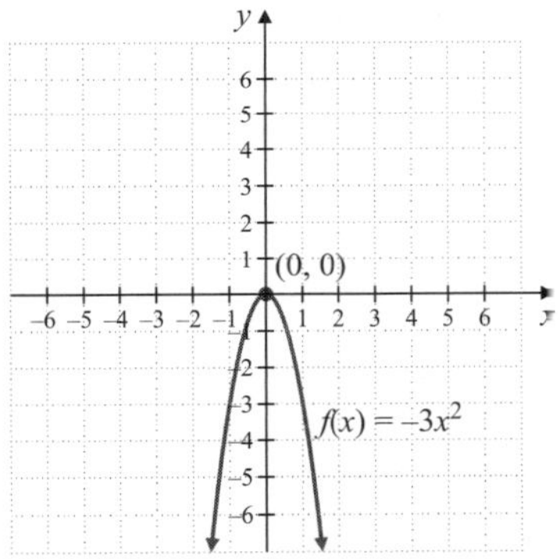

10. $f(x) = 2(x+3)^2 - 4$

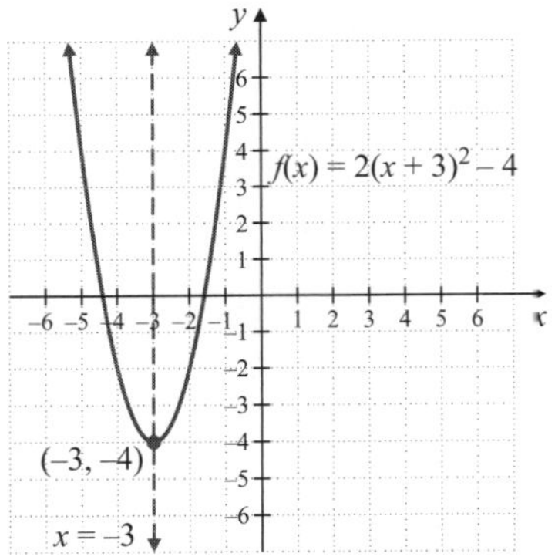

Calculator Explorations

1. $F(x)=\sqrt{x};\ G(x)=\sqrt{x}+1$

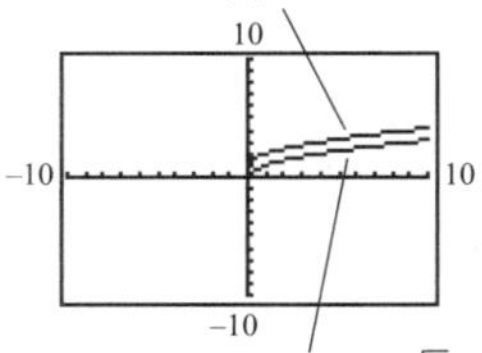

2. $g(x)=x^3,\ H(x)=x^3-2$

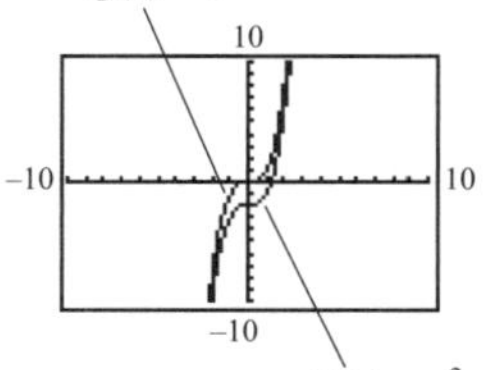

3. $H(x)=|x|;\ f(x)=|x-5|$

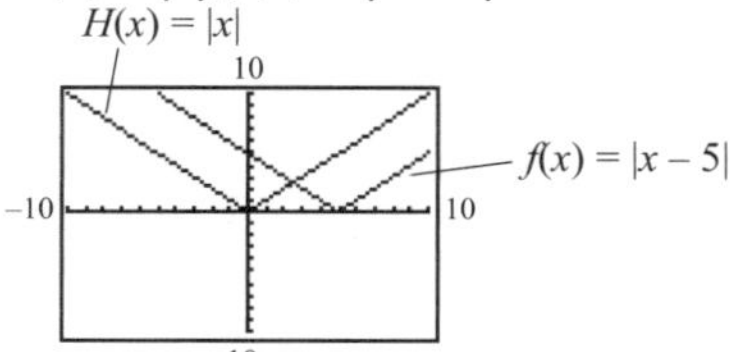

4. $h(x)=x^3+2;\ g(x)=(x-3)^3+2$

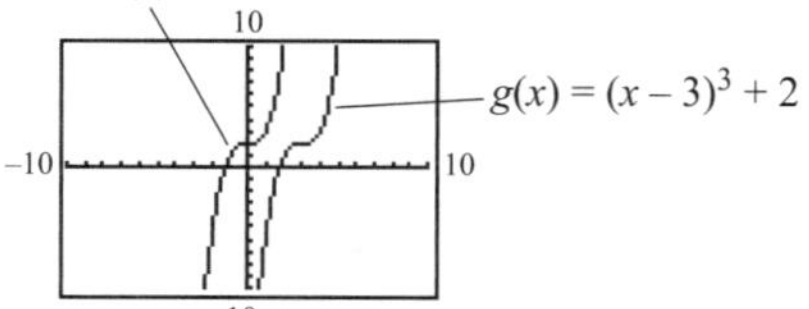

5. $f(x)=|x+4|;\ F(x)=|x+4|+3$

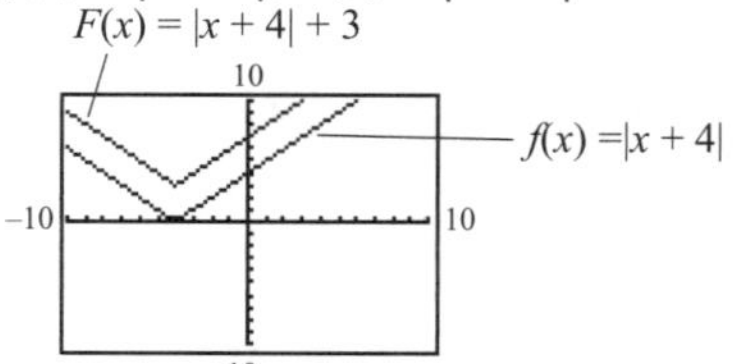

6. $G(x)=\sqrt{x}-2;\ \ g(x)=\sqrt{x-4}-2$

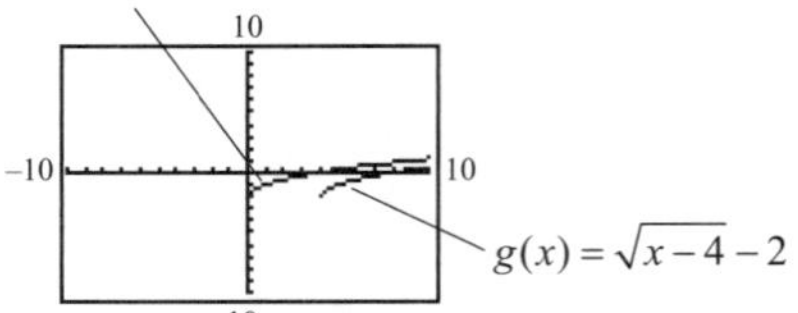

Mental Math

1. $f(x)=x^2$
vertex: (0, 0)

2. $f(x)=-5x^2$
vertex: (0, 0)

3. $g(x)=(x-2)^2$
vertex: (2, 0)

4. $g(x)=(x+5)^2$
vertex: (−5, 0)

5. $f(x)=2x^2+3$
vertex: (0, 3)

6. $h(x)=x^2-1$
vertex: (0, −1)

7. $g(x)=(x+1)^2+5$
vertex: (−1, 5)

8. $h(x)=(x-10)^2-7$
vertex: (10, −7)

Exercise Set 8.5

1. $f(x) = x^2 - 1$

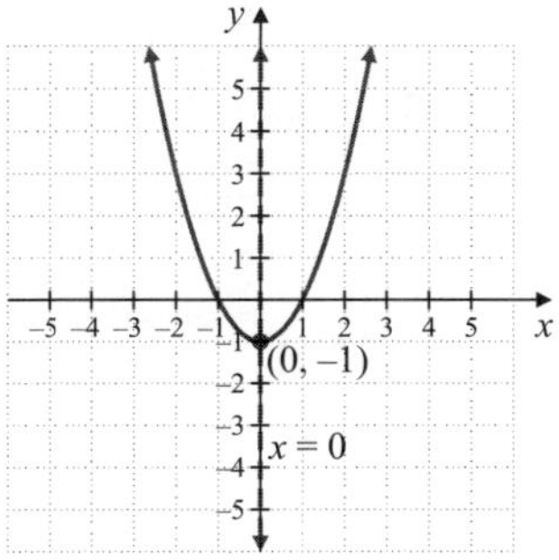

3. $f(x) = (x - 5)^2$

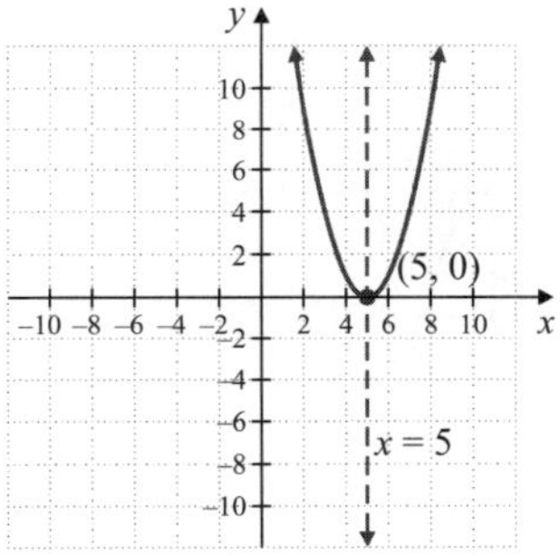

5. $h(x) = x^2 + 5$

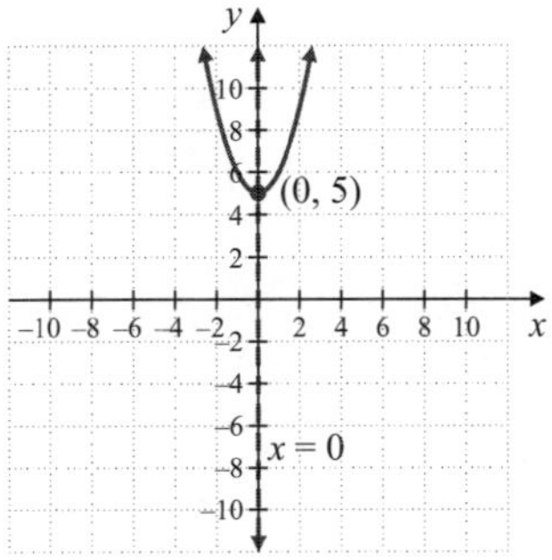

7. $h(x) = (x + 2)^2$

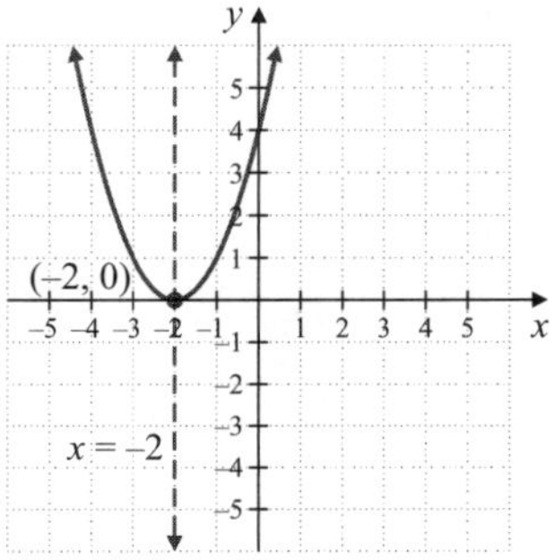

9. $g(x) = x^2 + 7$

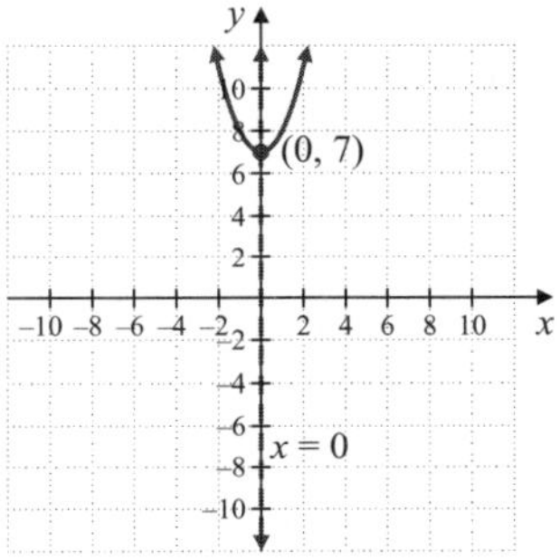

11. $G(x) = (x + 3)^2$

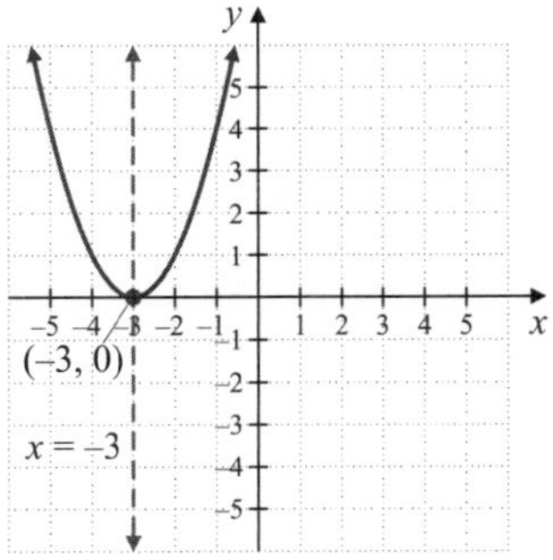

13. $f(x) = (x-2)^2 + 5$

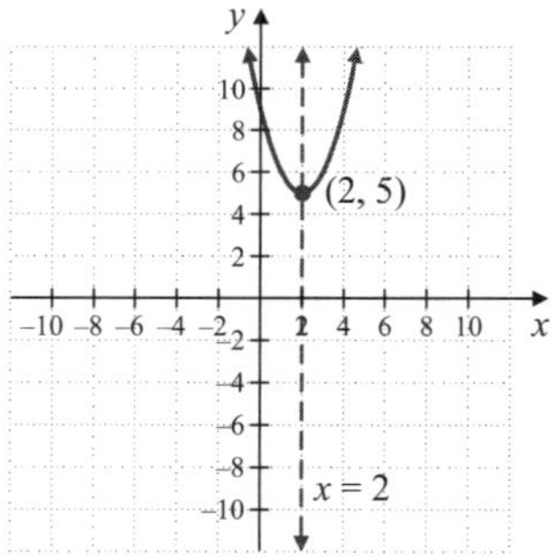

15. $h(x) = (x+1)^2 + 4$

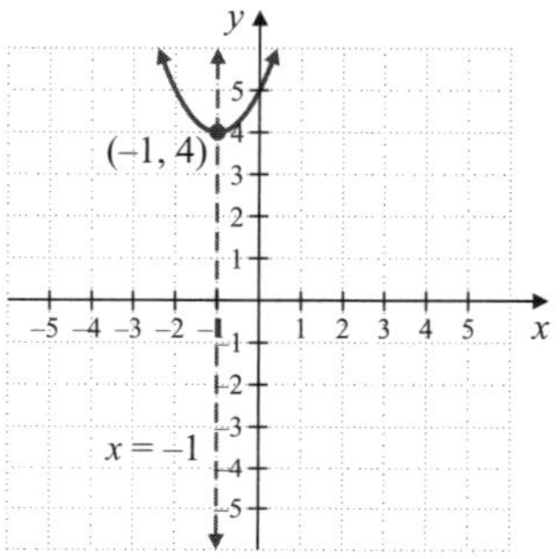

17. $g(x) = (x+2)^2 - 5$

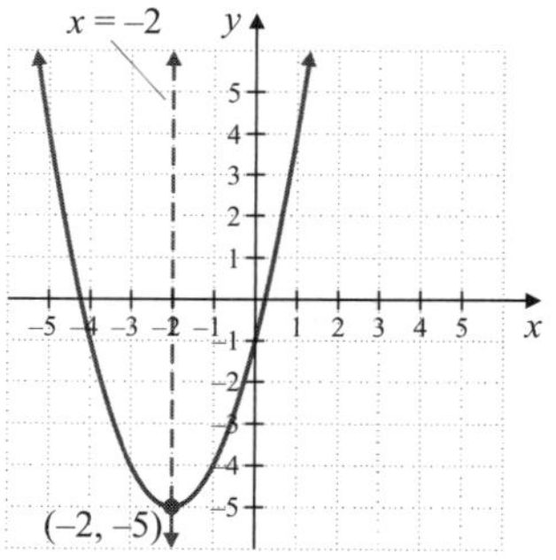

19. $h(x) = (x-3)^2 + 2$

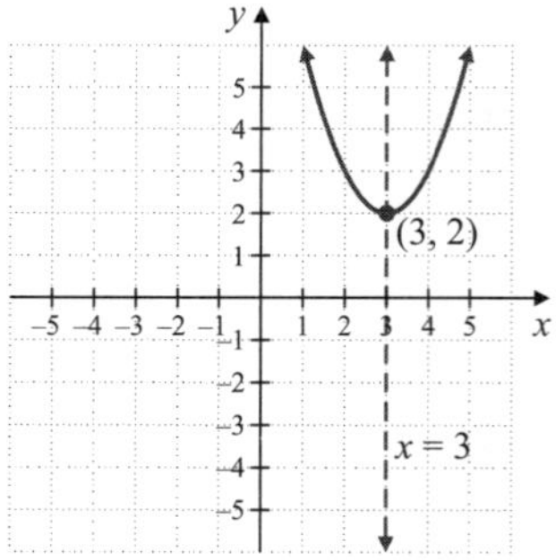

21. $g(x) = -x^2$

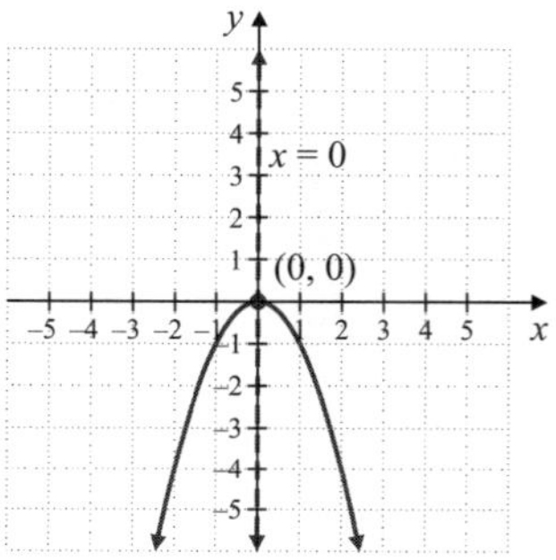

23. $h(x) = \frac{1}{3}x^2$

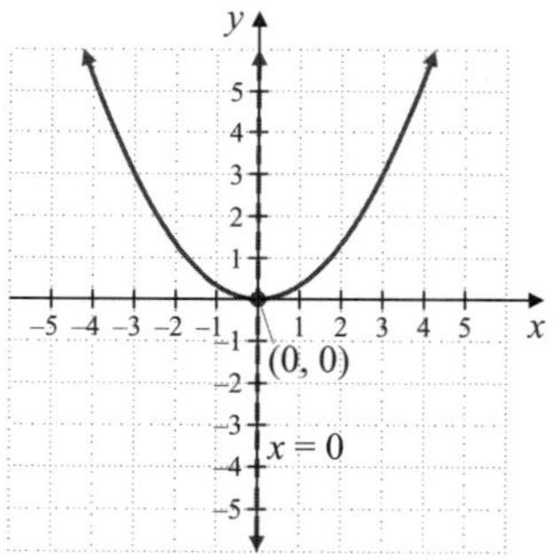

25. $H(x) = 2x^2$

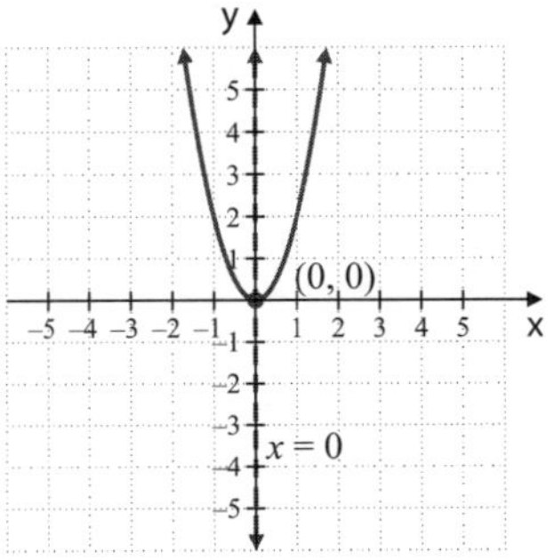

27. $F(x) = -4x^2$

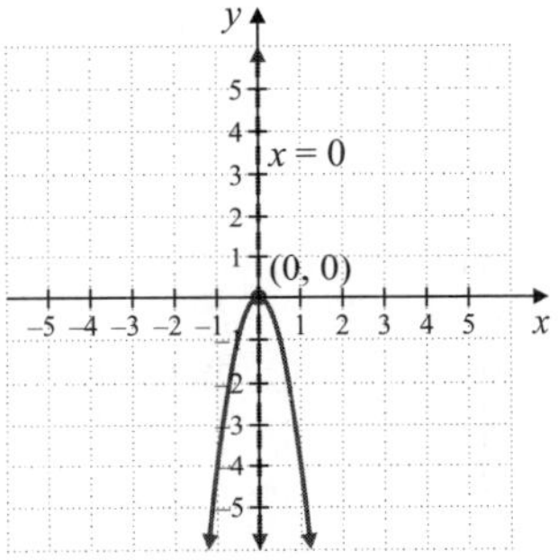

29. $f(x) = 10(x+4)^2 - 6$

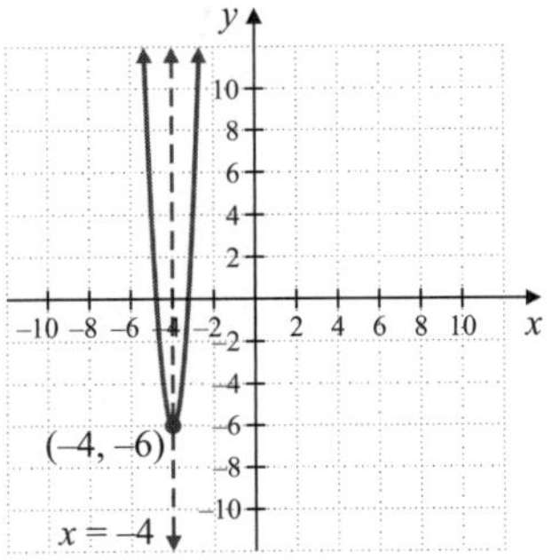

31. $h(x) = -3(x+3)^2 + 1$

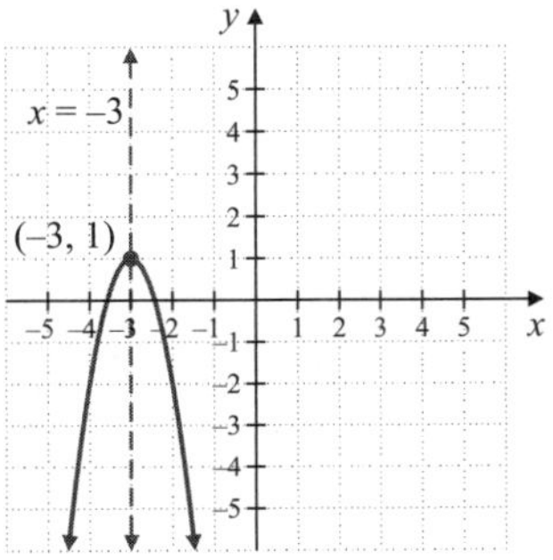

33. $H(x) = \frac{1}{2}(x-6)^2 - 3$

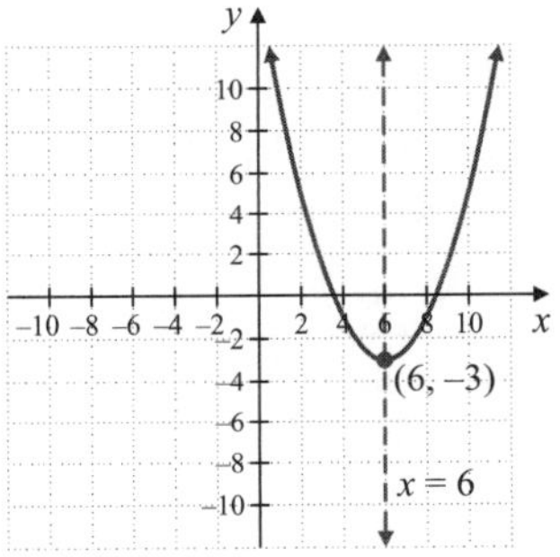

35. $f(x) = -(x-1)^2$

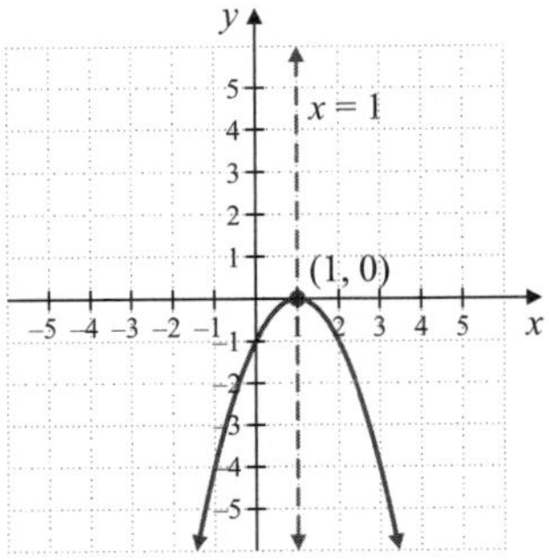

37. $F(x)=\left(x+\frac{1}{2}\right)^2-2$

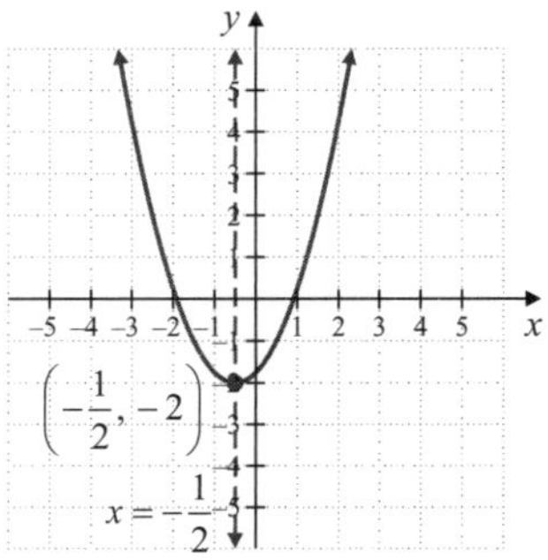

39. $F(x)=-x^2+2$

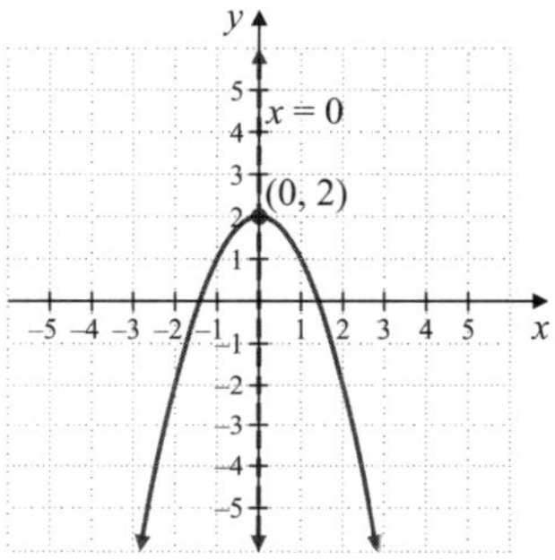

41. x^2+8x

$\left[\frac{1}{2}(8)\right]^2=4^2=16$

$x^2+8x+16$

43. z^2-16z

$\left[\frac{1}{2}(-16)\right]^2=(-8)^2=64$

$z^2-16z+64$

45. y^2+y

$\left[\frac{1}{2}(1)\right]^2=\left[\frac{1}{2}\right]^2=\frac{1}{4}$

$y^2+y+\frac{1}{4}$

47. $g(x)=5(x-2)^2+3$

49. $g(x)=5[x-(-3)]^2+6$

$g(x)=5(x+3)^2+6$

51. $y = f(x) + 1$
Shift the graph 1 unit up.

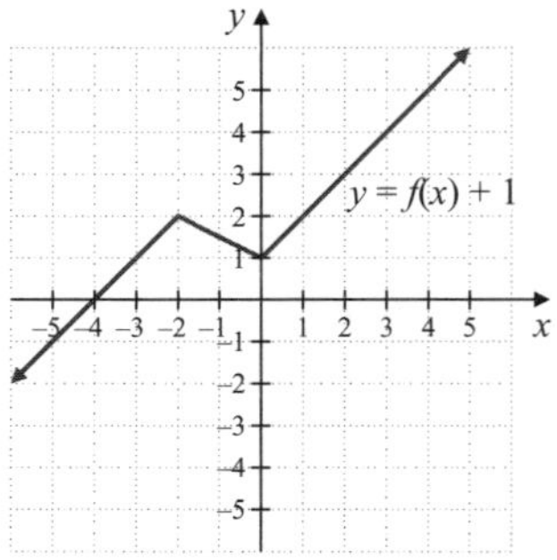

53. $y = f(x - 3)$
Shift the graph 3 units to the right.

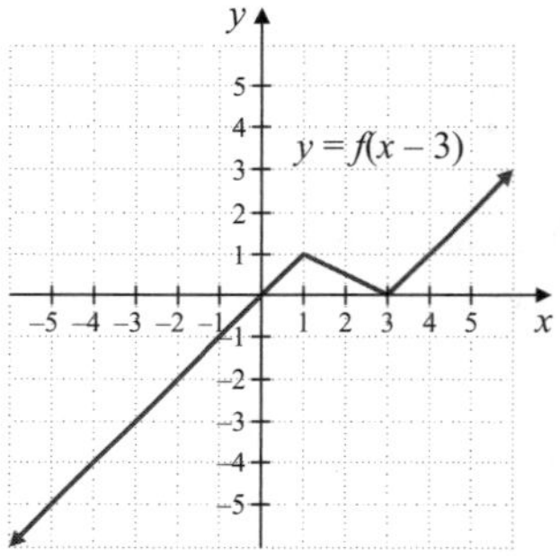

55. $y = f(x + 2) + 2$
Shift the graph 2 units up and 2 units to the left.

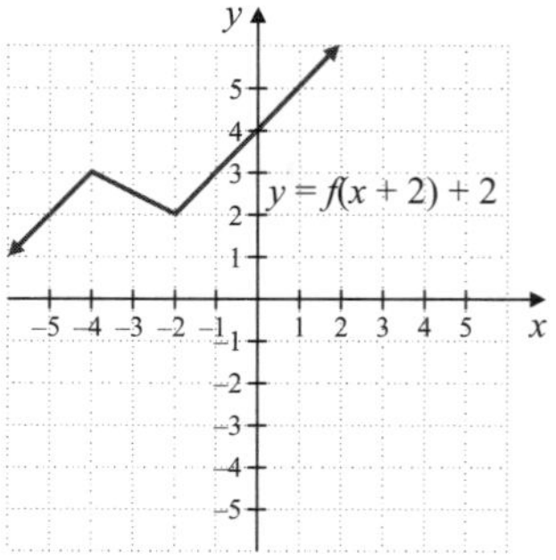

57. Since $a < 0$, the graph of $f(x)$ opens downward, with vertex (0.1, 3.6), which is choice c.

Section 8.6 Practice Problems

1. $f(x) = x^2 - 4x - 5$

$$y = x^2 - 4x - 5$$
$$y + 5 = x^2 - 4x$$
$$y + 5 + 4 = x^2 - 4x + 4$$
$$y + 9 = (x-2)^2$$
$$y = (x-2)^2 - 9$$

The vertex is (2, −9).

$0 = x^2 - 4x - 5$
$0 = (x-5)(x+1)$
$x - 5 = 0$ or $x + 1 = 0$
$x = 5$ $\quad$ $x = -1$

The x-intercepts are (−1, 0) and (5, 0).

$f(0) = 0^2 - 4(0) - 5 = -5$

The y-intercept is (0, −5).

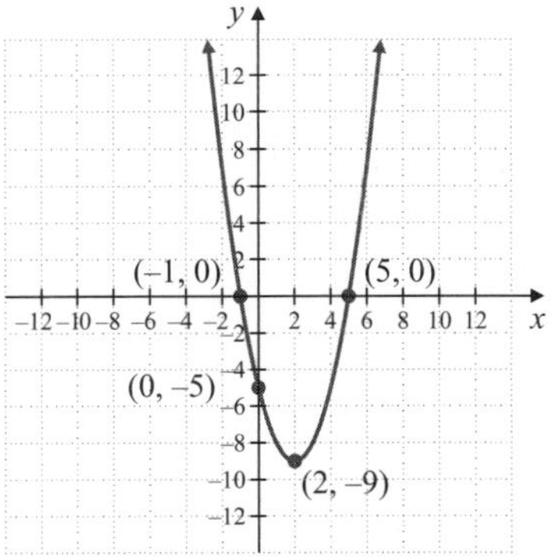

2. $f(x) = 2x^2 + 2x + 5$

$$y = 2x^2 + 2x + 5$$
$$y - 5 = 2x^2 + 2x$$
$$y - 5 = 2(x^2 + x)$$
$$y - 5 + 2\left(\frac{1}{4}\right) = 2\left(x^2 + x + \frac{1}{4}\right)$$
$$y - \frac{9}{2} = 2\left(x + \frac{1}{2}\right)^2$$
$$y = 2\left(x + \frac{1}{2}\right)^2 + \frac{9}{2}$$

The vertex is $\left(-\frac{1}{2}, \frac{9}{2}\right)$.

The parabola opens upward, since $a > 0$.

$f(0) = 2(0)^2 + 2(0) + 5 = 5$

The y-intercept is (0, 5).

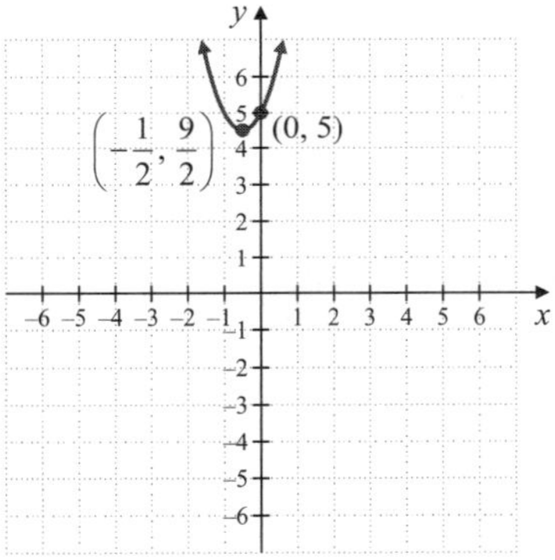

3. $f(x) = -x^2 - 2x + 8$

$$y = -x^2 - 2x + 8$$
$$y - 8 = -x^2 - 2x$$
$$y - 8 = -1(x^2 + 2x)$$
$$y - 8 - 1 = -1(x^2 + 2x + 1)$$
$$y - 9 = -1(x+1)^2$$
$$y = -1(x+1)^2 + 9$$

Since $a = -1$, the parabola opens downward with vertex (−1, 9).

$0 = -x^2 - 2x + 8$
$0 = -(x^2 + 2x - 8)$
$0 = -(x+4)(x-2)$

$x+4=0$ or $x-2=0$

$x=-4$ $\quad$ $x=2$

The x-intercepts are (–4, 0) and (2, 0).

$f(0)=0^2-2(0)-8=8$

The y-intercept is (0, 8).

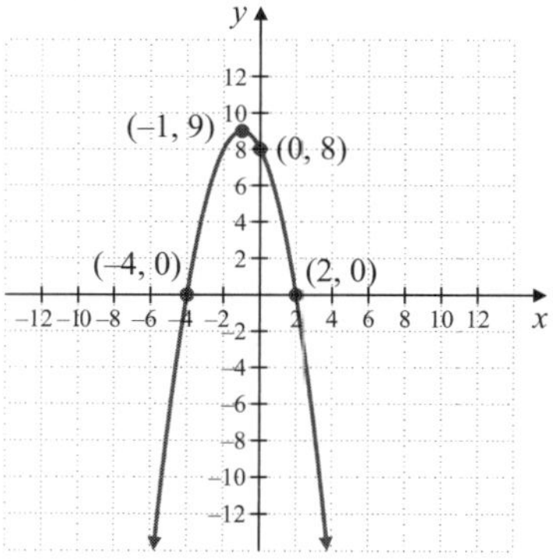

4. $f(x)=x^2-4x-5$

$a=1,\ b=-4,\ c=-5$

$\frac{-b}{2a}=\frac{-(-4)}{2(1)}=2$

$f(2)=2^2-4(2)-5=4-8-5=-9$

The vertex is (2, –9).

5. $f(t)=-16t^2+10t+100$

$a=-16,\ b=10,\ c=100$

$\frac{-b}{2a}=\frac{-10}{2(-16)}=\frac{5}{16}$

$$\begin{aligned}f\left(\frac{5}{16}\right)&=-16\left(\frac{5}{16}\right)^2+10\left(\frac{5}{16}\right)+100\\&=-16\left(\frac{25}{256}\right)+\frac{50}{16}+100\\&=-\frac{25}{16}+\frac{50}{16}+\frac{1600}{16}\\&=\frac{1625}{16}\\&=101\frac{9}{16}\end{aligned}$$

The vertex is $\left(\frac{5}{16}, 101\frac{9}{16}\right)$.

The object reaches its maximum height of $101\frac{9}{16}$ feet in $\frac{5}{16}$ second.

Exercise Set 8.6

1. $f(x)=x^2+8x+7$

$\frac{-b}{2a}=\frac{-8}{2(1)}=-4$ and

$f(-4)=(-4)^2+8(-4)+7=16-32+7=-9$

Thus, the vertex is (–4, –9).

3. $f(x)=-x^2+10x+5$

$\frac{-b}{2a}=\frac{-10}{2(-1)}=5$ and

$f(5)=-5^2+10(5)+5=-25+50+5=30$

Thus, the vertex is (5, 30).

5. $f(x)=5x^2-10x+3$

$\frac{-b}{2a}=\frac{-(-10)}{2(5)}=1$ and

$f(1)=5(1)^2-10(1)+3=5-10+3=-2$

Thus, the vertex is (1, –2).

7. $f(x)=-x^2+x+1$

$\frac{-b}{2a}=\frac{-1}{2(-1)}=\frac{1}{2}$ and

$f\left(\frac{1}{2}\right)=-\left(\frac{1}{2}\right)^2+\frac{1}{2}+1=-\frac{1}{4}+\frac{2}{4}+\frac{4}{4}=\frac{5}{4}$

Thus, the vertex is $\left(\frac{1}{2}, \frac{5}{4}\right)$.

9. $f(x)=x^2-4x+3$

$\frac{-b}{2a}=\frac{-(-4)}{2(1)}=2$

$f(2)=2^2-4(2)+3=-1$

The vertex is (2, –1), which is graph D.

11. $f(x) = x^2 - 2x - 3$

$\frac{-b}{2a} = \frac{-(-2)}{2(1)} = 1$

$f(1) = 1^2 - 2(1) - 3 = -4$

The vertex is (1, –4), which is graph B.

13. $f(x) = x^2 + 4x - 5$

$\frac{-b}{2a} = \frac{-4}{2(1)} = -2$ and

$f(-2) = (-2)^2 + 4(-2) - 5 = 4 - 8 - 5 = -9$

Thus, the vertex is (–2, –9).
The graph opens upward since $a > 0$.

$x^2 + 4x - 5 = 0$

$(x+5)(x-1) = 0$

$x + 5 = 0$ or $x - 1 = 0$

$x = -5$ or $x = 1$

The x-intercepts are (–5, 0) and (1, 0).
$f(0) = -5$, so (0, –5) is the y-intercept.

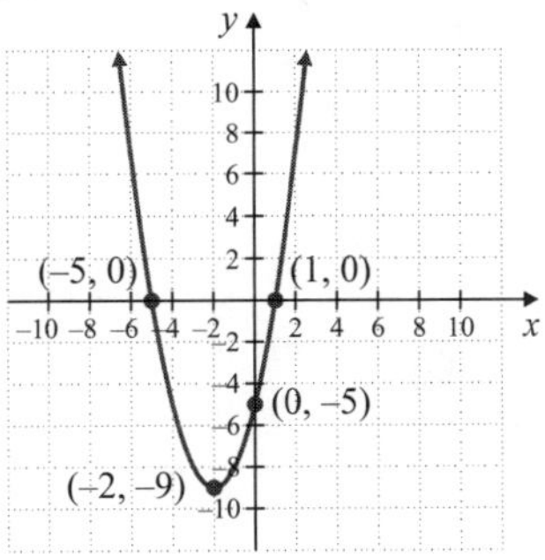

15. $f(x) = -x^2 + 2x - 1$

$\frac{-b}{2a} = \frac{-2}{2(-1)} = 1$ and

$f(1) = -1^2 + 2(1) - 1 = -1 + 2 - 1 = 0$

Thus, the vertex is (1, 0).
The graph opens downward, since $a < 0$.

$-x^2 + 2x - 1 = 0$

$-1(x^2 - 2x + 1) = 0$

$-1(x-1)^2 = 0$

$x - 1 = 0$

$x = 1$

The x-intercept is (1, 0).
$f(0) = -1$, so (0, –1) is the y-intercept.

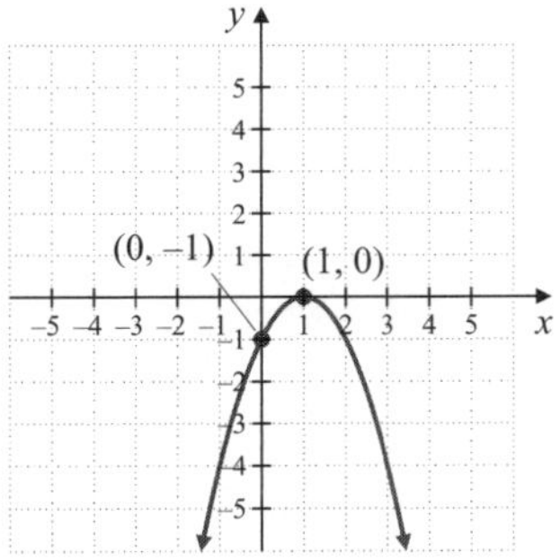

17. $f(x) = x^2 - 4$

$\frac{-b}{2a} = \frac{-0}{2(1)} = 0$ and $f(0) = -4$

Thus, the vertex is (0, –4).
The graph opens upward since $a > 0$.

$x^2 - 4 = 0$

$(x+2)(x-2) = 0$

$x + 2 = 0$ or $x - 2 = 0$

$x = -2$ or $x = 2$

The x-intercepts are (–2, 0) and (2, 0).
$f(0) = -4$, so (0, –4) is the y-intercept.

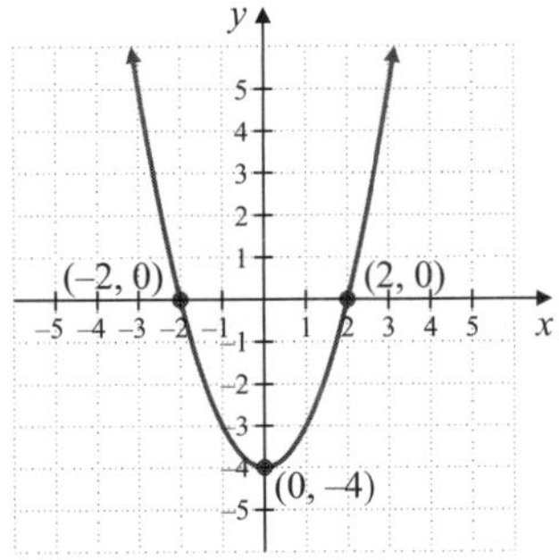

19. $f(x) = 4x^2 + 4x - 3$

$\frac{-b}{2a} = \frac{-4}{2(4)} = \frac{-4}{8} = -\frac{1}{2}$ and

$f\left(-\frac{1}{2}\right) = 4\left(-\frac{1}{2}\right)^2 + 4\left(-\frac{1}{2}\right) - 3 = 1 - 2 - 3 = -4$

Thus, the vertex is $\left(-\frac{1}{2}, -4\right)$.

The graph opens upward since $a > 0$.

$$4x^2 + 4x - 3 = 0$$
$$(2x-1)(2x+3) = 0$$
$$2x - 1 = 0 \quad \text{or} \quad 2x + 3 = 0$$
$$x = \frac{1}{2} \quad \text{or} \quad x = -\frac{3}{2}$$

The x-intercepts are $\left(\frac{1}{2}, 0\right)$ and $\left(-\frac{3}{2}, 0\right)$.

$f(0) = -3$, so $(0, -3)$ is the y-intercept.

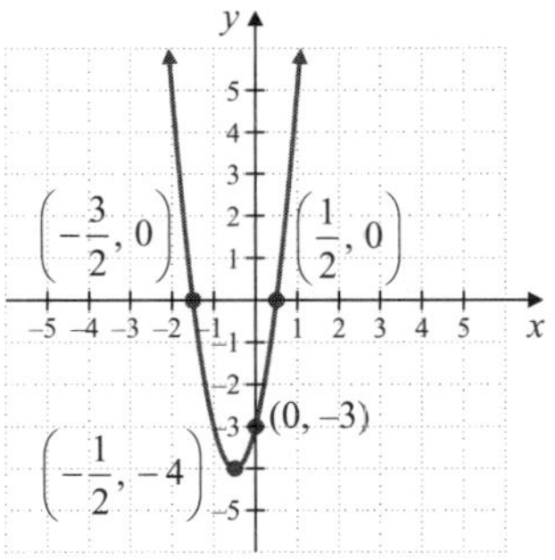

21. $f(x) = \frac{1}{2}x^2 + 4x + \frac{15}{2}$

$$\frac{-b}{2a} = \frac{-4}{2\left(\frac{1}{2}\right)} = -4$$

$$f(-4) = \frac{1}{2}(-4)^2 + 4(-4) + \frac{15}{2} = -\frac{1}{2}$$

Thus, the vertex is $\left(-4, -\frac{1}{2}\right)$.

The graph opens upward since $a > 0$.

$$\frac{1}{2}x^2 + 4x + \frac{15}{2} = 0$$
$$x^2 + 8x + 15 = 0$$
$$(x+5)(x+3) = 0$$

The x-intercepts are $(-5, 0)$ and $(-3, 0)$.

$f(0) = \frac{15}{2}$, so $\left(0, \frac{15}{2}\right)$ is the y-intercept.

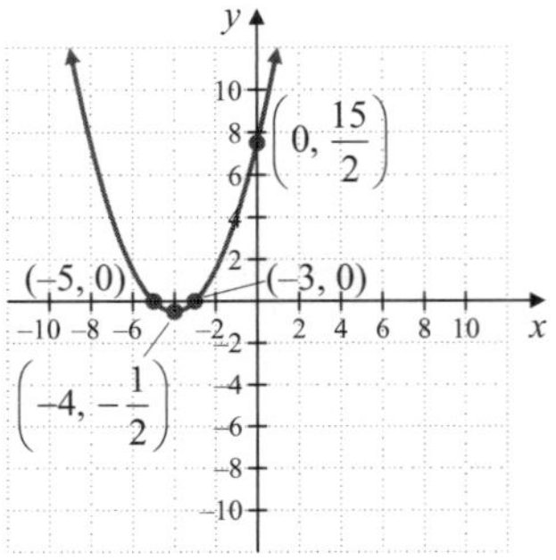

23. $f(x) = x^2 - 4x + 5$

$$\frac{-b}{2a} = \frac{-(-4)}{2(1)} = 2 \text{ and}$$

$$f(2) = 2^2 - 4(2) + 5 = 4 - 8 + 5 = 1$$

Thus, the vertex is $(2, 1)$.
The graph opens upward since $a > 0$. Since the graph opens upward and the vertex is in quadrant I, there are no x-intercepts.
$f(0) = 5$, so $(0, 5)$ is the y-intercept.

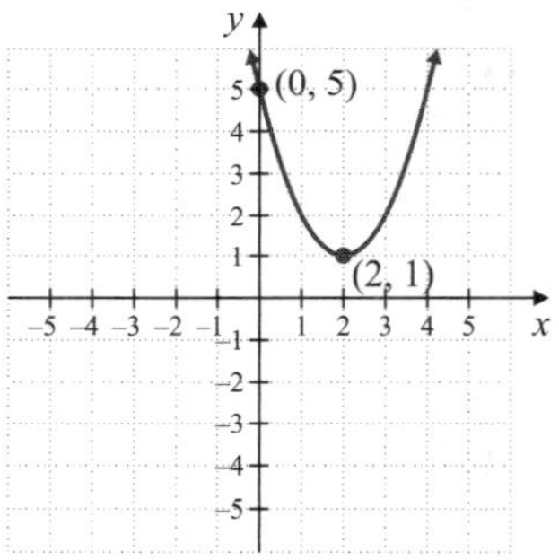

25. $f(x) = 2x^2 + 4x + 5$

$$\frac{-b}{2a} = \frac{-4}{2(2)} = -1$$

$$f(-1) = 2(-1)^2 + 4(-1) + 5 = 3$$

Thus, the vertex is $(-1, 3)$.
The graph opens upward since $a > 0$.
Since the graph opens upward and the vertex is in quadrant II, there are no x-intercepts.
$f(0) = 5$, so $(0, 5)$ is the y-intercept.

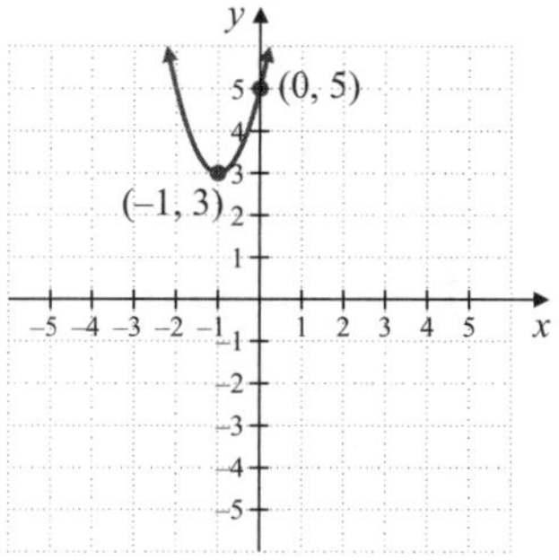

27. $f(x) = -2x^2 + 12x$

$$\frac{-b}{2a} = \frac{-12}{2(-2)} = 3$$

$f(3) = -2(3)^2 + 12(3) = -18 + 36 = 18$

Thus, the vertex is (3, 18).

The graph opens downward, since $a < 0$.

$-2x^2 + 12x = 0$

$-2x(x - 6) = 0$

$-2x = 0$ or $x - 6 = 0$

$x = 0$ or $x = 6$

The x-intercepts are (0, 0) and (6, 0).

$f(0) = 0$, so (0, 0) is the y-intercept.

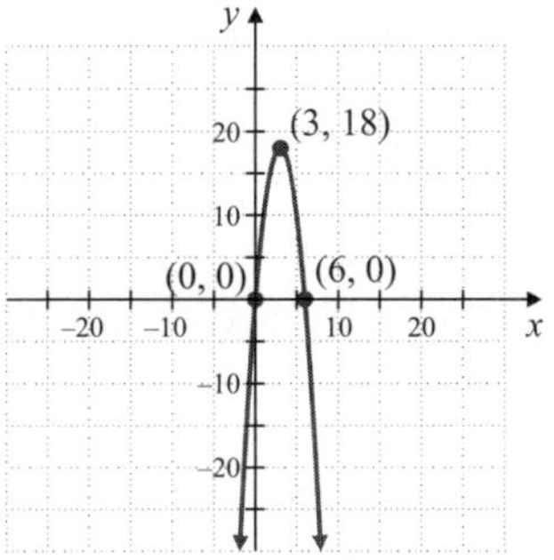

29. $h(t) = -16t^2 + 96t$

$\frac{-b}{2a} = \frac{-96}{2(-16)} = 3$ and

$h(3) = -16(3)^2 + 96(3) = -144 + 288 = 144$

The maximum height of the projectile is 144 feet.

31. a. $C(x) = 2x^2 - 800x + 92{,}000$

$$\frac{-b}{2a} = \frac{-(-800)}{2(2)} = \frac{800}{4} = 200$$

Thus, 200 bicycles must be manufactured to minimize the cost.

b.
$$\begin{aligned} C(200) &= 2(200)^2 - 800(200) + 92{,}000 \\ &= 80{,}000 - 160{,}000 + 92{,}000 \\ &= 12{,}000 \end{aligned}$$

The minimum cost is \$12,000.

33. Let x = one number.

$60 - x$ = other number

$f(x) = x(60 - x)$

$f(x) = 60x - x^2$

$f(x) = -x^2 + 60x$

$f(x) = -1(x^2 - 60x)$

$f(x) = -1(x^2 - 60x + 900) + 900$

$f(x) = -(x - 30)^2 + 900$

The maximum will occur at the vertex which is (30, 900). The numbers are 30 and 30.

35. Let x = one number.

$x + 10$ = other number

$f(x) = x(x + 10)$

$f(x) = x^2 + 10x$

$f(x) = (x^2 + 10x + 25) - 25$

$f(x) = (x + 5)^2 - 25$

The minimum will occur at the vertex, which is (−5, −25). The numbers are −5 and 5.

37. Let x = the width.
$40 - x$ = the length
$f(x) = x(40 - x)$
$f(x) = 40x - x^2$
$f(x) = -x^2 + 40x$
$f(x) = -1(x^2 - 40x)$
$f(x) = -1(x^2 - 40x + 400) + 400$
$f(x) = -(x - 20)^2 + 400$
The maximum will occur at the vertex which is (20, 400). The width is 20 units and the length is 20 units.

39. $f(x) = x^2 + 2$
$\frac{-b}{2a} = \frac{-0}{2(1)} = 0$
$f(0) = 2$
The vertex is (0, 2).

41. $g(x) = (x + 2)^2$
The vertex is (–2, 0).

43. $f(x) = (x + 5)^2 + 2$
The vertex is (–5, 2).

45. $f(x) = 3(x - 4)^2 + 1$
The vertex is (4, 1).

47. Since the coefficient of x^2 is positive, the graph of the parabola opens upward, so the function has a minimum value.

49. Since the coefficient of x^2 is negative, the graph of the parabola opens downward, so the function has a maximum value.

51. $f(x) = x^2 + 10x + 15$

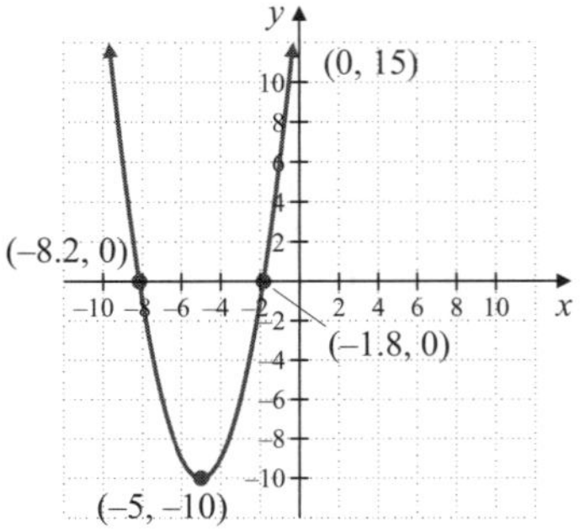

53. $f(x) = \frac{1}{2}x^2 + 4x + \frac{15}{2}$

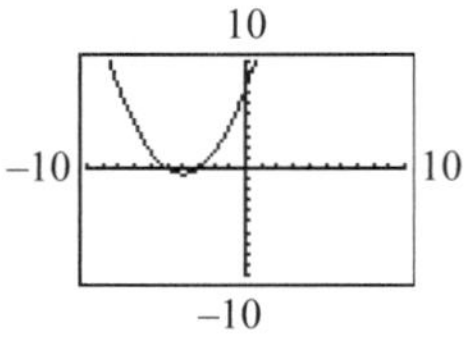

55. $f(x) = 2.3x^2 - 6.1x + 3.2$
$\frac{-b}{2a} = \frac{-(-6.1)}{2(2.3)} \approx 1.33$
$f(1.33) = 2.3(1.33)^2 - 6.1(1.33) + 3.2$
≈ -0.84
The minimum value of the function is approximately –0.84.

57. **a.** The function will have a maximum value; answers may vary.

b. $p(x) = -18x^2 + 298x + 988$
$a = -18$, $b = 298$, $c = 988$
$\frac{-b}{2a} = \frac{-298}{2(-18)} = \frac{149}{18} \approx 8.3$
The number of prison and jail inmates will be at a maximum about 8.3 years after 1995, or in 2003.

c. $p\left(\frac{149}{18}\right)$

$$= -18\left(\frac{149}{18}\right)^2 + 298\left(\frac{149}{18}\right) + 988$$
$$\approx 2221$$

The maximum number of inmates is predicted to be about 2220 thousand, or 2,220,000.

Chapter 8 Vocabulary Check

1. The <u>discriminant</u> helps us find the number and type of solutions of a quadratic equation.

2. If $a^2 = b$, then $a = \underline{\pm\sqrt{b}}$.

3. The graph of $f(x) = ax^2 + bx + c$, where a is not 0, is a parabola whose vertex has an x-value of $\underline{\frac{-b}{2a}}$.

4. A <u>quadratic inequality</u> is an inequality that can be written so that one side is a quadratic expression and the other side is 0.

5. The process of writing a quadratic equation so that one side is a perfect square trinomial is called <u>completing the square</u>.

6. The graph of $f(x) = x^2 + k$ has vertex $\underline{(0, k)}$.

7. The graph of $f(x) = (x-h)^2$ has vertex $\underline{(h, 0)}$.

8. The graph of $f(x) = (x-h)^2 + k$ has vertex $\underline{(h, k)}$.

9. The formula $x = \frac{-b \pm \sqrt{b^2 - 4ac}}{2a}$ is called the <u>quadratic formula</u>.

10. A <u>quadratic</u> equation is one that can be written in the form $ax^2 + bx + c = 0$, where a, b, and c are real numbers and a is not 0.

Chapter 8 Review

1. $x^2 - 15x + 14 = 0$

$$(x-14)(x-1) = 0$$
$$x - 14 = 0 \quad \text{or} \quad x - 1 = 0$$
$$x = 14 \qquad\qquad x = 1$$

The solution set is {14, 1}.

2. $7a^2 = 29a + 30$

$$7a^2 - 29a - 30 = 0$$
$$(7a+6)(a-5) = 0$$
$$7a + 6 = 0 \quad \text{or} \quad a - 5 = 0$$
$$a = -\frac{6}{7} \qquad\qquad a = 5$$

The solution set is $\left\{-\frac{6}{7}, 5\right\}$.

3. $4m^2 = 196$

$$m^2 = 49$$
$$m = \pm\sqrt{49}$$
$$m = \pm 7$$

The solution set is {−7, 7}.

4. $(5x-2)^2 = 2$

$$\sqrt{(5x-2)^2} = \pm\sqrt{2}$$
$$5x - 2 = \pm\sqrt{2}$$
$$5x = 2 \pm \sqrt{2}$$
$$x = \frac{2 \pm \sqrt{2}}{5}$$

The solution set is $\left\{\frac{2-\sqrt{2}}{5}, \frac{2+\sqrt{2}}{5}\right\}$.

5.
$$z^2+3z+1=0$$
$$z^2+3z=-1$$
$$z^2+3z+\left(\frac{3}{2}\right)^2=-1+\frac{9}{4}$$
$$\left(z+\frac{3}{2}\right)^2=\frac{5}{4}$$
$$z+\frac{3}{2}=\pm\sqrt{\frac{5}{4}}$$
$$z+\frac{3}{2}=\pm\frac{\sqrt{5}}{2}$$
$$z=-\frac{3}{2}\pm\frac{\sqrt{5}}{2}$$
The solution set is $\left\{\frac{-3-\sqrt{5}}{2}, \frac{-3+\sqrt{5}}{2}\right\}$.

6.
$$(2x+1)^2=x$$
$$4x^2+4x+1=x$$
$$4x^2+3x+1=0$$
$$x^2+\frac{3}{4}x+\frac{1}{4}=0$$
$$x^2+\frac{3}{4}x=-\frac{1}{4}$$
$$x^2+\frac{3}{4}x+\left(\frac{\frac{3}{4}}{2}\right)^2=-\frac{1}{4}+\frac{9}{64}$$
$$\left(x+\frac{3}{8}\right)^2=-\frac{7}{64}$$
$$x+\frac{3}{8}=\pm\sqrt{-\frac{7}{64}}$$
$$x+\frac{3}{8}=\pm\frac{i\sqrt{7}}{8}$$
$$x=-\frac{3}{8}\pm\frac{i\sqrt{7}}{8}$$
The solution set is $\left\{\frac{-3-i\sqrt{7}}{8}, \frac{-3+i\sqrt{7}}{8}\right\}$.

7. Find r when $A = 2717$, $P = 2500$, and $t = 2$.
$$A=P(1+r)^2$$
$$2717=2500(1+r)^2$$
$$\frac{2717}{2500}=(1+r)^2$$
$$\pm\sqrt{\frac{2717}{2500}}=1+r$$
$$-1\pm\sqrt{\frac{2717}{2500}}=r$$
Reject $-1-\sqrt{\frac{2717}{2500}}$ since the rate cannot be negative. The rate is
$-1+\sqrt{\frac{2717}{2500}}\approx 0.0425=4.25\%$.

8. Let x be the distance each ship traveled.
$$x^2+x^2=(150)^2$$
$$2x^2=22{,}500$$
$$x^2=11{,}250$$
$$x=\pm\sqrt{11{,}250}$$
$$x=\pm 75\sqrt{2}$$
Reject $-75\sqrt{2}$ since distance cannot be negative. Each ship traveled $75\sqrt{2}\approx 106.1$ miles.

9. Since the discriminant is negative, there are two complex but not real solutions.

10. Since the discriminant is positive, there are two real solutions.

11. Since the discriminant is positive, there are two real solutions.

12. Since the discriminant is 0, there is one real solution.

13. $x^2-16x+64=0$

$a=1,\ b=-16,\ c=64$

$$x=\frac{-b\pm\sqrt{b^2-4ac}}{2a}$$
$$=\frac{-(-16)\pm\sqrt{(-16)^2-4(1)(64)}}{2(1)}$$
$$=\frac{16\pm\sqrt{256-256}}{2}$$
$$=\frac{16\pm\sqrt{0}}{2}$$
$$=8$$

The solution set is {8}.

14. $x^2+5x=0$

$x^2+5x+0=0$

$a=1,\ b=5,\ c=0$

$$x=\frac{-b\pm\sqrt{b^2-4ac}}{2a}$$
$$=\frac{-5\pm\sqrt{5^2-4(1)(0)}}{2(1)}$$
$$=\frac{-5\pm\sqrt{25}}{2}$$
$$=\frac{-5\pm 5}{2}$$

$x=\frac{-5-5}{2}=-5$ or $x=\frac{-5+5}{2}=0$

The solution set is {–5, 0}.

15. $2x^2+3x=5$

$2x^2+3x-5=0$

$a=2,\ b=3,\ c=-5$

$$x=\frac{-b\pm\sqrt{b^2-4ac}}{2a}$$
$$=\frac{-3\pm\sqrt{3^2-4(2)(-5)}}{2(2)}$$
$$=\frac{-3\pm\sqrt{9+40}}{4}$$
$$=\frac{-3\pm\sqrt{49}}{4}$$
$$=\frac{-3\pm 7}{4}$$

$x=\frac{-3-7}{4}=-\frac{5}{2}$ or $x=\frac{-3+7}{4}=1$

The solution set is $\left\{-\frac{5}{2}, 1\right\}$.

16. $6x^2+7=5x$

$6x^2-5x+7=0$

$a=6,\ b=-5,\ c=7$

$$x=\frac{-b\pm\sqrt{b^2-4ac}}{2a}$$
$$=\frac{-(-5)\pm\sqrt{(-5)^2-4(6)(7)}}{2(6)}$$
$$=\frac{5\pm\sqrt{25-168}}{12}$$
$$=\frac{5\pm\sqrt{-143}}{12}$$
$$=\frac{5\pm i\sqrt{143}}{12}$$

The solution set is $\left\{\frac{5-i\sqrt{143}}{12}, \frac{5+i\sqrt{143}}{12}\right\}$.

17. $9a^2+4=2a$

$9a^2-2a+4=0$

$a=9,\ b=-2,\ c=4$

$$a=\frac{-b\pm\sqrt{b^2-4ac}}{2a}$$
$$=\frac{-(-2)\pm\sqrt{(-2)^2-4(9)(4)}}{2(9)}$$
$$=\frac{2\pm\sqrt{4-144}}{18}$$
$$=\frac{2\pm\sqrt{-140}}{18}$$
$$=\frac{2\pm 2i\sqrt{35}}{18}$$
$$=\frac{1\pm i\sqrt{35}}{9}$$

The solution set is $\left\{\frac{1-i\sqrt{35}}{9}, \frac{1+i\sqrt{35}}{9}\right\}$.

18. $(2x-3)^2=x$

$4x^2-12x+9=x$

$4x^2-13x+9=0$

$a=4,\ b=-13,\ c=9$

$$x=\frac{-b\pm\sqrt{b^2-4ac}}{2a}$$
$$=\frac{-(-13)\pm\sqrt{(-13)^2-4(4)(9)}}{2(4)}$$
$$=\frac{13\pm\sqrt{169-144}}{8}$$
$$=\frac{13\pm\sqrt{25}}{8}$$
$$=\frac{13\pm 5}{8}$$

$x=\frac{13-5}{8}=1$ or $x=\frac{13+5}{8}=\frac{9}{4}$

The solution set is $\left\{1, \frac{9}{4}\right\}$.

19. a. Let $t=1$.

$d(t)=-16t^2+30t+6$

$d(1)=-16(1)^2+30(1)+6=20$

The hat is 20 feet above the ground after 1 second.

b. Let $d(t)=0$ and find t.

$0=-16t^2+30t+6$

$a=-16,\ b=30,\ c=6$

$$b=\frac{-b\pm\sqrt{b^2-4ac}}{2a}$$
$$=\frac{-30\pm\sqrt{(30)^2-4(-16)(6)}}{2(-16)}$$
$$=\frac{-30\pm\sqrt{900+384}}{-32}$$
$$=\frac{-30\pm\sqrt{1284}}{-32}$$
$$=\frac{-30\pm 2\sqrt{321}}{-32}$$
$$=\frac{-2\left(15\pm\sqrt{321}\right)}{-2(16)}$$
$$=\frac{15\pm\sqrt{321}}{16}$$

Reject $\frac{15-\sqrt{321}}{6}$ since time cannot be negative. It will take $\frac{15+\sqrt{321}}{16}\approx 2.1$ seconds to hit the ground.

20. Let x be the length of each leg. Then $x+6$ is the length of the hypotenuse.

$x^2+x^2=(x+6)^2$

$2x^2=x^2+12x+36$

$x^2-12x-36=0$

$a=1,\ b=-12,\ c=-36$

$$x = \frac{-b \pm \sqrt{b^2 - 4ac}}{2a}$$
$$= \frac{-(-12) \pm \sqrt{(-12)^2 - 4(1)(-36)}}{2(1)}$$
$$= \frac{12 \pm \sqrt{144 + 144}}{2}$$
$$= \frac{12 \pm \sqrt{288}}{2}$$
$$= \frac{12 \pm 12\sqrt{2}}{2}$$
$$= 6 \pm 6\sqrt{2}$$

Reject $6 - 6\sqrt{2}$ since length cannot be negative. The length of each leg is $\left(6 + 6\sqrt{2}\right)$ centimeters.

21.
$$x^3 = 27$$
$$x^3 - 27 = 0$$
$$x^3 - 3^3 = 0$$
$$(x-3)(x^2 + 3x + 9) = 0$$
$$x - 3 = 0 \quad \text{or} \quad x^2 + 3x + 9 = 0$$
$$x = 3$$
$$a = 1, b = 3, c = 9$$
$$x = \frac{-b \pm \sqrt{b^2 - 4ac}}{2a}$$
$$= \frac{-3 \pm \sqrt{3^2 - 4(1)(9)}}{2(1)}$$
$$= \frac{-3 \pm \sqrt{9 - 36}}{2}$$
$$= \frac{-3 \pm \sqrt{-27}}{2}$$
$$= \frac{-3 \pm 3i\sqrt{3}}{2}$$

The solution set is $\left\{3, \frac{-3 - 3i\sqrt{3}}{2}, \frac{-3 + 3i\sqrt{3}}{2}\right\}$.

22.
$$y^3 = -64$$
$$y^3 + 64 = 0$$
$$y^3 + 4^3 = 0$$
$$(y+4)(y^2 - 4y + 16) = 0$$
$$y + 4 = 0 \quad \text{or} \quad y^2 - 4y + 16 = 0$$
$$y = -4$$
$$a = 1, b = -4, c = 16$$
$$y = \frac{-b \pm \sqrt{b^2 - 4ac}}{2a}$$
$$= \frac{-(-4) \pm \sqrt{(-4)^2 - 4(1)(16)}}{2(1)}$$
$$= \frac{4 \pm \sqrt{16 - 64}}{2}$$
$$= \frac{4 \pm \sqrt{-48}}{2}$$
$$= \frac{4 \pm 4i\sqrt{3}}{2}$$
$$= 2 \pm 2i\sqrt{3}$$

The solution set is $\left\{-4, 2 - 2i\sqrt{3}, 2 + 2i\sqrt{3}\right\}$.

23. The LCD is $x(x - 2)$.
$$\frac{5}{x} + \frac{6}{x-2} = 3$$
$$x(x-2)\left(\frac{5}{x}\right) + x(x-2)\left(\frac{6}{x-2}\right) = x(x-2)(3)$$
$$5(x-2) + 6x = 3x(x-2)$$
$$5x - 10 + 6x = 3x^2 - 6x$$
$$0 = 3x^2 - 17x + 10$$
$$0 = (3x-2)(x-5)$$
$$3x - 2 = 0 \quad \text{or} \quad x - 5 = 0$$
$$x = \frac{2}{3} \qquad x = 5$$

The solution set is $\left\{\frac{2}{3}, 5\right\}$.

24.
$$x^4 - 21x^2 - 100 = 0$$
$$(x^2 - 25)(x^2 + 4) = 0$$
$$(x+5)(x-5)(x^2+4) = 0$$
$$x+5=0 \quad \text{or} \quad x-5=0 \quad \text{or}$$
$$x=-5 \qquad x=5$$
$$x^2+4=0$$
$$x^2=-4$$
$$x=\pm\sqrt{-4}$$
$$x=\pm 2i$$
The solution set is $\{-5, 5, -2i, 2i\}$.

25. Let $y = x + 3$.
$$5(x+3)^2 - 19(x+3) = 4$$
$$5(x+3)^2 - 19(x+3) - 4 = 0$$
$$5y^2 - 19y - 4 = 0$$
$$(5y+1)(y-4) = 0$$
$$5y+1=0 \quad \text{or} \quad y-4=0$$
$$y=-\frac{1}{5} \qquad y=4$$
$$x+3=-\frac{1}{5} \qquad x+3=4$$
$$x=-3-\frac{1}{5} \qquad x=1$$
$$x=-\frac{16}{5}$$
The solution set is $\left\{-\frac{16}{5}, 1\right\}$.

26. Let $y = x^{1/3}$.
$$x^{2/3} - 6x^{1/3} + 5 = 0$$
$$(x^{1/3})^2 - 6x^{1/3} + 5 = 0$$
$$y^2 - 6y + 5 = 0$$
$$(y-1)(y-5) = 0$$
$$y-1=0 \quad \text{or} \quad y-5=0$$
$$y=1 \qquad y=5$$
$$x^{1/3}=1 \qquad x^{1/3}=5$$
$$x=1^3 \qquad x=5^3$$
$$x=1 \qquad x=125$$
The solution set is $\{1, 125\}$.

27.
$$a^6 - a^2 = a^4 - 1$$
$$a^6 - a^4 - a^2 + 1 = 0$$
$$a^4(a^2-1) - 1(a^2-1) = 0$$
$$(a^2-1)(a^4-1) = 0$$
$$(a+1)(a-1)(a^2-1)(a^2+1) = 0$$
$$(a+1)(a-1)(a+1)(a-1)(a^2+1) = 0$$
The solution set is $\{-1, 1, -i, i\}$.

28. Let $x = \frac{1}{y}$.
$$y^{-2} + y^{-1} = 20$$
$$\frac{1}{y^2} + \frac{1}{y} - 20 = 0$$
$$x^2 + x - 20 = 0$$
$$(x+5)(x-4) = 0$$
$$x+5=0 \quad \text{or} \quad x-4=0$$
$$x=-5 \qquad x=4$$
$$\frac{1}{y}=-5 \qquad \frac{1}{y}=4$$
$$y=-\frac{1}{5} \qquad y=\frac{1}{4}$$
The solution set is $\left\{-\frac{1}{5}, \frac{1}{4}\right\}$.

29. Let x be the time it takes Jerome to work alone. Then $x - 1$ is the time it takes Tim to work alone.

	Time	Part in 1 hour
Jerome	x	$\frac{1}{x}$
Tim	$x-1$	$\frac{1}{x-1}$
Together	5	$\frac{1}{5}$

The LCD is $5x(x-1)$.

$$\frac{1}{x}+\frac{1}{x-1}=\frac{1}{5}$$
$$5x(x-1)\left(\frac{1}{x}\right)+5x(x-1)\left(\frac{1}{x-1}\right)=5x(x-1)\left(\frac{1}{5}\right)$$
$$5(x-1)+5x=x(x-1)$$
$$5x-5+5x=x^2-x$$
$$0=x^2-11x+5$$

$a=1, b=-11, c=5$

$$x=\frac{-b\pm\sqrt{b^2-4ac}}{2a}$$
$$=\frac{-(-11)\pm\sqrt{(-11)^2-4(1)(5)}}{2(1)}$$
$$=\frac{11\pm\sqrt{101}}{2}$$

$x=\frac{11-\sqrt{101}}{2}\approx 0.5$ or $x=\frac{11+\sqrt{101}}{2}\approx 10.5$

Reject 0.5, since time cannot be negative and $0.5-1=-0.5$. It takes Jerome 10.5 hours to work alone and Tim $10.5-1=9.5$ hours to work alone.

30. Let x be the negative number.

$$x-\frac{1}{x}=-\frac{24}{5}$$
$$5x\left(x-\frac{1}{x}\right)=5x\left(-\frac{24}{5}\right)$$
$$5x^2-5=-24x$$
$$5x^2+24x-5=0$$
$$(5x-1)(x+5)=0$$

$5x-1=0$ or $x+5=0$

$x=\frac{1}{5}$ or $x=-5$

Reject $\frac{1}{5}$ since it is not a negative number. The number is -5.

31.
$$2x^2-50\le 0$$
$$2(x^2-25)\le 0$$
$$2(x+5)(x-5)\le 0$$
$$2(x+5)(x-5)=0$$

$x+5=0$ or $x-5=0$

$x=-5$ or $x=5$

Region	Test Point Value	$2(x+5)(x-5) \le 0$	Result
$(-\infty, -5)$	-6	$2(-1)(-11) \le 0$	False
$(-5, 5)$	0	$2(5)(-5) \le 0$	True
$(5, \infty)$	6	$2(11)(1) \le 0$	False

The solution set is $[-5, 5]$.

32.

$$\frac{1}{4}x^2 < \frac{1}{16}$$
$$4x^2 < 1$$
$$4x^2 - 1 < 0$$
$$(2x+1)(2x-1) < 0$$
$$(2x+1)(2x-1) = 0$$
$$2x+1=0 \quad \text{or} \quad 2x-1=0$$
$$x = -\frac{1}{2} \quad \text{or} \quad x = \frac{1}{2}$$

Region	Test Point Value	$(2x+1)(2x-1) < 0$	Result
$\left(-\infty, -\frac{1}{2}\right)$	-1	$(-1)(-3) < 0$	False
$\left(-\frac{1}{2}, \frac{1}{2}\right)$	0	$(1)(-1) < 0$	True
$\left(\frac{1}{2}, \infty\right)$	1	$(3)(1) < 0$	False

The solution set is $\left(-\frac{1}{2}, \frac{1}{2}\right)$.

33.

$$(x^2-16)(x^2-1) > 0$$
$$(x+4)(x-4)(x+1)(x-1) > 0$$
$$(x+4)(x-4)(x+1)(x-1) = 0$$
$$x+4=0 \quad \text{or} \quad x-4=0 \quad \text{or} \quad x+1=0 \quad \text{or} \quad x-1=0$$
$$x=-4 \quad \text{or} \quad x=4 \quad \text{or} \quad x=-1 \quad \text{or} \quad x=1$$

Region	Test Point Value	$(x+4)(x-4)(x+1)(x-1)>0$	Result
$(-\infty, -4)$	-5	$(-1)(-9)(-4)(-6)>0$	True
$(-4, -1)$	-2	$(2)(-6)(-1)(-3)>0$	False
$(-1, 1)$	0	$(4)(-4)(1)(-1)>0$	True
$(1, 4)$	2	$(6)(-2)(3)(1)>0$	False
$(4, \infty)$	5	$(9)(1)(6)(4)>0$	True

The solution set is $(-\infty, -4) \cup (-1, 1) \cup (4, \infty)$.

34. $\frac{x-5}{x-6}<0$

$x-6=0$

$x=6$ makes the denominator 0.

$$\frac{x-5}{x-6}=0$$

$$(x-6)\left(\frac{x-5}{x-6}\right)=(x-6)(0)$$

$$x-5=0$$

$$x=5$$

Region	Test Point Value	$\frac{x-5}{x-6}<0$	Result
$(-\infty, 5)$	0	$\frac{-5}{-6}<0$	False
$(5, 6)$	5.5	$\frac{0.5}{-0.5}<0$	True
$(6, \infty)$	7	$\frac{2}{1}<0$	False

The solution set is $(5, 6)$.

35. $\frac{(4x+3)(x-5)}{x(x+6)} > 0$

$x(x + 6) = 0$

$x = 0$ and $x = -6$ make the denominator 0.

$$\frac{(4x+3)(x-5)}{x(x+6)} = 0$$

$$x(x+6)\left(\frac{(4x+3)(x-5)}{x(x+6)}\right) = x(x+6)(0)$$

$$(4x+3)(x-5) = 0$$

$4x+3=0$ or $x-5=0$

$x = -\frac{3}{4}$ $\quad x = 5$

Region	Test Point Value	$\frac{(4x+3)(x-5)}{x(x+6)} > 0$	Result
$(-\infty, -6)$	-7	$\frac{(-25)(-12)}{(-7)(-1)} > 0$	True
$\left(-6, -\frac{3}{4}\right)$	-1	$\frac{(-1)(-6)}{(-1)(5)} > 0$	False
$\left(-\frac{3}{4}, 0\right)$	-0.5	$\frac{(1)(-5.5)}{(-0.5)(5.5)} > 0$	True
$(0, 5)$	1	$\frac{(7)(-4)}{(1)(7)} > 0$	False
$(5, \infty)$	6	$\frac{(27)(1)}{(6)(12)} > 0$	True

The solution set is $(-\infty, -6) \cup \left(-\frac{3}{4}, 0\right) \cup (5, \infty)$.

36. $(x + 5)(x - 6)(x + 2) \le 0$

$(x + 5)(x - 6)(x + 2) = 0$

$x+5=0$ or $x-6=0$ or $x+2=0$

$x=-5$ $\quad x=6$ $\quad x=-2$

Region	Test Point Value	$(x + 5)(x - 6)(x + 2) \le 0$	Result
$(-\infty, -5)$	-6	$(-1)(-12)(-4) \le 0$	True
$(-5, -2)$	-3	$(2)(-9)(-1) \le 0$	False
$(-2, 6)$	0	$(5)(-6)(2) \le 0$	True
$(6, \infty)$	7	$(12)(1)(9) \le 0$	False

The solution set is $(-\infty, -5] \cup [-2, 6]$.

37. $x^3 + 3x^2 - 25x - 75 > 0$

$x^2(x+3) - 25(x+3) > 0$

$(x+3)(x^2 - 25) > 0$

$(x+3)(x+5)(x-5) > 0$

$(x+3)(x+5)(x-5) = 0$

$x+3=0$ or $x+5=0$ or $x-5=0$

$x=-3$ $x=-5$ $x=5$

Region	Test Point Value	$(x + 3)(x + 5)(x - 5) > 0$	Result
$(-\infty, -5)$	-6	$(-3)(-1)(-11) > 0$	False
$(-5, -3)$	-4	$(-1)(1)(-9) > 0$	True
$(-3, 5)$	0	$(3)(5)(-5) > 0$	False
$(5, \infty)$	6	$(9)(11)(1) > 0$	True

The solution set is $(-5, -3) \cup (5, \infty)$.

38. $\dfrac{x^2+4}{3x} \le 1$

$3x = 0$

$x = 0$ makes the denominator 0.

$$\frac{x^2+4}{3x} = 1$$

$$3x\left(\frac{x^2+4}{3x}\right) = 3x(1)$$

$$x^2 + 4 = 3x$$

$$x^2 - 3x + 4 = 0$$

$a = 1, b = -3, c = 4$

$$x = \frac{-b \pm \sqrt{b^2 - 4ac}}{2a}$$

$$= \frac{-(-3) \pm \sqrt{(-3)^2 - 4(1)(4)}}{2(1)}$$

$$= \frac{3 \pm \sqrt{-7}}{2}$$

The related equation has no real number solutions.

Region	Test Point Value	$\frac{x^2+4}{3x} \le 1$	Result
$(-\infty, 0)$	-1	$\frac{5}{-3} \le 1$	True
$(0, \infty)$	1	$\frac{5}{3} \le 1$	False

The solution set is $(-\infty, 0)$.

39. $\dfrac{(5x+6)(x-3)}{x(6x-5)} < 0$

$x(6x-5) = 0$

$x = 0$ and $x = \dfrac{5}{6}$ make the denominator 0.

$$\frac{(5x+6)(x-3)}{x(6x-5)} = 0$$

$$x(6x-5)\left(\frac{(5x+6)(x-3)}{x(6x-5)}\right) = x(6x-5)(0)$$

$$(5x+6)(x-3) = 0$$

$5x+6=0$ or $x-3=0$

$x = -\dfrac{6}{5}$ $\quad$ $x = 3$

Region	Test Point Value	$\frac{(5x+6)(x-3)}{x(6x-5)} < 0$	Result
$\left(-\infty, -\frac{6}{5}\right)$	-2	$\frac{(-4)(-5)}{(-2)(-17)} < 0$	False
$\left(-\frac{6}{5}, 0\right)$	-1	$\frac{(1)(-4)}{(-1)(-11)} < 0$	True
$\left(0, \frac{5}{6}\right)$	0.5	$\frac{(8.5)(-2.5)}{(0.5)(-2)} < 0$	False
$\left(\frac{5}{6}, 3\right)$	1	$\frac{(11)(-2)}{(1)(1)} < 0$	True
$(3, \infty)$	4	$\frac{(26)(1)}{(4)(19)} < 0$	False

The solution set is $\left(-\dfrac{6}{5}, 0\right) \cup \left(\dfrac{5}{6}, 3\right)$.

40. $f(x) = x^2 - 4$

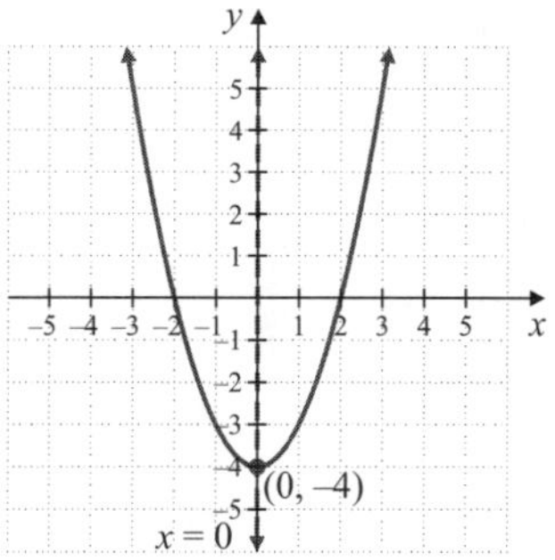

41. $g(x) = x^2 + 7$

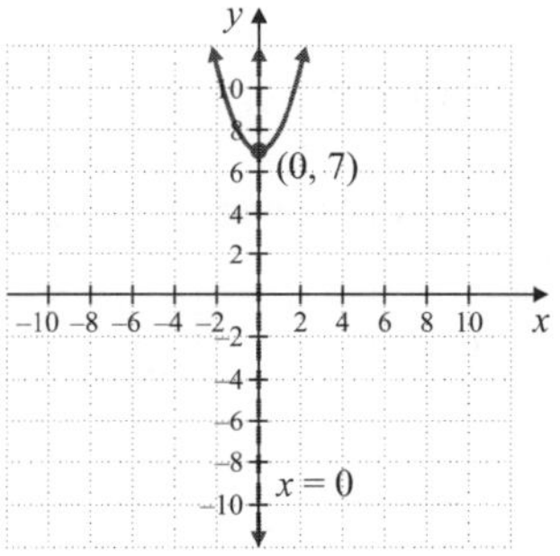

42. $H(x) = 2x^2$

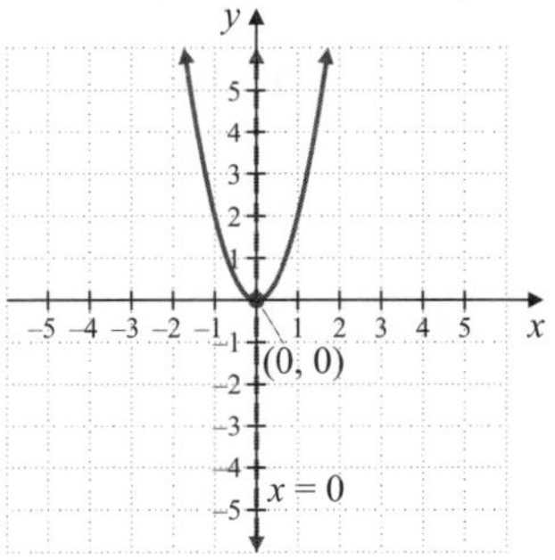

43. $h(x) = -\frac{1}{3}x^2$

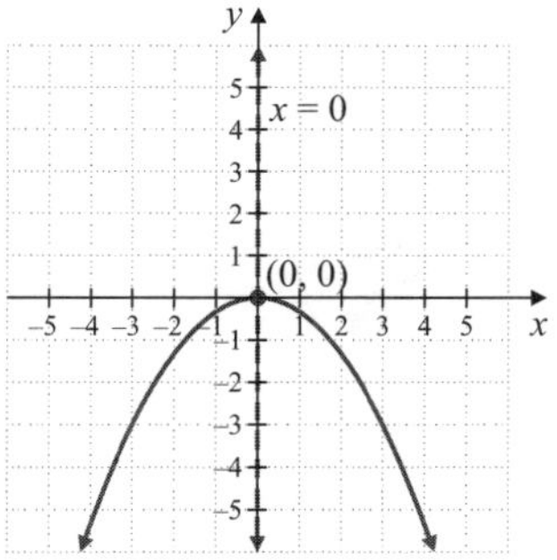

44. $F(x) = (x-1)^2$

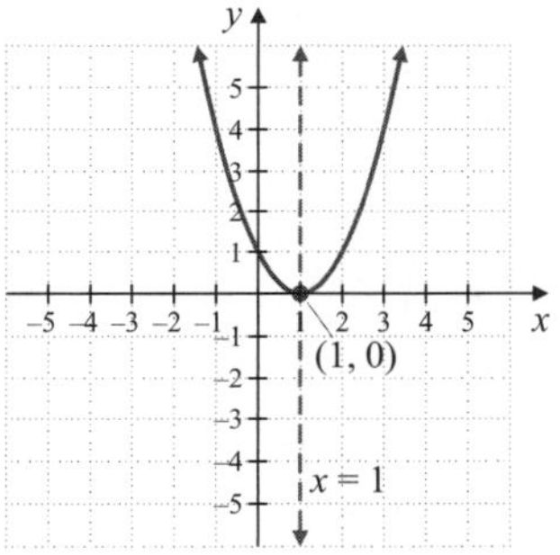

45. $G(x) = (x+5)^2$

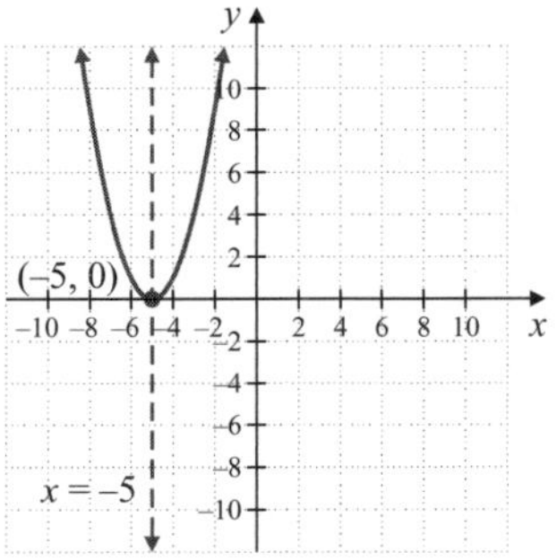

46. $f(x) = (x-4)^2 - 2$

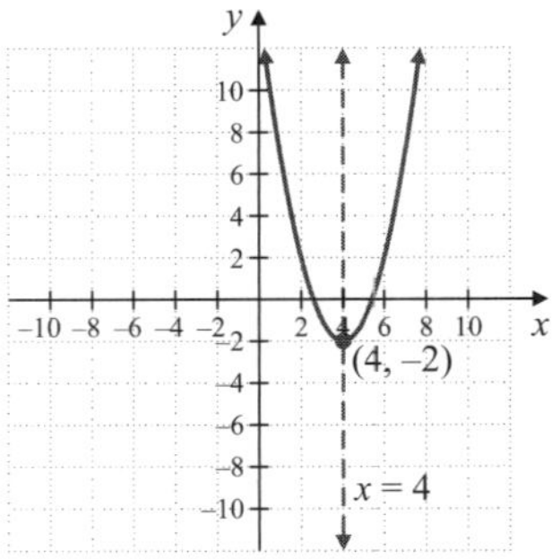

47. $f(x) = -3(x-1)^2 + 1$

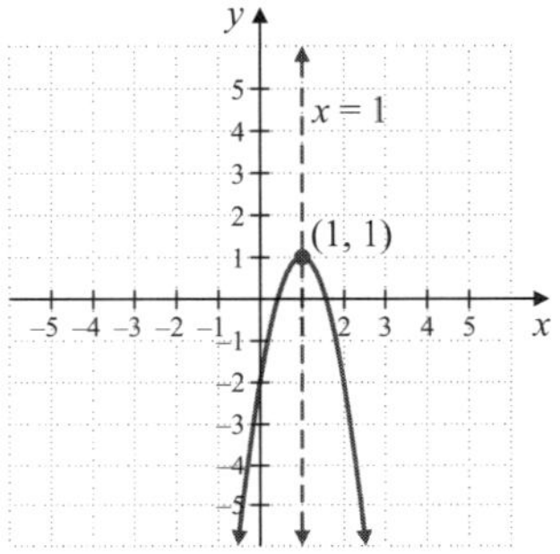

48. $f(x) = x^2 + 10x + 25$

$\dfrac{-b}{2a} = \dfrac{-10}{2(1)} = -5$

$f(-5) = (-5)^2 + 10(-5) + 25 = 0$

Thus, the vertex is (−5, 0). The graph opens upward, since $a > 0$.

$x^2 + 10x + 25 = 0$

$(x+5)^2 = 0$

$x + 5 = 0$

$x = -5$

The x-intercept is (−5, 0).
$f(0) = 25$, so (0, 25) is the y-intercept.

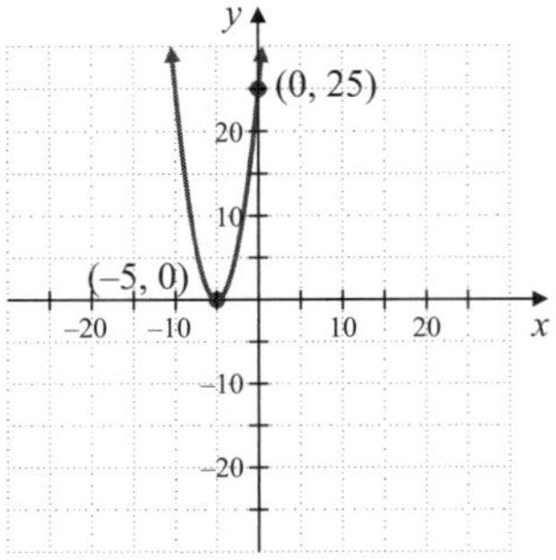

49. $f(x) = -x^2 + 6x - 9$

$\dfrac{-b}{2a} = \dfrac{-6}{2(-1)} = 3$

$f(3) = -3^2 + 6(3) - 9 = 0$

Thus, the vertex is (3, 0).
The graph opens downward, since $a < 0$.

$-x^2 + 6x - 9 = 0$

$-1(x^2 - 6x + 9) = 0$

$-1(x-3)^2 = 0$

$x - 3 = 0$

$x = 3$

The x-intercept is (3, 0).

$f(0) = -0^2 + 6(0) - 9 = -9$

The y-intercept is (0, −9).

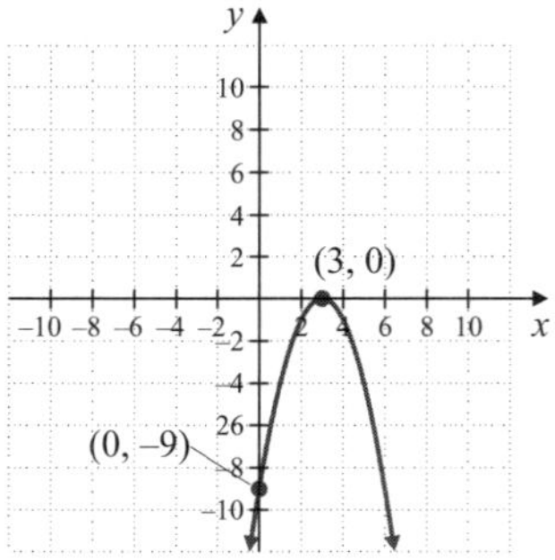

50. $f(x) = 4x^2 - 1$

$\frac{-b}{2a} = \frac{-0}{2(4)} = 0$

$f(0) = -1$, so the vertex is $(0, -1)$.
The graph opens upward, since $a > 0$.

$4x^2 - 1 = 0$

$(2x+1)(2x-1) = 0$

$2x+1 = 0$ or $2x-1 = 0$

$x = -\frac{1}{2}$ or $x = \frac{1}{2}$

The x-intercepts are $\left(-\frac{1}{2}, 0\right)$ and $\left(\frac{1}{2}, 0\right)$.

$f(0) = -1$, so the y-intercept is $(0, -1)$.

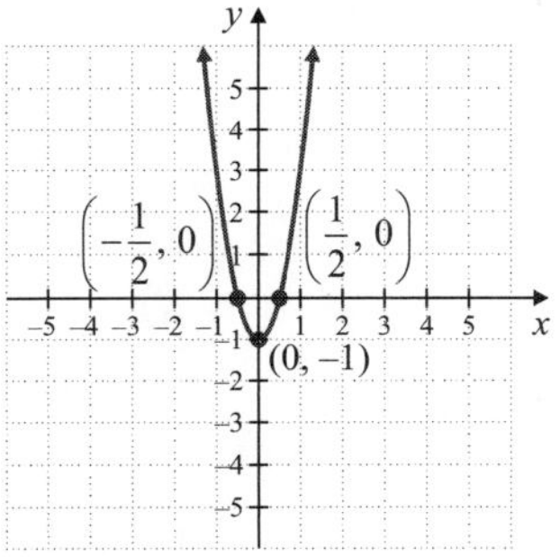

51. $f(x) = -5x^2 + 5$

$\frac{-b}{2a} = \frac{-0}{2(-5)} = 0$

$f(0) = 5$, so the vertex is $(0, 5)$.
The graph opens downward, since $a < 0$.

$-5x^2 + 5 = 0$

$-5(x^2 - 1) = 0$

$x^2 - 1 = 0$

$(x+1)(x-1) = 0$

$x = -1$ or $x = 1$

The x-intercepts are $(-1, 0)$ and $(1, 0)$.
$f(0) = 5$, so the y-intercept is $(0, 5)$.

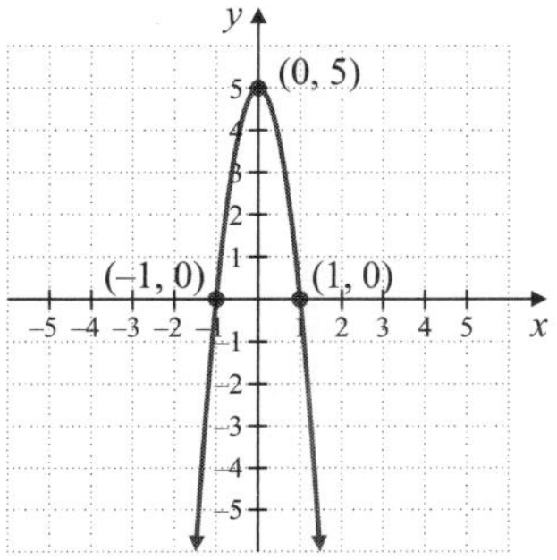

52. $f(x) = -3x^2 - 5x + 4$

$\frac{-b}{2a} = \frac{-(-5)}{2(-3)} = -\frac{5}{6}$

$f\left(-\frac{5}{6}\right) = -3\left(-\frac{5}{6}\right)^2 - 5\left(-\frac{5}{6}\right) + 4$

$= -\frac{25}{12} + \frac{50}{12} + \frac{48}{12}$

$= \frac{73}{12}$

The vertex is $\left(-\frac{5}{6}, \frac{73}{12}\right)$.

The graph opens downward, since $a < 0$.

$-3x^2 - 5x + 4 = 0$

$a = -3, b = -5, c = 4$

$x = \frac{-b \pm \sqrt{b^2 - 4ac}}{2a}$

$= \frac{-(-5) \pm \sqrt{(-5)^2 - 4(-3)(4)}}{2(-3)}$

$= \frac{5 \pm \sqrt{25 + 48}}{-6}$

$= \frac{5 \pm \sqrt{73}}{-6}$

$x \approx -2.3$ or $x \approx 0.6$

The x-intercepts are $(-2.3, 0)$ and $(0.6, 0)$.
$f(0) = 4$, so the y-intercept is $(0, 4)$.

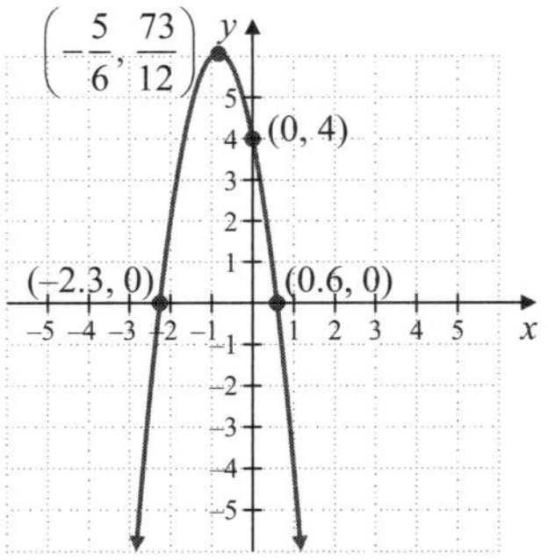

53. a. Let $h(t) = 350$ and solve for t.

$$h(t) = -16t^2 + 120t + 300$$
$$350 = -16t^2 + 120t + 300$$
$$0 = -16t^2 + 120t - 50$$
$$a = -16,\ b = 120,\ c = -50$$
$$t = \frac{-b \pm \sqrt{b^2 - 4ac}}{2a}$$
$$= \frac{-120 \pm \sqrt{(120)^2 - 4(-16)(-50)}}{2(-16)}$$
$$= \frac{-120 \pm \sqrt{14,400 - 3200}}{-32}$$
$$= \frac{-120 \pm \sqrt{11,200}}{-32}$$
$$x \approx 0.4 \quad \text{or} \quad x \approx 7.1$$

The object will reach a height of 350 feet after 0.4 second and 7.1 seconds.

b. answers may vary

54. Let x = one number.
Then $420 - x$ = other number.

$$f(x) = x(420 - x)$$
$$f(x) = 420x - x^2$$
$$f(x) = -1(x^2 - 420x)$$
$$f(x) = -1(x^2 - 420x + (-210)^2) + (-210)^2$$
$$f(x) = -1(x - 210)^2 + 44,100$$

The maximum will occur at the vertex, (210, −44,100). The numbers are 210 and 210.

55.
$$x^2 - x - 30 = 0$$
$$(x + 5)(x - 6) = 0$$
$$x + 5 = 0 \quad \text{or} \quad x - 6 = 0$$
$$x = -5 \qquad\qquad x = 6$$

The solution set is {−5, 6}.

56.
$$10x^2 = 3x + 4$$
$$10x^2 - 3x - 4 = 0$$
$$(5x - 4)(2x + 1) = 0$$
$$5x - 4 = 0 \quad \text{or} \quad 2x + 1 = 0$$
$$5x = 4 \qquad\qquad 2x = -1$$
$$x = \frac{4}{5} \qquad\qquad x = -\frac{1}{2}$$

The solution set is $\left\{-\frac{1}{2}, \frac{4}{5}\right\}$.

57.
$$9y^2 = 36$$
$$y^2 = 4$$
$$y = \pm\sqrt{4}$$
$$y = \pm 2$$

The solution set is {−2, 2}.

58.
$$(9n + 1)^2 = 9$$
$$\sqrt{(9n + 1)^2} = \pm\sqrt{9}$$
$$9n + 1 = \pm 3$$
$$9n = -1 \pm 3$$
$$n = \frac{-1 \pm 3}{9}$$
$$n = \frac{-1 - 3}{9} \quad \text{or} \quad n = \frac{-1 + 3}{9}$$
$$n = -\frac{4}{9} \qquad\qquad n = \frac{2}{9}$$

The solution set is $\left\{-\frac{4}{9}, \frac{2}{9}\right\}$.

59. $x^2 + x + 7 = 0$

$a = 1, b = 1, c = 7$

$$x = \frac{-b \pm \sqrt{b^2 - 4ac}}{2a}$$
$$= \frac{-1 \pm \sqrt{1^2 - 4(1)(7)}}{2(1)}$$
$$= \frac{-1 \pm \sqrt{1 - 28}}{2}$$
$$= \frac{-1 \pm \sqrt{-27}}{2}$$
$$= \frac{-1 \pm 3i\sqrt{3}}{2}$$

The solution set is $\left\{\frac{-1 - 3i\sqrt{3}}{2}, \frac{-1 + 3i\sqrt{3}}{2}\right\}$.

60. $(3x - 4)^2 = 10x$

$9x^2 - 24x + 16 = 10x$

$9x^2 - 34x + 16 = 0$

$a = 9, b = -34, c = 16$

$$x = \frac{-b \pm \sqrt{b^2 - 4ac}}{2a}$$
$$= \frac{-(-34) \pm \sqrt{(-34)^2 - 4(9)(16)}}{2(9)}$$
$$= \frac{34 \pm \sqrt{1156 - 576}}{18}$$
$$= \frac{34 \pm \sqrt{580}}{18}$$
$$= \frac{34 \pm 2\sqrt{145}}{18}$$
$$= \frac{17 \pm \sqrt{145}}{9}$$

The solution set is

$\left\{\frac{17 - \sqrt{145}}{9}, \frac{17 + \sqrt{145}}{9}\right\}$.

61. $x^2 + 11 = 0$

$x^2 = -11$

$x = \pm\sqrt{-11}$

$x = \pm i\sqrt{11}$

The solution set is $\left\{-i\sqrt{11}, i\sqrt{11}\right\}$.

62. $(5a - 2)^2 - a = 0$

$25a^2 - 20a + 4 - a = 0$

$25a^2 - 21a + 4 = 0$

$a = 25, b = -21, c = 4$

$$a = \frac{-b \pm \sqrt{b^2 - 4ac}}{2a}$$
$$= \frac{-(-21) \pm \sqrt{(-21)^2 - 4(25)(4)}}{2(25)}$$
$$= \frac{21 \pm \sqrt{441 - 400}}{50}$$
$$= \frac{21 \pm \sqrt{41}}{50}$$

The solution set is $\left\{\frac{21 - \sqrt{41}}{50}, \frac{21 + \sqrt{41}}{50}\right\}$.

63. $\frac{7}{8} = \frac{8}{x^2}$

$7x^2 = 64$

$x^2 = \frac{64}{7}$

$x = \pm\sqrt{\frac{64}{7}}$

$x = \pm\frac{\sqrt{64}}{\sqrt{7}}$

$x = \pm\frac{8}{\sqrt{7}}$

$x = \pm\frac{8\sqrt{7}}{7}$

The solution set is $\left\{-\frac{8\sqrt{7}}{7}, \frac{8\sqrt{7}}{7}\right\}$.

64. Let $y = x^{1/3}$.

$$x^{2/3} - 6x^{1/3} = -8$$
$$(x^{1/3})^2 - 6x^{1/3} + 8 = 0$$
$$y^2 - 6y + 8 = 0$$
$$(y-4)(y-2) = 0$$
$$y - 4 = 0 \quad \text{or} \quad y - 2 = 0$$
$$y = 4 \qquad y = 2$$
$$x^{1/3} = 4 \qquad x^{1/3} = 2$$
$$x = 4^3 \qquad x = 2^3$$
$$x = 64 \qquad x = 8$$

The solution set is $\{8, 64\}$.

65. $(2x - 3)(4x + 5) \geq 0$

$(2x - 3)(4x + 5) = 0$

$$2x - 3 = 0 \quad \text{or} \quad 4x + 5 = 0$$
$$2x = 3 \qquad 4x = -5$$
$$x = \frac{3}{2} \qquad x = -\frac{5}{4}$$

Region	Test Point Value	$(2x - 3)(4x + 5) \geq 0$	Result
$\left(-\infty, -\frac{5}{4}\right)$	-2	$(-7)(-3) \geq 0$	True
$\left(-\frac{5}{4}, \frac{3}{2}\right)$	0	$(-3)(5) \geq 0$	False
$\left(\frac{3}{2}, \infty\right)$	2	$(1)(13) \geq 0$	True

The solution set is $\left(-\infty, -\frac{5}{4}\right] \cup \left[\frac{3}{2}, \infty\right)$.

66. $\frac{x(x+5)}{4x-3} \geq 0$

$4x-3=0$

$x=\frac{3}{4}$ makes the denominator 0.

$$\frac{x(x+5)}{4x-3}=0$$
$$(4x-3)\left(\frac{x(x+5)}{4x-3}\right)=(4x-3)(0)$$
$$x(x+5)=0$$
$$x=0 \quad \text{or} \quad x+5=0$$
$$x=-5$$

Region	Test Point Value	$\frac{x(x+5)}{4x-3} \geq 0$	Result
$(-\infty, -5)$	-6	$\frac{(-6)(-1)}{(-27)} \geq 0$	False
$(-5, 0)$	-1	$\frac{(-1)(4)}{(-7)} \geq 0$	True
$\left(0, \frac{3}{4}\right)$	0.5	$\frac{(0.5)(5.5)}{(-1)} \geq 0$	False
$\left(\frac{3}{4}, \infty\right)$	1	$\frac{(1)(6)}{(1)} \geq 0$	True

The solution set is $[-5, 0] \cup \left(\frac{3}{4}, \infty\right)$.

67. $\frac{3}{x-2} > 2$

$x-2=0$

$x=2$ makes the denominator 0.

$$\frac{3}{x-2}=2$$
$$(x-2)\left(\frac{3}{x-2}\right)=(x-2)(2)$$
$$3=2x-4$$
$$7=2x$$
$$\frac{7}{2}=x$$

Region	Test Point Value	$\frac{3}{x-2} > 2$	Result
$(-\infty, 2)$	0	$\frac{3}{-2} > 2$	False
$\left(2, \frac{7}{2}\right)$	3	$\frac{3}{1} > 2$	True
$\left(\frac{7}{2}, \infty\right)$	4	$\frac{3}{2} > 2$	False

The solution set is $\left(2, \frac{7}{2}\right)$.

68. a. Let $x = 2000 - 1960 = 40$.

$$y = 26x^2 - 136x + 917$$
$$y = 26(40)^2 - 136(40) + 917$$
$$y = 37{,}077$$

The estimated number of passengers in 2000 is 37,077,000.

b. Let $y = 60{,}000$ and solve for x.

$$60{,}000 = 26x^2 - 136x + 917$$
$$0 = 26x^2 - 136x - 59{,}083$$

$a = 26$, $b = -136$, $c = -59{,}083$

$$x = \frac{-b \pm \sqrt{b^2 - 4ac}}{2a}$$
$$= \frac{-(-136) \pm \sqrt{(-136)^2 - 4(26)(-59{,}083)}}{2(26)}$$
$$= \frac{136 \pm \sqrt{6{,}163{,}128}}{52}$$

$x \approx -45$ or $x \approx 50$

Discard -45 since time cannot be negative. The number of passengers will reach 60,000,000 in $1960 + 50 = 2010$.

Chapter 8 Test

1. $5x^2 - 2x = 7$
$5x^2 - 2x - 7 = 0$
$(5x - 7)(x + 1) = 0$
$5x - 7 = 0$ or $x + 1 = 0$
$5x = 7$ or $x = -1$
$x = \frac{7}{5}$ or $x = -1$
The solution set is $\left\{\frac{7}{5}, -1\right\}$.

2. $(x+1)^2 = 10$
$\sqrt{(x+1)^2} = \pm\sqrt{10}$
$x + 1 = \pm\sqrt{10}$
$x = -1 \pm \sqrt{10}$
The solution set is $\left\{-1-\sqrt{10}, -1+\sqrt{10}\right\}$.

3. $m^2 - m + 8 = 0$
$a = 1, b = -1, c = 8$
$$m = \frac{-b \pm \sqrt{b^2 - 4ac}}{2a} = \frac{-(-1) \pm \sqrt{(-1)^2 - 4(1)(8)}}{2(1)} = \frac{1 \pm \sqrt{-31}}{2} = \frac{1 \pm i\sqrt{31}}{2}$$
The solution set is $\left\{\frac{1-i\sqrt{31}}{2}, \frac{1+i\sqrt{31}}{2}\right\}$.

4. $u^2 - 6u + 2 = 0$
$a = 1, b = -6, c = 2$
$$u = \frac{-b \pm \sqrt{b^2 - 4ac}}{2a} = \frac{-(-6) \pm \sqrt{(-6)^2 - 4(1)(2)}}{2(1)} = \frac{6 \pm \sqrt{36-8}}{2} = \frac{6 \pm \sqrt{28}}{2} = \frac{6 \pm 2\sqrt{7}}{2} = 3 \pm \sqrt{7}$$
The solution set is $\left\{3-\sqrt{7}, 3+\sqrt{7}\right\}$.

5. $7x^2 + 8x + 1 = 0$
$(7x + 1)(x + 1) = 0$
$7x + 1 = 0$ or $x + 1 = 0$
$7x = -1$ or $x = -1$
$x = -\frac{1}{7}$ or $x = -1$
The solution set is $\left\{-\frac{1}{7}, -1\right\}$.

6. $y^2 - 3y = 5$
$y^2 - 3y - 5 = 0$
$a = 1, b = -3, c = -5$
$$y = \frac{-b \pm \sqrt{b^2 - 4ac}}{2a} = \frac{-(-3) \pm \sqrt{(-3)^2 - 4(1)(-5)}}{2(1)} = \frac{3 \pm \sqrt{9+20}}{2} = \frac{3 \pm \sqrt{29}}{2}$$
The solution set is $\left\{\frac{3-\sqrt{29}}{2}, \frac{3+\sqrt{29}}{2}\right\}$.

7. The LCD is $(x + 2)(x - 2)$.

$$\frac{4}{x+2}+\frac{2x}{x-2}=\frac{6}{x^2-4}$$

$$(x+2)(x-2)\left(\frac{4}{x+2}\right)+(x+2)(x-2)\left(\frac{2x}{x-2}\right)=(x+2)(x-2)\left(\frac{6}{x^2-4}\right)$$

$$4(x-2)+2x(x+2)=6$$

$$4x-8+2x^2+4x=6$$

$$2x^2+8x-14=0$$

$$2(x^2+4x-7)=0$$

$a = 1, b = 4, c = -7$

$$\begin{aligned}x&=\frac{-b\pm\sqrt{b^2-4ac}}{2a}\\&=\frac{-4\pm\sqrt{4^2-4(1)(-7)}}{2(1)}\\&=\frac{-4\pm\sqrt{16+28}}{2}\\&=\frac{-4\pm\sqrt{44}}{2}\\&=\frac{-4\pm2\sqrt{11}}{2}\\&=-2\pm\sqrt{11}\end{aligned}$$

The solution set is $\{-2-\sqrt{11}, -2+\sqrt{11}\}$.

8. Let $y = x^2$.

$$x^4-8x^2-9=0$$

$$y^2-8y-9=0$$

$$(y-9)(y+1)=0$$

$y-9=0$ or $y+1=0$

$y=9$ $\quad$ $y=-1$

$x^2=9$ $\quad$ $x^2=-1$

$x=\pm\sqrt{9}$ $\quad$ $x=\pm\sqrt{-1}$

$x=\pm3$ $\quad$ $x=\pm i$

The solution set is $\{-3, 3, -i, i\}$.

9.
$$x^6+1=x^4+x^2$$
$$x^6-x^4-x^2+1=0$$
$$x^4(x^2-1)-(x^2-1)=0$$
$$(x^4-1)(x^2-1)=0$$
$$(x^2+1)(x^2-1)(x^2-1)=0$$
$$(x^2+1)(x^2-1)^2=0$$
$$(x^2+1)[(x+1)(x-1)]^2=0$$
$$(x^2+1)(x+1)^2(x-1)^2=0$$

$x^2+1=0$ or $(x+1)^2=0$ or $(x-1)^2=0$

$x^2=-1$ or $x+1=0$ or $x-1=0$

$x=\pm\sqrt{-1}=\pm i$ or $x=-1$ or $x=1$

The solution set is $\{1, -1, i, -i\}$.

10. Let $y = (x + 1)$.
$$(x+1)^2-15(x+1)+56=0$$
$$y^2-15y+56=0$$
$$(y-8)(y-7)=0$$

$y-8=0$ or $y-7=0$

$y=8$ $\quad y=7$

$x+1=8$ $\quad x+1=7$

$x=7$ $\quad x=6$

The solution set is $\{6, 7\}$.

11.
$$x^2-6x=-2$$
$$x^2-6x+\left(\frac{-6}{2}\right)^2=-2+9$$
$$(x-3)^2=7$$
$$\sqrt{(x-3)^2}=\pm\sqrt{7}$$
$$x-3=\pm\sqrt{7}$$
$$x=3\pm\sqrt{7}$$

The solution set is $\left\{3-\sqrt{7}, 3+\sqrt{7}\right\}$.

12.

$$2a^2+5=4a$$
$$2a^2-4a=-5$$
$$a^2-2a=-\frac{5}{2}$$
$$a^2-2a+\left(\frac{-2}{2}\right)^2=-\frac{5}{2}+1$$
$$(a-1)^2=-\frac{3}{2}$$
$$\sqrt{(a-1)^2}=\pm\sqrt{-\frac{3}{2}}$$
$$a-1=\pm\frac{i\sqrt{3}}{\sqrt{2}}$$
$$a=1\pm\frac{i\sqrt{6}}{2}$$

The solution set is $\left\{\frac{2-i\sqrt{6}}{2}, \frac{2+i\sqrt{6}}{2}\right\}$.

13.

$$2x^2-7x>15$$
$$2x^2-7x-15>0$$
$$(2x+3)(x-5)>0$$
$$(2x+3)(x-5)=0$$
$$2x+3=0 \quad \text{or} \quad x-5=0$$
$$x=-\frac{3}{2} \quad \text{or} \quad x=5$$

Region	Test Point Value	$(2x+3)(x-5)>0$	Result
$\left(-\infty, -\frac{3}{2}\right)$	−2	$(-1)(-7)>0$	True
$\left(-\frac{3}{2}, 5\right)$	0	$(3)(-5)>0$	False
$(5, \infty)$	6	$(15)(1)>0$	True

The solution set is $\left(-\infty, -\frac{3}{2}\right)\cup(5, \infty)$.

14.

$$(x^2-16)(x^2-25)\ge 0$$
$$(x+4)(x-4)(x+5)(x-5)\ge 0$$
$$(x+4)(x-4)(x+5)(x-5)=0$$
$$x+4=0 \quad \text{or} \quad x-4=0 \quad \text{or} \quad x+5=0 \quad \text{or} \quad x-5=0$$
$$x=-4 \qquad x=4 \qquad x=-5 \qquad x=5$$

Region	Test Point Value	$(x + 4)(x - 4)(x + 5)(x - 5) \geq 0$	Result
$(-\infty, -5)$	-6	$(-2)(-10)(-1)(-11) \geq 0$	True
$(-5, -4)$	-4.5	$(-0.5)(-8.5)(0.5)(-9.5) \geq 0$	False
$(-4, 4)$	0	$(4)(-4)(5)(-5) \geq 0$	True
$(4, 5)$	4.5	$(8.5)(0.5)(9.5)(-0.5) \geq 0$	False
$(5, \infty)$	6	$(10)(2)(11)(1) \geq 0$	True

The solution set is $(-\infty, -5] \cup [-4, 4] \cup [5, \infty)$.

15. $\frac{5}{x+3} < 1$

$x+3=0$

$x=-3$ makes the denominator 0.

$$\frac{5}{x+3}=1$$

$$(x+3)\left(\frac{5}{x+3}\right)=(x+3)(1)$$

$$5=x+3$$

$$2=x$$

Region	Test Point Value	$\frac{5}{x+3}<1$	Result
$(-\infty, -3)$	-4	$\frac{5}{-1}<1$	True
$(-3, 2)$	0	$\frac{5}{3}<1$	False
$(2, \infty)$	3	$\frac{5}{6}<1$	True

The solution set is $(-\infty, -3) \cup (2, \infty)$.

16. $\frac{7x-14}{x^2-9} \le 0$

$$x^2-9=0$$
$$(x+3)(x-3)=0$$
$$x+3=0 \quad \text{or} \quad x-3=0$$
$x=-3$ and $x=3$ make the denominator 0.
$$\frac{7x-14}{x^2-9}=0$$
$$(x^2-9)\left(\frac{7x-14}{x^2-9}\right)=(x^2-9)(0)$$
$$7x-14=0$$
$$7x=14$$
$$x=2$$

Region	Test Point Value	$\frac{7x-14}{x^2-9} \le 0$	Result
$(-\infty, -3)$	-4	$\frac{-42}{7} \le 0$	True
$(-3, 2)$	0	$\frac{-14}{-9} \le 0$	False
$(2, 3)$	2.5	$\frac{3.5}{-2.75} \le 0$	True
$(3, \infty)$	4	$\frac{14}{7} \le 0$	False

The solution set is $(-\infty, -3) \cup [2, 3)$.

17. $f(x)=3x^2$

vertex: (0, 0)

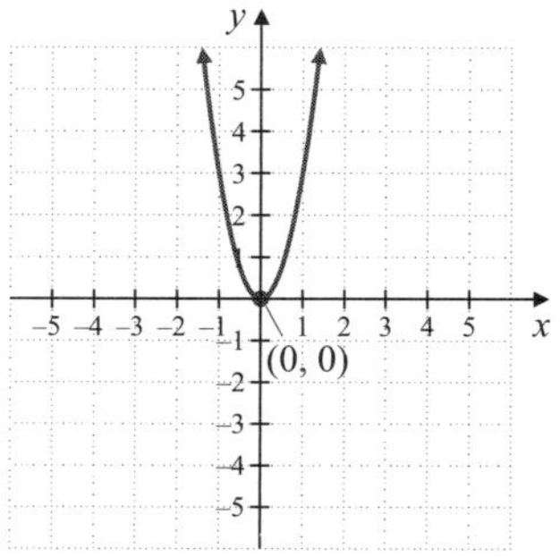

18. $G(x) = -2(x-1)^2 + 5$

vertex: (1, 5)

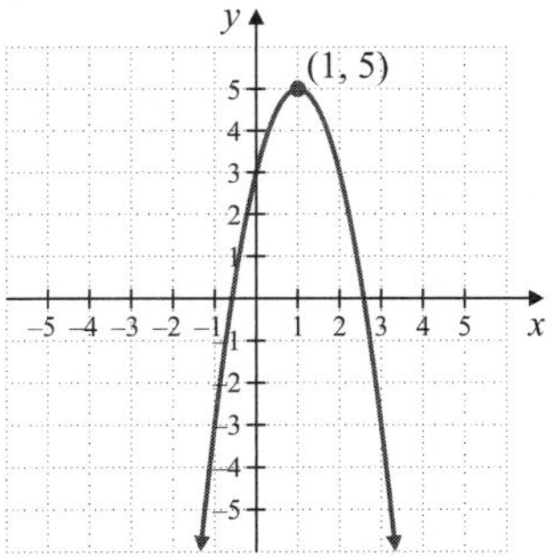

19. $h(x) = x^2 - 4x + 4$

$$\frac{-b}{2a} = \frac{-(-4)}{2(1)} = 2$$

$h(2) = 2^2 - 4(2) + 4 = 0$

Thus, the vertex is (2, 0).

The graph opens upward, since $a > 0$.

$$x^2 - 4x + 4 = 0$$
$$(x-2)^2 = 0$$
$$x - 2 = 0$$
$$x = 2$$

The x-intercept is (2, 0).

$h(0) = 4$, so the y-intercept is (0, 4).

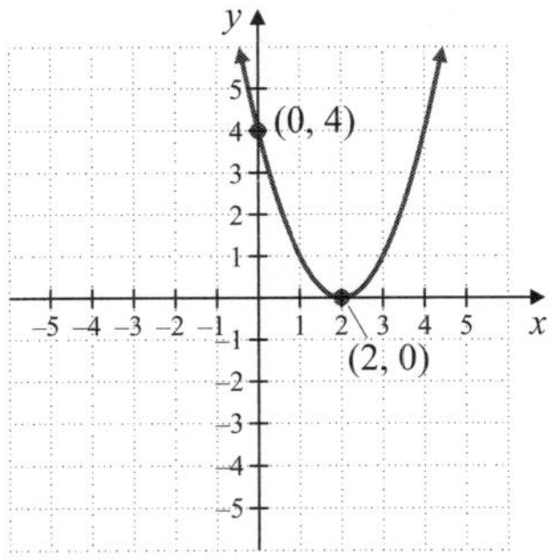

20. $F(x) = 2x^2 - 8x + 9$

$$\frac{-b}{2a} = \frac{-(-8)}{2(2)} = 2$$

$F(2) = 2(2)^2 - 8(2) + 9 = 1$

Thus, the vertex is (2, 1).

The graph opens upward, since $a > 0$.

Since the vertex is in quadrant I and the graph opens upward, there are no x-intercepts.

$F(0) = 9$, so (0, 9) is the y-intercept.

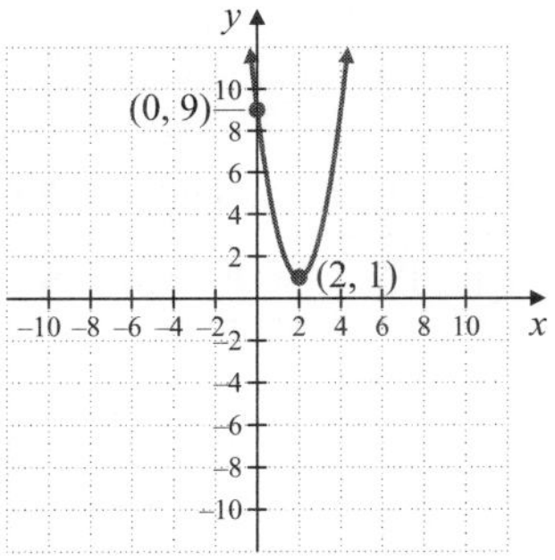

21. Use the Pythagorean theorem.

$$x^2 + (x+8)^2 = 20^2$$
$$x^2 + x^2 + 16x + 64 = 400$$
$$2x^2 + 16x - 336 = 0$$
$$x^2 + 8x - 168 = 0$$

$a = 1$, $b = 8$, $c = -168$

$$x = \frac{-b \pm \sqrt{b^2 - 4ac}}{2a}$$
$$x = \frac{-8 \pm \sqrt{8^2 - 4(1)(-168)}}{2(1)}$$
$$= \frac{-8 \pm \sqrt{64 + 672}}{2}$$
$$= \frac{-8 \pm \sqrt{736}}{2}$$
$$= \frac{-8 \pm 4\sqrt{46}}{2}$$
$$= \frac{2\left(-4 \pm 2\sqrt{46}\right)}{2}$$
$$= -4 \pm 2\sqrt{46}$$

Since the length cannot be negative, reject $-4 - 2\sqrt{46}$.

$x = -4 + 2\sqrt{46}$

$x + 8 = -4 + 2\sqrt{46} + 8 = 4 + 2\sqrt{46}$

The distance saved is $x + (x + 8) - 20$.

$$\begin{aligned} x+(x+8)-20 &= -4+2\sqrt{46}+\left(4+2\sqrt{46}\right)-20 \\ &= 4\sqrt{46}-20 \\ &\approx 7.1 \end{aligned}$$

About 7 feet of walking distance is saved by cutting across the lawn.

22. Let x be the time it takes Sandy to paint the room alone. Then $x - 2$ is the time it takes Dave alone.

	Time	Part in 1 hour
Sandy	x	$\frac{1}{x}$
Dave	$x - 2$	$\frac{1}{x-2}$
Together	4	$\frac{1}{4}$

The LCD is $4x(x - 2)$.

$$\begin{aligned} \frac{1}{x}+\frac{1}{x-2} &= \frac{1}{4} \\ 4x(x-2)\left(\frac{1}{x}\right)+4x(x-2)\left(\frac{1}{x-2}\right) &= 4x(x-2)\left(\frac{1}{4}\right) \\ 4(x-2)+4x &= x(x-2) \\ 4x-8+4x &= x^2-2x \\ 0 &= x^2-10x+8 \end{aligned}$$

$a = 1, b = -10, c = 8$

$$\begin{aligned} x &= \frac{-b\pm\sqrt{b^2-4ac}}{2a} \\ &= \frac{-(-10)\pm\sqrt{(-10)^2-4(1)(8)}}{2(1)} \\ &= \frac{10\pm\sqrt{68}}{2} \\ &= \frac{10\pm2\sqrt{17}}{2} \\ &= 5\pm\sqrt{17} \end{aligned}$$

$x = 5-\sqrt{17} \approx 0.88$ or $x = 5+\sqrt{17} \approx 9.12$

Discard 0.88 since 0.88 – 2 is negative. It takes Sandy $\left(5+\sqrt{17}\right)$ hr ≈ 9.12 hr to paint the room alone.

23. a. $s(t) = -16t^2 + 32t + 256$

$-\frac{b}{2a} = -\frac{32}{2(-16)} = 1$

$s(1) = -16(1)^2 + 32(1) + 256 = 272$

The maximum height occurs at the vertex, (1, 272). The stone's maximum height is 272 feet.

b. $0 = -16t^2 + 32t + 256$

$a = -16, b = 32, c = 256$

$$t = \frac{-b \pm \sqrt{b^2 - 4ac}}{2a}$$
$$= \frac{-32 \pm \sqrt{(32)^2 - 4(-16)(256)}}{2(-16)}$$
$$= \frac{-32 \pm \sqrt{17,408}}{-32}$$
$$= \frac{-32 \pm 32\sqrt{17}}{-32}$$
$$= 1 \pm \sqrt{17}$$

$t \approx -3.12$ or $t \approx 5.12$

Disregard –3.12 since time cannot be negative. The stone hits the water in about 5.12 seconds.

Cumulative Review Chapters 1–8

1. $730,000 = 7.3 \times 10^5$

2. $2.068 \times 10^{-3} = 0.002068$

3. $|2x - 1| + 5 = 6$

$|2x - 1| = 1$

$2x - 1 = 1$ or $2x - 1 = -1$

$2x = 2$ or $2x = 0$

$x = 1$ or $x = 0$

The solution set is {0, 1}.

4. $|3x - 2| = -5$

The absolute value of a number is always nonnegative. Therefore, the solution is ∅.

5. $x = y^2$

In $x = y^2$, if $y = 3$, then $x = 9$. Also, if $y = -3$, then $x = 9$. Since the x-value 9 corresponds to two y-values, 3 and –3, $x = y^2$ is not a function.

6. $f(x) = x^2 + x - 12$

$\frac{-b}{2a} = \frac{-1}{2(1)} = -\frac{1}{2}$

$$f\left(-\frac{1}{2}\right) = \left(-\frac{1}{2}\right)^2 - \frac{1}{2} - 12$$
$$= \frac{1}{4} - \frac{2}{4} - \frac{48}{4}$$
$$= -\frac{49}{4}$$

Thus, the vertex is $\left(-\frac{1}{2}, -\frac{49}{4}\right)$.

$x^2 + x - 12 = 0$

$(x + 4)(x - 3) = 0$

$x + 4 = 0$ or $x - 3 = 0$

$x = -4$ $\quad$ $x = 3$

The x-intercepts are (–4, 0) and (3, 0).

$f(0) = 0^2 + 0 - 12 = -12$

The y-intercept is (0, –12).

7. $\begin{cases} 3x + \frac{y}{2} = 2 \\ 6x + y = 5 \end{cases}$

Multiply the first equation by –2 and add the result to the second equation.

$$\begin{aligned} -6x - y &= -4 \\ 6x + y &= 5 \\ \hline 0 &= 1 \end{aligned}$$

Since the resulting equation is false, the system has no solution. The solution set is { } or ∅.

8. $\begin{cases} -6x+y=5 \\ 4x-2y=6 \end{cases}$

Multiply the first equation by 2 and add the result to the second equation.

$$\begin{array}{r} -12x+2y=10 \\ \underline{4x-2y=6} \\ -8x \quad =16 \\ x=-2 \end{array}$$

Replace x with -2 in the first equation.

$$-6(-2)+y=5$$
$$12+y=5$$
$$y=-7$$

The solution set is $\{(-2, -7)\}$.

9. $$\begin{array}{r} 10x^3-7x^2 \\ \underline{-(4x^3-3x^2+2)} \end{array} \rightarrow \begin{array}{r} 10x^3-7x^2 \\ \underline{+(-4x^3+3x^2-2)} \\ 6x^3-4x^2-2 \end{array}$$

10. $\dfrac{x^2-4x+4}{(2-x)}=\dfrac{(x-2)(x-2)}{-1(x-2)}=\dfrac{(x-2)}{-1}=2-x$

11. $(2x-7)(3x-4)$
$=2x(3x)+2x(-4)+(-7)(3x)+(-7)(-4)$
$=6x^2-8x-21x+28$
$=6x^2-29x+28$

12. $(4a-3)(7a-2)$
$=4a(7a)+(4a)(-2)+(-3)(7a)+(-3)(-2)$
$=28a^2-8a-21a+6$
$=28a^2-29a+6$

13. $(3x^2+y)(5x^2-2y)$
$=15x^4-6x^2y+5x^2y-2y^2$
$=15x^4-x^2y-2y^2$

14. $(2a+b)(3a-5b)=6a^2-10ab+3ab-5b^2$
$=6a^2-7ab-5b^2$

15. $$\begin{array}{r} 3x^2+2x+3 \\ x^2+0x-1\overline{\big)\,3x^4+2x^3+0x^2-8x+6} \\ \underline{3x^4+0x^3-3x^2} \\ 2x^3+3x^2-8x \\ \underline{2x^3+0x^2-2x} \\ 3x^2-6x+6 \\ \underline{3x^2+0x-3} \\ -6x+9 \end{array}$$

$$\frac{3x^4+2x^3-8x+6}{x^2-1}=3x^2+2x+3+\frac{-6x+9}{x^2-1}$$

16. a. $(a^{-2}bc^3)^{-3}=a^6b^{-3}c^{-9}=\dfrac{a^6}{b^3c^9}$

b. $\left(\dfrac{a^{-4}b^2}{c^3}\right)^{-2}=\dfrac{a^8b^{-4}}{c^{-6}}=\dfrac{a^8c^6}{b^4}$

c. $\left(\dfrac{3a^8b^2}{12a^5b^5}\right)^{-2}=\left(\dfrac{a^3}{4b^3}\right)^{-2}$
$=\dfrac{a^{-6}}{4^{-2}b^{-6}}$
$=\dfrac{16b^6}{a^6}$

17. $6x^2+3x^3=3x^2\cdot 2+3x^2\cdot x=3x^2(2+x)$

18. $9x^3+27x^2-15x$
$=3x\cdot 3x^2+3x\cdot 9x+3x(-5)$
$=3x(3x^2+9x-5)$

19. Let $x = a + 3$.
$2(a+3)^2-5(a+3)-7$
$=2x^2-5x-7$
$=(2x-7)(x+1)$
$=[2(a+3)-7][(a+3)+1]$
$=(2a+6-7)(a+3+1)$
$=(2a-1)(a+4)$

20. $2x(3y-2)-5(3y-2)=(3y-2)(2x-5)$

21. $$\begin{aligned}3a^2x-12abx+12b^2x &= 3x(a^2-4ab+4b^2)\\ &= 3x(a-2b)^2\end{aligned}$$

22. $$\begin{aligned}2xy+6x-y-3 &= 2x(y+3)-1(y+3)\\ &= (y+3)(2x-1)\end{aligned}$$

23. $$\begin{aligned}2x^2 &= \frac{17}{3}x+1\\ 3(2x^2) &= 3\left(\frac{17}{3}x+1\right)\\ 6x^2 &= 17x+3\end{aligned}$$

$$6x^2-17x-3=0$$
$$(6x+1)(x-3)=0$$
$$6x+1=0 \quad \text{or} \quad x-3=0$$
$$6x=-1 \qquad x=3$$
$$x=-\frac{1}{6}$$

The solution set is $\left\{-\frac{1}{6}, 3\right\}$.

24. $$\begin{aligned}2(a^2+2)-8 &= -2a(a-2)-5\\ 2a^2+4-8 &= -2a^2+4a-5\\ 4a^2-4a+1 &= 0\\ (2a-1)^2 &= 0\\ 2a-1 &= 0\\ 2a &= 1\\ a &= \frac{1}{2}\end{aligned}$$

The solution set is $\left\{\frac{1}{2}\right\}$.

25. $$\begin{aligned}\frac{3n+1}{2n}\cdot\frac{2n-4}{3n^2-2n-1} &= \frac{3n+1}{2n}\cdot\frac{2(n-2)}{(3n+1)(n-1)}\\ &= \frac{(3n+1)\cdot 2(n-2)}{2n(3n+1)(n-1)}\\ &= \frac{n-2}{n(n-1)}\end{aligned}$$

26. $$\begin{aligned}&\left(\sqrt{3}-4\right)\left(2\sqrt{3}+2\right)\\ &= \sqrt{3}\cdot 2\sqrt{3}+2\sqrt{3}-4\cdot 2\sqrt{3}-4\cdot 2\\ &= 2\cdot 3+2\sqrt{3}-8\sqrt{3}-8\\ &= 6-6\sqrt{3}-8\\ &= -2-6\sqrt{3}\end{aligned}$$

27. $$\begin{aligned}&\frac{x^3-1}{-3x+3}\cdot\frac{15x^2}{x^2+x+1}\\ &= \frac{(x-1)(x^2+x+1)}{-3(x-1)}\cdot\frac{15x^2}{x^2+x+1}\\ &= \frac{(x-1)(x^2+x+1)\cdot 3\cdot 5x^2}{-1\cdot 3(x-1)(x^2+x+1)}\\ &= \frac{5x^2}{-1}\\ &= -5x^2\end{aligned}$$

28. $$\begin{aligned}\frac{3}{x^2-4}-\frac{1-x}{x^2-4} &= \frac{3-(1-x)}{x^2-4}\\ &= \frac{3-1+x}{x^2-4}\\ &= \frac{2+x}{(x+2)(x-2)}\\ &= \frac{x+2}{(x+2)(x-2)}\\ &= \frac{1}{x-2}\end{aligned}$$

29. The LCD is $3x^3$.
$$\begin{aligned}\frac{2}{x^2}+\frac{5}{3x^3} &= \frac{2\cdot 3x}{x^2\cdot 3x}+\frac{5}{3x^3}\\ &= \frac{6x}{3x^3}+\frac{5}{3x^3}\\ &= \frac{6x+5}{3x^3}\end{aligned}$$

30. $a^2 - 6a + 8 = (a-4)(a-2)$

$16 - a^2 = (4-a)(4+a)$ or $-1(a-4)(a+4)$

The LCD is $-1(a-4)(a-2)(a+4)$.

$$\frac{a+1}{a^2-6a+8} - \frac{3}{16-a^2} = \frac{(-1)(a+4)}{(-1)(a+4)} \cdot \left(\frac{a+1}{a^2-6a+8}\right) - \frac{(a-2)}{(a-2)} \cdot \left(\frac{3}{16-a^2}\right)$$

$$= \frac{-(a+4)(a+1) - 3(a-2)}{-1(a-4)(a-2)(a+4)}$$

$$= \frac{-a^2 - 5a - 4 - 3a + 6}{-1(a-4)(a-2)(a+4)}$$

$$= \frac{-a^2 - 8a + 2}{-1(a-4)(a-2)(a+4)}$$

$$= \frac{-1(a^2 + 8a - 2)}{-1(a-4)(a-2)(a+4)}$$

$$= \frac{a^2 + 8a - 2}{(a-4)(a-2)(a+4)}$$

31. $$\frac{\frac{x}{y^2} + \frac{1}{y}}{\frac{y}{x^2} + \frac{1}{x}} = \frac{\left(\frac{x}{y^2} + \frac{1}{y}\right) x^2 y^2}{\left(\frac{y}{x^2} + \frac{1}{x}\right) x^2 y^2}$$

$$= \frac{\frac{x}{y^2} \cdot x^2 y^2 + \frac{1}{y} \cdot x^2 y^2}{\frac{y}{x^2} \cdot x^2 y^2 + \frac{1}{x} \cdot x^2 y^2}$$

$$= \frac{x^3 + x^2 y}{y^3 + xy^2}$$

$$= \frac{x^2(x+y)}{y^2(y+x)}$$

$$= \frac{x^2}{y^2}$$

32. $$\frac{(2a)^{-1}+b^{-1}}{a^{-1}+(2b)^{-1}}=\frac{\frac{1}{2a}+\frac{1}{b}}{\frac{1}{a}+\frac{1}{2b}}$$
$$=\frac{\left(\frac{1}{2a}+\frac{1}{b}\right)\cdot 2ab}{\left(\frac{1}{a}+\frac{1}{2b}\right)\cdot 2ab}$$
$$=\frac{\frac{1}{2a}\cdot 2ab+\frac{1}{b}\cdot 2ab}{\frac{1}{a}\cdot 2ab+\frac{1}{2b}\cdot 2ab}$$
$$=\frac{b+2a}{2b+a}$$
$$=\frac{2a+b}{a+2b}$$

33. The LCD is $(x - 3)(x + 3)$.

$$\frac{2x}{x-3}+\frac{6-2x}{x^2-9}=\frac{x}{x+3}$$
$$(x-3)(x+3)\left(\frac{2x}{x-3}\right)+(x-3)(x+3)\left(\frac{6-2x}{(x+3)(x-3)}\right)=(x+3)(x-3)\cdot\frac{x}{x+3}$$
$$2x(x+3)+(6-2x)=x(x-3)$$
$$2x^2+6x+6-2x=x^2-3x$$
$$x^2+7x+6=0$$
$$(x+6)(x+1)=0$$

$x+6=0$ or $x+1=0$

$x=-6$ $\quad x=-1$

The solution set is $\{-6, -1\}$.

34. The LCD is $2(x + 3)(x + 2)$.

$$\frac{x+3}{x^2+5x+6}=\frac{3}{2x+4}-\frac{1}{x+3}$$
$$2(x+3)(x+2)\left(\frac{x+3}{(x+3)(x+2)}\right)=2(x+3)(x+2)\cdot\frac{3}{2(x+2)}-2(x+3)(x+2)\cdot\frac{1}{x+3}$$
$$2(x+3)=3(x+3)-2(x+2)$$
$$2x+6=3x+9-2x-4$$
$$2x+6=x+5$$
$$x=-1$$

The solution set is $\{-1\}$.

35. The LCD is xyz.

$$\frac{1}{x}+\frac{1}{y}=\frac{1}{z}$$
$$xyz\left(\frac{1}{x}+\frac{1}{y}\right)=xyz\left(\frac{1}{z}\right)$$
$$xyz\left(\frac{1}{x}\right)+xyz\left(\frac{1}{y}\right)=xyz\left(\frac{1}{z}\right)$$
$$yz+xz=xy$$
$$yz=xy-xz$$
$$yz=x(y-z)$$
$$\frac{yz}{y-z}=x \quad \text{or} \quad x=\frac{yz}{y-z}$$

36.

$$\begin{array}{r|l} & x^2-6x+8 \\ x+3 & x^3-3x^2-10x+24 \\ & \underline{x^3+3x^2} \\ & -6x^2-10x \\ & \underline{-6x^2-18x} \\ & 8x+24 \\ & \underline{8x+24} \\ & 0 \end{array}$$

$$\frac{x^3-3x^2-10x+24}{x+3}=x^2-6x+8$$

37. $S=krh$

38.
$$y=\frac{k}{x}$$
$$8=\frac{k}{24}$$
$$192=k$$
$$y=\frac{192}{x}$$

39. $\sqrt[3]{-64}=-4$ because $(-4)^3=-64$.

40. $\sqrt[3]{24a^7b^9}=\sqrt[3]{(2a^2b^3)^3\cdot 3a}=2a^2b^3\sqrt[3]{3a}$

41. $\sqrt[3]{\frac{8}{125}}=\frac{2}{5}$ because $\left(\frac{2}{5}\right)^3=\frac{8}{125}$.

42. $\frac{2\sqrt{16}}{\sqrt{9x}}=\frac{2\cdot 4}{3\sqrt{x}}=\frac{8}{3\sqrt{x}}\cdot\frac{\sqrt{x}}{\sqrt{x}}=\frac{8\sqrt{x}}{3x}$

43. $\sqrt[3]{4}\cdot\sqrt[3]{2}=\sqrt[3]{4\cdot 2}=\sqrt[3]{8}=2$

44.
$$\left(\sqrt{5}-x\right)^2=\left(\sqrt{5}\right)^2-2\cdot\sqrt{5}\cdot x+x^2$$
$$=5-2x\sqrt{5}+x^2$$

45. $\sqrt{\frac{2}{a}}\cdot\sqrt{\frac{b}{3}}=\sqrt{\frac{2}{a}\cdot\frac{b}{3}}=\sqrt{\frac{2b}{3a}}$

46. $\left(\sqrt{a}+b\right)\left(\sqrt{a}-b\right)=\left(\sqrt{a}\right)^2-b^2=a-b^2$

47.
$$\sqrt[3]{54}-5\sqrt[3]{16}+\sqrt[3]{2}$$
$$=\sqrt[3]{27}\cdot\sqrt[3]{2}-5\cdot\sqrt[3]{8}\cdot\sqrt[3]{2}+\sqrt[3]{2}$$
$$=3\cdot\sqrt[3]{2}-5\cdot 2\cdot\sqrt[3]{2}+\sqrt[3]{2}$$
$$=3\sqrt[3]{2}-10\sqrt[3]{2}+\sqrt[3]{2}$$
$$=-6\sqrt[3]{2}$$

48. $\sqrt{x-2} = \sqrt{4x+1} - 3$

$\left(\sqrt{x-2}\right)^2 = \left(\sqrt{4x+1} - 3\right)^2$

$x - 2 = 4x + 1 - 6\sqrt{4x+1} + 9$

$6\sqrt{4x+1} = 3x + 12$

$\sqrt{4x+1} = \frac{1}{2}x + 2$

$\left(\sqrt{4x+1}\right)^2 = \left(\frac{1}{2}x + 2\right)^2$

$4x + 1 = \frac{1}{4}x^2 + 2x + 4$

$0 = \frac{1}{4}x^2 - 2x + 3$

$0 = \left(\frac{1}{2}x - 3\right)\left(\frac{1}{2}x - 1\right)$

$\frac{1}{2}x - 3 = 0$ or $\frac{1}{2}x - 1 = 0$

$\frac{1}{2}x = 3$ or $\frac{1}{2}x = 1$

$x = 6$ or $x = 2$

The solution set is {2, 6}.

49. $\sqrt[3]{\frac{7x}{8}} + 2\sqrt[3]{7x} = \frac{\sqrt[3]{7x}}{\sqrt[3]{8}} + 2\sqrt[3]{7x}$

$= \frac{\sqrt[3]{7x}}{2} + 2\sqrt[3]{7x}$

$= \frac{\sqrt[3]{7x}}{2} + \frac{2\sqrt[3]{7x} \cdot 2}{2}$

$= \frac{\sqrt[3]{7x}}{2} + \frac{4\sqrt[3]{7x}}{2}$

$= \frac{5\sqrt[3]{7x}}{2}$

50. $\sqrt[4]{48x^9} = \sqrt[4]{16x^8 \cdot 3x}$

$= \sqrt[4]{16x^8} \cdot \sqrt[4]{3x}$

$= 2x^2\sqrt[4]{3x}$

51. $p^2 + 2p = 4$

$p^2 + 2p + \left(\frac{2}{2}\right)^2 = 4 + 1$

$(p+1)^2 = 5$

$\sqrt{(p+1)^2} = \pm\sqrt{5}$

$p + 1 = \pm\sqrt{5}$

$p = -1 \pm \sqrt{5}$

The solution set is $\left\{-1-\sqrt{5}, -1+\sqrt{5}\right\}$.

52. $(y-1)^2 = 24$

$\sqrt{(y-1)^2} = \pm\sqrt{24}$

$y - 1 = \pm\sqrt{24}$

$y - 1 = \pm 2\sqrt{6}$

$y = 1 \pm 2\sqrt{6}$

The solution set is $\left\{1-2\sqrt{6}, 1+2\sqrt{6}\right\}$.

53. $2x^2 - 4x = 3$

$2x^2 - 4x - 3 = 0$

$a = 2, b = -4, c = -3$

$x = \frac{-b \pm \sqrt{b^2 - 4ac}}{2a}$

$= \frac{-(-4) \pm \sqrt{(-4)^2 - 4(2)(-3)}}{2(2)}$

$= \frac{4 \pm \sqrt{16+24}}{4}$

$= \frac{4 \pm \sqrt{40}}{4}$

$= \frac{4 \pm 2\sqrt{10}}{4}$

$= \frac{2\left(2 \pm \sqrt{10}\right)}{2 \cdot 2}$

$= \frac{2 \pm \sqrt{10}}{2}$

The solution set is $\left\{\frac{2-\sqrt{10}}{2}, \frac{2+\sqrt{10}}{2}\right\}$.

54.
$$m^2 = 4m + 8$$
$$m^2 - 4m - 8 = 0$$
$$a = 1, b = -4, c = -8$$
$$m = \frac{-b \pm \sqrt{b^2 - 4ac}}{2a}$$
$$= \frac{-(-4) \pm \sqrt{(-4)^2 - 4(1)(-8)}}{2(1)}$$
$$= \frac{4 \pm \sqrt{16 + 32}}{2}$$
$$= \frac{4 \pm \sqrt{48}}{2}$$
$$= \frac{4 \pm 4\sqrt{3}}{2}$$
$$= \frac{2\left(2 \pm 2\sqrt{3}\right)}{2}$$
$$= 2 \pm 2\sqrt{3}$$

The solution set is $\left\{2 - 2\sqrt{3},\ 2 + 2\sqrt{3}\right\}$.

55.
$$h = -16t^2 + 12t + 200$$
$$0 = -16t^2 + 12t + 200$$
$$0 = 4t^2 - 3t - 50$$
$$a = 4, b = -3, c = -50$$
$$t = \frac{-b \pm \sqrt{b^2 - 4ac}}{2a}$$
$$= \frac{-(-3) \pm \sqrt{(-3)^2 - 4(4)(-50)}}{2(4)}$$
$$= \frac{3 \pm \sqrt{9 + 800}}{8}$$
$$= \frac{3 \pm \sqrt{809}}{8}$$

Reject $\frac{3 - \sqrt{809}}{8}$ since time won't be negative. The object hits the ground in exactly $\frac{3 + \sqrt{809}}{8}$ seconds or approximately 3.9 seconds.

56. Let x be the time it takes them together.

	Time	Part in 1 hour
Mr. Briley	24	$\frac{1}{24}$
Son	40	$\frac{1}{40}$
Together	x	$\frac{1}{x}$

The LCD is $120x$.
$$\frac{1}{24} + \frac{1}{40} = \frac{1}{x}$$
$$120x\left(\frac{1}{24}\right) + 120x\left(\frac{1}{40}\right) = 120x\left(\frac{1}{x}\right)$$
$$5x + 3x = 120$$
$$8x = 120$$
$$x = 15$$

It takes them 15 hours working together.

Chapter 9

Section 9.1 Practice Problems

1. a. $$\begin{aligned}(f+g)(x) &= f(x)+g(x)\\ &= (x+3)+(3x-1)\\ &= 4x+2\end{aligned}$$

b. $$\begin{aligned}(f-g)(x) &= f(x)-g(x)\\ &= (x+3)-(3x-1)\\ &= x+3-3x+1\\ &= -2x+4\end{aligned}$$

c. $$\begin{aligned}(f\cdot g)(x) &= f(x)\cdot g(x)\\ &= (x+3)(3x-1)\\ &= 3x^2+8x-3\end{aligned}$$

d. $\left(\frac{f}{g}\right)(x)=\frac{f(x)}{g(x)}=\frac{x+3}{3x-1}$, where $x\neq\frac{1}{3}$.

2. a. $(f\circ g)(3)=f(g(3))=f(7)=7^2=49$
$$\begin{aligned}(g\circ f)(3) &= g(f(3))\\ &= g(9)\\ &= 2(9)+1\\ &= 19\end{aligned}$$

b. $$\begin{aligned}(f\circ g)(x) &= f(g(x))\\ &= f(2x+1)\\ &= (2x+1)^2\\ &= 4x^2+4x+1\end{aligned}$$
$(g\circ f)(x)=g(f(x))=g(x^2)=2x^2+1$

3. a. $$\begin{aligned}(f\circ g)(x) &= f(g(x))\\ &= f(x+1)\\ &= \sqrt{x+1}\end{aligned}$$

b. $(g\circ f)(x)=g(f(x))=g\left(\sqrt{x}\right)=\sqrt{x}+1$

4. a. $$\begin{aligned}F(x) &= |x+5|\\ &= |(x+5)|\\ &= |g(x)|\\ &= h(g(x))\\ &= (h\circ g)(x)\end{aligned}$$

b. $$\begin{aligned}G(x) &= 2x+5\\ &= (2x)+5\\ &= f(x)+5\\ &= g(f(x))\\ &= (g\circ f)(x)\end{aligned}$$

Exercise Set 9.1

1. a. $(f+g)(x)=(x-7)+(2x+1)=3x-6$

b. $$\begin{aligned}(f-g)(x) &= (x-7)-(2x+1)\\ &= x-7-2x-1\\ &= -x-8\end{aligned}$$

c. $$\begin{aligned}(f\cdot g)(x) &= (x-7)(2x+1)\\ &= 2x^2-13x-7\end{aligned}$$

d. $\left(\frac{f}{g}\right)(x)=\frac{x-7}{2x+1}$, where $x\neq-\frac{1}{2}$

3. a. $(f+g)(x)=(x^2+1)+5x=x^2+5x+1$

b. $(f-g)(x)=(x^2+1)-5x=x^2-5x+1$

c. $(f\cdot g)(x)=(x^2+1)(5x)=5x^3+5x$

d. $\left(\frac{f}{g}\right)(x)=\frac{x^2+1}{5x}$, where $x\neq 0$.

5. a. $(f+g)(x)=\sqrt{x}+(x+5)=\sqrt{x}+x+5$

b. $(f-g)(x)=\sqrt{x}-(x+5)=\sqrt{x}-x-5$

c. $(f\cdot g)(x)=\sqrt{x}(x+5)=x\sqrt{x}+5\sqrt{x}$

d. $\left(\frac{f}{g}\right)(x)=\frac{\sqrt{x}}{x+5}$, where $x \neq -5$

7. a. $(f+g)(x)=(-3x)+(5x^2)=5x^2-3x$

b. $(f-g)(x)=(-3x)-(5x^2)=-5x^2-3x$

c. $(f\cdot g)(x)=-3x\cdot 5x^2=-15x^3$

d. $\left(\frac{f}{g}\right)(x)=\frac{-3x}{5x^2}=-\frac{3}{5x}$, where $x \neq 0$.

9. $\begin{aligned}(f\circ g)(2)&=f(g(2))\\&=f(-4)\\&=(-4)^2-6(-4)+2\\&=16+24+2\\&=42\end{aligned}$

11. $\begin{aligned}(g\circ f)(-1)&=g(f(-1))\\&=g(9)\\&=-2(9)\\&=-18\end{aligned}$

13. $(g\circ h)(0)=g(h(0))=g(0)=-2(0)=0$

15. $\begin{aligned}(f\circ g)(x)&=f(g(x))\\&=f(5x)\\&=(5x)^2+1\\&=25x^2+1\end{aligned}$

$\begin{aligned}(g\circ f)(x)&=g(f(x))\\&=g(x^2+1)\\&=5(x^2+1)\\&=5x^2+5\end{aligned}$

17. $\begin{aligned}(f\circ g)(x)&=f(g(x))\\&=f(x+7)\\&=2(x+7)-3\\&=2x+14-3\\&=2x+11\end{aligned}$

$\begin{aligned}(g\circ f)(x)&=g(f(x))\\&=g(2x-3)\\&=(2x-3)+7\\&=2x+4\end{aligned}$

19. $\begin{aligned}(f\circ g)(x)&=f(g(x))\\&=f(-2x)\\&=(-2x)^3+(-2x)-2\\&=-8x^3-2x-2\end{aligned}$

$\begin{aligned}(g\circ f)(x)&=g(f(x))\\&=g(x^3+x-2)\\&=-2(x^3+x-2)\\&=-2x^3-2x+4\end{aligned}$

21. $(f\circ g)(x)=f(g(x))=f(10x-3)=|10x-3|$

$(g\circ f)(x)=g(f(x))=g(|x|)=10|x|-3$

23. $\begin{aligned}(f\circ g)(x)&=f(g(x))\\&=f(-5x+2)\\&=\sqrt{-5x+2}\end{aligned}$

$\begin{aligned}(g\circ f)(x)&=g(f(x))\\&=g\left(\sqrt{x}\right)\\&=-5\sqrt{x}+2\end{aligned}$

25. $\begin{aligned}H(x)&=\sqrt{x^2+2}\\&=\sqrt{(x^2+2)}\\&=\sqrt{h(x)}\\&=g(h(x))\\&=(g\circ h)(x)\end{aligned}$

27. $\begin{aligned}F(x)&=9x^2+2\\&=(3x)^2+2\\&=[f(x)]^2+2\\&=h(f(x))\\&=(h\circ f)(x)\end{aligned}$

29. $G(x) = 3\sqrt{x}$
$= 3 \cdot g(x)$
$= f(g(x))$
$= (f \circ g)(x)$

31. answers may vary

33. answers may vary

35. answers may vary

37. $x = y + 2$
$x - 2 = y$
$y = x - 2$

39. $x = 3y$
$\frac{x}{3} = y$
$y = \frac{x}{3}$

41. $x = -2y - 7$
$x + 7 = -2y$
$\frac{x+7}{-2} = y$
$y = -\frac{x+7}{2}$

43. $P(x) = R(x) - C(x)$

45. answers may vary

Section 9.2 Practice Problems

1. $f = \{(7, 3), (-1, 1), (5, 0), (4, -2)\}$
The function f is one-to-one since each y-value corresponds to only one x-value.

2. $g = \{(-3, 2), (6, 3), (2, 14), (-6, 2)\}$
The function g is not one-to-one since the y-value 2 in $(-3, 2)$ and $(-6, 2)$ corresponds to two different x-values.

3. $h = \{(0, 0), (1, 2), (3, 4), (5, 6)\}$
The function h is one-to-one since each y-value corresponds to only one x-value.

4. The table does not describe a one-to-one function since the output 7 corresponds to two different inputs, Mississippi and Utah.

5. The graph does not describe a one-to-one function since the y-value -2 corresponds to two different x-values, -3 and -2.

6. a. The graph does not represent a one-to-one function because it does not pass the vertical line test.

b. The graph does not represent a one-to-one function because it does not pass the vertical line test.

c. The graph represents a one-to-one function because it passes the vertical and horizontal line tests.

d. The graph represents a one-to-one function because it passes the vertical and horizontal line tests.

e. The graph does not represent a one-to-one function because it does not pass the horizontal line test.

7. $f = \{(2, -4), (-1, 13), (0, 0), (-7, -8)\}$
$f^{-1} = \{(-4, 2), (13, -1), (0, 0), (-8, -7)\}$

8. $f(x) = x - 6$
$y = x - 6$
$x = y - 6$
$x + 6 = y$
$f^{-1}(x) = x + 6$

9. $f(x) = 2x + 3$

$y = 2x + 3$

$x = 2y + 3$

$x - 3 = 2y$

$\frac{x-3}{2} = y$

$f^{-1}(x) = \frac{x-3}{2}$

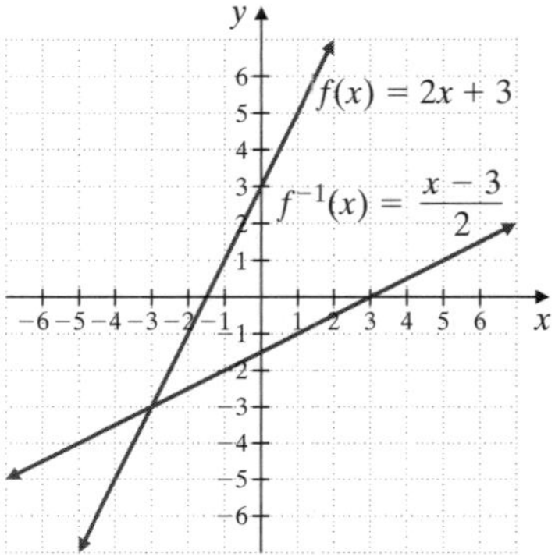

10. a.

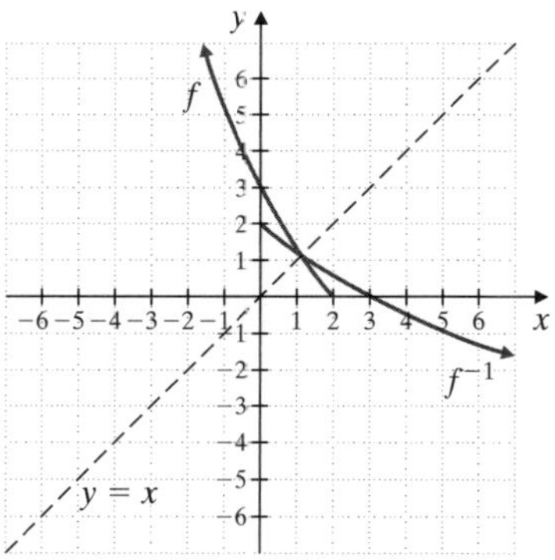

b.

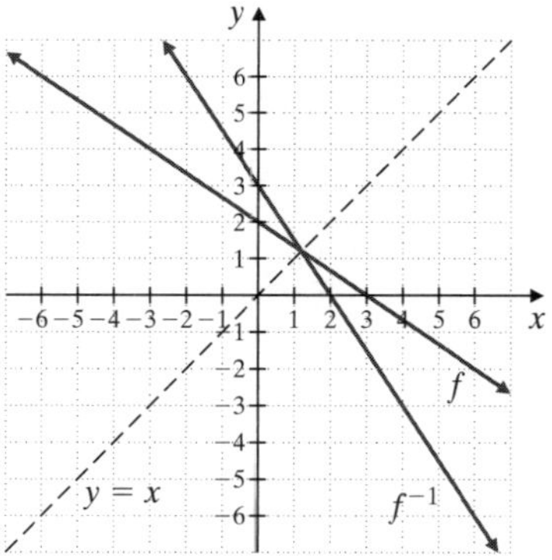

Exercise Set 9.2

1. Since the y-value 7 corresponds to two x-values, 3 and 6, the function is not one-to-one.

3. $h = \{(10, 10)\}$ is a one-to-one function.
$h^{-1}\{(10, 10)\}$

5. $f = \{(11, 12), (4, 3), (3, 4), (6, 6)\}$ is a one-to-one function.
$f^{-1} = \{(12, 11), (3, 4), (4, 3), (6, 6)\}$

7. The given function is not one-to-one since the output 5.5 corresponds to two different inputs. July and October. A similar statement could be made about the output 5.4.

9. The given function is one-to-one.

Rank in Population (Input)	1	47	14	24	36
State (Output)	California	Alaska	Indiana	Louisiana	New Mexico

11. $f(x) = x^3 + 2$

a. $f(1) = 1^3 + 2 = 3$

b. $f^{-1}(3) = 1$

13. $f(x) = x^3 + 2$

a. $f(-1) = (-1)^3 + 2 = 1$

b. $f^{-1}(1) = -1$

15. The graph represents a one-to-one function because it passes the vertical and horizontal line tests.

17. The graph does not represent a one-to-one function because it does not pass the horizontal line test.

19. The graph represents a one-to-one function because it passes the vertical and horizontal line tests.

21. The graph does not represent a one-to-one function because it does not pass the horizontal line test.

23. $f(x) = x + 4$

$$\begin{aligned} y &= x + 4 \\ x &= y + 4 \\ y &= x - 4 \end{aligned}$$

$f^{-1}(x) = x - 4$

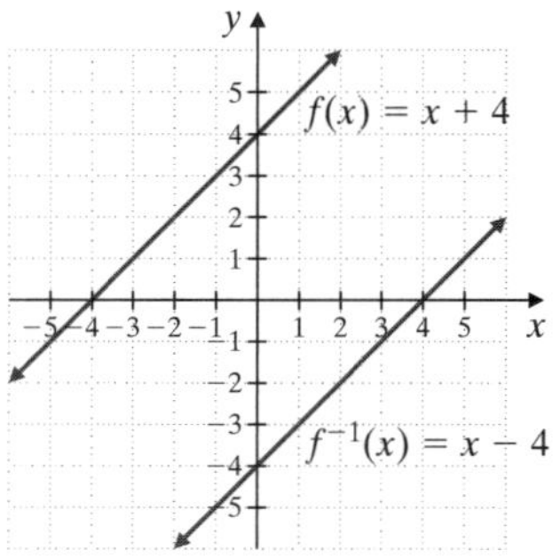

25.
$$f(x) = 2x - 3$$
$$y = 2x - 3$$
$$x = 2y - 3$$
$$2y = x + 3$$
$$y = \frac{x+3}{2}$$
$$f^{-1}(x) = \frac{x+3}{2}$$

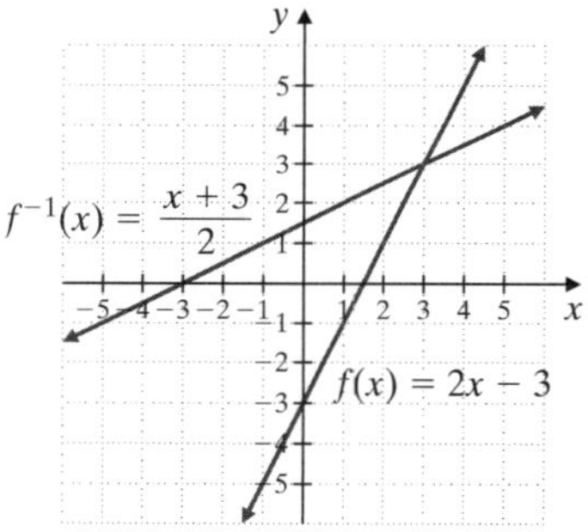

27.
$$f(x) = \frac{1}{2}x - 1$$
$$y = \frac{1}{2}x - 1$$
$$x = \frac{1}{2}y - 1$$
$$\frac{1}{2}y = x + 1$$
$$y = 2(x+1)$$
$$y = 2x + 2$$
$$f^{-1}(x) = 2x + 2$$

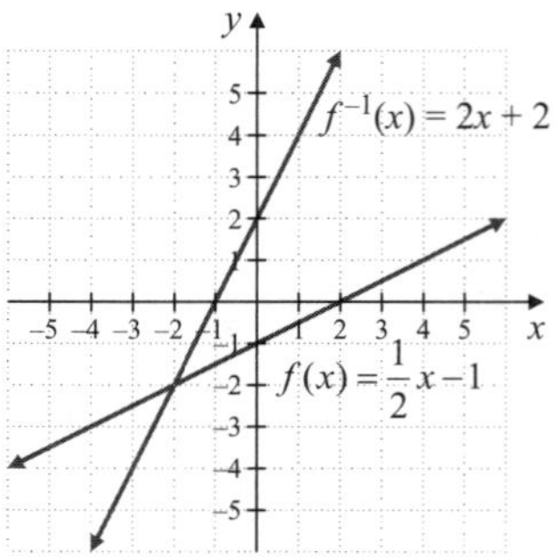

29.
$$f(x) = x^3$$
$$y = x^3$$
$$x = y^3$$
$$y = \sqrt[3]{x}$$
$$f^{-1}(x) = \sqrt[3]{x}$$

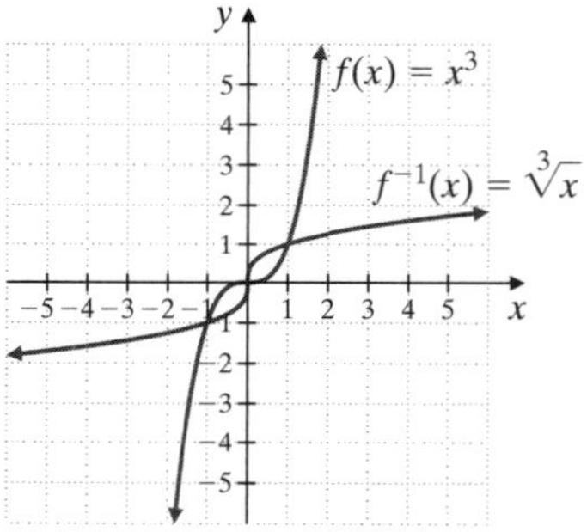

31.
$$f(x) = \frac{x-2}{5}$$
$$y = \frac{x-2}{5}$$
$$x = \frac{y-2}{5}$$
$$5x = y - 2$$
$$5x + 2 = y$$
$$f^{-1}(x) = 5x + 2$$

33.
$$f(x) = \sqrt[3]{x}$$
$$y = \sqrt[3]{x}$$
$$x = \sqrt[3]{y}$$
$$x^3 = y$$
$$f^{-1}(x) = x^3$$

35. $f(x)=\dfrac{5}{3x+1}$

$y=\dfrac{5}{3x+1}$

$x=\dfrac{5}{3y+1}$

$3y+1=\dfrac{5}{x}$

$3y=\dfrac{5}{x}-1$

$y=\dfrac{\frac{5}{x}-1}{3}$

$f^{-1}(x)=\dfrac{\frac{5}{x}-1}{3}$

37. $f(x)=(x+2)^3$

$y=(x+2)^3$

$x=(y+2)^3$

$\sqrt[3]{x}=y+2$

$\sqrt[3]{x}-2=y$

$f^{-1}(x)=\sqrt[3]{x}-2$

39.

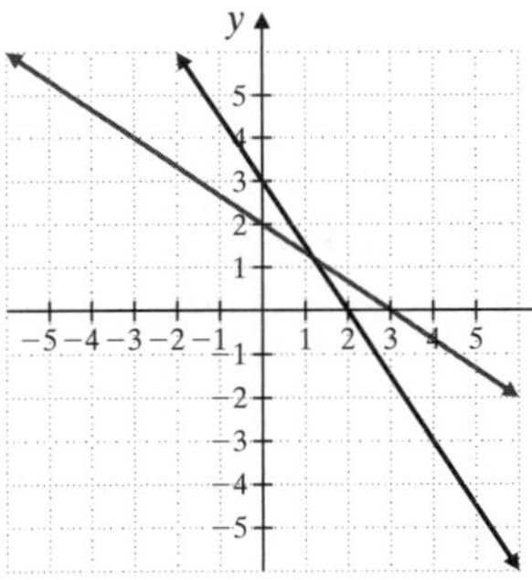

41.

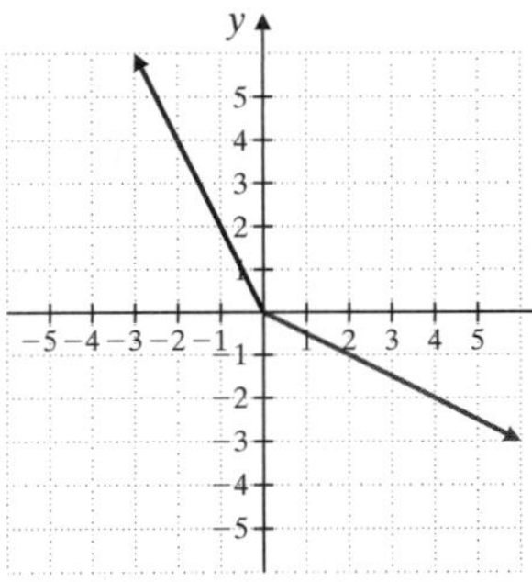

43. $25^{1/2}=\sqrt{25}=5$

45. $16^{3/4}=(16^{1/4})^3=(2)^3=8$

47. $9^{-3/2}=\dfrac{1}{9^{3/2}}=\dfrac{1}{(9^{1/2})^3}=\dfrac{1}{3^3}=\dfrac{1}{27}$

49. $f(x)=3^x$

$f(2)=3^2=9$

51. $f(x)=3^x$

$f\left(\dfrac{1}{2}\right)=3^{1/2}\approx 1.73$

53. a. The ordered pair corresponding to $f(2)$ is (2, 9).

b. One ordered pair that we know is a solution of f^{-1} is (9, 2).

55. a. $\left(-2,\dfrac{1}{4}\right),\left(-1,\dfrac{1}{2}\right)$, (0, 1), (1, 2), (2, 5)

b. $\left(\dfrac{1}{4},-2\right),\left(\dfrac{1}{2},-1\right)$, (1, 0), (2, 1), (5, 2)

c–d.

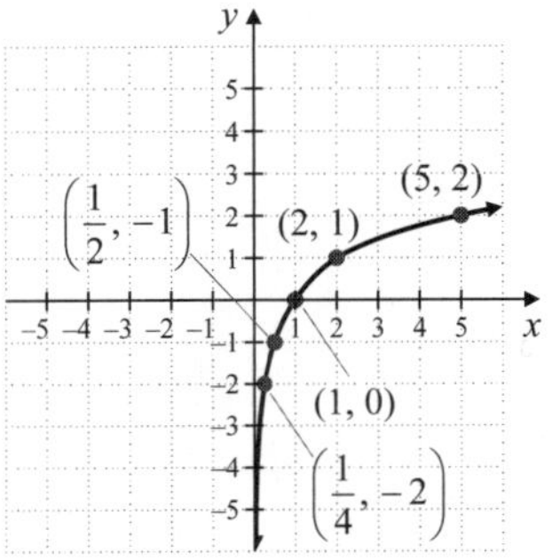

57. answers may vary

59. $f(x) = 3x + 1$

$y = 3x + 1$

$x = 3y + 1$

$x - 1 = 3y$

$\frac{x-1}{3} = y$

$f^{-1}(x) = \frac{x-1}{3}$

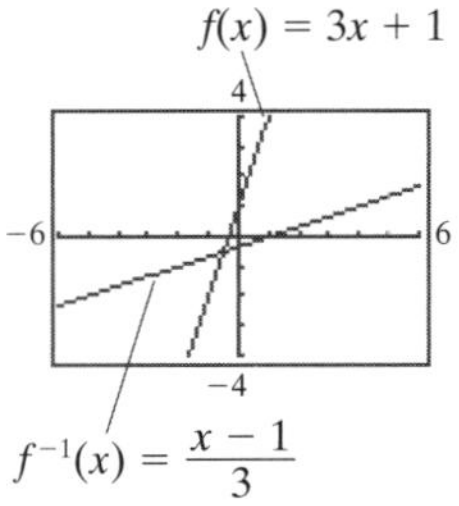

61. $f(x) = \sqrt[3]{x+3}$

$y = \sqrt[3]{x+3}$

$x = \sqrt[3]{y+3}$

$x^3 = \left(\sqrt[3]{y+3}\right)^3$

$x^3 = y + 3$

$x^3 - 3 = y$

$f^{-1}(x) = x^3 - 3$

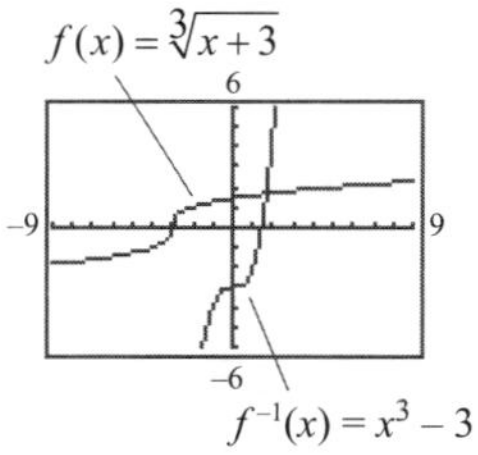

Section 9.3 Practice Problems

1.

x	$f(x) = 6^x$
−2	$6^{-2} = \frac{1}{36}$
−1	$6^{-1} = \frac{1}{6}$
0	$6^0 = 1$
1	$6^1 = 6$
2	$6^2 = 36$

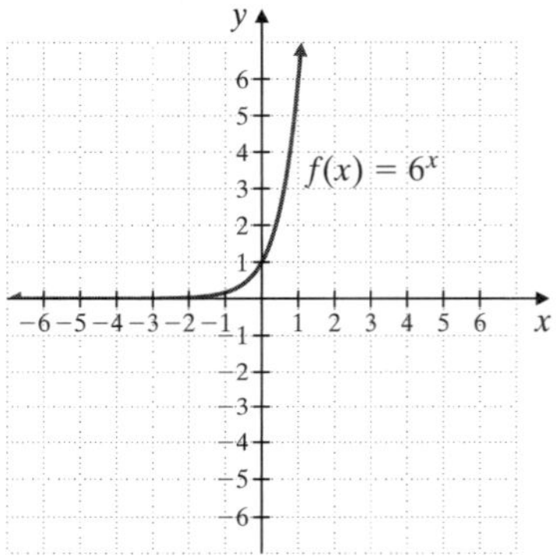

2.

x	$f(x) = \left(\frac{1}{5}\right)^x$
−2	$\left(\frac{1}{5}\right)^{-2} = 25$
−1	$\left(\frac{1}{5}\right)^{-1} = 5$
0	$\left(\frac{1}{5}\right)^0 = 1$
1	$\left(\frac{1}{5}\right)^1 = \frac{1}{5}$
2	$\left(\frac{1}{5}\right)^2 = \frac{1}{25}$

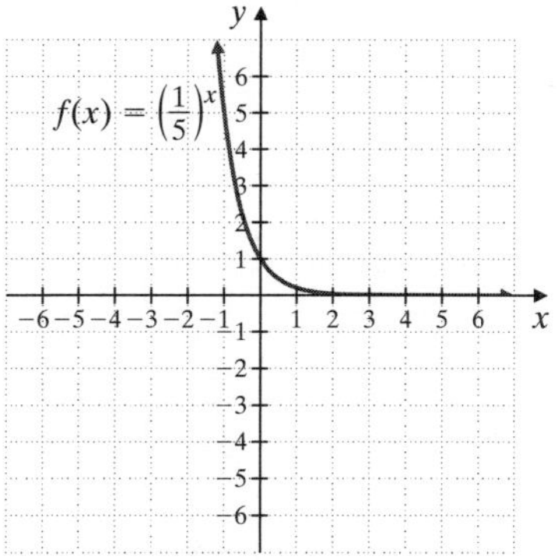

3.

x	$f(x) = 2^{x-1}$
-2	$2^{-3} = \frac{1}{8}$
-1	$2^{-2} = \frac{1}{4}$
0	$2^{-1} = \frac{1}{2}$
1	$2^0 = 1$
2	$2^1 = 2$

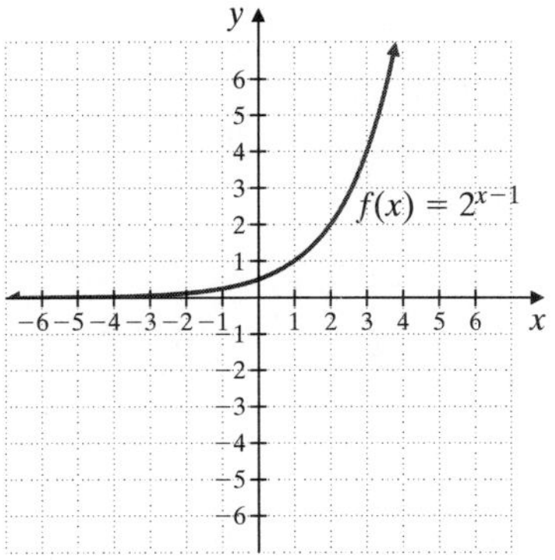

4. $5^x = 125$

$5^x = 5^3$

$x = 3$

The solution set is $\{3\}$.

5. $4^x = 8$

$(2^2)^x = 2^3$

$2^{2x} = 2^3$

$2x = 3$

$x = \frac{3}{2}$

The solution set is $\left\{\frac{3}{2}\right\}$.

6. $9^{x-1} = 27^x$

$(3^2)^{x-1} = (3^3)^x$

$3^{2x-2} = 3^{3x}$

$2x - 2 = 3x$

$-2 = x$

The solution set is $\{-2\}$.

7. a. $y = 100(2.7)^{-0.1t}$

$y = 100(2.7)^{-0.1(30)}$

$y \approx 5.08$

The expected percent of radioactive material is approximately 5.08%.

b. $A = P\left(1 + \frac{r}{n}\right)^{nt}$

$A = 23{,}000\left(1 + \frac{0.12}{4}\right)^{4 \cdot 6}$

$A = 23{,}000(1.03)^{24}$

$A = 46{,}754.26$

The amount owed is \$46,754.26.

Calculator Explorations

1. $Y_1 = 100(2.7)^{-0.1x}$

Use TRACE to find y when $x = 2$.
After 2 days, there is about 81.98% of radioactive material.

2. $Y_1 = 100(2.7)^{-0.1x}$

Use TRACE to find y when $x = 10$.
After 10 days, there is about 37.04% of radioactive material.

3. $Y_1 = 100(2.7)^{-0.1x}$
Use TRACE to find y when $x = 15$.
After 15 days, there is about 22.54% of radioactive material.

4. $Y_1 = 100(2.7)^{-0.1x}$
Use TRACE to find y when $x = 25$.
After 25 days, there is about 8.35% of radioactive material.

Exercise Set 9.3

1.

x	$y = 5^x$
−2	$5^{-2} = \frac{1}{5^2} = \frac{1}{25}$
−1	$5^{-1} = \frac{1}{5}$
0	$5^0 = 1$
1	$5^1 = 5$
2	$5^2 = 25$

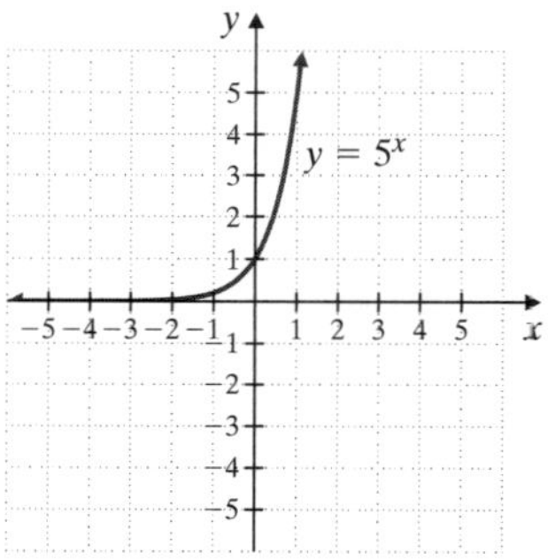

3.

x	$y = 1 + 2^x$
−2	$1 + 2^{-2} = 1 + \frac{1}{4} = \frac{5}{4}$
−1	$1 + 2^{-1} = 1 + \frac{1}{2} = \frac{3}{2}$
0	$1 + 2^0 = 1 + 1 = 2$
1	$1 + 2^1 = 1 + 2 = 3$
2	$1 + 2^2 = 1 + 4 = 5$

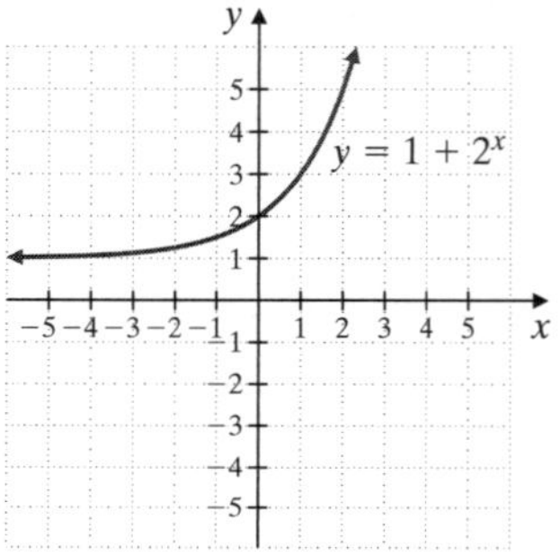

5.

x	$y = \left(\frac{1}{4}\right)^x$
−2	$\left(\frac{1}{4}\right)^{-2} = 4^2 = 16$
−1	$\left(\frac{1}{4}\right)^{-1} = 4^1 = 4$
0	$\left(\frac{1}{4}\right)^0 = 1$
1	$\left(\frac{1}{4}\right)^1 = \frac{1}{4}$
2	$\left(\frac{1}{4}\right)^2 = \frac{1}{16}$

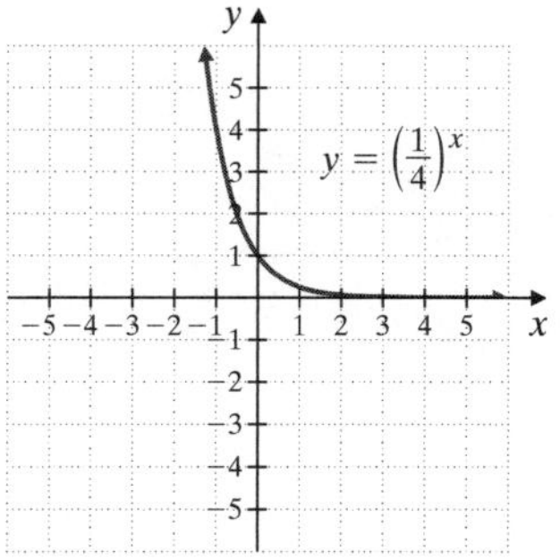

7.

x	$y = \left(\frac{1}{2}\right)^x - 2$
-2	$\left(\frac{1}{2}\right)^{-2} - 2 = 4 - 2 = 2$
-1	$\left(\frac{1}{2}\right)^{-1} - 2 = 2 - 2 = 0$
0	$\left(\frac{1}{2}\right)^{0} - 2 = 1 - 2 = -1$
1	$\left(\frac{1}{2}\right)^{1} - 2 = \frac{1}{2} - 2 = -\frac{3}{2}$
2	$\left(\frac{1}{2}\right)^{2} - 2 = \frac{1}{4} - 2 = -\frac{7}{4}$

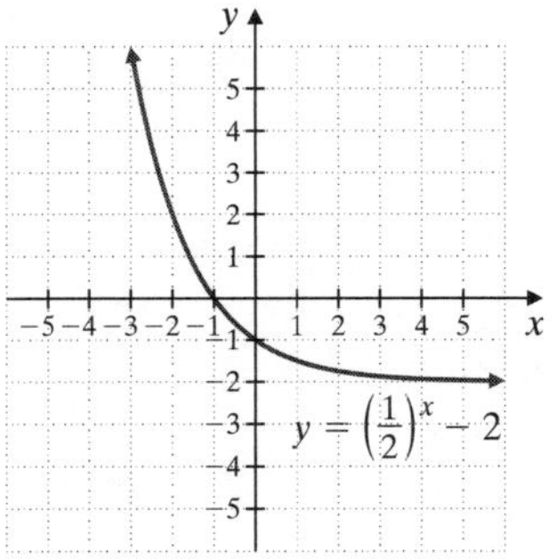

9.

x	$y = -2^x$
-2	$-2^{-2} = -\left(\frac{1}{2^2}\right) = -\frac{1}{4}$
-1	$-2^{-1} = -\left(\frac{1}{2^1}\right) = -\frac{1}{2}$
0	$-2^0 = -1$
1	$-2^1 = -2$
2	$-2^2 = -4$

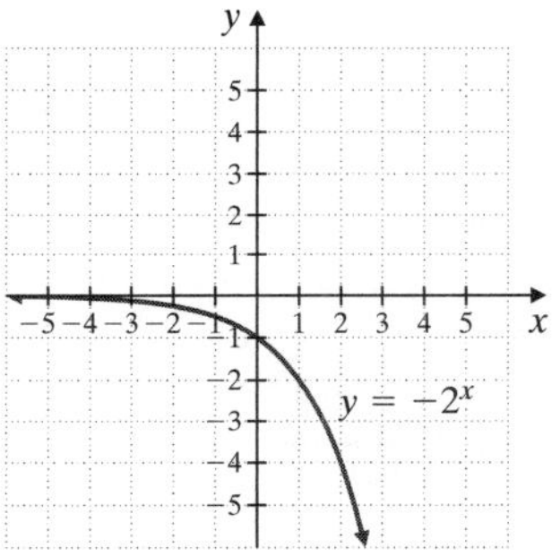

11.

x	$y = 3^x - 2$
-2	$3^{-2} - 2 = \frac{1}{9} - 2 = -\frac{17}{9}$
-1	$3^{-1} - 2 = \frac{1}{3} - 2 = -\frac{5}{3}$
0	$3^0 - 2 = 1 - 2 = -1$
1	$3^1 - 2 = 3 - 2 = 1$
2	$3^2 - 2 = 9 - 2 = 7$

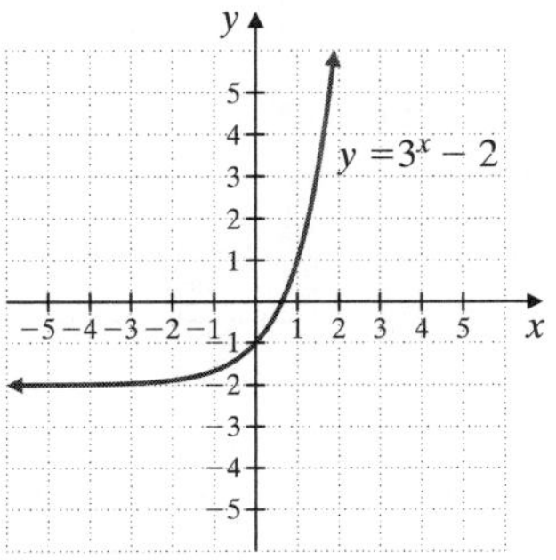

13.

x	$y=-\left(\frac{1}{4}\right)^x$
−2	$-\left(\frac{1}{4}\right)^{-2}=-(4)^2=-16$
−1	$-\left(\frac{1}{4}\right)^{-1}=-(4)^1=-4$
0	$-\left(\frac{1}{4}\right)^0=-1$
1	$-\left(\frac{1}{4}\right)^1=-\frac{1}{4}$
2	$-\left(\frac{1}{4}\right)^2=-\frac{1}{16}$

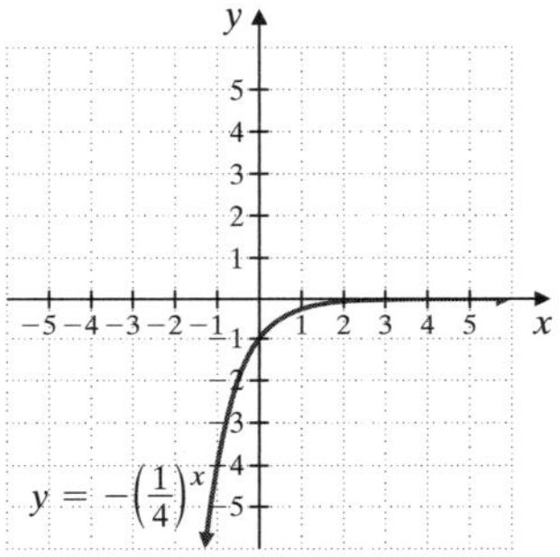

15.

x	$y=\left(\frac{1}{3}\right)^x+1$
−2	$\left(\frac{1}{3}\right)^{-2}+1=3^2+1=10$
−1	$\left(\frac{1}{3}\right)^{-1}+1=3+1=4$
0	$\left(\frac{1}{3}\right)^0+1=1+1=2$
1	$\left(\frac{1}{3}\right)^1+1=\frac{4}{3}$
2	$\left(\frac{1}{3}\right)^2+1=\frac{1}{9}+1=\frac{10}{9}$

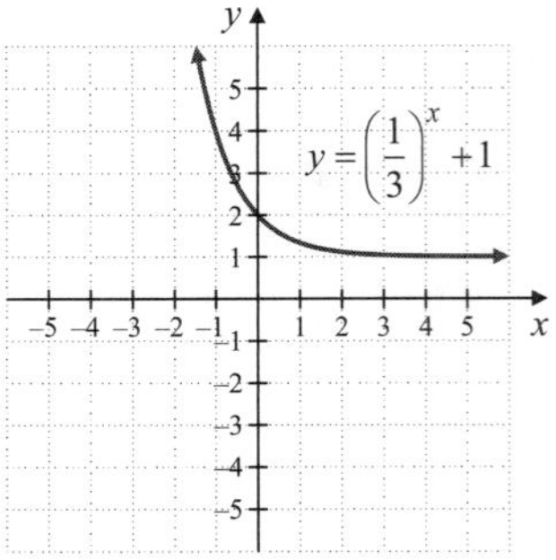

17.

x	$f(x)=2^{x-2}$
−1	$2^{-1-2}=2^{-3}=\frac{1}{2^3}=\frac{1}{8}$
0	$2^{0-2}=2^{-2}=\frac{1}{2^2}=\frac{1}{4}$
1	$2^{1-2}=2^{-1}=\frac{1}{2}$
2	$2^{2-2}=2^0=1$
3	$2^{3-2}=2^1=2$
4	$2^{4-2}=2^2=4$

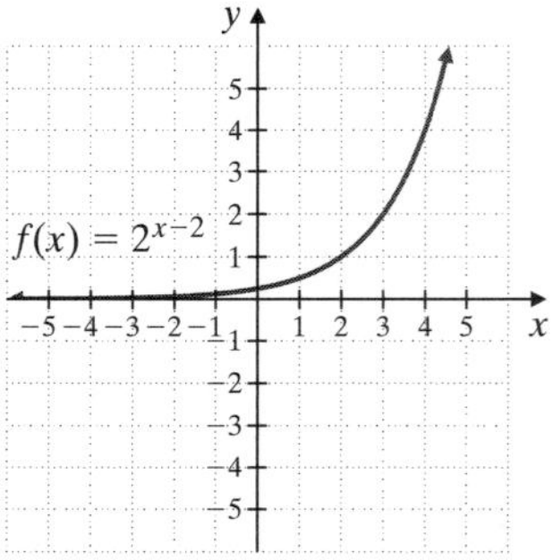

19.

x	$F(x)=5^{x+1}$
-2	$5^{-2+1}=5^{-1}=\frac{1}{5}$
-1	$5^{-1+1}=5^{0}=1$
0	$5^{0+1}=5^{1}=5$
1	$5^{1+1}=5^{2}=25$
2	$5^{2+1}=5^{3}=125$

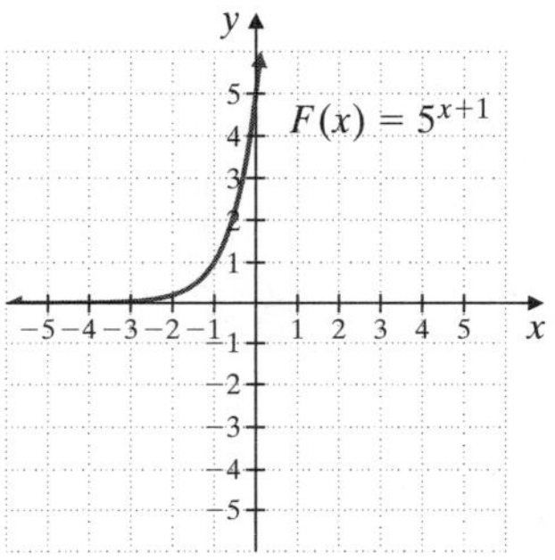

21. $3^x=27$
$3^x=3^3$
$x=3$
The solution set is $\{3\}$.

23. $16^x=8$
$(2^4)^x=2^3$
$2^{4x}=2^3$
$4x=3$
$x=\frac{3}{4}$
The solution set is $\left\{\frac{3}{4}\right\}$.

25. $32^{2x-3}=2$
$(2^5)^{2x-3}=2^1$
$10x-15=1$
$10x=16$
$x=\frac{8}{5}$
The solution set is $\left\{\frac{8}{5}\right\}$.

27. $\frac{1}{4}=2^{3x}$
$2^{-2}=2^{3x}$
$-2=3x$
$-\frac{2}{3}=x$
The solution set is $\left\{-\frac{2}{3}\right\}$.

29. $9^x=27$
$(3^2)^x=3^3$
$3^{2x}=3^3$
$2x=3$
$x=\frac{3}{2}$
The solution set is $\left\{\frac{3}{2}\right\}$.

31. $27^{x+1}=9$
$(3^3)^{x+1}=3^2$
$3^{3x+3}=3^2$
$3x+3=2$
$3x=-1$
$x=-\frac{1}{3}$
The solution set is $\left\{-\frac{1}{3}\right\}$.

33. $81^{x-1} = 27^{2x}$
$(3^4)^{x-1} = (3^3)^{2x}$
$3^{4x-4} = 3^{6x}$
$4x - 4 = 6x$
$-4 = 2x$
$-2 = x$
The solution set is $\{-2\}$.

35. $\left(\frac{1}{8}\right)^x = 16^{1-x}$
$(2^{-3})^x = (2^4)^{1-x}$
$2^{-3x} = 2^{4-4x}$
$-3x = 4 - 4x$
$x = 4$
The solution set is $\{4\}$.

37. $y = 30(2.7)^{-0.004t}$
$y = 30(2.7)^{-0.004(50)} = 30(2.7)^{-0.2} \approx 24.6$
There will be about 24.6 pounds left after 50 days.

39. a. $p(h) = 760(2.7)^{-0.145h}$
$p(2) = 760(2.7)^{-0.145(2)} \approx 569.8$
It will have an atmospheric pressure of about 569.8 millibars at 2 km.

b. $p(30.48) = 760(2.7)^{-0.145(30.48)} \approx 9.4$
It will have an atmospheric pressure of about 9.4 millibars at 30.48 km.

c. answers may vary

41. a. 2003 is 1 year after 2002, so $t = 1$.
$y = 44(1.29)^t$
$y = 44(1.29)^1 = 44(1.29) = 56.76$
The retail revenues from Internet shopping were predicted to be \$56.8 billion in 2003.

b. 2009 is 7 years after 2002, so $t = 7$.
$y = 44(1.29)^t$
$y = 44(1.29)^7 \approx 261.57$
The retail revenues from Internet shopping are predicted to be \$261.6 billion in 2009.

43. 2010 is 13 years after 1997, so $x = 13$.
$y = 52.47(1.196)^x$
$y = 52.47(1.196)^{13} \approx 537.5$
The prediction for 2010 is 537.5 million cellular phone users.

45. Use $P = 6000$, $t = 3$, $n = 12$, and $r = 8\% = 0.08$.
$A = P\left(1 + \frac{r}{n}\right)^{nt}$
$A = 6000\left(1 + \frac{0.08}{12}\right)^{(12)(3)}$
$= 6000\left(1 + \frac{0.08}{12}\right)^{36}$
≈ 7621.42
Erica owes \$7621.42.

47. $5x - 2 = 18$
$5x = 20$
$x = 4$
The solution set is $\{4\}$.

49. $3x - 4 = 3(x + 1)$
$3x - 4 = 3x + 3$
$-4 = 3$ False
Since the equation $-4 = 3$ is false, the equation has no solution. The solution set is $\varnothing$.

51. Since there are no variables in the exponent in $f(x) = 1.5x^2$, it is not an exponential function.

53. Since there are no variables in the exponent in $h(x) = \left(\frac{1}{2}x\right)^2$, it is not an exponential function.

55. $f(x) = \left(\frac{1}{2}\right)^x$

$f(1) = \left(\frac{1}{2}\right)^1 = \frac{1}{2}$

This is graph C.

57. $f(x) = \left(\frac{1}{4}\right)^x$

$f(1) = \left(\frac{1}{4}\right)^1 = \frac{1}{4}$

This is graph D.

59. answers may vary

61. The result is the same, 24.6 pounds.

63. $y = 30(2.7)^{-0.004t}$

$y = 30(2.7)^{-0.004(120)}$

$y \approx 18.62$

After 120 days, the estimate is 18.62 pounds left.

Section 9.4 Practice Problems

1. $\log_7 49 = 2$

$7^2 = 49$

2. $\log_8 \frac{1}{8} = -1$

$8^{-1} = \frac{1}{8}$

3. $\log_3 \sqrt{3} = \frac{1}{2}$

$3^{1/2} = \sqrt{3}$

4. $3^4 = 81$

$\log_3 81 = 4$

5. $2^{-3} = \frac{1}{8}$

$\log_2 \frac{1}{8} = -3$

6. $7^{1/3} = \sqrt[3]{7}$

$\log_7 \sqrt[3]{7} = \frac{1}{3}$

7. a. $\log_5 125 = 3$ because $5^3 = 125$.

b. $\log_7 \frac{1}{49} = -2$ because $7^{-2} = \frac{1}{49}$.

c. $\log_{100} 10 = \frac{1}{2}$ because $100^{1/2} = 10$.

8. $\log_2 x = 4$

$2^4 = x$

$16 = x$

The solution set is $\{16\}$.

9. $\log_x 9 = 2$

$x^2 = 9$

$x = 3$

The solution set is $\{3\}$.

10. $\log_2 1 = x$

$2^x = 1$

$x = 0$

The solution set is $\{0\}$.

11. a. $\log_6 6^3 = 3$

b. $\log_{11} 11^{-4} = -4$

c. $7^{\log_7 13} = 13$

d. $3^{\log_3 10} = 10$

12. $y = \log_4 x$ or $4^y = x$

$x = 4^y$	y
$\frac{1}{4}$	-1
1	0
4	1
16	2

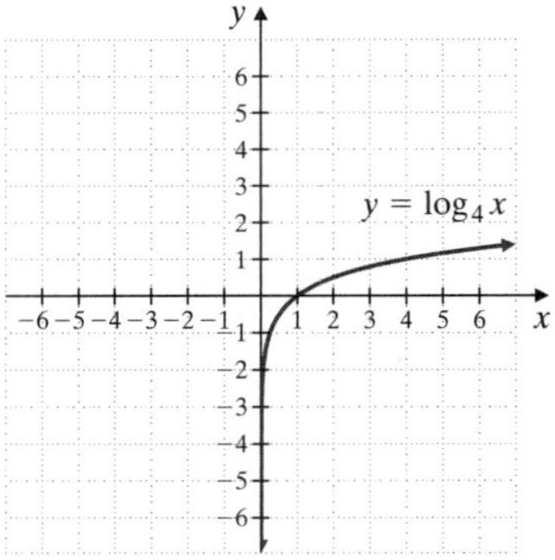

13. $f(x) = \log_{1/2} x$

$y = \log_{1/2} x$

$x = \left(\frac{1}{2}\right)^y$

$x = \left(\frac{1}{2}\right)^y$	y
4	-2
2	-1
1	0
$\frac{1}{2}$	1
$\frac{1}{4}$	2

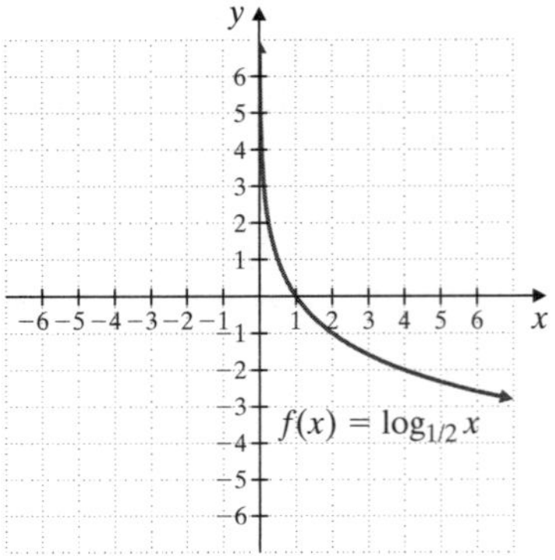

Exercise Set 9.4

1. $\log_6 36 = 2$

$6^2 = 36$

3. $\log_3 \frac{1}{27} = -3$

$3^{-3} = \frac{1}{27}$

5. $\log_{10} 1000 = 3$

$10^3 = 1000$

7. $\log_e x = 4$

$e^4 = x$

9. $\log_e \frac{1}{e^2} = -2$

$e^{-2} = \frac{1}{e^2}$

11. $\log_7 \sqrt{7} = \frac{1}{2}$

$7^{1/2} = \sqrt{7}$

13. $\log_{0.7} 0.343 = 3$

$0.7^3 = 0.343$

15. $\log_3 \frac{1}{81} = -4$

$3^{-4} = \frac{1}{81}$

17. $2^4 = 16$
$\log_2 16 = 4$

19. $10^2 = 100$
$\log_{10} 100 = 2$

21. $e^3 = x$
$\log_e x = 3$

23. $10^{-1} = \frac{1}{10}$
$\log_{10} \frac{1}{10} = -1$

25. $4^{-2} = \frac{1}{16}$
$\log_4 \frac{1}{16} = -2$

27. $5^{1/2} = \sqrt{5}$
$\log_5 \sqrt{5} = \frac{1}{2}$

29. $\log_2 8 = 3$ because $2^3 = 8$.

31. $\log_3 \frac{1}{9} = -2$ because $3^{-2} = \frac{1}{9}$.

33. $\log_{25} 5 = \frac{1}{2}$ because $25^{1/2} = 5$.

35. $\log_{1/2} 2 = -1$ because $\left(\frac{1}{2}\right)^{-1} = 2$.

37. $\log_6 1 = 0$ because $6^0 = 1$.

39. $\log_{10} 100 = 2$ because $10^2 = 100$.

41. $\log_3 81 = 4$ because $3^4 = 81$.

43. $\log_4 \frac{1}{64} = -3$ because $4^{-3} = \frac{1}{64}$.

45. $\log_3 9 = x$
$3^x = 9$
$3^x = 3^2$
$x = 2$
The solution set is $\{2\}$.

47. $\log_3 x = 4$
$3^4 = x$
$81 = x$
The solution set is $\{81\}$.

49. $\log_x 49 = 2$
$x^2 = 49$
$x^2 = 7^2$
$x = 7$
The solution set is $\{7\}$.

51. $\log_2 \frac{1}{8} = x$
$2^x = \frac{1}{8}$
$2^x = 2^{-3}$
$x = -3$
The solution set is $\{-3\}$.

53. $\log_3 \frac{1}{27} = x$
$3^x = \frac{1}{27}$
$3^x = \frac{1}{3^3}$
$3^x = 3^{-3}$
$x = -3$
The solution set is $\{-3\}$.

55. $\log_8 x = \frac{1}{3}$
$8^{1/3} = x$
$2 = x$
The solution set is $\{2\}$.

57. $\log_4 16 = x$
$$4^x = 16$$
$$4^x = 4^2$$
$$x = 2$$
The solution set is {2}.

59. $\log_{3/4} x = 3$
$$\left(\frac{3}{4}\right)^3 = x$$
$$\frac{27}{64} = x$$
The solution set is $\left\{\frac{27}{64}\right\}$.

61. $\log_x 100 = 2$
$$x^2 = 100$$
$$x^2 = 10^2$$
$$x = 10$$
The solution set is {10}.

63. $\log_2 2^4 = x$
$$2^x = 2^4$$
$$x = 4$$
The solution set is {4}.

65. $3^{\log_3 5} = x$
$$5 = x$$
The solution set is {5}.

67. $\log_x \frac{1}{7} = \frac{1}{2}$
$$x^{1/2} = \frac{1}{7}$$
$$(x^{1/2})^2 = \left(\frac{1}{7}\right)^2$$
$$x = \frac{1}{49}$$
The solution set is $\left\{\frac{1}{49}\right\}$.

69. $\log_5 5^3 = 3$

71. $2^{\log_2 3} = 3$

73. $\log_9 9 = 1$

75. $y = \log_3 x$
$$3^y = x$$

$x = 3^y$	y
$3^{-2} = \frac{1}{9}$	−2
$3^{-1} = \frac{1}{3}$	−1
$3^0 = 1$	0
$3^1 = 3$	1
$3^2 = 9$	2

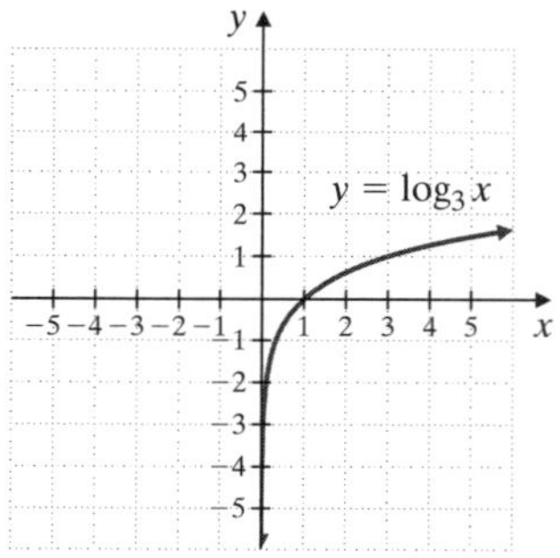

77. $f(x) = \log_{1/4} x$ or $y = \log_{1/4} x$
$$\left(\frac{1}{4}\right)^y = x$$

$x = \left(\frac{1}{4}\right)^y$	y
$\left(\frac{1}{4}\right)^{-1} = 4^1 = 4$	−1
$\left(\frac{1}{4}\right)^0 = 1$	0
$\left(\frac{1}{4}\right)^1 = \frac{1}{4}$	1

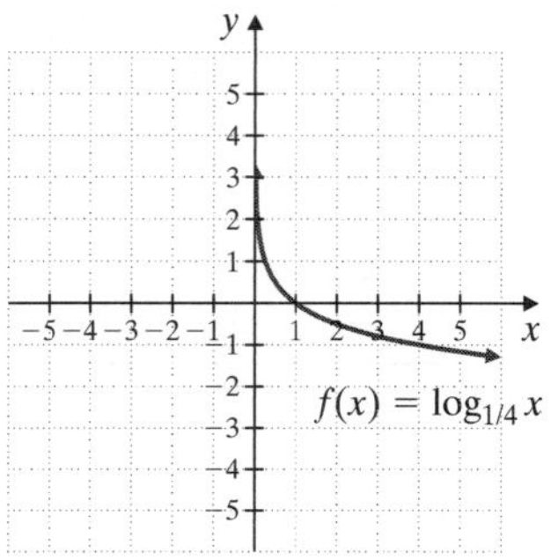

79. $f(x) = \log_5 x$ or $y = \log_5 x$

$5^y = x$

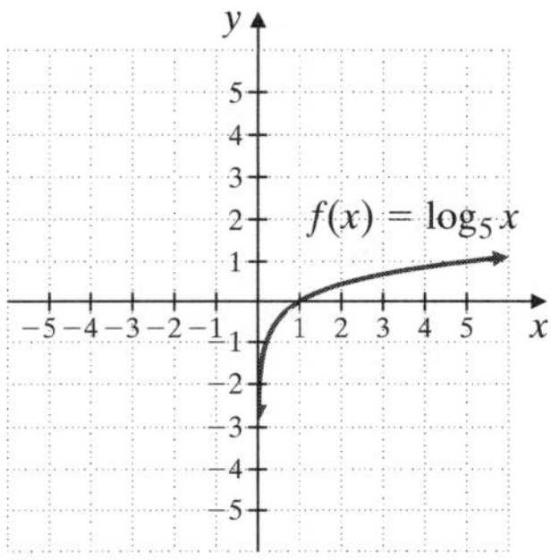

81. $f(x) = \log_{1/6} x$ or $y = \log_{1/6} x$

$\left(\frac{1}{6}\right)^y = x$

$x = \left(\frac{1}{6}\right)^y$	y
$\left(\frac{1}{6}\right)^{-1} = 6^1 = 6$	-1
$\left(\frac{1}{6}\right)^0 = 1$	0
$\left(\frac{1}{6}\right)^1 = \frac{1}{6}$	1

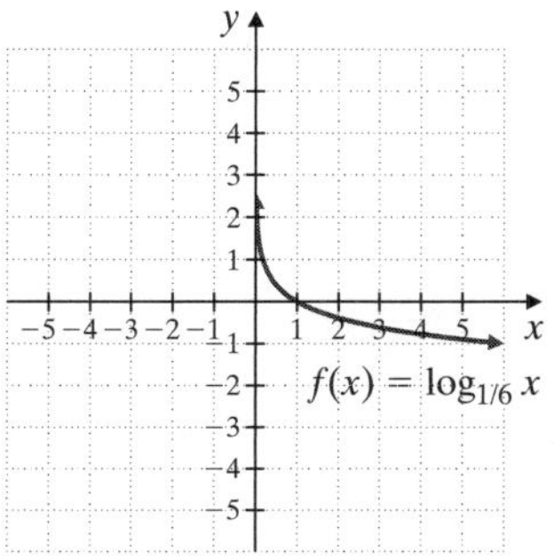

83. $\frac{x+3}{3+x} = \frac{x+3}{x+3} = 1$

85. $\frac{x^2 - 8x + 16}{2x - 8} = \frac{(x-4)(x-4)}{2(x-4)} = \frac{x-4}{2}$

87. **a.** $g(2) = 25$

b. Since (2, 25) is a solution of $g(x)$, then (25, 2) is a solution of $f(x)$.

c. $f(25) = 2$

89. answers may vary

91. $y = 4^x$; $y = \log_4 x$

x	$y = 4^x$
-2	$\frac{1}{16}$
-1	$\frac{1}{4}$
0	1
1	4
2	16

x	$y = \log_4 x$
$\frac{1}{16}$	−2
$\frac{1}{4}$	−1
1	0
4	1
16	2

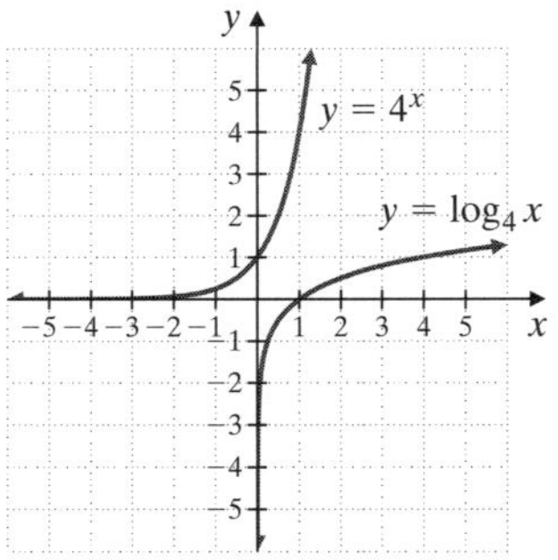

93. $y = \left(\frac{1}{3}\right)^x$; $y = \log_{1/3} x$

x	$y = \left(\frac{1}{3}\right)^x$
−2	9
−1	3
0	1
1	$\frac{1}{3}$
2	$\frac{1}{9}$

x	$y = \log_{1/3} x$
9	−2
3	−1
1	0
$\frac{1}{3}$	1
$\frac{1}{9}$	2

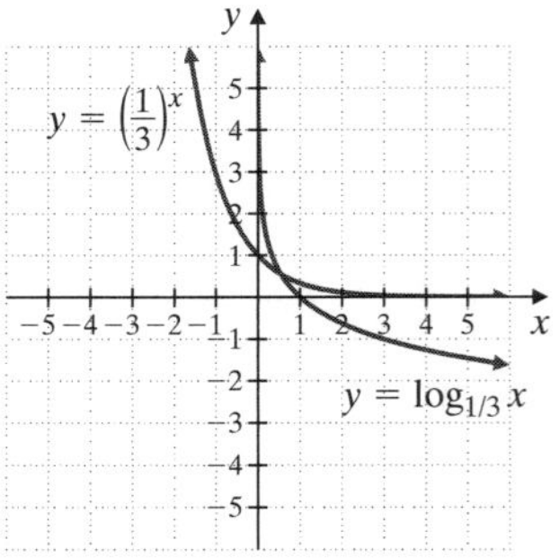

95. answers may vary

97. Use $H = 8$.

$$\log_{10}(1-k) = \frac{-0.3}{H}$$
$$\log_{10}(1-k) = \frac{-0.3}{8}$$
$$\log_{10}(1-k) = -0.0375$$
$$1-k = 10^{-0.0375}$$
$$1-10^{-0.0375} = k$$
$$0.0827 \approx k$$

The decay rate is 0.0827.

Section 9.5 Practice Problems

1. $\log_2 7 + \log_2 5 = \log_2(7 \cdot 5) = \log_2 35$

2. $\log_3 x + \log_3(x-9) = \log_3[x \cdot (x-9)]$
$= \log_3(x^2 - 9x)$

3. $\log_7 40 - \log_7 8 = \log_7 \frac{40}{8} = \log_7 5$

4. $\log_3(x^3+4)-\log_3(x^2+2)=\log_3\frac{x^3+4}{x^2+2}$

5. $\log_3 x^5 = 5\log_3 x$

6. $\log_7 \sqrt[3]{4} = \log_7 4^{1/3} = \frac{1}{3}\log_7 4$

7. $3\log_4 2 + 2\log_4 5 = \log_4 2^3 + \log_4 5^2$
$= \log_4 8 + \log_4 25$
$= \log_4(8\cdot 25)$
$= \log_4 200$

8. $5\log_2(2x-1)-\log_2 x$
$= \log_2(2x-1)^5 - \log_2 x$
$= \log_2\frac{(2x-1)^5}{x}$

9. $\log_7\frac{6\cdot 2}{5} = \log_7(6\cdot 2)-\log_7 5$
$= \log_7 6 + \log_7 2 - \log_7 5$

10. $\log_3\frac{x^4}{y^3} = \log_3 x^4 - \log_3 y^3$
$= 4\log_3 x - 3\log_3 y$

11. $\log_b 28 = \log_b(4\cdot 7)$
$= \log_b 4 + \log_b 7$
$= 0.86 + 1.21$
$= 2.07$

12. $\log_b 49 = \log_b 7^2$
$= 2\log_b 7$
$= 2(1.21)$
$= 2.42$

13. $\log_b \sqrt[3]{4} = \log_b 4^{1/3}$
$= \frac{1}{3}\log_b 4$
$= \frac{1}{3}(0.86)$
$= 0.28\overline{6}$

Exercise Set 9.5

1. $\log_5 2 + \log_5 7 = \log_5(2\cdot 7) = \log_5 14$

3. $\log_4 9 + \log_4 x = \log_4 9x$

5. $\log_6 x + \log_6(x+1) = \log_6[x\cdot(x+1)]$
$= \log_6(x^2+x)$

7. $\log_{10} 5 + \log_{10} 2 + \log_{10}(x^2+2)$
$= \log_{10}[5\cdot 2\cdot(x^2+2)]$
$= \log_{10}[10(x^2+2)]$
$= \log_{10}(10x^2+20)$

9. $\log_5 12 - \log_5 4 = \log_5\frac{12}{4} = \log_5 3$

11. $\log_3 8 - \log_3 2 = \log_3\frac{8}{2} = \log_3 4$

13. $\log_2 x - \log_2 y = \log_2\frac{x}{y}$

15. $\log_2(x^2+6)-\log_2(x^2+1)=\log_2\frac{x^2+6}{x^2+1}$

17. $\log_3 x^2 = 2\log_3 x$

19. $\log_4 5^{-1} = -1\log_4 5 = -\log_4 5$

21. $\log_5\sqrt{y} = \log_5 y^{1/2} = \frac{1}{2}\log_5 y$

23. $\log_2 5 + \log_2 x^3 = \log_2 5x^3$

25. $3\log_4 2 + \log_4 6 = \log_4 2^3 + \log_4 6$
$= \log_4 8 + \log_4 6$
$= \log_4(8\cdot 6)$
$= \log_4 48$

27. $3\log_5 x + 6\log_5 z = \log_5 x^3 - \log_5 z^6$
$= \log_5 x^3 z^6$

29. $\log_4 2 + \log_4 10 - \log_4 5$
$= \log_4(2\cdot 10) - \log_4 5$
$= \log_4 20 - \log_4 5$
$= \log_4 \frac{20}{5}$
$= \log_4 4$
$= 1$

31. $\log_7 6 + \log_7 3 - \log_7 4 = \log_7(6\cdot 3) - \log_7 4$
$= \log_7 18 - \log_7 4$
$= \log_7 \frac{18}{4}$
$= \log_7 \frac{9}{2}$

33. $\log_{10} x - \log_{10}(x+1) + \log_{10}(x^2 - 2)$
$= \log_{10} \frac{x}{x+1} + \log_{10}(x^2 - 2)$
$= \log_{10} \frac{x(x^2-2)}{x+1}$
$= \log_{10} \frac{x^3 - 2x}{x+1}$

35. $3\log_2 x + \frac{1}{2}\log_2 x - 2\log_2(x+1)$
$= \log_2 x^3 + \log_2 x^{1/2} - \log_2(x+1)^2$
$= \log_2(x^3 \cdot x^{1/2}) - \log_2(x+1)^2$
$= \log_2 x^{7/2} - \log_2(x+1)^2$
$= \log_2 \frac{x^{7/2}}{(x+1)^2}$

37. $2\log_8 x - \frac{2}{3}\log_8 x + 4\log_8 x$
$= \log_8 x^2 - \log_8 x^{2/3} + \log_8 x^4$
$= \log_8 \frac{x^2}{x^{2/3}} + \log_8 x^4$
$= \log_8 \frac{x^2 \cdot x^4}{x^{2/3}}$
$= \log_8 x^{2+4-2/3}$
$= \log_8 x^{16/3}$

39. $\log_3 \frac{4y}{5} = \log_3 4y - \log_3 5$
$= \log_3 4 + \log_3 y - \log_3 5$

41. $\log_4 \frac{2}{9z} = \log_4 2 - \log_4 9z$
$= \log_4 2 - (\log_4 9 + \log_4 z)$
$= \log_4 2 - \log_4 9 - \log_4 z$

43. $\log_2 \frac{x^3}{y} = \log_2 x^3 - \log_2 y$
$= 3\log_2 x - \log_2 y$

45. $\log_b \sqrt{7x} = \log_b (7x)^{1/2}$
$= \frac{1}{2}\log_b 7x$
$= \frac{1}{2}(\log_b 7 + \log_b x)$
$= \frac{1}{2}\log_b 7 + \frac{1}{2}\log_b x$

47. $\log_6 x^4 y^5 = \log_6 x^4 + \log_6 y^5$
$= 4\log_6 x + 5\log_6 y$

49. $\log_5 x^3(x+1) = \log_5 x^3 + \log_5(x+1)$
$= 3\log_5 x + \log_5(x+1)$

51. $\log_6 \frac{x^2}{x+3} = \log_6 x^2 - \log_6(x+3)$
$= 2\log_6 x - \log_6(x+3)$

53. $\log_b \frac{5}{3} = \log_b 5 - \log_b 3 = 0.7 - 0.5 = 0.2$

55. $\log_b 15 = \log_b(3 \cdot 5)$
$= \log_b 3 + \log_b 5$
$= 0.5 + 0.7$
$= 1.2$

57. $\log_b \sqrt{5} = \log_b 5^{1/2}$
$= \frac{1}{2}\log_b 5$
$= \frac{1}{2}(0.7)$
$= 0.35$

59. $\log_b 8 = \log_b 2^3 = 3\log_b 2 = 3(0.43) = 1.29$

61. $\log_b \frac{3}{9} = \log_b \frac{1}{3}$
$= \log_b 3^{-1}$
$= -1 \cdot \log_b 3$
$= -1 \cdot (0.68)$
$= -0.68$

63. $\log_b \sqrt{\frac{2}{3}} = \log_b \left(\frac{2}{3}\right)^{1/2}$
$= \frac{1}{2}\log_b \frac{2}{3}$
$= \frac{1}{2}(\log_b 2 - \log_b 3)$
$= \frac{1}{2}(0.43 - 0.68)$
$= -0.125$

65.

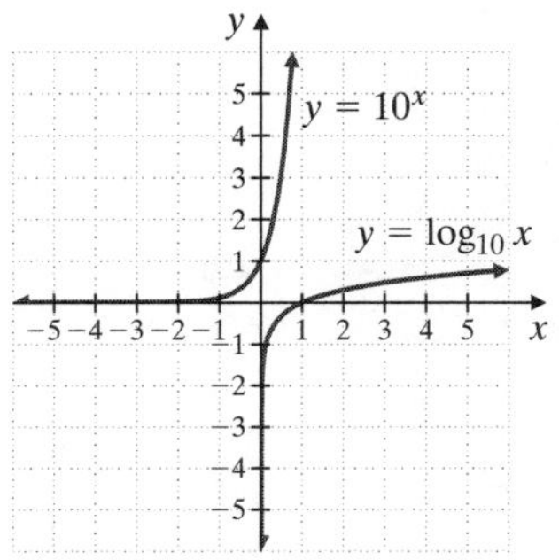

67. $\log_{10} \frac{1}{10} = \log_{10} 10^{-1} = -1$

69. $\log_7 \sqrt{7} = \log_7 7^{1/2} = \frac{1}{2}\log_7 7 = \frac{1}{2}(1) = \frac{1}{2}$

71. $\log_9 \frac{21}{3} = \log_9 21 - \log_9 3$ or

$\log_9 \frac{21}{3} = \log_9 7$

The choices are a and d.

73. $\log_3 x + \log_3 y = \log_3 xy \neq \log_3(x + y)$
The statement is false.

75. $\log_7 \frac{14}{8} = \log_7 14 - \log_7 8$
The statement is true.

77. $\log_3 24 = \log_3(6 \cdot 4)$
$= \log_3 6 + \log_3 4$
$\neq (\log_3 6) \cdot (\log_3 4)$
The statement is false.

Integrated Review

1. $(f + g)(x) = f(x) + g(x)$
$= (x - 6) + (x^2 + 1)$
$= x^2 + x - 5$

2. $(f - g)(x) = f(x) - g(x)$
$= (x - 6) - (x^2 + 1)$
$= x - 6 - x^2 - 1$
$= -x^2 + x - 7$

3. $(f \cdot g)(x) = f(x)g(x)$
$= (x - 6)(x^2 + 1)$
$= x^3 + x - 6x^2 - 6$
$= x^3 - 6x^2 + x - 6$

4. $\left(\frac{f}{g}\right)(x) = \frac{f(x)}{g(x)} = \frac{x - 6}{x^2 + 1}$

5. $(f \circ g)(x) = f(g(x)) = f(3x-1) = \sqrt{3x-1}$

6. $(g \circ f)(x) = g(f(x)) = g\left(\sqrt{x}\right) = 3\sqrt{x} - 1$

7. $f = \{(-2, 6), (4, 8), (2, -6), (3, 3)\}$ is a one-to-one function since each y-value corresponds to only one x-value.

$f^{-1} = \{(6, -2), (8, 4), (-6, 2), (3, 3)\}$

8. $g = \{(4, 2), (-1, 3), (5, 3), (7, 1)\}$ is not a one-to-one function since the y-value 3 corresponds to two different x-values, -1 and 5.

9. The graph does not represent a one-to-one function because it does not pass the horizontal line test.

10. The graph represents a one-to-one function because it passes the vertical and horizontal line tests.

11. The graph does not represent a one-to-one function because it does not pass the horizontal line test.

12. $f(x) = 3x$

$y = 3x$

$x = 3y$

$\frac{x}{3} = y$

$f^{-1}(x) = \frac{x}{3}$

13. $f(x) = x + 4$

$y = x + 4$

$x = y + 4$

$x - 4 = y$

$f^{-1}(x) = x - 4$

14. $f(x) = 5x - 1$

$y = 5x - 1$

$x = 5y - 1$

$x + 1 = 5y$

$\frac{x+1}{5} = y$

$f^{-1}(x) = \frac{x+1}{5}$

15. $f(x) = 3x + 2$

$y = 3x + 2$

$x = 3y + 2$

$x - 2 = 3y$

$\frac{x-2}{3} = y$

$f^{-1}(x) = \frac{x-2}{3}$

16.

x	$y = \left(\frac{1}{2}\right)^x$
-2	$\left(\frac{1}{2}\right)^{-2} = 4$
-1	$\left(\frac{1}{2}\right)^{-1} = 2$
0	$\left(\frac{1}{2}\right)^{0} = 1$
1	$\left(\frac{1}{2}\right)^{1} = \frac{1}{2}$
2	$\left(\frac{1}{2}\right)^{2} = \frac{1}{4}$

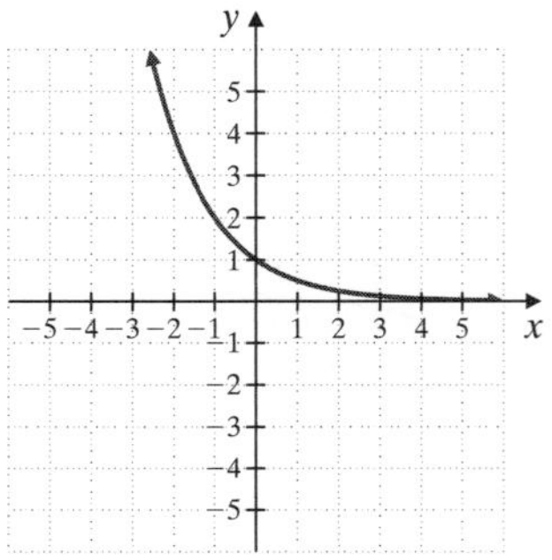

17.

x	$y = 2^x + 1$
-2	$2^{-2} + 1 = \frac{1}{4} + 1 = \frac{5}{4}$
-1	$2^{-1} + 1 = \frac{1}{2} + 1 = \frac{3}{2}$
0	$2^0 + 1 = 1 + 1 = 2$
1	$2^1 + 1 = 2 + 1 = 3$
2	$2^2 + 1 = 4 + 1 = 5$

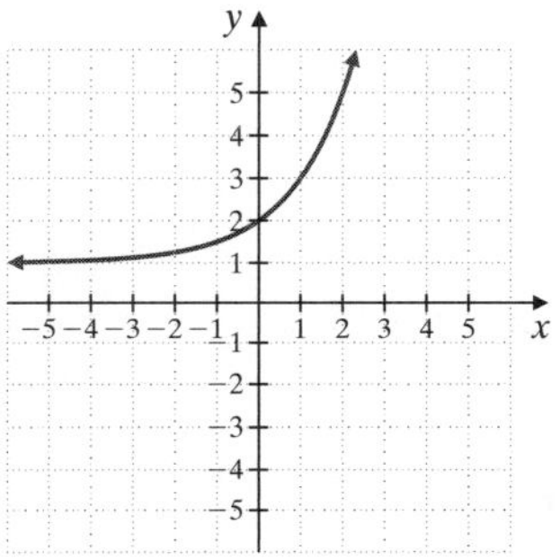

18. $y = \log_3 x$

$3^y = x$

$x = 3^y$	y
$3^{-2} = \frac{1}{9}$	-2
$3^{-1} = \frac{1}{3}$	-1
$3^0 = 1$	0
$3^1 = 3$	1
$3^2 = 9$	2

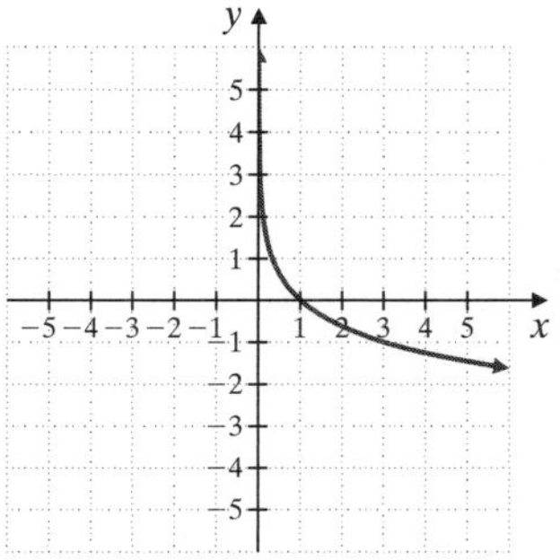

19. $y = \log_{1/3} x$

$\left(\frac{1}{3}\right)^y = x$

$x = \left(\frac{1}{3}\right)^y$	y
$\left(\frac{1}{3}\right)^{-2} = 9$	-2
$\left(\frac{1}{3}\right)^{-1} = 3$	-1
$\left(\frac{1}{3}\right)^0 = 1$	0
$\left(\frac{1}{3}\right)^1 = \frac{1}{3}$	1
$\left(\frac{1}{3}\right)^2 = \frac{1}{9}$	2

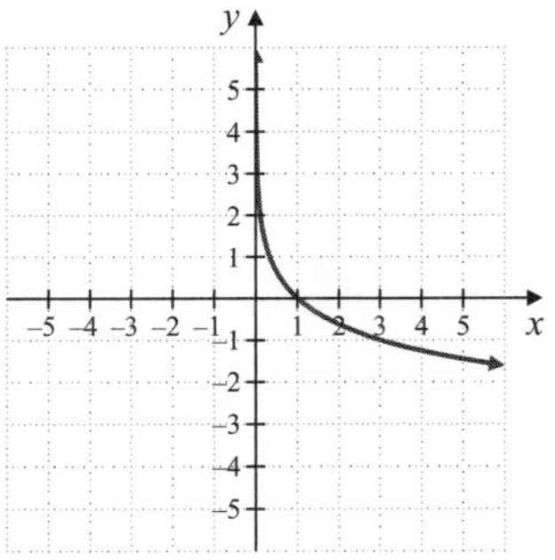

20. $2^x = 8$
$2^x = 2^3$
$x = 3$
The solution set is {3}.

21. $9 = 3^{x-5}$
$3^2 = 3^{x-5}$
$2 = x - 5$
$7 = x$
The solution set is {7}.

22. $4^{x-1} = 8^{x+2}$
$(2^2)^{x-1} = (2^3)^{x+2}$
$2^{2x-2} = 2^{3x+6}$
$2x - 2 = 3x + 6$
$-2 = x + 6$
$-8 = x$
The solution set is {−8}.

23. $25^x = 125^{x-1}$
$(5^2)^x = (5^3)^{x-1}$
$5^{2x} = 5^{3x-3}$
$2x = 3x - 3$
$-x = -3$
$x = 3$
The solution set is {3}.

24. $\log_4 16 = x$
$4^x = 16$
$4^x = 4^2$
$x = 2$
The solution set is {2}.

25. $\log_{49} 7 = x$
$49^x = 7$
$(7^2)^x = 7$
$7^{2x} = 7^1$
$2x = 1$
$x = \frac{1}{2}$
The solution set is $\left\{\frac{1}{2}\right\}$.

26. $\log_2 x = 5$
$2^5 = x$
$32 = x$
The solution set is {32}.

27. $\log_x 64 = 3$
$x^3 = 64$
$x^3 = 4^3$
$x = 4$
The solution set is {4}.

28. $\log_x \frac{1}{125} = -3$
$x^{-3} = \frac{1}{125}$
$x^{-3} = 5^{-3}$
$x = 5$
The solution set is {5}.

29. $\log_3 x = -2$
$3^{-2} = x$
$\frac{1}{9} = x$
The solution set is $\left\{\frac{1}{9}\right\}$.

30. $\log_2 x + \log_2 14 = \log_2(x \cdot 14) = \log_2 14x$

31. $x\log_2 5 + \log_2 8 = \log_2 5^x + \log_2 8$
$= \log_2 5^x \cdot 8$

32. $3\log_5 x - 5\log_5 y = \log_5 x^3 - \log_5 y^5$
$= \log_5 \frac{x^3}{y^5}$

33. $9\log_5 x + 3\log_5 y = \log_5 x^9 + \log_5 y^3$
$= \log_5 x^9 y^3$

34. $\log_2 x + \log_2(x-3) - \log_2(x^2+4)$
$= \log_2 x(x-3) - \log_2(x^2+4)$
$= \log_2 \dfrac{x(x-3)}{x^2+4}$
$= \log_2 \dfrac{x^2-3x}{x^2+4}$

35. $\log_3 y - \log_3(y+2) + \log_3(y^3+11)$
$= \log_3 \dfrac{y}{y+2} + \log_3(y^3+11)$
$= \log_3 \left(\dfrac{y}{y+2}\right)\cdot(y^3+11)$
$= \log_3 \dfrac{y^4+11y}{y+2}$

36. $\log_7 \dfrac{9x^2}{y} = \log_7 9x^2 - \log_7 y$
$= \log_7 9 + \log_7 x^2 - \log_7 y$
$= \log_7 9 + 2\log_7 x - \log_7 y$

37. $\log_6 \dfrac{5y}{z^2} = \log_6 5y - \log_6 z^2$
$= \log_6 5 + \log_6 y - \log_6 z^2$
$= \log_6 5 + \log_6 y - 2\log_6 z$

Section 9.6 Practice Problems

1. $\log 21 \approx 1.3222$

2. $\log 1000 = \log 10^3 = 3$

3. $\log \dfrac{1}{100} = \log 10^{-2} = -2$

4. $\log 10{,}000 = \log 10^4 = 4$

5. $\log \sqrt[3]{10} = \log 10^{1/3} = \dfrac{1}{3}$

6. $\log x = 2.9$
$10^{2.9} = x$
$x \approx 794.3282$

7. $\ln 11 \approx 2.3979$

8. $\ln e^9 = 9$

9. $\ln \sqrt[3]{e} = \ln e^{1/3} = \dfrac{1}{3}$

10. $\ln 7x = 10$
$e^{10} = 7x$
$\dfrac{e^{10}}{7} = x$
$x \approx 3146.6380$

11. $A = Pe^{rt}$, where $P = 1200$, $r = 0.08$, $t = 3$.
$A = 1200e^{0.08(3)}$
$A = 1200e^{0.24}$
$A \approx 1525.50$
The total amount of money owed is approximately \$1525.50.

12. $\log_7 5 = \dfrac{\log 5}{\log 7} \approx \dfrac{0.69897}{0.84510} \approx 0.8271$

Exercise Set 9.6

1. $\log 8 \approx 0.9031$

3. $\log 2.31 \approx 0.3636$

5. $\ln 2 \approx 0.6931$

7. $\ln 0.0716 \approx -2.6367$

9. $\log 12.6 \approx 1.1004$

11. $\ln 5 = 1.6094$

13. $\log 41.5 \approx 1.6180$

15. $\log 100 = \log 10^2 = 2$

17. $\log\frac{1}{1000} = \log 10^{-3} = -3$

19. $\ln e^2 = 2$

21. $\ln\sqrt[4]{e} = \ln e^{1/4} = \frac{1}{4}$

23. $\log 10^3 = 3$

25. $\ln e^{3.1} = 3.1$

27. $\log 0.0001 = \log 10^{-4} = -4$

29. $\ln\sqrt{e} = \ln e^{1/2} = \frac{1}{2}$

31. $\log x = 1.3$
$x = 10^{1.3} \approx 19.9526$
The solution set is $\{10^{1.3}\}$ or approximately $\{19.9526\}$.

33. $\ln x = 1.4$
$x = e^{1.4} \approx 4.0552$
The solution set is $\{e^{1.4}\}$ or approximately $\{4.0552\}$.

35. $\log x = 2.3$
$x = 10^{2.3} \approx 199.5262$
The solution set is $\{10^{2.3}\}$ or approximately $\{199.5262\}$.

37. $\ln x = -2.3$
$x = e^{-2.3} \approx 0.1003$
The solution set is $\{e^{-2.3}\}$ or approximately $\{0.1003\}$.

39. $\log 2x = 1.1$
$2x = 10^{1.1}$
$x = \frac{10^{1.1}}{2} \approx 6.2946$
The solution set is $\left\{\frac{10^{1.1}}{2}\right\}$ or approximately $\{6.2946\}$.

41. $\ln 4x = 0.18$
$4x = e^{0.18}$
$x = \frac{1}{4}e^{0.18} \approx 0.2993$
The solution set is $\left\{\frac{1}{4}e^{0.18}\right\}$ or approximately $\{0.2993\}$.

43. $\ln(3x-4) = 2.3$
$3x - 4 = e^{2.3}$
$3x = e^{2.3} + 4$
$x = \frac{e^{2.3}+4}{3} \approx 4.6581$
The solution set is $\left\{\frac{e^{2.3}+4}{3}\right\}$ or approximately $\{4.6581\}$.

45. $\log(2x+1) = -0.5$
$2x + 1 = 10^{-0.5}$
$2x = 10^{-0.5} - 1$
$x = \frac{10^{-0.5}-1}{2} \approx -0.3419$
The solution set is $\left\{\frac{10^{-0.5}-1}{2}\right\}$ or approximately $\{-0.3419\}$.

47. Use $A = Pe^{rt}$ with $P = 1400$, $r = 8\% = 0.08$, and $t = 12$.

$$\begin{aligned} A &= Pe^{rt} \\ &= 1400e^{0.08(12)} \\ &= 1400e^{0.96} \\ &\approx 3656.38 \end{aligned}$$

She has \$3656.38.

49. Use $A = Pe^{rt}$ with $P = 2000$, $r = 6\% = 0.06$, and $t = 4$.

$$\begin{aligned} A &= Pe^{rt} \\ &= 2000e^{0.06(4)} \\ &= 2000e^{0.24} \\ &\approx 2542.50 \end{aligned}$$

She owes \$2542.50.

51. $\log_2 3 = \dfrac{\log 3}{\log 2} \approx \dfrac{0.4771}{0.3010} \approx 1.5850$

53. $\log_8 6 = \dfrac{\log 6}{\log 8} \approx \dfrac{0.7782}{0.9031} \approx 0.8617$

55. $\log_4 9 = \dfrac{\log 9}{\log 4} \approx \dfrac{0.9542}{0.6021} \approx 1.5850$

57.
$$\begin{aligned} \log_3 \frac{1}{6} &= \log_3 6^{-1} \\ &= -\log_3 6 \\ &= -\frac{\ln 6}{\ln 3} \\ &\approx -\frac{1.79176}{1.09861} \\ &\approx -1.6309 \end{aligned}$$

59. $\log_{1/2} 5 = \dfrac{\log 5}{\log \frac{1}{2}} \approx \dfrac{0.69897}{-0.30103} \approx -2.3219$

61.
$$\begin{aligned} 6x - 3(2 - 5x) &= 6 \\ 6x - 6 + 15x &= 6 \\ 21x &= 12 \\ x &= \frac{12}{21} \\ x &= \frac{4}{7} \end{aligned}$$

The solution set is $\left\{\frac{4}{7}\right\}$.

63.
$$\begin{aligned} 2x + 3y &= 6x \\ 3y &= 4x \\ \frac{3y}{4} &= x \quad \text{or} \quad x = \frac{3y}{4} \end{aligned}$$

65.
$$\begin{aligned} x^2 + 7x &= -6 \\ x^2 + 7x + 6 &= 0 \\ (x + 6)(x + 1) &= 0 \\ x + 6 = 0 \quad &\text{or} \quad x + 1 = 0 \\ x = -6 \quad & \quad\quad\;\; x = -1 \end{aligned}$$

The solution set is $\{-6, -1\}$.

67. answers may vary

69. $f(x) = e^x$

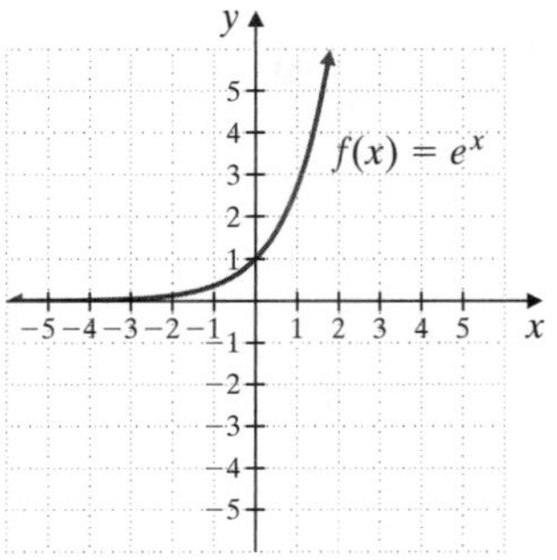

71. $f(x) = \ln x$

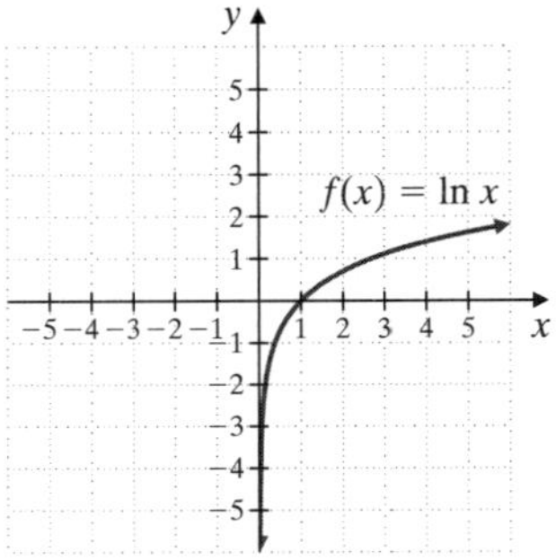

73. answers may vary

75. $R = \log\left(\frac{a}{T}\right) + B = \log\left(\frac{200}{1.6}\right) + 2.1 \approx 4.2$

The magnitude is 4.2.

77. $R = \log\left(\frac{a}{T}\right) + B = \log\left(\frac{400}{2.6}\right) + 3.1 \approx 5.3$

The magnitude is 5.3.

Section 9.7 Practice Problems

1.
$$2^x = 5$$
$$\log 2^x = \log 5$$
$$x \log 2 = \log 5$$
$$x = \frac{\log 5}{\log 2} \approx 2.3219$$

The solution set is $\left\{\frac{\log 5}{\log 2}\right\}$ or approximately $\{2.3219\}$.

2. $\log_3(x+5) = 2$
$$3^2 = x+5$$
$$9 = x+5$$
$$4 = x$$

The solution set is $\{4\}$.

3. $\log_6 x + \log_6(x+1) = 1$
$$\log_6[x(x+1)] = 1$$
$$\log_6(x^2+x) = 1$$
$$6^1 = x^2 + x$$
$$0 = x^2 + x - 6$$
$$0 = (x+3)(x-2)$$
$$0 = x+3 \quad \text{or} \quad 0 = x-2$$
$$-3 = x \qquad\qquad 2 = x$$

Because the logarithm of a negative number is undefined, –3 is rejected. The solution set is $\{2\}$.

4. $\log(x+1) - \log x = 1$
$$\log\frac{x+1}{x} = 1$$
$$10^1 = \frac{x+1}{x}$$
$$10x = x+1$$
$$9x = 1$$
$$x = \frac{1}{9}$$

The solution set is $\left\{\frac{1}{9}\right\}$.

5. Let $5000 = y_0$ and $8 = t$.
$$y = y_0 e^{0.15t}$$
$$y = 5000e^{0.15(8)}$$
$$= 5000e^{1.2}$$
$$\approx 16{,}600$$

There were approximately 16,600 lemmings in 8 months.

6. Let $P = 1000$, $A = 2000$, $r = 6\% = 0.06$, and $n = 4$.

$$A = P\left(1+\frac{r}{n}\right)^{nt}$$
$$2000 = 1000\left(1+\frac{0.06}{4}\right)^{4t}$$
$$2 = (1.015)^{4t}$$
$$\log 2 = \log 1.015^{4t}$$
$$\log 2 = 4t \log 1.015$$
$$\frac{\log 2}{4\log 1.015} = t$$
$$11.64 \approx t$$

It takes about $11\frac{3}{4}$ years for the investment to double.

Calculator Explorations

1. $$A = P\left(1+\frac{r}{n}\right)^{nt}$$
$$6000 = 5000\left(1+\frac{0.05}{4}\right)^{4t}$$
$$Y_1 = 5000\left(1+\frac{0.05}{4}\right)^{4x}$$
$$Y_2 = 6000$$
The intersection is at $x \approx 3.67$. It takes about 3.67 years or 3 years and 8 months.

2. $$A = P\left(1+\frac{r}{n}\right)^{nt}$$
$$2000 = 1000\left(1+\frac{0.045}{365}\right)^{365t}$$
$$Y_1 = 1000\left(1+\frac{0.045}{365}\right)^{365x}$$
$$Y_2 = 2000$$
The intersection is at $x \approx 15.40$. It takes about 15.40 years or 15 years and 5 months.

3. $$A = P\left(1+\frac{r}{n}\right)^{nt}$$
$$40,000 = 10,000\left(1+\frac{0.06}{12}\right)^{12t}$$
$$Y_1 = 10,000\left(1+\frac{0.06}{12}\right)^{12x}$$
$$Y_2 = 40,000$$
The intersection is at $x \approx 23.16$. It takes about 23.16 years or 23 years and 2 months.

4. $$A = P\left(1+\frac{r}{n}\right)^{nt}$$
$$800 = 500\left(1+\frac{0.04}{2}\right)^{2t}$$
$$Y_1 = 500\left(1+\frac{0.04}{2}\right)^{2x}$$
$$Y_2 = 800$$
The intersection is at $x \approx 11.87$. It takes about 11.87 years or 11 years and 11 months.

Exercise Set 9.7

1. $$3^x = 6$$
$$\log 3^x = \log 6$$
$$x \log 3 = \log 6$$
$$\frac{\log 6}{\log 3} \approx 1.6309$$
The solution set is $\left\{\frac{\log 6}{\log 3}\right\}$ or approximately $\{1.6309\}$.

3. $$9^x = 5$$
$$\log 9^x = \log 5$$
$$x \log 9 = \log 5$$
$$x = \frac{\log 5}{\log 9} \approx 0.7325$$
The solution set is $\left\{\frac{\log 5}{\log 9}\right\}$ or approximately $\{0.7325\}$.

5.
$$\begin{aligned} 3^{2x} &= 3.8 \\ \log 3^{2x} &= \log 3.8 \\ 2x \log 3 &= \log 3.8 \\ 2x &= \frac{\log 3.8}{\log 3} \\ x &= \frac{\log 3.8}{2\log 3} \approx 0.6076 \end{aligned}$$

The solution set is $\left\{\frac{\log 3.8}{2\log 3}\right\}$ or approximately $\{0.6076\}$.

7.
$$\begin{aligned} e^{6x} &= 5 \\ \ln e^{6x} &= \ln 5 \\ 6x &= \ln 5 \\ x &= \frac{\ln 5}{6} \approx 0.2682 \end{aligned}$$

The solution set is $\left\{\frac{\ln 5}{6}\right\}$ or approximately $\{0.2682\}$.

9.
$$\begin{aligned} 2^{x-3} &= 5 \\ \log 2^{x-3} &= \log 5 \\ (x-3)\log 2 &= \log 5 \\ x-3 &= \frac{\log 5}{\log 2} \\ x &= 3+\frac{\log 5}{\log 2} \approx 5.3219 \end{aligned}$$

The solution set is $\left\{3+\frac{\log 5}{\log 2}\right\}$ or approximately $\{5.3219\}$.

11.
$$\begin{aligned} 4^{x+7} &= 3 \\ \log 4^{x+7} &= \log 3 \\ (x+7)\log 4 &= \log 3 \\ x+7 &= \frac{\log 3}{\log 4} \\ x &= \frac{\log 3}{\log 4} - 7 \approx -6.2075 \end{aligned}$$

The solution set is $\left\{\frac{\log 3}{\log 4}-7\right\}$ or approximately $\{-6.2075\}$.

13.
$$\begin{aligned} 7^{3x-4} &= 11 \\ \log 7^{3x-4} &= \log 11 \\ (3x-4)\log 7 &= \log 11 \\ 3x-4 &= \frac{\log 11}{\log 7} \\ 3x &= 4+\frac{\log 11}{\log 7} \\ x &= \frac{1}{3}\left(4+\frac{\log 11}{\log 7}\right) \approx 1.7441 \end{aligned}$$

The solution set is $\left\{\frac{1}{3}\left(4+\frac{\log 11}{\log 7}\right)\right\}$ or approximately $\{1.7441\}$.

15.
$$\begin{aligned} \log_2(x+5) &= 4 \\ x+5 &= 2^4 \\ x+5 &= 16 \\ x &= 11 \end{aligned}$$

The solution set is $\{11\}$.

17.
$$\begin{aligned} \log_4 2 + \log_4 x &= 0 \\ \log_4(2x) &= 0 \\ 2x &= 4^0 \\ 2x &= 1 \\ x &= \frac{1}{2} \end{aligned}$$

The solution set is $\left\{\frac{1}{2}\right\}$.

19. $\log_2 6 - \log_2 x = 3$

$$\log_2 \frac{6}{x} = 3$$
$$\frac{6}{x} = 2^3$$
$$\frac{6}{x} = 8$$
$$\frac{6}{8} = x$$
$$\frac{3}{4} = x$$

The solution set is $\left\{\frac{3}{4}\right\}$.

21. $\log_2(x^2 + x) = 1$

$$x^2 + x = 2^1$$
$$x^2 + x = 2$$
$$x^2 + x - 2 = 0$$
$$(x+2)(x-1) = 0$$
$$x + 2 = 0 \quad \text{or} \quad x - 1 = 0$$
$$x = -2 \qquad\qquad x = 1$$

Both solutions check, so the solution set is $\{-2, 1\}$.

23. $\log_4 x + \log_4(x+6) = 2$

$$\log_4[x(x+6)] = 2$$
$$x(x+6) = 4^2$$
$$x^2 + 6x = 16$$
$$x^2 + 6x - 16 = 0$$
$$(x+8)(x-2) = 0$$
$$x + 8 = 0 \quad \text{or} \quad x - 2 = 0$$
$$x = -8 \qquad\qquad x = 2$$

The value of $x = -8$ does not check, so the solution set is $\{2\}$.

25. $\log_5(x+3) - \log_5 x = 2$

$$\log_5 \frac{x+3}{x} = 2$$
$$\frac{x+3}{x} = 5^2$$
$$\frac{x+3}{x} = 25$$
$$x + 3 = 25x$$
$$3 = 24x$$
$$\frac{3}{24} = x$$
$$\frac{1}{8} = x$$

The solution set is $\left\{\frac{1}{8}\right\}$.

27. $\log_4(x^2 - 3x) = 1$

$$x^2 - 3x = 4^1$$
$$x^2 - 3x = 4$$
$$x^2 - 3x - 4 = 0$$
$$(x-4)(x+1) = 0$$
$$x - 4 = 0 \quad \text{or} \quad x + 1 = 0$$
$$x = 4 \qquad\qquad x = -1$$

Both solutions check, so the solution set is $\{4, -1\}$.

29. $\log_2 x + \log_2(3x+1) = 1$

$$\log_2[x(3x+1)] = 1$$
$$x(3x+1) = 2^1$$
$$3x^2 + x = 2$$
$$3x^2 + x - 2 = 0$$
$$(3x-2)(x+1) = 0$$
$$3x - 2 = 0 \quad \text{or} \quad x + 1 = 0$$
$$3x = 2 \qquad\qquad x = -1$$
$$x = \frac{2}{3}$$

The value $x = -1$ does not check, so the solution set is $\left\{\frac{2}{3}\right\}$.

31. Let $y_0 = 83$ and $t = 5$.

$y = y_0 e^{0.043t}$

$y = 83e^{0.043(5)} = 83e^{0.215} \approx 103$

The population is estimated to be 103 wolves in 5 years.

33. 2010 is 6 years after 2004, so $t = 6$.

$y = y_0 e^{0.025t}$

$y = 6{,}191{,}000e^{0.025(6)}$

$= 6{,}191{,}000e^{0.15}$

$\approx 7{,}192{,}916$

The population in 2010 is estimated to be 7,192,916 people.

35. Let $y = 9{,}000{,}000$ and $y_0 = 10{,}032{,}000$.

$$y = y_0 e^{-0.004t}$$

$$9{,}000{,}000 = 10{,}032{,}000e^{-0.004t}$$

$$\frac{9{,}000{,}000}{10{,}032{,}000} = e^{-0.004t}$$

$$\ln\frac{9{,}000{,}000}{10{,}032{,}000} = \ln e^{-0.004t}$$

$$\ln\frac{9{,}000{,}000}{10{,}032{,}000} = -0.004t$$

$$\frac{1}{-0.004}\ln\frac{9{,}000{,}000}{10{,}032{,}000} = t$$

$$t \approx 27.1$$

The population will decline to 9,000,000 in about 27.1 years.

37. $A = P\left(1+\frac{r}{n}\right)^{nt}$, $P = 600$,

$A = 2(600) = 1200$, $r = 7\% = 0.07$, and $n = 12$.

$$1200 = 600\left(1+\frac{0.07}{12}\right)^{12t}$$

$$2 = \left(1+\frac{0.07}{12}\right)^{12t}$$

$$\log 2 = \log\left(1+\frac{0.07}{12}\right)^{12t}$$

$$\log 2 = 12t\log\left(1+\frac{0.07}{12}\right)$$

$$\frac{\log 2}{12\log\left(1+\frac{0.07}{12}\right)} = t$$

$$9.9 \approx t$$

It would take approximately 9.9 years for the \$600 to double.

39. $A = P\left(1+\frac{r}{n}\right)^{nt}$

$A = 200 + 1200 = 1400$, $P = 1200$, $r = 9\% = 0.09$, and $n = 4$.

$$1400 = 1200\left(1+\frac{0.09}{4}\right)^{4t}$$

$$\frac{1400}{1200} = \left(1+\frac{0.09}{4}\right)^{4t}$$

$$\log\frac{1400}{1200} = \log\left(1+\frac{0.09}{4}\right)^{4t}$$

$$\log\frac{1400}{1200} = 4t\log\left(1+\frac{0.09}{4}\right)$$

$$\frac{\log\frac{1400}{1200}}{4\log\left(1+\frac{0.09}{4}\right)} = t$$

$$1.7 \approx t$$

It will take approximately 1.7 years for the investment to earn \$200.

41. $A = P\left(1+\frac{r}{n}\right)^{nt}$, $P = 1000$,
$A = 2(1000) = 2000$, $r = 8\% = 0.08$, and $n = 2$.

$$2000 = 1000\left(1+\frac{0.08}{2}\right)^{2t}$$
$$2 = (1.04)^{2t}$$
$$\log 2 = \log 1.04^{2t}$$
$$\log 2 = 2t \log 1.04$$
$$\frac{\log 2}{2\log 1.04} = t$$
$$8.8 \approx t$$

It would take approximately 8.8 years for the \$1000 to double.

43. $w = 0.00185h^{2.67}$, $w = 85$

$$85 = 0.00185h^{2.67}$$
$$\frac{85}{0.00185} = h^{2.67}$$
$$\log\left(\frac{85}{0.00185}\right) = \log h^{2.67}$$
$$\log\left(\frac{85}{0.00185}\right) = 2.67\log h$$
$$\frac{1}{2.67}\log\left(\frac{85}{0.00185}\right) = \log h$$
$$1.7462 \approx \log h$$
$$10^{1.7462} = h$$
$$55.7 \approx h$$

The expected height is 55.7 inches.

45. $P = 14.7e^{-0.21x}$, $x = 1$

$P = 14.7e^{-0.21(1)} = 14.7e^{-0.21} \approx 11.9$

The average atmospheric pressure in Denver is approximately 11.9 pounds per square inch.

47. $P = 14.7e^{-0.21x}$, $P = 7.5$

$$7.5 = 14.7e^{-0.21x}$$
$$\frac{7.5}{14.7} = e^{-0.21x}$$
$$\ln\frac{7.5}{14.7} = \ln e^{-0.21x}$$
$$\ln\frac{7.5}{14.7} = -0.21x$$
$$\frac{1}{-0.21}\ln\frac{7.5}{14.7} = x$$
$$3.2 \approx x$$

The elevation is estimated at 3.2 miles.

49. $t = \frac{1}{c}\ln\left(\frac{A}{A-N}\right)$

$$t = \frac{1}{0.09}\ln\left(\frac{75}{75-50}\right)$$
$$t = \frac{1}{0.09}\ln(3)$$
$$t \approx 12.21$$

It will take about 12 weeks.

51. $t = \frac{1}{c}\ln\left(\frac{A}{A-N}\right)$

$$t = \frac{1}{0.07}\ln\left(\frac{210}{210-150}\right)$$
$$t \approx 18$$

It will take about 18 weeks.

53. $\frac{x^2 - y + 2z}{3x} = \frac{(-2)^2 - (0) + 2(3)}{3(-2)}$

$$= \frac{4+6}{-6}$$
$$= \frac{10}{-6}$$
$$= -\frac{5}{3}$$

55. $\frac{3z - 4x + y}{x + 2z} = \frac{3(3) - 4(-2) + 0}{-2 + 2(3)} = \frac{9+8}{-2+6} = \frac{17}{4}$

57. Use $y = y_0e^{kt}$ with $y_0 = 5,130,632$ and $y = 5,880,811$ when $t = 3$.

$$5,880,811 = 5,130,632e^{3k}$$
$$\frac{5,880,811}{5,130,632} = e^{3k}$$
$$\ln\frac{5,880,811}{5,130,632} = 3k$$
$$k = \frac{1}{3}\ln\frac{5,880,811}{5,130,632} \approx 0.04549$$

The annual rate of population growth is 4.5%.

59. answers may vary

61. {6.93}

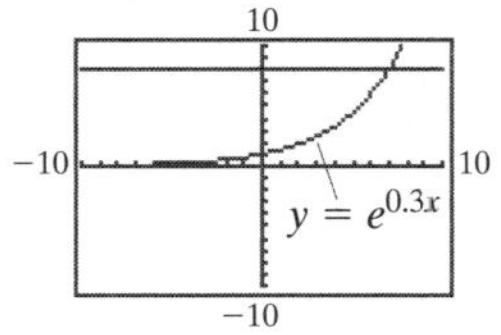

The Bigger Picture

1.
$$8^x = 2^{x-3}$$
$$2^{3x} = 2^{x-3}$$
$$3x = x - 3$$
$$2x = -3$$
$$x = -\frac{3}{2}$$

The solution set is $\left\{-\frac{3}{2}\right\}$.

2.
$$11^x = 5$$
$$\log 11^x = \log 5$$
$$x\log 11 = \log 5$$
$$x = \frac{\log 5}{\log 11} \approx 0.6712$$

The solution set is $\left\{\frac{\log 5}{\log 11} \approx 0.6712\right\}$.

3.
$$-7x + 3 \le -5x + 13$$
$$-7x + 5x \le 13 - 3$$
$$-2x \le 10$$
$$x \ge -5$$

The solution set is $[-5, \infty)$.

4.
$$-7 \le 3x + 6 \le 0$$
$$-13 \le 3x \le -6$$
$$\frac{-13}{3} \le x \le \frac{-6}{3}$$
$$-\frac{13}{3} \le x \le -2$$

The solution set is $\left[-\frac{13}{3}, -2\right]$.

5.
$$|5y + 3| < 3$$
$$-3 < 5y + 3 < 3$$
$$-6 < 5y < 0$$
$$-\frac{6}{5} < y < 0$$

The solution set is $\left(-\frac{6}{5}, 0\right)$.

6. $(x - 6)(5x + 1) = 0$

$5x + 1 = 0$ or $x - 6 = 0$

$x = -\frac{1}{5}$ $\quad$ $x = 6$

The solution set is $\left\{6, -\frac{1}{5}\right\}$.

7.
$$\log_{13} 8 + \log_{13}(x - 1) = 1$$
$$\log_{13}[8(x - 1)] = 1$$
$$8(x - 1) = 13^1$$
$$8x - 8 = 13$$
$$8x = 21$$
$$x = \frac{21}{8}$$

The solution set is $\left\{\frac{21}{8}\right\}$.

8. $\left|\frac{3x-1}{4}\right| = 2$

$\frac{3x-1}{4} = -2$ or $\frac{3x-1}{4} = 2$

$3x-1 = -8$ $3x-1 = 8$

$3x = -7$ $3x = 9$

$x = -\frac{7}{3}$ $x = 3$

The solution set is $\left\{-\frac{7}{3}, 3\right\}$.

9. $|7x + 1| > -2$
Since the absolute value of a number is always nonnegative, the solution set is $(-\infty, \infty)$.

10. $x^2 = 4$

$x = \pm\sqrt{4}$

$x = \pm 2$

The solution set is $\{-2, 2\}$.

11. $(x+5)^2 = 3$

$x+5 = \pm\sqrt{3}$

$x = -5 \pm \sqrt{3}$

The solution set is $\left\{-5-\sqrt{3}, -5+\sqrt{3}\right\}$.

12. $\log_7(4x^2 - 27x) = 1$

$4x^2 - 27x = 7^1$

$4x^2 - 27x = 7$

$4x^2 - 27x - 7 = 0$

$(4x+1)(x-7) = 0$

$4x+1 = 0$ or $x-7 = 0$

$x = -\frac{1}{4}$ $x = 7$

The solution set is $\left\{-\frac{1}{4}, 7\right\}$.

Chapter 9 Vocabulary Check

1. For a one-to-one function, we can find its inverse function by switching the coordinates of the ordered pairs of the function.

2. The composition of functions f and g is $(f \circ g)(x) = f(g(x))$.

3. A function of the form $f(x) = b^x$ is called an exponential function if $b > 0$ b is not 1, and x is a real number.

4. The graphs of f and f^{-1} are symmetric about the line $y = x$.

5. Natural logarithms are logarithms to base e.

6. Common logarithms are logarithms to base 10.

7. To see whether a graph is the graph of a one-to-one function, apply the vertical line test to see whether it is a function, and then apply the horizontal line test to see whether it is a one-to-one function.

8. A logarithmic function is a function that can be defined by $f(x) = \log_b x$ where x is a positive real number, b is a constant positive real number, and b is not 1.

Chapter 9 Review

1. $(f+g)(x) = f(x) + g(x)$

$= (x-5) + (2x+1)$

$= 3x - 4$

2. $(f-g)(x) = f(x) - g(x)$

$= (x-5) - (2x+1)$

$= x - 5 - 2x - 1$

$= -x - 6$

3. $(f \cdot g)(x) = f(x) \cdot g(x)$

$= (x-5)(2x+1)$

$= 2x^2 - 9x - 5$

4. $\left(\frac{g}{f}\right)(x) = \frac{2x+1}{x-5}$, where $x \neq 5$.

5. $$\begin{aligned}(f \circ g)(x) &= f(g(x)) \\ &= f(x+1) \\ &= (x+1)^2 - 2 \\ &= x^2 + 2x + 1 - 2 \\ &= x^2 + 2x - 1\end{aligned}$$

6. $$\begin{aligned}(g \circ f)(x) &= g(f(x)) \\ &= g(x^2 - 2) \\ &= x^2 - 2 + 1 \\ &= x^2 - 1\end{aligned}$$

7. $$\begin{aligned}(h \circ g)(2) &= h(g(2)) \\ &= h(3) \\ &= 3^3 - 3^2 \\ &= 27 - 9 \\ &= 18\end{aligned}$$

8. $$\begin{aligned}(f \circ f)(x) &= f(f(x)) \\ &= f(x^2 - 2) \\ &= (x^2 - 2)^2 - 2 \\ &= x^4 - 4x^2 + 4 - 2 \\ &= x^4 - 4x^2 + 2\end{aligned}$$

9. $(f \circ g)(-1) = f(g(-1)) = f(0) = 0^2 - 2 = -2$

10. $$\begin{aligned}(h \circ h)(2) &= h(h(2)) \\ &= h(4) \\ &= 4^3 - 4^2 \\ &= 64 - 16 \\ &= 48\end{aligned}$$

11. $h = \{(-9, 14), (6, 8), (-11, 12), (15, 15)\}$ is a one-to-one function since each *y*-value corresponds to only one *x*-value.
$h^{-1} = \{(14, -9), (8, 6), (12, -11), (15, 15)\}$

12. $f = \{(-5, 5), (0, 4), (13, 5), (11, -6)\}$ is not a one-to-one function because the *y*-value 5 corresponds to two *x*-values, −5 and 13.

13. The function is one-to-one since each output corresponds to only one input.

Rank in Automobile Thefts (Input)	2	4	1	3
U.S. Region (Output)	West	Midwest	South	Northeast

14. The function is not one-to-one because the output 4 corresponds to three different inputs, square, parallelogram, and rectangle.

15. The graph does not represent a one-to-one function because it does not pass the horizontal line test.

16. The graph does not represent a one-to-one function because it does not pass the horizontal line test.

17. The graph does not represent a one-to-one function because it does not pass the horizontal line test.

18. The graph represents a one-to-one function because it passes the vertical and horizontal line tests.

19.
$$\begin{aligned} f(x) &= 6x+11 \\ y &= 6x+11 \\ x &= 6y+11 \\ x-11 &= 6y \\ \frac{x-11}{6} &= y \\ f^{-1}(x) &= \frac{x-11}{6} \end{aligned}$$

20.
$$\begin{aligned} f(x) &= 12x \\ y &= 12x \\ x &= 12y \\ \frac{x}{12} &= y \\ f^{-1}(x) &= \frac{x}{12} \end{aligned}$$

21.
$$\begin{aligned} f(x) &= 3x-5 \\ y &= 3x-5 \\ x &= 3y-5 \\ x+5 &= 3y \\ \frac{x+5}{3} &= y \\ f^{-1}(x) &= \frac{x+5}{3} \end{aligned}$$

22. $f(x) = 2x + 1$
$y = 2x + 1$
$x = 2y + 1$
$x - 1 = 2y$
$\frac{x-1}{2} = y$
$f^{-1}(x) = \frac{x-1}{2}$

23. $f(x) = -2x + 3$
$y = -2x + 3$
$x = -2y + 3$
$x - 3 = -2y$
$-\frac{x-3}{2} = y$
$f^{-1}(x) = -\frac{x-3}{2}$

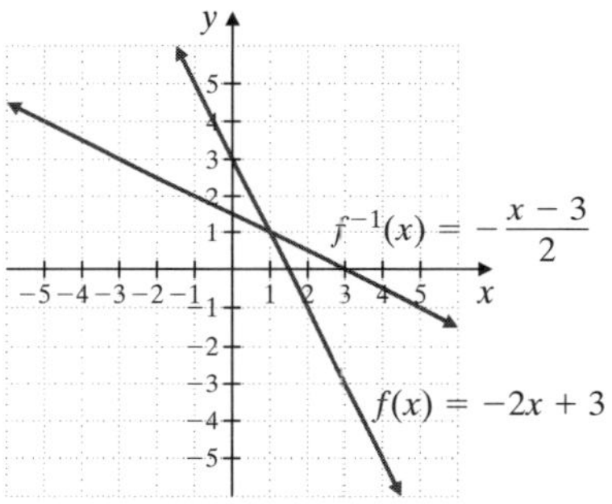

24. $f(x) = 5x - 5$
$y = 5x - 5$
$x = 5y - 5$
$x + 5 = 5y$
$\frac{x+5}{5} = y$
$f^{-1}(x) = \frac{x+5}{5}$

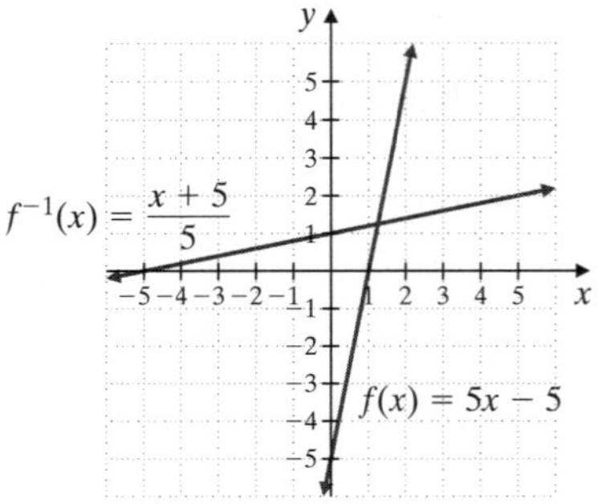

25. $4^x = 64$
$4^x = 4^3$
$x = 3$
The solution set is $\{3\}$.

26. $2^{3x} = \frac{1}{16}$
$2^{3x} = 2^{-4}$
$3x = -4$
$x = -\frac{4}{3}$
The solution set is $\left\{-\frac{4}{3}\right\}$.

27. $9^{x+1} = 243$
$(3^2)^{x+1} = 3^5$
$3^{2x+2} = 3^5$
$2x + 2 = 5$
$2x = 3$
$x = \frac{3}{2}$
The solution set is $\left\{\frac{3}{2}\right\}$.

28.

x	$y = 3^x$
-2	$3^{-2} = \frac{1}{3^2} = \frac{1}{9}$
-1	$3^{-1} = \frac{1}{3}$
0	$3^0 = 1$
1	$3^1 = 3$
2	$3^2 = 9$

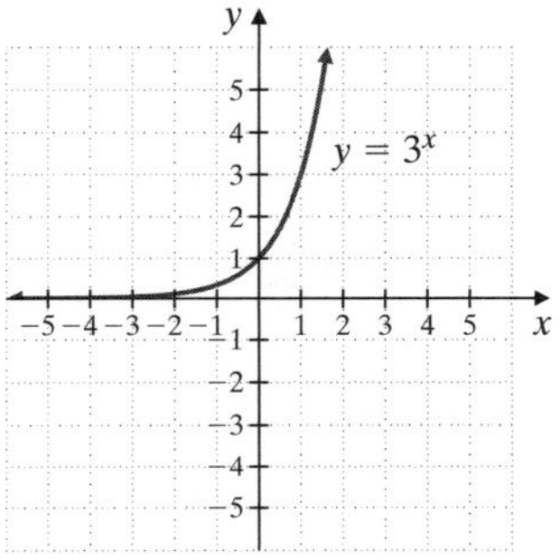

29.

x	$y = \left(\frac{1}{3}\right)^x$
-2	$\left(\frac{1}{3}\right)^{-2} = 3^2 = 9$
-1	$\left(\frac{1}{3}\right)^{-1} = 3$
0	$\left(\frac{1}{3}\right)^0 = 1$
1	$\left(\frac{1}{3}\right)^1 = \frac{1}{3}$
2	$\left(\frac{1}{3}\right)^2 = \frac{1}{9}$

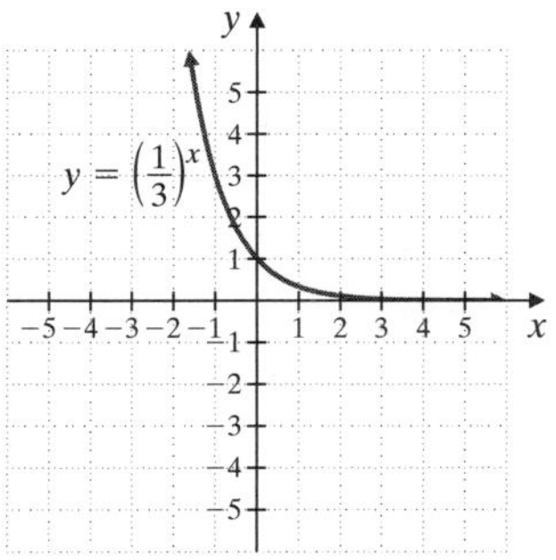

30.

x	$y = 2^{x-4}$
-2	$2^{-2-4} = 2^{-6} = \frac{1}{2^6} = \frac{1}{64}$
-1	$2^{-1-4} = 2^{-5} = \frac{1}{2^5} = \frac{1}{32}$
0	$2^{0-4} = 2^{-4} = \frac{1}{2^4} = \frac{1}{16}$
1	$2^{1-4} = 2^{-3} = \frac{1}{2^3} = \frac{1}{8}$
2	$2^{2-4} = 2^{-2} = \frac{1}{2^2} = \frac{1}{4}$

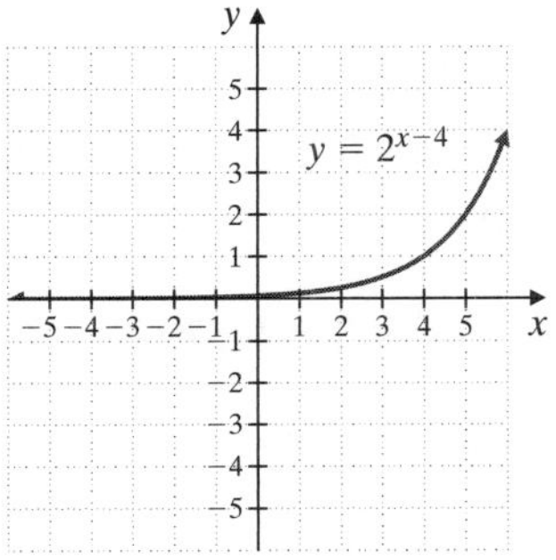

31.

x	$y = 2^x + 4$
-2	$2^{-2} + 4 = \frac{1}{4} + 4 = \frac{17}{4}$
-1	$2^{-1} + 4 = \frac{1}{2} + 4 = \frac{9}{2}$
0	$2^0 + 4 = 1 + 4 = 5$
1	$2^1 + 4 = 2 + 4 = 6$
2	$2^2 + 4 = 4 + 4 = 8$

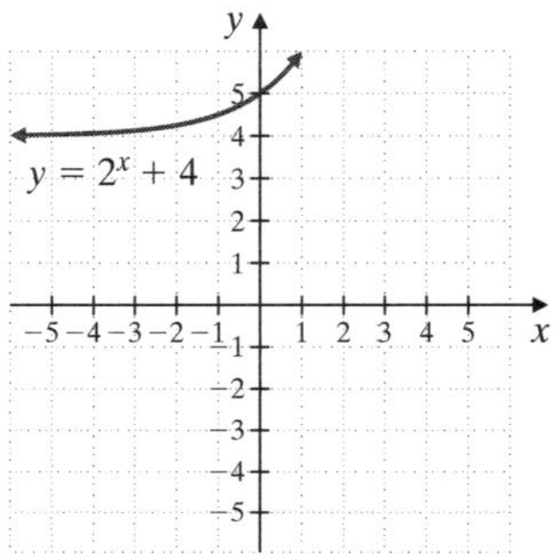

32. Let $P = 1600$, $r = 0.09$, $n = 2$, and $t = 7$.

$$A = P\left(1 + \frac{r}{n}\right)^{nt}$$
$$A = 1600\left(1 + \frac{0.09}{2}\right)^{2 \cdot 7}$$
$$A = 1600(1.045)^{14}$$
$$A \approx 2963.11$$

The annual accrued is \$2963.11.

33. Let $P = 800$, $r = 0.07$, $n = 4$, and $t = 5$.

$$A = P\left(1 + \frac{r}{n}\right)^{nt}$$
$$A = 800\left(1 + \frac{0.07}{4}\right)^{4 \cdot 5}$$
$$A = 800(1.0175)^{20}$$
$$A \approx 1131.82$$

The value will be \$1131.82.

34.
$$49 = 7^2$$
$$\log_7 49 = 2$$

35.
$$2^{-4} = \frac{1}{16}$$
$$\log_2 \frac{1}{16} = -4$$

36.
$$\log_{1/2} 16 = -4$$
$$\left(\frac{1}{2}\right)^{-4} = 16$$

37.
$$\log_{0.4} 0.064 = 3$$
$$0.4^3 = 0.064$$

38.
$$\log_4 x = -3$$
$$4^{-3} = x$$
$$\frac{1}{4^3} = x$$
$$\frac{1}{64} = x$$

The solution set is $\left\{\frac{1}{64}\right\}$.

39.
$$\log_3 x = 2$$
$$3^2 = x$$
$$9 = x$$

The solution set is $\{9\}$.

40.
$$\log_3 1 = x$$
$$3^x = 1$$
$$3^x = 3^0$$
$$x = 0$$

The solution set is $\{0\}$.

41.
$$\log_x 64 = 2$$
$$x^2 = 64$$
$$x^2 = 8^2$$
$$x = 8$$

The solution set is $\{8\}$.

42.
$$\log_4 4^5 = x$$
$$4^x = 4^5$$
$$x = 5$$

The solution set is $\{5\}$.

43. $\log_7 7^{-2} = x$

$7^x = 7^{-2}$

$x = -2$

The solution set is $\{-2\}$.

44. $5^{\log_5 4} = x$

$4 = x$

The solution set is $\{4\}$.

45. $2^{\log_2 9} = x$

$9 = x$

The solution set is $\{9\}$.

46. $\log_3(2x+5) = 2$

$3^2 = 2x+5$

$9 = 2x+5$

$4 = 2x$

$2 = x$

The solution set is $\{2\}$.

47. $\log_8(x^2+7x) = 1$

$8^1 = x^2+7x$

$8 = x^2+7x$

$0 = x^2+7x-8$

$(x+8)(x-1) = 0$

$x+8 = 0$ or $x-1 = 0$

$x = -8$ $\quad$ $x = 1$

The solution set is $\{-8, 1\}$.

48. $y = 2^x$; $y = \log_2 x$

x	$y = 2^x$
-2	$2^{-2} = \frac{1}{4}$
-1	$2^{-1} = \frac{1}{2}$
0	$2^0 = 1$
1	$2^1 = 2$
2	$2^2 = 4$

x	$y = \log_2 x$
$\frac{1}{4}$	-2
$\frac{1}{2}$	-1
1	0
2	1
4	2

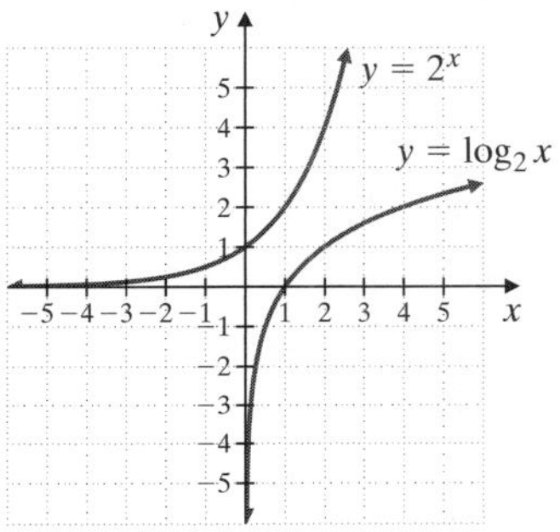

49. $y = \left(\frac{1}{2}\right)^x$; $y = \log_{1/2} x$

x	$y = \left(\frac{1}{2}\right)^x$
-2	$\left(\frac{1}{2}\right)^{-2} = 4$
-1	$\left(\frac{1}{2}\right)^{-1} = 2$
0	$\left(\frac{1}{2}\right)^0 = 1$
1	$\left(\frac{1}{2}\right)^1 = \frac{1}{2}$
2	$\left(\frac{1}{2}\right)^2 = \frac{1}{4}$

x	$y = \log_{1/2} x$
4	−2
2	−1
1	0
$\frac{1}{2}$	1
$\frac{1}{4}$	2

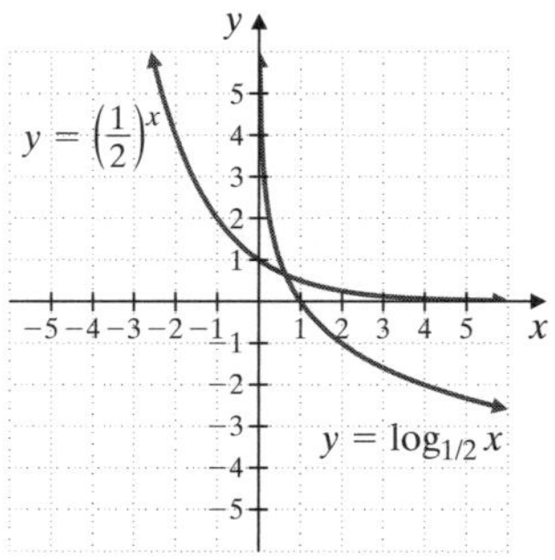

50. $\log_3 8 + \log_3 4 = \log_3 8 \cdot 4 = \log_3 32$

51. $\log_2 6 + \log_2 3 = \log_2 6 \cdot 3 = \log_2 18$

52. $\log_7 15 - \log_7 20 = \log_7 \frac{15}{20} = \log_7 \frac{3}{4}$

53. $\log_e 18 - \log_e 12 = \log_e \frac{18}{12} = \log_e \frac{3}{2}$

54. $\log_{11} 8 + \log_{11} 3 - \log_{11} 6$
$= \log_{11} 8 \cdot 3 - \log_{11} 6$
$= \log_{11} 24 - \log_{11} 6$
$= \log_{11} \frac{24}{6}$
$= \log_{11} 4$

55. $\log_5 14 + \log_5 3 - \log_5 21$
$= \log_5 14 \cdot 3 - \log_5 21$
$= \log_5 42 - \log_5 21$
$= \log_5 \frac{42}{21}$
$= \log_5 2$

56. $2\log_5 x - 2\log_5(x+1) + \log_5 x$
$= \log_5 x^2 - \log_5(x+1)^2 + \log_5 x$
$= \log_5 \frac{x^2}{(x+1)^2} + \log_5 x$
$= \log_5 \frac{x^2}{(x+1)^2} \cdot x$
$= \log_5 \frac{x^3}{(x+1)^2}$

57. $4\log_3 x - \log_3 x + \log_3(x+2)$
$= \log_3 x^4 - \log_3 x + \log_3(x+2)$
$= \log_3 \frac{x^4}{x} + \log_3(x+2)$
$= \log_3 x^3 + \log_3(x+2)$
$= \log_3 x^3(x+2)$
$= \log_3(x^4 + 2x^3)$

58. $\log_3 \frac{x^3}{x+2} = \log_3 x^3 - \log_3(x+2)$
$= 3\log_3 x - \log_3(x+2)$

59. $\log_4 \frac{x+5}{x^2} = \log_4(x+5) - \log_4 x^2$
$= \log_4(x+5) - 2\log_4 x$

60. $\log_2 \frac{3x^2 y}{z}$
$= \log_2 3x^2 y - \log_2 z$
$= \log_2 3 + \log_2 x^2 + \log_2 y - \log_2 z$
$= \log_2 3 + 2\log_2 x + \log_2 y - \log_2 z$

61. $\log_7 \frac{yz^3}{x} = \log_7 yz^3 - \log_7 x$
$= \log_7 y + \log_7 z^3 - \log_7 x$
$= \log_7 y + 3\log_7 z - \log_7 x$

62. $\log_b 50 = \log_b 2\cdot 25$
$= \log_b 2 + \log_b 25$
$= \log_b 2 + \log_b 5^2$
$= \log_b 2 + 2\log_b 5$
$= 0.36 + 2(0.83)$
$= 2.02$

63. $\log_b \frac{4}{5} = \log_b 4 - \log_b 5$
$= \log_b 2^2 - \log_b 5$
$= 2\log_b 2 - \log_b 5$
$= 2(0.36) - 0.83$
$= -0.11$

64. $\log 3.6 \approx 0.5563$

65. $\log 0.15 \approx -0.8239$

66. $\ln 1.25 \approx 0.2231$

67. $\ln 4.63 \approx 1.5326$

68. $\log 1000 = \log 10^3 = 3$

69. $\log \frac{1}{10} = \log 10^{-1} = -1$

70. $\ln \frac{1}{e} = \ln e^{-1} = -1$

71. $\ln e^4 = 4$

72. $\log 2x = 2$
$2x = 10^2$
$2x = 100$
$x = 50$
The solution set is $\{50\}$.

73. $\ln(3x) = 1.6$
$3x = e^{1.6}$
$x = \frac{e^{1.6}}{3} \approx 1.6510$
The solution set is $\left\{\frac{e^{1.6}}{3}\right\}$ or approximately $\{1.6510\}$.

74. $\ln(2x-3) = -1$
$2x - 3 = e^{-1}$
$2x = e^{-1} + 3$
$x = \frac{e^{-1}+3}{2} \approx 1.6839$
The solution set is $\left\{\frac{e^{-1}+3}{2}\right\}$ or approximately $\{1.6839\}$.

75. $\ln(3x+1) = 2$
$3x + 1 = e^2$
$3x = e^2 - 1$
$x = \frac{e^2-1}{3} \approx 2.1297$
The solution set is $\left\{\frac{e^2-1}{3}\right\}$ or approximately $\{2.1297\}$.

76. $\log_5 1.6 = \frac{\log 1.6}{\log 5} \approx \frac{0.20412}{0.69897} \approx 0.2920$

77. $\log_3 4 = \frac{\log 4}{\log 3} \approx \frac{0.60206}{0.47712} \approx 1.2619$

78. Let $P = 1450$, $r = 6\% = 0.06$, and $t = 5$.
$A = Pe^{rt}$
$A = 1450e^{0.06(5)}$
$A \approx 1957.30$
The amount accrued is \$1957.30.

79. Let $P = 940$, $r = 11\% = 0.11$, and $t = 3$.

$$A = Pe^{rt}$$
$$A = 940e^{0.11(3)}$$
$$A \approx 1307.51$$

The amount grows to \$1307.51.

80.
$$7^x = 20$$
$$\log 7^x = \log 20$$
$$x \log 7 = \log 20$$
$$x = \frac{\log 20}{\log 7} \approx 1.5395$$

The solution set is $\left\{\frac{\log 20}{\log 7}\right\}$ or approximately $\{1.5395\}$.

81.
$$3^{2x} = 7$$
$$\log 3^{2x} = \log 7$$
$$2x \log 3 = \log 7$$
$$2x = \frac{\log 7}{\log 3}$$
$$x = \frac{\log 7}{2\log 3} \approx 0.8856$$

The solution set is $\left\{\frac{\log 7}{2\log 3}\right\}$ or approximately $\{0.8856\}$.

82.
$$3^{2x+1} = 6$$
$$\log 3^{2x+1} = \log 6$$
$$(2x+1)\log 3 = \log 6$$
$$2x+1 = \frac{\log 6}{\log 3}$$
$$2x = \frac{\log 6}{\log 3} - 1$$
$$x = \frac{1}{2}\left(\frac{\log 6}{\log 3} - 1\right) \approx 0.3155$$

The solution set is $\left\{\frac{1}{2}\left(\frac{\log 6}{\log 3} - 1\right)\right\}$ or approximately $\{0.3155\}$.

83.
$$8^{4x-2} = 3$$
$$\log 8^{4x-2} = \log 3$$
$$(4x-2)\log 8 = \log 3$$
$$4x - 2 = \frac{\log 3}{\log 8}$$
$$4x = \frac{\log 3}{\log 8} + 2$$
$$x = \frac{1}{4}\left(\frac{\log 3}{\log 8} + 2\right) \approx 0.6321$$

The solution set is $\left\{\frac{1}{4}\left(\frac{\log 3}{\log 8} + 2\right)\right\}$ or approximately $\{0.6321\}$.

84.
$$\log_5 2 + \log_5 x = 2$$
$$\log_5 2x = 2$$
$$2x = 5^2$$
$$2x = 25$$
$$x = \frac{25}{2}$$

The solution set is $\left\{\frac{25}{2}\right\}$.

85.
$$\log 5x - \log(x+1) = 4$$
$$\log \frac{5x}{x+1} = 4$$
$$\frac{5x}{x+1} = 10^4$$
$$\frac{5x}{x+1} = 10{,}000$$
$$5x = 10{,}000x + 10{,}000$$
$$-9995x = 10{,}000$$
$$x = -\frac{10{,}000}{9995}$$

Since the log of a negative number is undefined, there is no solution, $\varnothing$.

86. $\log_2 x + \log_2 2x - 3 = 1$

$$\log_2 2x^2 - 3 = 1$$
$$\log_2 2x^2 = 4$$
$$2x^2 = 2^4$$
$$2x^2 = 16$$
$$x^2 = 8$$
$$x = \pm\sqrt{8}$$
$$x = \pm 2\sqrt{2}$$

Since the log of a negative number is undefined, the solution set is $\{2\sqrt{2}\}$.

87. $\log_3(x^2 - 8x) = 2$

$$x^2 - 8x = 3^2$$
$$x^2 - 8x = 9$$
$$x^2 - 8x - 9 = 0$$
$$(x-9)(x+1) = 0$$
$$x - 9 = 0 \quad \text{or} \quad x + 1 = 0$$
$$x = 9 \qquad\qquad x = -1$$

The solution set is $\{9, -1\}$.

88. Let $y_0 = 155,000$, $k = 6\% = 0.06$, and $t = 4$.

$$y = y_0 e^{kt}$$
$$y = 155,000e^{0.06(4)}$$
$$y \approx 197,044$$

Therefore, 197,044 ducks are expected after 4 weeks.

89. Let $y_0 = 5,844,000$, $k = 2.7\% = 0.027$, and $t = 2010 - 2004 = 6$.

$$y = y_0 e^{kt}$$
$$y = 5,844,000e^{0.027(6)}$$
$$y \approx 6,871,727$$

The expected population is 6,871,727 in 2010.

90. Let $y_0 = 60,424,000$, $y = 65,000,000$, and $k = 0.4\% = 0.004$.

$$y = y_0 e^{kt}$$
$$65,000,000 = 60,424,000e^{0.004t}$$
$$\frac{65,000,000}{60,424,000} = e^{0.004t}$$
$$\ln\frac{65,000,000}{60,424,000} = 0.004t$$
$$\frac{1}{0.004}\ln\frac{65,000,000}{60,424,000} = t$$
$$18.3 \approx t$$

It will take about 18.3 years for the population to reach 65,000,000.

91. Let $y_0 = 19,913,000$, $y = 2(19,913,000) = 39,826,000$, and $k = 0.6\% = 0.006$.

$$y = y_0 e^{kt}$$
$$39,826,000 = 19,913,000e^{0.006t}$$
$$2 = e^{0.006t}$$
$$\ln 2 = 0.006t$$
$$\frac{\ln 2}{0.006} = t$$
$$115.5 \approx t$$

It will take about 115.5 years for the population to double.

92. Let $y_0 = 104,960,000$, $y = 2(104,960,000) = 209,920,000$, and $k = 2.1\% = 0.021$.

$$y = y_0 e^{kt}$$
$$209,920,000 = 104,960,000e^{0.021t}$$
$$2 = e^{0.021t}$$
$$\ln 2 = 0.021t$$
$$\frac{\ln 2}{0.021} = t$$
$$33.0 \approx t$$

It will take about 33.0 years for the population to double.

93. Let $P = 5000$, $A = 10{,}000$, $r = 8\% = 0.08$, and $n = 4$.

$$A = P\left(1+\frac{r}{n}\right)^{nt}$$
$$10{,}000 = 5000\left(1+\frac{0.08}{4}\right)^{4t}$$
$$2 = (1.02)^{4t}$$
$$\log 2 = \log 1.02^{4t}$$
$$\log 2 = 4t \log 1.02$$
$$\frac{\log 2}{4\log 1.02} = t$$
$$8.8 \approx t$$

It will take about 8.8 years for the investment to grow to \$10,000.

94. Let $P = 6000$, $A = 10{,}000$, $r = 6\% = 0.06$, and $n = 12$.

$$A = P\left(1+\frac{r}{n}\right)^{nt}$$
$$10{,}000 = 6000\left(1+\frac{0.06}{12}\right)^{12t}$$
$$\frac{10{,}000}{6000} = (1.005)^{12t}$$
$$\log\frac{10{,}000}{6000} = \log 1.005^{12t}$$
$$\log\frac{10{,}000}{6000} = 12t\log 1.005$$
$$\frac{\log\frac{10{,}000}{6000}}{12\log 1.005} = t$$
$$8.5 \approx t$$

It took about 8.5 years for the investment to grow to \$10,000.

95. $3^x = \frac{1}{9}$

$$3^x = 3^{-2}$$
$$x = -2$$

The solution set is $\{-2\}$.

96. $5^{2x} = 125$

$$5^{2x} = 5^3$$
$$2x = 3$$
$$x = \frac{3}{2}$$

The solution set is $\left\{\frac{3}{2}\right\}$.

97. $8^{3x-2} = 4$

$$(2^3)^{3x-2} = 2^2$$
$$2^{3(3x-2)} = 2^2$$
$$3(3x-2) = 2$$
$$9x - 6 = 2$$
$$9x = 8$$
$$x = \frac{8}{9}$$

The solution set is $\left\{\frac{8}{9}\right\}$.

98. $\log_4 64 = x$

$$4^x = 64$$
$$4^x = 4^3$$
$$x = 3$$

The solution set is $\{3\}$.

99. $\log_x 81 = 4$

$$x^4 = 81$$
$$x^4 = 3^4$$
$$x = 3$$

The solution set is $\{3\}$.

100. $\log_4(x^2 - 3x) = 1$

$$x^2 - 3x = 4^1$$
$$x^2 - 3x = 4$$
$$x^2 - 3x - 4 = 0$$
$$(x+1)(x-4) = 0$$
$$x+1 = 0 \quad \text{or} \quad x-4 = 0$$
$$x = -1 \qquad\qquad x = 4$$

The solution set is $\{-1, 4\}$.

101. $\log_2(3x-1)=4$

$$3x-1=2^4$$
$$3x-1=16$$
$$3x=17$$
$$x=\frac{17}{3}$$

The solution set is $\left\{\frac{17}{3}\right\}$.

102. $\ln x=-1.2$

$$x=e^{-1.2}$$

The solution set is $\{e^{-1.2}\}$.

103. $\log_3 x+\log_3 10=2$

$$\log_3 10x=2$$
$$10x=3^2$$
$$10x=9$$
$$x=\frac{9}{10}$$

The solution set is $\left\{\frac{9}{10}\right\}$.

104. $\ln 3x-\ln(x-3)=2$

$$\ln\frac{3x}{x-3}=2$$
$$\frac{3x}{x-3}=e^2$$
$$3x=e^2(x-3)$$
$$3x=e^2x-3e^2$$
$$3e^2=e^2x-3x$$
$$3e^2=x(e^2-3)$$
$$\frac{3e^2}{e^2-3}=x$$

The solution set is $\left\{\frac{3e^2}{e^2-3}\right\}$.

105. $\log_6 x-\log_6(4x+7)=1$

$$\log_6\frac{x}{4x+7}=1$$
$$\frac{x}{4x+7}=6^1$$
$$x=6(4x+7)$$
$$x=24x+42$$
$$-23x=42$$
$$x=-\frac{42}{23}$$

Since the log of a negative number is undefined, the solution set is $\varnothing$.

Chapter 9 Test

1. $(f\cdot g)(x)=f(x)\cdot g(x)$
$$=x(2x-3)$$
$$=2x^2-3x$$

2. $(f-g)(x)=f(x)-g(x)$
$$=x-(2x-3)$$
$$=x-2x+3$$
$$=-x+3 \text{ or } 3-x$$

3. $(f\circ h)(0)=f(h(0))=f(5)=5$

4. $(g\circ f)(x)=g(f(x))=g(x)=x-7$

5. $(g\circ h)(x)=g(h(x))$
$$=g(x^2-6x+5)$$
$$=x^2-6x+5-7$$
$$=x^2-6x-2$$

6. $f(x) = 7x - 14$
$y = 7x - 14$
$x = 7y - 14$
$x + 14 = 7y$
$\frac{x+14}{7} = y$
$f^{-1}(x) = \frac{x+14}{7}$

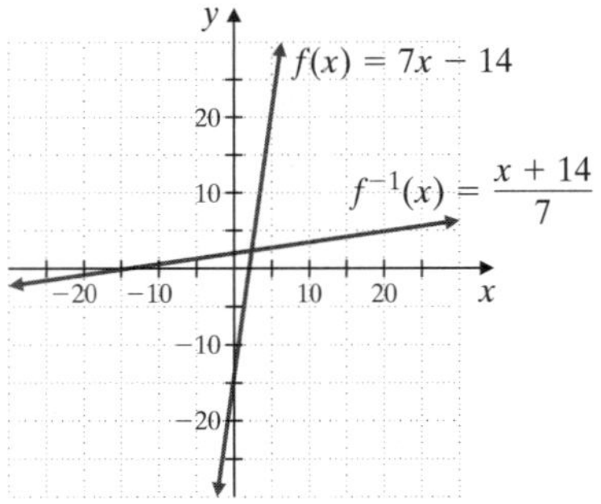

7. The graph represents a one-to-one function since it passes the horizontal and vertical line tests.

8. The graph does not represent a one-to-one function since it does not pass either the vertical or the horizontal line test.

9. $y = 6 - 2x$
$f(x) = -2x + 6$ so the function is one-to-one.
$y = -2x + 6$
$x = -2y + 6$
$x - 6 = -2y$
$\frac{x-6}{-2} = y$
$f^{-1}(x) = \frac{x-6}{-2}$ or $f^{-1}(x) = \frac{-x+6}{2}$

10. $f = \{(0, 0), (2, 3), (-1, 5)\}$
The function f is one-to-one since each y-value corresponds to only one x-value.
$f^{-1} = \{(0, 0), (3, 2), (5, -1)\}$

11. The table does not describe a one-to-one function since the output d corresponds to two different inputs, Dog and Desk. A similar argument could be made for output c.

12. $\log_3 6 + \log_3 4 = \log_3 6 \cdot 4 = \log_3 24$

13. $\log_5 x + 3\log_5 x - \log_5(x+1)$
$= 4\log_5 x - \log_5(x+1)$
$= \log_5 x^4 - \log_5(x+1)$
$= \log_5 \frac{x^4}{x+1}$

14. $\log_6 \frac{2x}{y^3} = \log_6 2x - \log_6 y^3$
$= \log_6 2 + \log_6 x - 3\log_6 y$

15. $\log_b \frac{3}{25} = \log_b 3 - \log_b 25$
$= \log_b 3 - \log_b 5^2$
$= \log_b 3 - 2\log_b 5$
$= 0.79 - 2(1.16)$
$= -1.53$

16. $\log_7 8 = \frac{\log 8}{\log 7} \approx \frac{0.90309}{0.84510} \approx 1.0686$

17. $8^{x-1} = \frac{1}{64}$
$8^{x-1} = 8^{-2}$
$x - 1 = -2$
$x = -1$
The solution set is $\{-1\}$.

18.
$$3^{2x+5} = 4$$
$$\log 3^{2x+5} = \log 4$$
$$(2x+5)\log 3 = \log 4$$
$$2x+5 = \frac{\log 4}{\log 3}$$
$$2x = \frac{\log 4}{\log 3} - 5$$
$$x = \frac{1}{2}\left(\frac{\log 4}{\log 3} - 5\right) \approx -1.8691$$

The solution set is $\left\{\frac{1}{2}\left(\frac{\log 4}{\log 3} - 5\right)\right\}$ or approximately $\{-1.8691\}$.

19. $\log_3 x = -2$
$$x = 3^{-2}$$
$$x = \frac{1}{9}$$

The solution set is $\left\{\frac{1}{9}\right\}$.

20. $\ln \sqrt{e} = x$
$$\ln e^{1/2} = x$$
$$\frac{1}{2} = x$$

The solution set is $\left\{\frac{1}{2}\right\}$.

21. $\log_8(3x-2) = 2$
$$3x - 2 = 8^2$$
$$3x - 2 = 64$$
$$3x = 66$$
$$x = \frac{66}{3}$$
$$x = 22$$

The solution set is $\{22\}$.

22. $\log_5 x + \log_5 3 = 2$
$$\log_5 3x = 2$$
$$3x = 5^2$$
$$3x = 25$$
$$x = \frac{25}{3}$$

The solution set is $\left\{\frac{25}{3}\right\}$.

23. $\log_4(x+1) - \log_4(x-2) = 3$
$$\log_4 \frac{x+1}{x-2} = 3$$
$$\frac{x+1}{x-2} = 4^3$$
$$\frac{x+1}{x-2} = 64$$
$$x+1 = 64(x-2)$$
$$x+1 = 64x - 128$$
$$129 = 63x$$
$$\frac{129}{63} = x$$
$$\frac{43}{21} = x$$

The solution set is $\left\{\frac{43}{21}\right\}$.

24. $\ln(3x+7) = 1.31$
$$3x + 7 = e^{1.31}$$
$$3x = e^{1.31} - 7$$
$$x = \frac{e^{1.31} - 7}{3}$$
$$x \approx -1.0979$$

The solution set is $\{-1.0979\}$.

25. $f(x) = \left(\frac{1}{2}\right)^x + 1$

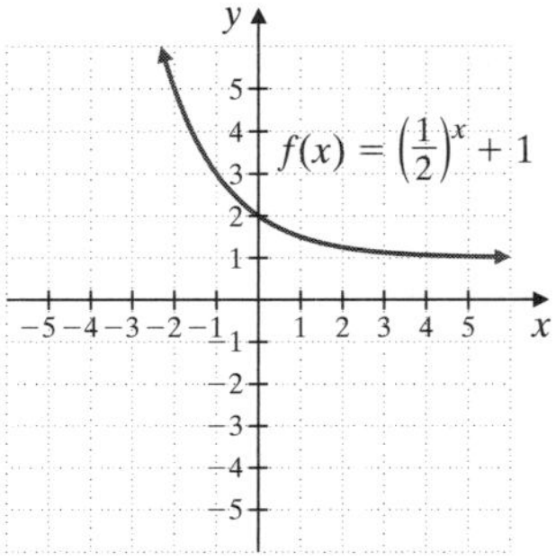

26. $y = 3^x$, $y = \log_3 x$

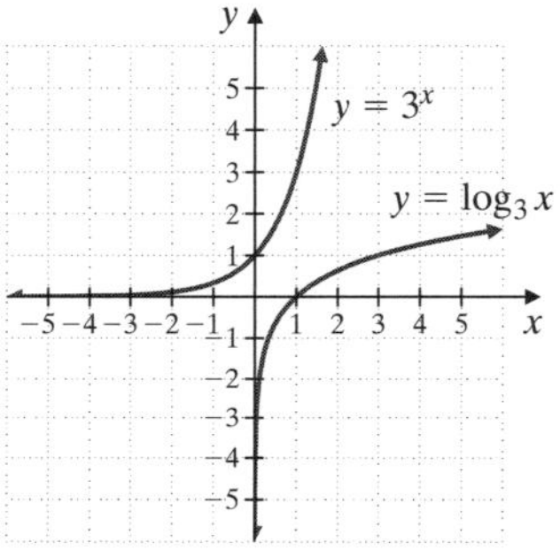

27. Let $P = 4000$, $r = 9\% = 0.09$, $n = 12$, and $t = 3$.

$$A = P\left(1 + \frac{r}{n}\right)^{nt}$$

$$A = 4000\left(1 + \frac{0.09}{12}\right)^{12(3)}$$

$$A \approx 5234.58$$

The amount will be \$5234.58 in 3 years.

28. Let $P = 2000$, $A = 3000$, $r = 7\% = 0.07$, and $n = 2$.

$$A = P\left(1 + \frac{r}{n}\right)^{nt}$$

$$3000 = 2000\left(1 + \frac{0.07}{2}\right)^{2t}$$

$$\frac{3}{2} = (1.035)^{2t}$$

$$\log\frac{3}{2} = \log(1.035)^{2t}$$

$$\log\frac{3}{2} = 2t \log 1.035$$

$$\frac{\log\frac{3}{2}}{2\log 1.035} = t$$

$$6 \approx t$$

It will take about 6 years for the money to grow to \$3000.

29. $y = y_0 e^{kt}$, $y_0 = 57,000$, $k = 2.6\% = 0.026$, and $t = 5$.

$y = 57,000e^{0.026(5)} = 57,000e^{0.13} \approx 64,913$

There will be approximately 64,913 prairie dogs 5 years from now.

30. $y = y_0 e^{kt}$, $y_0 = 400$, $y = 1000$, and $k = 6.2\% = 0.062$.

$$1000 = 400e^{0.062t}$$

$$\frac{5}{2} = e^{0.062t}$$

$$\ln\frac{5}{2} = 0.062t$$

$$\frac{\ln\frac{5}{2}}{0.062} = t$$

$$15 \approx t$$

It will take about 15 years to reach their goal.

31. Let $t = 0.5$.

$R(t) = 2.7^{-(1/3)t}$

$R(0.5) = 2.7^{-(1/3)(0.5)} \approx 0.85$

The probability is about 85%.

32. Let $t = 2$.

$R(t) = 2.7^{-(1/3)(2)} \approx 0.52$

The probability is about 52%.

Cumulative Review Chapters 1–9

1. $\sqrt[3]{27} = 3$, since $3^3 = 27$.

2. $\sqrt[3]{64x^7y^2} = \sqrt[3]{64x^6 \cdot xy^2} = 4x^2\sqrt[3]{xy^2}$

3. $\sqrt[4]{16} = 2$, since 2 is positive and $2^4 = 16$.

4. $\sqrt[4]{162a^4b^5} = \sqrt[4]{81a^4b^4 \cdot 2b} = 3ab\sqrt[4]{2b}$

5. $\begin{cases} x+y+z=180 \\ y=x+80 \\ z=x+10 \end{cases}$

Substitute $y = x + 80$ and $z = x + 10$ in the first equation.

$$\begin{aligned} x+y+z &= 180 \\ x+x+80+x+10 &= 180 \\ 3x+90 &= 180 \\ 3x &= 90 \\ x &= 30 \end{aligned}$$

Then $y = x + 80 = 30 + 80 = 110$, and $z = x + 10 = 30 + 10 = 40$. The ordered triple solution is (30, 110, 40). The angles measure 30°, 110°, and 40°.

6. $\begin{cases} x+y=180 \\ y=x-40 \end{cases}$

Use substitution.

$$\begin{aligned} x+y &= 180 \\ x+x-40 &= 180 \\ 2x-40 &= 180 \\ 2x &= 220 \\ x &= 110 \end{aligned}$$

Then $y = x - 40 = 110 - 40 = 70$.

7. $7x(x^2+5y)-(x^2+5y)$

$= 7x(x^2+5y)-1(x^2+5y)$

$= (x^2+5y)(7x-1)$

8. $$\frac{1}{3}(x-2)=\frac{1}{4}(x+1)$$
$$12\left[\frac{1}{3}(x-2)\right]=12\left[\frac{1}{4}(x+1)\right]$$
$$4(x-2)=3(x+1)$$
$$4x-8=3x+3$$
$$4x-3x=3+8$$
$$x=11$$
The solution set is {11}.

9. $$\frac{5k}{k^2-4}-\frac{2}{k^2+k-2}=\frac{5k}{(k+2)(k-2)}-\frac{2}{(k+2)(k-1)}$$
The LCD is $(k+2)(k-2)(k-1)$.
$$\frac{5k}{(k+2)(k-2)}-\frac{2}{(k+2)(k-1)}=\frac{5k\cdot(k-1)}{(k+2)(k-2)\cdot(k-1)}-\frac{2\cdot(k-2)}{(k+2)(k-1)\cdot(k-2)}$$
$$=\frac{5k^2-5k}{(k+2)(k-2)(k-1)}-\frac{2k-4}{(k+2)(k-2)(k-1)}$$
$$=\frac{5k^2-5k-2k+4}{(k+2)(k-2)(k-1)}$$
$$=\frac{5k^2-7k+4}{(k+2)(k-2)(k-1)}$$

10. $$\frac{5}{x-2}+\frac{3}{x^2+4x+4}-\frac{6}{x+2}=\frac{5}{x-2}+\frac{3}{(x+2)^2}-\frac{6}{x+2}$$
The LCD is $(x-2)(x+2)^2$.
$$\frac{5}{x-2}+\frac{3}{(x+2)^2}-\frac{6}{x+2}=\frac{5\cdot(x+2)^2}{(x-2)\cdot(x+2)^2}+\frac{3\cdot(x-2)}{(x+2)^2\cdot(x-2)}-\frac{6\cdot(x-2)(x+2)}{(x+2)\cdot(x-2)(x+2)}$$
$$=\frac{5(x^2+4x+4)+3(x-2)-6(x^2-4)}{(x-2)(x+2)^2}$$
$$=\frac{5x^2+20x+20+3x-6-6x^2+24}{(x-2)(x+2)^2}$$
$$=\frac{-x^2+23x+38}{(x-2)(x+2)^2}$$

11. $f(8)=\sqrt{8-4}=\sqrt{4}=2$

12. $f(28)=\sqrt{28-4}=\sqrt{24}=\sqrt{4\cdot 6}=2\sqrt{6}$

13. $g(-1)=\sqrt[3]{-1+2}=\sqrt[3]{1}=1$

14. $g(0)=\sqrt[3]{0+2}=\sqrt[3]{2}$

15. $x^{1/2}x^{1/3}=x^{\frac{1}{2}+\frac{1}{3}}=x^{\frac{3}{6}+\frac{2}{6}}=x^{5/6}$

16.
$$\begin{aligned}(a^{-2}b^{3}c^{-4})^{-2}&=a^{(-2)\cdot(-2)}b^{3(-2)}c^{(-4)(-2)}\\&=a^{4}b^{-6}c^{8}\\&=\frac{a^{4}c^{8}}{b^{6}}\end{aligned}$$

17.
$$\begin{aligned}\frac{(2x^{2/5})^{5}}{x^{2}}&=\frac{2^{5}(x^{2/5})^{5}}{x^{2}}\\&=\frac{32x^{2}}{x^{2}}\\&=32x^{2-2}\\&=32x^{0}\\&=32\cdot 1\\&=32\end{aligned}$$

18. $\left(\dfrac{3^{-2}}{x}\right)^{-3}=\dfrac{3^{(-2)(-3)}}{x^{-3}}=\dfrac{3^{6}}{x^{-3}}=729x^{3}$

19. $\sqrt{\dfrac{x}{9}}=\dfrac{\sqrt{x}}{\sqrt{9}}=\dfrac{\sqrt{x}}{3}$

20. $\dfrac{\sqrt{108a^{2}}}{3\sqrt{3}}=\dfrac{\sqrt{36a^{2}\cdot 3}}{3\sqrt{3}}=\dfrac{6a\sqrt{3}}{3\sqrt{3}}=2a$

21. $\sqrt[4]{\dfrac{3}{16y^{4}}}=\dfrac{\sqrt[4]{3}}{\sqrt[4]{16y^{4}}}=\dfrac{\sqrt[4]{3}}{2y}$

22.
$$\begin{aligned}\frac{3\sqrt[3]{81a^{5}b^{10}}}{\sqrt[3]{3b^{4}}}&=\frac{3\sqrt[3]{27a^{3}b^{9}\cdot 3a^{2}b}}{\sqrt[3]{b^{3}\cdot 3b}}\\&=\frac{3\cdot 3ab^{3}\sqrt[3]{3a^{2}b}}{b\sqrt[3]{3b}}\\&=\frac{9ab^{3}\sqrt[3]{3b}\cdot\sqrt[3]{a^{2}}}{b\cdot\sqrt[3]{3b}}\\&=9ab^{2}\sqrt[3]{a^{2}}\end{aligned}$$

23.
$$\begin{aligned}\left(\sqrt{2x}+5\right)\left(\sqrt{2x}-5\right)&=\left(\sqrt{2x}\right)^{2}-5^{2}\\&=2x-25\end{aligned}$$

24. $\left(\dfrac{1}{2}x+3\right)\left(\dfrac{1}{2}x-3\right)=\left(\dfrac{1}{2}x\right)^{2}-3^{2}=\dfrac{1}{4}x^{2}-9$

25.
$$\begin{aligned}\left(\sqrt{3}-1\right)^{2}&=\left(\sqrt{3}\right)^{2}-2\cdot\sqrt{3}\cdot 1+1^{2}\\&=3-2\sqrt{3}+1\\&=4-2\sqrt{3}\end{aligned}$$

26. a.
$$\begin{aligned}a^{1/4}(a^{3/4}-a^{8})&=a^{1/4}\cdot a^{3/4}-a^{1/4}\cdot a^{8}\\&=a^{\frac{1}{4}+\frac{3}{4}}-a^{\frac{1}{4}+8}\\&=a^{\frac{4}{4}}-a^{\frac{1}{4}+\frac{32}{4}}\\&=a-a^{\frac{33}{4}}\end{aligned}$$

b.
$$\begin{aligned}&(x^{1/2}-3)(x^{1/2}+5)\\&=(x^{1/2})^{2}+5x^{1/2}-3x^{1/2}-15\\&=x+2x^{1/2}-15\end{aligned}$$

27.
$$\begin{aligned}\frac{\sqrt{7}}{\sqrt{45}}&=\frac{\sqrt{7}}{\sqrt{9}\cdot\sqrt{5}}\\&=\frac{\sqrt{7}}{3\sqrt{5}}\\&=\frac{\sqrt{7}\cdot\sqrt{7}}{3\sqrt{5}\cdot\sqrt{7}}\\&=\frac{7}{3\sqrt{5\cdot 7}}\\&=\frac{7}{3\sqrt{35}}\end{aligned}$$

28. $\sqrt[3]{\frac{27}{m^4n^8}} = \frac{\sqrt[3]{27}}{\sqrt[3]{m^4n^8}}$

$= \frac{3}{\sqrt[3]{m^3n^6} \cdot \sqrt[3]{mn^2}}$

$= \frac{3}{mn^2\sqrt[3]{mn^2}}$

$= \frac{3 \cdot \sqrt[3]{m^2n}}{mn^2\sqrt[3]{mn^2} \cdot \sqrt[3]{m^2n}}$

$= \frac{3\sqrt[3]{m^2n}}{m^2n^3}$

29. $\sqrt{4-x} = x-2$

$\left(\sqrt{4-x}\right)^2 = (x-2)^2$

$4-x = x^2-4x+4$

$x^2-3x = 0$

$x(x-3) = 0$

$x = 0$ or $x-3 = 0$

$x = 3$

The proposed solution 3 checks, but 0 does not. The solution set is {3}.

30. $f(x) = -3x + 4$

The slope of this line is –3 ($y = mx + b$). The slope of a line perpendicular to it is $\frac{1}{3}$.

Using $m = \frac{1}{3}$ and $(x_1, y_1) = (-2, 6)$, write the equation of the line.

$y - y_1 = m(x - x_1)$

$y - 6 = \frac{1}{3}(x+2)$

$y = \frac{1}{3}(x+2)+6$

$3y = (x+2)+18$

$3y = x+20$

$y = \frac{x+20}{3}$

$f(x) = \frac{x+20}{3}$

31. $(2+3i)+(-3+2i) = (2-3)+(3+2)i$

$= -1+5i$

32. a. $i^8 = (i^4)^2 = 1^2 = 1$

b. $i^{21} = i^{20} \cdot i^1 = (i^4)^5 \cdot i^1 = 1^5 \cdot i = i$

c. $i^{42} = i^{40} \cdot i^2$

$= (i^4)^{10} \cdot i^2$

$= 1^{10} \cdot (-1)$

$= -1$

d. $i^{-13} = \frac{1}{i^{13}}$

$= \frac{1}{i^{12} \cdot i}$

$= \frac{1}{(i^4)^3 \cdot i}$

$= \frac{1}{1^3 \cdot i}$

$= \frac{1}{i}$

$= \frac{1}{i} \cdot \frac{i}{i}$

$= \frac{i}{i^2}$

$= \frac{i}{-1}$

$= -i$

33.
$$4x^2 - 24x + 41 = 0$$
$$x^2 - 6x + \frac{41}{4} = 0$$
$$x^2 - 6x = -\frac{41}{4}$$
$$x^2 - 6x + \left(\frac{-6}{2}\right)^2 = -\frac{41}{4} + \left(\frac{-6}{2}\right)^2$$
$$x^2 - 6x + 9 = -\frac{41}{4} + \frac{36}{4}$$
$$(x-3)^2 = -\frac{5}{4}$$
$$x - 3 = \pm\sqrt{-\frac{5}{4}}$$
$$x - 3 = \pm\frac{i\sqrt{5}}{2}$$
$$x = 3 \pm \frac{i\sqrt{5}}{2}$$
$$x = \frac{6}{2} \pm \frac{i\sqrt{5}}{2}$$
$$x = \frac{6 \pm i\sqrt{5}}{2}$$

The solution set is $\left\{\frac{6-i\sqrt{5}}{2}, \frac{6+i\sqrt{5}}{2}\right\}$.

34.
$$4x^2+8x-1=0$$
$$x^2+2x-\frac{1}{4}=0$$
$$x^2+2x=\frac{1}{4}$$
$$x^2+2x+\left(\frac{2}{2}\right)^2=\frac{1}{4}+\left(\frac{2}{2}\right)^2$$
$$x^2+2x+1=\frac{1}{4}+\frac{4}{4}$$
$$(x+1)^2=\frac{5}{4}$$
$$x+1=\pm\sqrt{\frac{5}{4}}$$
$$x+1=\pm\frac{\sqrt{5}}{2}$$
$$x=-1\pm\frac{\sqrt{5}}{2}$$
$$x=\frac{-2\pm\sqrt{5}}{2}$$
The solution set is $\left\{\frac{-2-\sqrt{5}}{2}, \frac{-2+\sqrt{5}}{2}\right\}$.

35.
$$\frac{1}{4}m^2-m+\frac{1}{2}=0$$
$$4\left(\frac{1}{4}m^2-m+\frac{1}{2}\right)=4\cdot 0$$
$$m^2-4m+2=0$$
$a=1$, $b=-4$, $c=2$
$$m=\frac{-b\pm\sqrt{b^2-4ac}}{2a}$$
$$m=\frac{-(-4)\pm\sqrt{(-4)^2-4(1)(2)}}{2(1)}$$
$$m=\frac{4\pm\sqrt{16-8}}{2}=\frac{4\pm\sqrt{8}}{2}=\frac{4\pm2\sqrt{2}}{2}=2\pm\sqrt{2}$$
The solution set is $\{2-\sqrt{2}, 2+\sqrt{2}\}$.

36. $\left(x-\frac{1}{2}\right)^2 = \frac{x}{2}$

$x^2 - x + \frac{1}{4} = \frac{x}{2}$

$4x^2 - 4x + 1 = 2x$

$4x^2 - 6x + 1 = 0$

$a = 4, b = -6, c = 1$

$$x = \frac{-b \pm \sqrt{b^2 - 4ac}}{2a}$$

$$= \frac{-(-6) \pm \sqrt{(-6)^2 - 4(4)(1)}}{2(4)}$$

$$= \frac{6 \pm \sqrt{36-16}}{8}$$

$$= \frac{6 \pm \sqrt{20}}{8}$$

$$= \frac{6 \pm 2\sqrt{5}}{8}$$

$$= \frac{3 \pm \sqrt{5}}{4}$$

The solution set is $\left\{\frac{3-\sqrt{5}}{4}, \frac{3+\sqrt{5}}{4}\right\}$.

37. Let $m = x^{1/3}$.

$x^{2/3} - 5x^{1/3} + 6 = 0$

$(x^{1/3})^2 - 5x^{1/3} + 6 = 0$

$m^2 - 5m + 6 = 0$

$(m-3)(m-2) = 0$

$m - 3 = 0$ or $m - 2 = 0$

$m = 3$ $\quad$ $m = 2$

Replace m with $x^{1/3}$.

$x^{1/3} = 3$ $\quad$ $x^{1/3} = 2$

$x = 3^3 = 27$ $\quad$ $x = 2^3 = 8$

The solution set is $\{8, 27\}$.

38. $\frac{1}{a+5} = \frac{1}{3a+6} - \frac{a+2}{a^2+7a+10}$

$\frac{1}{a+5} = \frac{1}{3(a+2)} - \frac{a+2}{(a+2)(a+5)}$

The LCD is $3(a + 2)(a + 5)$.

$$3(a+2)(a+5)\cdot\frac{1}{a+5}=3(a+2)(a+5)\cdot\frac{1}{3(a+2)}-3(a+2)(a+5)\cdot\frac{a+2}{(a+2)(a+5)}$$
$$3(a+2)=(a+5)-3(a+2)$$
$$3a+6=a+5-3a-6$$
$$5a=-7$$
$$a=-\frac{7}{5}$$

The solution set is $\left\{-\frac{7}{5}\right\}$.

39. $\frac{5}{x+1}<-2$

$x+1=0$

$x=-1$ makes the denominator 0.

Now solve $\frac{5}{x+1}=-2$.

$$(x+1)\cdot\frac{5}{x+1}=(x+1)\cdot(-2)$$
$$5=-2x-2$$
$$7=-2x$$
$$-\frac{7}{2}=x$$

Region	Test Point Value	$\frac{5}{x+1}<-2$	Result
$\left(-\infty, -\frac{7}{2}\right)$	−4	$\frac{5}{-3}<-2$	False
$\left(-\frac{7}{2}, -1\right)$	−2	$\frac{5}{-1}<-2$	True
$(-1, \infty)$	0	$\frac{5}{1}<-2$	False

The solution set is $\left(-\frac{7}{2}, -1\right)$.

40. Let x = length of unknown side.

$$8^2=x^2+4^2$$
$$64=x^2+16$$
$$48=x^2$$
$$\pm\sqrt{48}=x$$
$$\pm4\sqrt{3}=x$$

Discard $-4\sqrt{3}$ since length cannot be negative. The length is $4\sqrt{3}$ inches.

41.
$$\begin{aligned} f(x) &= 3x^2+3x+1 \\ y &= 3x^2+3x+1 \\ y-1 &= 3x^2+3x \\ y-1 &= 3(x^2+x) \\ y-1+3\left(\frac{1}{4}\right) &= 3\left(x^2+x+\frac{1}{4}\right) \\ y-\frac{1}{4} &= 3\left(x+\frac{1}{2}\right)^2 \\ y &= 3\left(x+\frac{1}{2}\right)^2+\frac{1}{4} \end{aligned}$$

$$f(x) = 3\left(x+\frac{1}{2}\right)^2+\frac{1}{4}$$

Then $a = 3$, $h = -\frac{1}{2}$, and $k = \frac{1}{4}$. The parabola opens upward and the vertex is $\left(-\frac{1}{2}, \frac{1}{4}\right)$.

Let $x = 0$.

$f(0) = 3(0)^2+3(0)+1 = 1$

The y-intercept is (0, 1).

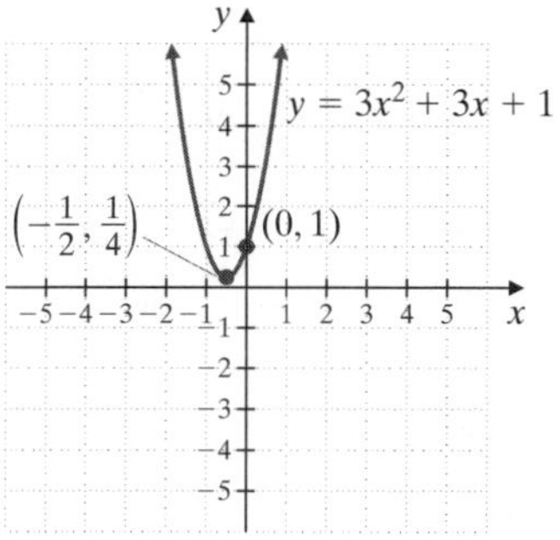

42. $\dfrac{x^3-8}{x-2} = \dfrac{(x-2)(x^2+2x+4)}{x-2} = x^2+2x+4$

43. a. $(f \circ g)(x) = f(g(x)) = f(x-2) = |x-2|$

b. $(g \circ f)(x) = g(f(x)) = g(|x|) = |x|-2$

44. a. $\dfrac{\frac{a}{5}}{\frac{a-1}{10}} = \dfrac{a}{5} \div \dfrac{a-1}{10} = \dfrac{a}{5} \cdot \dfrac{10}{a-1} = \dfrac{2a}{a-1}$

b.
$$\begin{aligned} &\frac{\frac{3}{2+a}+\frac{6}{2-a}}{\frac{5}{a+2}-\frac{1}{a-2}} \\ &= \frac{\frac{3(a-2)}{(a+2)(a-2)}+\frac{-6(a+2)}{(a+2)(a-2)}}{\frac{5(a-2)}{(a+2)(a-2)}-\frac{1(a+2)}{(a+2)(a-2)}} \\ &= \frac{\frac{3(a-2)-6(a+2)}{(a+2)(a-2)}}{\frac{5(a-2)-(a+2)}{(a+2)(a-2)}} \\ &= \frac{3(a-2)-6(a+2)}{(a+2)(a-2)} \cdot \frac{(a+2)(a-2)}{5(a-2)-(a+2)} \\ &= \frac{3(a-2)-6(a+2)}{5(a-2)-(a+2)} \\ &= \frac{3a-6-6a-12}{5a-10-a-2} \\ &= \frac{-3a-18}{4a-12} \\ &= \frac{-3(a+6)}{4(a-3)} \end{aligned}$$

c. $\dfrac{x^{-1}+y^{-1}}{xy} = \dfrac{\frac{1}{x}+\frac{1}{y}}{xy} = \dfrac{xy\left(\frac{1}{x}+\frac{1}{y}\right)}{xy(xy)} = \dfrac{y+x}{x^2y^2}$

45.
$$\begin{aligned} f(x) &= 3x-5 \\ y &= 3x-5 \\ x &= 3y-5 \\ x+5 &= 3y \\ \frac{x+5}{3} &= y \\ f^{-1}(x) &= \frac{x+5}{3} \end{aligned}$$

46.

2	8	−12	−7
		16	8
	8	4	1

$(8x^2-12x-7) \div (x-2) = 8x+4+\dfrac{1}{x-2}$

47.
$$\begin{aligned}4^{x+3} &= 8^x\\ (2^2)^{x+3} &= (2^3)^x\\ 2^{2x+6} &= 2^{3x}\\ 2x+6 &= 3x\\ 6 &= x\end{aligned}$$
The solution set is $\{6\}$.

48.
$$\begin{aligned}y &= kx\\ \frac{1}{2} &= k(12)\\ \frac{1}{24} &= k\\ y &= \frac{1}{24}x\end{aligned}$$

49.
$$\begin{aligned}\log_x 25 &= 2\\ 25 &= x^2\\ 5^2 &= x^2\\ 5 &= x\end{aligned}$$
The solution set is $\{5\}$.

50. a.
$$\begin{aligned}\frac{\sqrt{20}}{3}+\frac{\sqrt{5}}{4} &= \frac{2\sqrt{5}}{3}+\frac{\sqrt{5}}{4}\\ &= \frac{4\cdot 2\sqrt{5}}{4\cdot 3}+\frac{3\cdot\sqrt{5}}{3\cdot 4}\\ &= \frac{8\sqrt{5}}{12}+\frac{3\sqrt{5}}{12}\\ &= \frac{8\sqrt{5}+3\sqrt{5}}{12}\\ &= \frac{11\sqrt{5}}{12}\end{aligned}$$

b.
$$\begin{aligned}\sqrt[3]{\frac{24x}{27}}-\frac{\sqrt[3]{3x}}{2} &= \frac{\sqrt[3]{24x}}{\sqrt[3]{27}}-\frac{\sqrt[3]{3x}}{2}\\ &= \frac{\sqrt[3]{8\cdot 3x}}{\sqrt[3]{27}}-\frac{\sqrt[3]{3x}}{2}\\ &= \frac{\sqrt[3]{2^3}\cdot\sqrt[3]{3x}}{\sqrt[3]{3^3}}-\frac{\sqrt[3]{3x}}{2}\\ &= \frac{2\sqrt[3]{3x}}{3}-\frac{\sqrt[3]{3x}}{2}\\ &= \frac{2\sqrt[3]{3x}\cdot 2}{3\cdot 2}-\frac{\sqrt[3]{3x}\cdot 3}{2\cdot 3}\\ &= \frac{4\sqrt[3]{3x}}{6}-\frac{3\sqrt[3]{3x}}{6}\\ &= \frac{4\sqrt[3]{3x}-3\sqrt[3]{3x}}{6}\\ &= \frac{\sqrt[3]{3x}}{6}\end{aligned}$$

51. Let $P = 1600$, $r = 9\% = 0.09$, and $t = 5$.
$$\begin{aligned}A &= Pe^{rt}\\ A &= 1600e^{0.09(5)}\\ &= 1600e^{0.45}\\ &\approx 2509.30\end{aligned}$$
The amount owed is \$2509.30.

52.
$$\begin{aligned}7\log_3 x+9\log_3 y &= \log_3 x^7+\log_3 y^9\\ &= \log_3 x^7y^9\end{aligned}$$

53.
$$\begin{aligned}\log(x+2)-\log x &= 2\\ \log\frac{x+2}{x} &= 2\\ \frac{x+2}{x} &= 10^2\\ \frac{x+2}{x} &= 100\\ x+2 &= 100x\\ 2 &= 99x\\ \frac{2}{99} &= x\end{aligned}$$
The solution set is $\left\{\frac{2}{99}\right\}$.

54. $\log_7(8x-6)=2$

$$\begin{aligned} 8x-6 &= 7^2 \\ 8x-6 &= 49 \\ 8x &= 55 \\ x &= \frac{55}{8} \end{aligned}$$

The solution set is $\left\{\frac{55}{8}\right\}$.

Chapter 10

Section 10.1 Practice Problems

1. $x = 4y^2$

$x = 4(y-0)^2 + 0$

$V(0, 0)$

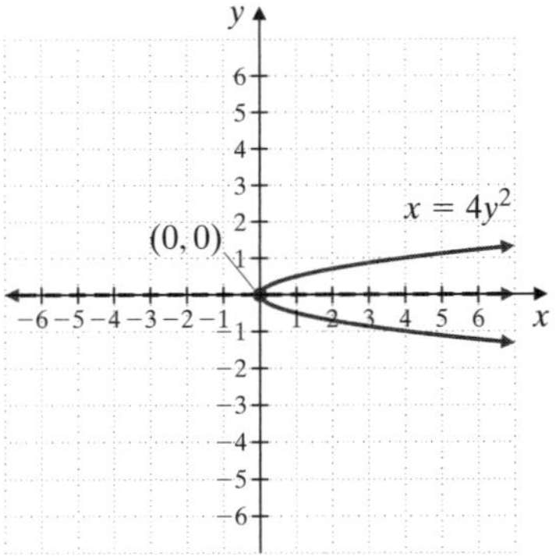

2. $x = -2(y-3)^2 + 1$

$V(1, 3)$

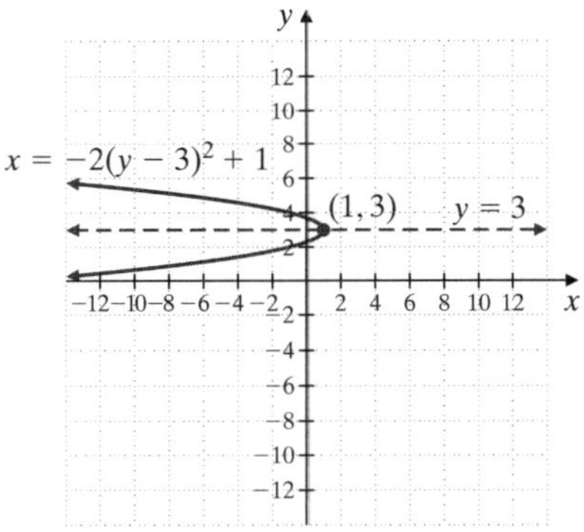

3.

$$\begin{aligned} y &= -x^2 - 4x + 12 \\ y - 12 &= -1(x^2 + 4x) \\ y - 12 - 4 &= -1(x^2 + 4x + 4) \\ y - 16 &= -1(x+2)^2 \\ y &= -1(x+2)^2 + 16 \end{aligned}$$

$V(-2, 16)$

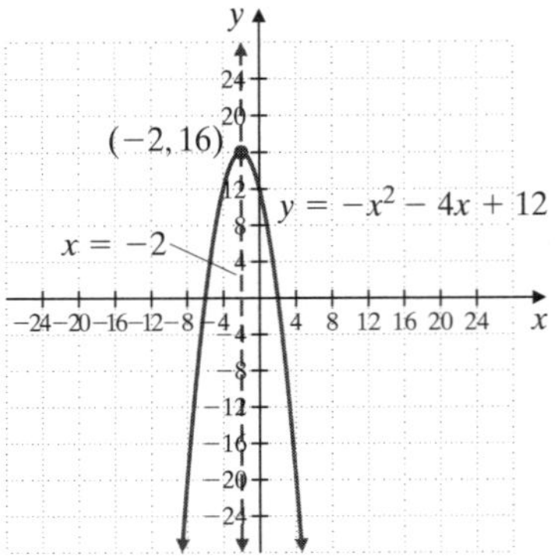

4. $x = 3y^2 + 12y + 13$

$$y = \frac{-b}{2a} = \frac{-12}{2(3)} = -2$$

$$x = 3(-2)^2 + 12(-2) + 13 = 1$$

$V(1, -2)$; axis of symmetry $y = -2$; opens to the right since $a > 0$.

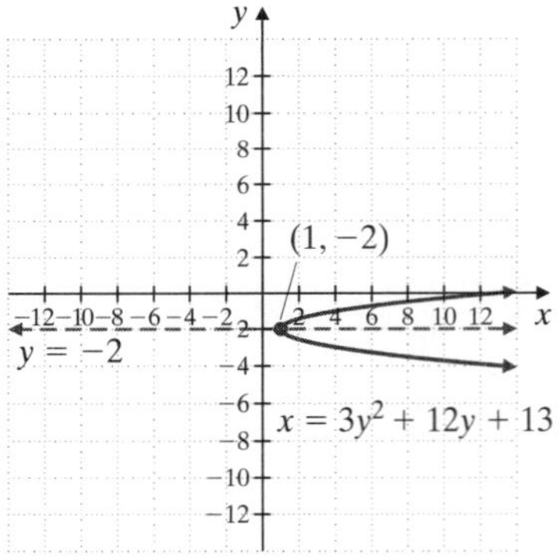

5. $$x^2 + y^2 = 36$$
$$(x-0)^2 + (y-0)^2 = 6^2$$
$C(0, 0)$ and $r = 6$

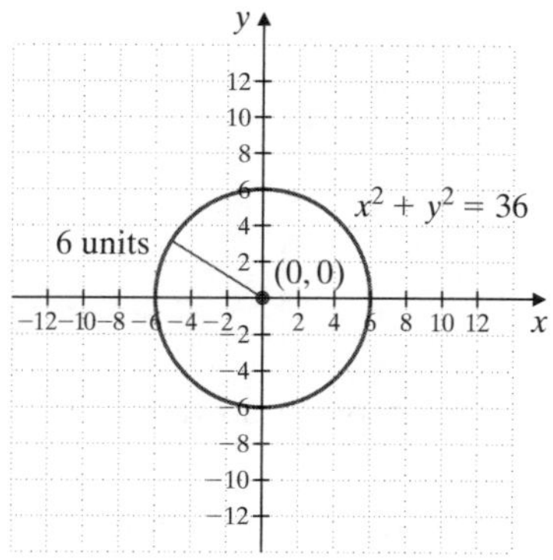

6. $$x^2 + (y+2)^2 = 6$$
$$(x-0)^2 + (y-(-2))^2 = \left(\sqrt{6}\right)^2$$
$C(0, -2)$ and $r = \sqrt{6}$

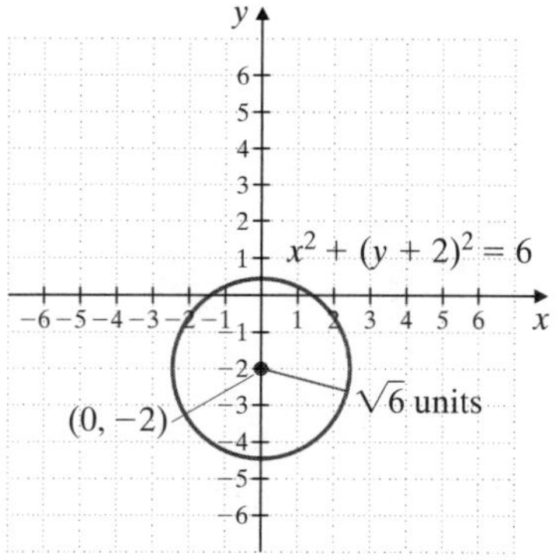

7. $C(2, -5)$, $r = 7$
$$(x-h)^2 + (y-k)^2 = r^2$$
$$(x-2)^2 + (y-(-5))^2 = 7^2$$
$$(x-2)^2 + (y+5)^2 = 49$$

8. $$x^2 + y^2 - 2x + 6y = 6$$
$$(x^2 - 2x) + (y^2 + 6y) = 6$$
$$(x^2 - 2x + 1) + (y^2 + 6y + 9) = 6 + 1 + 9$$
$$(x-1)^2 + (y+3)^2 = 16$$
$C(1, -3)$ and $r = 4$

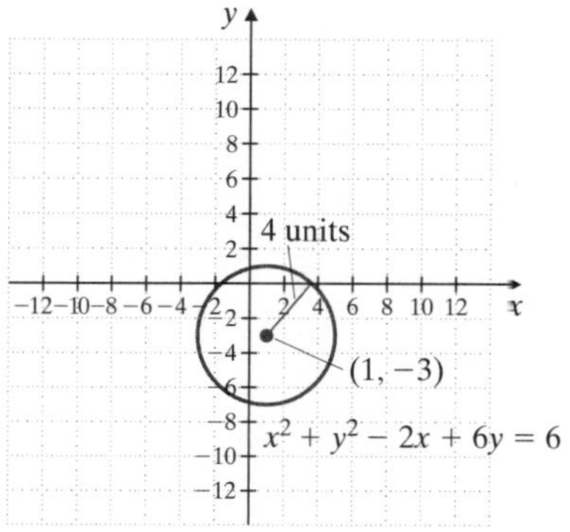

Calculator Explorations

1. $x^2 + y^2 = 55$
$$y^2 = 55 - x^2$$
$$y = \pm\sqrt{55 - x^2}$$
$$Y_1 = \sqrt{55 - x^2}$$
$$Y_2 = -\sqrt{55 - x^2}$$

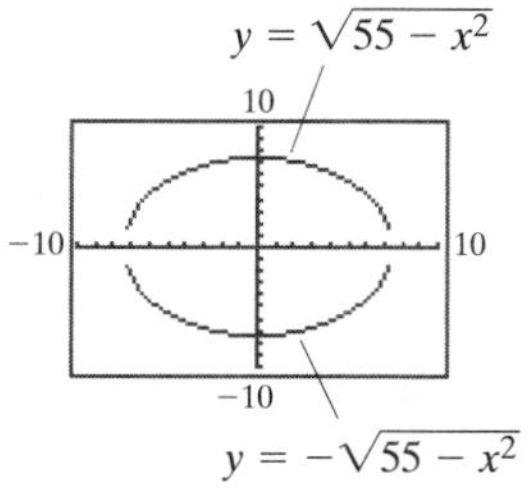

2. $x^2 + y^2 = 20$
$$y^2 = 20 - x^2$$
$$y = \pm\sqrt{20 - x^2}$$
$$Y_1 = \sqrt{20 - x^2}$$
$$Y_2 = -\sqrt{20 - x^2}$$

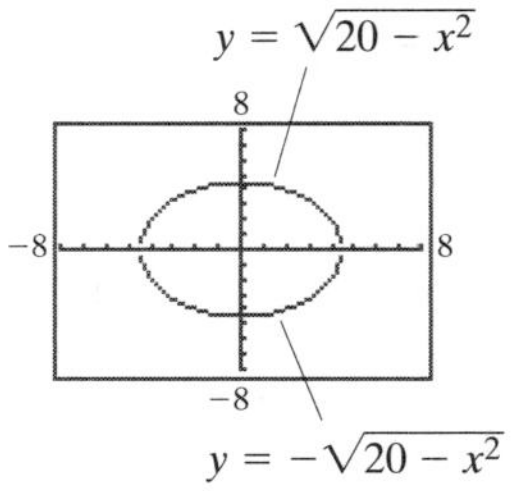

3. $7x^2 + 7y^2 - 89 = 0$

$$7x^2 + 7y^2 = 89$$

$$x^2 + y^2 = \frac{89}{7}$$

$$y^2 = \frac{89}{7} - x^2$$

$$y = \pm\sqrt{\frac{89}{7} - x^2}$$

$$Y_1 = \sqrt{\frac{89}{7} - x^2}$$

$$Y_2 = -\sqrt{\frac{89}{7} - x^2}$$

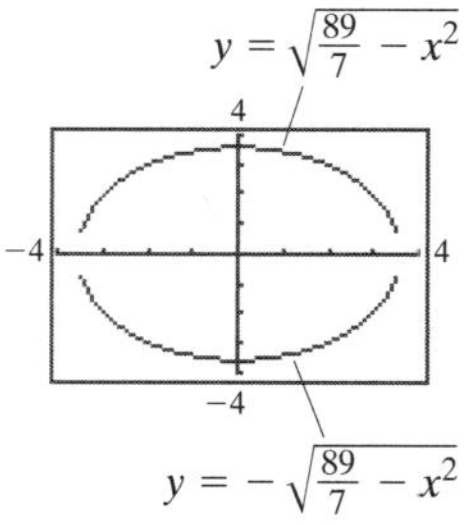

4. $3x^2 + 3y^2 - 35 = 0$

$$3x^2 + 3y^2 = 35$$

$$x^2 + y^2 = \frac{35}{3}$$

$$y^2 = \frac{35}{3} - x^2$$

$$y = \pm\sqrt{\frac{35}{3} - x^2}$$

$$Y_1 = \sqrt{\frac{35}{3} - x^2}$$

$$Y_2 = -\sqrt{\frac{35}{3} - x^2}$$

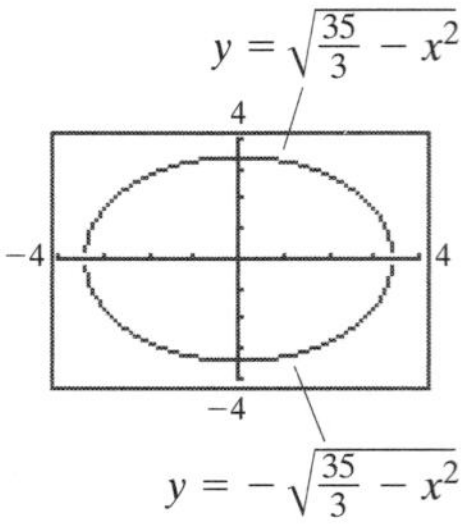

Mental Math

1. $y = x^2 - 7x + 5$ opens upward.

2. $y = -x^2 + 16$ opens downward.

3. $x = -y^2 - y + 2$ opens to the left.

4. $x = 3y^2 + 2y - 5$ opens to the right.

5. $y = -x^2 + 2x + 1$ opens downward.

6. $x = -y^2 + 2y - 6$ opens to the left.

Exercise Set 10.1

1. $x = 3y^2$

   $x = 3(y - 0)^2 + 0$

   $V(0, 0)$; opens to the right

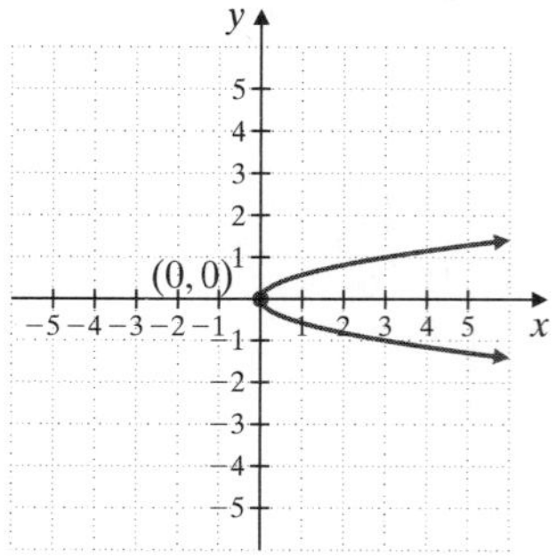

3. $x = -2y^2$

$x = -2(y-0)^2 + 0$

$V(0, 0)$; opens to the left

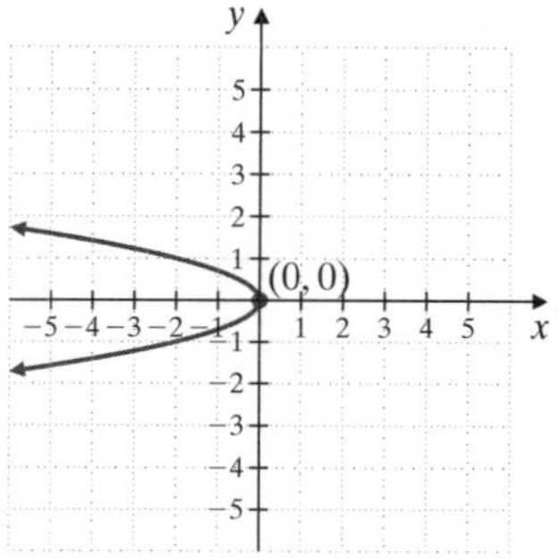

5. $y = -4x^2 = -4(x-0)^2 + 0$

$V(0, 0)$; opens downward

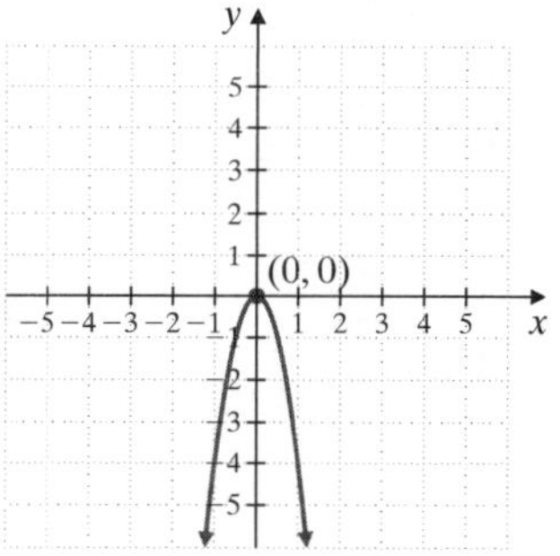

7. $x = (y-2)^2 + 3$

$V(3, 2)$; opens to the right

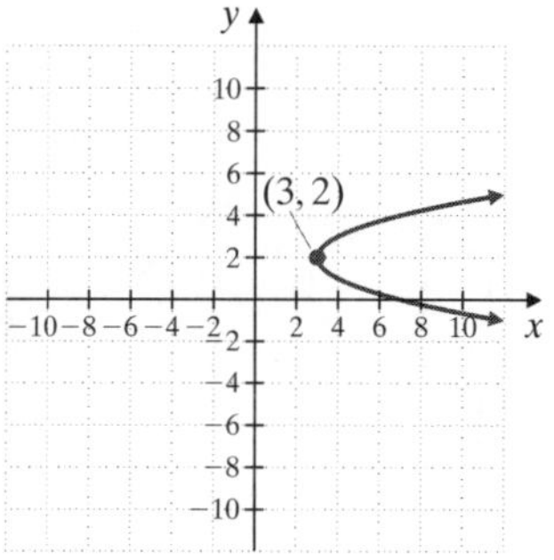

9. $y = -3(x-1)^2 + 5$

$V(1, 5)$; opens downward

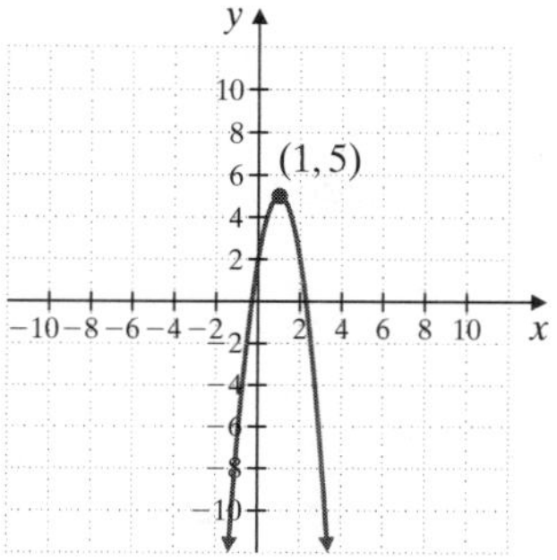

11. $x = y^2 + 6y + 8$

$x = y^2 + 6y + 9 + 8 - 9$

$x = (y+3)^2 - 1$

$V(-1, -3)$; opens to the right

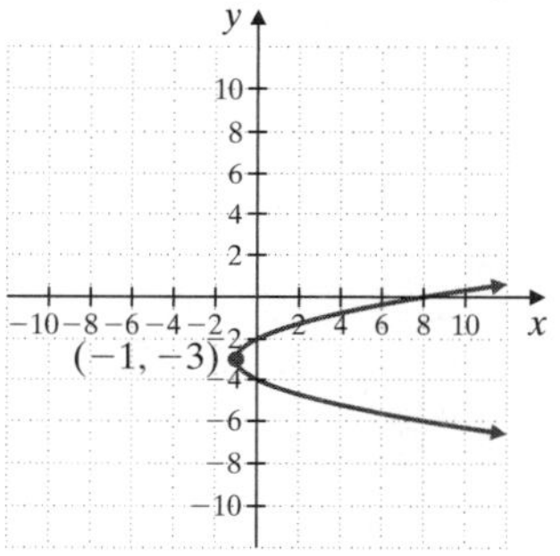

13. $y = x^2 + 10x + 20$

$y = x^2 + 10x + 25 + 20 - 25$

$y = (x+5)^2 - 5$

$V(-5, -5)$; opens upward

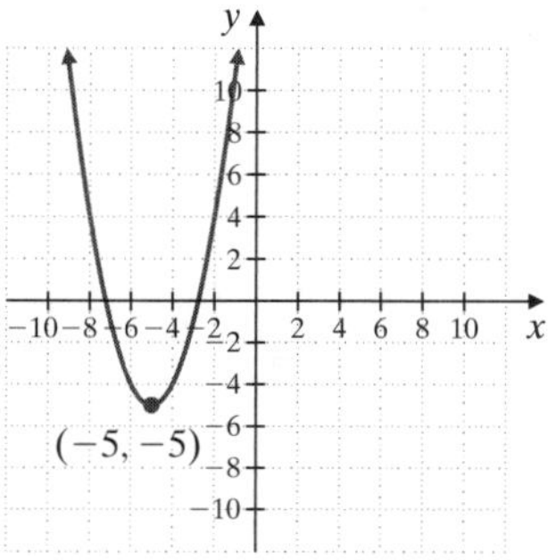

15. $x = -2y^2 + 4y + 6$

$x = -2(y^2 - 2y) + 6$

$x = -2(y^2 - 2y + 1) + 6 + 2$

$x = -2(y-1)^2 + 8$

$V(8, 1)$; opens to the left

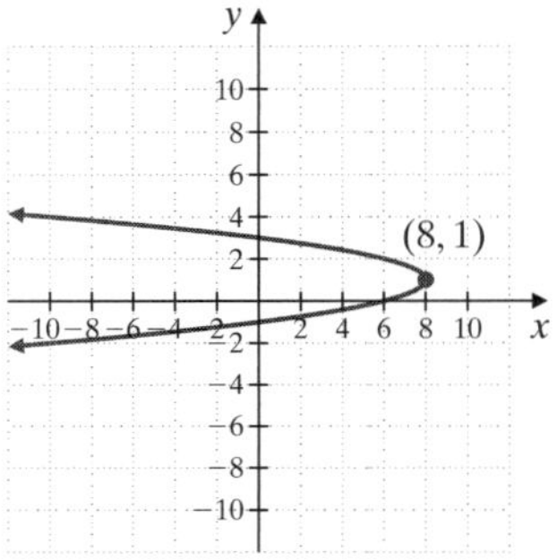

17. $x^2 + y^2 = 9$

$(x-0)^2 + (y-0)^2 = 3^2$

$C(0, 0)$ and $r = 3$

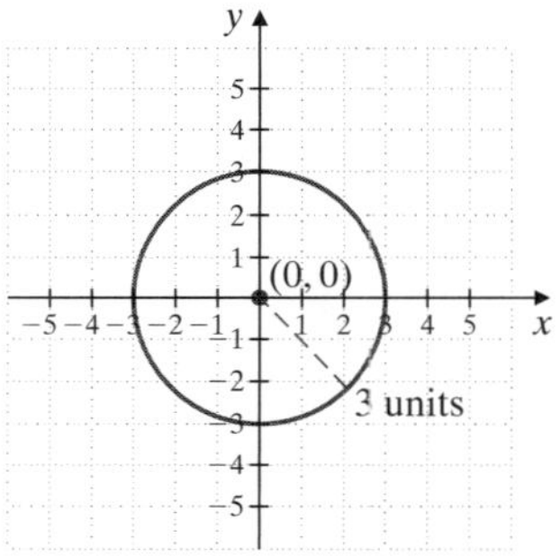

19. $x^2 + (y-2)^2 = 1$

$(x-0)^2 + (y-2)^2 = 1^2$

$C(0, 2)$ and $r = 1$

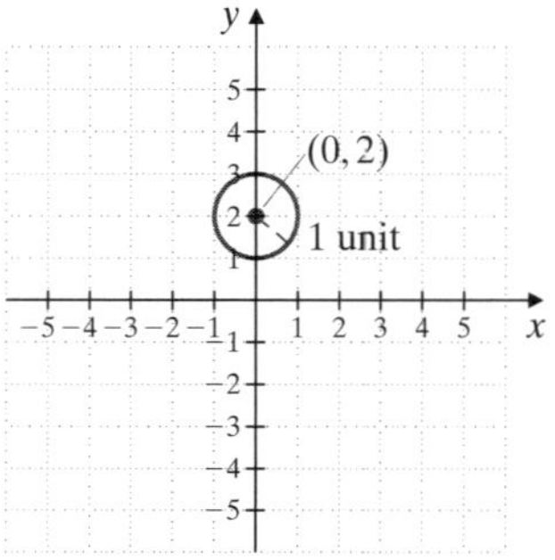

21. $(x-5)^2 + (y+2)^2 = 1$

$(x-5)^2 + (y+2)^2 = 1^2$

$C(5, -2)$ and $r = 1$

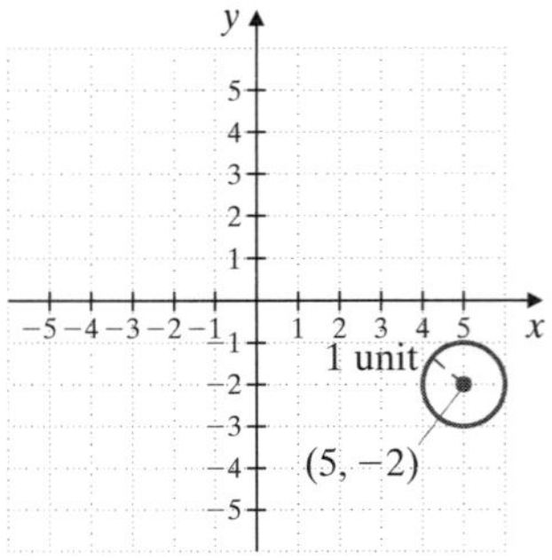

23. $x^2 + y^2 + 6y = 0$

$x^2 + y^2 + 6y + 9 = 0 + 9$

$(x-0)^2 + (y+3)^2 = 3^2$

$C(0, -3)$ and $r = 3$

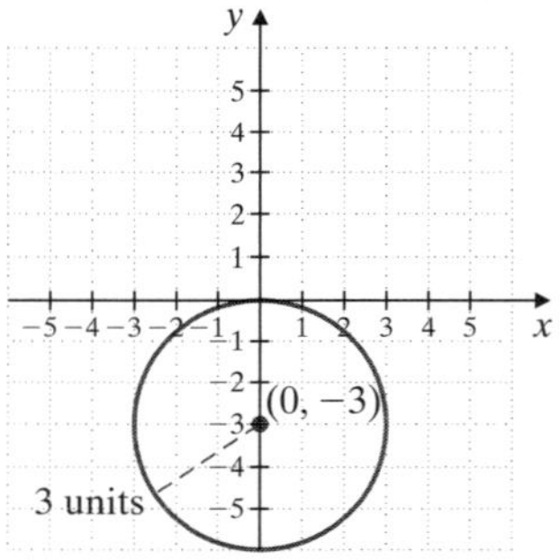

25. $x^2 + y^2 + 2x - 4y = 4$

$x^2 + 2x + 1 + y^2 - 4y + 4 = 4 + 1 + 4$

$(x+1)^2 + (y-2)^2 = 9$

$(x+1)^2 + (y-2)^2 = 3^2$

$C(-1, 2)$ and $r = 3$

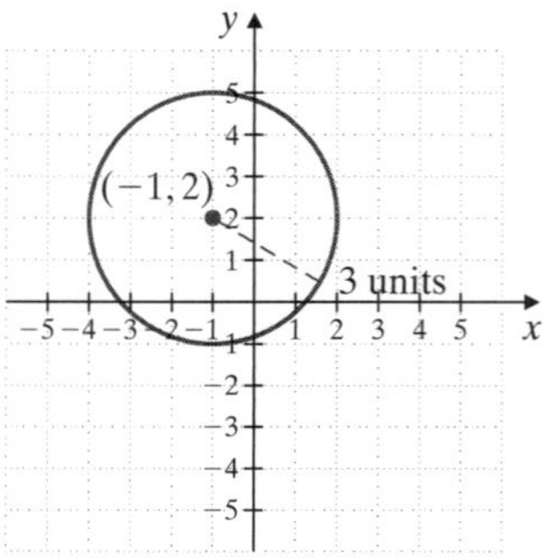

27.
$$x^2+y^2-4x-8y-2=0$$
$$x^2-4x+4+y^2-8y+16=2+4+16$$
$$(x-2)^2+(y-4)^2=22$$
$$(x-2)^2+(y-4)^2=\left(\sqrt{22}\right)^2$$

$C(2, 4)$ and $r=\sqrt{22}$

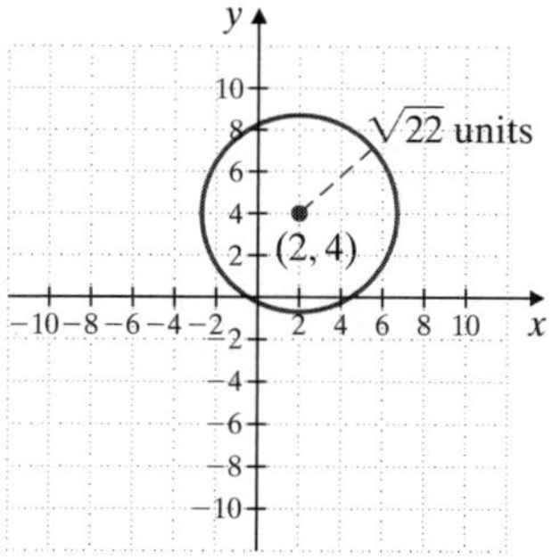

29.
$$3x^2+3y^2=75$$
$$x^2+y^2=25$$
$$(x-0)^2+(y-0)^2=5^2$$

$C(0, 0)$ and $r = 5$

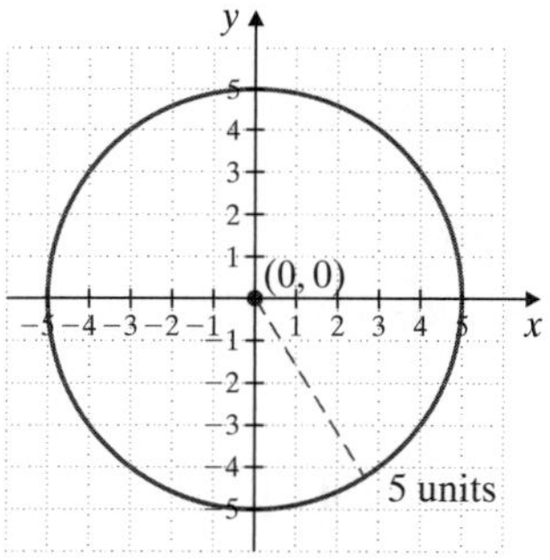

31.
$$4(x+1)^2+4(y-3)^2=12$$
$$(x+1)^2+(y-3)^2=3$$
$$(x+1)^2+(y-3)^2=\left(\sqrt{3}\right)^2$$

$C(-1, 3)$ and $r=\sqrt{3}$

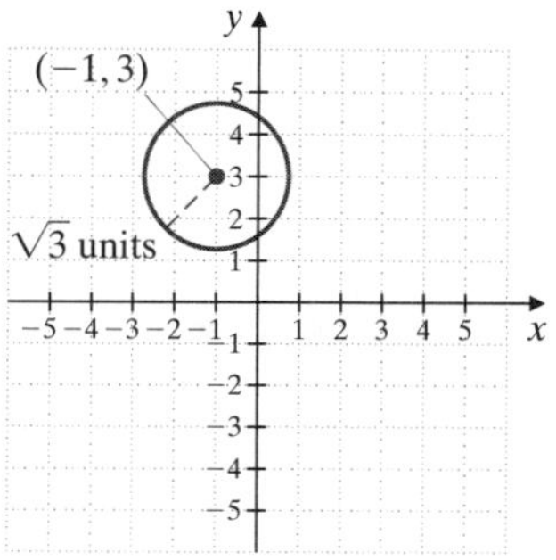

33. $C(2, 3)$; $r = 6$
$$(x-2)^2+(y-3)^2=6^2$$
$$(x-2)^2+(y-3)^2=36$$

35. $C(0, 0)$; $r=\sqrt{3}$
$$(x-0)^2+(y-0)^2=\left(\sqrt{3}\right)^2$$
$$x^2+y^2=3$$

37. $C(-5, 4)$; $r=3\sqrt{5}$
$$[x-(-5)]^2+(y-4)^2=\left(3\sqrt{5}\right)^2$$
$$(x+5)^2+(y-4)^2=45$$

39. $y = 2x + 5$

x-intercept: Let $y = 0$.
$$0=2x+5$$
$$-\frac{5}{2}=x$$
$$\left(-\frac{5}{2}, 0\right)$$

y-intercept: Let $x = 0$.

$y = 2(0) + 5$

$y = 5$

$(0, 5)$

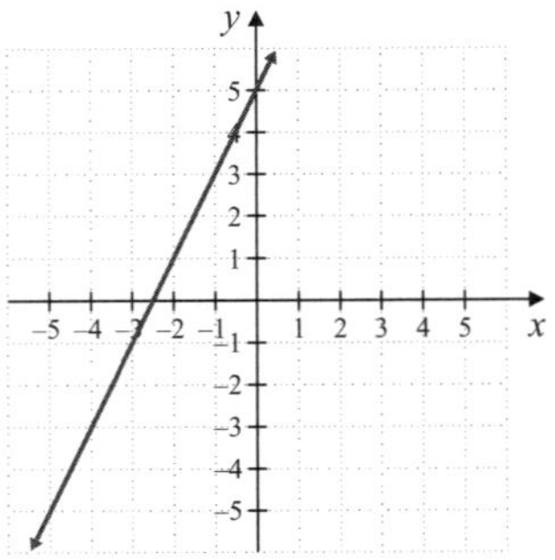

41. $y = 3$

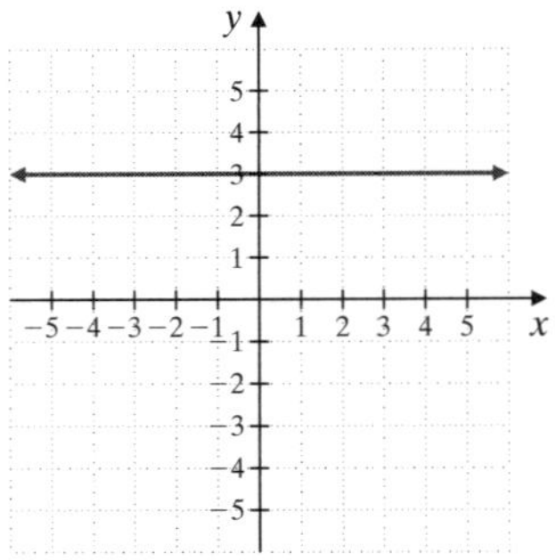

43. $\frac{1}{\sqrt{3}} = \frac{1}{\sqrt{3}} \cdot \frac{\sqrt{3}}{\sqrt{3}} = \frac{\sqrt{3}}{3}$

45. $\frac{4\sqrt{7}}{\sqrt{6}} = \frac{4\sqrt{7}}{\sqrt{6}} \cdot \frac{\sqrt{6}}{\sqrt{6}}$

$= \frac{4\sqrt{42}}{6}$

$= \frac{2\sqrt{42}}{3}$

47. a. The radius is one-half of the diameter.

$r = \frac{d}{2} = \frac{33}{2} = 16.5$ meters

b. $C = 2\pi r = 2\pi(16.5) \approx 103.67$ meters

c. $\frac{103.67}{30} \approx 3.5$ meters apart

d. From the drawing, the center is at (0, 16.5).

e. $(x-0)^2 + (y-16.5)^2 = (16.5)^2$

$x^2 + (y-16.5)^2 = (16.5)^2$

49. a. The radius was one-half of the diameter, or 125 feet.

b. $264 - 250 = 14$
The wheel is 14 feet above the ground.

c. $125 + 14 = 139$
The center of the wheel is 139 feet from the ground.

d. From the drawing, the center is at (0, 139).

e. $(x-0)^2 + (y-139)^2 = 125^2$

$x^2 + (y-139)^2 = 125^2$

51. answers may vary

53. Using A as (0, 0), point B is at (3, 1) and point C is at (19, 13).

$d(B, C) = \sqrt{(x_2 - x_1)^2 + (y_2 - y_1)^2}$

$= \sqrt{(19-3)^2 + (13-1)^2}$

$= \sqrt{16^2 + 12^2}$

$= \sqrt{256 + 144}$

$= \sqrt{400}$

$= 20$

The distance across the lake is 20 meters.

55. $y = a(x-h)^2 + k$

$y = a(x-0)^2 + 40$

Set $y = 0$ and $x = 50$ to find a.

$0 = a(50)^2 + 40$

$\frac{-40}{50^2} = a = -\frac{2}{125}$

$y = -\frac{2}{125}(x-0)^2 + 40$

$y = -\frac{2}{125}x^2 + 40$

Section 10.2 Practice Problems

1. $\frac{x^2}{9}+\frac{y^2}{4}=1$

$\frac{x^2}{3^2}+\frac{y^2}{2^2}=1$

$C(0, 0)$

x-intercepts: $(-3, 0)$, $(3, 0)$

y-intercepts $(0, -2)$, $(0, 2)$

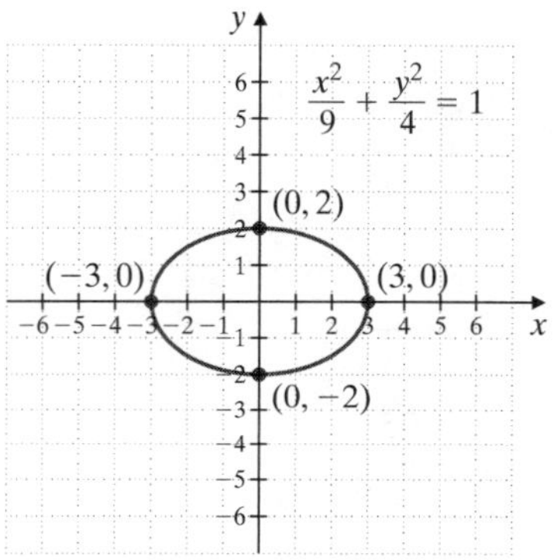

2. $4x^2+36y^2=144$

$\frac{x^2}{36}+\frac{y^2}{4}=1$

$\frac{x^2}{6^2}+\frac{y^2}{2^2}=1$

$C(0, 0)$

x-intercepts: $(-6, 0)$, $(6, 0)$

y-intercepts: $(0, -2)$, $(0, 2)$

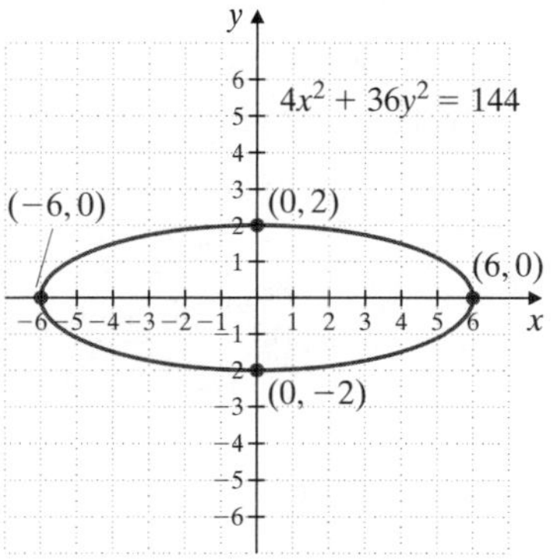

3. $\frac{(x-1)^2}{9}+\frac{(y-3)^2}{16}=1$

$\frac{(x-1)^2}{3^2}+\frac{(y-3)^2}{4^2}=1$

$C(1, 3)$

other points:

$(1, 3+4)$, or $(1, 7)$

$(1, 3-4)$, or $(1, -1)$

$(1+3, 3)$, or $(4, 3)$

$(1-3, 3)$, or $(-2, 3)$

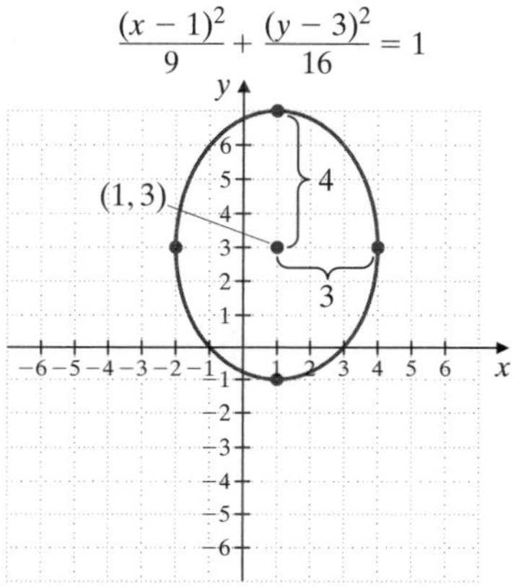

4. $\frac{x^2}{9}-\frac{y^2}{4}=1$

$\frac{x^2}{3^2}-\frac{y^2}{2^2}=1$

$a=3$, $b=2$

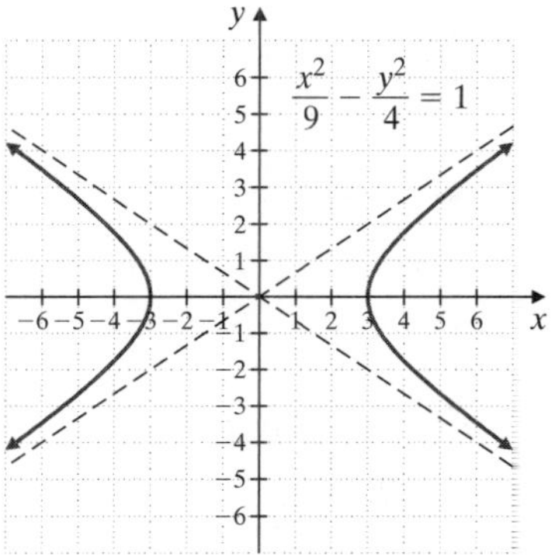

5. $9y^2 - 16x^2 = 144$

$$\frac{y^2}{16} - \frac{x^2}{9} = 1$$

$$\frac{y^2}{4^2} - \frac{x^2}{3^2} = 1$$

$a = 3, b = 4$

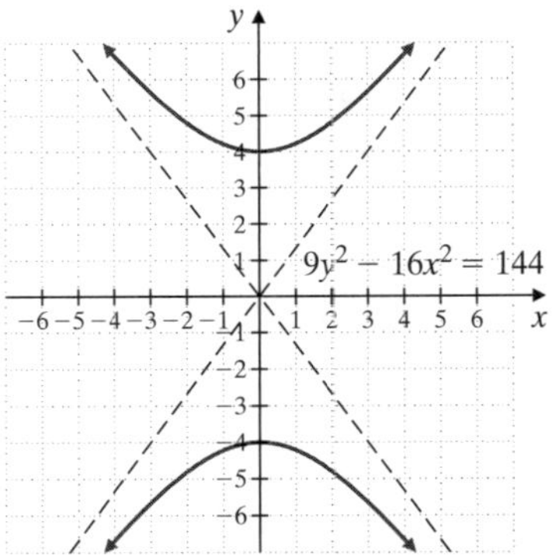

Calculator Explorations

1. $10x^2 + y^2 = 32$

$$y^2 = 32 - 10x^2$$

$$y = \pm\sqrt{32 - 10x^2}$$

$$Y_1 = \sqrt{32 - 10x^2}$$

$$Y_2 = -\sqrt{32 - 10x^2}$$

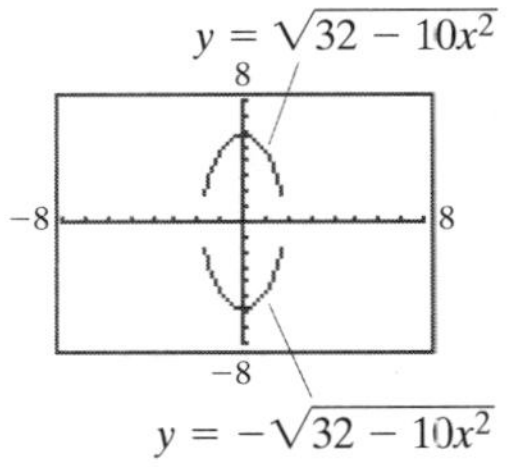

2. $20x^2 + 5y^2 = 100$

$$5y^2 = 100 - 20x^2$$

$$y^2 = 20 - 4x^2$$

$$y = \pm\sqrt{20 - 4x^2}$$

$$Y_1 = \sqrt{20 - 4x^2}$$

$$Y_2 = -\sqrt{20 - 4x^2}$$

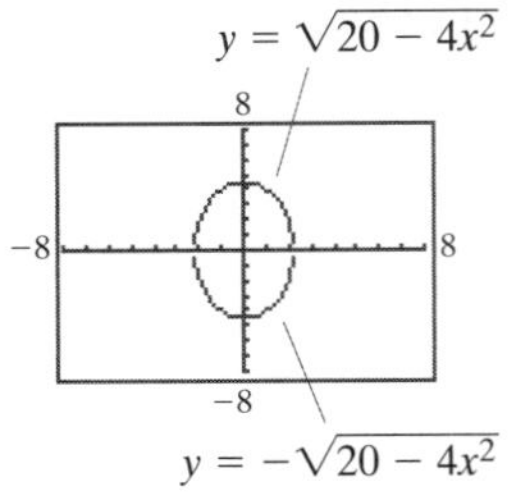

3. $7.3x^2 + 15.5y^2 = 95.2$

$$15.5y^2 = 95.2 - 7.3x^2$$

$$y^2 = \frac{95.2 - 7.3x^2}{15.5}$$

$$y = \pm\sqrt{\frac{95.2 - 7.3x^2}{15.5}}$$

$$Y_1 = \sqrt{\frac{95.2 - 7.3x^2}{15.5}}$$

$$Y_2 = -\sqrt{\frac{95.2 - 7.3x^2}{15.5}}$$

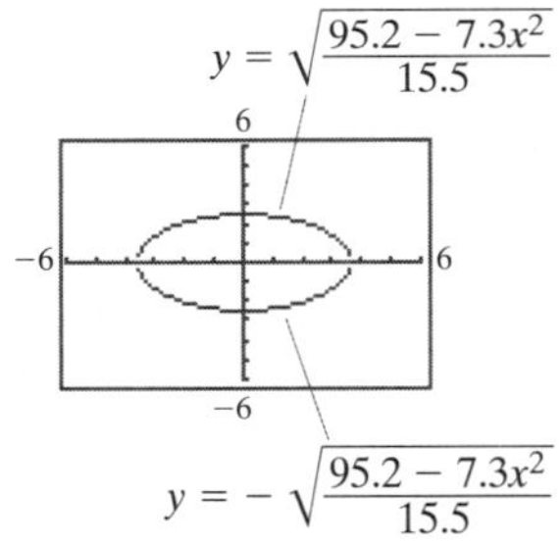

4. $18.8x^2 + 36.1y^2 = 205.8$

$$36.1y^2 = 205.8 - 18.8x^2$$

$$y^2 = \frac{205.8 - 18.8x^2}{36.1}$$

$$y = \pm\sqrt{\frac{205.8 - 18.8x^2}{36.1}}$$

$$Y_1 = \sqrt{\frac{205.8 - 18.8x^2}{36.1}}$$

$$Y_2 = -\sqrt{\frac{205.8 - 18.8x^2}{36.1}}$$

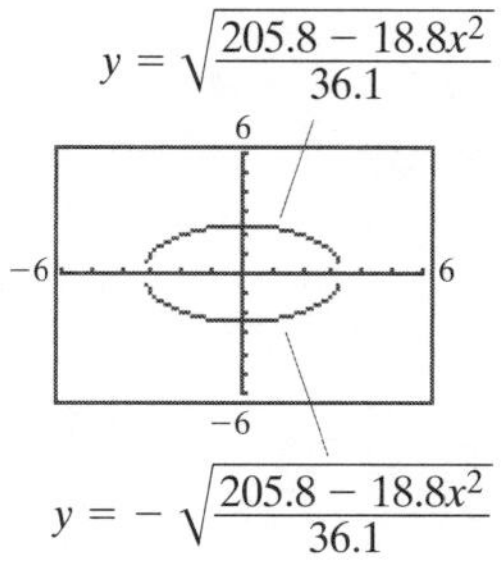

Mental Math

1. $\frac{x^2}{16} + \frac{y^2}{4} = 1$
ellipse

2. $\frac{x^2}{16} - \frac{y^2}{4} = 1$
hyperbola

3. $x^2 - 5y^2 = 3$
hyperbola

4. $-x^2 + 5y^2 = 3$
hyperbola

5. $-\frac{y^2}{25} + \frac{x^2}{36} = 1$
hyperbola

6. $\frac{y^2}{25} + \frac{x^2}{36} = 1$
ellipse

Exercise Set 10.2

1. $\frac{x^2}{4} + \frac{y^2}{25} = 1$
$\frac{x^2}{2^2} + \frac{y^2}{5^2} = 1$
$C(0, 0)$
x-intercepts: (−2, 0), (2, 0)
y-intercepts: (0, −5), (0, 5)

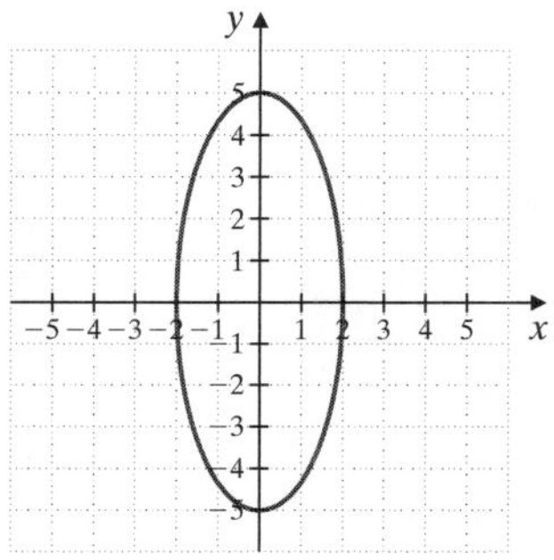

3. $\frac{x^2}{9} + y^2 = 1$
$\frac{x^2}{3^2} + \frac{y^2}{1^2} = 1$
$C(0, 0)$
x-intercepts: (−3, 0), (3, 0)
y-intercepts: (0, −1), (0, 1)

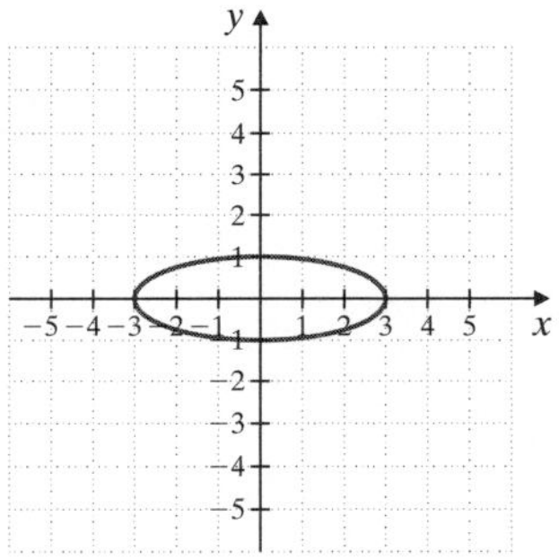

5. $9x^2 + 4y^2 = 36$
$\frac{x^2}{4} + \frac{y^2}{9} = 1$
$\frac{x^2}{2^2} + \frac{y^2}{3^2} = 1$
$C(0, 0)$
x-intercepts: (−2, 0), (2, 0)
y-intercepts: (0, −3), (0, 3)

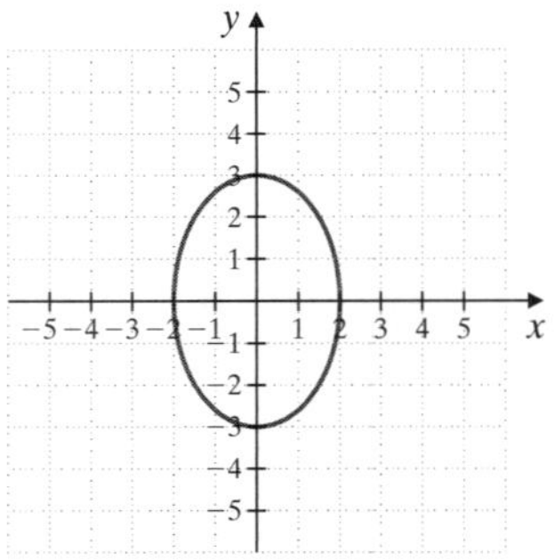

7. $4x^2 + 25y^2 = 100$

$$\frac{x^2}{25} + \frac{y^2}{4} = 1$$

$$\frac{x^2}{5^2} + \frac{y^2}{2^2} = 1$$

$C(0, 0)$
x-intercepts: (–5, 0), (5, 0)
y-intercepts: (0, –2), (0, 2)

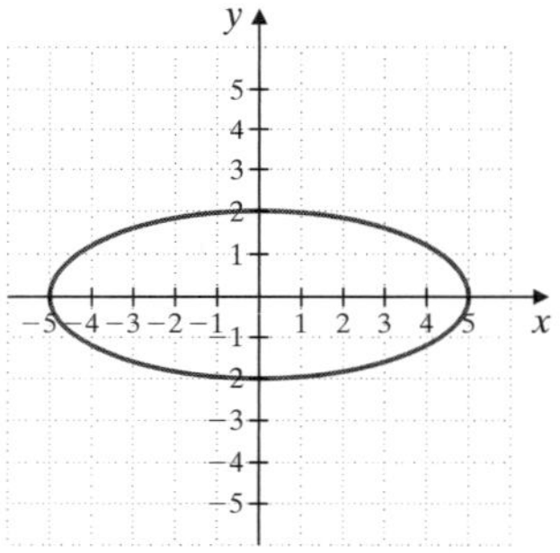

9. $\frac{(x+1)^2}{36} + \frac{(y-2)^2}{49} = 1$

$$\frac{(x+1)^2}{6^2} + \frac{(y-2)^2}{7^2} = 1$$

$C(-1, 2)$
other points:
(–1 – 6, 2), or (–7, 2)
(–1 + 6, 2), or (5, 2)
(–1, 2 – 7), or (–1, –5)
(–1, 2 + 7), or (–1, 9)

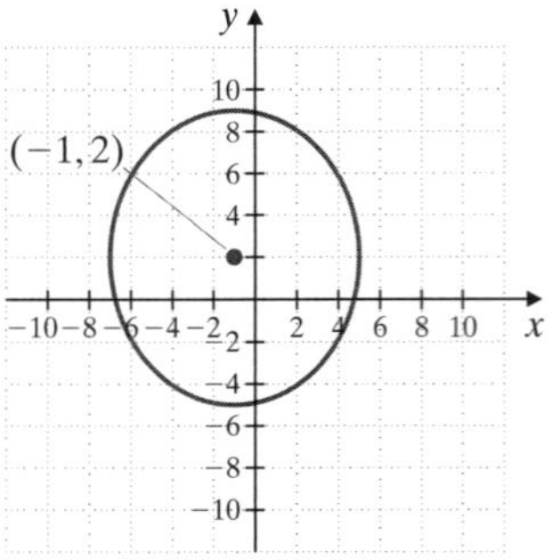

11. $\frac{(x-1)^2}{4} + \frac{(y-1)^2}{25} = 1$

$$\frac{(x-1)^2}{2^2} + \frac{(y-1)^2}{5^2} = 1$$

$C(1, 1)$
other points:
(1 – 2, 1), or (–1, 1)
(1 + 2, 1), or (3, 1)
(1, 1 – 5), or (1, –4)
(1, 1 + 5), or (1, 6)

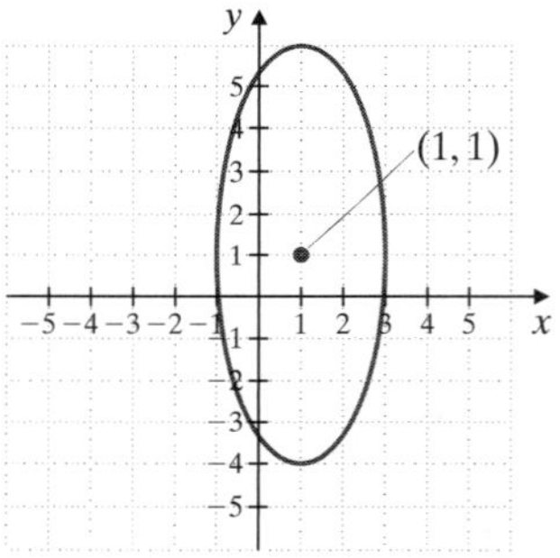

13. $\frac{x^2}{4} - \frac{y^2}{9} = 1$

$$\frac{x^2}{2^2} - \frac{y^2}{3^2} = 1$$

$a = 2, b = 3$

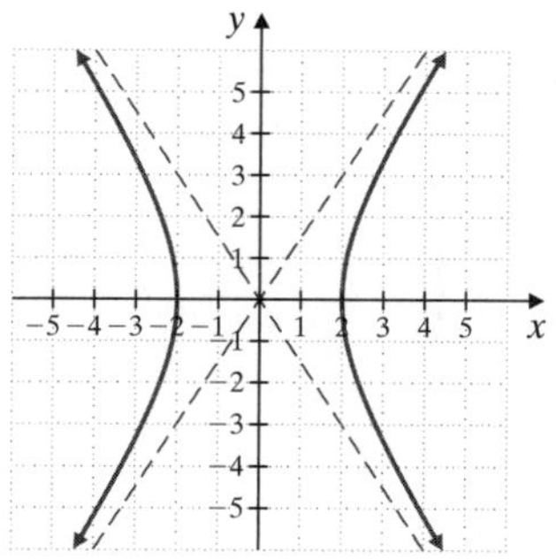

15. $\dfrac{y^2}{25} - \dfrac{x^2}{16} = 1$

$\dfrac{y^2}{5^2} - \dfrac{x^2}{4^2} = 1$

$a = 4,\ b = 5$

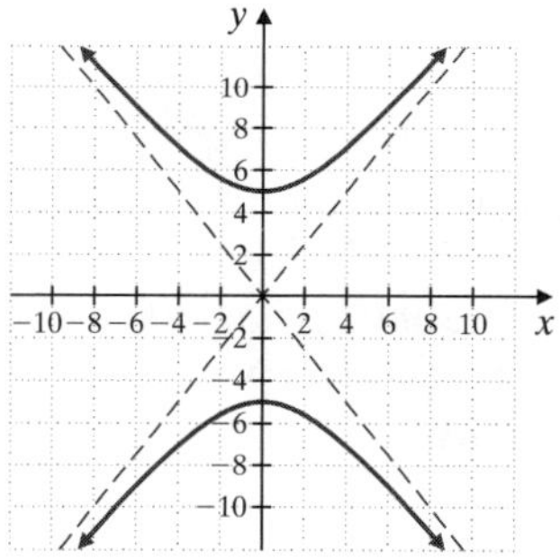

17. $x^2 - 4y^2 = 16$

$\dfrac{x^2}{16} - \dfrac{y^2}{4} = 1$

$\dfrac{x^2}{4^2} - \dfrac{y^2}{2^2} = 1$

$a = 4,\ b = 2$

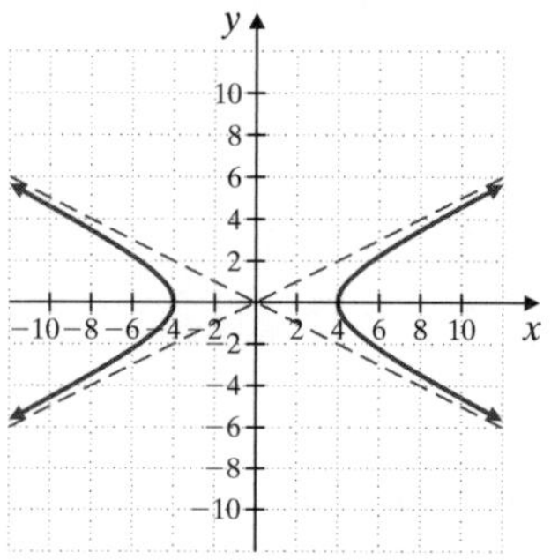

19. $16y^2 - x^2 = 16$

$y^2 - \dfrac{x^2}{16} = 1$

$\dfrac{y^2}{1^2} - \dfrac{x^2}{4^2} = 1$

$a = 4,\ b = 1$

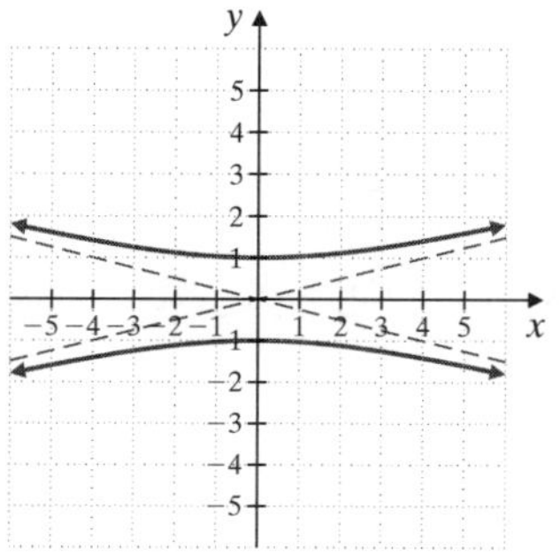

21. $\dfrac{y^2}{36} = 1 - x^2$

$x^2 + \dfrac{y^2}{36} = 1$

$\dfrac{x^2}{1^2} + \dfrac{y^2}{6^2} = 1$

$C(0, 0)$

x-intercepts: $(-1, 0)$, $(1, 0)$

y-intercepts: $(0, -6)$, $(0, 6)$

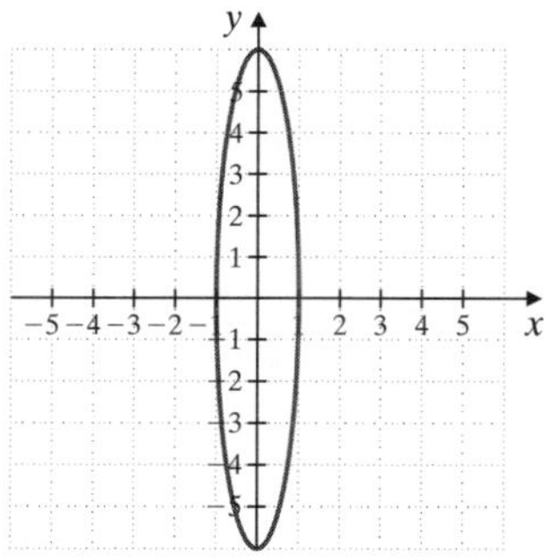

23. $4(x-1)^2 + 9(y+2)^2 = 36$

$\dfrac{(x-1)^2}{9} + \dfrac{(y+2)^2}{4} = 1$

$\dfrac{(x-1)^2}{3^2} + \dfrac{(y+2)^2}{2^2} = 1$

$C(1, -2)$

other points:
$(1-3, -2)$, or $(-2, -2)$
$(1+3, -2)$, or $(4, -2)$
$(1, -2-2)$, or $(1, -4)$
$(1, -2+2)$, or $(1, 0)$

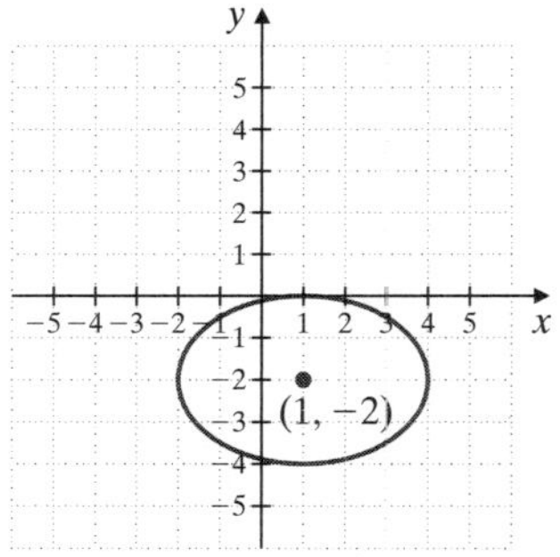

25. $8x^2+2y^2=32$

$$\frac{8x^2}{32}+\frac{2y^2}{32}=1$$
$$\frac{x^2}{4}+\frac{y^2}{16}=1$$
$$\frac{x^2}{2^2}+\frac{y^2}{4^2}=1$$

$C(0, 0)$
x-intercepts: $(-2, 0)$, $(2, 0)$
y-intercepts: $(0, -4)$, $(0, 4)$

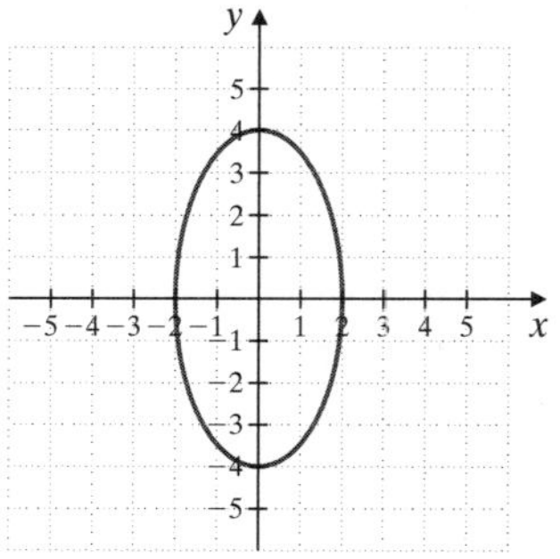

27. $25x^2-y^2=25$

$$x^2-\frac{y^2}{25}=1$$
$$\frac{x^2}{1^2}-\frac{y^2}{5^2}=1$$

$a=1, b=5$

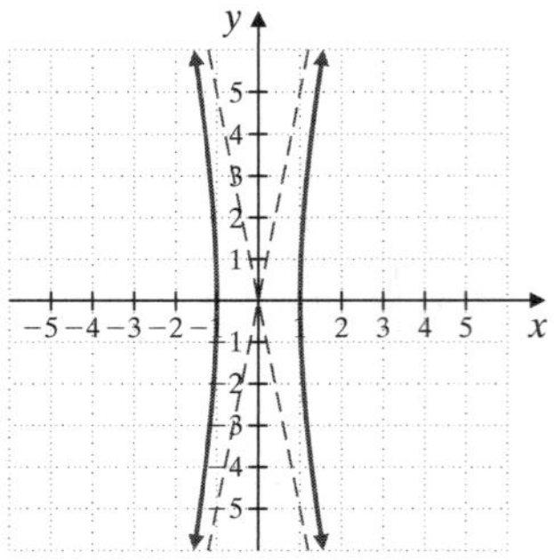

29. $(2x^3)(-4x^2)=2(-4)x^{3+2}=-8x^5$

31. $-5x^2+x^2=(-5+1)x^2=-4x^2$

33. $\frac{x^2}{16}+\frac{y^2}{25}=1$

$$\frac{x^2}{4^2}+\frac{y^2}{5^2}=1$$

The x-intercepts are at $(4, 0)$ and $(-4, 0)$. The
y-intercepts are at $(0, 5)$ and $(0, -5)$. The y-intercepts are farther apart, by $10-8=2$ units.

35. $4x^2+y^2=16$

$$\frac{x^2}{4}+\frac{y^2}{16}=1$$
$$\frac{x^2}{2^2}+\frac{y^2}{4^2}=1$$

The x-intercepts are at $(-2, 0)$ and $(2, 0)$. The y-intercepts are at $(0, -4)$ and $(0, 4)$. The y-intercepts are farther apart, by $8-4=4$ units.

37. $x^2+y^2=25$

$$\frac{x^2}{25}+\frac{y^2}{25}=\frac{25}{25}$$
$$\frac{x^2}{25}+\frac{y^2}{25}=1$$

This resembles an ellipse. This type of conic section is a circle when $a=b$.

39. A: $a^2 = 36,\ b^2 = 13$

B: $a^2 = 4, b^2 = 4$

C: $a^2 = 25,\ b^2 = 16$

D: $a^2 = 39,\ b^2 = 25$

E: $a^2 = 17,\ b^2 = 81$

F: $a^2 = 36,\ b^2 = 36$

G: $a^2 = 16,\ b^2 = 65$

H: $a^2 = 144,\ b^2 = 140$

41. A: $d = 6$
B: $d = 2$
C: $d = 5$
D: $d = 5$
E: $d = 9$
F: $d = 6$
G: $d = 4$
H: $d = 12$

43. The values of e are greater than 0 and less than 1.

45. The values of e are greater than 1.

47. answers may vary

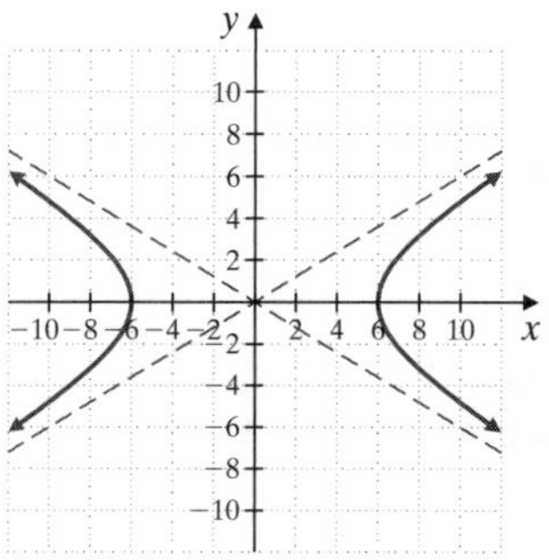

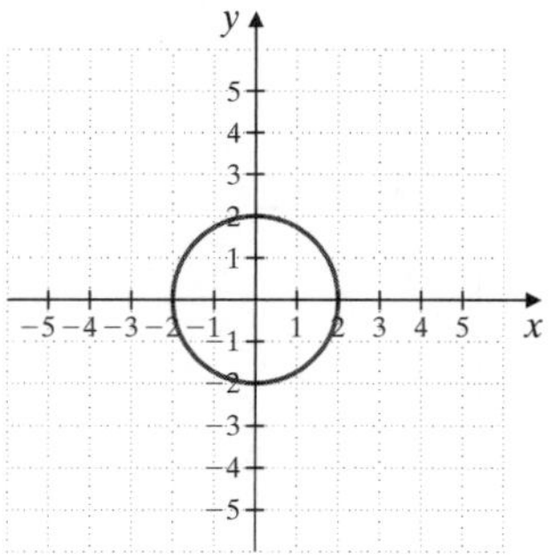

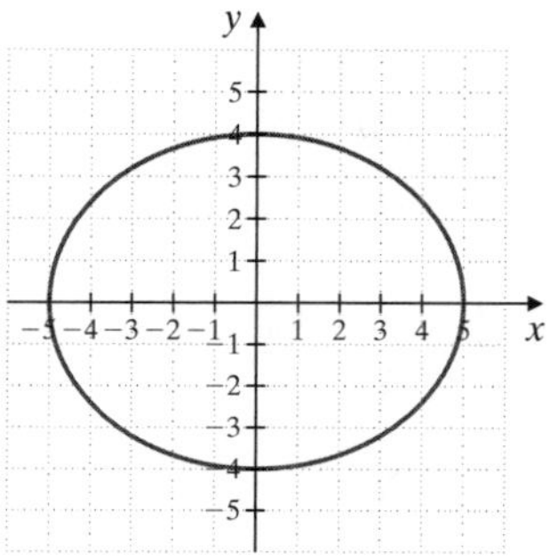

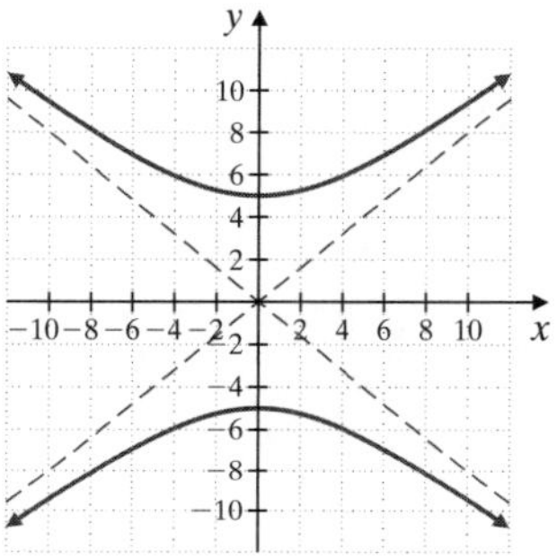

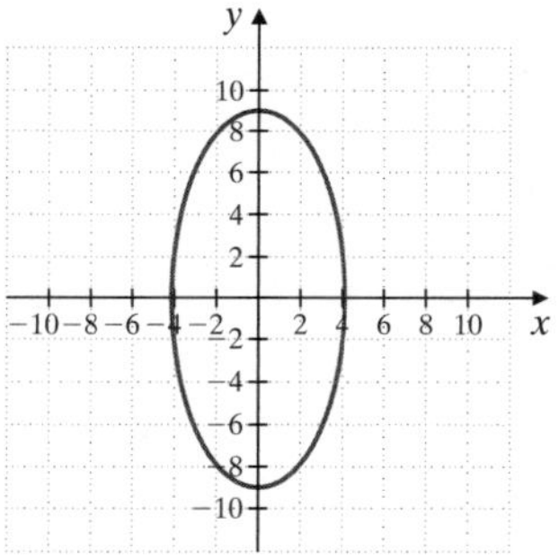

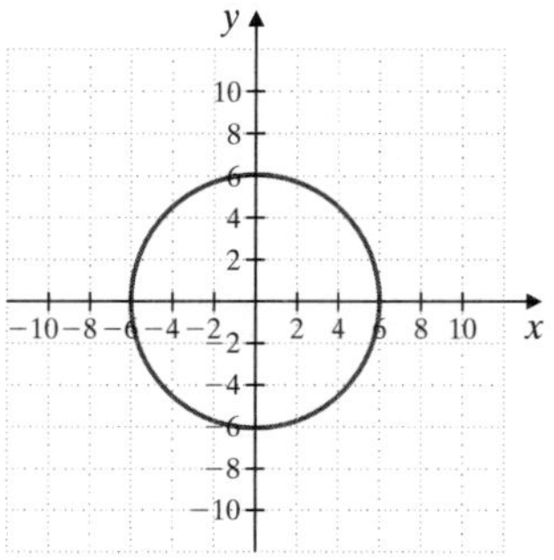

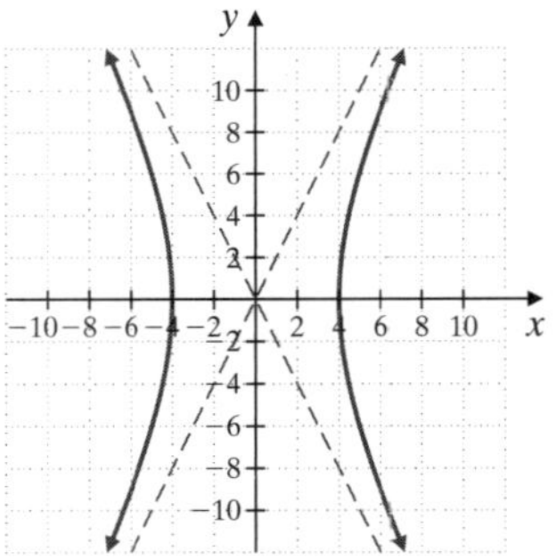

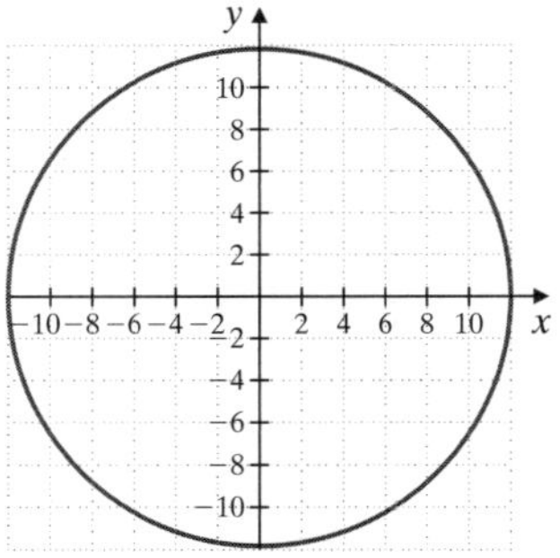

Integrated Review

1. $(x-7)^2+(y-2)^2=4$
$(x-7)^2+(y-2)^2=2^2$
circle
$C(7, 2)$ and $r=2$

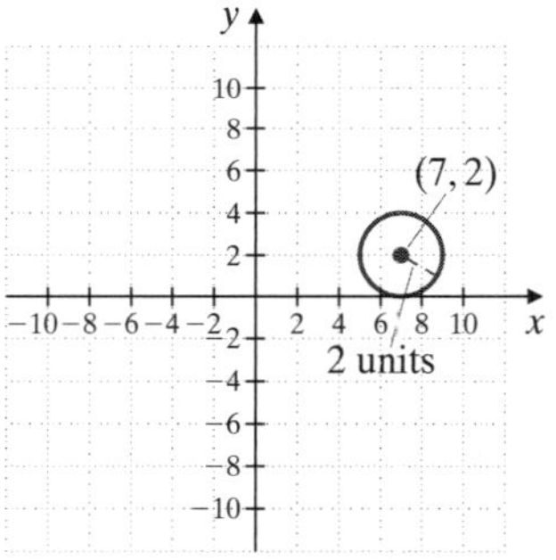

2. $y=x^2+4$
$y=(x-0)^2+4$
parabola
$V(0, 4)$; opens upward

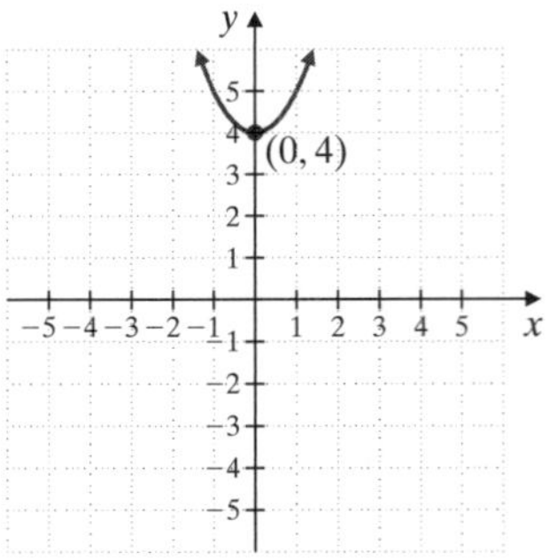

3. $y=x^2+12x+36$
$y=(x+6)^2+0$
parabola
$V(-6, 0)$; opens upward

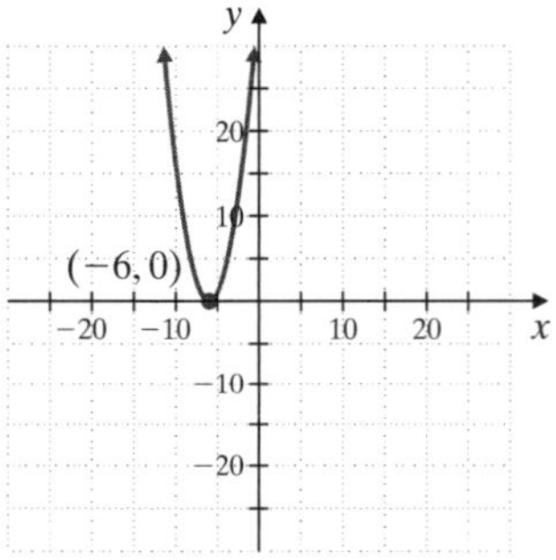

4. $\frac{x^2}{4}+\frac{y^2}{9}=1$
$\frac{x^2}{2^2}+\frac{y^2}{3^2}=1$
ellipse
$C(0, 0)$
x-intercepts: $(-2, 0)$, $(2, 0)$
y-intercepts: $(0, -3)$, $(0, 3)$

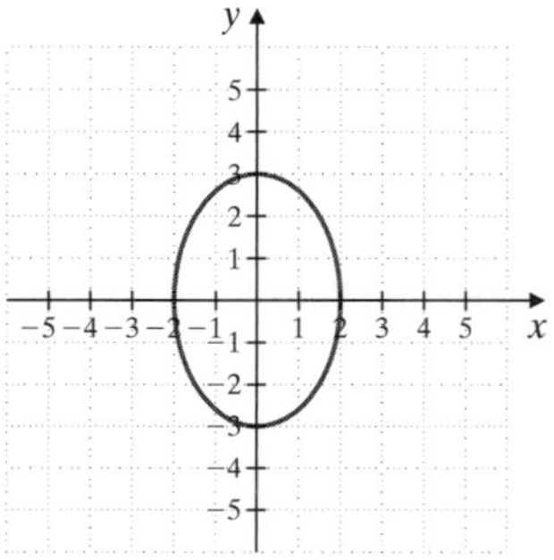

5. $\frac{y^2}{9} - \frac{x^2}{9} = 1$

$\frac{y^2}{3^2} - \frac{x^2}{3^2} = 1$

hyperbola

$a = 3$, $b = 3$

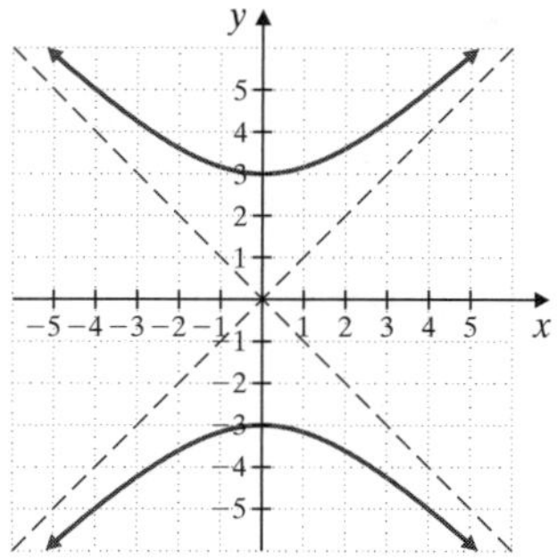

6. $\frac{x^2}{16} - \frac{y^2}{4} = 1$

$\frac{x^2}{4^2} - \frac{y^2}{2^2} = 1$

hyperbola

$a = 4$, $b = 2$

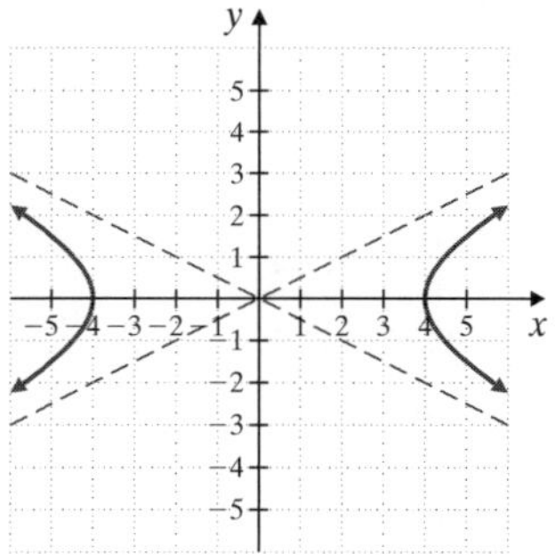

7. $2x^2 + 8y^2 = 32$

$\frac{x^2}{16} + \frac{y^2}{4} = 1$

$\frac{x^2}{4^2} + \frac{y^2}{2^2} = 1$

ellipse

$C(0, 0)$

x-intercepts: $(-4, 0)$, $(4, 0)$

y-intercepts: $(0, -2)$, $(0, 2)$

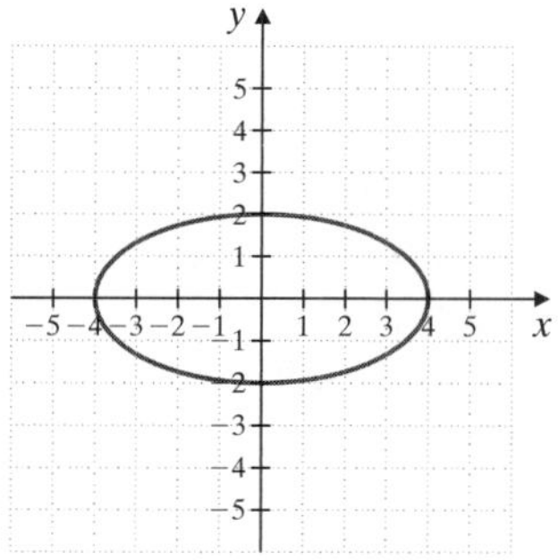

8. $x^2 + y^2 = 16$

$(x - 0)^2 + (y - 0)^2 = 4^2$

circle

$C(0, 0)$ and $r = 4$

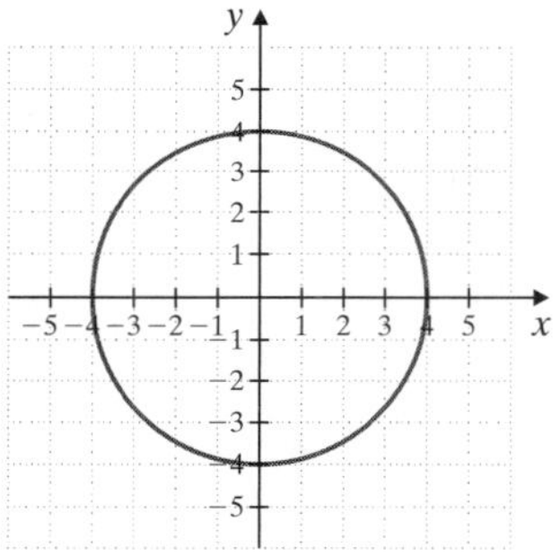

9. $x = y^2 + 4y - 1$

$x = y^2 + 4y + 4 - 1 - 4$

$x = (y + 2)^2 - 5$

parabola

$V(-5, -2)$; opens to the right

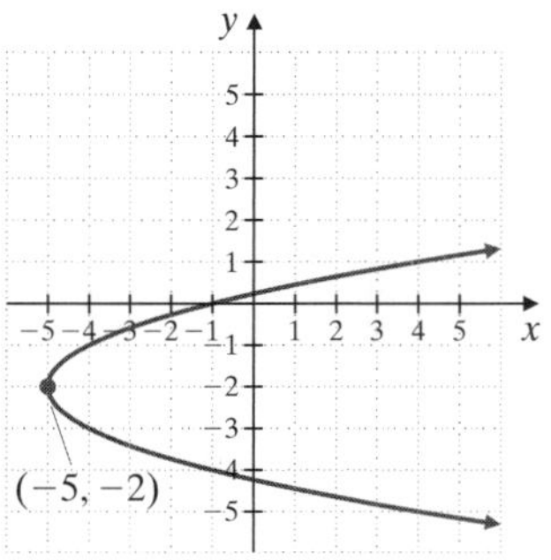

10. $x = -y^2 + 6y$

$x = -1(y^2 - 6y)$

$x = -1(y^2 - 6y + 9) + 9$

$x = -1(y - 3)^2 + 9$

parabola

$V(9, 3)$; opens to the left

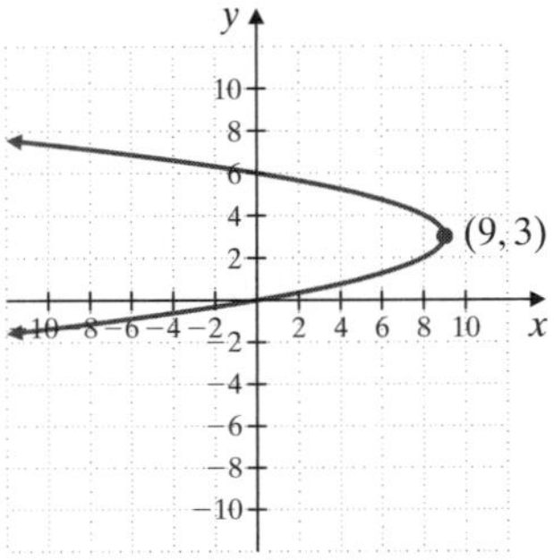

11. $9x^2 - 4y^2 = 36$

$\frac{x^2}{4} - \frac{y^2}{9} = 1$

$\frac{x^2}{2^2} - \frac{y^2}{3^2} = 1$

hyperbola

$a = 2$, $b = 3$

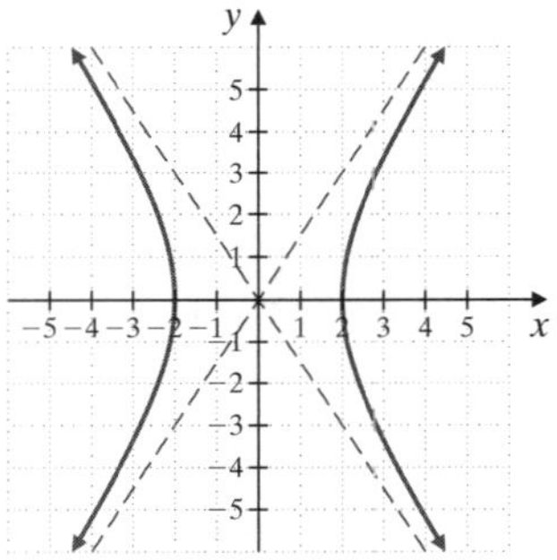

12. $18x^2 + 8y^2 = 72$

$\frac{x^2}{4} + \frac{y^2}{9} = 1$

$\frac{x^2}{2^2} + \frac{y^2}{3^2} = 1$

ellipse

$C(0, 0)$

x-intercepts: $(-2, 0)$, $(2, 0)$

y-intercepts: $(0, -3)$, $(0, 3)$

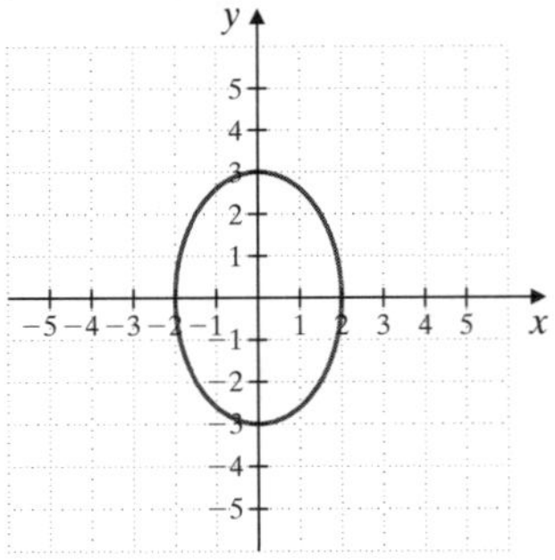

13. $\frac{(x-1)^2}{49} + \frac{(y+2)^2}{25} = 1$

$\frac{(x-1)^2}{7^2} + \frac{(y+2)^2}{5^2} = 1$

ellipse

$C(1, -2)$

other points:

$(1 - 7, -2)$, or $(-6, -2)$

$(1 + 7, -2)$, or $(8, -2)$

$(1, -2 - 5)$, or $(1, -7)$

$(1, -2 + 5)$, or $(1, 3)$

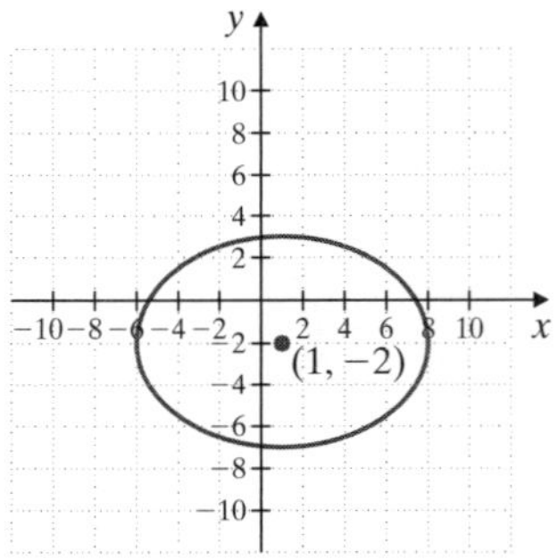

14. $y^2 = x^2 + 16$

$y^2 - x^2 = 16$

$\frac{y^2}{16} - \frac{x^2}{16} = 1$

$\frac{y^2}{4^2} - \frac{x^2}{4^2} = 1$

hyperbola

$a = 4$, $b = 4$

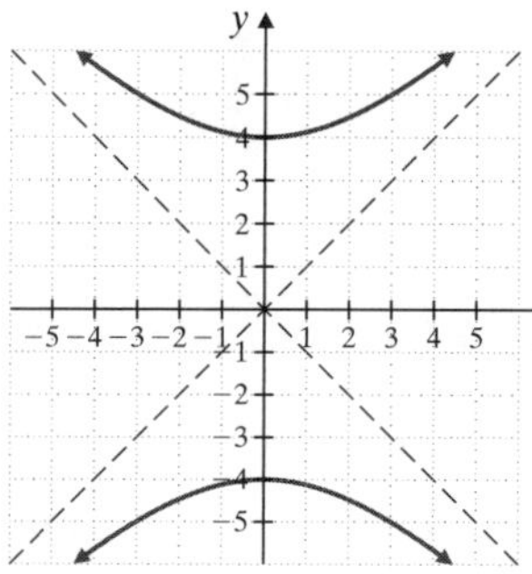

15. $\left(x+\frac{1}{2}\right)^2+\left(y-\frac{1}{2}\right)^2=1$

circle

$C\left(-\frac{1}{2}, \frac{1}{2}\right)$ and $r = 1$

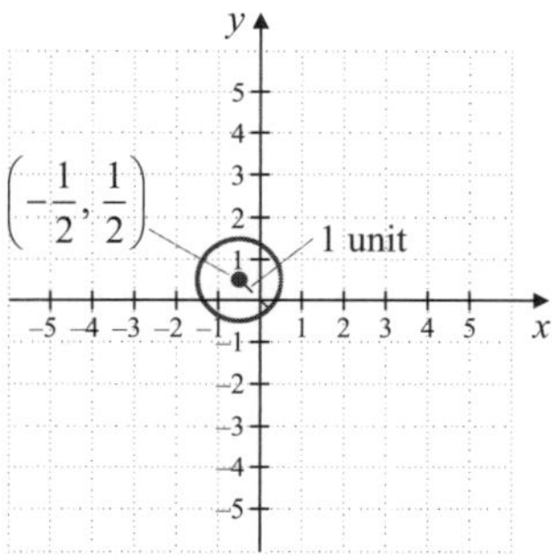

Section 10.3 Practice Problems

1. $\begin{cases} x^2 - 2y = 5 \\ x + y = -1 \end{cases}$

Solve equation 2 for *y*.

$y = -1 - x$

Substitute.

$$x^2 - 2(-1-x) = 5$$
$$x^2 + 2 + 2x = 5$$
$$x^2 + 2x - 3 = 0$$
$$(x+3)(x-1) = 0$$
$$x+3=0 \quad \text{or} \quad x-1=0$$
$$x=-3 \qquad x=1$$

Let $x = -3$. Let $x = 1$.

$y = -1 - (-3)$ $y = -1 - 1$

$y = 2$ $y = -2$

The solution set is $\{(-3, 2), (1, -2)\}$.

2. $\begin{cases} y = \sqrt{x} \\ x^2 + y^2 = 30 \end{cases}$

Substitute.

$$x^2 + y^2 = 30$$
$$x^2 + \left(\sqrt{x}\right)^2 = 30$$
$$x^2 + x = 30$$
$$x^2 + x - 30 = 0$$
$$(x+6)(x-5) = 0$$
$$x = -6 \quad \text{or} \quad x = 5$$

Let $x = -6$. Let $x = 5$.

$y = \sqrt{-6}$ $y = \sqrt{5}$

Not a real number

The solution set is $\left\{\left(5, \sqrt{5}\right)\right\}$.

3. $\begin{cases} x^2 + y^2 = 1 \\ x + y = 4 \end{cases}$

Solve equation 2 for *x*.

$x = 4 - y$

Substitute.

$$(4-y)^2 + y^2 = 1$$
$$16 - 8y + y^2 + y^2 = 1$$
$$2y^2 - 8y + 15 = 0$$

$a = 2$, $b = -8$, $c = 15$

$$y = \frac{-(-8) \pm \sqrt{(-8)^2 - 4(2)(15)}}{2(2)}$$
$$y = \frac{8 \pm \sqrt{64-120}}{4}$$
$$y = \frac{8 \pm \sqrt{-56}}{4}$$

Since $\sqrt{-56}$ is not a real number, there is no solution.

4. $\begin{cases} x^2+3y^2=21 \\ x^2-y^2=1 \end{cases}$

Multiply equation 2 by −1. Add the results.

$$\begin{aligned} x^2+3y^2&=21 \\ -x^2+y^2&=-1 \\ \hline 4y^2&=20 \\ y^2&=5 \\ y&=\pm\sqrt{5} \end{aligned}$$

Let $y=-\sqrt{5}$.

$$\begin{aligned} x^2-y^2&=1 \\ x^2-\left(-\sqrt{5}\right)^2&=1 \\ x^2-5&=1 \\ x^2&=6 \\ x&=\pm\sqrt{6} \end{aligned}$$

Let $y=\sqrt{5}$.

$$\begin{aligned} x^2-y^2&=1 \\ x^2-\left(\sqrt{5}\right)^2&=1 \\ x^2-5&=1 \\ x^2&=6 \\ x&=\pm\sqrt{6} \end{aligned}$$

The solution set is $\left\{\left(\sqrt{6}, \sqrt{5}\right), \left(\sqrt{6}, -\sqrt{5}\right), \left(-\sqrt{6}, \sqrt{5}\right), \left(-\sqrt{6}, -\sqrt{5}\right)\right\}$.

Exercise Set 10.3

1. $\begin{cases} x^2+y^2=25 \\ 4x+3y=0 \end{cases}$

Solve equation 2 for *y*.

$$\begin{aligned} 3y&=-4x \\ y&=\frac{-4x}{3} \end{aligned}$$

Substitute.

$$\begin{aligned} x^2+\left(-\frac{4x}{3}\right)^2&=25 \\ x^2+\frac{16x^2}{9}&=25 \\ \frac{25}{9}x^2&=25 \\ \frac{x^2}{9}&=1 \\ x^2&=9 \\ x&=\pm\sqrt{9}=\pm 3 \end{aligned}$$

$x = 3$: $y = -\frac{4}{3}(3) = -4$

$x = -3$: $y = -\frac{4}{3}(-3) = 4$

The solution set is $\{(3, -4), (-3, 4)\}$.

3. $\begin{cases} x^2 + 4y^2 = 10 \\ y = x \end{cases}$

Substitute.

$$x^2 + 4y^2 = 10$$
$$x^2 + 4x^2 = 10$$
$$5x^2 = 10$$
$$x^2 = 2$$
$$x = \pm\sqrt{2}$$

$x = -\sqrt{2}$: $y = -\sqrt{2}$

$x = \sqrt{2}$: $y = \sqrt{2}$

The solution set is $\left\{\left(-\sqrt{2}, -\sqrt{2}\right), \left(\sqrt{2}, \sqrt{2}\right)\right\}$.

5. $\begin{cases} y^2 = 4 - x \\ x - 2y = 4 \end{cases}$

Solve equation 2 for x.

$$x - 2y = 4$$
$$x = 2y + 4$$

Substitute.

$$y^2 = 4 - (2y + 4)$$
$$y^2 = -2y$$
$$y^2 + 2y = 0$$
$$y(y + 2) = 0$$

$y = 0$ or $y + 2 = 0$

$y = -2$

$y = 0$: $x - 2(0) = 4$

$x = 4$

$y = -2$: $x - 2(-2) = 4$

$x + 4 = 4$

$x = 0$

The solution set is $\{(4, 0), (0, -2)\}$.

7. $\begin{cases} x^2 + y^2 = 9 \\ 16x^2 - 4y^2 = 64 \end{cases}$

Multiply equation 1 by 4. Add the results.

$$\begin{aligned} 4x^2 + 4y^2 &= 36 \\ \underline{16x^2 - 4y^2} &\underline{= 64} \\ 20x^2 \qquad &= 100 \end{aligned}$$
$$x^2 = 5$$
$$x = \pm\sqrt{5}$$

$x = -\sqrt{5}$: $\left(-\sqrt{5}\right)^2 + y^2 = 9$

$$5 + y^2 = 9$$
$$y^2 = 4$$
$$y = \pm 2$$

$x = \sqrt{5}$: $\left(\sqrt{5}\right)^2 + y^2 = 9$

$$5 + y^2 = 9$$
$$y^2 = 4$$
$$y = \pm 2$$

The solution set is $\left\{\left(-\sqrt{5}, -2\right), \left(-\sqrt{5}, 2\right), \left(\sqrt{5}, -2\right), \left(\sqrt{5}, 2\right)\right\}$.

9. $\begin{cases} x^2 + 2y^2 = 2 \\ x - y = 2 \end{cases}$

Solve equation 2 for x.

$x = y + 2$

Substitute.

$$(y + 2)^2 + 2y^2 = 2$$
$$y^2 + 4y + 4 + 2y^2 = 2$$
$$3y^2 + 4y + 4 = 2$$
$$3y^2 + 4y + 2 = 0$$

$$\begin{aligned} b^2 - 4ac &= 4^2 - 4(3)(2) \\ &= 16 - 24 \\ &= -8 < 0 \end{aligned}$$

Therefore, no real solutions exist. The solution set is $\varnothing$.

11. $\begin{cases} y = x^2 - 3 \\ 4x - y = 6 \end{cases}$

Substitute.

$4x-(x^2-3)=6$
$4x-x^2+3=6$
$0=x^2-4x+3$
$0=(x-3)(x-1)$
$x-3=0$ or $x-1=0$
$x=3$ $x=1$
$x=3$: $y=3^2-3=6$
$x=1$: $y=1^2-3=-2$
The solution set is $\{(1, -2), (3, 6)\}$.

13. $\begin{cases} y=x^2 \\ 3x+y=10 \end{cases}$

Substitute.
$3x+x^2=10$
$x^2+3x-10=0$
$(x+5)(x-2)=0$
$x+5=0$ or $x-2=0$
$x=-5$ or $x=2$
$x=-5$: $y=(-5)^2=25$
$x=2$: $y=2^2=4$
The solution set is $\{(-5, 25), (2, 4)\}$.

15. $\begin{cases} y=2x^2+1 \\ x+y=-1 \end{cases}$

Substitute.
$x+2x^2+1=-1$
$2x^2+x+2=0$
$a=2, b=1, c=2$

$$x=\frac{-1\pm\sqrt{1^2-4(2)(2)}}{2(2)}=\frac{-1\pm\sqrt{-15}}{4}$$

Since $\sqrt{-15}$ is not a real number, there is no solution. The solution set is $\varnothing$.

17. $\begin{cases} y=x^2-4 \\ y=x^2-4x \end{cases}$

Substitute.
$x^2-4=x^2-4x$
$-4=-4x$
$x=1$
$y=1^2-4=-3$
The solution set is $\{(1, -3)\}$.

19. $\begin{cases} 2x^2+3y^2=14 \\ -x^2+y^2=3 \end{cases}$

Multiply equation 2 by 2 and add the results.

$$\begin{array}{r} 2x^2+3y^2=14 \\ \underline{-2x^2+2y^2=6} \\ 5y^2=20 \end{array}$$

$y^2=4$
$y=\pm\sqrt{4}$
$y=\pm 2$
$y=-2$: $-x^2+(-2)^2=3$
$x^2=1$
$x=\pm 1$
$y=2$: $-x^2+2^2=3$
$x^2=1$
$x=\pm 1$
The solution set is
$\{(-1, -2), (-1, 2), (1, -2), (1, 2)\}$.

21. $\begin{cases} x^2+y^2=1 \\ x^2+(y+3)^2=4 \end{cases}$

Subtract equation 1 from equation 2.
$(y+3)^2-y^2=3$
$y^2+6y+9-y^2=3$
$6y+9=3$
$6y=-6$
$y=-1$
Substitute back.
$x^2+(-1)^2=1$
$x^2+1=1$
$x^2=0$
$x=0$
The solution set is $\{(0, -1)\}$.

23. $\begin{cases} y = x^2 + 2 \\ y = -x^2 + 4 \end{cases}$

Substitute.

$$x^2 + 2 = -x^2 + 4$$
$$2x^2 = 2$$
$$x^2 = 1$$
$$x = \pm\sqrt{1}$$
$$x = \pm 1$$

$x = -1$: $y = (-1)^2 + 2 = 1 + 2 = 3$

$x = 1$: $y = 1^2 + 2 = 1 + 2 = 3$

The solution set is $\{(-1, 3), (1, 3)\}$.

25. $\begin{cases} 3x^2 + y^2 = 9 \\ 3x^2 - y^2 = 9 \end{cases}$

Subtract.

$$2y^2 = 0$$
$$y^2 = 0$$
$$y = 0$$

Substitute back.

$$3x^2 + 0 = 9$$
$$3x^2 = 9$$
$$x^2 = 3$$
$$x = \pm\sqrt{3}$$

The solution set is $\left\{\left(\sqrt{3}, 0\right), \left(-\sqrt{3}, 0\right)\right\}$.

27. $\begin{cases} x^2 + 3y^2 = 6 \\ x^2 - 3y^2 = 10 \end{cases}$

Add the equations.

$$2x^2 = 16$$
$$x^2 = 8$$
$$x = \pm\sqrt{8}$$
$$x = \pm 2\sqrt{2}$$

$x = -2\sqrt{2}$: $\left(-2\sqrt{2}\right)^2 + 3y^2 = 6$

$$8 + 3y^2 = 6$$
$$3y^2 = -2$$
$$y^2 = -\frac{2}{3}$$
$$y = \pm\sqrt{-\frac{2}{3}}$$

$x = 2\sqrt{2}$: $\left(2\sqrt{2}\right)^2 + 3y^2 = 6$

$$8 + 3y^2 = 6$$
$$3y^2 = -2$$
$$y^2 = -\frac{2}{3}$$
$$y = \pm\sqrt{-\frac{2}{3}}$$

Since $\sqrt{-\frac{2}{3}}$ is not a real number, there is no solution. The solution set is $\varnothing$.

29. $\begin{cases} x^2 + y^2 = 36 \\ y = \frac{1}{6}x^2 - 6 \end{cases}$

$$y + 6 = \frac{1}{6}x^2$$
$$x^2 = 6(y + 6)$$

Substitute.

$$6(y + 6) + y^2 = 36$$
$$6y + 36 + y^2 = 36$$
$$6y + y^2 = 0$$
$$y(6 + y) = 0$$

$y = 0$ or $6 + y = 0$

$y = -6$

$y = 0$: $x^2 + 0^2 = 36$

$$x^2 = 36$$
$$x = \pm 6$$

$$y = -6:\ x^2 + (-6)^2 = 36$$
$$x^2 + 36 = 36$$
$$x^2 = 0$$
$$x = 0$$

The solution set is {(6, 0), (–6, 0), (0, –6)}.

31. $\begin{cases} y = \sqrt{x} \\ x^2 + y^2 = 12 \end{cases}$

Substitute.

$$x^2 + \left(\sqrt{x}\right)^2 = 12$$
$$x^2 + x = 12$$
$$x^2 + x - 12 = 0$$
$$(x+4)(x-3) = 0$$
$$x + 4 = 0 \quad \text{or} \quad x - 3 = 0$$
$$x = -4 \qquad\qquad x = 3$$
$$x = -4:\ y = \sqrt{-4}$$
$$x = 3:\ y = \sqrt{3}$$

Since $\sqrt{-4}$ is not a real number, the solution set is $\left\{\left(3, \sqrt{3}\right)\right\}$.

33. $x > -3$

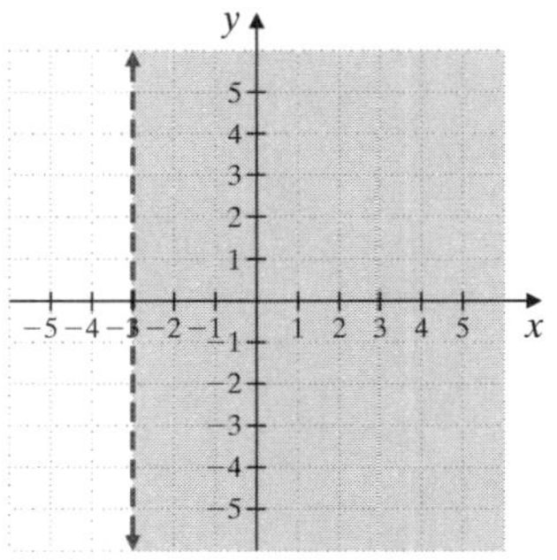

35. $y < 2x - 1$

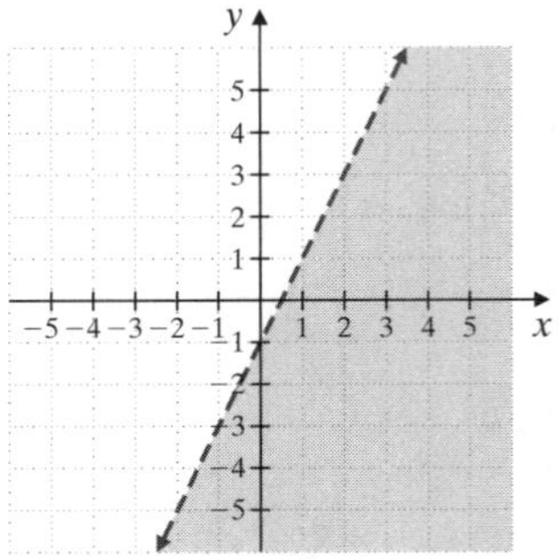

37. answers may vary

39. A circle and a parabola can intersect at 0, 1, 2, 3, or 4 points. Therefore, there are 0, 1, 2, 3, or 4 solutions.

41. $\begin{cases} x^2 + y^2 = 130 \\ x^2 - y^2 = 32 \end{cases}$

Add.

$$2x^2 = 162$$
$$x^2 = 81$$
$$x = \pm 9$$

Substitute back.

$$9^2 + y^2 = 130 \qquad (-9)^2 + y^2 = 130$$
$$y^2 = 49 \qquad y^2 = 49$$
$$y = \pm 7 \qquad y = \pm 7$$

The numbers are 9 and 7, 9 and –7, –9 and 7, or –9 and –7.

43. Let x = width and y = length.

$\begin{cases} xy = 285 \\ 2x + 2y = 68 \end{cases}$

Solve equation 2 for x.

$$2x + 2y = 68$$
$$x + y = 34$$
$$x = 34 - y$$

Substitute.

$$xy = 285$$
$$(34 - y)y = 285$$
$$34y - y^2 = 285$$
$$0 = y^2 - 34y + 285$$
$$0 = (y - 15)(y - 19)$$

$y-15=0$ or $y-19=0$
$y=15$ $\quad y=19$
$y = 15$: $x = 34 - 15 = 19$
$y = 19$: $x = 34 - 19 = 15$
The dimensions are 15 cm by 19 cm.

45. $\begin{cases} p=-0.01x^2-0.2x+9 \\ p=0.01x^2-0.1x+3 \end{cases}$

Substitute.

$$\begin{aligned} -0.01x^2-0.2x+9 &= 0.01x^2-0.1x+3 \\ 0 &= 0.02x^2+0.1x-6 \\ 0 &= x^2+5x-300 \\ 0 &= (x+20)(x-15) \end{aligned}$$

$x+20=0$ or $x-15=0$
$x=-20$ or $x=15$

Disregard the negative.

$p=-0.01(15)^2-0.2(15)+9$
$p=3.75$

15 thousand compact discs; price, $3.75

47. $\begin{cases} x^2+4y^2=10 \\ y=x \end{cases}$

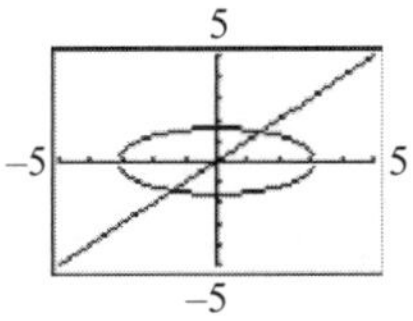

49. $\begin{cases} y=x^2+2 \\ y=-x^2+4 \end{cases}$

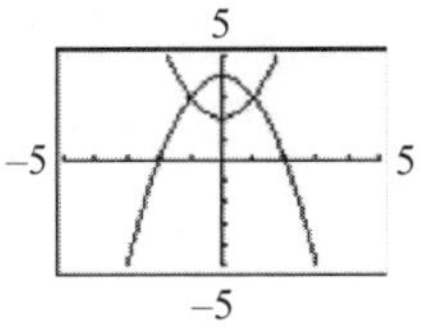

Section 10.4 Practice Problems

1. $\dfrac{x^2}{25}+\dfrac{y^2}{4}\le 1$

Graph $\dfrac{x^2}{5^2}+\dfrac{y^2}{2^2}=1.$

Then shade the region containing (0, 0),

since $\dfrac{0^2}{25}+\dfrac{0^2}{4}\le 1$ is true.

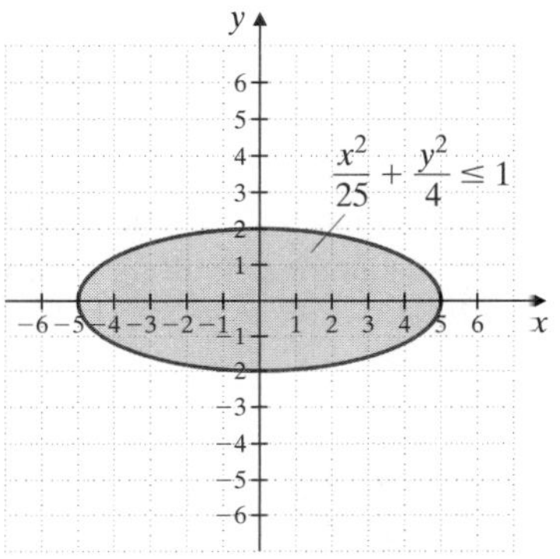

2. $9x^2>4y^2+144$

Graph $9x^2=4y^2+144$
or $9x^2-4y^2=144$
$\dfrac{x^2}{16}-\dfrac{y^2}{36}=1$

Shade the appropriate regions.

Test (−5, 0): $9(-5)^2>4(0)^2+144$ True

Test (0, 0): $9(0)^2>4(0)^2+144$ False

Test (5, 0): $9(5)^2>4(0)^2+144$ True

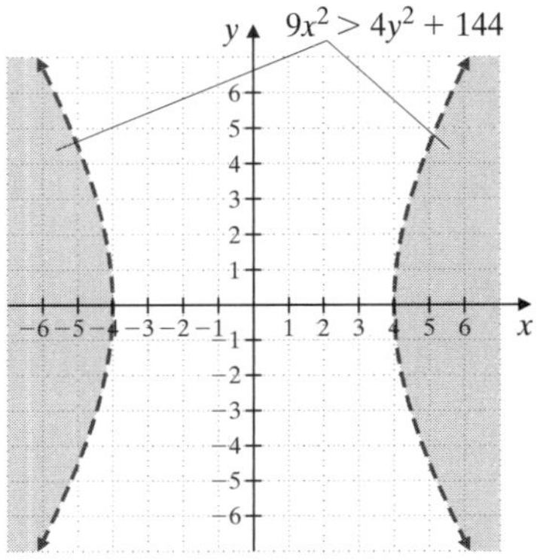

3. $\begin{cases} y \ge x^2 \\ y \le -4x + 2 \end{cases}$

Graph each inequality. The intersection is the darkest shaded region along with its boundary lines.

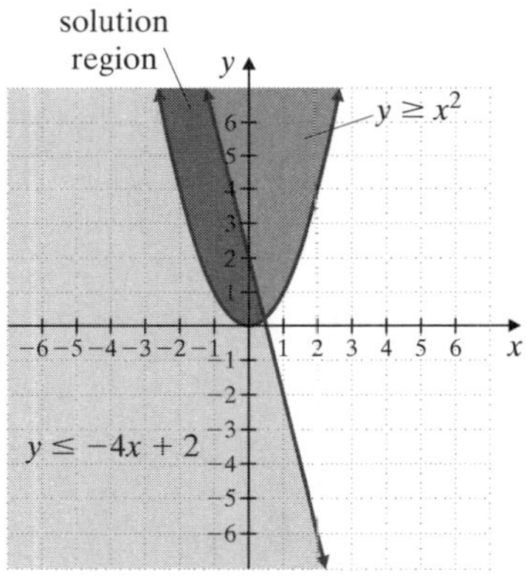

4. $\begin{cases} x^2 + y^2 < 9 \\ \dfrac{x^2}{9} - \dfrac{y^2}{4} < 1 \\ y > x - 2 \end{cases}$

Graph each inequality. The intersection region is shown as shaded.

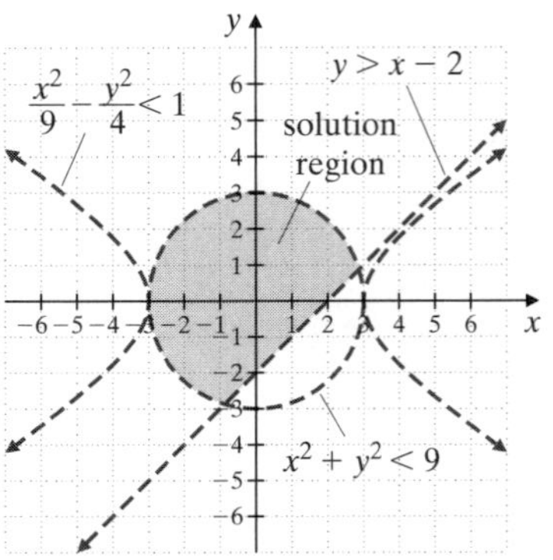

Exercise Set 10.4

1. $y < x^2$

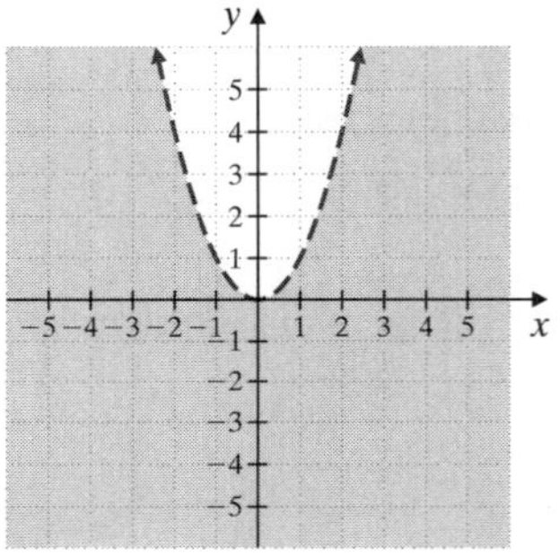

3. $x^2 + y^2 \ge 16$

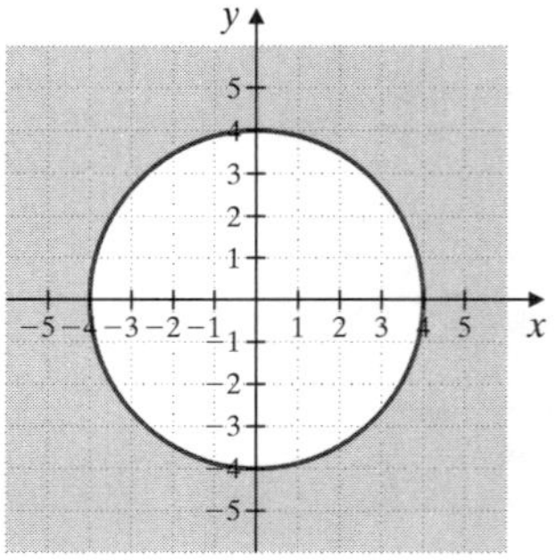

5. $\frac{x^2}{4} - y^2 < 1$

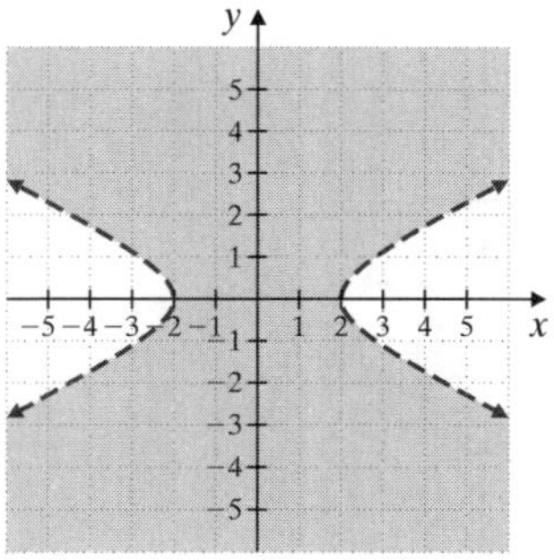

7. $y > (x-1)^2 - 3$

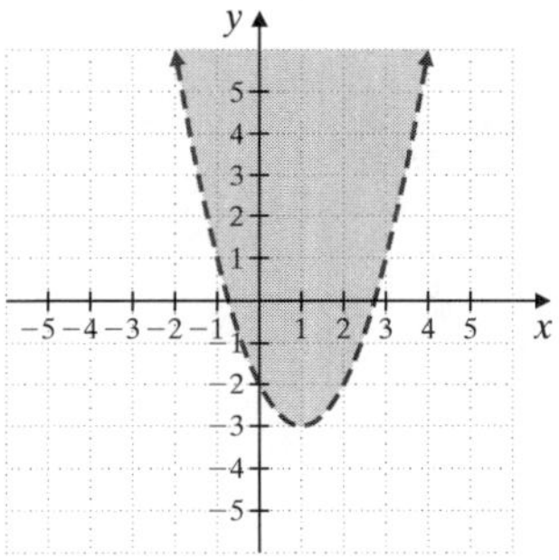

9. $x^2 + y^2 \le 9$

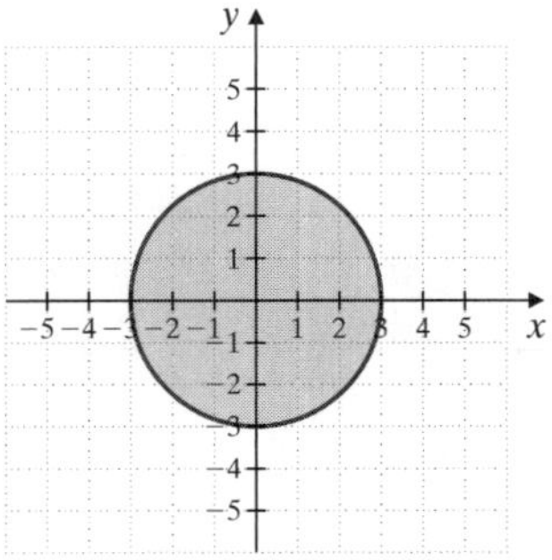

11. $y > -x^2 + 5$

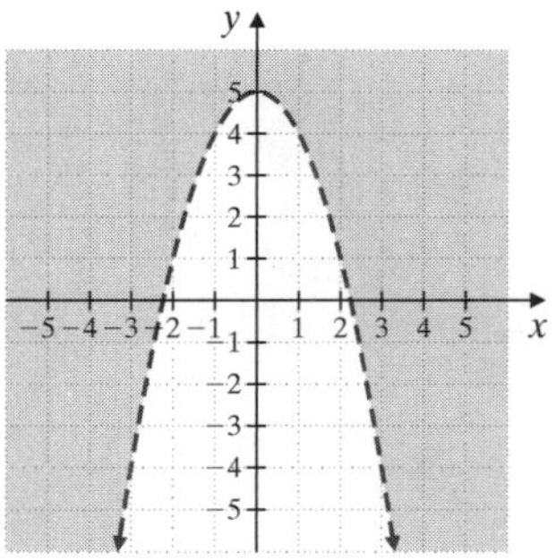

13. $\frac{x^2}{4} + \frac{y^2}{9} \le 1$

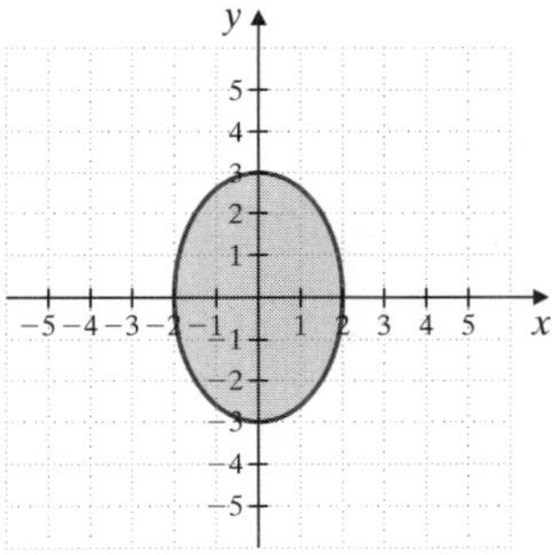

15. $\frac{y^2}{4} - x^2 \le 1$

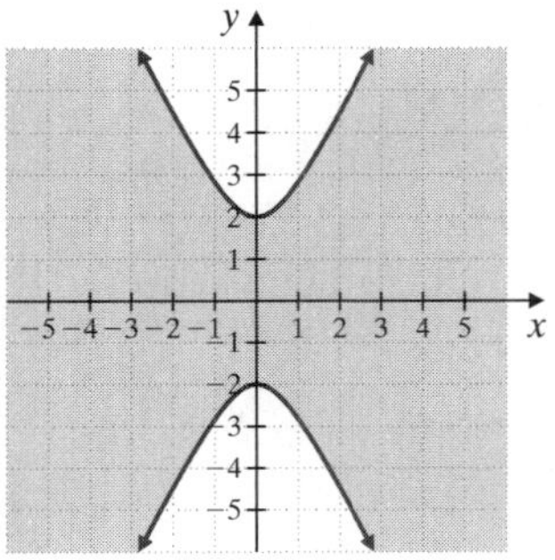

17. $y < (x-2)^2 + 1$

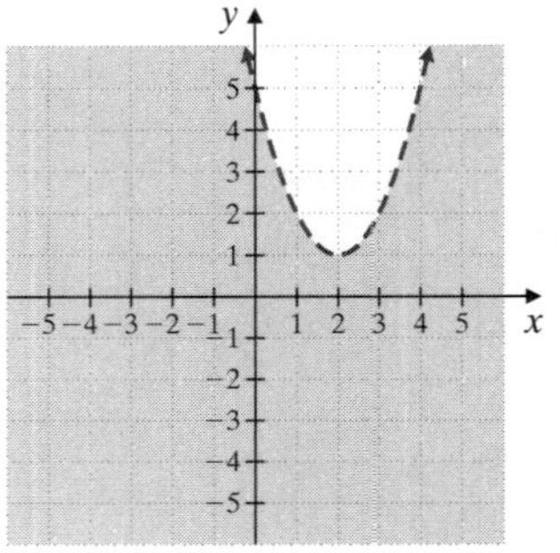

19. $y \le x^2 + x - 2$

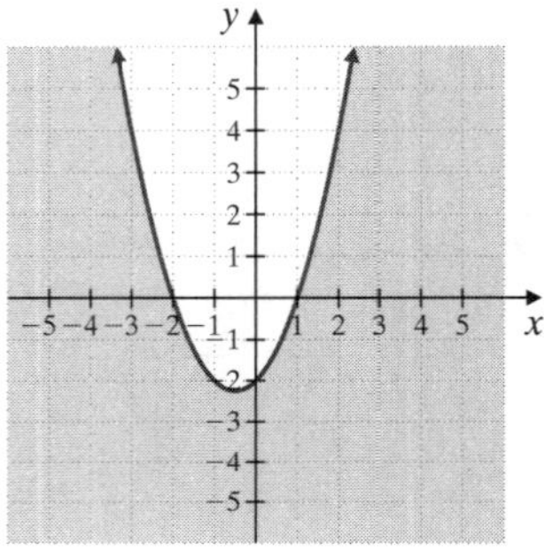

21. $\begin{cases} 4x + 3y \ge 12 \\ x^2 + y^2 < 16 \end{cases}$

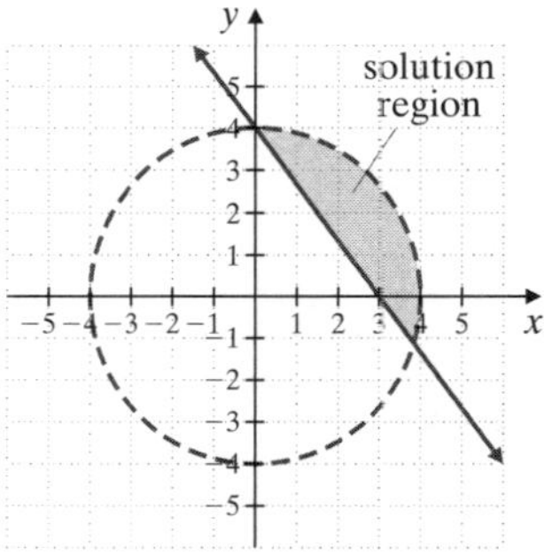

23. $\begin{cases} x^2 + y^2 \le 9 \\ x^2 + y^2 \ge 1 \end{cases}$

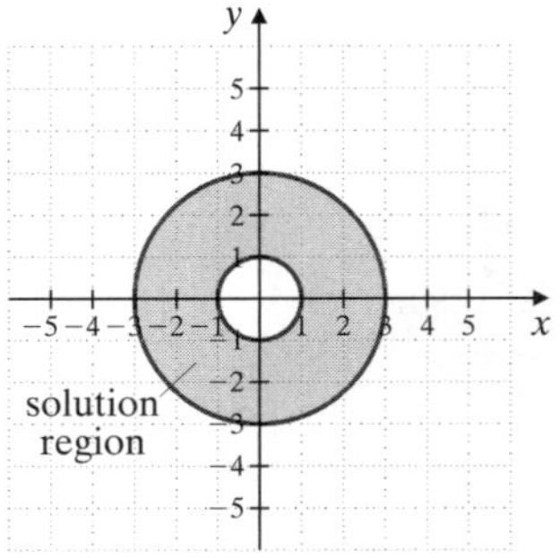

25. $\begin{cases} y > x^2 \\ y \ge 2x + 1 \end{cases}$

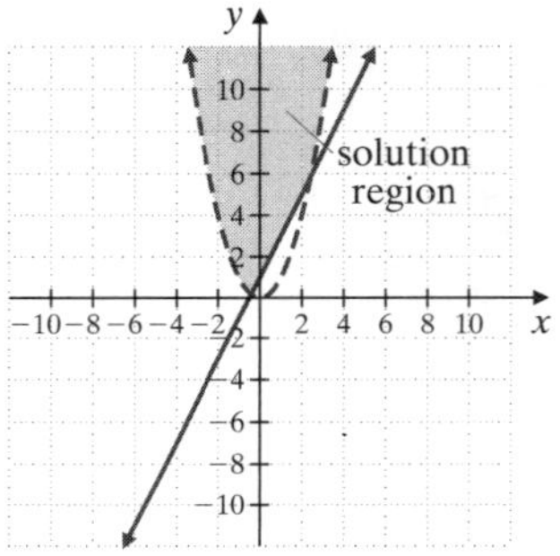

27. $\begin{cases} x^2 + y^2 > 9 \\ y > x^2 \end{cases}$

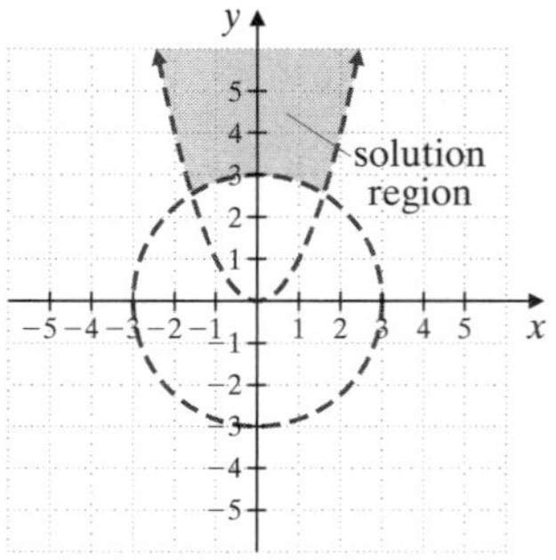

29. $\begin{cases} \dfrac{x^2}{4} + \dfrac{y^2}{9} \ge 1 \\ x^2 + y^2 \ge 4 \end{cases}$

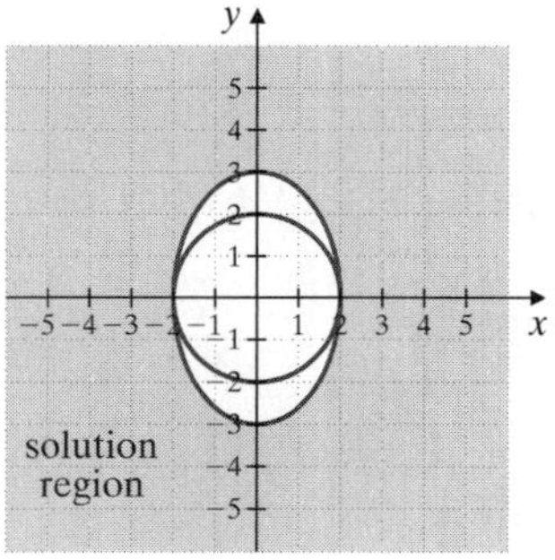

31. $\begin{cases} x^2 - y^2 \ge 1 \\ y \ge 0 \end{cases}$

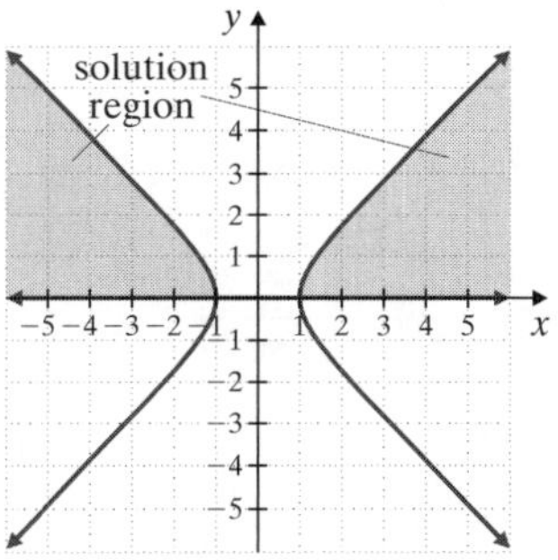

33. $\begin{cases} x + y \ge 1 \\ 2x + 3y < 1 \\ x > -3 \end{cases}$

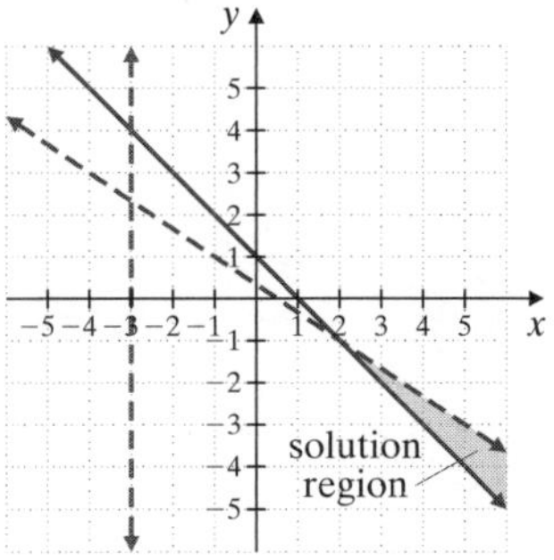

35. $\begin{cases} x^2 - y^2 < 1 \\ \dfrac{x^2}{16} + y^2 \le 1 \\ x \ge -2 \end{cases}$

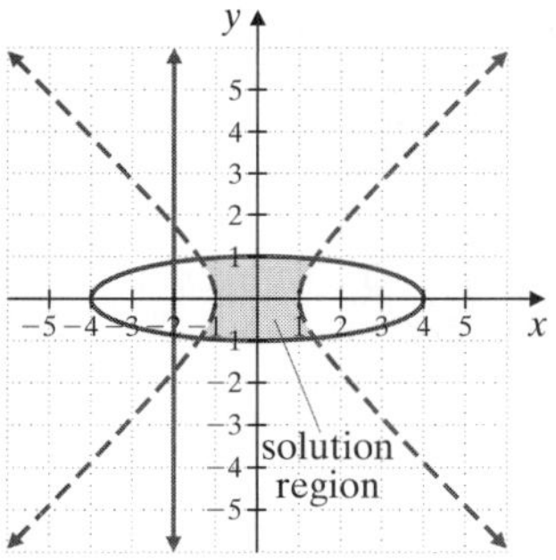

37. This is not the graph of a function because a vertical line can cross the graph in two places.

39. This is a function because it passes the vertical line test.

41. answers may vary

43. $\begin{cases} y \le x^2 \\ y \ge x + 2 \\ x \ge 0 \\ y \ge 0 \end{cases}$

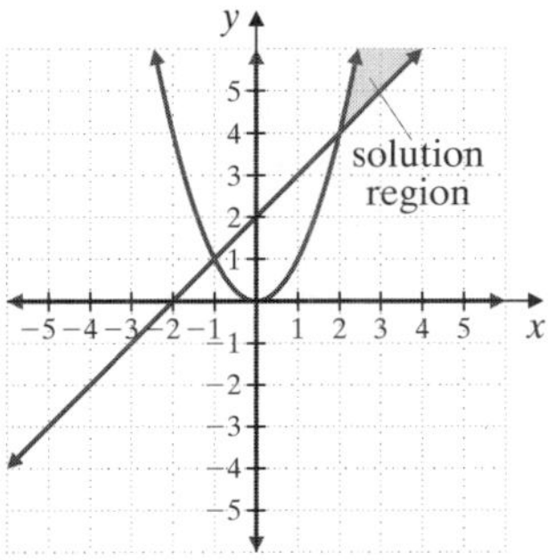

Chapter 10 Vocabulary Check

1. A(n) circle is the set of all points in a plane that are the same distance from a fixed point, called the center.

2. A <u>nonlinear system of equations</u> is a system of equations at least one of which is not linear.

3. A(n) <u>ellipse</u> is the set of points on a plane such that the sum of the distances of those points from two fixed points is a constant.

4. In a circle, the distance from the center to a point of the circle is called its <u>radius</u>.

5. A(n) <u>hyperbola</u> is the set of points in a plane such that the absolute value of the difference of the distance from two fixed points is constant.

Chapter 10 Review

1. $C(-4, 4)$, $r = 3$
$$[x-(-4)]^2 + (y-4)^2 = 3^2$$
$$(x+4)^2 + (y-4)^2 = 9$$

2. $C(-7, -9)$, $r = \sqrt{11}$
$$[x-(-7)]^2 + [y-(-9)]^2 = \left(\sqrt{11}\right)^2$$
$$(x+7)^2 + (y+9)^2 = 11$$

3. $C(5, 0)$, $d = 10$, $r = \frac{10}{2} = 5$
$$(x-5)^2 + (y-0)^2 = 5^2$$
$$(x-5)^2 + y^2 = 25$$

4. $C(0, 0)$, $d = 7$, $r = \frac{7}{2}$
$$(x-0)^2 + (y-0)^2 = \left(\frac{7}{2}\right)^2$$
$$x^2 + y^2 = \frac{49}{4}$$

5. $$x^2 + y^2 = 4$$
$$(x-0)^2 + (y-0)^2 = 2^2$$
$C(0, 0)$, $r = 2$

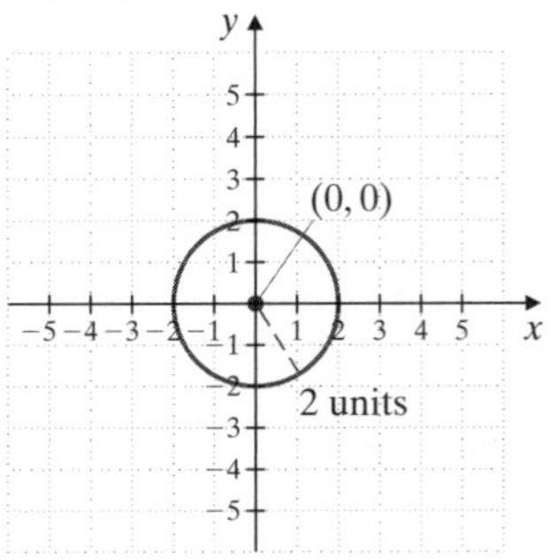

6. $x = 2(y-5)^2 + 4$
$V(4, 5)$
The parabola opens to the right.

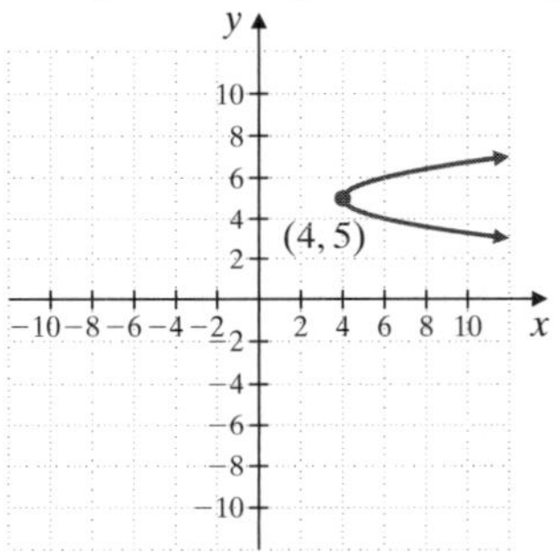

7. $x = -(y+2)^2 + 3$
$V(3, -2)$
The parabola opens to the left.

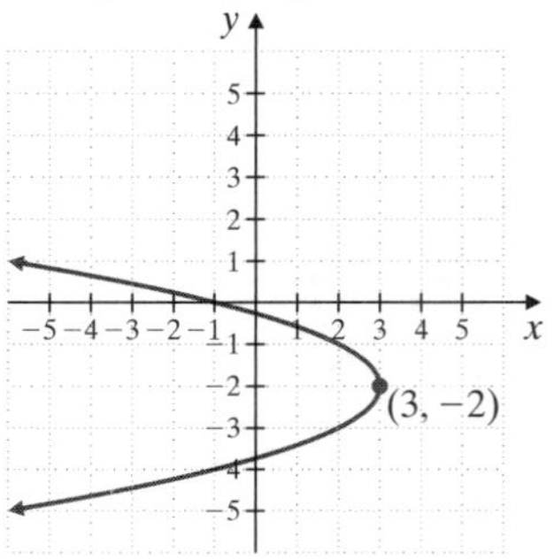

8. $(x-1)^2+(y-2)^2=4$

$(x-1)^2+(y-2)^2=2^2$

$C(1, 2), r=2$

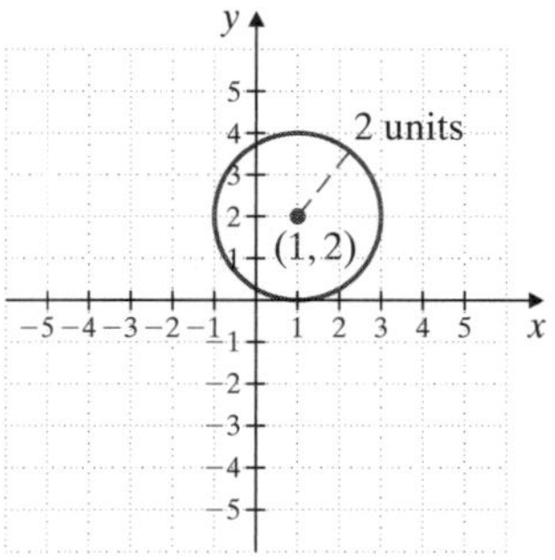

9. $y=-x^2+4x+10$

$y=-(x^2-4x)+10$

$y=-(x^2-4x+4)+10+4$

$y=-(x-2)^2+14$

$V(2, 14)$

The parabola opens downward.

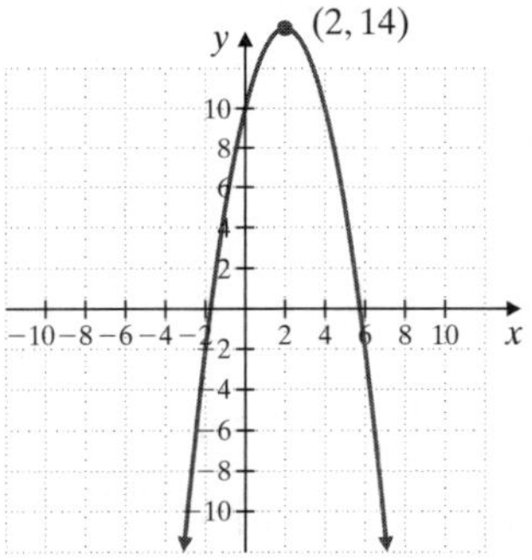

10. $x=-y^2-4y+6$

$x=-(y^2+4y)+6$

$x=-(y^2+4y+4)+6+4$

$x=-(y+2)^2+10$

$V(10, -2)$

The parabola opens to the left.

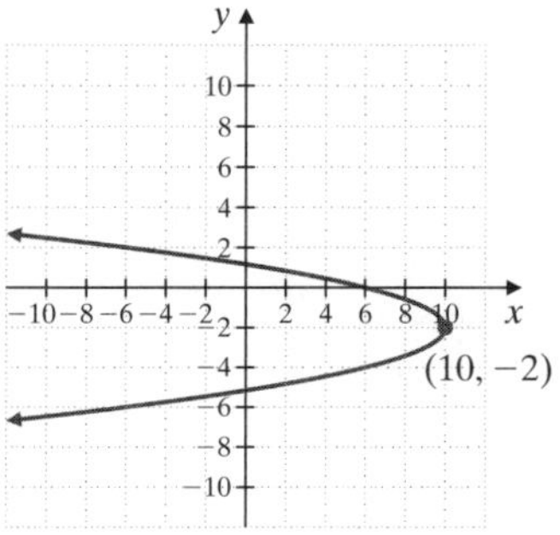

11. $x=\frac{1}{2}y^2+2y+1$

$x=\frac{1}{2}(y^2+4y)+1$

$x=\frac{1}{2}(y^2+4y+4)+1-2$

$x=\frac{1}{2}(y+2)^2-1$

$V(-1, -2)$

The parabola opens to the right.

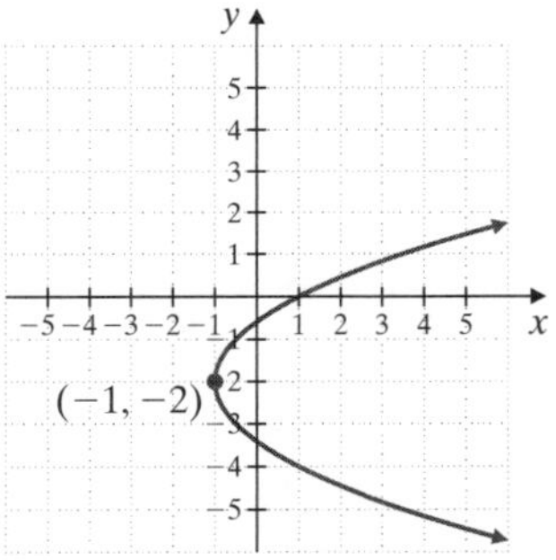

12. $y=-3x^2+\frac{1}{2}x+4$

$y=-3\left(x^2-\frac{1}{6}x\right)+4$

$y=-3\left(x^2-\frac{1}{6}x+\frac{1}{144}\right)+4+\frac{1}{48}$

$y=-3\left(x-\frac{1}{12}\right)^2+4\frac{1}{48}$

$V\left(\frac{1}{12}, 4\frac{1}{48}\right)$

The parabola opens downward.

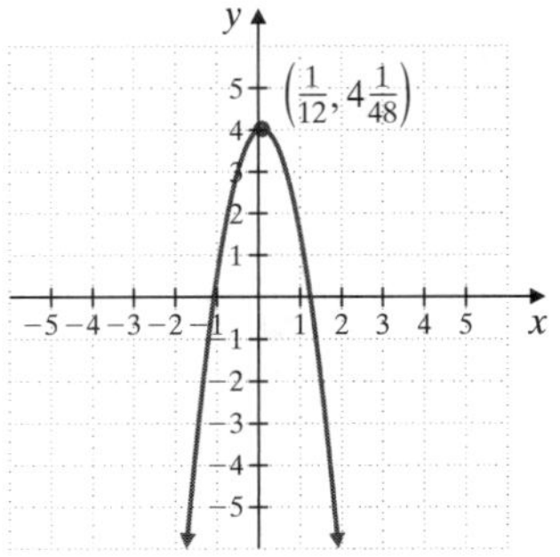

13.
$$x^2 + y^2 + 2x + y = \frac{3}{4}$$
$$(x^2 + 2x) + (y^2 + y) = \frac{3}{4}$$
$$(x^2 + 2x + 1) + \left(y^2 + y + \frac{1}{4}\right) = \frac{3}{4} + 1 + \frac{1}{4}$$
$$(x+1)^2 + \left(y + \frac{1}{2}\right)^2 = 2$$
$$(x+1)^2 + \left(y + \frac{1}{2}\right)^2 = \left(\sqrt{2}\right)^2$$
$$C\left(-1, -\frac{1}{2}\right),\ r = \sqrt{2}$$

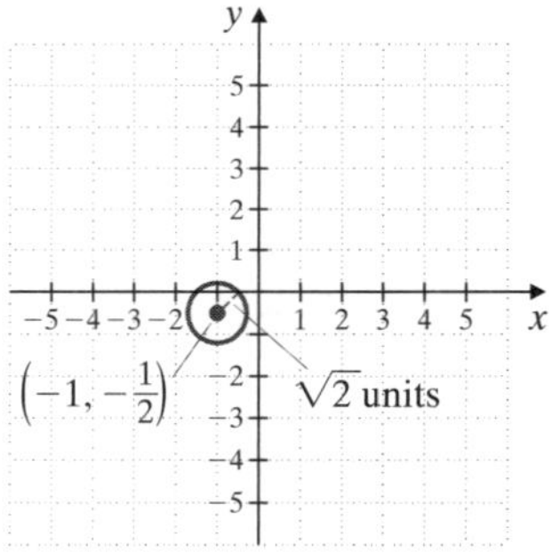

14.
$$x^2 + y^2 - 3y = \frac{7}{4}$$
$$x^2 + \left(y^2 - 3y + \frac{9}{4}\right) = \frac{7}{4} + \frac{9}{4}$$
$$x^2 + \left(y - \frac{3}{2}\right)^2 = \frac{16}{4}$$
$$(x-0)^2 + \left(y - \frac{3}{2}\right)^2 = 2^2$$
$$C\left(0, \frac{3}{2}\right),\ r = 2$$

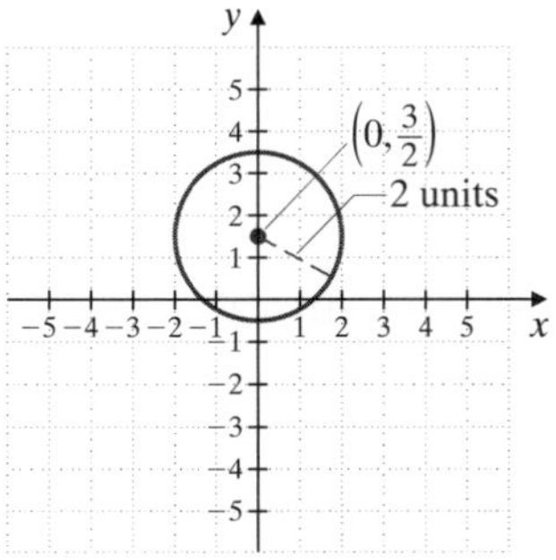

15.
$$4x^2 + 4y^2 + 16x + 8y = 1$$
$$x^2 + y^2 + 4x + 2y = \frac{1}{4}$$
$$(x^2 + 4x) + (y^2 + 2y) = \frac{1}{4}$$
$$(x^2 + 4x + 4) + (y^2 + 2y + 1) = \frac{1}{4} + 4 + 1$$
$$(x+2)^2 + (y+1)^2 = \frac{21}{4}$$
$$(x+2)^2 + (y+1)^2 = \left(\frac{\sqrt{21}}{2}\right)^2$$
$$C(-2, -1),\ r = \frac{\sqrt{21}}{2}$$

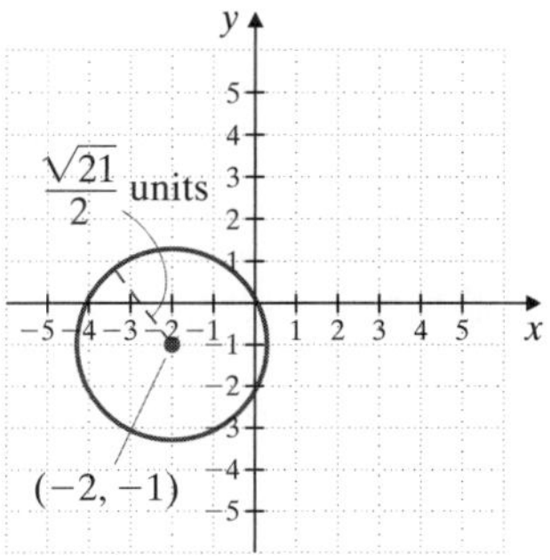

16. $x^2 + \frac{y^2}{4} = 1$

$\frac{x^2}{1^2} + \frac{y^2}{2^2} = 1$

$C(0, 0)$

x-intercepts: (−1, 0), (1, 0)

y-intercepts: (0, −2), (0, 2)

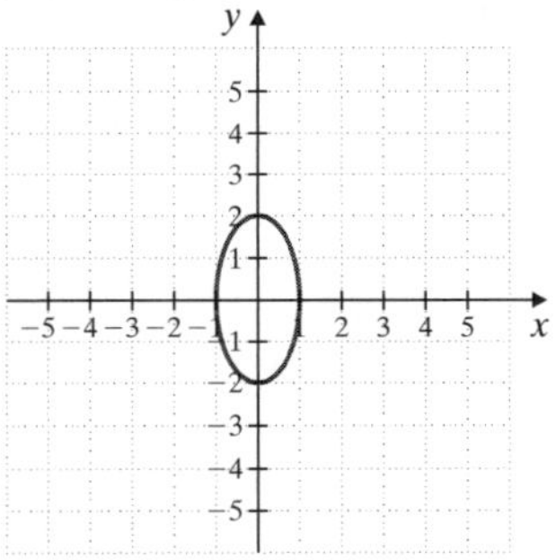

17. $x^2 - \frac{y^2}{4} = 1$

$\frac{x^2}{1^2} - \frac{y^2}{2^2} = 1$

$a = 1, b = 2$

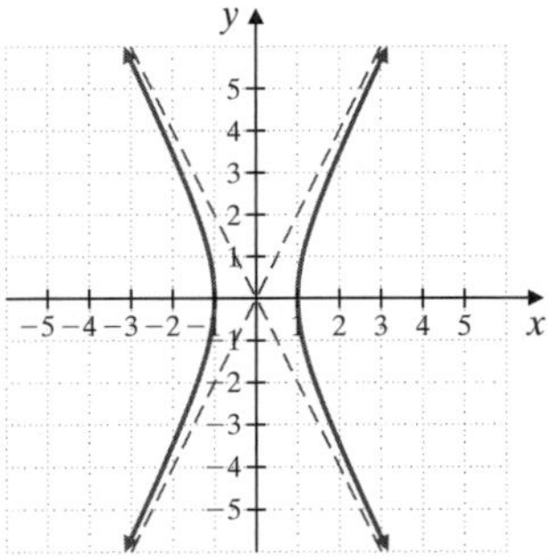

18. $-5x^2 + 25y^2 = 125$

$-\frac{x^2}{25} + \frac{y^2}{5} = 1$

$\frac{y^2}{\left(\sqrt{5}\right)^2} - \frac{x^2}{5^2} = 1$

$a = 5,\ b = \sqrt{5}$

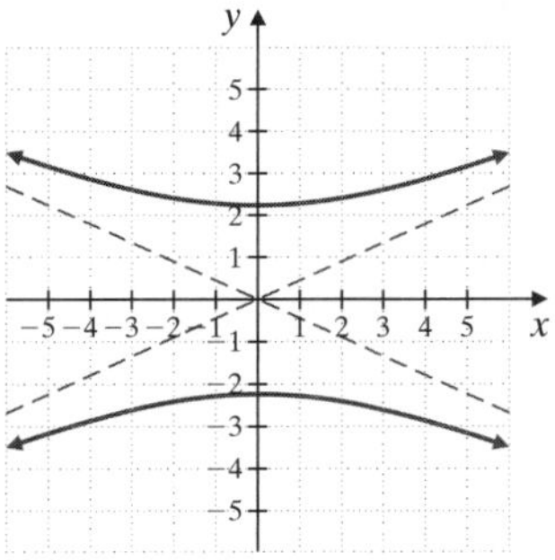

19. $4y^2 + 9x^2 = 36$

$\frac{y^2}{9} + \frac{x^2}{4} = 1$

$\frac{x^2}{2^2} + \frac{y^2}{3^2} = 1$

$C(0, 0)$

x-intercepts: (−2, 0), (2, 0)

y-intercepts: (0, −3), (0, 3)

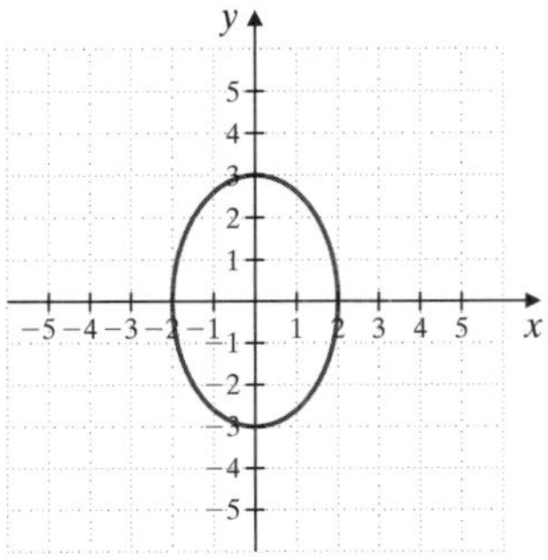

20. $\frac{(x+3)^2}{9} + \frac{(y-4)^2}{25} = 1$

$\frac{(x+3)^2}{3^2} + \frac{(y-4)^2}{5^2} = 1$

$C(-3, 4)$

other points:

(−3 + 3, 4), or (0, 4)

$(-3-3, 4)$, or $(-6, 4)$
$(-3, 4-5)$, or $(-3, -1)$
$(-3, 4+5)$, or $(-3, 9)$

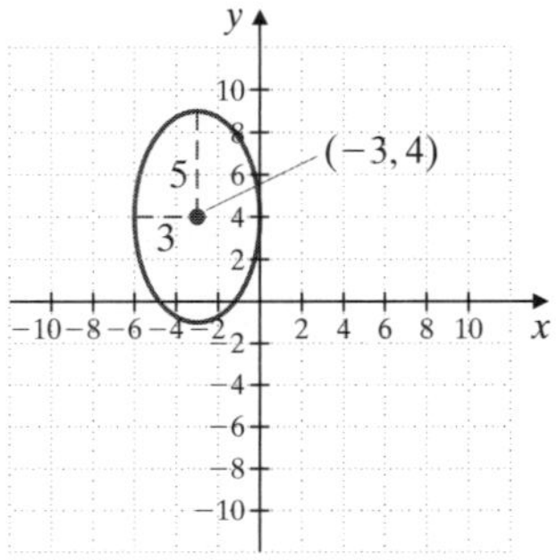

21. $x^2 - y^2 = 1$

$$\frac{x^2}{1^2} - \frac{y^2}{1^2} = 1$$

$a = 1, b = 1$

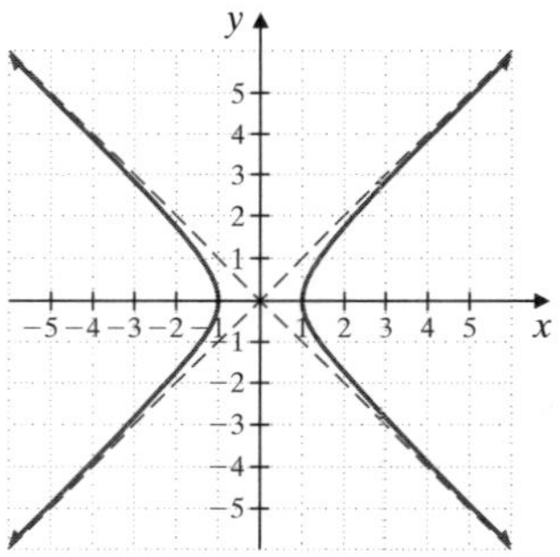

22. $36y^2 - 49x^2 = 1764$

$$\frac{y^2}{49} - \frac{x^2}{36} = 1$$

$$\frac{y^2}{7^2} - \frac{x^2}{6^2} = 1$$

$a = 6, b = 7$

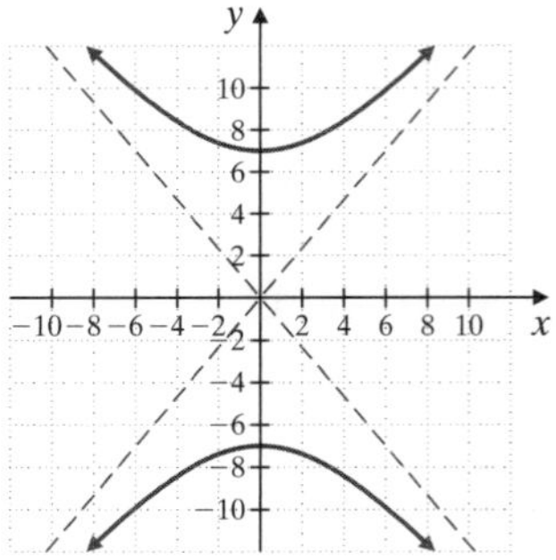

23. $y = x^2 + 9$

$V(0, 9)$

The parabola opens upward.

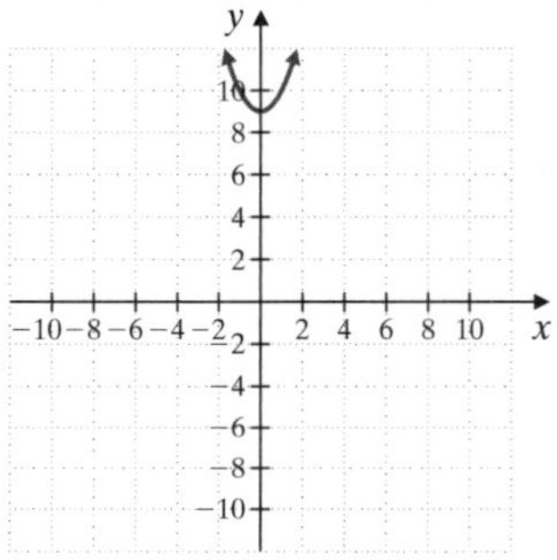

24. $x = 4y^2 - 16$

$V(-16, 0)$

The parabola opens to the right.

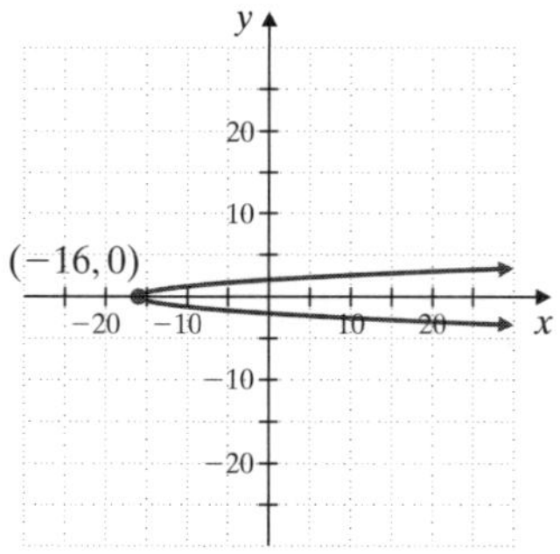

25. $y = x^2 + 4x + 6$

$y = (x^2 + 4x + 4) + 6 - 4$

$y = (x+2)^2 + 2$

$V(-2, 2)$

The parabola opens upward.

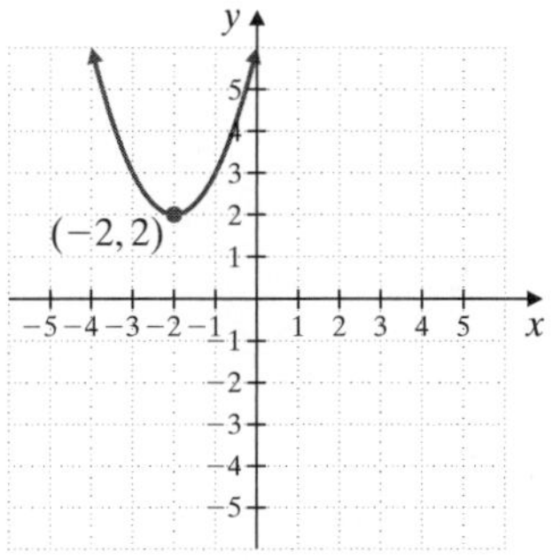

26.

$$y^2+2x^2=4x+6$$
$$y^2+2x^2-4x-6=0$$
$$y^2+2(x^2-2x)-6=0$$
$$y^2+2(x^2-2x+1)-6-2=0$$
$$y^2+2(x-1)^2-8=0$$
$$y^2+2(x-1)^2=8$$
$$\frac{y^2}{8}+\frac{(x-1)^2}{4}=1$$
$$\frac{y^2}{\left(\sqrt{8}\right)^2}+\frac{(x-1)^2}{2^2}=1$$

$C(1, 0)$

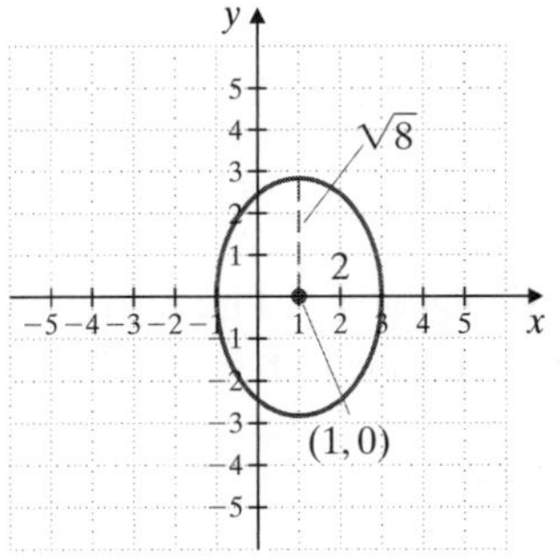

27. $x-4y=y^2$

$$x=y^2+4y$$
$$x=y^2+4y+4-4$$
$$x=(y+2)^2-4$$

$V(-4, -2)$

The parabola opens to the right.

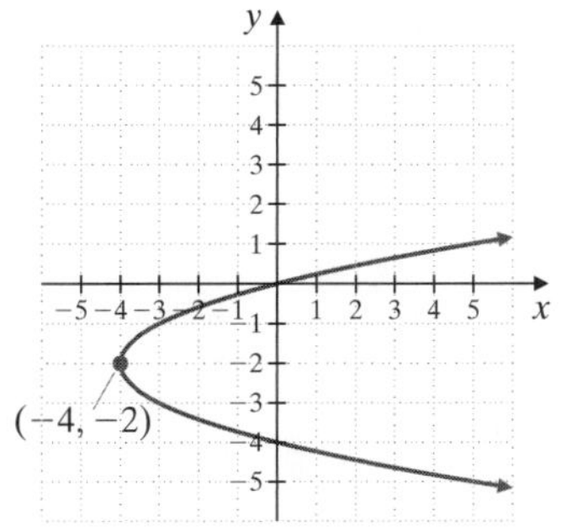

28. $x^2-4=y^2$

$$x^2-y^2=4$$
$$\frac{x^2}{4}-\frac{y^2}{4}=1$$
$$\frac{x^2}{2^2}-\frac{y^2}{2^2}=1$$

$a=2, b=2$

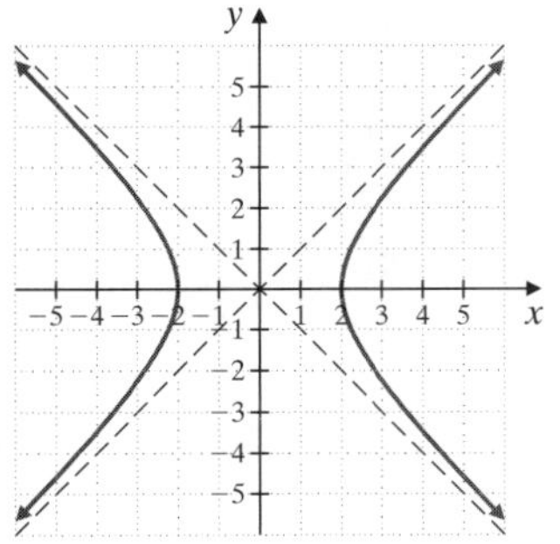

29. $x^2=4-y^2$

$$x^2+y^2=4$$
$$x^2+y^2=2^2$$

$C(0, 0), r=2$

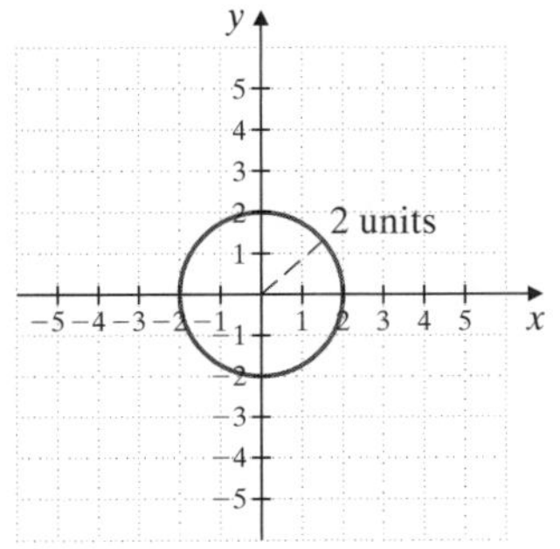

30. $36y^2 = 576 + 16x^2$

$36y^2 - 16x^2 = 576$

$\frac{y^2}{16} - \frac{x^2}{36} = 1$

$\frac{y^2}{4^2} - \frac{x^2}{6^2} = 1$

$a = 6, b = 4$

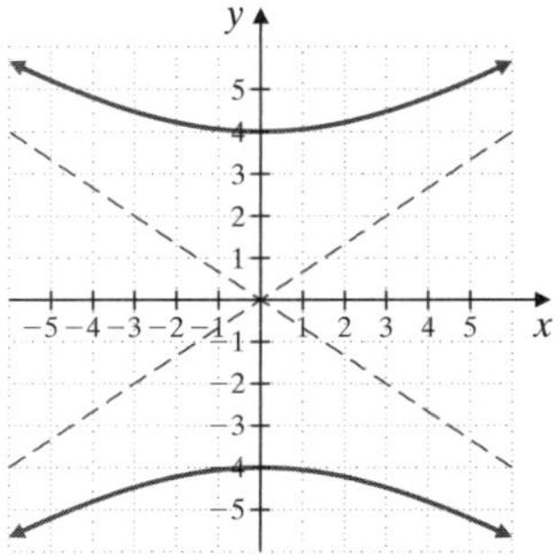

31. $3(x-7)^2 + 3(y+4)^2 = 1$

$(x-7)^2 + (y+4)^2 = \frac{1}{3}$

$(x-7)^2 + (y+4)^2 = \left(\sqrt{\frac{1}{3}}\right)^2$

$(x-7)^2 + (y+4)^2 = \left(\frac{\sqrt{3}}{3}\right)^2$

$C(7, -4)$

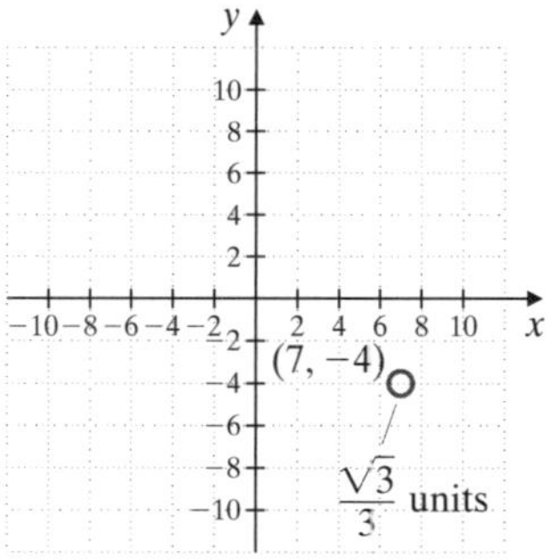

32. $\begin{cases} y = 2x - 4 \\ y^2 = 4x \end{cases}$

Substitute.

$y^2 = 4x$

$(2x-4)^2 = 4x$

$4x^2 - 16x + 16 = 4x$

$4x^2 - 16x + 16 = 4x$

$4x^2 - 20x + 16 = 0$

$4(x^2 - 5x + 4) = 0$

$4(x-4)(x-1) = 0$

$x - 4 = 0$ or $x - 1 = 0$

$x = 4 \qquad x = 1$

$x = 4$: $y = 2(4) - 4 = 4$

$x = 1$: $y = 2(1) - 4 = -2$

The solution set is $\{(1, -2), (4, 4)\}$.

33. $\begin{cases} x^2 + y^2 = 4 \\ x - y = 4 \end{cases}$

Solve equation 2 for x.

$x = 4 + y$

Substitute.

$x^2 + y^2 = 4$

$(4+y)^2 + y^2 = 4$

$16 + 8y + y^2 + y^2 = 4$

$2y^2 + 8y + 12 = 0$

$a = 2, b = 8, c = 12$

$y = \frac{-8 \pm \sqrt{8^2 - 4(2)(12)}}{2(2)} = \frac{-8 \pm \sqrt{-32}}{4}$

Since $\sqrt{-32}$ is not a real number, there is no solution. The solution set is $\varnothing$.

34. $\begin{cases} y = x + 2 \\ y = x^2 \end{cases}$

Substitute.

$x + 2 = x^2$

$0 = x^2 - x - 2$

$0 = (x-2)(x+1)$

$x - 2 = 0$ or $x + 1 = 0$

$x = 2 \qquad x = -1$

$x = 2$: $y = 2^2 = 4$

$x = -1$: $y = (-1)^2 = 1$

The solution set is $\{(-1, 1), (2, 4)\}$.

35. $\begin{cases} 4x - y^2 = 0 \\ 2x^2 + y^2 = 16 \end{cases}$

Add the equations.

$$2x^2 + 4x = 16$$
$$2x^2 + 4x - 16 = 0$$
$$2(x^2 + 2x - 8) = 0$$
$$2(x + 4)(x - 2) = 0$$
$$x + 4 = 0 \quad \text{or} \quad x - 2 = 0$$
$$x = -4 \qquad\qquad x = 2$$

$x = -4$: $4(-4) - y^2 = 0$
$$-16 = y^2$$
$$\pm\sqrt{-16} = y$$

$x = 2$: $4(2) - y^2 = 0$
$$8 = y^2$$
$$\pm\sqrt{8} = y$$
$$\pm 2\sqrt{2} = y$$

Since $\sqrt{-16}$ is not a real number, the solution set is $\left\{\left(2, 2\sqrt{2}\right), \left(2, -2\sqrt{2}\right)\right\}$.

36. $\begin{cases} x^2 + 4y^2 = 16 \\ x^2 + y^2 = 4 \end{cases}$

Multiply equation 2 by −1. Add the results.

$$\begin{array}{r} x^2 + 4y^2 = 16 \\ \underline{-x^2 - y^2 = -4} \\ 3y^2 = 12 \end{array}$$
$$y^2 = 4$$
$$y = \pm\sqrt{4}$$
$$y = \pm 2$$

$y = -2$: $x^2 + (-2)^2 = 4$
$$x^2 = 0$$
$$x = 0$$

$y = 2$: $x^2 + (2)^2 = 4$
$$x^2 = 0$$
$$x = 0$$

The solution set is $\{(0, 2), (0, -2)\}$.

37. $\begin{cases} x^2 + 2y = 9 \\ 5x - 2y = 5 \end{cases}$

Add the equations.

$$x^2 + 5x = 14$$
$$x^2 + 5x - 14 = 0$$
$$(x + 7)(x - 2) = 0$$
$$x + 7 = 0 \quad \text{or} \quad x - 2 = 0$$
$$x = -7 \qquad\qquad x = 2$$

$x = -7$: $5(-7) - 2y = 5$
$$-2y = 40$$
$$y = -20$$

$x = 2$: $5(2) - 2y = 5$
$$-2y = -5$$
$$y = \frac{5}{2}$$

The solution set is $\left\{\left(2, \frac{5}{2}\right), (-7, -20)\right\}$.

38. $\begin{cases} y = 3x^2 + 5x - 4 \\ y = 3x^2 - x + 2 \end{cases}$

Substitute.

$$3x^2 + 5x - 4 = 3x^2 - x + 2$$
$$5x - 4 = -x + 2$$
$$6x = 6$$
$$x = 1$$

$x = 1$: $y = 3(1)^2 + 5(1) - 4 = 4$

The solution set is $\{(1, 4)\}$.

39. $\begin{cases} x^2 - 3y^2 = 1 \\ 4x^2 + 5y^2 = 21 \end{cases}$

Solve equation 1 for x^2.

$$x^2 = 1 + 3y^2$$

Substitute.

$$4(1 + 3y^2) + 5y^2 = 21$$
$$4 + 12y^2 + 5y^2 = 21$$
$$17y^2 = 17$$
$$y^2 = 1$$
$$y = \pm 1$$

$y=-1$: $x^2=1+3(-1)^2$
$x^2=4$
$x=\pm 2$

$y=1$: $x^2=1+3(1)^2$
$x^2=4$
$x=\pm 2$

The solution set is $\{(-2,-1), (-2, 1), (2,-1), (2, 1)\}$.

40. Let x be the width and y be the length.

$$\begin{cases} xy=150 \\ 2x+2y=50 \end{cases}$$

Solve equation 2 for x.

$2x+2y=50$
$x+y=25$
$x=25-y$

Substitute.

$xy=150$
$(25-y)y=150$
$25y-y^2=150$
$0=y^2-25y+150$
$0=(y-15)(y-10)$

$y-15=0$ or $y-10=0$
$y=15$ $\quad y=10$

$y=15$: $x=25-15=10$
$y=10$: $x=25-10=15$

The dimensions are length = 15 feet and width = 10 feet.

41. An ellipse and a hyperbola can intersect at a maximum of 4 points. Therefore, there are a maximum of 4 real number solutions.

42. $y \le -x^2+3$

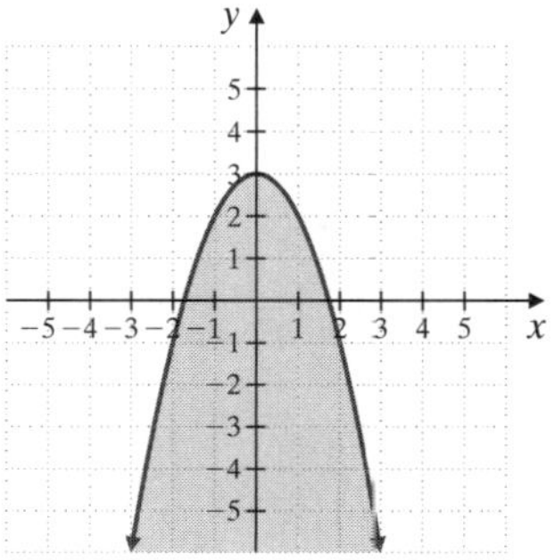

43. $x^2+y^2<9$

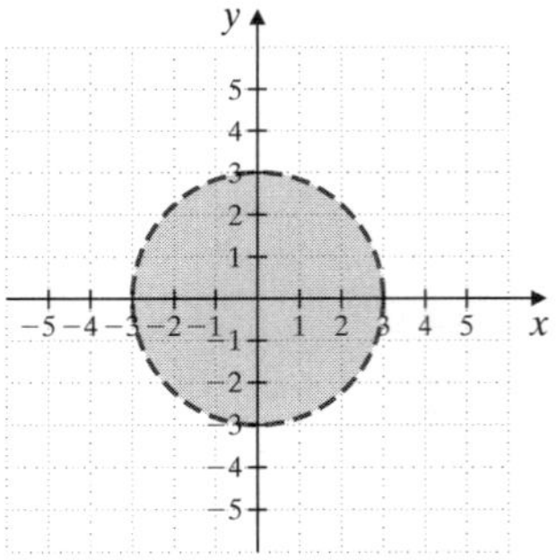

44. $\dfrac{x^2}{4}+\dfrac{y^2}{9}\ge 1$

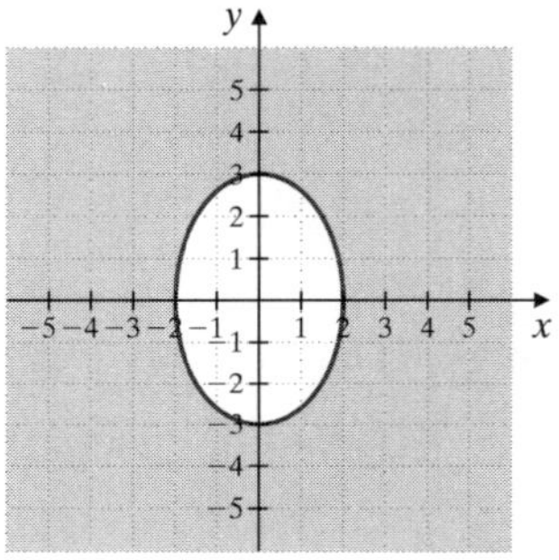

45. $\begin{cases} 2x \le 4 \\ x+y \ge 1 \end{cases}$

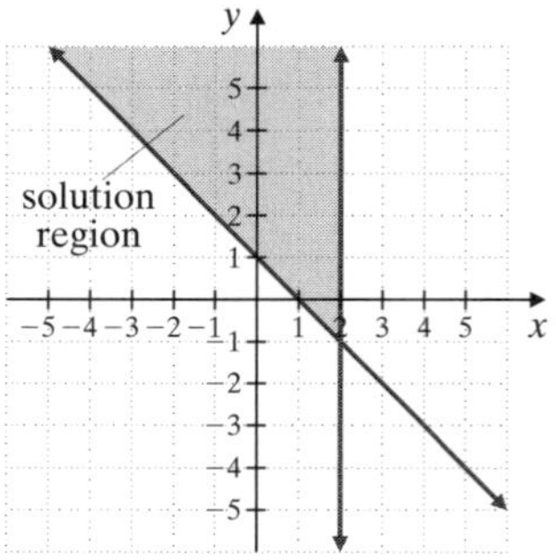

46. $\begin{cases} 3x+4y \le 12 \\ x-2y > 6 \end{cases}$

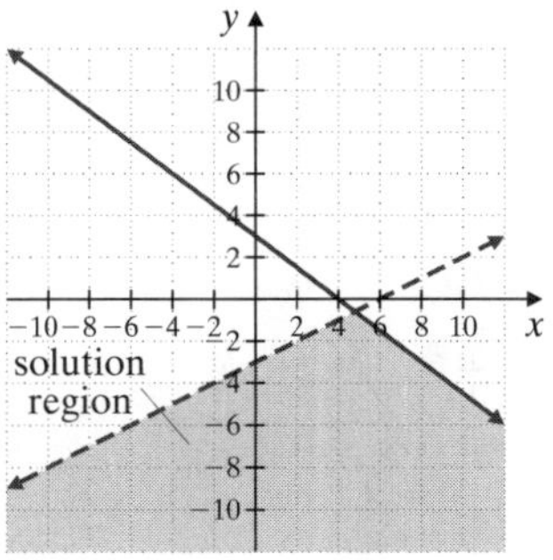

47. $\begin{cases} x^2+y^2 \le 16 \\ x^2+y^2 \ge 4 \end{cases}$

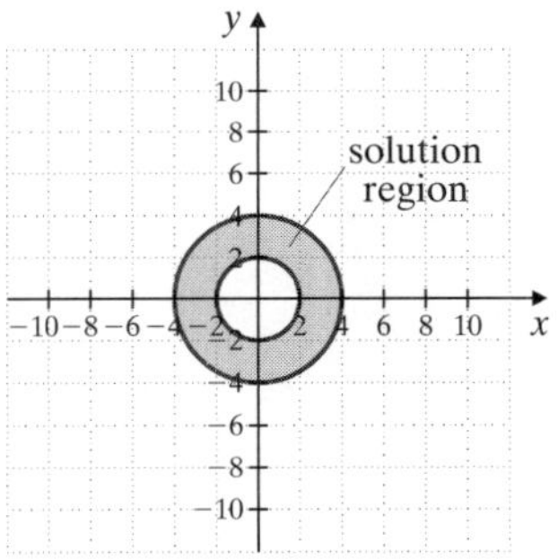

48. $\begin{cases} x^2+y^2 < 4 \\ x^2-y^2 \le 1 \end{cases}$

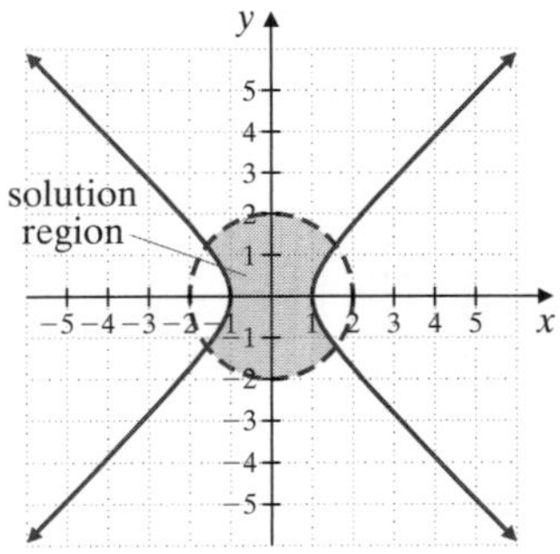

49. $\begin{cases} x^2+y^2 < 4 \\ y \ge x^2-1 \\ x \ge 0 \end{cases}$

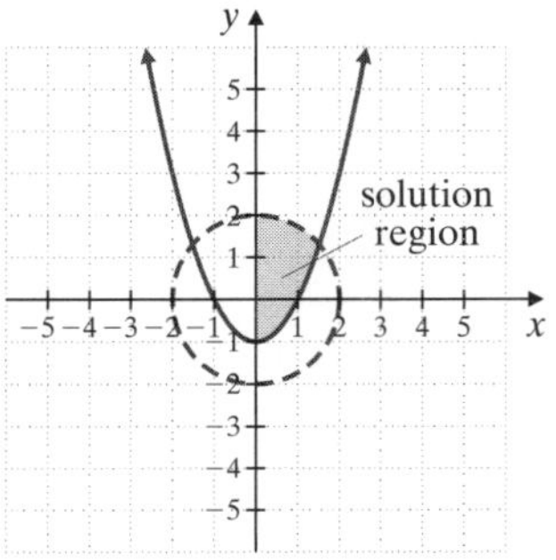

50. $C(-7, 8)$, $r = 5$

$$[x-(-7)]^2+(y-8)^2 = 5^2$$
$$(x+7)^2+(y-8)^2 = 25$$

51.
$$3x^2+6x+3y^2 = 9$$
$$x^2+2x+y^2 = 3$$
$$(x^2+2x+1)+y^2 = 3+1$$
$$(x+1)^2+y^2 = 4$$
$$(x+1)^2+(y-0)^2 = 2^2$$

$C(-1, 0)$, $r = 2$

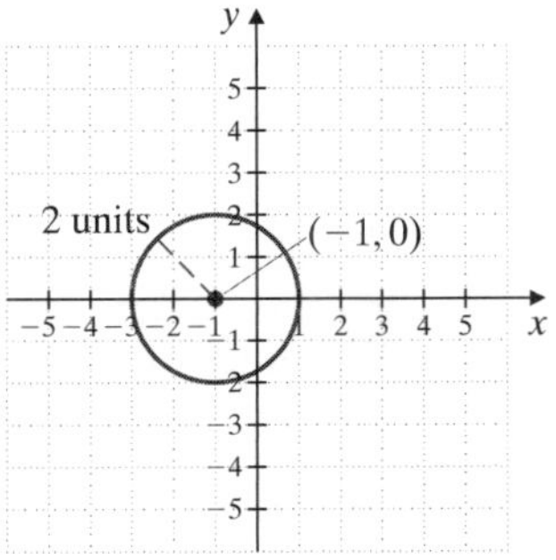

52. $y = x^2+6x+9$

$y = (x+3)^2$

$V(-3, 0)$

The parabola opens upward.

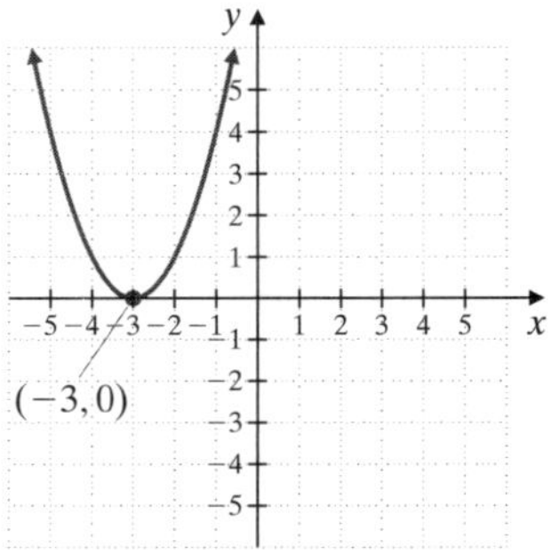

53. $x = y^2 + 6y + 9$

$x = (y+3)^2$

$V(0, -3)$

The parabola opens to the right.

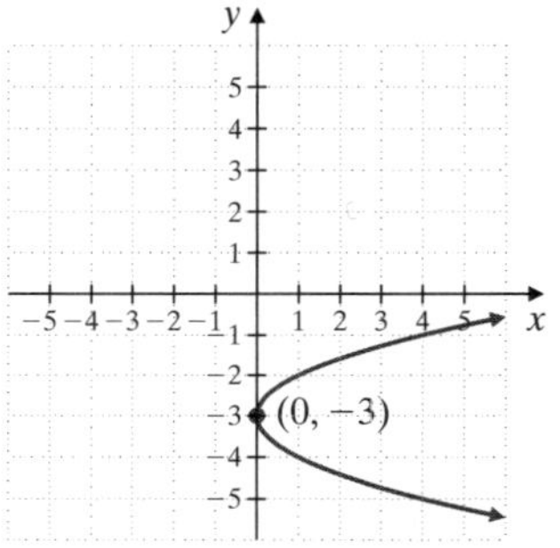

54. $\frac{y^2}{4} - \frac{x^2}{16} = 1$

$\frac{y^2}{2^2} - \frac{x^2}{4^2} = 1$

$a = 4, b = 2$

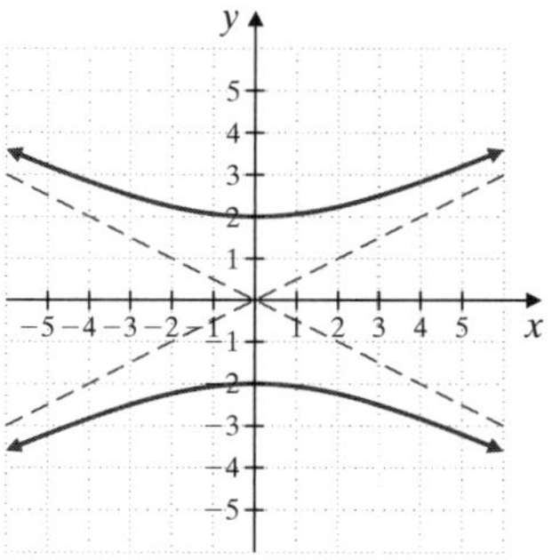

55. $\frac{y^2}{4} + \frac{x^2}{16} = 1$

$\frac{y^2}{2^2} + \frac{x^2}{4^2} = 1$

$C(0, 0)$

x-intercepts: $(-4, 0), (4, 0)$

y-intercepts: $(0, -2), (0, 2)$

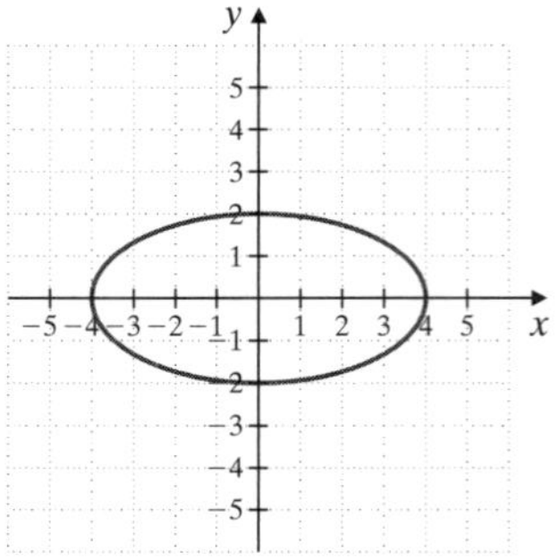

56. $\frac{(x-2)^2}{4} + (y-1)^2 = 1$

$\frac{(x-2)^2}{2^2} + \frac{(y-1)^2}{1^2} = 1$

$C(2, 1)$

other points:

$(2 - 2, 1)$, or $(0, 1)$

$(2 + 2, 1)$, or $(4, 1)$

$(2, 1 - 1)$, or $(2, 0)$

$(2, 1 + 1)$, or $(2, 2)$

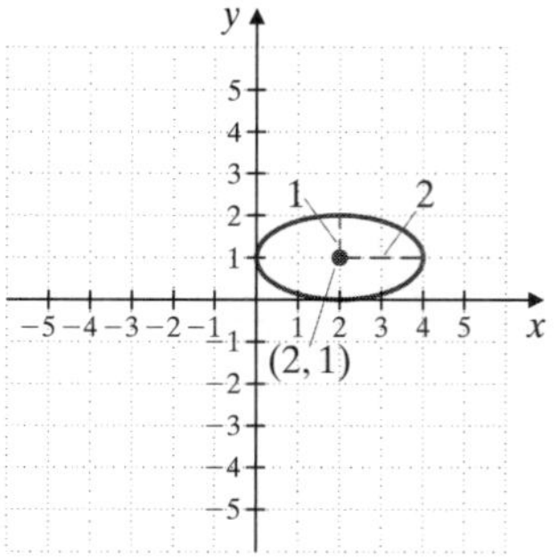

57.
$$y^2 = x^2 + 6$$
$$y^2 - x^2 = 6$$
$$\frac{y^2}{6} - \frac{x^2}{6} = 1$$
$$\frac{y^2}{\left(\sqrt{6}\right)^2} - \frac{x^2}{\left(\sqrt{6}\right)^2} = 1$$
$a = \sqrt{6}$, $b = \sqrt{6}$

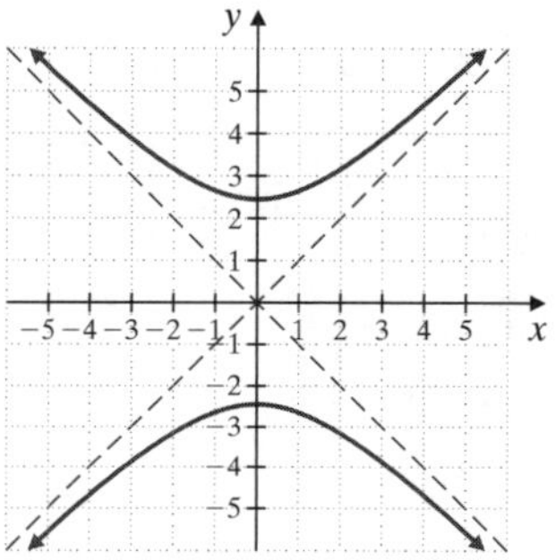

58.
$$y^2 + x^2 = 4x + 6$$
$$y^2 + (x^2 - 4x) = 6$$
$$y^2 + (x^2 - 4x + 4) = 6 + 4$$
$$y^2 + (x-2)^2 = 10$$
$$(x-2)^2 + y^2 = \left(\sqrt{10}\right)^2$$
$C(2, 0)$, $r = \sqrt{10}$

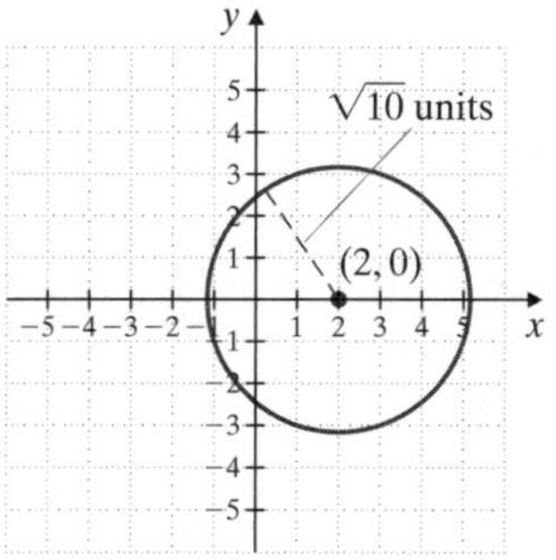

59.
$$x^2 + y^2 - 8y = 0$$
$$x^2 + (y^2 - 8y + 16) = 0 + 16$$
$$x^2 + (y-4)^2 = 16$$
$$(x-0)^2 + (y-4)^2 = 4^2$$
$C(0, 4)$, $r = 4$

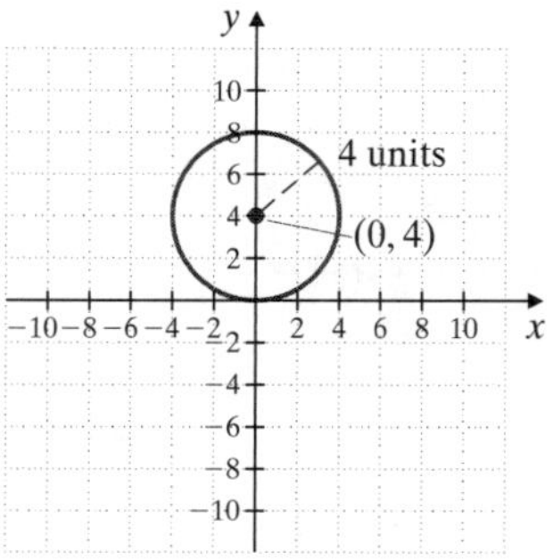

60. $6(x-2)^2 + 9(y+5)^2 = 36$
$$\frac{(x-2)^2}{6} + \frac{(y+5)^2}{4} = 1$$
$$\frac{(x-2)^2}{\left(\sqrt{6}\right)^2} + \frac{(y+5)^2}{2^2} = 1$$
$C(2, -5)$

other points:

$\left(2-\sqrt{6}, -5\right)$

$\left(2+\sqrt{6}, -5\right)$

$(2, -5 - 2)$, or $(2, -7)$

$(2, -5 + 2)$, or $(2, -3)$

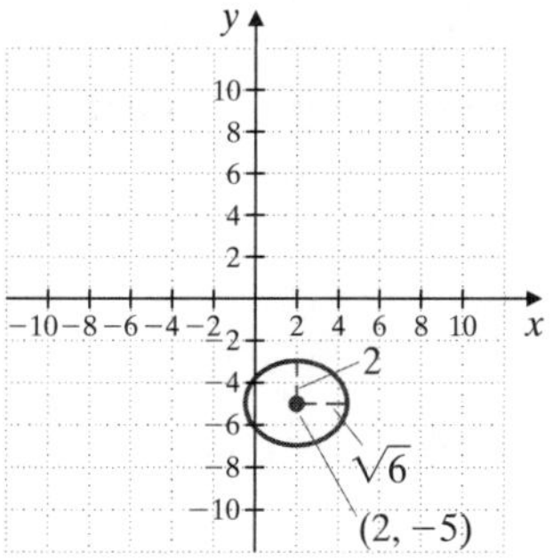

61. $\frac{x^2}{16} - \frac{y^2}{25} = 1$

$\frac{x^2}{4^2} - \frac{y^2}{5^2} = 1$

$a = 4, b = 5$

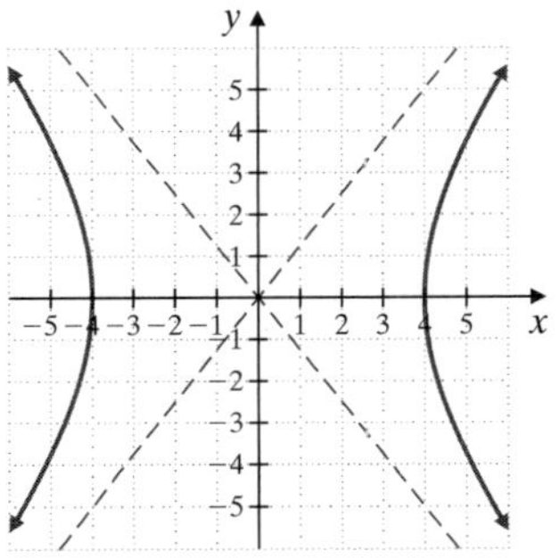

62. $\begin{cases} y = x^2 - 5x + 1 \\ y = -x + 6 \end{cases}$

Substitute.

$x^2 - 5x + 1 = -x + 6$

$x^2 - 4x - 5 = 0$

$(x-5)(x+1) = 0$

$x - 5 = 0$ or $x + 1 = 0$

$x = 5$ $\quad$ $x = -1$

$x = 5$: $y = -5 + 6 = 1$

$x = -1$: $y = -(-1) + 6 = 7$

The solution set is $\{(5, 1), (-1, 7)\}$.

63. $\begin{cases} x^2 + y^2 = 10 \\ 9x^2 + y^2 = 18 \end{cases}$

Multiply equation 1 by -1. Add the results.

$-x^2 - y^2 = -10$

$\underline{9x^2 + y^2 = 18}$

$8x^2 = 8$

$x^2 = 1$

$x = \pm\sqrt{1}$

$x = \pm 1$

$x = -1$: $(-1)^2 + y^2 = 10$

$y^2 = 9$

$y = \pm 3$

$x = 1$: $(1)^2 + y^2 = 10$

$y^2 = 9$

$y = \pm 3$

The problem set is $\{(-1, 3), (-1, -3), (1, 3), (1, -3)\}$.

64. $x^2 - y^2 < 1$

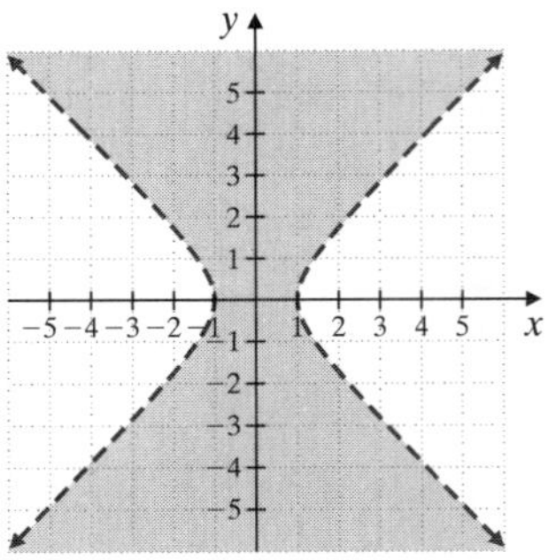

65. $\begin{cases} y > x^2 \\ x + y \geq 3 \end{cases}$

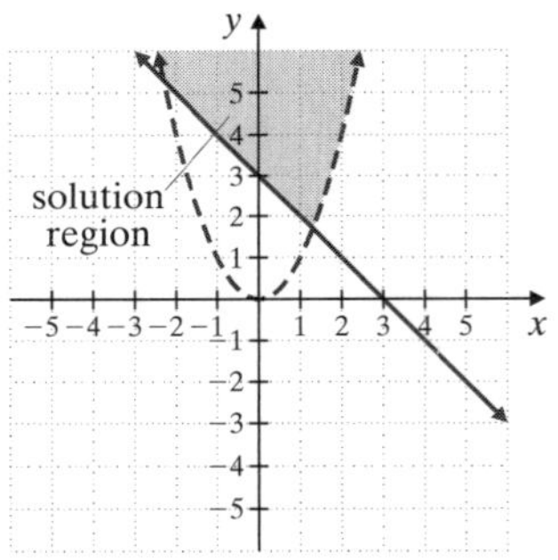

Chapter 10 Test

1. $x^2 + y^2 = 36$ or $(x-0)^2 + (y-0)^2 = 6^2$

Circle: $C(0, 0)$, $r = 6$

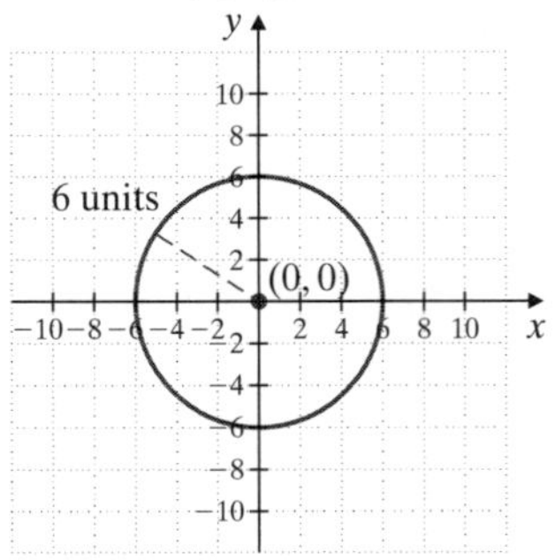

2. $x^2 - y^2 = 36$

$$\frac{x^2}{36} - \frac{y^2}{36} = 1$$

$$\frac{x^2}{6^2} - \frac{y^2}{6^2} = 1$$

Hyperbola: $a = 6$, $b = 6$

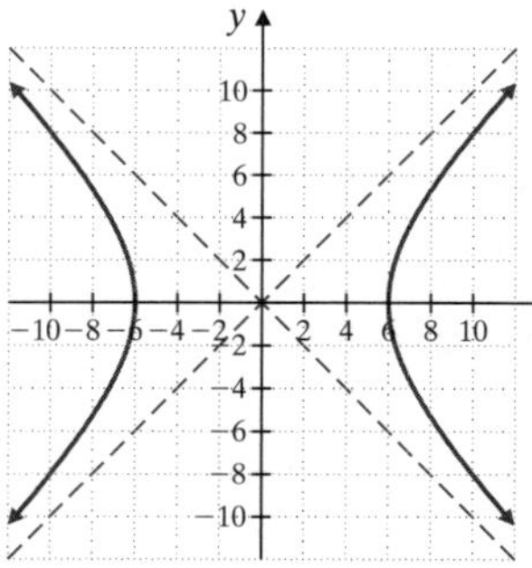

3. $16x^2 + 9y^2 = 144$

$$\frac{x^2}{9} + \frac{y^2}{16} = 1$$

$$\frac{x^2}{3^2} + \frac{y^2}{4^2} = 1$$

Ellipse:
$C(0, 0)$
x-intercepts: $(-3, 0)$, $(3, 0)$
y-intercepts: $(0, -4)$, $(0, 4)$

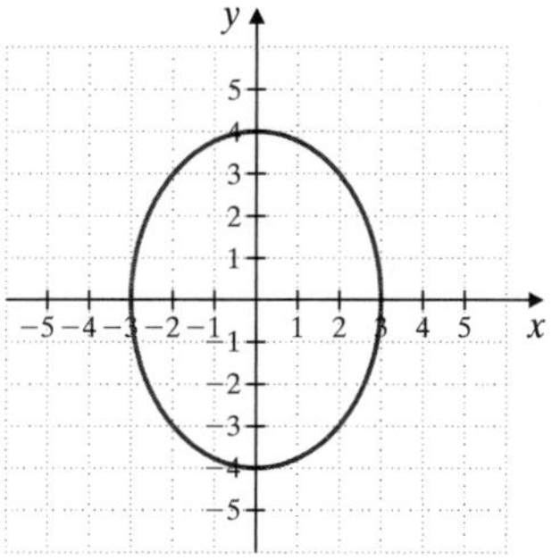

4. $y = x^2 - 8x + 16$

$y = (x - 4)^2$

Parabola: $V(4, 0)$
opens upward

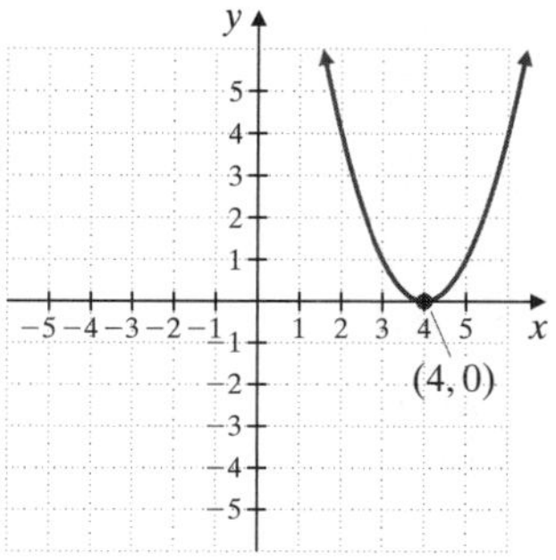

5.

$$x^2 + y^2 + 6x = 16$$
$$x^2 + 6x + y^2 = 16$$
$$(x^2 + 6x + 9) + y^2 = 16 + 9$$
$$(x + 3)^2 + y^2 = 5^2$$

Circle: $C(-3, 0)$, $r = 5$

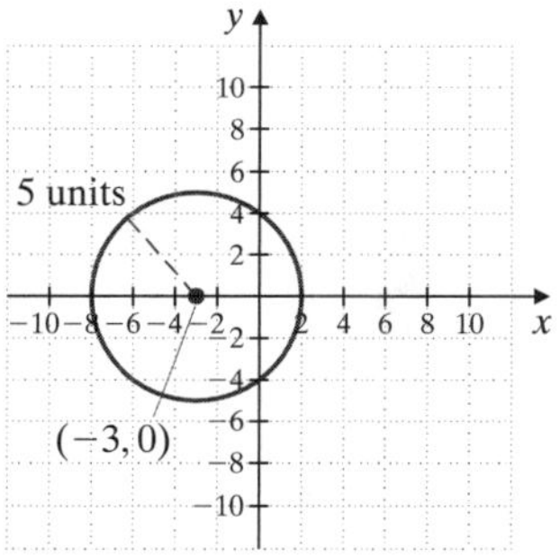

6. $x = y^2 + 8y - 3$

$x = (y^2 + 8y + 16) - 3 - 16$

$x = (y + 4)^2 - 19$

Parabola: $V(-19, -4)$,
opens to the right

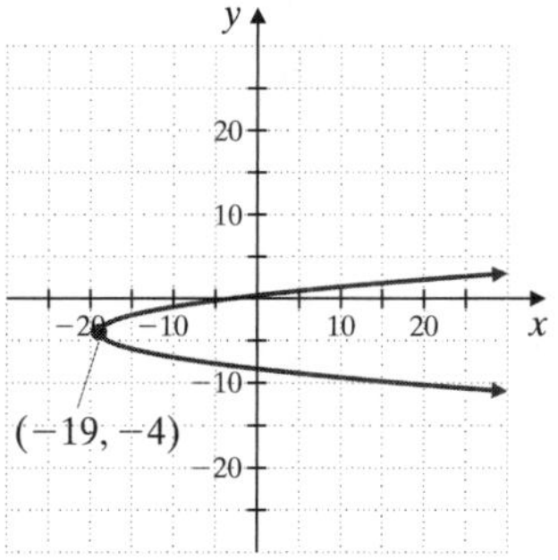

7. $\dfrac{(x-4)^2}{16}+\dfrac{(y-3)^2}{9}=1$

$\dfrac{(x-4)^2}{4^2}+\dfrac{(y-3)^2}{3^2}=1$

Ellipse: $C(4, 3)$
other points:
$(4 - 4, 3)$, or $(0, 3)$
$(4 + 4, 3)$, or $(8, 3)$
$(4, 3 - 3)$, or $(4, 0)$
$(4, 3 + 3)$, or $(4, 6)$

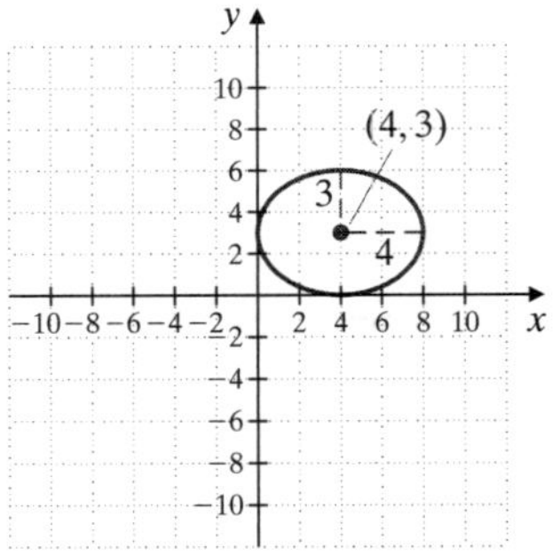

8. $y^2-x^2=1$

$\dfrac{y^2}{1^2}-\dfrac{x^2}{1^2}=1$

Hyperbola: $a = 1$, $b = 1$

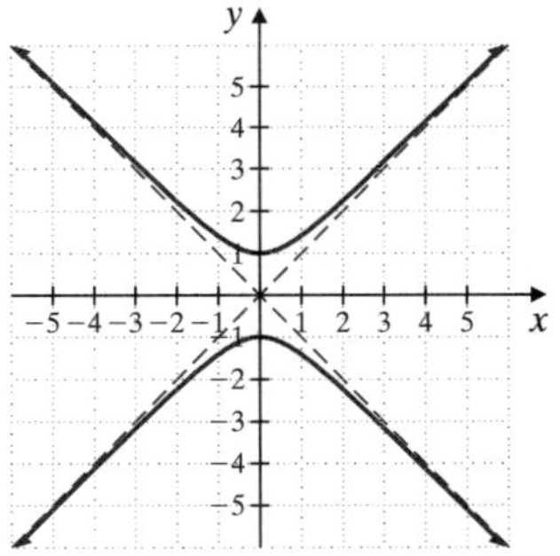

9. $\begin{cases} x^2+y^2=169 \\ 5x+12y=0 \end{cases}$

$12y=-5x$

$y=-\dfrac{5x}{12}$

Substitute.

$x^2+\left(-\dfrac{5x}{12}\right)^2=169$

$x^2+\dfrac{25x^2}{144}=169$

$\dfrac{169x^2}{144}=169$

$\dfrac{x^2}{144}=1$

$x^2=144$ so $x=\pm 12$.

Substitute back.

$x = 12$: $y=-\dfrac{5}{12}(12)=-5$

$x=-12$: $y=-\dfrac{5}{12}(-12)=5$

The solution set is $\{(12, -5), (-12, 5)\}$.

10. $\begin{cases} x^2+y^2=26 \\ x^2-2y^2=23 \end{cases}$

Multiply the second equation by -1. Add the results.

$$\begin{aligned} x^2+y^2&=26 \\ -x^2+2y^2&=-23 \\ \hline 3y^2&=3 \\ y^2&=1 \\ y&=\pm 1 \end{aligned}$$

$y=-1$: $x^2+(-1)^2=26$
$x^2=25$
$x=\pm 5$

$y = 1$: $x^2+1^2=26$
$x^2=25$
$x=\pm 5$

The solution set is
$\{(-5, -1), (-5, 1), (5, -1), (5, 1)\}$.

11. $\begin{cases} y=x^2-5x+6 \\ y=2x \end{cases}$

Substitute.

$x^2-5x+6=2x$
$x^2-7x+6=0$
$(x-6)(x-1)=0$
$x-6=0$ or $x-1=0$
$x=6$ $\quad x=1$
$x=6$: $y=2(6)=12$
$x=1$: $y=2(1)=2$
The solution set is $\{(6, 12), (1, 2)\}$.

12. $\begin{cases} x^2+4y^2=5 \\ y=x \end{cases}$

Substitute.
$x^2+4x^2=5$
$5x^2=5$
$x^2=1$
$x=\pm 1$
$x=-1$: $y=-1$
$x=1$: $y=1$
The solution set is $\{(-1, -1), (1, 1)\}$.

13. $\begin{cases} 2x+5y\ge 10 \\ y\ge x^2+1 \end{cases}$

First graph $\begin{cases} 2x+5y=10 \\ y=x^2+1 \end{cases}$ or

$\begin{cases} y=-\dfrac{2}{5}x+2 \\ y=1\cdot(x-0)^2+1 \end{cases}$

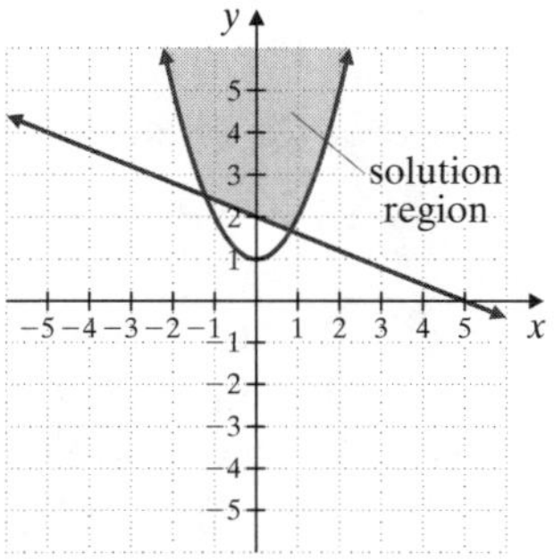

14. $\begin{cases} \dfrac{x^2}{4}+y^2\le 1 \\ x+y>1 \end{cases}$

First graph $\begin{cases} \dfrac{x^2}{2^2}+\dfrac{y^2}{1^2}=1 \\ x+y=1 \end{cases}$

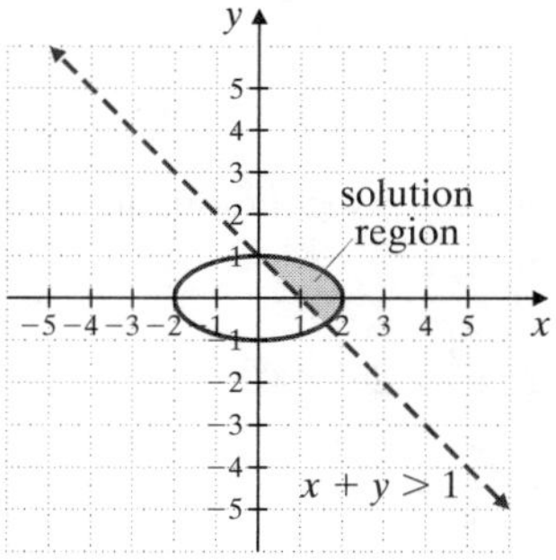

15. $\begin{cases} x^2+y^2>1 \\ \dfrac{x^2}{4}-y^2\ge 1 \end{cases}$

First graph $\begin{cases} x^2+y^2=1 \\ \dfrac{x^2}{2^2}-\dfrac{y^2}{1^2}=1 \end{cases}$

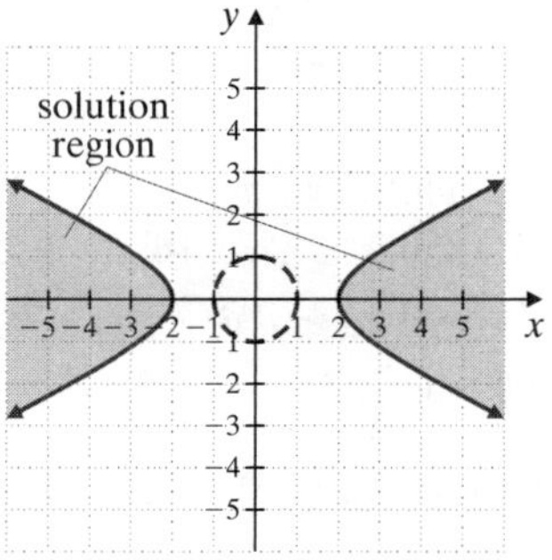

16. $\begin{cases} x^2+y^2\ge 4 \\ x^2+y^2<16 \\ y\ge 0 \end{cases}$

First graph $\begin{cases} x^2+y^2=2^2 \\ x^2+y^2=4^2 \\ y=0 \end{cases}$

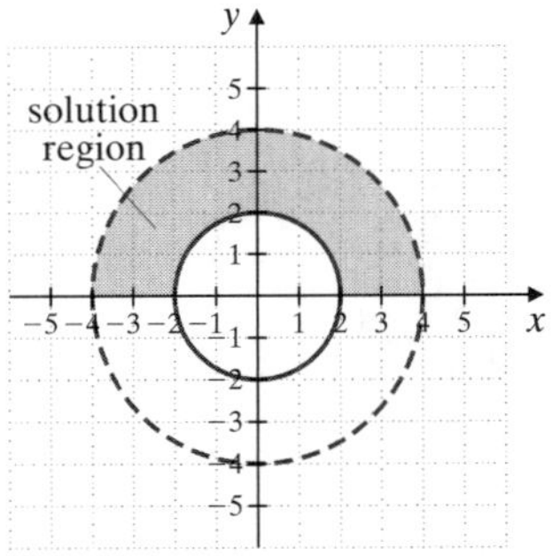

17. Graph B; vertex in second quadrant, opens to the right.

18. $100x^2 + 225y^2 = 22,500$
$$\frac{x^2}{225} + \frac{y^2}{100} = 1$$
$$\frac{x^2}{15^2} + \frac{y^2}{10^2} = 1$$
Height = 10 feet
Width = 2(15) = 30 feet

Cumulative Review Chapters 1–10

1. $(11x^3 - 12x^2 + x - 3) + (x^3 - 10x + 5)$
$= 11x^3 + x^3 - 12x^2 + x - 10x - 3 + 5$
$= 12x^3 - 12x^2 - 9x + 2$

2. $3x + 4 > 1$ and $2x - 5 \le 9$
$3x > -3$ and $2x \le 14$
$x > -1$ and $x \le 7$
The solution is (–1, 7].

3. $(x+3)(2x+5) = x(2x+5) + 3(2x+5)$
$= 2x^2 + 5x + 6x + 15$
$= 2x^2 + 11x + 15$

4. $m = \frac{y_2 - y_1}{x_2 - x_1} = \frac{-4-2}{1-3} = \frac{-6}{-2} = 3$

5.
–2	1	–2	–11	0	34
		–2	8	6	–12
	1	–4	–3	6	22

$(x^4 - 2x^3 - 11x^2 + 34) \div (x+2)$
$= x^3 - 4x^2 - 3x + 6 + \frac{22}{x+2}$

6.
	rate	time	distance = rate · time
North	r	2	$r(2)$
South	$r + 25$	2	$(r + 25)2$

The sum of their distances is 650 miles.
$2r + 2(r+25) = 650$
$2r + 2r + 50 = 650$
$4r = 600$
$r = 150$
$r + 25 = 175$
The planes are traveling at 150 mph and 175 mph.

7. $2(x - 5) + 3a(x - 5) = (x - 5)(2 + 3a)$

8. $3y^2 + 14y + 15 = (3y+5)(y+3)$

9. $x^2 - 12x + 35$
$(-5)(-7) = 35$ and $-5 - 7 = -12$
$x^2 - 12x + 35 = (x-5)(x-7)$

10. $20a^5 + 54a^4 + 10a^3 = 2a^3(10a^2 + 27a + 5)$
$= 2a^3(2a+5)(5a+1)$

11. $(x+3)^2 - 36 = (x+3)^2 - 6^2$
$= [(x+3) - 6][(x+3) + 6]$
$= (x + 3 - 6)(x + 3 + 6)$
$= (x-3)(x+9)$

12. Let $x = y - 3$.

$$(y-3)^2 - 2(y-3) - 8 = x^2 - 2x - 8$$
$$= (x-4)(x+2)$$

Replace x with $y - 3$.

$$(x-4)(x+2) = [(y-3)-4][(y-3)+2]$$
$$= (y-3-4)(y-3+2)$$
$$= (y-7)(y-1)$$

13. $2x^2 + 9x - 5 = 0$

$(2x-1)(x+5) = 0$

$2x - 1 = 0$ or $x + 5 = 0$

$x = \frac{1}{2}$ or $x = -5$

The solution set is $\left\{-5, \frac{1}{2}\right\}$.

14. The LCD is $(x + 3)(x - 3)$.

$$\frac{2}{x+3} = \frac{1}{x^2-9} - \frac{1}{x-3}$$
$$(x+3)(x-3)\left(\frac{2}{x+3}\right) = (x+3)(x-3)\cdot\frac{1}{(x+3)(x-3)} - (x+3)(x-3)\cdot\frac{1}{x-3}$$
$$2(x-3) = 1-(x+3)$$
$$2x-6 = 1-x-3$$
$$3x = 4$$
$$x = \frac{4}{3}$$

The solution set is $\left\{\frac{4}{3}\right\}$.

15. $\frac{2+x}{x+2} = \frac{x+2}{x+2} = 1$

16. $\frac{8x^3-1}{2x-1} = \frac{(2x-1)(4x^2+2x+1)}{2x-1} = 4x^2 + 2x + 1$

17. $\frac{2-x}{x-2} = \frac{-1(x-2)}{x-2} = -1$

18. $\dfrac{4x^9-25x^5y^2}{2x^7+5x^5y}=\dfrac{x^5(4x^4-25y^2)}{x^5(2x^2+5y)}$
$=\dfrac{x^5[(2x^2)^2-(5y)^2]}{x^5(2x^2+5y)}$
$=\dfrac{x^5(2x^2+5y)(2x^2-5y)}{x^5(2x^2+5y)}$
$=2x^2-5y$

19. $\dfrac{x^2}{x+7}-\dfrac{49}{x+7}=\dfrac{x^2-49}{x+7}=\dfrac{(x+7)(x-7)}{x+7}=x-7$

20. $\dfrac{2}{3a-15}-\dfrac{a}{25-a^2}$
$=\dfrac{2}{3(a-5)}-\dfrac{a}{(5-a)(5+a)}$
$=\dfrac{2}{3(a-5)}\cdot\dfrac{a+5}{a+5}-\dfrac{a}{-1(a-5)(a+5)}\cdot\dfrac{3}{3}$
$=\dfrac{2(a+5)}{3(a-5)(a+5)}+\dfrac{3a}{3(a-5)(a+5)}$
$=\dfrac{2(a+5)+3a}{3(a-5)(a+5)}$
$=\dfrac{2a+10+3a}{3(a-5)(a+5)}$
$=\dfrac{5a+10}{3(a-5)(a+5)}$
$=\dfrac{5(a+2)}{3(a-5)(a+5)}$

21. $\dfrac{x}{3y^2}-\dfrac{x+1}{3y^2}=\dfrac{x-(x+1)}{3y^2}=\dfrac{x-x-1}{3y^2}=-\dfrac{1}{3y^2}$

22. $\dfrac{x}{3x^2-13x-10}+\dfrac{9}{9x^2-4}=\dfrac{x}{(3x+2)(x-5)}+\dfrac{9}{(3x-2)(3x+2)}$

$$=\frac{x}{(3x+2)(x-5)}\cdot\frac{(3x-2)}{(3x-2)}+\frac{9}{(3x-2)(3x+2)}\cdot\frac{(x-5)}{(x-5)}$$
$$=\frac{x(3x-2)+9(x-5)}{(3x+2)(x-5)(3x-2)}$$
$$=\frac{3x^2-2x+9x-45}{(3x+2)(x-5)(3x-2)}$$
$$=\frac{3x^2+7x-45}{(3x+2)(x-5)(3x-2)}$$

23. $\dfrac{\frac{5x}{x+2}}{\frac{10}{x-2}}=\dfrac{5x}{x+2}\cdot\dfrac{x-2}{10}=\dfrac{5x(x-2)}{2\cdot 5(x+2)}=\dfrac{x(x-2)}{2(x+2)}$

24. $y=\dfrac{k}{x}$

$3=\dfrac{k}{\frac{2}{3}}$

$3=\dfrac{3k}{2}$

$6=3k$

$2=k$

$y=\dfrac{2}{x}$

25.
$$\frac{2x}{x-3}+\frac{6-2x}{x^2-9}=\frac{x}{x+3}$$
$$\frac{2x}{x-3}+\frac{-2(x-3)}{(x+3)(x-3)}=\frac{x}{x+3}$$
$$\frac{2x}{x-3}+\frac{-2}{x+3}=\frac{x}{x+3}$$
$$(x+3)(x-3)\left(\frac{2x}{x-3}\right)+(x+3)(x-3)\cdot\frac{-2}{x+3}=(x+3)(x-3)\cdot\frac{x}{x+3}$$
$$2x(x+3)-2(x-3)=x(x-3)$$
$$2x^2+6x-2x+6=x^2-3x$$
$$x^2+7x+6=0$$
$$(x+6)(x+1)=0$$

$x+6=0$ or $x+1=0$

$x=-6$ or $x=-1$

The solution set is $\{-6, -1\}$.

26. **a.** $\sqrt{5}\left(2+\sqrt{15}\right)=2\sqrt{5}+\sqrt{75}$
$=2\sqrt{5}+\sqrt{25\cdot 3}$
$=2\sqrt{5}+5\sqrt{3}$

b. $\left(\sqrt{3}-\sqrt{5}\right)\left(\sqrt{7}-1\right)$
$=\sqrt{21}-\sqrt{3}-\sqrt{35}+\sqrt{5}$

c. $\left(2\sqrt{5}-1\right)^2=\left(2\sqrt{5}\right)^2-2\cdot 2\sqrt{5}+1$
$=4\cdot 5-4\sqrt{5}+1$
$=20-4\sqrt{5}+1$
$=21-4\sqrt{5}$

d. $\left(3\sqrt{2}+5\right)\left(3\sqrt{2}-5\right)=\left(3\sqrt{2}\right)^2-5^2$
$=9\cdot 2-25$
$=18-25$
$=-7$

27. Let n be the number.

$$\frac{9-n}{19+n}=\frac{1}{3}$$
$$3(19+n)\cdot\frac{9-n}{19+n}=3(19+n)\cdot\frac{1}{3}$$
$$3(9-n)=19+n$$
$$27-3n=19+n$$
$$8=4n$$
$$2=n$$

The unknown number is 2.

28. $\frac{-2}{\sqrt{3}+3}=\frac{-2}{\sqrt{3}+3}\cdot\frac{\sqrt{3}-3}{\sqrt{3}-3}$
$=\frac{-2\left(\sqrt{3}-3\right)}{3-9}$
$=\frac{-2\left(\sqrt{3}-3\right)}{-6}$
$=\frac{\sqrt{3}-3}{3}$

29. $\sqrt{0}=0$ because $0^2=0$.

30. $\sqrt{\frac{x^7}{49}}=\frac{\sqrt{x^7}}{\sqrt{49}}=\frac{\sqrt{x^6\cdot x}}{7}=\frac{x^3\sqrt{x}}{7}$

31. $\sqrt{0.25}=\sqrt{(0.5)^2}=0.5$

32. $\sqrt{90a^5b^2}=\sqrt{9a^4b^2\cdot 10a}$
$=\sqrt{(3a^2b)^2\cdot 10a}$
$=3a^2b\sqrt{10a}$

33. $\sqrt[8]{x^4}=x^{4/8}=x^{1/2}=\sqrt{x}$

34. $\sqrt[4]{x^2y^4}=(x^2y^4)^{1/4}$
$=x^{2/4}y^{4/4}$
$=x^{1/2}y$
$=y\sqrt{x}$

35. $\sqrt[6]{r^2s^4}=(r^2s^4)^{1/6}$
$=r^{2/6}s^{4/6}$
$=r^{1/3}s^{2/3}$
$=(rs^2)^{1/3}$
$=\sqrt[3]{rs^2}$

36.
$$\sqrt{2x-3}=x-3$$
$$\left(\sqrt{2x-3}\right)^2=(x-3)^2$$
$$2x-3=x^2-6x+9$$
$$0=x^2-8x+12$$
$$0=(x-6)(x-2)$$
$x-6=0$ or $x-2=0$
$x=6$ or $x=2$

Discard $x = 2$ since it does not check. The solution set is $\{6\}$.

37. $\sqrt[3]{24}=\sqrt[3]{8\cdot 3}=\sqrt[3]{8}\cdot\sqrt[3]{3}=2\sqrt[3]{3}$

38. $\sqrt{144y^2}=\sqrt{(12y)^2}=12y$

39. $\sqrt[4]{32}=\sqrt[4]{16\cdot 2}=\sqrt[4]{16}\cdot\sqrt[4]{2}=2\sqrt[4]{2}$

40. $-\sqrt[3]{-27x^3} = -\sqrt[3]{(-3x)^3} = -(-3x) = 3x$

41. $\dfrac{2}{\sqrt{5}} = \dfrac{2}{\sqrt{5}} \cdot \dfrac{\sqrt{5}}{\sqrt{5}} = \dfrac{2\sqrt{5}}{5}$

42. a. $\dfrac{\sqrt{32}}{\sqrt{4}} = \sqrt{\dfrac{32}{4}} = \sqrt{8} = 2\sqrt{2}$

b. $\dfrac{\sqrt[3]{240y^2}}{5\sqrt[3]{3y^{-4}}} = \dfrac{1}{5}\sqrt[3]{\dfrac{240y^2}{3y^{-4}}}$

$= \dfrac{1}{5}\sqrt[3]{80y^6}$

$= \dfrac{2y^2\sqrt[3]{10}}{5}$

c. $\dfrac{\sqrt[5]{64x^9y^2}}{\sqrt[5]{2x^2y^{-8}}} = \sqrt[5]{\dfrac{64}{2}x^{9-2}y^{2-(-8)}}$

$= \sqrt[5]{32x^7y^{10}}$

$= 2xy^2\sqrt[5]{x^2}$

43. $\sqrt{-10x-1} + 3x = 0$

$\sqrt{-10x-1} = -3x$

$\left(\sqrt{-10x-1}\right)^2 = (-3x)^2$

$-10x - 1 = 9x^2$

$9x^2 + 10x + 1 = 0$

$(9x+1)(x+1) = 0$

$9x + 1 = 0$ or $x + 1 = 0$

$x = -\dfrac{1}{9}$ $\quad$ $x = -1$

The solution set is $\left\{-\dfrac{1}{9}, -1\right\}$.

44. $x^3 + 2x^2 - 4x \ge 8$

$x^3 + 2x^2 - 4x - 8 \ge 0$

$x^2(x+2) - 4(x+2) \ge 0$

$(x+2)(x^2-4) \ge 0$

$(x+2)(x+2)(x-2) \ge 0$

Region	Test Point Value	$(x+2)(x+2)(x-2) \ge 0$	Result
$(-\infty, -2)$	-3	$(-1)(-1)(-5) \ge 0$	False
$(-2, 2)$	0	$(2)(2)(-2) \ge 0$	False
$(2, \infty)$	3	$(5)(5)(1) \ge 0$	True

The solution set is $[2, \infty)$.

45. $$\begin{aligned}(2-5i)(4+i) &= 2(4+i)-5i(4+i)\\ &= 8+2i-20i-5i^2\\ &= 8+2i-20i-5(-1)\\ &= 8-18i+5\\ &= 13-18i\end{aligned}$$

46. $$\begin{aligned}f(x) &= \frac{x+1}{2}\\ y &= \frac{x+1}{2}\\ x &= \frac{y+1}{2}\\ 2x &= y+1\\ 2x-1 &= y\\ f^{-1}(x) &= 2x-1\end{aligned}$$

47. $$\begin{aligned}p^2+2p &= 4\\ p^2+2p+\left(\frac{2}{2}\right)^2 &= 4+\left(\frac{2}{2}\right)^2\\ p^2+2p+1 &= 4+1\\ (p+1)^2 &= 5\\ p+1 &= \pm\sqrt{5}\\ p &= -1\pm\sqrt{5}\end{aligned}$$

The solution set is $\left\{-1+\sqrt{5}, -1-\sqrt{5}\right\}$.

48. $f(x) = -x^2 - 6x + 4$

$a = -1, b = -6, c = 4$

$$x = \frac{-b}{2a} = \frac{-(-6)}{2(-1)} = -3$$

$$\begin{aligned} f(-3) &= -(-3)^2 - 6(-3) + 4 \\ &= -9 + 18 + 4 \\ &= 13 \end{aligned}$$

The maximum value is 13.

49. $2x^2 - 4x = 3$

$2x^2 - 4x - 3 = 0$

$a = 2, b = -4, c = -3$

$$\begin{aligned} x &= \frac{-b \pm \sqrt{b^2 - 4ac}}{2a} \\ &= \frac{-(-4) \pm \sqrt{(-4)^2 - 4(2)(-3)}}{2(2)} \\ &= \frac{4 \pm \sqrt{16 + 24}}{4} \\ &= \frac{4 \pm \sqrt{40}}{4} \\ &= \frac{4 \pm 2\sqrt{10}}{4} \\ &= \frac{2\left(2 \pm \sqrt{10}\right)}{2 \cdot 2} \\ &= \frac{2 \pm \sqrt{10}}{2} \end{aligned}$$

The solution set is $\left\{\frac{2-\sqrt{10}}{2}, \frac{2+\sqrt{10}}{2}\right\}$.

50. a. $64^x = 4$

$$\begin{aligned} (4^3)^x &= 4 \\ 4^{3x} &= 4^1 \\ 3x &= 1 \\ x &= \frac{1}{3} \end{aligned}$$

The solution set is $\left\{\frac{1}{3}\right\}$.

b.
$$\begin{aligned} 125^{x-3} &= 25 \\ (5^3)^{x-3} &= 5^2 \\ 5^{3x-9} &= 5^2 \\ 3x-9 &= 2 \\ 3x &= 11 \\ x &= \frac{11}{3} \end{aligned}$$

The solution set is $\left\{\frac{11}{3}\right\}$.

c.
$$\begin{aligned} \frac{1}{81} &= 3^{2x} \\ 3^{-4} &= 3^{2x} \\ -4 &= 2x \\ -2 &= x \end{aligned}$$

The solution set is $\{-2\}$.

51.
$$\begin{aligned} d &= \sqrt{(y_2-y_1)^2+(x_2-x_1)^2} \\ &= \sqrt{[-4-(-5)]^2+(1-2)^2} \\ &= \sqrt{1^2+(-1)^2} \\ &= \sqrt{2} \\ &\approx 1.414 \end{aligned}$$

52. The LCD is $(x-2)(x+2)$.

$$\frac{4}{x-2}-\frac{x}{x+2}=\frac{16}{x^2-4}$$
$$(x-2)(x+2)\left(\frac{4}{x-2}\right)-(x-2)(x+2)\left(\frac{x}{x+2}\right)=(x-2)(x+2)\left(\frac{16}{x^2-4}\right)$$
$$\begin{aligned} 4(x+2)-x(x-2) &= 16 \\ 4x+8-x^2+2x &= 16 \\ 0 &= x^2-6x+8 \\ 0 &= (x-4)(x-2) \end{aligned}$$

$x-4=0$ or $x-2=0$
$x=4$ or $x=2$

Discard $x = 2$ since $\frac{4}{2-2}=\frac{4}{0}$ is undefined. The solution set is $\{4\}$.

53. $\frac{x^2}{9}+\frac{y^2}{16}=1$

$\frac{x^2}{3^2}+\frac{y^2}{4^2}=1$

Ellipse: $C(0, 0)$

x-intercepts: $(-3, 0)$, $(3, 0)$

y-intercepts: $(0, -4)$, $(0, 4)$

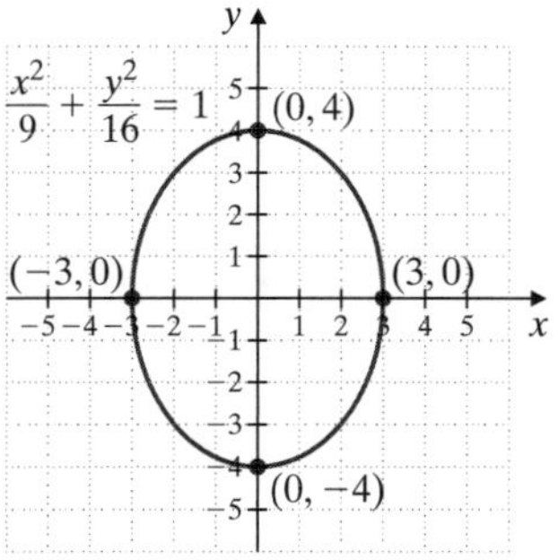

54. $f(x)=x^2-3x+2;\ g(x)=-3x+5$

a. $$\begin{aligned}(f\circ g)(x)&=f(g(x))\\&=f(-3x+5)\\&=(-3x+5)^2-3(-3x+5)+2\\&=9x^2-30x+25+9x-15+2\\&=9x^2-21x+12\end{aligned}$$

b. $$\begin{aligned}(f\circ g)(-2)&=f(g(-2))\\&=f(11)\\&=11^2-3(11)+2\\&=121-33+2\\&=90\end{aligned}$$

c. $$\begin{aligned}(g\circ f)(x)&=g(f(x))\\&=g(x^2-3x+2)\\&=-3(x^2-3x+2)+5\\&=-3x^2+9x-6+5\\&=-3x^2+9x-1\end{aligned}$$

d. $$\begin{aligned}(g\circ f)(5)&=g(f(5))\\&=g(12)\\&=-3(12)+5\\&=-31\end{aligned}$$

55. $\begin{cases}x^2+y^2=4\\ x+y=3\end{cases}$

Solve equation 2 for y.

$y=3-x$

Substitute.

$$\begin{aligned}x^2+(3-x)^2&=4\\x^2+9-6x+x^2&=4\\2x^2-6x+5&=0\end{aligned}$$

$a=2$, $b=-6$, $c=5$

$$x=\frac{-(-6)\pm\sqrt{(-6)^2-4(2)(5)}}{2(2)}$$

$$x=\frac{6\pm\sqrt{36-40}}{4}$$

$$x=\frac{6\pm\sqrt{-4}}{4}$$

Since $\sqrt{-4}$ is not a real number, there is no solution. The solution set is $\varnothing$.

56. $\begin{cases}x^2+y^2=36\\ y=x+6\end{cases}$

Substitute.

$$\begin{aligned}x^2+y^2&=36\\x^2+(x+6)^2&=36\\x^2+x^2+12x+36&=36\\2x^2+12x&=0\\2x(x+6)&=0\end{aligned}$$

$2x=0$ or $x+6=0$

$x=0$ or $x=-6$

$x=0$: $y=0+6=6$

$x=-6$: $y=-6+6=0$

The solution set is $\{(-6, 0), (0\ 6)\}$.

57. $\begin{cases} x^2 + 2y^2 = 10 \\ x^2 - y^2 = 1 \end{cases}$

Multiply the second equation by −1. Add the results.

$$\begin{aligned} x^2 + 2y^2 &= 10 \\ \underline{-x^2 + y^2} &\underline{= -1} \\ 3y^2 &= 9 \\ y^2 &= 3 \\ y &= \pm\sqrt{3} \end{aligned}$$

$y = -\sqrt{3}:$
$$\begin{aligned} x^2 - y^2 &= 1 \\ x^2 - \left(-\sqrt{3}\right)^2 &= 1 \\ x^2 - 3 &= 1 \\ x^2 &= 4 \\ x &= \pm 2 \end{aligned}$$

$y = \sqrt{3}:$
$$\begin{aligned} x^2 - \left(\sqrt{3}\right)^2 &= 1 \\ x^2 - 3 &= 1 \\ x^2 &= 4 \\ x &= \pm 2 \end{aligned}$$

The solution set is

$\left\{\left(2, \sqrt{3}\right), \left(-2, \sqrt{3}\right), \left(2, -\sqrt{3}\right), \left(-2, -\sqrt{3}\right)\right\}$.

Appendices

Appendix A Practice Problems

1. $\frac{1}{4} = \frac{1 \cdot 5}{4 \cdot 5} = \frac{5}{20}$

2. $\frac{20}{35} = \frac{2 \cdot 2 \cdot 5}{5 \cdot 7} = \frac{2 \cdot 2}{7} = \frac{4}{7}$

3. There are no common factors other than 1, so $\frac{7}{20}$ is already simplified.

4. $\frac{12}{40} = \frac{2 \cdot 2 \cdot 3}{2 \cdot 2 \cdot 2 \cdot 5} = \frac{3}{10}$

5. $\frac{4}{4} = 1$ since $4 \div 4 = 1$.

6. $\frac{9}{3} = 3$ since $9 \div 3 = 3$.

7. $\frac{10}{10} = 1$ since $10 \div 10 = 1$.

8. $\frac{5}{1} = 5$ since $5 \div 1 = 5$.

9. $\frac{3}{7} \cdot \frac{3}{5} = \frac{3 \cdot 3}{7 \cdot 5} = \frac{9}{35}$

10. $\frac{2}{9} \div \frac{3}{4} = \frac{2}{9} \cdot \frac{4}{3} = \frac{2 \cdot 4}{9 \cdot 3} = \frac{8}{27}$

11. $$\begin{aligned} \frac{8}{11} \div 24 &= \frac{8}{11} \div \frac{24}{1} \\ &= \frac{8}{11} \cdot \frac{1}{24} \\ &= \frac{2 \cdot 2 \cdot 2}{11 \cdot 2 \cdot 2 \cdot 2 \cdot 3} \\ &= \frac{1}{33} \end{aligned}$$

12. $\frac{5}{4} \div \frac{5}{8} = \frac{5}{4} \cdot \frac{8}{5} = \frac{5 \cdot 4 \cdot 2}{4 \cdot 5} = \frac{2}{1} = 2$

13. $\frac{2}{11} + \frac{5}{11} = \frac{2+5}{11} = \frac{7}{11}$

14. $\frac{1}{8} + \frac{3}{8} = \frac{1+3}{8} = \frac{4}{8} = \frac{1 \cdot 4}{2 \cdot 4} = \frac{1}{2}$

15. $\frac{13}{10} - \frac{3}{10} = \frac{13-3}{10} = \frac{10}{10} = 1$

16. $\frac{7}{6} - \frac{2}{6} = \frac{7-2}{6} = \frac{5}{6}$

17. $\frac{3}{8} + \frac{1}{20} = \frac{3 \cdot 5}{8 \cdot 5} + \frac{1 \cdot 2}{20 \cdot 2} = \frac{15}{40} + \frac{2}{40} = \frac{17}{40}$

18. $$\begin{aligned} \frac{8}{15} - \frac{1}{3} &= \frac{8}{15} - \frac{1 \cdot 5}{3 \cdot 5} \\ &= \frac{8}{15} - \frac{5}{15} \\ &= \frac{3}{15} \\ &= \frac{1 \cdot 3}{5 \cdot 3} \\ &= \frac{1}{5} \end{aligned}$$

19. $5\frac{1}{6} \cdot 4\frac{2}{5} = \frac{31}{6} \cdot \frac{22}{5} = \frac{682}{30} = 22\frac{22}{30} = 22\frac{11}{15}$

20. $7\frac{3}{8} - 6\frac{1}{4} = \frac{59}{8} - \frac{25}{4} = \frac{59}{8} - \frac{50}{8} = \frac{9}{8}$ or $1\frac{1}{8}$

21. $76\frac{1}{9}+35\frac{3}{4}=\frac{685}{9}+\frac{143}{4}$
$=\frac{2740}{36}+\frac{1287}{36}$
$=\frac{4027}{36}$
$=111\frac{31}{36}$

Appendix A Exercise Set

1. $\frac{7}{10}=\frac{7\cdot 3}{10\cdot 3}=\frac{21}{30}$

3. $\frac{2}{9}=\frac{2\cdot 2}{9\cdot 2}=\frac{4}{18}$

5. $\frac{4}{5}=\frac{4\cdot 4}{5\cdot 4}=\frac{16}{20}$

7. $\frac{2}{4}=\frac{1\cdot 2}{2\cdot 2}=\frac{1}{2}$

9. $\frac{10}{15}=\frac{2\cdot 5}{3\cdot 5}=\frac{2}{3}$

11. $\frac{3}{7}$ is already simplified.

13. $\frac{20}{20}=1$

15. $\frac{35}{7}=\frac{5\cdot 7}{7}=5$

17. $\frac{18}{30}=\frac{2\cdot 3\cdot 3}{2\cdot 3\cdot 5}=\frac{3}{5}$

19. $\frac{16}{20}=\frac{2\cdot 2\cdot 2\cdot 2}{2\cdot 2\cdot 5}=\frac{2\cdot 2}{5}=\frac{4}{5}$

21. $\frac{66}{48}=\frac{2\cdot 3\cdot 11}{2\cdot 2\cdot 2\cdot 2\cdot 3}=\frac{11}{2\cdot 2\cdot 2}=\frac{11}{8}$

23. $\frac{120}{244}=\frac{2\cdot 2\cdot 2\cdot 3\cdot 5}{2\cdot 2\cdot 61}=\frac{2\cdot 3\cdot 5}{61}=\frac{30}{61}$

25. $\frac{192}{264}=\frac{2\cdot 2\cdot 2\cdot 2\cdot 2\cdot 2\cdot 3}{2\cdot 2\cdot 2\cdot 3\cdot 11}=\frac{2\cdot 2\cdot 2}{11}=\frac{8}{11}$

27. $\frac{1}{2}\cdot\frac{3}{4}=\frac{1\cdot 3}{2\cdot 4}=\frac{3}{8}$

29. $\frac{2}{3}\cdot\frac{3}{4}=\frac{2\cdot 3}{3\cdot 2\cdot 2}=\frac{1}{2}$

31. $5\frac{1}{9}\cdot 3\frac{2}{3}=\frac{46}{9}\cdot\frac{11}{3}=\frac{506}{27}=18\frac{20}{27}$

33. $7\frac{2}{5}\div\frac{1}{5}=\frac{37}{5}\div\frac{1}{5}=\frac{37}{5}\cdot\frac{5}{1}=\frac{37}{1}=37$

35. $\frac{1}{2}\div\frac{7}{12}=\frac{1}{2}\cdot\frac{12}{7}=\frac{1\cdot 2\cdot 2\cdot 3}{2\cdot 7}=\frac{6}{7}$

37. $\frac{3}{4}\div\frac{1}{20}=\frac{3}{4}\cdot\frac{20}{1}=\frac{3\cdot 4\cdot 5}{4\cdot 1}=\frac{3\cdot 5}{1}=\frac{15}{1}=15$

39. $\frac{7}{10}\cdot\frac{5}{21}=\frac{7\cdot 5}{2\cdot 5\cdot 3\cdot 7}=\frac{1}{6}$

41. $\frac{9}{20}\div 12=\frac{9}{20}\div\frac{12}{1}$
$=\frac{9}{20}\cdot\frac{1}{12}$
$=\frac{3\cdot 3\cdot 1}{20\cdot 3\cdot 4}$
$=\frac{3}{20\cdot 4}$
$=\frac{3}{80}$

43. $4\frac{2}{11}\cdot 2\frac{1}{2}=\frac{46}{11}\cdot\frac{5}{2}=\frac{2\cdot 23\cdot 5}{11\cdot 2}=\frac{115}{11}=10\frac{5}{11}$

45. $8\frac{3}{5} \div 2\frac{9}{10} = \frac{43}{5} \div \frac{29}{10}$
$= \frac{43}{5} \cdot \frac{10}{29}$
$= \frac{43 \cdot 2 \cdot 5}{5 \cdot 29}$
$= \frac{43 \cdot 2}{29}$
$= \frac{86}{29}$
$= 2\frac{28}{29}$

47. $\frac{4}{5} + \frac{1}{5} = \frac{4+1}{5} = \frac{5}{5} = 1$

49. $\frac{4}{5} - \frac{1}{5} = \frac{4-1}{5} = \frac{3}{5}$

51. $\frac{23}{105} + \frac{4}{105} = \frac{23+4}{105} = \frac{27}{105} = \frac{3 \cdot 3 \cdot 3}{3 \cdot 5 \cdot 7} = \frac{9}{35}$

53. $\frac{17}{21} - \frac{10}{21} = \frac{17-10}{21} = \frac{7}{21} = \frac{7}{3 \cdot 7} = \frac{1}{3}$

55. $9\frac{7}{8} + 2\frac{3}{8} = \frac{79}{8} + \frac{19}{8} = \frac{98}{8} = 12\frac{2}{8} = 12\frac{1}{4}$

57. $5\frac{2}{5} - 3\frac{4}{5} = \frac{27}{5} - \frac{19}{5} = \frac{8}{5} = 1\frac{3}{5}$

59. $\frac{2}{3} + \frac{3}{7} = \frac{2 \cdot 7}{3 \cdot 7} + \frac{3 \cdot 3}{7 \cdot 3} = \frac{14}{21} + \frac{9}{21} = \frac{23}{21}$

61. $\frac{10}{3} - \frac{5}{21} = \frac{10 \cdot 7}{3 \cdot 7} - \frac{5}{21} = \frac{70}{21} - \frac{5}{21} = \frac{65}{21}$

63. $\frac{10}{21} + \frac{5}{21} = \frac{10+5}{21} = \frac{15}{21} = \frac{3 \cdot 5}{3 \cdot 7} = \frac{5}{7}$

65. $\frac{5}{22} - \frac{5}{33} = \frac{5 \cdot 3}{22 \cdot 3} - \frac{5 \cdot 2}{33 \cdot 2} = \frac{15}{66} - \frac{10}{66} = \frac{5}{66}$

67. $8\frac{11}{12} - 1\frac{5}{6} = \frac{107}{12} - \frac{11}{6} = \frac{107}{12} - \frac{22}{12} = \frac{85}{12} = 7\frac{1}{12}$

69.
$$\begin{array}{rr} 17\frac{2}{5} & 17\frac{6}{15} \\ +\ 30\frac{2}{3} & +\ 30\frac{10}{15} \\ \hline & 47\frac{16}{15} \end{array} = 47 + 1\frac{1}{15} = 48\frac{1}{15}$$

71. $\frac{12}{5} - 1 = \frac{12}{5} - \frac{5}{5} = \frac{7}{5}$

73. $\frac{2}{3} - \frac{5}{9} + \frac{5}{6} = \frac{2 \cdot 6}{3 \cdot 6} - \frac{5 \cdot 2}{9 \cdot 2} + \frac{5 \cdot 3}{6 \cdot 3}$
$= \frac{12}{18} - \frac{10}{18} + \frac{15}{18}$
$= \frac{12-10+15}{18}$
$= \frac{17}{18}$

75. answers may vary

77. $1 - \frac{3}{10} - \frac{5}{10} = \frac{10}{10} - \frac{3}{10} - \frac{5}{10}$
$= \frac{10-3-5}{10}$
$= \frac{2}{10}$
$= \frac{2}{2 \cdot 5}$
$= \frac{1}{5}$

79. $1 - \frac{1}{4} - \frac{3}{8} = \frac{8}{8} - \frac{2}{8} - \frac{3}{8} = \frac{8-2-3}{8} = \frac{3}{8}$

81.

$$\begin{array}{r} 237\frac{1}{6} \\ -\ 224\frac{5}{12} \\ \hline \end{array} \qquad \begin{array}{r} 237\frac{2}{12} \\ -\ 224\frac{5}{12} \\ \hline \end{array} \qquad \begin{array}{r} 236\frac{14}{12} \\ -\ 224\frac{5}{12} \\ \hline 12\frac{9}{12} = 12\frac{3}{4} \end{array}$$

The Olympic record throw is $12\frac{3}{4}$ feet longer.

Appendix B Practice Problems

1. a.
$$\begin{vmatrix} -3 & 6 \\ 2 & 1 \end{vmatrix} = ad - bc$$
$$= (-3)(1) - (6)(2)$$
$$= -3 - 12$$
$$= -15$$

b.
$$\begin{vmatrix} 4 & 5 \\ 0 & -5 \end{vmatrix} = ad - bc$$
$$= 4(-5) - (5)(0)$$
$$= -20 - 0$$
$$= -20$$

2.
$$\begin{cases} x - y = -4 \\ 2x + 3y = 2 \end{cases}$$
$$D = \begin{vmatrix} 1 & -1 \\ 2 & 3 \end{vmatrix} = 1(3) - (-1)(2) = 3 + 2 = 5$$
$$D_x = \begin{vmatrix} -4 & -1 \\ 2 & 3 \end{vmatrix}$$
$$= -4(3) - (-1)(2)$$
$$= -12 + 2$$
$$= -10$$
$$D_y = \begin{vmatrix} 1 & -4 \\ 2 & 2 \end{vmatrix} = 1(2) - (-4)(2) = 2 + 8 = 10$$
$$x = \frac{D_x}{D} = \frac{-10}{5} = -2$$
$$y = \frac{D_y}{D} = \frac{10}{5} = 2$$
The solution is (−2, 2).

3.
$$\begin{cases} 4x + y = 3 \\ 2x - 3y = -9 \end{cases}$$
$$D = \begin{vmatrix} 4 & 1 \\ 2 & -3 \end{vmatrix} = 4(-3) - (1)(2) = -12 - 2 = -14$$
$$D_x = \begin{vmatrix} 3 & 1 \\ -9 & -3 \end{vmatrix} = 3(-3) - (1)(-9) = -9 + 9 = 0$$
$$D_y = \begin{vmatrix} 4 & 3 \\ 2 & -9 \end{vmatrix}$$
$$= 4(-9) - (3)(2)$$
$$= -36 - 6$$
$$= -42$$
$$x = \frac{D_x}{D} = \frac{0}{-14} = 0$$
$$y = \frac{D_y}{D} = \frac{-42}{-14} = 3$$
The solution is (0, 3).

4. a.
$$\begin{vmatrix} 2 & 0 & 1 \\ -1 & 3 & 2 \\ 5 & 1 & 4 \end{vmatrix}$$
$$= 2 \cdot \begin{vmatrix} 3 & 2 \\ 1 & 4 \end{vmatrix} - (-1) \cdot \begin{vmatrix} 0 & 1 \\ 1 & 4 \end{vmatrix} + 5 \cdot \begin{vmatrix} 0 & 1 \\ 3 & 2 \end{vmatrix}$$
$$= 2(12 - 2) + 1(0 - 1) + 5(0 - 3)$$
$$= 20 + (-1) + (-15)$$
$$= 4$$

b.
$$\begin{vmatrix} 2 & 0 & 1 \\ -1 & 3 & 2 \\ 5 & 1 & 4 \end{vmatrix}$$
$$= 5 \cdot \begin{vmatrix} 0 & 1 \\ 3 & 2 \end{vmatrix} - 1 \cdot \begin{vmatrix} 2 & 1 \\ -1 & 2 \end{vmatrix} + 4 \cdot \begin{vmatrix} 2 & 0 \\ -1 & 3 \end{vmatrix}$$
$$= 5(0 - 3) - 1(4 + 1) + 4(6 - 0)$$
$$= -15 - 5 + 24$$
$$= 4$$

5.
$$\begin{cases} x + 2y - z = 3 \\ 2x - 3y + z = -9 \\ -x + y - 2z = 0 \end{cases}$$
Find D, D_x, D_y, and D_z by expanding by the minors of the first column.

$$D = \begin{vmatrix} 1 & 2 & -1 \\ 2 & -3 & 1 \\ -1 & 1 & -2 \end{vmatrix}$$
$$= 1 \cdot \begin{vmatrix} -3 & 1 \\ 1 & -2 \end{vmatrix} - 2 \cdot \begin{vmatrix} 2 & -1 \\ 1 & -2 \end{vmatrix} - 1 \cdot \begin{vmatrix} 2 & -1 \\ -3 & 1 \end{vmatrix}$$
$$= 1 \cdot (6-1) - 2 \cdot (-4+1) - 1 \cdot (2-3)$$
$$= 5 + 6 + 1$$
$$= 12$$

$$D_x = \begin{vmatrix} 3 & 2 & -1 \\ -9 & -3 & 1 \\ 0 & 1 & -2 \end{vmatrix}$$
$$= 3 \cdot \begin{vmatrix} -3 & 1 \\ 1 & -2 \end{vmatrix} - (-9) \cdot \begin{vmatrix} 2 & -1 \\ 1 & -2 \end{vmatrix} + 0 \cdot \begin{vmatrix} 2 & -1 \\ -3 & 1 \end{vmatrix}$$
$$= 3(6-1) + 9(-4+1) + 0$$
$$= 15 + (-27)$$
$$= -12$$

$$D_y = \begin{vmatrix} 1 & 3 & -1 \\ 2 & -9 & 1 \\ -1 & 0 & -2 \end{vmatrix}$$
$$= 1 \cdot \begin{vmatrix} -9 & 1 \\ 0 & -2 \end{vmatrix} - 2 \cdot \begin{vmatrix} 3 & -1 \\ 0 & -2 \end{vmatrix} + (-1) \cdot \begin{vmatrix} 3 & -1 \\ -9 & 1 \end{vmatrix}$$
$$= (18-0) - 2(-6-0) - 1(3-9)$$
$$= 18 + 12 + 6$$
$$= 36$$

$$D_z = \begin{vmatrix} 1 & 2 & 3 \\ 2 & -3 & -9 \\ -1 & 1 & 0 \end{vmatrix}$$
$$= 1 \cdot \begin{vmatrix} -3 & -9 \\ 1 & 0 \end{vmatrix} - 2 \cdot \begin{vmatrix} 2 & 3 \\ 1 & 0 \end{vmatrix} + (-1) \cdot \begin{vmatrix} 2 & 3 \\ -3 & -9 \end{vmatrix}$$
$$= (0+9) - 2(0-3) - 1(-18+9)$$
$$= 9 + 6 + 9$$
$$= 24$$

$$x = \frac{D_x}{D} = \frac{-12}{12} = -1$$
$$y = \frac{D_y}{D} = \frac{36}{12} = 3$$
$$z = \frac{D_z}{D} = \frac{24}{12} = 2$$

The solution is (−1, 3, 2).

Mental Math

1. $\begin{vmatrix} 7 & 2 \\ 0 & 8 \end{vmatrix} = 56 - 0 = 56$

2. $\begin{vmatrix} 6 & 0 \\ 1 & 2 \end{vmatrix} = 12 - 0 = 12$

3. $\begin{vmatrix} -4 & 2 \\ 0 & 8 \end{vmatrix} = -32 - 0 = -32$

4. $\begin{vmatrix} 5 & 0 \\ 3 & -5 \end{vmatrix} = -25 - 0 = -25$

5. $\begin{vmatrix} -2 & 0 \\ 3 & -10 \end{vmatrix} = 20 - 0 = 20$

6. $\begin{vmatrix} -1 & 4 \\ 0 & -18 \end{vmatrix} = 18 - 0 = 18$

Appendix B Exercise Set

1. $\begin{vmatrix} 3 & 5 \\ -1 & 7 \end{vmatrix} = ad - bc$

$$= 3(7) - (5)(-1)$$
$$= 21 - (-5)$$
$$= 21 + 5$$
$$= 26$$

3. $\begin{vmatrix} 9 & -2 \\ 4 & -3 \end{vmatrix} = ad - bc$

$$= 9(-3) - (-2)(4)$$
$$= -27 + 8$$
$$= -19$$

5. $\begin{vmatrix} -2 & 9 \\ 4 & -18 \end{vmatrix} = ad - bc$

$$= (-2)(-18) - 9(4)$$
$$= 36 - 36$$
$$= 0$$

7. $\begin{vmatrix} \frac{3}{4} & \frac{5}{2} \\ -\frac{1}{6} & \frac{7}{3} \end{vmatrix} = ad - bc$

$$= \frac{3}{4} \cdot \frac{7}{3} - \frac{5}{2} \cdot \left(-\frac{1}{6}\right)$$
$$= \frac{21}{12} + \frac{5}{12}$$
$$= \frac{26}{12}$$
$$= \frac{13}{6}$$

9. $\begin{cases} 2y - 4 = 0 \\ x + 2y = 5 \end{cases} \to \begin{cases} 0x + 2y = 4 \\ x + 2y = 5 \end{cases}$

$D = \begin{vmatrix} 0 & 2 \\ 1 & 2 \end{vmatrix} = 0(2) - (2)(1) = 0 - 2 = -2$

$D_x = \begin{vmatrix} 4 & 2 \\ 5 & 2 \end{vmatrix} = 4(2) - (2)(5) = 8 - 10 = -2$

$D_y = \begin{vmatrix} 0 & 4 \\ 1 & 5 \end{vmatrix} = 0(5) - 4(1) = 0 - 4 = -4$

$x = \frac{D_x}{D} = \frac{-2}{-2} = 1$

$y = \frac{D_y}{D} = \frac{-4}{-2} = 2$

The solution is (1, 2).

11. $\begin{cases} 3x + y = 1 \\ 2y = 2 - 6x \end{cases} \to \begin{cases} 3x + y = 1 \\ 6x + 2y = 2 \end{cases} \to \begin{cases} 3x + y = 1 \\ 3x + y = 1 \end{cases}$

The equations are the same. The system is dependent. The solution set is $\{(x, y) | 3x + y = 1\}$.

13. $\begin{cases} 5x - 2y = 27 \\ -3x + 5y = 18 \end{cases}$

$D = \begin{vmatrix} 5 & -2 \\ -3 & 5 \end{vmatrix} = 5(5) - (-2)(-3) = 25 - 6 = 19$

$D_x = \begin{vmatrix} 27 & -2 \\ 18 & 5 \end{vmatrix} = 27(5) - (-2)(18)$

$$= 135 - (-36)$$
$$= 135 + 36$$
$$= 171$$

$$D_y = \begin{vmatrix} 5 & 27 \\ -3 & 18 \end{vmatrix} = 5(18) - (27)(-3)$$
$$= 90 - (-81)$$
$$= 90 + 81$$
$$= 171$$

$$x = \frac{D_x}{D} = \frac{171}{19} = 9$$

$$y = \frac{D_y}{D} = \frac{171}{19} = 9$$

The solution is (9, 9).

15. $\begin{cases} 2x - 5y = 4 \\ x + 2y = -7 \end{cases}$

$$D = \begin{vmatrix} 2 & -5 \\ 1 & 2 \end{vmatrix} = 2(2) - (-5)(1) = 4 + 5 = 9$$

$$D_x = \begin{vmatrix} 4 & -5 \\ -7 & 2 \end{vmatrix} = 4(2) - (-5)(-7) = 8 - 35 = -27$$

$$D_y = \begin{vmatrix} 2 & 4 \\ 1 & -7 \end{vmatrix} = 2(-7) - 4(1) = -14 - 4 = -18$$

$$x = \frac{D_x}{D} = \frac{-27}{9} = -3$$

$$y = \frac{D_y}{D} = \frac{-18}{9} = -2$$

The solution is (–3, –2).

17. $\begin{cases} \frac{2}{3}x - \frac{3}{4}x = -1 \\ -\frac{1}{6}x + \frac{3}{4}y = \frac{5}{2} \end{cases}$

$$D = \begin{vmatrix} \frac{2}{3} & -\frac{3}{4} \\ -\frac{1}{6} & \frac{3}{4} \end{vmatrix} = \frac{2}{3}\left(\frac{3}{4}\right) - \left(-\frac{3}{4}\right)\left(-\frac{1}{6}\right)$$
$$= \frac{1}{2} - \frac{3}{24}$$
$$= \frac{4}{8} - \frac{1}{8}$$
$$= \frac{3}{8}$$

$$D_x = \begin{vmatrix} -1 & -\frac{3}{4} \\ \frac{5}{2} & \frac{3}{4} \end{vmatrix} = (-1)\left(\frac{3}{4}\right) - \left(-\frac{3}{4}\right)\left(\frac{5}{2}\right)$$
$$= -\frac{3}{4} - \left(-\frac{15}{8}\right)$$
$$= -\frac{6}{8} + \frac{15}{8}$$
$$= \frac{9}{8}$$

$$D_y = \begin{vmatrix} \frac{2}{3} & -1 \\ -\frac{1}{6} & \frac{5}{2} \end{vmatrix} = \frac{2}{3}\left(\frac{5}{2}\right) - (-1)\left(-\frac{1}{6}\right)$$
$$= \frac{5}{3} - \frac{1}{6}$$
$$= \frac{10}{6} - \frac{1}{6}$$
$$= \frac{9}{6}$$
$$= \frac{3}{2}$$

$$x = \frac{D_x}{D} = \frac{\frac{9}{8}}{\frac{3}{8}} = \frac{9}{8} \div \frac{3}{8} = \frac{9}{8} \cdot \frac{8}{3} = \frac{9}{3} = 3$$

$$y = \frac{D_y}{D} = \frac{\frac{3}{2}}{\frac{3}{8}} = \frac{3}{2} \div \frac{3}{8} = \frac{3}{2} \cdot \frac{8}{3} = \frac{8}{2} = 4$$

The solution is (3, 4).

19. Expand by the minors of the first column.

$$\begin{vmatrix} 2 & 1 & 0 \\ 0 & 5 & -3 \\ 4 & 0 & 2 \end{vmatrix} = 2\begin{vmatrix} 5 & -3 \\ 0 & 2 \end{vmatrix} - 0\begin{vmatrix} 1 & 0 \\ 0 & 2 \end{vmatrix} + 4\begin{vmatrix} 1 & 0 \\ 5 & -3 \end{vmatrix}$$
$$= 2(10 - 0) - 0 + 4(-3 - 0)$$
$$= 20 + (-12)$$
$$= 8$$

21. Expand by the minors of the third column.

$$\begin{vmatrix} 4 & -6 & 0 \\ -2 & 3 & 0 \\ 4 & -6 & 1 \end{vmatrix}$$
$$= 0\begin{vmatrix} -2 & 3 \\ 4 & -6 \end{vmatrix} - 0\begin{vmatrix} 4 & -6 \\ 4 & -6 \end{vmatrix} + 1\begin{vmatrix} 4 & -6 \\ -2 & 3 \end{vmatrix}$$
$$= 0 - 0 + 1(12 - 12)$$
$$= 0 - 0 + 1(0)$$
$$= 0$$

23. Expand by the minors of the first row.

$$\begin{vmatrix} 1 & 0 & 4 \\ 1 & -1 & 2 \\ 3 & 2 & 1 \end{vmatrix} = 1\begin{vmatrix} -1 & 2 \\ 2 & 1 \end{vmatrix} - 0\begin{vmatrix} 1 & 2 \\ 3 & 1 \end{vmatrix} + 4\begin{vmatrix} 1 & -1 \\ 3 & 2 \end{vmatrix}$$
$$= 1(-1-4) - 0 + 4(2+3)$$
$$= -5 - 0 + 20$$
$$= 15$$

25. Expand by the minors of the first row.

$$\begin{vmatrix} 3 & 6 & -3 \\ -1 & -2 & 3 \\ 4 & -1 & 6 \end{vmatrix}$$
$$= 3\begin{vmatrix} -2 & 3 \\ -1 & 6 \end{vmatrix} - 6\begin{vmatrix} -1 & 3 \\ 4 & 6 \end{vmatrix} + (-3)\begin{vmatrix} -1 & -2 \\ 4 & -1 \end{vmatrix}$$
$$= 3(-12 - (-3)) - 6(-6 - 12) - 3(1 - (-8))$$
$$= -27 + 108 - 27$$
$$= 54$$

27. $\begin{cases} 3x \quad + z = -1 \\ -x - 3y + z = 7 \\ \quad 3y + z = 5 \end{cases}$

$$D = \begin{vmatrix} 3 & 0 & 1 \\ -1 & -3 & 1 \\ 0 & 3 & 1 \end{vmatrix}$$
$$= 3\begin{vmatrix} -3 & 1 \\ 3 & 1 \end{vmatrix} - (-1)\begin{vmatrix} 0 & 1 \\ 3 & 1 \end{vmatrix} + 0\begin{vmatrix} 0 & 1 \\ -3 & 1 \end{vmatrix}$$
$$= 3(-3-3) + 1(0-3) + 0$$
$$= -18 + (-3)$$
$$= -21$$

$$D_x = \begin{vmatrix} -1 & 0 & 1 \\ 7 & -3 & 1 \\ 5 & 3 & 1 \end{vmatrix}$$
$$= -1\begin{vmatrix} -3 & 1 \\ 3 & 1 \end{vmatrix} - 0\begin{vmatrix} 7 & 1 \\ 5 & 1 \end{vmatrix} + 1\begin{vmatrix} 7 & -3 \\ 5 & 3 \end{vmatrix}$$
$$= -(-3-3) - 0 + (21+15)$$
$$= 6 + 36$$
$$= 42$$

$$D_y = \begin{vmatrix} 3 & -1 & 1 \\ -1 & 7 & 1 \\ 0 & 5 & 1 \end{vmatrix}$$
$$= 3\begin{vmatrix} 7 & 1 \\ 5 & 1 \end{vmatrix} - (-1)\begin{vmatrix} -1 & 1 \\ 5 & 1 \end{vmatrix} + 0\begin{vmatrix} -1 & 1 \\ 7 & 1 \end{vmatrix}$$
$$= 3(7-5) + 1(-1-5) + 0$$
$$= 6 - 6$$
$$= 0$$

$$D_z = \begin{vmatrix} 3 & 0 & -1 \\ -1 & -3 & 7 \\ 0 & 3 & 5 \end{vmatrix}$$
$$= 3\begin{vmatrix} -3 & 7 \\ 3 & 5 \end{vmatrix} - (-1)\begin{vmatrix} 0 & -1 \\ 3 & 5 \end{vmatrix} + 0\begin{vmatrix} 0 & -1 \\ -3 & 7 \end{vmatrix}$$
$$= 3(-15-21) + 1(0+3) + 0$$
$$= -108 + 3$$
$$= -105$$

$$x = \frac{D_x}{D} = \frac{42}{-21} = -2$$
$$y = \frac{D_y}{D} = \frac{0}{-21} = 0$$
$$z = \frac{D_z}{D} = \frac{-105}{-21} = 5$$

The solution is (−2, 0, 5).

29. $\begin{cases} x+y+z=8 \\ 2x-y-z=10 \\ x-2y+3z=22 \end{cases}$

$$D=\begin{vmatrix} 1 & 1 & 1 \\ 2 & -1 & -1 \\ 1 & -2 & 3 \end{vmatrix}$$
$$=1\begin{vmatrix} -1 & -1 \\ -2 & 3 \end{vmatrix}-1\begin{vmatrix} 2 & -1 \\ 1 & 3 \end{vmatrix}+1\begin{vmatrix} 2 & -1 \\ 1 & -2 \end{vmatrix}$$
$$=1(-3-2)-1(6-(-1))+1(-4-(-1))$$
$$=-5-7-3$$
$$=-15$$

$$D_x=\begin{vmatrix} 8 & 1 & 1 \\ 10 & -1 & -1 \\ 22 & -2 & 3 \end{vmatrix}$$
$$=8\begin{vmatrix} -1 & -1 \\ -2 & 3 \end{vmatrix}-10\begin{vmatrix} 1 & 1 \\ -2 & 3 \end{vmatrix}+22\begin{vmatrix} 1 & 1 \\ -1 & -1 \end{vmatrix}$$
$$=8(-3-2)-10(3-(-2))+22(-1-(-1))$$
$$=-40-50+0$$
$$=-90$$

$$D_y=\begin{vmatrix} 1 & 8 & 1 \\ 2 & 10 & -1 \\ 1 & 22 & 3 \end{vmatrix}$$
$$=-8\begin{vmatrix} 2 & -1 \\ 1 & 3 \end{vmatrix}+10\begin{vmatrix} 1 & 1 \\ 1 & 3 \end{vmatrix}-22\begin{vmatrix} 1 & 1 \\ 2 & -1 \end{vmatrix}$$
$$=-8(6-(-1))+10(3-1)-22(-1-2)$$
$$=-56+20+66$$
$$=30$$

$$D_z=\begin{vmatrix} 1 & 1 & 8 \\ 2 & -1 & 10 \\ 1 & -2 & 22 \end{vmatrix}$$
$$=8\begin{vmatrix} 2 & -1 \\ 1 & -2 \end{vmatrix}-10\begin{vmatrix} 1 & 1 \\ 1 & -2 \end{vmatrix}+22\begin{vmatrix} 1 & 1 \\ 2 & -1 \end{vmatrix}$$
$$=8(-4-(-1))-10(-2-1)+22(-1-2)$$
$$=-24+30-66$$
$$=-60$$

$$x=\frac{D_x}{D}=\frac{-90}{-15}=6$$

$$y=\frac{D_y}{D}=\frac{30}{-15}=-2$$

$$z=\frac{D_z}{D}=\frac{-60}{-15}=4$$

The solution is (6, –2, 4).

31. $\begin{cases} 2x+2y+z=1 \\ -x+y+2z=3 \\ x+2y+4z=0 \end{cases}$

$$D=\begin{vmatrix} 2 & 2 & 1 \\ -1 & 1 & 2 \\ 1 & 2 & 4 \end{vmatrix}$$
$$=2\begin{vmatrix} 1 & 2 \\ 2 & 4 \end{vmatrix}-2\begin{vmatrix} -1 & 2 \\ 1 & 4 \end{vmatrix}+1\begin{vmatrix} -1 & 1 \\ 1 & 2 \end{vmatrix}$$
$$=2(4-4)-2(-4-2)+1(-2-1)$$
$$=2(0)-2(-6)+1(-3)$$
$$=0+12-3$$
$$=9$$

$$D_x=\begin{vmatrix} 1 & 2 & 1 \\ 3 & 1 & 2 \\ 0 & 2 & 4 \end{vmatrix}$$
$$=1\begin{vmatrix} 1 & 2 \\ 2 & 4 \end{vmatrix}-3\begin{vmatrix} 2 & 1 \\ 2 & 4 \end{vmatrix}+0\begin{vmatrix} 2 & 1 \\ 1 & 2 \end{vmatrix}$$
$$=1(4-4)-3(8-2)+0$$
$$=0-3(6)+0$$
$$=-18$$

$$D_y=\begin{vmatrix} 2 & 1 & 1 \\ -1 & 3 & 2 \\ 1 & 0 & 4 \end{vmatrix}$$
$$=2\begin{vmatrix} 3 & 2 \\ 0 & 4 \end{vmatrix}-1\begin{vmatrix} -1 & 2 \\ 1 & 4 \end{vmatrix}+1\begin{vmatrix} -1 & 3 \\ 1 & 0 \end{vmatrix}$$
$$=2(12-0)-1(-4-2)+1(0-3)$$
$$=24+6+(-3)$$
$$=27$$

$$D_z=\begin{vmatrix} 2 & 2 & 1 \\ -1 & 1 & 3 \\ 1 & 2 & 0 \end{vmatrix}$$
$$=2\begin{vmatrix} 1 & 3 \\ 2 & 0 \end{vmatrix}-2\begin{vmatrix} -1 & 3 \\ 1 & 0 \end{vmatrix}+1\begin{vmatrix} -1 & 1 \\ 1 & 2 \end{vmatrix}$$
$$=2(0-6)-2(0-3)+1(-2-1)$$
$$=-12+6-3$$
$$=-9$$

$$x = \frac{D_x}{D} = \frac{-18}{9} = -2$$
$$y = \frac{D_y}{D} = \frac{27}{9} = 3$$
$$z = \frac{D_z}{D} = \frac{-9}{9} = -1$$

The solution is (−2, 3, −1).

33. $\begin{cases} x - 2y + z = -5 \\ 3y + 2z = 4 \\ 3x - y = -2 \end{cases}$

$$D = \begin{vmatrix} 1 & -2 & 1 \\ 0 & 3 & 2 \\ 3 & -1 & 0 \end{vmatrix}$$
$$= 1\begin{vmatrix} 3 & 2 \\ -1 & 0 \end{vmatrix} - (-2)\begin{vmatrix} 0 & 2 \\ 3 & 0 \end{vmatrix} + 1\begin{vmatrix} 0 & 3 \\ 3 & -1 \end{vmatrix}$$
$$= 1(0-(-2)) + 2(0-6) + 1(0-9)$$
$$= 2 - 12 - 9$$
$$= -19$$

$$D_x = \begin{vmatrix} -5 & -2 & 1 \\ 4 & 3 & 2 \\ -2 & -1 & 0 \end{vmatrix}$$
$$= -5\begin{vmatrix} 3 & 2 \\ -1 & 0 \end{vmatrix} - 4\begin{vmatrix} -2 & 1 \\ -1 & 0 \end{vmatrix} + (-2)\begin{vmatrix} -2 & 1 \\ 3 & 2 \end{vmatrix}$$
$$= -5(0-(-2)) - 4(0-(-1)) - 2(-4-3)$$
$$= -10 - 4 + 14$$
$$= 0$$

$$D_y = \begin{vmatrix} 1 & -5 & 1 \\ 0 & 4 & 2 \\ 3 & -2 & 0 \end{vmatrix}$$
$$= -(-5)\begin{vmatrix} 0 & 2 \\ 3 & 0 \end{vmatrix} + 4\begin{vmatrix} 1 & 1 \\ 3 & 0 \end{vmatrix} - (-2)\begin{vmatrix} 1 & 1 \\ 0 & 2 \end{vmatrix}$$
$$= 5(0-6) + 4(0-3) + 2(2-0)$$
$$= -30 - 12 + 4$$
$$= -38$$

$$D_z = \begin{vmatrix} 1 & -2 & -5 \\ 0 & 3 & 4 \\ 3 & -1 & -2 \end{vmatrix}$$
$$= -5\begin{vmatrix} 0 & 3 \\ 3 & -1 \end{vmatrix} - 4\begin{vmatrix} 1 & -2 \\ 3 & -1 \end{vmatrix} + (-2)\begin{vmatrix} 1 & -2 \\ 0 & 3 \end{vmatrix}$$
$$= -5(0-9) - 4(-1-(-6)) - 2(3-0)$$
$$= 45 - 20 - 6$$
$$= 19$$

$$x = \frac{D_x}{D} = \frac{0}{-19} = 0$$
$$y = \frac{D_y}{D} = \frac{-38}{-19} = 2$$
$$z = \frac{D_z}{D} = \frac{19}{-19} = -1$$

The solution is (0, 2, −1).

35. $\begin{vmatrix} 1 & x \\ 2 & 7 \end{vmatrix} = -3$

$$(1)(7) - x(2) = -3$$
$$7 - 2x = -3$$
$$-2x = -10$$
$$x = 5$$

37. 0; answers may vary

Appendix D Exercise Set

1. Since the x- and y-values are between −10 and 10, the ordered pairs will lie within a standard viewing rectangle.

3. Since the x-value of ordered pair (−11, 0) is less than −10, not all ordered pairs will lie within a standard viewing rectangle.

5. answers may vary

7. answers may vary

9. answers may vary

11. Xmin = −12 Ymin = −12
Xmax = 12 Ymax = 12
$\text{Xscl} = \frac{6}{5}$ $\text{Yscl} = \frac{6}{5}$

13. Xmin = –9 Ymin = –12
Xmax = 9 Ymax = 12
Xscl = 1 Yscl = 2

15. Xmin = –10 Ymin = –25
Xmax = 10 Ymax = 25
Xscl = 2 Yscl = 5

17. Xmin = –5 Ymin = –15
Xmax = 5 Ymax = 15
Xscl = 1 Yscl = 3

19. Xmin = –20 Ymin = –30
Xmax = 30 Ymax = 50
Xscl = 5 Yscl = 10

21. Setting A:

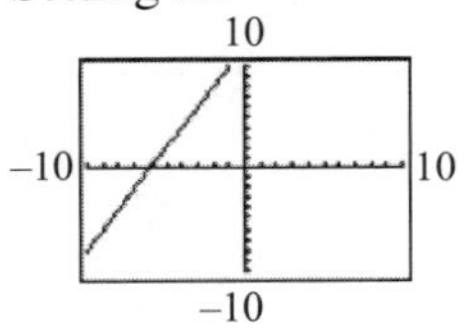

Setting B:

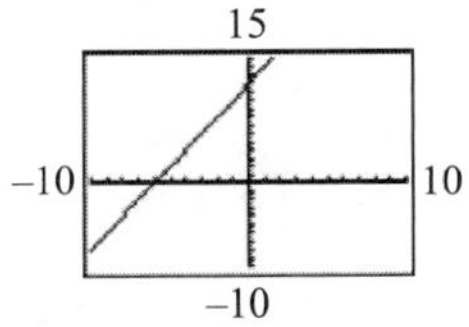

Setting B shows all the intercepts.

23. Setting A:

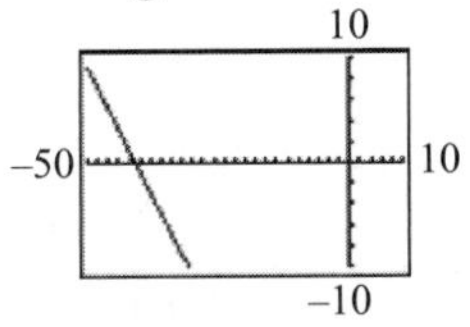

Setting B:

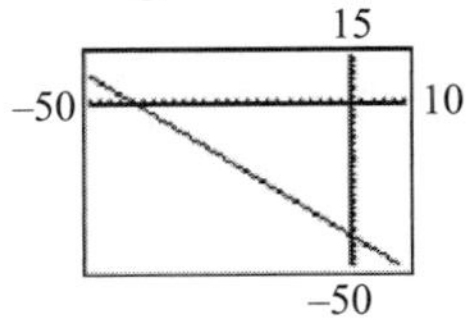

Setting B shows all the intercepts.

25. Setting A:

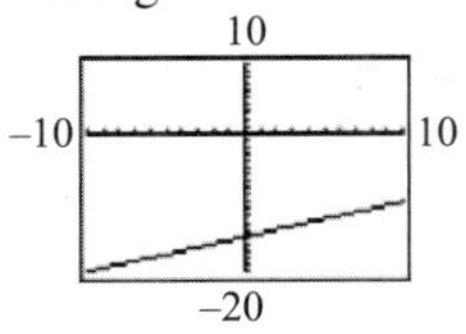

Setting B:

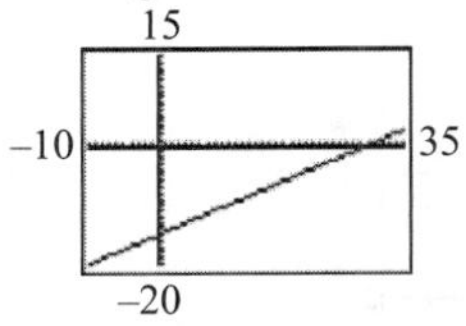

Setting B shows all the intercepts.

27. $3x = 5y$ or $y = \dfrac{3x}{5}$

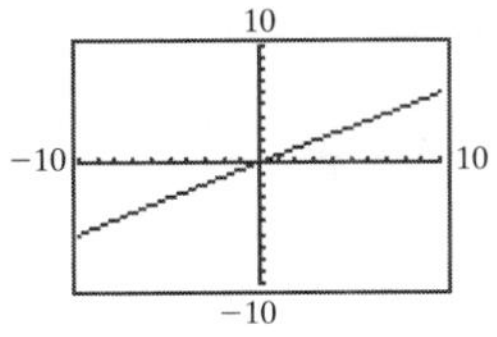

29. $9x - 5y = 30$ or $y = \dfrac{9x - 30}{5}$

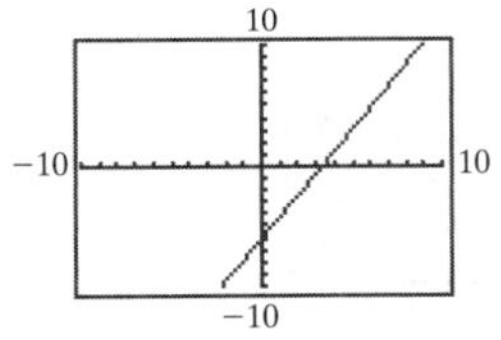

31. $y = -7$

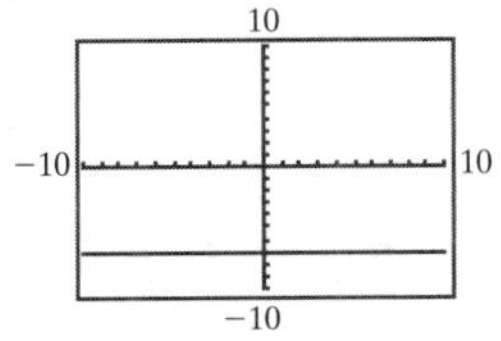

33. $x + 10y = -5$ or $y = \dfrac{-5 - x}{10}$

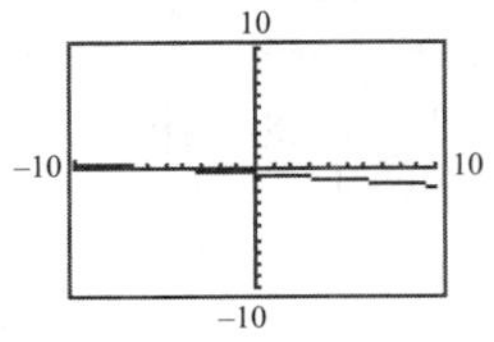

35. $y = \sqrt{x}$

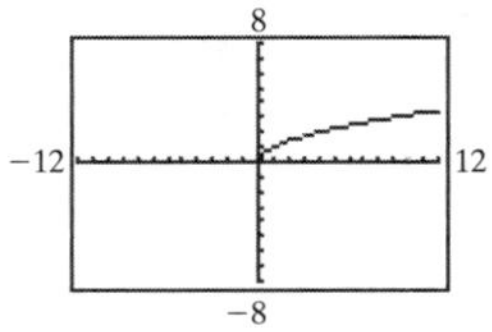

37. $y = x^2 + 2x + 1$

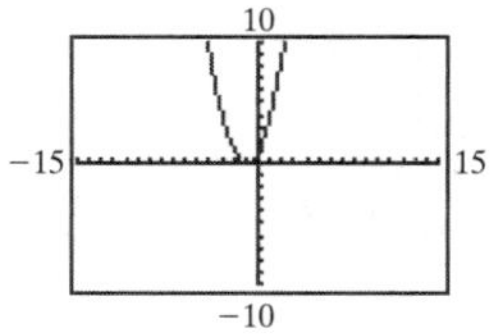

39. $y = |x|$

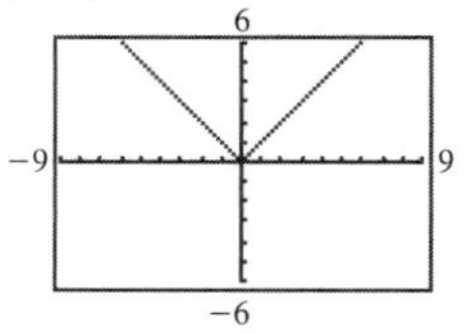

41. $x + 2y = 30$ or $y = \frac{30 - x}{2}$

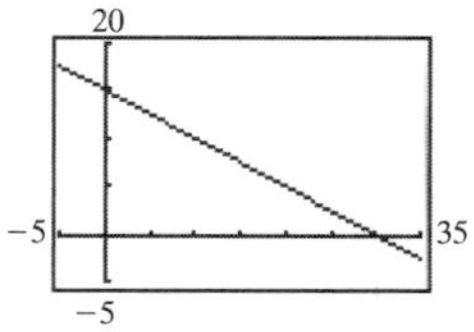

Appendix E Practice Problems

1.

x	$h(x) = 4\|x\|$	$g(x) = \frac{1}{5}\|x\|$
−2	$4\|-2\| = 4 \cdot 2 = 8$	$\frac{1}{5}\|-2\| = \frac{1}{5} \cdot 2 = \frac{2}{5}$
−1	$4\|-1\| = 4 \cdot 1 = 4$	$\frac{1}{5}\|-1\| = \frac{1}{5} \cdot 1 = \frac{1}{5}$
0	$4\|0\| = 4 \cdot 0 = 0$	$\frac{1}{5}\|0\| = \frac{1}{5} \cdot 0 = 0$
1	$4\|1\| = 4 \cdot 1 = 4$	$\frac{1}{5}\|1\| = \frac{1}{5} \cdot 1 = \frac{1}{5}$
2	$4\|2\| = 4 \cdot 2 = 8$	$\frac{1}{5}\|2\| = \frac{1}{5} \cdot 2 = \frac{2}{5}$

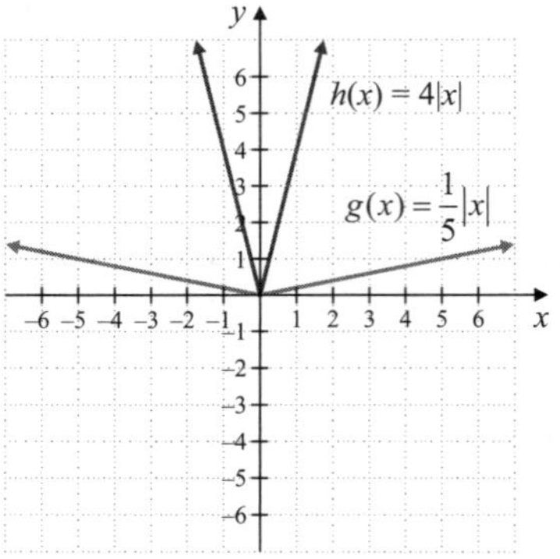

2.

x	$f(x) = -\frac{1}{2}\|x+1\| + 3$
−5	$f(-5) = -\frac{1}{2}\|-5+1\| + 3 = 1$
1	$f(1) = -\frac{1}{2}\|1+1\| + 3 = 2$
3	$f(3) = -\frac{1}{2}\|3+1\| + 3 = 1$

vertex = (−1, 3)
The graph opens down since $a < 0$.
The graph is wider than $y = |x|$ since $\left|-\frac{1}{2}\right| = \frac{1}{2} < 1$.

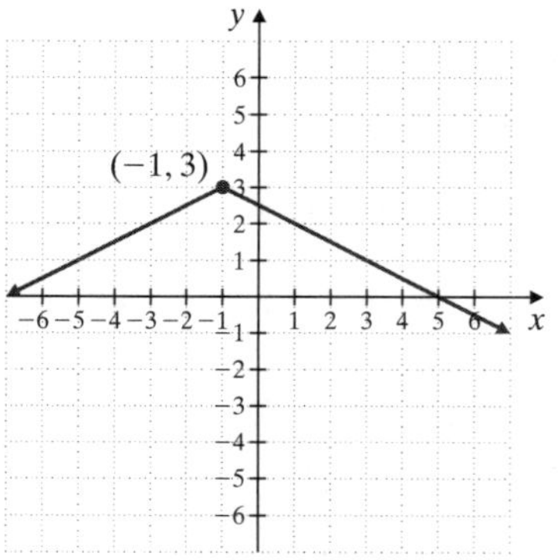

Appendix E Exercise Set

1. $f(x) = 3|x| = 3|x - 0| + 0$
The vertex is (0, 0), the graph opens up, and the graph is narrower than $y = |x|$.

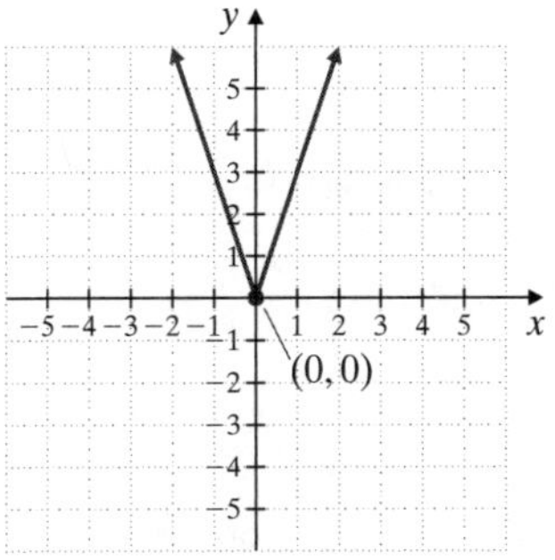

3. $f(x) = \frac{1}{4}|x| = \frac{1}{4}|x - 0| + 0$
The vertex is (0, 0), the graph opens up, and the graph is wider than $y = |x|$.

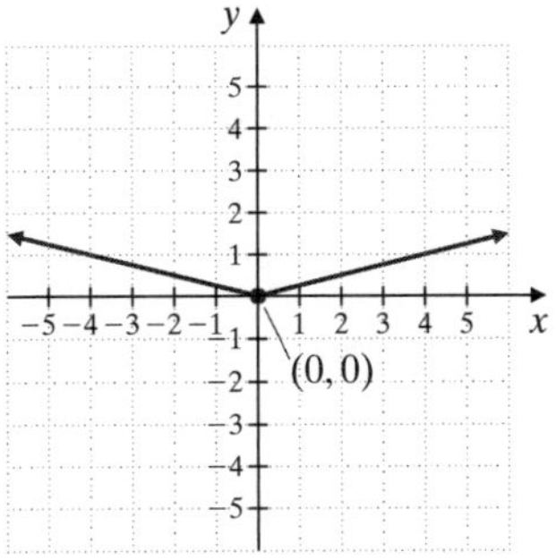

5. $g(x) = 2|x| + 3 = 2|x - 0| + 3$
The vertex is (0, 3), the graph opens up, and the graph is narrower than $y = |x|$.

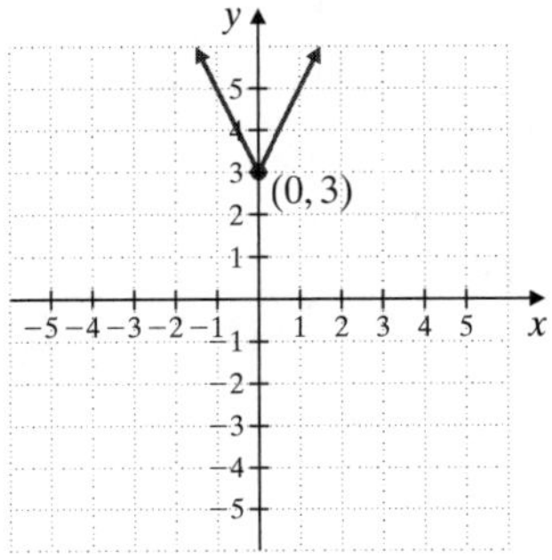

7. $h(x) = -\frac{1}{2}|x| = -\frac{1}{2}|x - 0| + 0$
The vertex is (0, 0), the graph opens down, and the graph is wider than $y = |x|$.

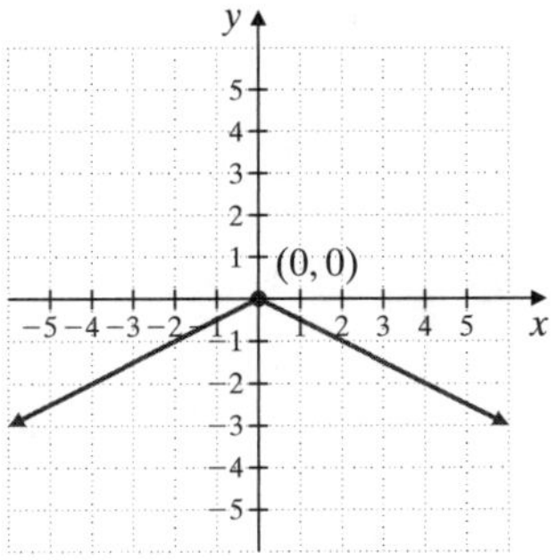

9. $f(x) = 4|x - 1| = 4|x - 1| + 0$
The vertex is (1, 0), the graph opens up, and the graph is narrower than $y = |x|$.

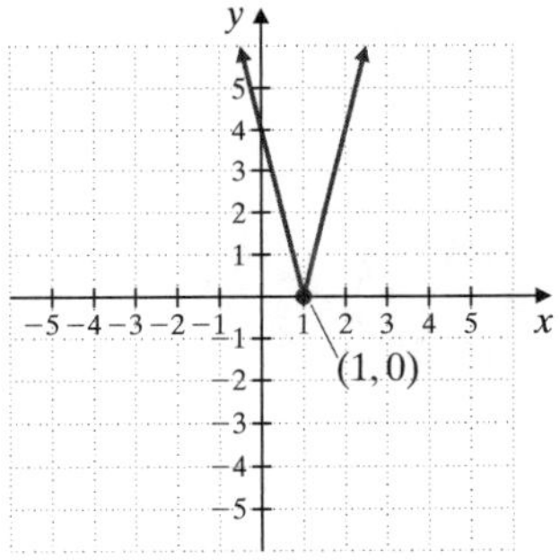

11. $g(x) = -\frac{1}{3}|x| - 2 = -\frac{1}{3}|x - 0| - 2$
The vertex is (0, −2), the graph opens down, and the graph is wider than $y = |x|$.

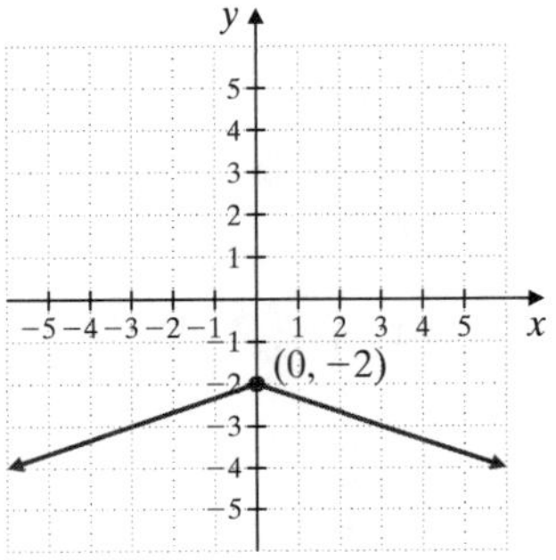

13. $f(x) = -2|x-3| + 4$
The vertex is (3, 4), the graph opens down, and the graph is narrower than $y = |x|$.

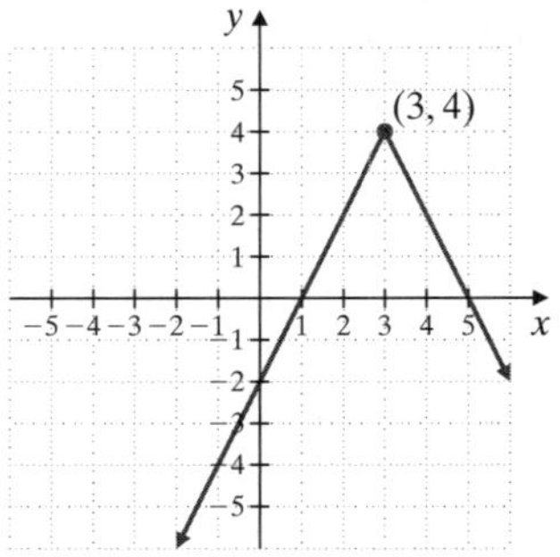

15. $f(x) = \frac{2}{3}|x+2| - 5$

The vertex is (–2, –5), the graph opens up, and the graph is wider than $y = |x|$.

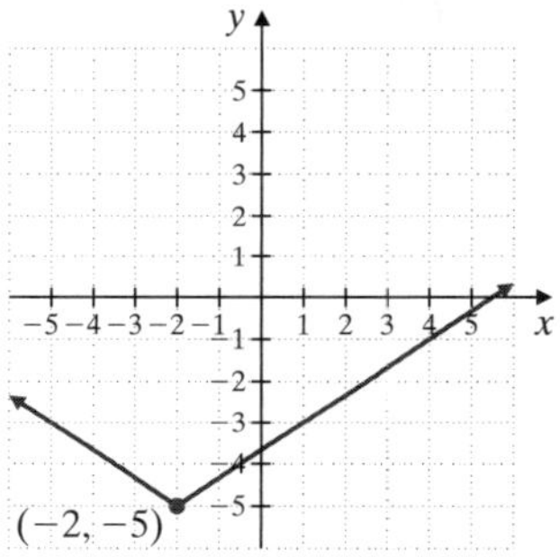

Appendix F Exercise Set

1. $90° - 19° = 71°$

3. $90° - 70.8° = 19.2°$

5. $90° - 11\frac{1}{4}° = 78\frac{3}{4}°$

7. $180° - 150° = 30°$

9. $180° - 30.2° = 149.8°$

11. $180° - 79\frac{1}{2}° = 100\frac{1}{2}°$

13. $m\angle 1 = m\angle 5 = m\angle 7 = 110°$
$180° - 110° = 70°$
$m\angle 2 = m\angle 3 = m\angle 4 = m\angle 6 = 70°$

15. $180° - 11° - 79° = 90°$

17. $180° - 25° - 65° = 90°$

19. $180° - 30° - 60° = 90°$

21. $90° - 45° = 45°$ and $90°$

23. $90° - 17° = 73°$ and $90°$

25. $90° - 39\frac{3}{4}° = 50\frac{1}{4}°$ and $90°$

27. $\frac{12}{4} = \frac{18}{x}$
$12x = 4(18)$
$12x = 72$
$x = 6$

29. $\frac{6}{9} = \frac{3}{x}$
$6x = 27$
$x = 4.5$

31. $a^2 + b^2 = c^2$
$6^2 + 8^2 = c^2$
$36 + 64 = c^2$
$100 = c^2$
$10 = c$

33. $a^2 + b^2 = c^2$
$a^2 + 5^2 = 13^2$
$a^2 + 25 = 169$
$a^2 = 144$
$a = 12$